Ornithology

Ornithology

SECOND EDITION

Frank B. Gill

The Academy of Natural Sciences of Philadelphia
and The University of Pennsylvania

Bird portraits by James E. Coe
Consultant for bird portraiture — Guy Tudor

W.H. Freeman and Company
New York

Cover photograph: Bohemian Waxwings by Art Wolfe

Library of Congress Cataloging-in-Publication Data

Gill, Frank B.
Ornithology / by Frank B. Gill.—2d ed.
p. cm.
Includes bibliographical references (p.) and index.
ISBN 0-7167-2415-4
1. Ornithology. I. Title. 60565128
QL673.G515 1994
598–dc20 94-12109
CIP

Printed in the United States of America

Third Printing

3 4 5 6 7 8 9 0 HL 9 9 8 7 6 5

To my grandfather, Frank Rockingham Downing,
who was the first of many to share with me
his knowledge and love of birds

Contents

Chapter 3: Systematics 45

PART II: FORM AND FUNCTION

Chapter 4: Feathers 65

Chapter 5: Flight 93

Chapter 6: Physiology 115

Chapter 7: Feeding 147

PART III: BEHAVIOR AND COMMUNICATION

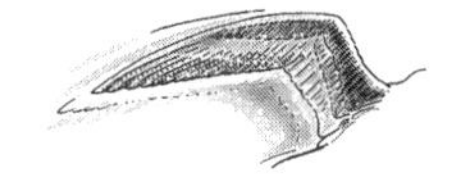

Chapter 8: Brains and Senses 177

Chapter 9: Visual Communication 203

Chapter 13: Navigation 309

Chapter 14: Social Behavior 327

PART V: REPRODUCTION AND DEVELOPMENT

Chapter 15: Reproduction 349

Chapter 16: Nests and Incubation 375

Chapter 17: Mates 403

Chapter 18: Growth and Development 425

Chapter 19: Parental Care 453

PART VI POPULATION DYNAMICS AND CONSERVATION

Chapter 20: Demography 481

Chapter 21: Populations 507

Chapter 22: Species 527

Preface

As in its original edition, this revision of *Ornithology* introduces the biology of birds from a contemporary ornithological perspective. The book is designed for undergraduate students, but I have always had in mind, as well, bird enthusiasts young and old, who simply want to know more about birds. Ornithology invites the participation of persons with great diversity of backgrounds and interests. I hope this book will be useful to similarly diverse readers, and, to this end, I have tried to share the excitement that birds and knowledge of them provide me. Throughout I have emphasized the effects of evolution in birds, especially the integration of morphological, behavioral, and physiological adaptations. I give much attention to recent discoveries and to exciting prospects for study in the future. I have avoided theory for theory's sake and have stressed discovery rather than the mathematical models that may guide discovery.

My original intentions when I undertook the revision of *Ornithology* were relatively modest. I planned to incorporate all corrections that had been brought to my attention, to lightly revise and update the content, and to add a chapter on conservation in response to the growing interest in that topic. Once begun, however, compelling revisions tended to multiply in number and scope, limited only by time constraints and editorial deadlines. Consequently, each chapter has been revised and updated with new information or perspective.

This second edition of *Ornithology* includes much that will be familiar to those who have read the original edition. It also includes some substantial changes. In addition to the new final chapter on conservation (Chapter 24), I have reorganized the text in several other important ways. Key elements of the original chapters on speciation and geography have been incorporated into Chapters 1 and 2 to provide a better introductory foundation for the material that follows. Speciation (Chapter 22) has been moved forward to follow discussion of populations (Chapter 21), because speciation is inherently a population process, the products of which—species—become fundamental units of ecological communities (Chapter 23). Also redistributed to other chapters (Chapters 10 and 18) in this second edition is the material from the original Chapter 11, previously dedicated to learning and the development of behavior. I have moved mating systems (Chapter 17), which originally started the section on reproduction and development (Part V), to a later position between nests and incubation (Chapter 16) and growth and development (Chapter 18). This change reflects our increasing awareness that the varied mating systems of birds may not reflect so much the ecology of social behavior (Chapter 14) as issues of genetic parentage and parental care of chicks. Finally, I integrated the intriguing topics of brood parasitism and cooperative breeding into a single chapter on parental care (Chapter 19).

New to this edition are boxes that highlight interesting examples pertinent to the general text discussion and in some cases, that isolate technical information from the main flow of the text. For purely practical reasons, it was often easier to add exciting, new information as a box than to integrate it fully into the text itself.

In an attempt to make the text accessible to readers who are not specialists, technical terms are defined clearly when they are first used; the enhanced, comprehensive index also refers the reader to the first use of a term and its definition.

The first edition of *Ornithology* included an appendix that listed the scientific names of each bird species mentioned in the text. I have eliminated this appendix from the second edition, adopting instead the comprehensive list of species of the birds of the world by Charles G. Sibley and Burt L. Monroe, Jr. (1990), with only a few exceptions. I give precedence to current, familiar names still used in the sixth edition (and supplements) of the American Ornithologists' Union's *Checklist of North American Birds.* For example, I use Eared Grebe, not Black-necked Grebe, and European Starling, not Common Starling. I also retain the names Rock Dove for the feral city form of Rock Pigeon and Common Canary for the caged, domesticated form of the Island Canary. The English names of bird species are always capitalized to leave no doubt that, for example, a Brown Booby is *Sula leucogaster,* not just a booby that is brown. Following K. C. Parkes (1993), "chicken" or "domestic fowl" refers to domesticated forms of the Red Junglefowl; "turkey" or "domestic turkey," to domesticated forms of *Meleagris gallopavo;* "pigeon" or "homing pigeon," to domesticated forms of *Columba livia;* and "Japanese Quail," to laboratory strains of *Coturnix,* which remain of uncertain taxonomic status.

Those who teach a one-semester course in ornithology will doubtless have to choose among the many available topics discussed in *Ornithology, Second Edition.* To allow for differences in the outlines of courses, I have packaged the chapters in short topical units. As a supplement to *Ornithology*, I am still able to recommend *The Birder's Handbook* by Paul Ehrlich and his colleagues (published by Simon and Schuster, Inc.), which provides students with information on the biology of the species of birds they learn to identify on field trips. Also available for lectures are slide sets covering the birds of the world, bird behavior, familiar eastern North American birds, eastern wood warblers, and others. These can be ordered directly from VIREO, c/o The Academy of Natural Sciences, 19th Street and the Parkway, Philadelphia, Pennsylvania 19103.

This edition has benefited greatly from comments by many colleagues and students. The next edition will also. I sincerely invite creative suggestions, corrections of errors, updates, and reprints containing interesting, new ornithological information.

Acknowledgments

IN ADDITION TO THE MANY FRIENDS who contributed to the first edition of *Ornithology,* acknowledged in the preface to that edition, I thank with greatest appreciation those who read chapters at my request, sent their comments to the editors or to me directly, or patiently responded to my queries. Among the many who helped substantially were C. Ankney, R. Curry, S. Drennan, F. Gehlbach, D. Kroodsma, D. Ligon, R. Moldenhauer, R. Montgomerie, J.P. Myers, K.C. Parkes, A. Poole, R. Ridgely, R. Roth, S. Senner, F. Sheldon, B. Slikas, N. Steele, P. Stettenheim, and N. Udvardy. Special thanks are due enthusiastically to R. Rose, R. Zink, and especially to P. Ryan, who read critically every chapter, sometimes more than once, and who greatly improved the book's clarity and content. The artistic contributions of Guy Tudor and Jim Coe continue to grace this edition. My family—Diana, James, and Frances—contributed both indirectly with their patience and love as well as directly with editorial and computer expertise. Equally appreciated has been the advice of my three primary mentors in ornithology—namely, Wesley E. Lanyon, Robert W. Storer, and Harrison B. Tordoff—who continue as trusted consultants and valued friends.

Ornithology: A Short History

In my hand I held the most remarkable of all living things, a creature of astounding abilities that elude our understanding, of extraordinary, even bizarre senses, of stamina and endurance far surpassing anything else in the animal world. Yet my captive measured a mere five inches in length and weighed less than half an ounce, about the weight of a fifty-cent piece. I held that truly awesome enigma, a bird. (Fisher, 1979)

WITH NO OTHER ANIMAL has our relationship been so constant, so varied, so enriched by symbol, myth, art, and science, and so contradictory as has our relationship with birds. Since earliest records of humankind, birds have served as symbols of peace and war, as subjects of art, as objects for study and for sport. Birds and their eggs range from the most exotic to the commonplace. Their command of our imagination is not surprising, for they are astonishing creatures, most notably for their versatility, their diversity, their flight, and their song.

Birds are conspicuous and are found everywhere: Snowy Owls in the Arctic Circle, Black-bellied Sandgrouse in the deserts of the Middle East, the White-winged Diuca-Finch at the highest elevations of the Peruvian Andes, and Emperor Penguins hundreds of meters beneath Antarctic seas. Huge eagles and bright parrots course over the rain forests of the world, and bustards, plovers, and larks stride and scurry across the arid plains.

These highly mobile creatures are travelers of the long distance and the short. Some birds, like the Nicobar Pigeon in Indonesia, move incessantly from island to island, whereas others are master navigators, traveling phenomenal distances. The Sooty Shearwater migrates from islands off Australia to the coasts of California and Oregon, the Arctic Tern from New England to Antarctica, and the Rufous Hummingbird from Alaska to Mexico.

And birds please the eye. Little in nature is more extravagant than the Twelve-wired Bird-of-Paradise, more subtly beautiful than the Evening Grosbeak, more stylish than the Horned Sungem hummingbird, or more improbable than the Javan Frogmouth.

Birds as Cultural and Religious Symbols

All these qualities seem to have provoked wonder and a sense of mystery since the dawn of human existence. Indeed, in almost every primitive culture birds were divine messengers and agents: To understand their language was to understand the gods. To interpret the meaning of the flight of birds was to foretell the future. Our words *augury* and *auspice* literally mean "bird talk" and "bird view." By the time Greek lyric poetry was flourishing (fifth and fourth centuries BC), the words for bird and omen were almost synonymous, and a person seldom undertook an act of consequence without benefit of augury and auspice. This practice still prevails in Southeast Asia and the Western Pacific.

As symbols of ideology and inspiration, birds have figured largely in many religions and in most cultures. The dove was a symbol of motherhood in Mesopotamia and was especially associated with Aphrodite, the Greek goddess of love. For the Phoenicians, Syrians, and Greeks, the dove was the voice of oracles; and in Islam it is said to call the faithful to prayer. In Christianity it represents the Holy Spirit and is associated with the Virgin Mary. Bearing an olive branch in its bill, the dove continues to be a potent symbol of peace. In contrast, the dove was a messenger of war in early Japanese culture.

The eagle appeared as a symbol in Western civilization as early as 3000 BC in the Sumerian city of Lugash. In Greek mythology the eagle was the messenger of Zeus. At least since Roman times, the symbolic eagle in Europe was the Golden Eagle, and that species also was the war symbol of many North American Indians at the time of early English settlement. In 1782 the Bald Eagle became the symbol of the fledgling United States.

Less common than the eagle, but prevalent in myth and legend, is the raven. As Apollo's messenger, the raven reported a nymph's infidelity and, as a consequence, Apollo changed the bird's color from white to black. After 40 days Noah sent forth both a dove and a raven to discover whether the floodwaters had receded. The faithless raven, according to some accounts, did not return and so earned Noah's curse and, once again, a color change from white to black. The belief in the raven's color change appears in a Greenland Eskimo legend in which the Snowy Owl, long the raven's best friend, poured sooty lamp oil over him in the heat of a disagreement.

In other legends the raven plays a more favorable role. North American Indian folklore described the raven's generosity in sharing its food with men stranded by floodwaters. Norse sailors, like Hindu sailors half the world away, carried ravens, which they released to lead them to land. Two ravens are said to have guided Alexander the Great through a duststorm on his long journey across the Egyptian desert to consult the prophet at the Temple of Ammon.

Diversity of Human Interest in Birds

Not only is our association with birds as old as human society, it is characterized by the diversity of our interest in them. We can do no more here than give a few examples of that diversity and, by way of those examples, come finally to the rich and varied science of ornithology.

The earliest records indicate that eggs have always been part of the human diet. The domesticated chicken, a form of the Red Junglefowl, existed in India before 3000 BC and was known in China by 1500 BC and in Greece by 700 BC. Mallard ducks and geese were domesticated in the Far East nearly 1000 years before Christ, and domestication of the turkey in Mexico appears to be very ancient. The Romans developed large-scale breeding and raising of poultry for food, but the practice on that scale disappeared after the fall of the Roman Empire and did not reappear in Europe until the nineteenth century.

The first American poultry exhibition was held in Boston in 1849, and in 1873 the American Poultry Association (APA) was founded, the oldest association of livestock breeders and growers in the United States. In 1905, the APA published the *American Standard of Perfection*. Now in its 1983 edition, the book is a wonderfully informative and entertaining illustrated guide to the ideal characteristics of more than 100 domestic fowl, ducks, geese, and turkeys. Presently there are at least 37 different food breeds of chickens and at least 24 ornamental breeds.

The pigeon has had a dual role as a carrier and as a prized food. There were ancient pigeon posts in Babylon, and the bird was used as a carrier in early Egyptian dynasties. The use of carrier pigeons as messengers was commonplace in Roman times and continued into the twentieth century until the invention of the radio, telegraph, and telephone.

Falconry is enjoying a modest renaissance. Originating perhaps as long as 4000 years ago, the sport flourished in Europe in the Middle Ages, and the Crusaders introduced Islamic techniques that increased and refined European falconry. After a sharp decline of Peregrine Falcons and several small accipiters in Europe and North America in the 1960s, breeding and release programs arose, and now the ancient sport, with its historical tradition of studying and protecting birds of prey, is being revived.

Use of feathers as ornamentation was widespread among North and South American peoples, in Africa, and in the Western Pacific from the earliest known times. The elaborate feather capes of the Hawaiian kings and the feather mosaics of the Mayas and Aztecs were works of high art. Among native North Americans, particular uses of feathers as badges of rank and status were common. Feather clothing was also common for protection from weather, much as goose down is widely used today.

Birds have always been influential in the arts. The earliest piece of English secular music of which we know, "Sumer is Icumen in," is a canon for four voices and the words are those of the thirteenth-century lyric in which the cuckoo welcomes summer with its song. The cuckoo, nightingale, and quail are heard in Beethoven's Sixth Symphony. The eighteenth-century composer Boccherini wrote a string quartet called "The Aviary," perhaps the first complex composition in which a number of birds are imitated. The composers Maurice Ravel and Béla Bartók used bird songs in their works for orchestra, voice, and piano.

Birds as subject and as metaphor are found frequently in opera. Wagner wrote an aria about owls, ravens, jackdaws, and magpies for *Die Meistersinger*. In Puccini's *Madama Butterfly* a character sings of a robin, and in *La Bohème*, another sings of swallows. In what is probably the most popular aria in the most popular opera of all time, the "Habanera" in Bizet's *Carmen*, the opening words are "Love is a rebel bird that no one is able to tame."

An interesting confluence of the name of a musician—in this instance, the nickname—and the name of the music brought together one of the most memorable of American jazz musicians and one of the most memorable tunes: Charlie "Bird" Parker and "Ornithology."

The role of birds in painting and sculpture is impressively large. Birds appear in paleolithic cave paintings in France and Spain as early as 14,000 BC, and as neolithic cave paintings in eastern Turkey, 8000 years later. In Egyptian tombs at Thebes, very accurate bird paintings appear before 2000 BC. In Knossos, on Crete, a famous Minoan fresco of a partridge and a hoopoe survives from about 1800 BC. Among the most vibrant and brilliantly colored Roman mosaics are those of birds, from Pompeii.

A remarkable work is an assemblage of bird species in a thirteenth-century illuminated manuscript of the Book of Revelation. Hieronymus Bosch's *Garden of Delights* (about 1500) is filled with birds. Among twentieth-century artists, Matisse and Picasso showed recurring interest in birds, and Brancusi's sleek birds in both chrome and stone are memorable.

Birds are ubiquitous in literature. For its perfect matching of avian and human characteristics, Aristophanes' comedy *The Birds*, has been described as an "ornithomorphic view of man." Birds are prominent enough in Shakespeare's plays and poems to have led the scholar James Harting to write an entire book on the subject, *The Ornithology of Shakespeare*, first published in 1871.

Some lyric poets were excellent ornithologists, notably the seventeenth-century Englishmen Michael Drayton and Andrew Marvell, whose descriptions of birds are very precise. More recently, Shelley's skylark, Keats's nightingale, and Yeats's swan have become the best known birds in English literature.

Beginning as early as the fifteenth century, books with numerous bird illustrations began to appear. Bird illustrations continued over the centuries, with the Englishmen Mark Catesby and Thomas Bewick (both in the eighteenth century) and the American John James Audubon, who published his four-volume *The Birds of America* (1827–1830) among the most prominent. By the turn of the twentieth century, a great flourishing that continues to this day was underway and served as an impetus to the rise of modern ornithology and field guidebooks.

Among the finest illustrators of the early twentieth century were Bruno Liljefors of Sweden, Archibald Thorburn of England, and Louis Agassiz Fuertes of the United States. Fuertes, with his unerring eye and his faultless sense of the salient characteristics of any bird, is believed by some to have made his birds more dazzlingly alive than any other painter.

Early Beginnings of Ornithology

With all the disparate appeal of birds, it is little wonder that some human beings have chosen to study them, a study that has evolved into the modern science of ornithology. Aristotle's fourth-century BC *History of Animals* is the first effort we know of in Western culture to account systematically for what we observe in nature, and the writing reflects the first organized scientific research. Birds figure prominently in all of Aristotle's work in natural history. Alexander of Myndos, in the first century AD, wrote a three-volume work on animals, two of which are about birds. Only fragments survive in quotation. Pliny the Elder (AD 23–79) produced an elaborate natural history encyclopedia in 37 volumes, all of which survive. He summarized the work of some 500 ancient authors and offered his own critical point of view. Aelian (AD 170–235), a Roman who wrote in Greek, devoted much attention to birds in his *On the Characteristics of Animals.*

Until the Renaissance, our knowledge of the natural history of birds depended largely on these and other Greek and Roman writers. They told us much that was reliable, but they also left us with many wrong notions. The quotations from Alexander's work reflect close and accurate observation, but Aelian was steadfastly uncritical of his sources and perpetuated two remarkably wrong notions about the behavior of cranes: one, that they flew against the wind and swallowed a stone for ballast so as not to be swept off course; the other, that they posted sentinels at night, requiring them to stand on one foot while holding a stone in the other, thereby ensuring that if the sentinel fell asleep, it would drop the stone and be awakened by the noise.

A major step toward modern ornithology was the growth of field observation in the eighteenth century. In 1789, Gilbert White, an English clergyman, published a natural history of his parish, gathered over 40 years' time. His observations of birds were marvelously precise and beautifully expressed. But he also asked incisive questions about the basic biology of birds, about species, ecological niches, physiology, and migration. Many of his curiosities still pertain to research in ornithology.

Ornithology Today

Contemporary ornithology has benefited from years of careful field observation by devoted amateurs as well as by professional ornithologists. Our knowledge of avian life histories and populations is more complete than that of most other classes of animals. Owing in part to this wealth of information and in part to the attributes of birds, birds have increasingly become the subject in primary biological studies. By the middle of the 1980s, birds provided more textbook examples of biological phenomena than any other class of vertebrates. Birds have been central to work on the formation of species and have been used in some of the most detailed molecular analyses of phylogeny. Perhaps the greatest contribution of bird studies has been to population and community ecology, but their contribution to evolutionary ecology and to the discovery of new connections between animal behavior and ecology is not far behind. Birds are particularly well suited for the study of mating systems and strategies. They are similarly useful in investigation of the roles of kinship in evolution and of altruism.

The rules governing communication and physiological mechanisms that connect communication to behavior have been elucidated by bird studies. Because only humans, parrots, and songbirds can imitate sounds, birds are immensely important to the investigation of the interplay between inheritance and learning, mediated by the central nervous system. Because birds in captivity continue partly to maintain their natural behavior, and to some degree because they are long-lived, they are useful in the study of effects of natural stimuli on physiology and behavior. The same characteristics also make it possible to study the environmental control of reproduction and the role of circadian and circannual rhythms.

Birds are, more obviously perhaps, ideal subjects for the study of adaptation to extreme conditions and unusual niches, of navigation, and of energetics of flight. And at the very origin of ethology, the work of Niko Tinbergen with gulls and of Konrad Lorenz with ducks and geese provide classic examples of the attempt to understand the evolution of behavior. Their studies earned them a Nobel Prize in Physiology or Medicine. In cell biology and medicine, the discovery of B vitamins and their roles in nutrition came from studies of chickens, which readily reveal dietary deficiency. Albert Szent-Györgyi won a Nobel Prize in Physiology or Medicine for the elucidation of the Krebs cycle from studies of pigeon breast muscle, as did Payton Rous for studies of avian sarcoma that linked viruses to cancer for the first time. Recent discoveries of regenerating hair cells of a chick's ear offer promise for the treatment of human deafness (Corwin and Cotanche 1988).

Birds are, of course, sensitive indicators of environmental change, witness Rachel Carson's account in *Silent Spring* or warning of the El Niño phenomenon, well before meterologists announce it, by the failure of seabird reproduction that results from changes in food and climate. We are only now beginning to understand that, besides urbanization and modern modes of transportation, the practices of agriculture and forestry, especially mechanized deforestation, have profound effects on habitats and populations. Because of their migration, birds more than other animals help us to understand the global nature of these effects. Study of birds is also helping us to understand the detrimental effects of our introducing into the environment domestic animals, rats, nonnative species, and oil and chemical pollution.

A few stories of extinction in modern times are well known; those of the Passenger Pigeon and the Great Auk are dramatic. Less well known is an extraordinary story of a battle won by the birds. It occurred in Western Australia in 1932 and is known as the Emu War. At the time, it attracted much attention and was covered by the press. It seems that some 20,000 emus threatened wheat fields. Soldiers employing machine guns and artillery spent a month attacking the birds. The birds, in the words of Dominick Serventy of the Australian Wildlife Research Office, "apparently adopted guerilla tactics and split into small units. This made the use of military equipment uneconomic." After the soldiers withdrew, fences were built to separate the emus from the grain.

A satisfactory outcome indeed. Perhaps through basic research and thoughtful practice, we will come to a more harmonious relationship with those astonishing creatures that have charmed and fascinated us throughout our own existence.

Further Readings

John Kastner's *A World of Watchers* (Sierra Club, 1988) is a delightful ramble through the world of birdwatching and almost everything connected to it. Kastner and Miriam T. Gross are the editors of *The Bird Illustrated from 1550–1900* (Abrams, 1988), a lovely set of annotated reproductions of paintings and drawings from the collection of the New York Public Library, where Gross is a librarian.

Two books of wide-ranging interest are *A Book of Birds* by Mary Priestley (1937, but out of print) and Joseph Wood Krutch's *A Treasury of Bird Love* (Doubleday, 1962).

Peggy Munsterberg wrote a fine introduction to *The Penguin Book of Bird Poetry* (Penguin, 1984); and John R.T. Pollard's graceful *Birds in Greek Life and Myth* (Thames and Hudson, 1977) is the most complete account of its subject written entirely in English.

In a fine essay celebrating the centennial birthday of the American Ornithologists' Union, Ernst Mayr (1984) reviewed *The Contributions of Ornithology to Biology*, with a sequel from different perspectives by Masakazu Konishi and three eminent colleagues (1989).

PART I

Origins

C H A P T E R 1

The Diversity of Birds

MILLIONS OF YEARS AGO, a small, bipedal reptile lived among the dinosaurs. Over time its stiff scales evolved into soft feathers. Feathered insulation enhanced its control of a high body temperature, thereby increasing its activity and endurance. Its leaps led to powerful, graceful flight. Mastery of flight opened a world of ecological opportunities. A new group of vertebrates—the Class Aves—evolved.

This chapter previews the major features of bird diversity, their form and function, their scientific classification, and their geography. Chapter 2 examines the evolutionary history of birds from the first well-known fossil bird, *Archaeopteryx lithographica.*

Basic Characteristics of Birds

Birds are two-legged (bipedal) vertebrates—animals with backbones—the animal group that also includes mammals, amphibians, reptiles, and bony fishes. Despite their diversity of form, birds are a well-defined group of bipedal vertebrates. They are distinguished from other vertebrates by feathers, which are unique modifications of the outer skin. Compared with the scales of reptiles, feathers are filamentous, soft in texture, flexible, lightweight structures (Figure 1–1). No comparable structures exist in other vertebrates. Dead structures that wear easily and must be replaced regularly, feathers are essential for both temperature regulation and flight. They insulate the body and help maintain high body temperature. Lightweight and strong, the long feathers of the wing generate lift and thrust for flight.

All birds have bills, a distinctive attribute that facilitates instant recognition. The avian bill varies greatly in form and function but is always toothless and covered with a horny sheath. The avian bill has no exact

Figure 1–1 Snowy Egret with display plumes. (Courtesy of A. Cruickshank/VIREO)

parallel among other extant vertebrates; it is approximated only by the snout of the duck-billed platypus, a strange, egg-laying mammal of Australia.

Birds are feathered flying machines (Figure 1–2). Their wings and their ability to fly are familiar attributes but, unlike feathers, are not diagnostic features; bats and flying insects also have wings. The entire avian body is structured for flight. Bird bones, for example, typically are lightweight structures, being spongy, strutted, or hollow. The skeleton generally is strengthened and reinforced through fusions of bones of the hands, head, pelvis, and feet. Horizontal, backward-curved projections on the ribs—called uncinate processes—overlap other ribs and so strengthen the walls of the thorax. The furcula—or wishbone—compresses and rebounds like a powerful spring in rhythm to the beat of the wings. The wing itself is a highly modified forelimb that, with a few remarkable exceptions, is nearly incapable of functions other than flight. Fused hand bones support and maneuver the large and powerful primary flight feathers.

On land, a bird's center of gravity must remain over and between its feet, particularly when it perches, squats, or rises (Figure 1–3). The equal length of the two main leg bones—tibiotarsus and tarsometatarsus—of long-legged birds ensures this relationship. Foot-propelled diving birds such as loons have sacrificed balance on land for their considerable swimming abilities. For efficient propulsion, they have powerful legs situated at the rear of a streamlined body (Figure 1–4).

Arboreal—tree-dwelling—species, which constitute the majority of birds, have feet that tightly grip branches. Among the features of such feet are long tendons that pass around the backside of the ankle joints. When

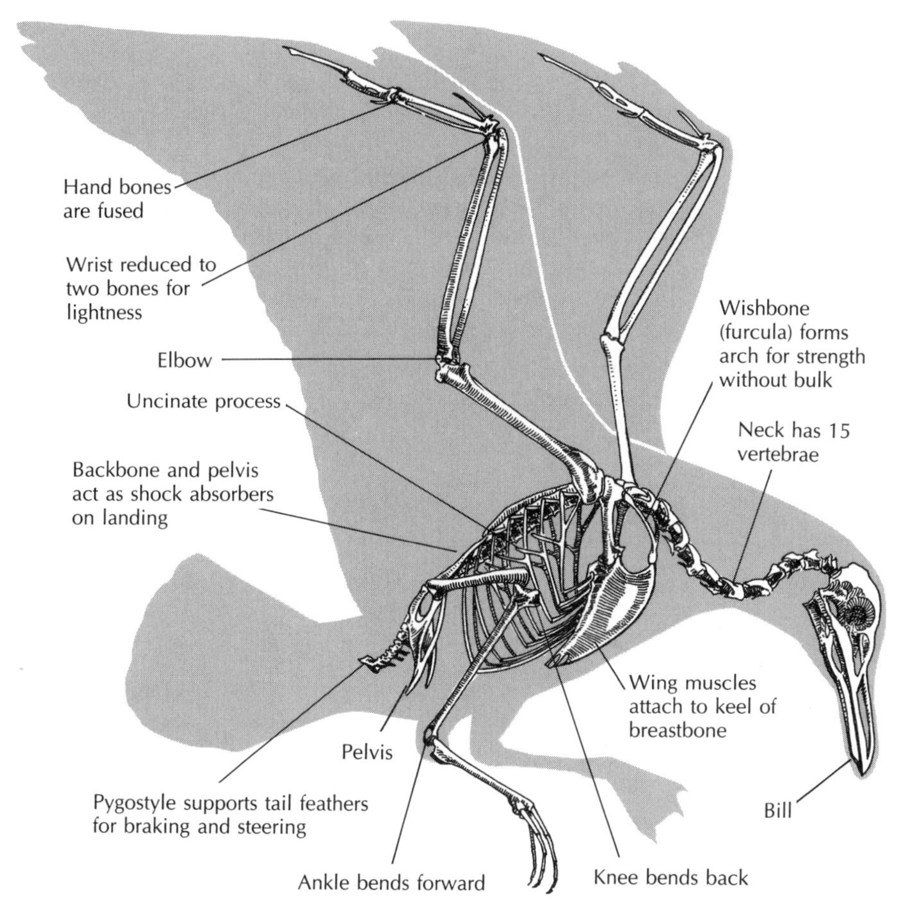

Figure 1–2 Adaptations for flight. Supporting the large wings of the Herring Gull is a strong but lightweight skeleton. An enlarged, keeled sternum houses and anchors the large breast muscles that empower the wings. The bones of the hand and wrist, which support and maneuver the large and powerful primary flight feathers, are reduced in number and fused for extra strength. Similarly, the pygostyle, made of fused tail vertebrae, supports and controls the tail feathers, which are used for braking and steering. Strengthening the body skeleton are fusions of the pelvic bones and associated vertebrae, plus horizontal rib projections, called uncinate processes. The furcula, or wishbone, compresses and rebounds like a powerful spring in rhythm to the beat of the wings. (After Pasquier 1983; drawing by Biruta Akerbergs)

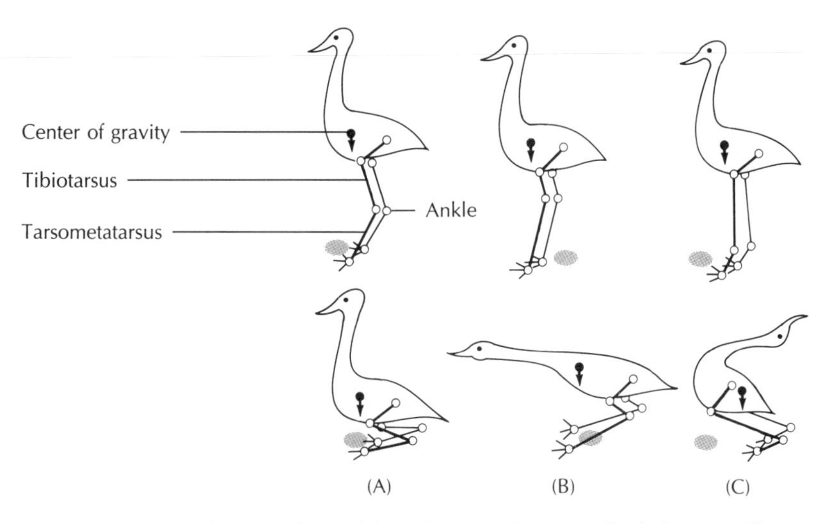

Figure 1–3 (A) Leg bones of equal lengths contribute to the balance of long-legged birds. When a bird crouches to incubate its egg, for example, leg bones of different lengths (B and C) would displace the center of gravity. What appears at first glance to be a backward-bending knee joint is really the ankle joint. In birds, the foot bones (three tarsals) are fused both to one another and to the metatarsals, thereby creating a long, strong, single leg element, the tarsometatarsus, which enables birds to walk on their toes rather than on the whole foot. (After Storer 1971a)

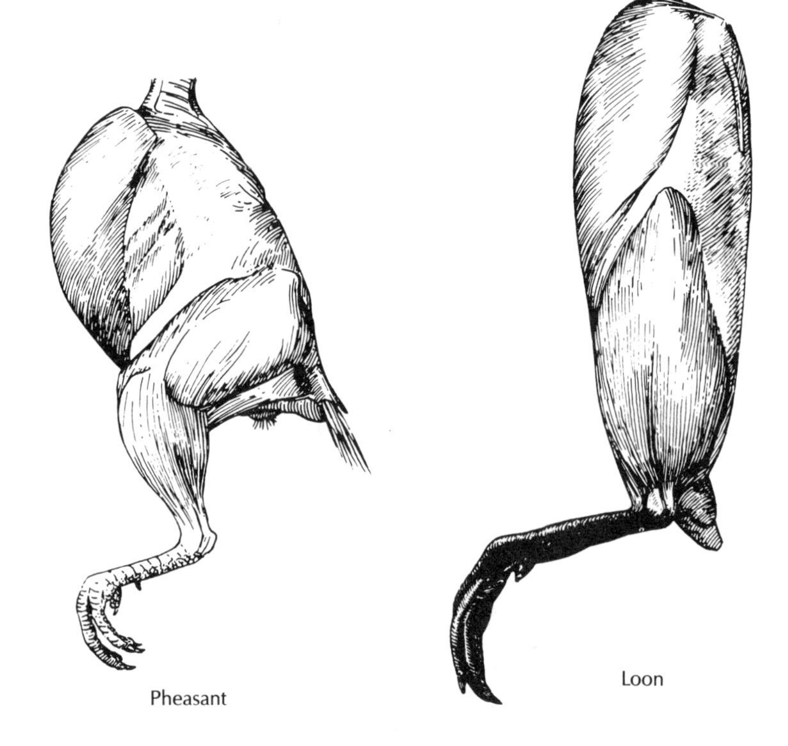

Figure 1–4 The body of a loon, which is streamlined for diving, compared with that of a pheasant. Note also the loon's highly developed hindlimb musculature. (From Storer 1971a)

a bird bends its joints to squat, the tendons automatically flex, locking the toes around the branch (Figure 1–5). When a bird stands, the tension relaxes and the toes open. The foot of the songbirds—Order Passeriformes—is perhaps the most advanced in this respect. A special system of ridges and pads between the tendons that flex the toes and the insides of the toe pads acts as a natural locking mechanism and permits birds to sleep while perching. The large, opposable single rear toe—hallux—which enhances the ability of a bird to grip a branch, is unusual among vertebrates.

Avian physiology accommodates the extreme metabolic demands of flight and temperature regulation. The red fibers of avian flight muscles have an extraordinary capacity for sustained work and can also produce heat by shivering (Chapter 5). Birds maintain high body temperatures (40° to 44°C) over a wide range of ambient temperatures. The circulatory and respiratory systems of birds include a powerful four-chambered heart and efficient, flow-through lungs, which deliver fuel and remove both waste and heat produced by metabolic activities.

The reproductive systems of birds also are unusual. Birds produce large, richly provisioned external eggs, the most elaborate reproductive cells of any animal. No species bear live young like those produced by other classes of vertebrates. Nurturing the growth of the embryos in the egg and of the young after they hatch requires dedicated parental care. Most birds form monogamous pair bonds, some for life; but many, it turns out, engage in additional sexual liaisons. Indeed, the eggs in one nest may be of mixed paternities and even maternities. Mating systems, spacing behavior, competition, and cooperation reflect varied solutions to the challenges of successful reproduction.

Birds have large, well-developed brains, 6 to 11 times larger than those of like-sized reptiles. Bird brains and primate brains exhibit functional lateralization, with left hemispheric dominance associated with learning and innovation in vocal repertoires. Substantial learning by birds guides the mastery of complex motor tasks, social behavior, and vocalizations.

Highly developed neural systems and acute senses mediate feats of communication and navigation. Birds, particularly the songbirds, have the greatest sound-producing capabilities of all vertebrates. Mammals are constrained in their vocalizations by the position, structural simplicity, and neural innervation of the larynx, which is the analogue of the syrinx—the song-producing organ of birds. Birds can navigate by using patterns of the Earth's magnetism, celestial cues, and perhaps polarized light. The highly developed color vision of birds reaches into the near-ultraviolet range of the spectrum. The broad hearing range of birds encompasses infrasounds—sounds below the hearing range of humans.

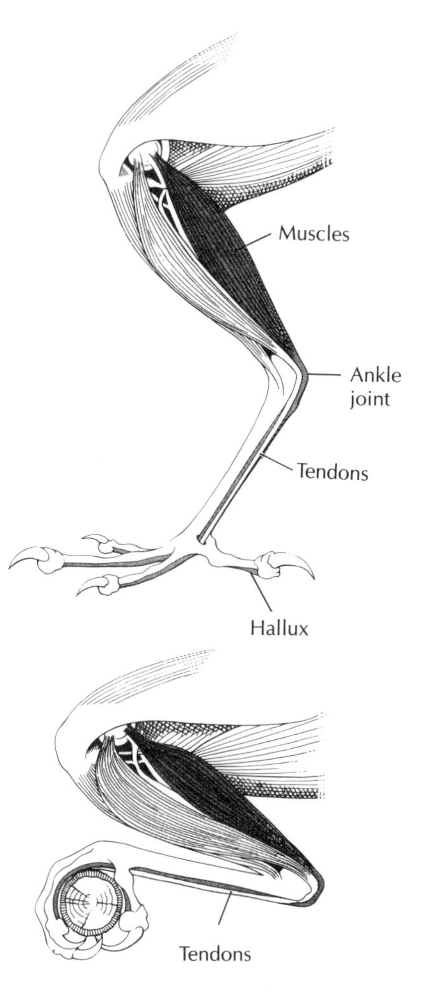

Figure 1–5 When a perching bird squats, the leg tendons, which are located on the rear side of the ankle, automatically cause the toes to grip. (From Wilson 1980, with permission of Scientific American, Inc.)

Adaptive Radiation of Form and Function

The variety of birds is the grand result of millions of years of evolutionary change and adaptation. Roughly 300 billion birds of roughly 10,000 species now inhabit the Earth. Yet this number is only a small fraction of the number of species that have existed since the Age of Dinosaurs. The earliest known bird is *Archaeopteryx lithographica.* Its fossilized remains are 155 million years old. It had feathers and probably could fly after a fashion. Responding to ecological opportunities, subsequent birds diversified in form and function. From the fundamental anatomy of their common ancestor evolved perching songbirds such as robins; nocturnal forest hunters such as owls; aquatic divers such as penguins; oceanic mariners such as albatrosses; shoreline waders such as plovers; and large, flightless ground birds such as the Ostrich (Figure 1–6). Birds range in size from only 2 grams (hummingbird) to 100,000 grams (Ostrich).

The diversity of birds reflects the evolution of additional varied species adapted to different ecologies and behaviors, a phenomenon called adaptive radiation. Bill sizes and bill shapes change in relation to the types of

Figure 1–6 Birds have evolved along major lines, each adapted to a particular mode of life. (From *Evolution of Vertebrates* by E.H. Colbert. Copyright 1955 John Wiley & Sons, Inc.; reprinted by permission of John Wiley & Sons, Inc.)

food eaten. Leg lengths change in relation to habits of perching or terrestrial locomotion, and wing shapes change in relation to patterns of flight. For example, from a single ancestral species of shorebird evolved aerial pirates such as skuas and plunging divers such as terns, as well as a host of

Figure 1–7 A classic example of adaptive radiation: Hawaiian honeycreepers (Family Drepanididae) have evolved bills that range from thin warblerlike bills to long sicklelike bills to heavy grosbeaklike bills. (From Raikow 1976; drawing by H. Douglas Pratt)

waders, including sandpipers, plovers, turnstones, stilts, oystercatchers, snipes, woodcocks, curlews, and godwits, each with characteristic leg lengths and bill lengths, shapes, and curvatures. As varied as the habitats they occupy, shorebirds also include aerial pratincoles, deep-water divers such as auks, and the grouselike seedsnipes of South American moorlands. All these related species are included in the Order Charadriiformes.

The varied diets of modern birds generally include green leaves, buds, fruits, nectar, seeds, invertebrates of all sizes, and vertebrates of many kinds, including carrion. Fruits, seeds, and insects nourish the majority of birds, especially the passerine land birds, whose adaptive radiation was coupled to those of flowering plants and their associated insects. Few birds are specialized herbivores; apparently mammals have usurped most of the terrestrial grazing and browsing niches. In the absence of mammals on New Zealand, numerous species of flightless, herbivorous moas evolved. The long, complex digestive tracts of herbivores and the weight of slowly digesting plant matter also may not be compatible with flight.

Corresponding to a diversity of diets is a diversity of bills. The variety of bill forms that can evolve during the process of adaptive radiation is seen in the Hawaiian honeycreepers, which apparently evolved from a flock of small finches that strayed out over the Pacific Ocean from Asia or North America millions of years ago. The finches made landfall on one of the Hawaiian islands, then flourished and spread throughout the archipelago. Isolated populations changed in genetic composition and appearance, at first imperceptibly and then conspicuously. Subtle changes in bill shapes and bill sizes led to a proliferation of bill types and their related feeding behaviors, from heavy grosbeaklike bills for cracking large legume seeds to long sicklelike bills for sipping nectar from flowers or probing bark crevices for insects (Figure 1–7).

Different modes of locomotion further expand the ecological opportunities of birds. Shorebirds, as we have mentioned, include aerial, wading, and diving species. Birds soar through the sky, scurry and stride across the land, hop agilely from branch to branch, hitch up tree trunks, and swim powerfully to great depths in the sea. The combination of forelimbs adapted for flight and hindlimbs for bipedal locomotion gives birds a tremendous range of ecological options.

There are specialized flying birds, as well as specialized swimmers, runners, waders, climbers, and perchers. Wing shapes and modes of flight range from the long, narrow wings of the albatross, adapted for soaring over the oceans, to the short, round wings of wrens, adapted for agile fluttering through dense vegetation. At another extreme are the adaptations of wing-propelled diving birds, such as penguins, which use their flipperlike wings to move underwater.

Like the structures of bills and wings, the anatomy of feet and legs reflects different life styles (Figure 1–8). At one extreme are the long, powerful legs of wading and cursorial—running—birds such as storks and Ostriches. At the other extreme are the tiny feet and short legs of

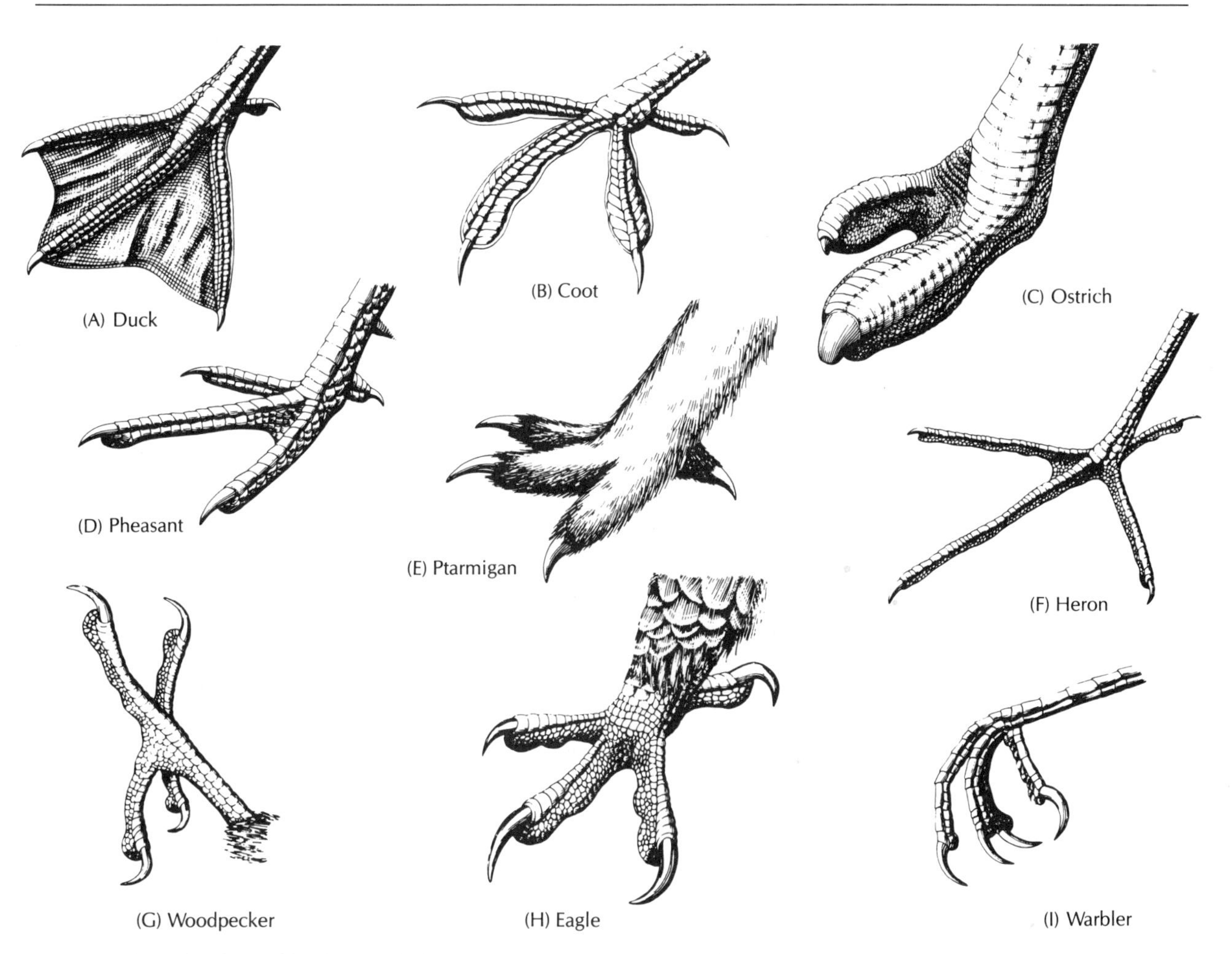

Figure 1–8 The feet of birds reveal their ecological habits: Water birds have (A) webbed or (B) lobed toes for swimming; terrestrial birds have toes specialized for (C) running, (D) scratching in dirt, (E) walking on snow, or (F) wading. Other land birds have feet designed for (G) climbing, (H) holding prey, or (I) perching. (From Wilson 1980, with permission of Scientific American, Inc.)

specialized aerial species such as swifts. The long toes of herons and jacanas spread the bird's weight over a large surface area and facilitate walking on soft surfaces. Sandgrouse scurry on soft desert sands, and ptarmigan can walk on snow by virtue of snowshoelike adaptations of their feet. Lobes on the toes of coots and webbing between the toes of ducks aid swimming. Climbing birds such as woodpeckers have large, sharply curved claws; nuthatches climb downward by gripping a tree's bark with a large claw on the hind toe.

We accept that the close fit between form and function portrayed so vividly by the diversity of birds reflects evolutionary adaptation through natural selection. Natural selection, as set forth by Charles Darwin in 1859 and confirmed subsequently through experiment and independent observation, is simply the predictable relative increase of individuals with advantageous traits. Healthy individuals leave more offspring than do sickly individuals. Camouflaged chicks are more likely to escape predation and to reproduce themselves than are boldly colored chicks. To be favored by natural selection, however, traits need not be dramatically better. Subtle or slight advantages in egg quality, camouflage, or agility increase in prevalence in a local population. Adaptation by natural selection is a process without plan or purpose, a process that gradually transforms the appearances and abilities of organisms. The elucidation of this process, of its limitations as well as its mechanisms, is the unifying theme of modern biology.

Bird Names and Classifications

The Class Aves comprises almost 10,000 recognized species of varied forms descended from one another through the process of adaptation by natural selection. A logical system of names for each species is an essential prerequisite for the scientific study of the biology of birds. The nonscientific names of birds, however, tend to vary with locale. The American Goldfinch, for example, is also called the yellow-bird, thistle-bird, wild canary, and beet-bird (Figure 1–9). Each human culture employs its local bird names. Ornithologists require standardized names to communicate efficiently and without ambiguity. The science of naming and classifying birds according to standardized rules is called taxonomy, and the scientists who do this work are taxonomists. A taxon (pl. *taxa*) is any group of

Figure 1–9 The American Goldfinch has many local names, such as wild canary, yellow-bird, thistle-bird, and beet-bird. (Courtesy of A. Cruickshank/VIREO)

animals that is recognized in a classification. The Class Aves is a taxon that includes all species of birds.

The standard rules of taxonomy are based on the system of nomenclature developed in 1758 by Carolus Linnaeus, a Swedish botanist. Linnaeus assigned two latinized names to each species: The first denotes the genus—a group of similar species; the second denotes the species. Thus, the American Goldfinch is known formally as *Carduelis tristis*, which is a taxon that includes all populations of that species. This combination of names is unique; no other bird—indeed, no other animal—may have the same pair of names. Prior to Linnaeus, names were not standardized in length but consisted instead of a string of descriptive Latin words. The Great Black-backed Gull, for example, was once *Larus maximus ex alba et nigro feu caeruleo nigricante varius* (Willoughby and Ray 1676). Now it is simply *Larus marinus*.

In addition to their Latin scientific names, birds have standard English names. The Checklist Committee of the American Ornithologists' Union establishes and regularly revises the list of valid names, both English and Latin, for all bird species in North America. Comprehensive lists of all the birds of the world are presented by Clements (1991), Howard and Moore (1991), and Sibley and Monroe (1990). Reflecting the growing popular interest in birds, ornithologists in America are more prone to use English names for species than are their colleagues who study mammals (mammalogists), insects (entomologists), or plants (botanists).

If we examine the world of birds, we can see the possibility of constructing a hierarchy—or ranking—of differences. A cursory survey of North American birds will distinguish woodpeckers from owls. Less obvious are the differences between the Downy Woodpecker, the Red-bellied Woodpecker, and the Northern Flicker, or the differences between the Great Horned Owl, the Barred Owl, and the Eastern Screech-Owl. Recognition of the subtle differences between Downy and Hairy Woodpeckers (Figure 1–10) or between the Eastern Screech-Owl and the Whiskered Screech-Owl requires even more expertise.

Related taxa—that is, those sharing a common evolutionary history or genealogy, as do the species of woodpeckers or owls, or birds themselves—constitute a lineage. Ornithologists classify the diverse species of modern birds into 29 different major lineages—or orders. Owls and woodpeckers thus are classified in different orders, Strigiformes and Piciformes, respectively. Note that each order ends in *-formes*. In turn, each of the 29 orders comprises a hierarchical set of families and genera. All woodpeckers are classified in the same order and also the same family, the Picidae. The very similar, closely related Downy Woodpecker and Hairy Woodpecker are classified in the genus *Picoides*, but the less closely related Northern Flicker is classified in the genus *Colaptes*, along with other species of flickers (Table 1–1). The distinguishing features of genera, families, and orders reflect the evolutionary adaptations of birds—anatomical, physiological, ecological, and behavioral—which are the main themes of this text. To take your first major step toward appreciating the diversity

TABLE 1–1
Classification of three species of woodpeckers

Taxon	Downy Woodpecker	Hairy Woodpecker	Northern Flicker
Class	Aves	Aves	Aves
Order	Piciformes	Piciformes	Piciformes
Family	Picidae	Picidae	Picidae
Genus	*Picoides*	*Picoides*	*Colaptes*
Species	*pubescens*	*villosus*	*auratus*

Figure 1–10 Three species of woodpeckers: (A) Downy Woodpecker, (B) Hairy Woodpecker, (C) Northern Flicker. The Downy Woodpecker and the Hairy Woodpecker are more closely related to each other than either is to the Northern Flicker.

of the Class Aves, review the illustrations of the major groups of the birds of the world, presented in the Appendix.

The current classification of living birds is a hierarchical arrangement of roughly 29 orders, 187 families, over 2000 genera, and over 9600 species (Table 1–2). The arrangements of species into taxonomic categories and the numbers of species themselves are not fixed, because a classification is more than an authoritative basis for orderly communication about birds. It is also a set of working hypotheses about the relationships, similarities, and differences among birds. Different taxonomic concepts and new information prompt regular changes in the official classification of birds.

TABLE 1–2
The orders of the birds of the world

Order	Number of taxa			Members
	Families	Genera	Species	
Tinamiformes	1	9	47	Tinamous
Rheiformes	1	1	2	Rheas
Struthioniformes	1	1	1	Ostrich
Casuariiformes	2	2	4	Cassowaries, Emu
Dinornithiformes	1	1	3	Kiwis
Podicipediformes	1	6	21	Grebes
Sphenisciformes	1	6	17	Penguins
Procellariiformes	4	24	115	Tube-nosed seabirds: petrels, shearwaters, albatrosses, storm-petrels, diving-petrels
Pelecaniformes	6	8	67	Water birds with totipalmate feet: cormorants, pelicans, anhingas, boobies, gannets, frigatebirds, tropicbirds
Anseriformes	2	48	161	Waterfowl: ducks, geese, swans, screamers of South America
Phoenicopteriformes	1	1	5	Flamingos
Ciconiiformes	5	42	120	Long-legged wading birds: storks, herons, ibises, spoonbills, Hammerkop, Shoebill
Falconiformes	5	81	311	Raptors: falcons, caracaras, hawks, eagles, Old World vultures, kites, Osprey, Secretary-bird, and New World vultures
Galliformes	5	70	258	Gallinaceous birds: grouse, quails, pheasants, chickens, curassows, guans, chachalacas, guineafowl, and moundbuilders; excludes Hoatzin (see Cuculiformes)

(continued on next page)

TABLE 1–2 *(continued)*
The orders of the birds of the world

Order	Number of taxa: Families	Genera	Species	Members
Gruiformes	11	55	213	Diverse terrestrial and marsh birds: rails, coots, sungrebes, cranes, Sunbittern, Kagu, Limpkin, seriemas, bustards, buttonquails, trumpeters, roatelos of Madagascar
Charadriiformes	19	85	366	Shorebirds and their relatives: sandpipers, plovers, phalaropes, stilts, jacanas, painted-snipes, pratincoles, gulls and terns, seed-snipes, sheathbills, skimmers, skuas, auks, sandgrouse
Gaviiformes	1	1	5	Loons
Columbiformes	1	40	310	Pigeons, doves
Psittaciformes	3	80	358	Parrots, macaws, lories, cockatoos
Coliiformes	1	2	6	Mousebirds
Musophagiformes	1	5	23	Turacos and plaintain-eaters
Cuculiformes	2	30	143	Cuckoos, Hoatzin
Strigiformes	2	25	178	Owls, barn-owls
Caprimulgiformes	5	20	113	Nightjars, potoos, frogmouths, Oilbird, owlet-frogmouths
Apodiformes	3	128	422	Swifts, crested swifts, hummingbirds
Trogoniformes	1	6	39	Trogons, quetzals
Coraciiformes	10	46	218	Kingfishers and allies: todies, motmots, bee-eaters, rollers, Courol, Hoopoe, woodhoo-poes, hornbills
Piciformes	8	66	410	Woodpeckers and allies: wrynecks, piculets, barbets, toucans, honeyguides, jacamars, puffbirds
Passeriformes[a]	83	1161	5712	Perching birds, songbirds, passerines
Totals	187	2050	9648	

a. The values given for numbers of genera and species of Passeriformes are larger than the sum of the numbers appearing in the corresponding tables in the Appendix because the values in this table include various birds that have not been assigned to specific families and, hence, are not included in the Appendix tables.

After Bock and Farrand 1980; Sibley and Monroe 1990.

Biogeography

Biogeography is the study of the geographical distributions of plants and animals. For over a century, biogeographers have divided the Earth into six major faunal regions corresponding roughly to the major continental areas (Figure 1–11). Each faunal region has its characteristic birds:

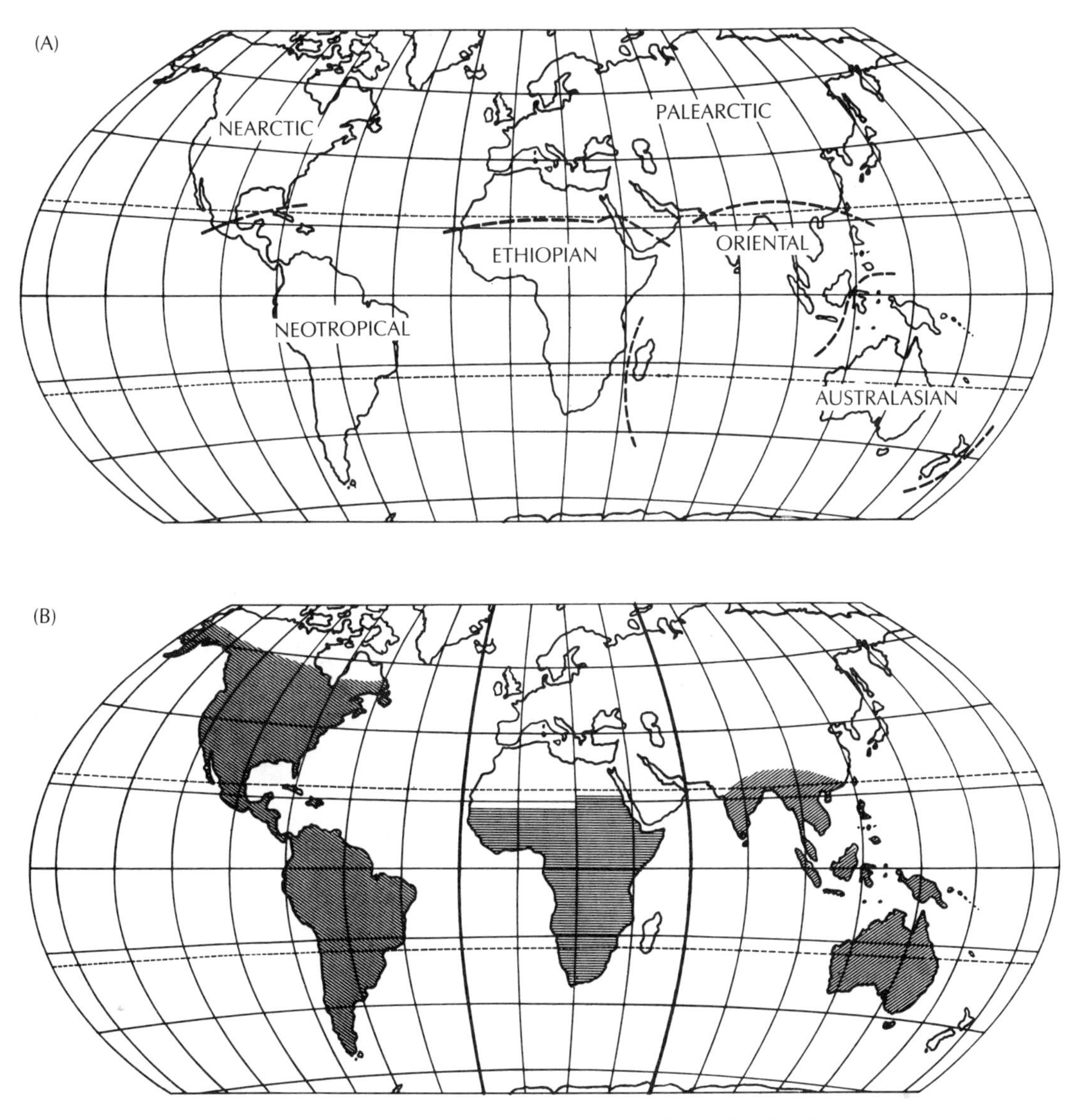

Figure 1–11 (A) The six major zoogeographical regions and (B) the distributions of three avian families: tyrant-flycatchers, limited to the New World (Neotropical and Nearctic regions); turacos, limited to Africa south of the Sahara (Ethiopian region); and wood-swallows, limited to the Oriental and Australasian regions. (From Thompson 1964)

TABLE 1–3
Avian specialties of the major biogeographic regions

Regions	Endemic nonpasserine families	Representative family radiations
Nearctic and Palearctic (Holarctic): all of North America, Mexico, and the West Indies, plus all of Europe, northern Asia south to the Himalayas, and Africa north of the Sahara	Loons (Gaviidae) Auks (Alcidae)	Accentors (Prunellidae) [Palearctic] Buntings (Emberizidae) Carduelline finches (Fringillidae) Wood warblers (Parulidae) [Nearctic] Old World warblers (Sylviidae) [Palearctic]
Neotropical: all of South America plus Central America south of the Isthmus of Tehauntepec, Mexico	Rheas (Rheidae) Tinamous (Tinamidae) Curassows (Cracidae) Trumpeters (Psophiidae) Sunbittern (Eurypygidae) Seriemas (Cariamidae) Limpkin (Aramidae) Oilbird (Steatornithidae) Hoatzin (Opisthocomidae) Motmots (Momotidae) Jacamars (Galbulidae) Puffbirds (Bucconidae) Toucans (Ramphastidae)	Hummingbirds (Trochilidae) Tyrant-flycatchers (Tyrannidae) Tanagers (Thraupidae) Typical antbirds (Formicariidae) Ovenbirds (Furnariidae) Woodcreepers (Dendrocolaptidae) Manakins (Pipridae) Cotingas (Cotingidae)
Ethiopian: Africa south of the Sahara	Ostrich (Struthionidae) Secretarybird (Sagittariidae) Guineafowl (Numididae) Roatelos (Mesoenatidae) (M)[a] Turacos (Musophagidae) Mousebirds (Coliidae) Groundrollers (Brachypteraciidae) (M) Courol (Leptosomatidae) (M) Woodhoopoes (Phoeniculidae)	Larks (Alaudidae) Sunbirds (Nectariniidae) Weavers (Ploceidae)
Oriental: Southeast Asia from the Himalayas to northern Indonesia	None	Leafbirds (Irenidae) Pheasants (Phasianidae) Broadbills (Eurylaimidae) Pittas (Pittidae) Babblers (Timaliidae) Flowerpeckers (Dicaeidae)

TABLE 1–3 *(continued)*

Regions	Endemic nonpasserine families	Representative family radiations
Australasian: Australia and New Guinea from Lombok south plus the islands of the southwest Pacific.	Emus (Dromiceidae) Cassowaries (Casuariidae) Kiwis (Apterygidae) Kagu (Rhynochetidae) Cockatoos (Cacatuidae) Lories (Loriidae) Owlet-nightjars (Aegothelidae)	Birds-of-paradise (Paradisaeidae) Whistlers (Pachycephalidae) Honeyeaters (Meliphagidae) Monarch-flycatchers (Monarchidae) Australian warblers (Acanthizidae)

a. (M), Madagascar only.

so-called endemic taxa or species, which are found nowhere else, and other birds that represent major adaptive radiations of more widespread taxa (Table 1–3). Waxwings and loons are restricted to North America and Eurasia, the Nearctic and Palearctic regions, respectively. The birds that are endemic to Africa, or the Ethiopian region, include Ostriches, mousebirds, and turacos. Australia and New Guinea, the Australasian region, have Emus, honeyeaters, and birds-of-paradise. South America, the Neotropical region, has toucans, tanagers, and trumpeters.

Most avifaunas—regional assemblages of bird species—are mixtures of species of various ages and origins. The history of bird distributions can be viewed as a series of waves of adaptive radiations, moving north, south, east, and west. New groups of birds replaced older ones and in turn produced complex mosaics of ancient, recent, and new colonists from different regions. The birds of North America include old and new colonists from Asia and South America, remnants of ancient avifaunas, plus diverse species groups that evolved only on that continent, for example, the colorful wood warblers. On the other side of the globe, most of the birds of Madagascar are derived from African ancestors, but 5 percent came from India. Early colonists of lands not subject to regular invasion by members of other faunas diversified locally in response to the ecological opportunities available to them. The diverse species of vanga shrikes on Madagascar and of honeycreepers on the Hawaiian islands are two such groups.

Summary

Characterized as vertebrates with feathers, birds have distinctive bills, maintain high body temperatures, produce large external eggs, and have elaborate parental behavior and extraordinary vocal abilities. The anatomy and physiology of most birds are adapted for flight.

The diversity of birds reflects millions of years of divergence and adaptation by natural selection. The process of adaptive radiation is well illustrated by the members of the Order Charadriiformes, which include terrestrial waders, aerial plungers, and wing-propelled divers. The radiation of Hawaiian honeycreepers illustrates the way in which bill forms can evolve in relation to feeding habits.

The classification of the kinds of birds of the world helps ornithologists to communicate with one another and serves as a tool in the continuing study of avian evolutionary relationships. Formal taxonomic classifications comprise a hierarchical series of inclusive categories that reflect the relationships among lineages. Orders, families, and genera are the principal taxonomic groupings of birds, all of which belong to the Class Aves.

The classification of birds is organized as a hierarchy of taxa—or groupings of species and populations of the Class Aves—comprising 29 orders, 187 families, over 2000 genera, and over 9600 species. These numbers, which reflect our knowledge and biases in 1994, will change with future research, changes in taxonomic philosophy, and discoveries of new taxa.

The birds of the world comprise geographical assemblages of species; these large groups are called avifaunas. The six major avifaunas of the world are the Nearctic (North America), Neotropical (Central and South America), Palearctic (Europe and Asia), Ethiopian (Africa), Oriental (Southeast Asia), and Australasian (Australia and New Guinea). Each region has its characteristic birds.

Further Readings

Page, J., and E.S. Morton. 1989. Lords of the Air. Washington, D.C.: Smithsonian Books. *An overview of the fundamentals of ornithology for general audiences; with photographs.*

Perrins, C.M., and A.L.A. Middleton. 1985. The Encyclopedia of Birds. New York: Facts on File. *A comprehensive, colorful volume of the biology of the birds of the world.*

Storer, R.W. 1971. Adaptive radiation of birds. Avian Biology 1: 149–188. *An elegant summary of the diversification of birds.*

C H A P T E R 2

History

THE DIVERSITY OF LIFE is the result of three evolutionary processes: phyletic evolution, which is the gradual change of a single lineage; speciation, the splitting of one phyletic lineage into two or more; and extinction, the termination of a lineage (see Figure 3–1). If we had a complete record of life on Earth, it would be possible to reconstruct accurately the historical patterns of speciation and phyletic evolution. Extinctions, however, fragment the historical record; they erase the connections between related lineages. The challenge of reconstructing the history of life belongs to a field of scholarly endeavor called systematics. Systematists are scientists who study and hypothesize evolutionary relationships among organisms through the comparison of fossils, preserved specimens, behavior, and, increasingly, the genetic code—DNA—of life itself.

The history of birds written in the geological record is brief compared with the histories of fishes and reptiles, but the ancestors of birds still reach back in time to the Mesozoic era, more than 150 million years ago. Avian history starts with the transformation of reptile ancestors into a feathered flying bird. This chapter first examines the reptilian features of birds, to prepare for a discussion of *Archaeopteryx lithographica*, a famous fossil creature that combined features of both reptiles and birds. Also presented are hypotheses about the evolution of feathers from scales, the specific reptilian ancestors of birds, and the evolution of avian flight. A fossil bird, *Sinornis santensis*, discovered in 1987 takes us one step closer to modern perching birds with full flight capabilities. This chapter then concludes with a brief review of the main stages of diversification of modern birds and an introduction to the process of geographical speciation that has been responsible for the diversity of species. Chapter 3 (Systematics) examines how systematists reconstruct evolutionary relationships among species.

Reptilian Features of Birds

Birds evolved from reptiles (Figure 2–1). Thomas H. Huxley, the great evolutionary biologist of the nineteenth century, asserted that birds were "merely glorified reptiles" and accordingly classified them together in the taxonomic category Sauropsida (Huxley 1867). Indeed, birds and modern reptiles share many characteristics (Figure 2–2). The skulls of both articulate with the first neck vertebra by means of a single ball-and-socket device—the occipital condyle; mammals, which evolved from a different line of reptiles, have two of these. Birds and modern reptiles have a simple middle ear that has only one ear bone—the stapes; mammals have three middle ear bones. The lower jaws—or mandibles—of both birds and modern reptiles are composed of five or six bones on each side; mammals have only one mandibular bone. The ankles of both birds and modern reptiles are sited in the tarsal bones (Figures 1–2 and 1–3), not between the long lower leg bone—or tibia—and tarsi as in mammals. The scales on the legs of birds are similar in structure to the body scales of modern reptiles.

Both birds and modern reptiles lay a yolked, polar egg in which the embryo develops by shallow divisions of the cytoplasm on the surface of the egg. Female birds and some female reptiles have the ZW sex chromosome combination and are referred to as heterogametic (Chapter 15); males are the heterogametic sex among mammals. Birds and reptiles both have nucleated red blood cells, whereas the red blood cells of mammals lack nuclei.

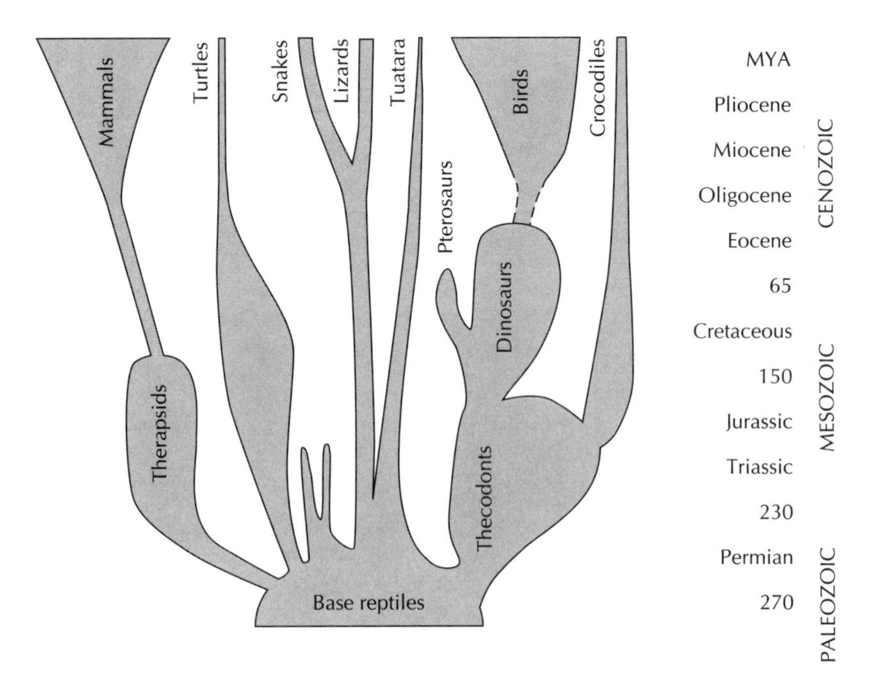

Figure 2–1 A simplified family tree of the vertebrates. Birds, dinosaurs and pterosaurs, and crocodiles evolved from one group of reptiles, the thecodonts; mammals evolved from another, the therapsids. Other groups of early reptiles gave rise to turtles, snakes and lizards, and the iguanalike tuatara now found only on little islands near New Zealand.

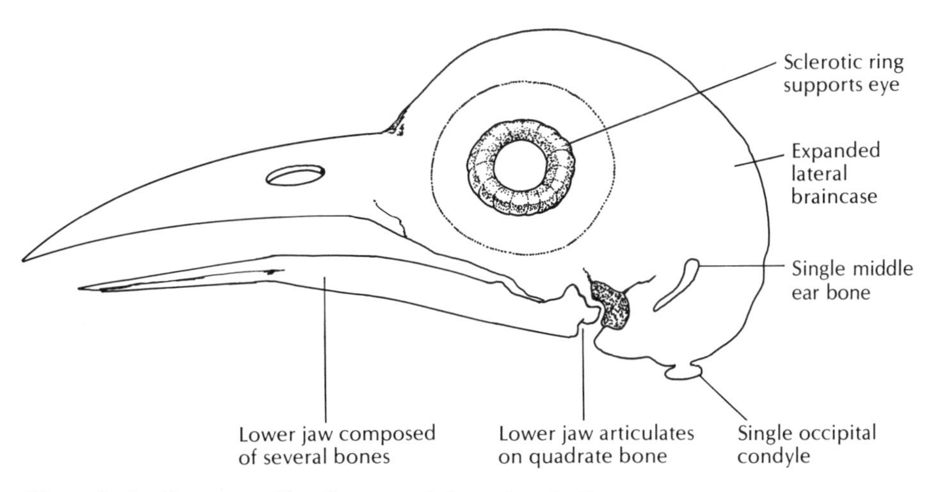

Figure 2–2 Some reptilian features of the avian skull.

Archaeopteryx: The Link Between Birds and Reptiles

The similarities between birds and reptiles leave no doubt of their evolutionary relationship. Yet we are not content with that. We want to know which reptiles gave rise to birds, and how the transformation proceeded. For this knowledge we must turn to the fossil record. There, an extinct creature, *Archaeopteryx lithographica*, reveals much about the origin of birds. The fossils of *Archaeopteryx* are the missing links that help paleontologists to reconstruct the evolution of the Class Aves.

Fine-grained limestone deposits in central Europe contain a record of creatures that occupied that region during the Age of Dinosaurs—in the late Jurassic period, 135 to 155 million years ago (Table 2–1). At that time, central Europe was tropical, sporting palmlike plants. Great, warm seas and lagoons covered parts of the European continent. The coastal lagoons attracted pterodactyls—flying reptiles—some as small as sparrows and others as large as eagles, which flew on batlike wings made of stretched skin. Sometimes they perished in the lagoons where gentle fossilization in the fine calcareous sediments preserved their features in exquisite anatomical detail. Also preserved in the same lagoons were the remains of the feathered reptile, now called *Archaeopteryx*.

The first evidence of the origin of birds was an impression of just a single feather found in a Bavarian quarry, from which Jurassic limestone was mined for lithographic slabs. The impression was brought to the attention of Hermann von Meyer of Munich in 1861. A complete skeleton of a small reptilelike animal with feathers also was found and brought to von Meyer's attention just a few months later. He named the fossil creature *Archaeopteryx* (*archios*, ancient; *pteryx*, wing) *lithographica*. The

TABLE 2–1
Geological time scale

Era	Period	Epoch	Million years before present
	Quaternary	Recent	0.01
		Pleistocene	
			1.5–3.5
Cenozoic (age of birds and mammals)	Tertiary	Pliocene	
			7
		Miocene	
			26
		Oligocene	
			37–38
		Eocene	
			53–54
		Paleocene	
			65
Mesozoic (age of reptiles)	Cretaceous	Late	
			100
		Early	
			135
	Jurassic	Late	
			155
		Middle	
			170
		Early	
			180–190
	Triassic		230

From Feduccia 1980.

discovery of a second complete specimen of *Archaeopteryx* in another quarry near Eichstätt, Bavaria, followed in 1877 (Figure 2–3). The first specimen is now in the British Museum; the second is in the Humboldt Museum für Naturkunde in Berlin. The specimen in the Humboldt Museum "may well be the most important natural history specimen in existence, perhaps comparable in value to the Rosetta Stone" (Feduccia 1980, p. 19). It is fully articulated, revealing details of the wing bones, flight feathers, and the pairs of feathers attached to each vertebra of its long tail. These feathers are indistinguishable from modern feathers.

Three of the other four fossils of *Archaeopteryx* are of poorer quality, so poor that one was misidentified (ironically, by Hermann von Meyer) as a pterodactyl. Another one found near Eichstätt in 1951 was misidentified as a small dinosaur, *Compsognathus*, until 1973 when Franz Mayr of Eichstätt noticed slight feather impressions. The last found specimen of

Archaeopteryx is a well-preserved one that was discovered in 1987 in a private collection in Solnhofen, Germany (Wellnhofer 1988). It is the largest of all known specimens.

Archaeopteryx was a crow-sized, bipedal "reptile" with a blunt snout and many small, reptilian teeth. It bore feathers on both wings and tail, and probably also over most of its body, like modern birds. It possessed, however, numerous reptilian features (Figure 2–4). Like the modern guans (Cracidae), it may have been a strong-running, terrestrial "bird" that could leap into trees, jump among large branches, and make short flights between trees. Paleontologists debate whether *Archaeopteryx* was

Figure 2–3 This fully articulated skeleton of *Archaeopteryx lithographica* was found in 1877 near Eichstätt, Bavaria. (Courtesy of J. Ostrom)

actually warm-blooded—or endothermic—like modern birds (Houck et al. 1990; Ruben 1991).

Paleontologists agree, however, that *Archaeopteryx* was a semiarboreal creature capable of gliding and weak flapping but not of long, sustained flights. It had strongly curved claws, like those of perching and climbing modern birds, not the long flattened claws of ground birds (Feduccia

Figure 2–4 Skeletal features of (A) the reptilelike *Archaeopteryx* and (B) a modern bird, the domestic pigeon. In modern birds, (1) the braincase is expanded and the head bones are fused; (2) the separate hand bones of reptiles are also fused into fewer rigid elements in birds; (3) the separate pelvic bones of reptiles are fused into a single, sturdy structure in modern birds; (4) the many tail vertebrae of *Archaeopteryx* are reduced in number and partially fused into a pygostyle in modern birds; (5) the tiny, cartilaginous sternum of *Archaeopteryx* has expanded in modern birds to a large, keeled bony structure for the attachment of flight muscles; and (6) the typical reptile rib cage is strengthened with horizontal uncinate processes. (From *Evolution of Vertebrates* by E.H. Colbert. Copyright 1955 John Wiley & Sons, Inc.; reprinted by permission of John Wiley & Sons, Inc.)

1993a). *Archaeopteryx* probably could not launch itself into the air from the ground because it lacked the principal (supracoracoideus) muscles that lift the wing in a rapid recovery stroke. Even so, these deficiencies do not preclude powered flight after, for example, jumping from a tree branch (Olson and Feduccia 1979). Indications of *Archaeopteryx*'s flight capability include its large furcula, which probably anchored strong pectoralis muscles, and the acute angle of its scapula, which supported dorsal elevator muscles that helped to lift the wings, as they do in modern flying birds. Also, the vanes of *Archaeopteryx*'s primary wing feathers were asymmetrical, a characteristic shared by nearly all flying birds and most pronounced in strong fliers (Feduccia and Tordoff 1979). In flightless birds, these vanes are symmetrical (Figure 2–5).

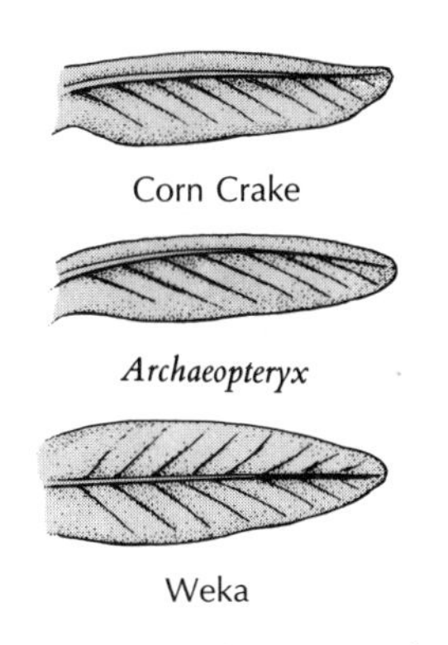

Figure 2–5 The vanes of the primaries of *Archaeopteryx* were asymmetrical like those in modern flying birds, such as the Corn Crake, a kind of rail, not symmetrical like those of flightless birds, such as the Weka, a flightless rail of New Zealand. The asymmetry has an aerodynamic function and presumably evolved in relation to flight in this primitive bird. (After Feduccia and Tordoff 1979)

The discovery of *Archaeopteryx* linked the evolution of birds directly to reptiles. Some biologists think these specimens may still be the best fossil proof of the process of evolution (Ostrom 1976). *Archaeopteryx* was a reptile with several of the characteristics of birds. It was a timely discovery of an animal that was intermediate between two higher taxonomic categories, a transition from ancestral to descendant stocks. It seemed that Darwin's prediction of intermediate evolutionary links in *On the Origin of Species by Means of Natural Selection* (1859), published only two years prior to the discovery of the first two fossils, had been fulfilled.

The intermediate morphology of *Archaeopteryx* quickly moved into the center of the debate between opponents and supporters of evolution by natural selection. Creationists, defending their views of the separate and unchanging appearances of birds and reptiles, insisted that Darwinists were misinterpreting the apparent intermediacy of *Archaeopteryx*. On the other side of the debate, Thomas H. Huxley, Darwin's most eloquent champion, was convinced of the link between birds and birdlike reptiles and soon converted leading American paleontologists. Charles Marsh of Yale University was one of these converts. He wrote:

> The classes of Birds and Reptiles, as now living, are separated by a gulf so profound that a few years since it was cited by the opponents of evolution as the most important break in the animal series, and one which that doctrine could not bridge over. Since then, as Huxley has clearly shown, this gap has been virtually filled by the discovery of bird-like Reptiles and reptilian Birds. *Compsognathus* and *Archaeopteryx* of the Old World . . . are the stepping stones by which the evolutionist of to-day leads the doubting brother across the shallow remnant of the gulf, once thought impassable. (Marsh 1877, p. 352; Feduccia 1980, p. 15)

Archaeopteryx contributed to the acceptance of Darwin's theory of evolution as well as to our understanding of the origin of birds. Still unanswered, however, were many key questions. What more does *Archaeopteryx* reveal about the ancestry of birds? Do its features suggest which reptiles were the ancestors of birds?

Evolution of Feathers

Archaeopteryx was a feathered reptile. We presume that feathers evolved from scales of some kind (Waterman 1977), but we are not certain exactly what advantages promoted the evolution of feathers from scales. One hypothesis is that feathers first aided flight (Parkes 1966; Feduccia 1993b). Slightly elongated and then frayed scales on the trailing edges of the forelimbs possibly enhanced either primitive gliding or parachuting. As gliding abilities improved and steering requirements increased, the hypothesis continues, so did the elaboration of feathers on the wings and tail. Reptile experts favor a different, physiological hypothesis (Regal 1975). They suggest that scales first became more featherlike as temperature regulation devices, particularly as heat shields that enabled the ancestors of birds to be more active in hot, sunny habitats. Modern lizards

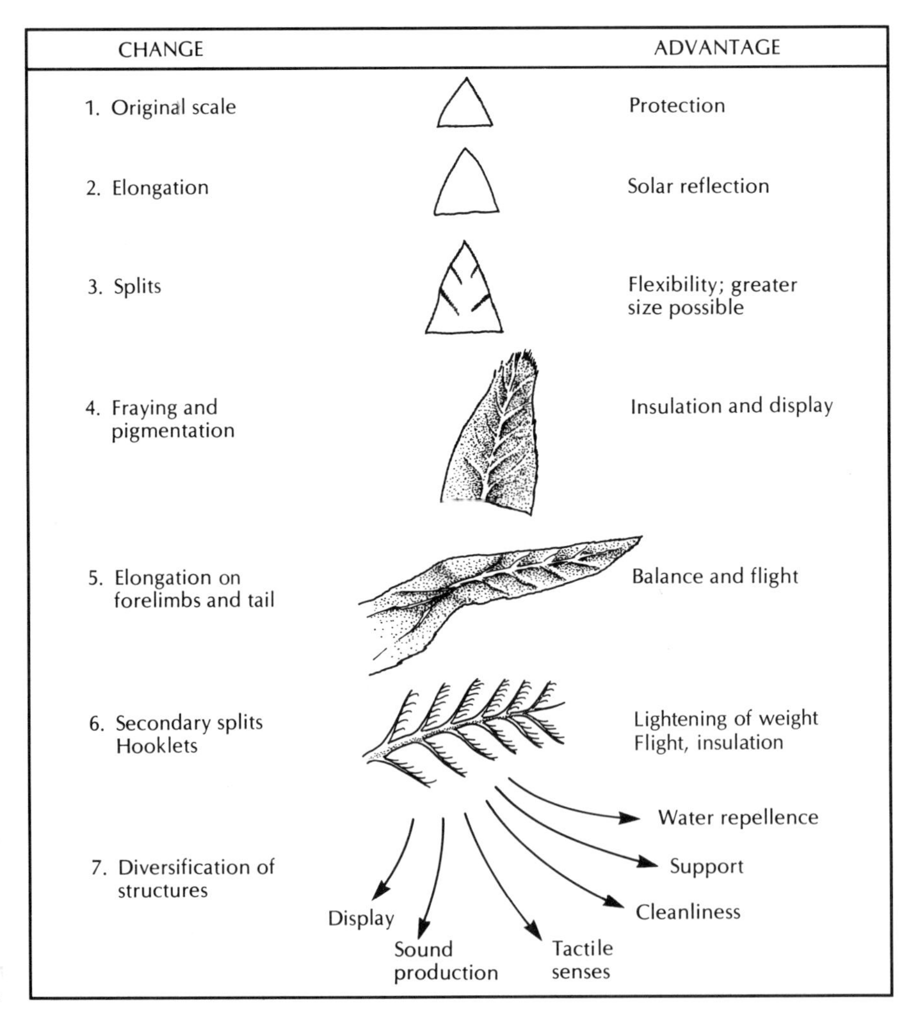

Figure 2–6 Hypothetical steps in the evolution of feathers from scales. Elongation and splitting of reptilian scales aided reflection of solar heat and permitted larger, flexible scales. Increased fraying and pigmentation of the large scales contributed to insulation and to displays. Elongation of feathers on the forelimbs and tail improved balance on extended leaps and, ultimately, flight. Secondary splits that led to the evolution of branches with interlocking hooklets gave rise to the modern light feather structure that aids flight and insulation. In addition, this versatile structure is easily modified for special purposes, including sound production, tactile sensation, support, and water repellency.

that live in hot, sunny climates tend to have large scales that reduce heat loads. Featherlike fraying of a scale's edges would increase its flexibility and effectiveness as a heat shield during the day and as insulation at night. Once lengthened and frayed, the advantages of the novel, flexible, lightweight structure of the protofeather coupled with its usefulness in thermoregulation and primitive flight apparently catalyzed the rapid evolution of definitive feather structure in the ancestors of birds (Figure 2–6).

Reptilian Ancestors of Birds

There is little doubt that birds evolved from some line of Mesozoic reptiles. Which line is still a matter of debate. One possibility is that birds evolved directly from a group of reptiles called thecodonts. Another more likely possibility is that birds evolved from small theropod dinosaurs (Figure 2–7).

The thecodontian theory of the origin of birds looks to a large group of primitive reptiles that prevailed in the early years of the Mesozoic era. Among these were the lightly built thecodonts, a diverse group of reptiles that gave rise to dinosaurs of different sorts, some of which were arboreal;

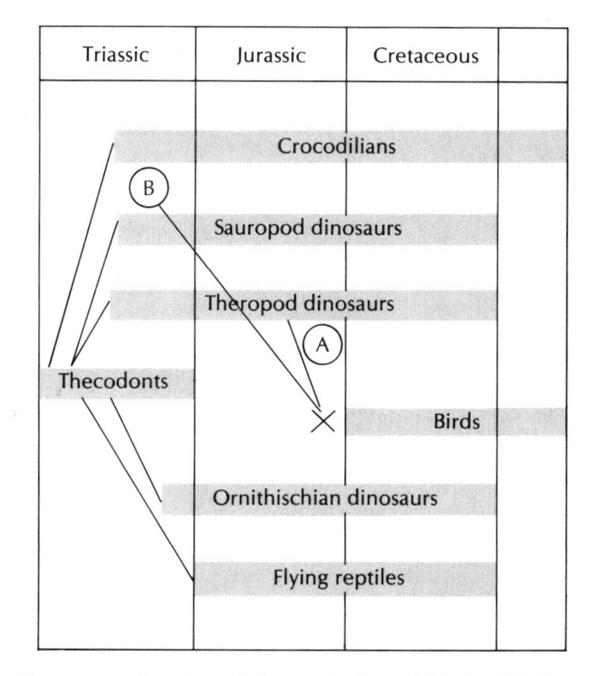

Figure 2–7 Alternative theories of the evolution of birds. (A) Some experts believe that birds evolved from small theropod dinosaurs. (B) Other experts believe that birds evolved directly from the thecodont ancestors of dinosaurs and crocodiles. (After Ostrom 1975)

to flying reptiles called pterosaurs; and to crocodiles. Some thecodonts, such as *Longisquama*, had elongated scales that seemed like the natural precursors of feathers (Figure 2–8B).

The early works of Maximillian Fürbringer (1888), Robert Broom (1913), and especially Gerhard Heilmann (1927) favored the theory of thecodontian ancestry of birds. More recently, Samuel Tarsitano and Max Hecht (1980) and Larry Martin (1983) also found support for this theory. Fourteen characters seem to unite modern birds and crocodilian thecodonts. This theory suggests that birds split early from the main stem of crocodilian evolution by the middle Triassic period. If so, critics ask, why is there a gap of 90 million years in the fossil record between crocodilian thecodonts and *Archaeopteryx*?

Potentially closing that gap is a controversial new fossil named *Protoavis* from the late Triassic period, 75 million years before *Archaeopteryx* (Chatterchee 1991). Excavated in western Texas in 1961, *Protoavis* made headlines because of its importance to the competing theories about the origin of birds. Its heralded avian features, however, have not been verified. The poor condition of *Protoavis* leads most experts to doubt that it will ever be proved to be part bird (Monastersky 1991).

The alternative, dinosaur theory of the origin of birds is attractive because dinosaurs and *Archaeopteryx* coexist in the fossil record. Although we usually think first of the large, spectacular species, dinosaurs varied greatly in size and habits. One group of dinosaurs that evolved from the thecodont reptiles, the theropod dinosaurs, included not only large carnivores such as *Tyrannosaurus rex* but also many small ones—called coelurosaurs—that were close in size to modern iguanas. These were agile, lightly built, bipedal, little dinosaurs with many small, sharp teeth. They probably chased small vertebrates and large insects. Some may even have been warm-blooded. These small dinosaurs figure prominently in Michael Crichton's book *Jurassic Park*.

The hypothesis that birds evolved from small theropod dinosaurs goes back to the early debates that followed the discovery of fossil *Archaeopteryx*. Thomas H. Huxley (1868) was particularly impressed by the similarities between *Archaeopteryx* and *Compsognathus*, a small dinosaur preserved in the same limestone deposits (Figure 2–8A). Recent studies by John Ostrom support Huxley's theory of a dinosaur ancestry of birds. *Archaeopteryx* and small dinosaurs share 23 of 42 specialized skeletal features of the hand, vertebrae, humerus and ulna, pectoral arch, hindlimb, and pelvis. Ostrom noted:

> It has repeatedly been observed that the *Archaeopteryx* specimens are very birdlike, but also possess a number of reptilian features. . . . The actual fact is that these specimens are not particularly like modern birds at all. If feather impressions had not been preserved in the London and Berlin specimens, they would never have been identified as birds. Instead, they would unquestionably have been labelled as coelurosaurian dinosaurs. (Ostrom 1975, p. 61)

(A)

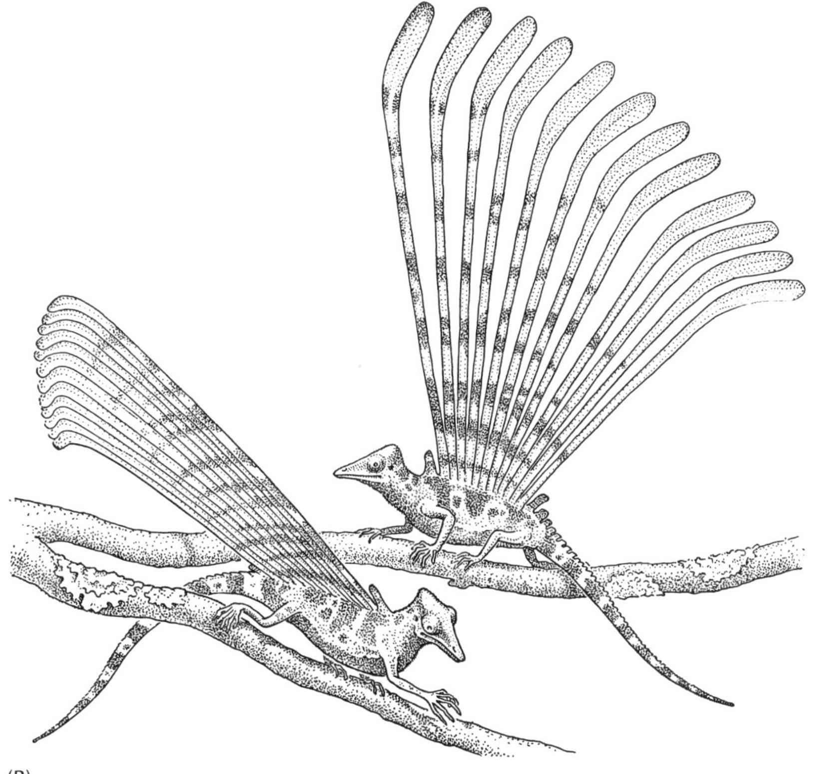

(B)

Figure 2–8 Two possible relatives of birds. (A) *Compsognathus* was a small theropod dinosaur that was preserved in the same limestone deposits as *Archaeopteryx*. (B) *Longisquama* was a lightly built, arboreal thecodont reptile with elongated scales. Whether a theropod or a thecodont reptile was the ancestor of birds remains uncertain. (A from Heilmann 1927; B from Bakker 1975, with permission of Scientific American, Inc.)

Supporting this conclusion also is the discovery that juvenile dinosaurs and birds share a distinctive anatomical feature of their leg bones: growth plates with unique cellular structures (Barreto et al. 1993). These growth plates are disks of cartilage that are present near the ends of long bones and are responsible for rapid rates of leg bone elongation. Broader surveys of the distribution of "dinosaur" growth plates are now needed to confirm their value as characters that link birds to dinosaurs.

Evolution of Avian Flight

Whatever the ancestor of *Archaeopteryx*, the major features of modern birds would evolve next. In particular, the evolution of modern, powered flight overhauled both the aerodynamic structures of the body and the physiology that provided energy. These changes opened the door for the diversification of birds. The evolution of avian flight and how well *Archaeopteryx* could fly have long been debated among ornithologists. One wonders, for example, what caused the forelimbs of the reptilian

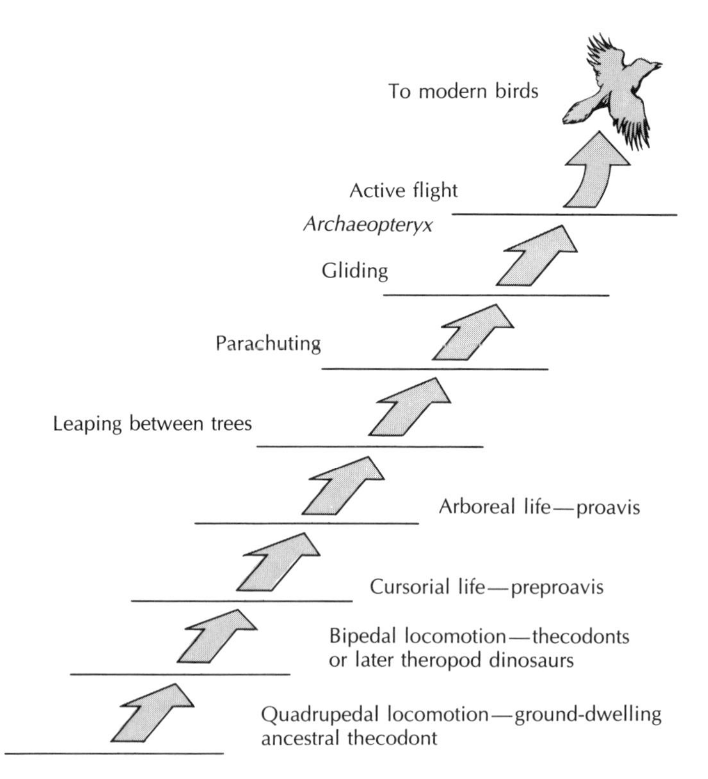

Figure 2–9 The arboreal theory of the evolution of avian flight suggests that, after evolving bipedal locomotion, the reptilian ancestors of birds became arboreal and leaped between trees. Active flight evolved from earlier stages of parachuting and gliding flight that enhanced the leaping abilities of the ancestors of *Archaeopteryx*. (After Feduccia 1980; adapted from Bock 1965)

ancestor of *Archaeopteryx* to evolve into protowings in the first place. Two theories exist: an arboreal theory and a cursorial theory.

The arboreal theory proposes that the evolution of flight started with parachuting and gliding from elevated perches. Most drawings and models of *Archaeopteryx* depict an arboreal reptile clambering around trees, grasping branches with clawed fingers. The arboreal theory proposes that extensions of the bones of the forelimb enhanced by elongated (flight) feathers enabled the ancestor of *Archaeopteryx* to parachute and glide between trees. The arboreal theory has been the favored theory for many years (Bock 1965; Feduccia 1980) (Figure 2–9).

The cursorial theory proposes that forelimbs first elongated because they heightened leaping ability in a small bipedal theropod dinosaur that ran and jumped to catch insects in its jaws. Extensions of the forelimbs would help to control and extend its leaps (Caple et al. 1983, 1984). Elongation of three extensions of the body—such as two wings and a tail—would not enable flight at first but would have a dramatic effect on the control of the body's roll and pitch. Elongation of these radii would lead to greater feeding volumes, enhanced maneuverability, and higher velocities of running and jumping. The cursorial theory provides an initial series of adaptive steps based on the path of a projectile—called trajectory ballistics—rather than the aerodynamics of true flight. The flight capabilities of modern birds seem, in this light, a logical extension of the first small jumps by little dinosaurs. Protowings, increased arboreal habits, and gliding with feeble flapping as proposed for the life style of *Archaeopteryx* would be the next logical evolutionary steps.

After *Archaeopteryx*

The fossils of *Archaeopteryx* are our principal evidence of the transition from reptiles to birds. The early fossil record of the Class Aves provided few additional clues of the transition to modern birds until the discovery in 1987 of another amazing fossil bird; this one was found in China and dates from the early Cretaceous period, 140 million years ago (Sereno and Chenggang 1992). *Sinornis santensis* was a toothed, sparrow-sized bird with many features of theropod dinosaurs and *Archaeopteryx*, a pattern suggesting that it occupied a basal position in the evolutionary tree of birds (Figure 2–10). *Sinornis* also exhibited features intermediate between *Archaeopteryx* and modern perching birds. Advances over the semiarboreal capabilities of *Archaeopteryx* include strengthening and modifications of the hand, forearm, and pectoral girdle for flight functions; the ability to raise the wings high over the body and also to fold them; a large pygostyle for support of a tail fan, which improves steering and braking in flight; and a perching foot with an opposable rear toe, the hallux. These features suggest that avian flight and perching abilities evolved in small-bodied birds that followed *Archaeopteryx*.

Better known than *Sinornis* is a side branch of toothed seabirds—*Hesperornis*, *Ichthyornis*, and their relatives in the extinct Order

Figure 2–10 *Sinornis santensis*, a new fossil bird from China, which is intermediate between *Archaeopteryx* and modern perching birds. (After Sereno and Chenggang 1992)

Hesperornithiformes—which thrived in the late Cretaceous period. Some of these seabirds superficially resembled modern loons. Their size ranged from that of a small chicken to that of a large penguin. The largest was *Hesperornis regalis*, 1 to 2 meters in length. All 13 known species of divers were flightless, with vestigial wing bones. They had large, powerful, lobed feet and inhabited the Cretaceous seas that covered the central portions of the North American continent. Flying above the same shallow seas were toothed, ternlike birds (*Ichthyornis*), of which six species are known from the fossil record. Both groups of toothed Cretaceous birds had well-formed, unserrated, reptilian teeth with constricted bases and expanded roots set into distinct sockets in their bony jaws. None of these lineages has survived. These toothed birds disappeared along with dinosaurs in the cataclysmic period of mass extinction that marked the end of the Mesozoic era. Among the few survivors were the ancestors of toothless modern birds.

Modern Birds

The Tertiary period that followed the Mesozoic era produced some huge carnivorous, semimodern birds that temporarily occupied some of the niches left vacant by bipedal dinosaurs. Two-meter-tall diatrymas with powerful legs, clawed toes, massive horse-sized skulls, and tearing, eaglelike beaks must have terrorized many lesser creatures before becoming extinct in the Eocene epoch (Figure 2–11). In the Eocene, long-legged vulturelike birds (*Neocathartes*) lived in Wyoming beside shorebirds with ducklike heads (*Presbyornis*). From the Oligocene to the Pliocene epochs, 12 known species of phorusrhacids—predatory birds 2 to 3 meters tall, with powerful, rapacious bills—ranged throughout South America and north to Florida. As recently as the last ice age, huge vulturelike teratorns dominated the skies. One teratorn with a 4-meter wingspan was abundant in southern California. Another known from caves in Nevada had a wingspan of 5 to 6 meters, and yet another, recently discovered in Argentina, had an 8-meter wingspan. They were the size of small planes! These enormous birds symbolize some of the extremes of past avian achievements.

The orders of modern birds diverged from one another approximately 60 million years ago, or early in the Tertiary period. Specialized water birds, such as loons, auks, gulls, ducks, cranes, and petrels, invaded aquatic niches during the Eocene epoch, 37 to 54 million years ago.

Figure 2–11 Large flightless birds flourished during the Tertiary period. These included diatrymas. (From Heilmann 1927)

Primitive woodpeckers and their relatives also appeared during the early Eocene and became the predominant perching birds during the Miocene epoch. Relatives of rollers, kingfishers, and hornbills diversified in the Oligocene epoch. In the Miocene, the rapid evolution of flowering plants and insects opened new niches for insect-eating, fruit-eating, and nectar-feeding birds; this ecological diversity resulted in an explosive radiation of songbirds (Regal 1977). By the end of the Tertiary, 5 to 10 million years ago, birds had diversified into a broad range of forms that included many modern genera.

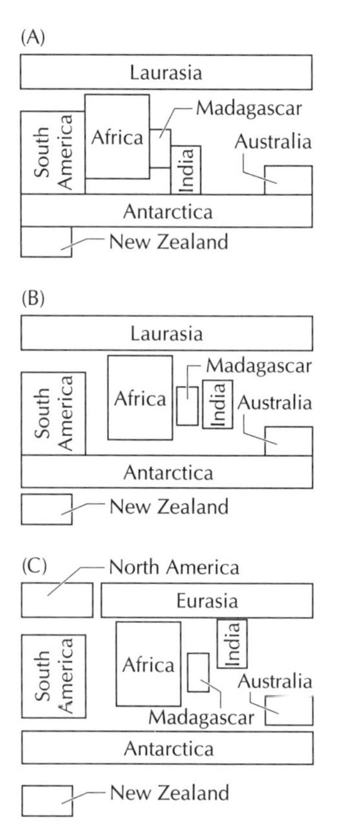

Figure 2–12 Schematic diagram of past configurations of the continents. (A) Once combined into two supercontinents, a northern Laurasia and a southern Gondwanaland, the continents drifted apart during the (B) Cretaceous periods and (C) Eocene epoch of the early Tertiary into the present configuration. (From Cracraft 1974b)

The History of Avifaunas

The early diversification of birds took place on a very different Earth; neither the arrangement of the continental landmasses nor the climates resembled those of today. Through much of the Tertiary period, the world's climates were warm from pole to pole; there was no striking polar gradient from frigid to hot as there is today. For example, during the late Eocene and early Oligocene epochs (see Table 2–1 for the geological time scale), subtropical to tropical climates with abundant precipitation and no frost prevailed in the far north of both North America and Eurasia (Cracraft 1973a). The floras of Great Britain and of western Europe in the early Eocene resembled the modern rain forests of Southeast Asia. Tropical birds—trogons, parrots, hornbills, barbets, broadbills, and mousebirds—once lived in central Europe. Alligators and large tortoises lived on Ellesmere Island above the Arctic Circle.

Joel Cracraft (1974b) postulates that the adaptive radiation of modern songbirds took place primarily on the landmass of the far north during the early Cenozoic era and then spread southward into Africa, Australasia, and North America. Passerines in each major area then underwent their own adaptive radiations.

The arrangements of continents and their connections have changed over the course of avian evolution (Figure 2–12). The modern continents have been moving apart since the late Jurassic and Cretaceous periods. Much of the major reorganization during the Mesozoic era of the single great landmass known as Pangaea, with Laurasia in the north and Gondwanaland in the south, preceded the evolution of modern bird taxa. Nevertheless, modern orders of birds were present during the next stages of continental drift. Brazil had separated from Africa by the late Cretaceous, and India had split from Antarctica and was moving north to collide with Asia. From the late Cretaceous period (65 to 70 million years ago) of the Mesozoic era and into the Paleocene epoch of the Cenozoic era, the northern landmasses (North America, Europe, and Asia) were broadly connected as a single continent, called Laurasia. A continuous land connection existed between northeastern Asia (modern Siberia) and northwestern North America (modern Alaska) during much of the Tertiary period.

Throughout the Tertiary period, birds moved through the tropical–subtropical or warm temperate forests that covered Eurasia and North America. At first, movement was easy across a broad North Atlantic land bridge. After the separation of North America from Europe during the Eocene epoch, the Bering land bridge became the main corridor for faunal exchange. Wet tropical forests also covered most of South America during the Tertiary period. The rising of the Andes mountains in the late Tertiary period created a cooler, drier climate that favored the grasslands of Argentina and the coastal deserts of modern Chile and Peru. Even Antarctica supported lush temperate forests of southern beech, conifers, and palms during the Eocene epoch and into the Miocene epoch.

Coupled with favorable climates, the Gondwanaland association of the modern Southern Hemisphere continents fostered the exchange of taxa among Africa, South America, and Australia. Dispersal between Australia and South America was once possible via the reasonable climates and forests of Antarctica. Ancestral fowl—moundbuilders in Australia, cracids in South America, and guineafowl in Africa—appear to have originated in the main parts of Gondwanaland (Cracraft 1973a). Radiations of pheasants, partridge, New World quail, and grouse in North America came after the northward expansion of ancestral groups into Laurasia and the separation of Laurasia into North America and Eurasia in the Eocene. Representatives of these radiations on the continents of the Northern Hemisphere subsequently moved southward into Africa and the East Indies.

At the close of the Tertiary period, the Earth's climates cooled, especially in the polar regions, and became more strongly seasonal. During the Pleistocene epoch that followed, starting 2 to 3.5 million years ago, climatic changes and glaciers drastically altered the distributions of birds throughout the world. Alternating dry–cool and wet–warm climates split the geographical ranges of birds and promoted both speciation and extinction. Tropical birds became restricted to equatorial latitudes and, even there, were sometimes concentrated in limited refuges as the alpine zone extended to lower elevations and as grasslands expanded during dry, glacial periods. In Africa, montane conditions extended to as low as 500 meters above sea level during glaciations, reducing the lowland forests to less than half of their area today (Mayr and O'Hara 1986). Later warm periods reversed this pattern, thereby restricting montane species to isolated mountaintops.

New geographical or habitat connections accompany changing global climates, permitting fusions of isolated avifaunas. The birds of Central America, for example, include North and South American species. South American birds expanded into Central America with the northward movement of tropical rain forests in the late Pleistocene epoch. There was a reciprocal southward movement of some North American birds into the rain forests of South America.

The islands of the West Indies have been continuously colonized, despite the area's substantial open-water barriers (Bond 1948, 1963).

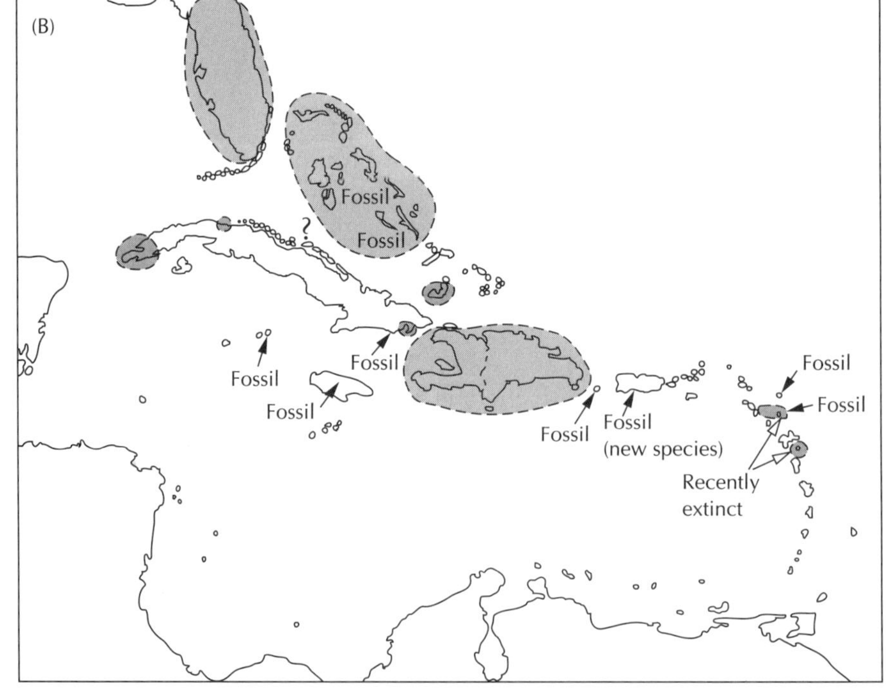

Figure 2–13 Fossils reveal that Burrowing Owls (A) were once more widely distributed in the West Indies. (Courtesy of A. Cruickshank/VIREO) (B) Their current distribution is indicated by shading. (From Pregill and Olson 1981)

Both ancient and modern contributions are apparent among the birds of this avifauna. The original elements of the modern West Indies avifauna came primarily from Central America, followed more recently by species from South America. Thirty-three species of South American land birds entered the West Indies via the Lesser Antilles. Fifteen of these remain on the fringes of the West Indies, not having moved past Grenada and St. Vincent. Thirteen have moved to the middle of the central part of the Lesser Antilles, but only three have reached eastern Puerto Rico.

The climates and principal habitats of the West Indies have changed in recent times. Dry climates prevailed in the West Indies during the Pleistocene epoch, enabling species in grasslands, savannahs, and desert habitats to be widely distributed (Pregill and Olson 1981). Increases in annual rainfall during the last 10,000 years have caused many of these birds to go extinct. Today, only relict—remnant—populations of Burrowing Owls, for example, exist (Figure 2–13). Fossils indicate they were once widespread in the West Indies. Caracaras, thick-knees, and the Bahama Mockingbird of the dry country once were widely distributed in the West Indies, but they too exist now only as isolated relicts.

Speciation

The evolutionary legacy of the earliest birds is believed to include roughly 100,000 species, of which only 1 in 10 is now with us. Behind this legacy lies the process of speciation, the multiplication of species through division of one species into two or more as a result of the genetic divergence of isolated populations. Geographical separation of populations reduces the exchange of genes, thereby allowing independent

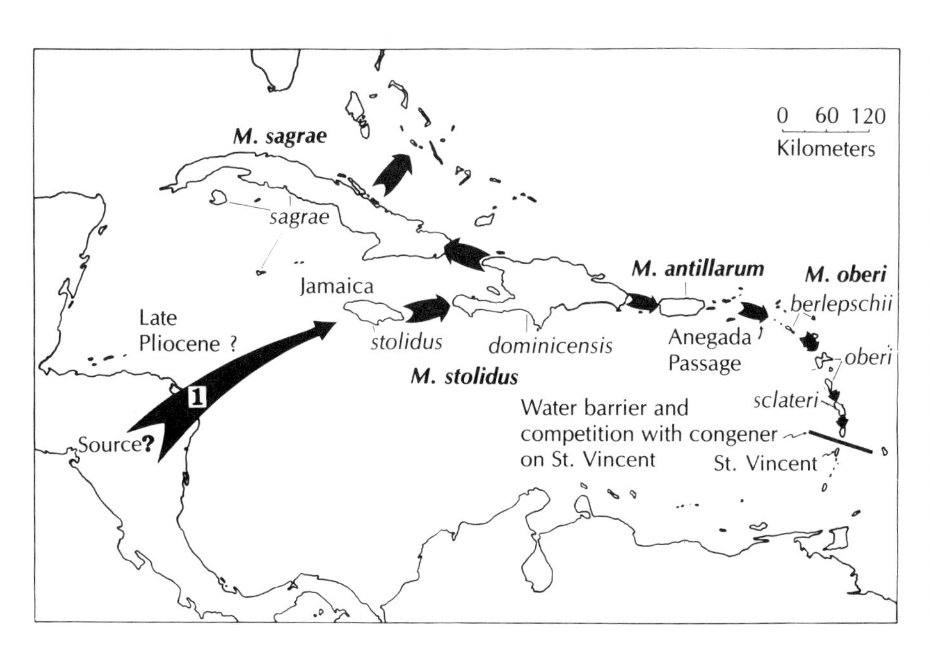

Figure 2–14 Historical invasions account for *Myiarchus* flycatchers now found in the West Indies. An invasion of Jamaica from Central America in the late Pliocene, perhaps, was followed by the spread and differentiation of four related species (Latin names in boldface): Stolid Flycatcher (*M. stolidus*), La Sagra's Flycatcher (*M. sagrae*), Puerto Rican Flycatcher (*M. antillarum*), and Lesser Antillean Flycatcher (*M. oberi*). The subspecies names (in lightface) of divergent island populations of each of these species are also indicated. (From Lanyon 1967)

divergence and enabling speciation. Most species of birds evolve as geographical isolates.

Bird populations become isolated in two principal ways. First, pioneering individuals may colonize an oceanic island, for example, that is separated from their main population on the mainland or other islands. For instance, flycatchers of the genus *Myiarchus*, which includes the Great Crested and Ash-throated Flycatchers of the continental United States, repeatedly invaded the West Indies. The first invasion from Middle America in the late Tertiary led to the development of the Stolid Flycatcher on Jamaica and Hispaniola (Figure 2–14). Subsequent expansions of the Stolid Flycatcher from Hispaniola, Wesley Lanyon (1967) hypothesized, led to the evolution of two other species, the La Sagra's Flycatcher and the Puerto Rican Flycatcher in the Greater Antilles and, more recently, to the Lesser Antillean Flycatcher in the Lesser Antilles.

Classic examples of divergence and speciation come from remote islands such as the Galápagos and Hawaiian archipelagos. The birds on the Channel islands off the coast of southern California also are distinct, as are the kingfishers on small satellite islands off the coast of New Guinea. Islands of special habitats, such as desert oases or subalpine mountain forests, may set a similar stage for divergence and speciation of the populations that occupy them.

Alternatively, fragmentation of habitats that were once continuous may also isolate bird populations. Some ornithologists believe that the dry cold climates of the Pleistocene epoch, for example, shrank the great Amazonian rain forests into much smaller fragments surrounded by grasslands. Restricted to these forest refuges, toucans, manakins, and flycatchers were among the many kinds of birds that underwent speciation (Haffer 1969, 1974) (Figure 2–15). In North America also, the distributions of closely related bird species, such as the Rose-breasted versus Black-headed Grosbeak, reflect past glacial fragmentation of their ranges into eastern and western components (Mengel 1970).

Remnant populations are one of the major consequences of historical habitat changes. Ostriches, now restricted to Africa, once roamed throughout Asia. Todies, which are tiny, colorful relatives of kingfishers, are presently found only on the Greater Antilles of the West Indies but once lived also in Wyoming and France (Olson 1976, 1985). The extremely localized distributions of many birds of New Guinea are most likely a result of gradual extinction throughout most of the former range and of persistence of the species in a few favorable sites (Diamond 1973, 1975). The Obscure Berrypecker, for example, now lives in only two localities at opposite ends of New Guinea (Figure 2–16). Widely separated areas may consequently share peculiar taxa. The African River-Martin inhabits the Congo River basin, whereas the closely related White-eyed River-Martin inhabits only Thailand; no related species are found between these locations. The scrub jays of Florida are separated from other populations in the southwestern United States by over 3000 kilometers.

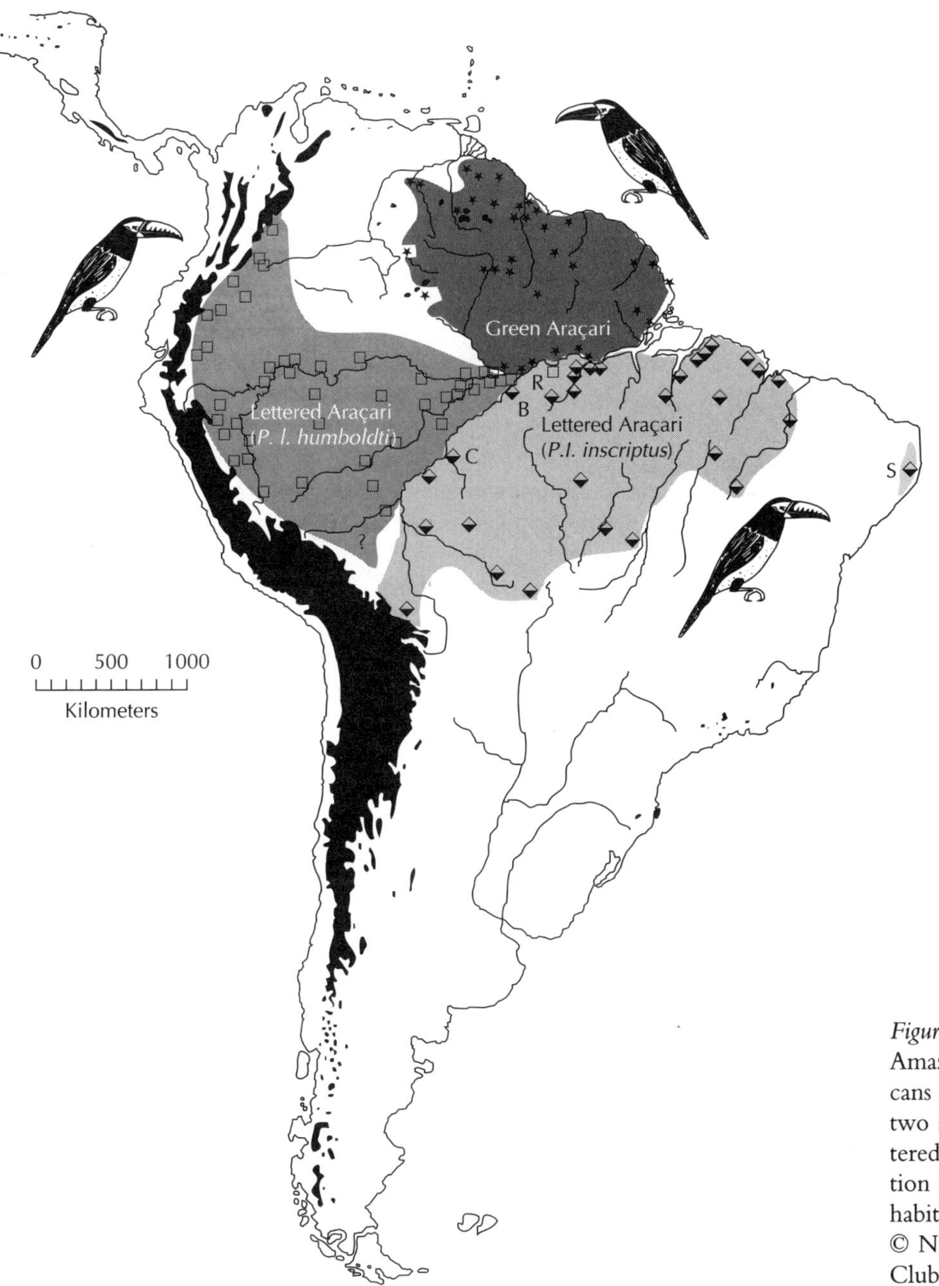

Figure 2–15 The ranges in Amazonia of three small toucans (the Green Araçari and two subspecies of the Lettered Araçari) reflect past isolation in refuges of wet forest habitats. (After Haffer 1974. © Nuttall Ornithological Club)

Although the general patterns of geographical speciation in birds are well known, the details of the process of speciation are not. Slow adaptive divergence of fragments of large populations or rapid genetic reorganization in small populations appear to be the primary modes of speciation.

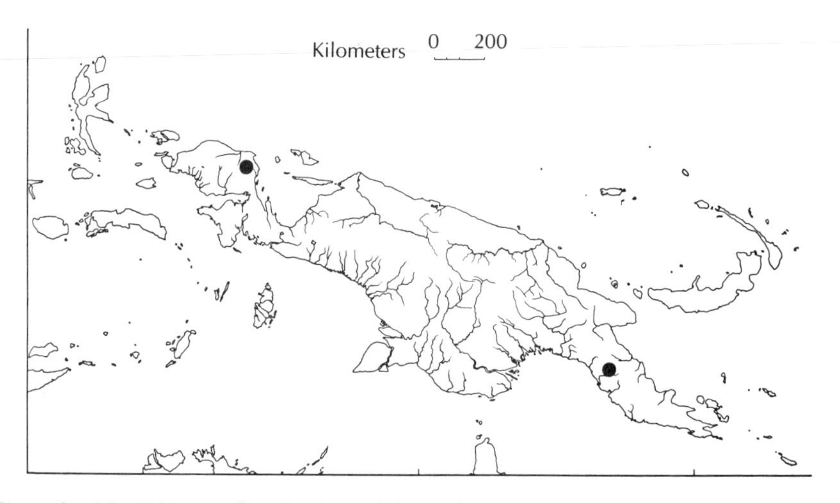

Figure 2–16 Disjunct distribution of the Obscure Berrypecker, which inhabits two sites (black circles) at the opposite ends of New Guinea. (Adapted from Diamond 1975)

The process of speciation, which is discussed in detail in Chapter 22, may involve behavioral transformation as well as genetic transformation. Still to be resolved is the relative importance of ecological and social adaptations, as well as the timing and nature of the genetic changes related to them.

Summary

Birds evolved from small, bipedal reptiles more than 150 million years ago in the Mesozoic era. Birds share with reptiles many anatomical features that distinguish them from mammals, including a single occipital condyle on the back of the skull, a single middle ear bone, and nucleated red blood cells.

Archaeopteryx lithographica, one of the most important fossils of all time, was a crow-sized, toothed, bipedal reptile with two essential avian features: feathers and a furcula ("wishbone"). It could clamber around trees and could fly. Known from five specimens preserved in fine limestone deposited in the late Jurassic period in Bavaria, it represents an evolutionary link between birds and reptiles. Because of its timely discovery, *Archaeopteryx* fostered acceptance of Darwin's theory of evolution. Exactly which group of reptiles was the ancestor of *Archaeopteryx*, and hence which gave rise to birds, remains uncertain. Small theropod dinosaurs are the most likely candidates.

A new fossil species, *Sinornis santensis*, found in China in 1987 illustrates further changes toward a modern perching bird. *Sinornis santensis* retained many primitive features that place it in a basal position in the evolution of birds. Its advanced features, however, suggest that avian flight and perching abilities evolved in small-bodied birds.

Once established as a new group of vertebrates, birds diversified in both form and function. A variety of toothed birds established themselves in the late Cretaceous period but then went extinct. The modern orders of birds diverged from one another 60 million years ago, near the beginning of the Tertiary period. Water birds multiplied in variety in the Eocene epoch, as did land birds in the Miocene epoch. The characteristics that distinguish modern species evolved during the last 2 million years, although some living species seem to have changed little from their fossil ancestors tens of millions of years old.

Bird species multiply through the isolation of populations, which acquire new behavioral, ecological, and genetic characteristics. Geographical isolation usually results either from the colonization of a remote place, such as an island, or through fragmentation of habitats as a result of climatic changes.

Avifaunas are the grand result of millions of years of evolution, adaptive radiation, dispersal, and extinction of avian taxa with varied ecological roles. Throughout the history of avian evolution, neither the world's climates nor the arrangement of the continents were as we know them today. Many modern birds occupy only remnants of their original distributions.

Further Readings

Cracraft, J. 1973. Continental drift, paleoclimatology, and the evolution and biogeography of birds. J. Zool. (Lond.) 169: 455–545. *A detailed review of the modern perspective.*

Cracraft, J. 1986. The origin and early diversification of birds. Paleobiology 12: 383–399. *An analysis of the relationships among the earliest fossil birds.*

Feduccia, J.A. 1980. The Age of Birds. Cambridge, Mass.: Harvard University Press. *A well-illustrated overview of the evolution of birds, with particular attention to the fossil record.*

Gauthier, J. 1986. Saurischian monophyly and the origin of birds. *In* The Origin of Birds and the Evolution of Flight, K. Padian, Ed., pp. 1–55. Memoirs of the California Academy of Sciences, No. 8. *A detailed analysis of the phylogenetic relationships of reptiles and birds.*

Hecht, M.K., J.R. Ostrom, G. Viohl, and P. Wellenhofer. 1985. The Beginnings of Birds. Willibaldsburg: Freunde des Jura-Museums Eichstätt. *An important series of papers presented at the International Archaeopteryx Conference, Eichstätt, 1984.*

Olson, S.L. 1985. The fossil record of birds. Avian Biology 8: 79–238. *The most current, thorough compendium.*

Vuilleumier, F. 1975. Zoogeography. Avian Biology 5: 421–495. *A detailed review of the classic perspective.*

CHAPTER 3

Systematics

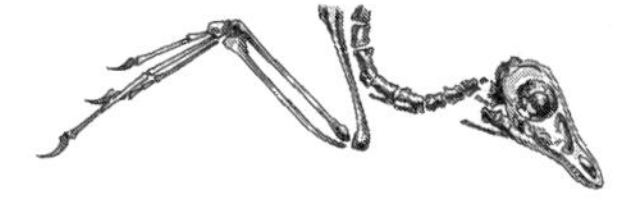

COMPARATIVE BIOLOGICAL SCIENCES such as ornithology help us to understand the evolution of diversity. Closely related species share immediate common ancestors, which, in turn, share earlier common ancestors. The tree of genealogical relationships among species—their phylogeny—provides a foundation for taxonomic classification and also a framework for understanding the evolution of behavior, ecology, and morphology.

This chapter presents an overview of avian systematics—the scientific study of evolutionary relationships of birds. First is a summary of the nature of species, the fundamental units of biological classification. Then follow a review of the relationship between phylogeny and formal classification and a review of the morphological, behavioral, and genetic attributes of birds that provide clues to evolutionary history. The long-standing controversy concerning the evolutionary relationships of flamingos provides an example of the kinds of problems that confront ornithologists. For example, convergence—the independent evolution of similar adaptations in unrelated organisms—can mislead scientists who attempt to determine which species have a common ancestor. The chapter concludes with an introduction to some of the primary methodologies of systematics, including cladistics—the study of evolutionary branching sequences—and biochemical genetics—the study of changes in an organism's DNA, which contains the genetic code of life.

Species

Species are the fundamental units of biological classification. They have characteristic sizes, shapes, songs, and colors, as well as ecological niches and geographical ranges. Species may interact ecologically, but they cannot exchange genes or novel genetic-based adaptations. According to the prevailing biological species concept, species evolve independently of one another. By definition, *"Species are groups of interbreeding natural populations that are reproductively isolated from other such groups"* (Mayr 1970, p. 12).

The criteria involved in the definition of biological species are the reproductive compatibility of individuals and the potential for blending of differences between two populations.

Concerns about the practical application of the biological species concept prompt some ornithologists to question its working merits and to recommend new approaches to the study of speciation and geographical variation in birds (McKitrick and Zink 1988; Cracraft 1989). Chapter 22 considers both the process of speciation and current debates about the species concept. Here we review the ways that ornithologists classify species into natural evolutionary groups.

The number of avian species recognized by ornithologists continually changes. A list of 19,000 species in the early 1900s shrank to 8600 in 1940, partly because better information revealed some "species" to be variations due to age or sex. Then the species concept itself broadened to incorporate variant populations once classified separately. Reassessments of isolated populations that differ slightly in size or color reduced the official species list even more. In recent years, however, taxonomic revisions, new discoveries, and a growing tendency to elevate the status of isolated populations have increased the total number of bird species recognized by ornithologists from 8600 to over 9600, and this number is likely to continue to increase (Sibley and Monroe 1990).

Classification and Phylogeny

The process of naming and classifying birds is an ancient and continuing one that includes clustering similar species into diagnosable taxa and ranking these taxa on the basis of their hypothesized sequences of evolutionary relationship. New philosophical and technological advances have enabled ornithologists to reassess previous hypotheses and sometimes to resolve previous dilemmas.

Francis Willoughby and John Ray's *Ornithologiae*, published in 1676, was the first formal classification of birds. The authors of this seminal work attempted to arrange all birds then known into a logical, hierarchical classification—the "cornerstone of modern systematic ornithology" (Zimmer 1926). Willoughby and Ray divided birds first into two obvious groups, Landfowl and Waterfowl, and then into additional subgroups. Nearly a century later, Carolus Linnaeus (1758) used this elementary classification in his *Systema Naturae*, which became the model for subsequent classifications. Important as they were, these early efforts tended to classify according to superficial similarities such as adaptations to aquatic versus terrestrial habitats, rather than according to an overarching concept of evolutionary relationship.

Darwin's theory of evolution by natural selection transformed the philosophical basis of systematics. His concept of evolutionary relationships provided a sound theoretical and rational basis for finding evidence of common ancestries. In his classic work *On the Origin of Species by Means of*

Natural Selection (1859), Charles Darwin reflected on the hierarchy of similarity due to evolutionary relationships:

> I believe that the arrangement of the groups within each class, in due subordination and relation to each other, must be strictly genealogical in order to be natural; but that the amount of difference in the several branches or groups, though allied in the same degree in blood to their common progenitor, may differ greatly, being due to the different degrees of modification which they have undergone; and this is expressed by the forms being ranked under different genera, families, sections, or orders. (Darwin 1859, p. 420)

The birth of a convincing avian classification took decades but was finally manifest in the work of Hans Friedrich Gadow (Gadow 1892, 1893). Based on the integrated assessment of 40 anatomical characters, Gadow's classification of birds was a milestone in ornithological systematics, one that still prevails. Modern classifications of the birds of the world derive from Gadow's work—including those of Erwin Stresemann (1927–1934), Alexander Wetmore (1960), Ernst Mayr and Dean Amadon (1951), Robert Storer (1971b), and James Peters in the 15-volume *Checklist of the Birds of the World* (1931–1986). Challenging these traditional classifications is a new one by Charles Sibley and Jon Ahlquist (1990), based on comparisons of the DNA of over 1700 species.

Prevailing classifications of birds attempt to portray the evolutionary relationships of the various lineages (Chapter 2) as proposed by Darwin.

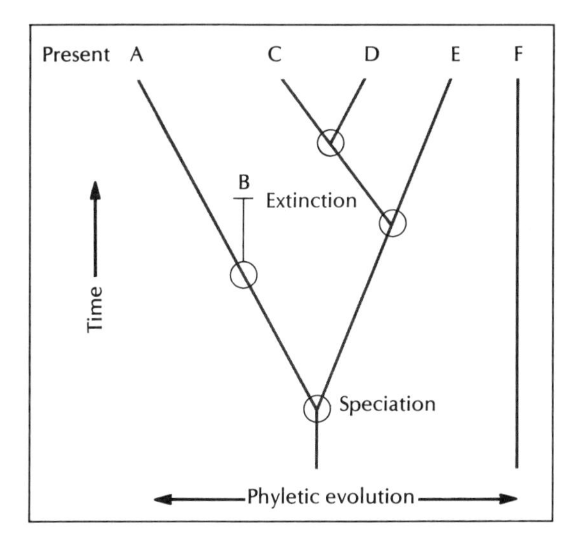

Figure 3–1 Diversification of evolutionary lineages includes speciation (circled nodes), the splitting of lineages; extinction, the loss of lineages; and phyletic evolution, the gradual change of a lineage with time. Clusters of similar, related taxa, such as C, D, and E, present in modern times, result from these changes. Taxon B went extinct. Taxon F is not related to the other taxa, which shared a recent common ancestor.

Theoretically, each taxon (Chapter 1) is monophyletic, that is, it contains a set of birds—called clades—related by evolutionary descent from a common ancestor. A hierarchical organization of taxa reflects relative closeness or distance of the evolutionary relationships among those taxa. Of the hypothetical taxa in Figure 3–1, five (A, B, C, D, and E) should be placed in the same higher taxon, the same family perhaps. Taxon F probably is not related and thus belongs in a different family. If F had evolved from the ancestor of A–E at some much earlier date, it might be assigned to the same order. Taxa C, D, and E would be grouped together in the same genus because they form a cluster of similar taxa. Because taxon B became extinct, an outcome resulting in a gap between taxon A and taxon C, taxon A might be recognized as a genus of its own.

Taxonomic Characters

Reconstruction of the evolutionary history of birds requires the analysis of characters that are shared as a result of common ancestry. Conservative characters—those that do not change easily in the course of ecological

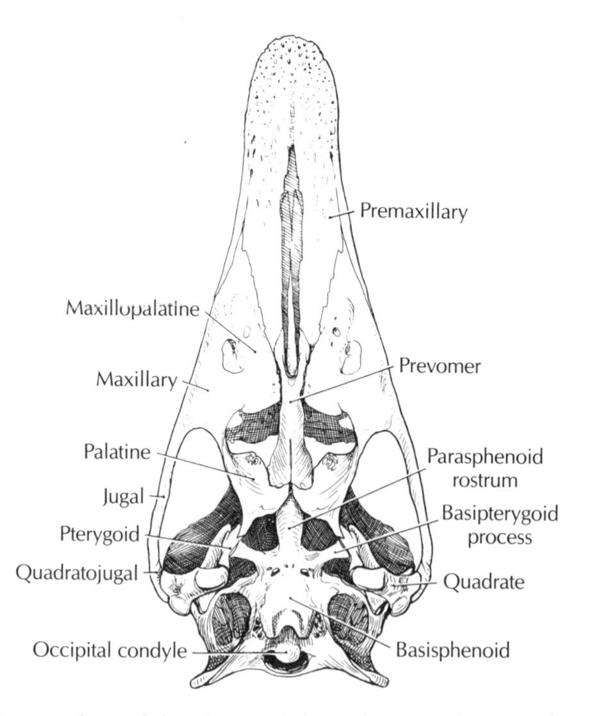

Figure 3–2 Bony palate of the Greater Rhea, showing the complex arrangement of bones that represent the unique paleognathous palate of ratite birds. Other orders of birds have different arrangements of the elements of the bony palate. (From *Fundamentals of Ornithology* by J. Van Tyne and A.J. Berger. Copyright 1976 John Wiley & Sons, Inc.; reprinted by permission of John Wiley & Sons, Inc.)

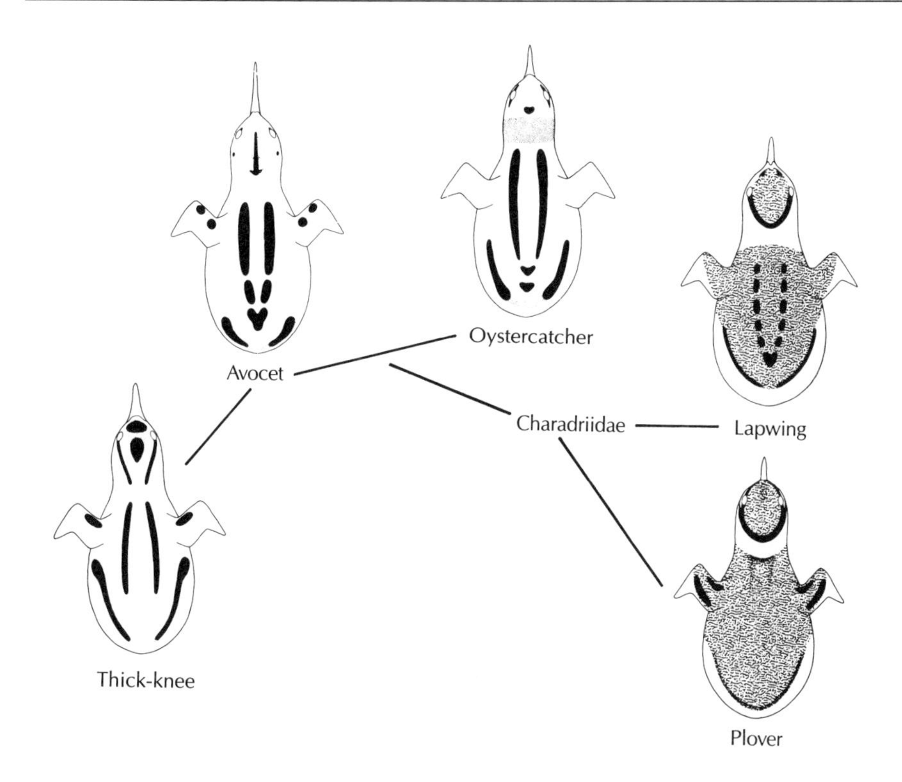

Figure 3–3 Plumage color patterns of downy young shorebirds provide clues to their evolutionary relationships (represented by branching lines). (Adapted from Jehl 1968)

adaptation—are of the greatest value because they retain clues to potential ancestors.

Darwin's champion, Thomas H. Huxley, helped to lay the foundations of evolutionary systematics in birds with his study of the arrangement of the bones of the avian bony palate, which forms a partition between the nasal cavities and the mouth (Huxley 1867) (Figure 3–2). In response to Darwin and Huxley, anatomists searched for meaningful taxonomic characters, features that suggest common ancestry of groups of species. In addition to the bony palate, some of the most important features include the form of the nostrils; the structure of the leg muscles and tendons of the feet; the arrangement of toes; the form, size, arrangement, and number of scales (scutes) on the tarsus; and the presence or absence of the fifth secondary flight feather on the wings. Ornithologists soon learned, however, that no single conservative attribute of birds provided a comprehensive solution. The choice of taxonomic characters varied and so did their application from one group of birds to another. Behavior yielded clues to evolutionary relationships among some birds, and so did plumage patterns of downy young (Figure 3–3), calls and morphology of vocal apparatus, and biochemical assays of proteins and the genes themselves.

Unique characters may define related groups of species, that is, those with a common ancestor. Songbirds, the members of the Order Passeriformes, for example, have several unique characters. They have a preen gland with a unique nipple structure (Figure 4–8) and unique sperm (Figure 15–4). They also have a specialized perching foot with a large hallux (rear-directed toe), uniquely arranged deep tendons, and simplified foot muscles that facilitate perching at the expense of more delicate toe movements (Raikow 1982). These features indicate that members of the Order Passeriformes evolved from a common ancestor, that is, they are monophyletic.

Taxonomic characters must be homologous structures, that is, structures that are shared by two organisms and that "can be traced phylogenetically to the same feature in the immediate common ancestor of both organisms" (Bock 1973, p. 386). Particularly interesting for phylogenetic studies are conservative, homologous characters that exist in both their original and their changed states. For example, the flipperlike wings of penguins evolved from the typical avian wings of their petrel ancestors. In

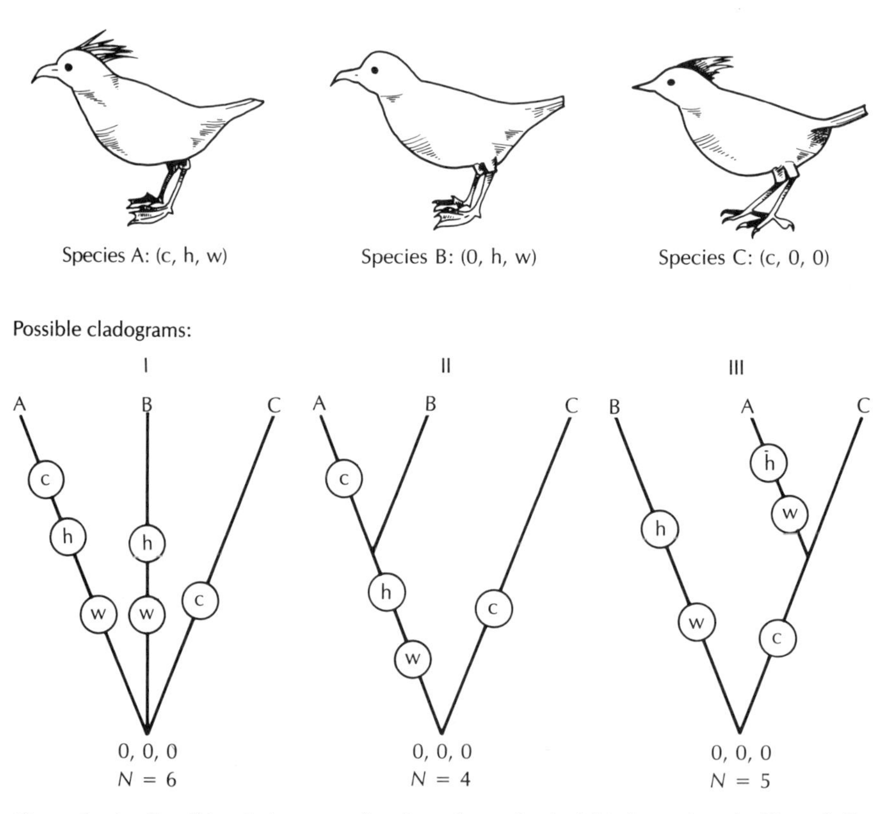

Figure 3–4 Possible cladograms for three hypothetical bird species, A, B, and C, that have different combinations of three derived characters: (c) crest, (h) hooked bill, and (w) webbed feet. Primitive character states (no crest, unhooked bill, or unwebbed feet) are denoted by 0. The changes from primitive (0) to derived character states (c, h, or w) are indicated for the evolution of each lineage (A, B, or C) from the common ancestor (0, 0, 0). The center cladogram ($N = 4$) is the most parsimonious, requiring fewest total changes to account for the distribution of derived characters among the three species; it also has the advantage that it postulates no convergence between species A and species B.

this case, the wings of petrels represent the ancestral—or primitive—character state whereas the flipperlike wings of penguins represent the advanced—or derived—character state. For another example, hooked bills represent the derived character state if they evolved from unhooked bills, the primitive character state.

If two species share a derived character state, we can hypothesize that they have a common ancestor with the same derived character state. The flipperlike wings of penguin species presumably reflect their common ancestry. We can then draw simple hypothetical branching sequences—or cladograms—based on the characters of extant species and their hypothetical ancestors. The construction of cladograms based on patterns of shared derived character states is called cladistic character analysis. Figure 3–4 shows three hypothetical birds that have different feet, crests, and bills. We assume that the cladogram with the fewest evolutionary changes, the most parsimonious one, illustrates the most plausible phylogeny. In this case, cladogram II, which assumes a common ancestor for A and B that looked like B, is most plausible.

One example of an avian relationship based on primitive and derived character states is that proposed by Alan Feduccia (1977) in a study of the stapes (middle ear bone) of perching birds (see Chapter 8). The woodhoopoes and hoopoes share a unique derived character state, an anvil stapes, which supports the traditional hypothesis that the two families are closely related (Figure 3–5). Most coraciiform birds—the kingfishers and their allies (Order Coraciiformes)—have the primitive, column-shaped stapes.

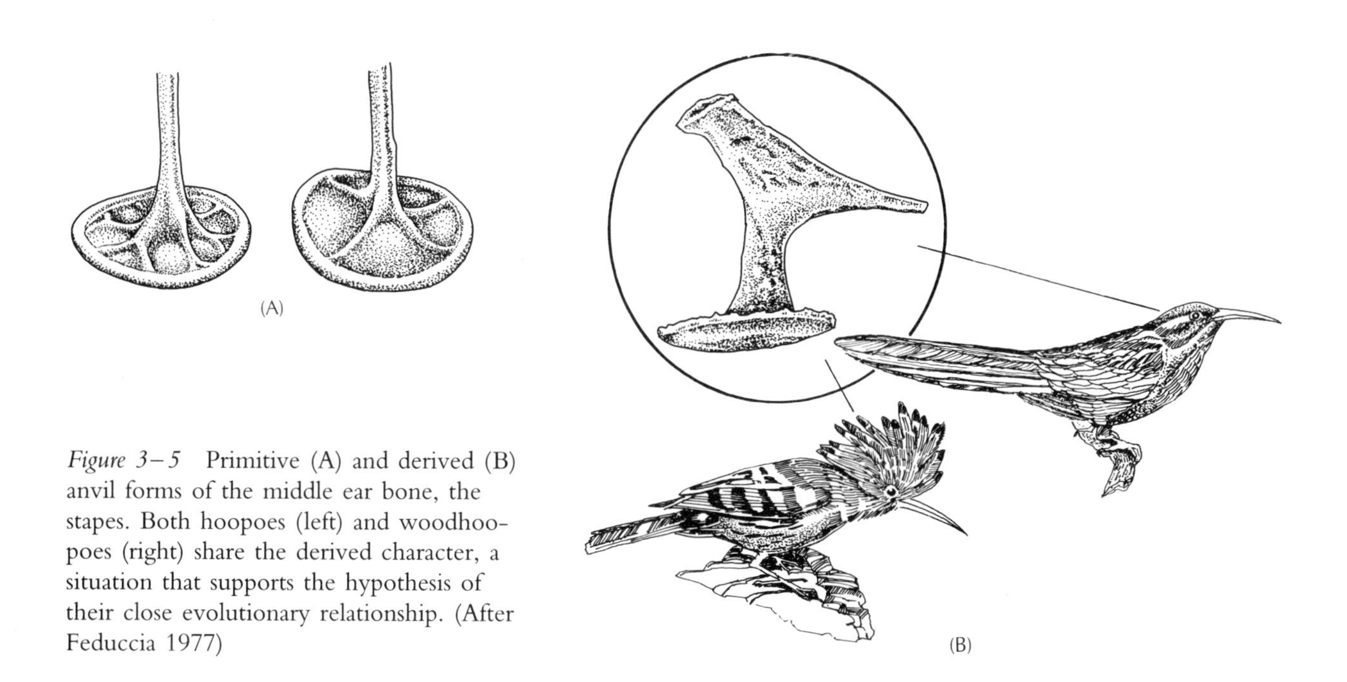

Figure 3–5 Primitive (A) and derived (B) anvil forms of the middle ear bone, the stapes. Both hoopoes (left) and woodhoopoes (right) share the derived character, a situation that supports the hypothesis of their close evolutionary relationship. (After Feduccia 1977)

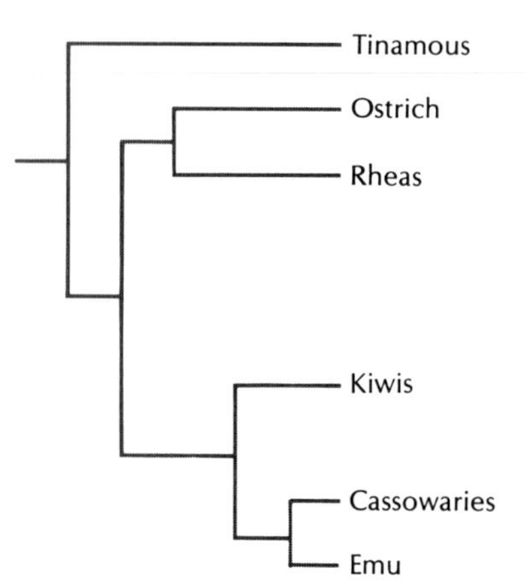

Figure 3–6 A phylogeny of ratite birds. Tinamous, which are the only extant flying members of this group, represent the most primitive forms. The flightless forms either evolved from flightless ancestors and dispersed throughout the southern continents prior to the breakup of Gondwanaland 75 to 80 million years ago or evolved flightlessness separately from immigrant northern flighted forms.

Ratite Relationships

The first major cladistic analysis in ornithology was Joel Cracraft's work (1974a) on the relationships of the large, flightless birds called ratites. Ostriches, rheas, Emus, and cassowaries are all ratites. The tinamous of South America also belong to this assemblage, even though they can fly, because they possess the characters that unite the ratites as a monophyletic group, including a unique (paleognathous) palate structure (see Figure 3–2), an unusual segmented bill covering in the downy chick, similar configurations of the openings in the pelvis called the ilioischiatic fenestra, similar DNA nucleotide sequences, similar chromosome arrangements, and similar amino acid sequences of an eye lens protein. For his study, Cracraft distinguished between the primitive and derived states of characters of the skeleton. Primitive character states were those that are also present in unrelated fowllike birds of the Order Galliformes, and derived character states were those unique to ratites. Cladistic analyses of the characters suggested that tinamous separated from the main ratite lineage very early, that rheas are most closely related to Ostriches, and that Emus are more closely related to Ostriches than are moas and kiwis.

This pioneering study, however, did not establish the phylogeny of the ratite birds once and for all. Anthony Bledsoe (1988) reevaluated Cracraft's data, disagreed with some of his decisions about which character states were derived, and made allowances for evolutionary reversals of character states. The revised analysis of 83 postcranial skeletal characters suggests that cassowaries and Emus are more closely related to kiwis than to Ostriches (Figure 3–6). A recent biochemical study supports this conclusion (Cooper et al. 1992). Further still, Peter Houde and Storrs Olson (1981) suggested that modern ratite birds may be relicts of an early radiation of flying palaeognathous birds in the Northern Hemisphere. The modern ratites may be united only through diverse and ancient relatives.

Convergence

Characters that change easily in the course of adaptation provide little information about ancestry and are sometimes misleading. Adaptation to similar ecological roles causes unrelated species of birds to become superficially similar in details of appearance, morphology, and behavior.

The meadowlarks of North American grasslands and the longclaws of the African grasslands are classic cases of convergence in color pattern. Both live in open grasslands and are about the same size and shape. Both have streaked brown backs, bright yellow underparts with a black V on the neck, and white outer tail feathers. But meadowlarks are related to the Red-winged Blackbird and other members of the New World Family Icteridae because they all have conical bill shapes and distinctive jaw muscles, and lack bristles at the base of the bill. Longclaws are classified as relatives of the pipits and wagtails in the Family Motacillidae, on the basis of their slender notched bills, configuration of jaw muscles, and presence of bristles at the base of the bill.

Another classic case of morphological convergence is that of the northern ocean auklets, such as the Dovekie, and the southern ocean diving-petrels. Both are compact, black-and-white seabirds that use their wings to propel themselves underwater to capture marine crustaceans. Dovekies are related to gulls and other members of the Order Charadriiformes; diving-petrels are related to albatrosses and other members of the Order Procellariiformes. All procellariiform birds have distinctive derived skeletons and tubular nostrils.

One could make a long list of instances of morphological convergence. Tyrant-flycatchers share ecological and morphological attributes with shrikes, wheatears, tits, warblers, pipits, or thrushes (Keast 1972). Australian land birds share so many morphological attributes with shrikes, flycatchers, and small insect-eating warblers that they were misclassified with superficially similar European and Asian birds.

Convergence can sometimes be unveiled by the study of fine anatomical details. Generally, the more complex the character, the less likely it is

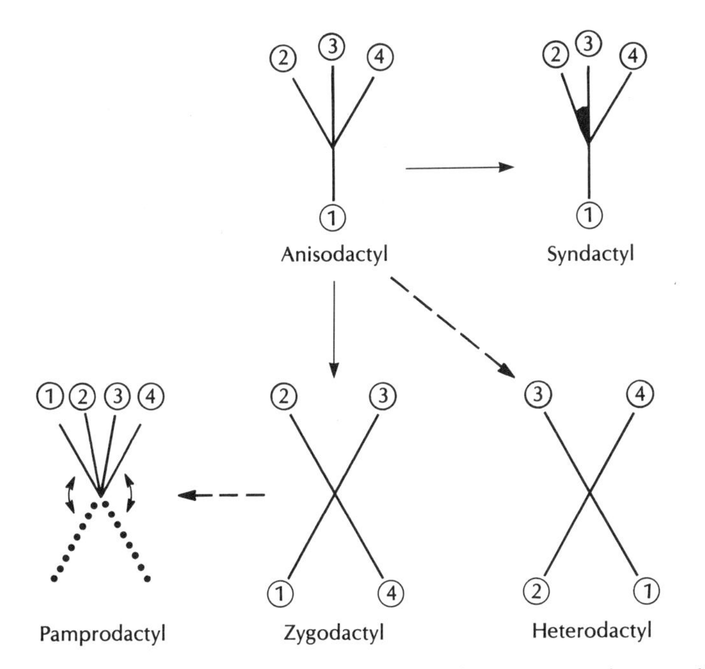

Figure 3–7 Toe arrangements of perching birds. Alternatives to the prevalent (anisodactyl) arrangement of three toes in front and the hallux (the first digit) pointing to the rear have evolved several times (solid arrows). The syndactyl foot, in which the bases of toes 2 and 3 are fused, characterizes the Coraciiformes. The zygodactyl arrangement, with two forward-pointing toes and two rear-pointing toes, has been achieved in different ways nine times during the evolution of birds. In trogons, toe 2, not toe 4, is rear-directed (heterodactyl). In the pamprodactyl foot, the positions of toes 1 and 4 are not fixed; all four toes may point to the front. Dashed arrows indicate uncertain derivations.

that anatomical details will be precisely the same. The details of avian foot structure reveal how unrelated birds evolved similar, but not identical, arrangements of the four toes (Bock and Miller 1959).

Although most perching birds have anisodactyl feet, with three forward toes and one rear toe (Figure 3–7), at least nine groups, including woodpeckers and their allies, most parrots, cuckoos, owls, the Osprey, turacos, mousebirds, Courol (Leptosomatidae), and some swifts, have zygodactyl feet, with two forward and two rear toes. Different orientations of the working surfaces (condyles) of cuckoo toe bones versus woodpecker toe bones, for example, indicate that these unrelated birds have evolved the zygodactyl foot arrangement in different ways.

Another group of birds, the trogons, appear to have the zygodactyl toe arrangement, but their second toe, not their fourth, is directed backward, forming what is called the "heterodactyl" toe arrangement. Still other toe configurations are possible (Figure 3–7). The "syndactyl foot" with two or three toes fused basally characterizes the Order Coraciiformes; and the "pamprodactyl" foot with all four toes directed forward characterizes the mousebirds (Order Coliiformes) and some swifts (Order Apodiformes).

Flamingo Relationships

Many of the current debates in avian systematics center on the possibility that similarities result not from ancestral relationship but from convergence. This possibility is, in part, the reason behind the historically troubled classification of flamingos (Table 3–1). Gadow (1893) classified flamingos as a subgroup (Suborder Phoenicopteri) of the storks, Order Ciconiiformes. For years, however, ornithologists have wondered whether flamingos are really ducks with long storklike legs or storks with ducklike, filter-feeding bills. Evidence for both conclusions seemed strong. Lacking firm resolution, taxonomists have given flamingos an order of their own (Phoenicopteriformes) between the Orders Anseriformes and Ciconiiformes (Storer 1971b).

Adding to the historical debate is the proposal by Storrs Olson and Alan Feduccia (1980a) that flamingos may not be related to either ducks or storks. They suggest instead that shorebirds, specifically stilts, are the modern relatives of flamingos. The past affiliations with ducks and storks, they assert, are based solely on "traditions and misconceptions." The musculature of flamingos is like that of the Banded Stilt of Australia, and both possess a unique leg muscle. Other characteristics, including the flamingo's distinctive life history, natal down, skeletal structure, and behavior, also resemble those of the Banded Stilt. Olson and Feduccia recommend that the taxonomic status of the flamingos be reduced to the Family Phoenicopteridae placed immediately after the Family Recurvirostridae, putative ancestors of the flamingos in the Order Charadriiformes. One wonders, however, whether the similarities between flamingos and the Banded Stilt reflect convergence. Both, after all, are adapted to an unusual environment—the shores of shallow, saline lakes.

TABLE 3–1
Different classifications of flamingos

Taxon	Gadow 1893	Storer 1971b	Olson and Feduccia 1980a
Order	Ciconiiformes	Ciconiiformes	Ciconiiformes
Suborder	**Phoenicopteri**		
Order		**Phoenicopteriformes**	
Order	Anseriformes	Anseriformes	Anseriformes
Order	Charadriiformes	Charadriiformes	Charadriiformes
Family		Recurvirostridae	Recurvirostridae
Family			**Phoenicopteridae**

These classifications reflect scientists' uncertainty of the evolutionary relationships of flamingos. Traditionally allied to storks (Ciconiiformes), either as a suborder (Phoenicopteri) or as a separate order (Phoenicopteriformes), Olson and Feduccia suggested that they are really shorebirds that merit only family status (Phoenicopteridae) in the order Charadriiformes.
Taxon representing flamingos appears in boldface.

Current Revolutions in Avian Systematics

Despite a major effort in the first half of the twentieth century, ornithologists failed to discover phylogenetic clues that seriously challenged the prevailing arrangements of the higher categories of birds. The possibilities of convergence loomed large. The relationships of the flamingos remained unresolved. Erwin Stresemann, a great German ornithologist, said,

> But as far as the problem of the relationship of the orders of birds is concerned, so many distinguished investigators have labored in this field in vain, that little hope is left for spectacular breakthroughs. . . . Science ends where comparative morphology, comparative physiology, comparative ethology have failed us after neary [*sic*] 200 years of efforts. The rest is silence. (Stresemann 1959, p. 277)

Two major revolutions that started in the 1970s have infused new vigor into the analysis of evolutionary relationships among birds. One revolution was conceptual, the other was technological. The conceptual revolution—cladistic character analysis—enables ornithologists to separate shared primitive characters from shared derived characters. The latter provide the best clues to evolutionary relationships. Cladistic character analysis also allows rigorous separation of convergent characters from characters shared as a result of common ancestry. The other revolution was the advent of new biochemical technologies for the laboratory comparisons of DNA (nucleotide sequences) of species.

Rapidly increasing knowledge of DNA structure and of the genetic control of protein synthesis enables testing of hypotheses about evolutionary relationships based on morphological characters. There is little reason to suspect that convergence would operate in the same manner on both proteins or nucleic acids and morphological features, such as bills. Because different genes or genomes tend to evolve at different rates, some (those that evolve fast) serve best in comparisons of recently evolved species and others (those that evolve slowly) serve best in comparisons of taxa such as families and orders that separated long ago. The best estimates of phylogenetic relationships lie finally with concordance of phylogenies produced by different methods (Bledsoe and Raikow 1990). Confidence increases when different characters or analytical assumptions yield the same result. In general, biochemical studies tend to corroborate previous morphological evidence of relationships. Sometimes, however, biochemical analyses challenge traditional views, reveal overlooked cases of convergence, and suggest unsuspected relationships among taxa.

The analysis of egg white proteins by Charles Sibley (1970), for example, produced startling information about the relationships of some odd birds, each of which was assigned to its own (monotypic) family. For example, the skulking Wrenthrush of highland bamboo habitats in Central America was long thought to be a thrush (Turdidae), but study of its

Figure 3–8 Electrophoretic studies of proteins suggested that the Hoatzin is related to *Guira* cuckoos, not to guans, as had been thought. (From original by E. Poole, courtesy of Academy of Natural Sciences, Philadelphia)

proteins revealed that it was a wood warbler (Parulidae) (Sibley 1968). Subsequent analyses of the vocal apparatus (Ames 1975), hindlimb muscles (Raikow 1978), and life history (Hunt 1971) support the protein-based hypothesis. Support from characters as disparate as protein structure, muscle arrangements, and behavior is strong support indeed. For another example, the Hoatzin, a strange bird of the Amazon forests, may be related to *Guira* cuckoos (Order Cuculiformes), not guans (Order Galliformes) as once thought (Sibley and Ahlquist 1973) (Figure 3–8).

The biochemical analysis of egg white proteins has been largely superseded by the analysis of enzymes found in the body tissues of adult birds (Barrowclough 1983; Evans 1987). Enzymes with easily interpreted patterns of inheritance—called allozymes—have served a variety of phylogenetic studies, including those of wood warblers (Barrowclough and Corbin 1978) and grassland sparrows (Zink and Avise 1990). However, allozyme comparisons are rapidly yielding to more direct comparisons of DNA, as the technology of nucleic acid chemistry becomes increasingly accessible to biologists as well as biochemists (Quinn and White 1987).

The technique called DNA–DNA hybridization played a major role in the revitalization of avian systematics (Box 3–1). Charles Sibley and his colleague Jon Ahlquist (1990) assembled thousands of samples of DNA from birds around the world, representing species from all but three families. They tested more than 25,000 DNA hybrids in an unprecedented effort to revise the entire class of organisms by a singular molecular technique. Their radical new phylogeny and classification of the birds of the world based on DNA–DNA hybridization challenges many traditions. The resulting work, *Phylogeny and Classification of Birds: A Study in Molecular Evolution*, is a milestone in ornithology by virtue of its herculean scope and pioneering methodology. It remains to be seen, however, whether their results will be corroborated and confirmed. The work is a controversial one because of potentially serious flaws in the design of the experiments and in the technical analysis of the data (Gill and Sheldon 1991; Lanyon 1992a).

Among the results of Sibley and Ahlquist's DNA hybridization experiments that have received acclaim are revelations of convergence among ecological equivalents. Many of the Australian songbirds, for example, had been previously classified with morphologically similar Asian and European forms. DNA experiments showed that diverse Australian songbirds are more closely related to one another than to morphologically similar forms. The adaptive radiation of Australian songbirds parallels the extraordinary radiations of marsupial mammals and eucalyptus plants on that continent. The results of DNA hybridization also support the hypothesis (Ligon 1967) that New World vultures, such as Turkey Vultures and Andean Condors, are related to storks (Ciconiiformes), not raptors (Accipitriformes), as previously thought. Storks and New World vultures share a variety of skeletal and behavioral characters, including the unusual habit of defecating on their legs as a means of increasing heat loss through evaporation of liquid excreta.

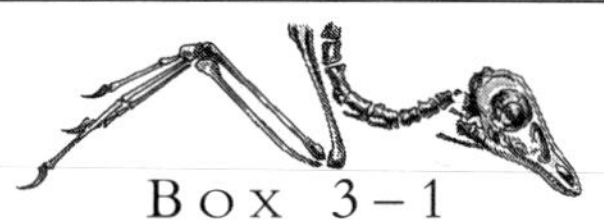

Box 3–1
DNA hybridization compares total genetic divergence

DNA hybridization estimates the amount of genetic change in the entire genetic code that has taken place since the time two groups diverged from their most recent common ancestor. The estimates of genetic change between species are then used to cluster most-similar pairs of species together. To build a phylogeny, the estimates of genetic change are converted to standardized "distances" that form the basis of a hierarchical branching diagram—or evolutionary tree.

DNA is composed of four primary units called nucleotides. The four different nucleotides each have a unique base—adenine (A), cytosine (C), thymine (T), or guanine (G). The linear sequence of the nucleotides forms the genetic basis of life. Hydrogen bonds between certain pairs of bases—guanine and cytosine, adenine and thymine—hold the double-stranded DNA molecule together. However, these bonds can be broken by high temperatures.

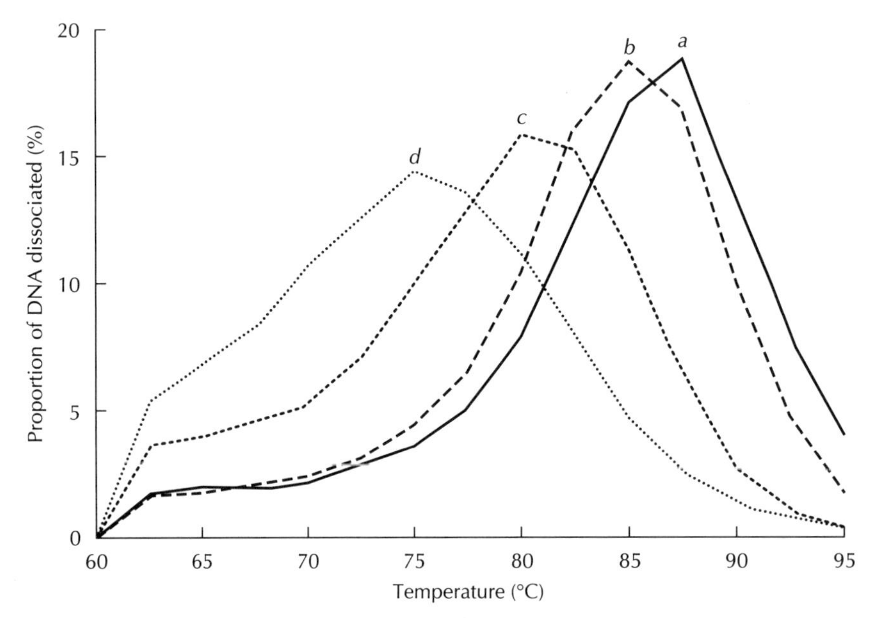

Results of a DNA hybridization experiment. The four curves are examples of heron DNA thermal disassociation profiles, which show the proportion of the DNA that disassociates at each temperature. Curve *a* is the melting profile of the control, the homoduplex, in which Great Blue Heron DNA is hybridized with itself. It separates at a higher temperature than do the other hybrid DNAs because the best possible match obtains when DNA is hybridized with itself. The other three profiles of hybrid DNAs of different species—or heteroduplexes—demonstrate decreasing similarities between Great Blue Heron DNA and the DNAs of the Great Egret (curve *b*), the American Bittern (curve *c*), and the Glossy Ibis (curve *d*).

The difference between the modal or peak melting temperatures of the homoduplex profile and the heteroduplex profiles (ΔT_{mode}) is used to estimate the genetic distance. Here, the distance between the Great Blue Heron and the Great Egret is 86.5°C − 85.0°C = 1.5. The genetic distance between the Great Blue Heron and the American Bittern or the Glossy Ibis is 5.5 or 10.5, respectively.

Fragments of avian DNA from two species form double-stranded hybrid complexes when they are heated and separated into single strands and then allowed to reassociate under special laboratory conditions. The hybridized DNA complexes of two samples from a single species are stable and separate only at high temperatures, but hybrid DNA complexes of distantly related species, such as a penguin and a warbler, have few sequences in common and readily dissociate, even at low temperatures. Genomic similarity—or the number of bases in a specific nucleotide sequence that two species have in common—is reflected in the degree of thermal stability of the DNA–DNA hybrid molecule. Each 1 percent increase in the match between the paired sequences of a hybrid complex requires a 1°C increase in temperature to separate them (see figure on page 58).

Fred Sheldon's (1987a, b) DNA hybridization experiments serve to illustrate this approach to biochemical systematics (see figures). The experiments demonstrate that the Great Blue Heron is genetically more similar to the Great Egret than to the American Bittern and more similar to both these species than to the Glossy Ibis. These studies also revealed different rates of genetic change among different lineages of herons.

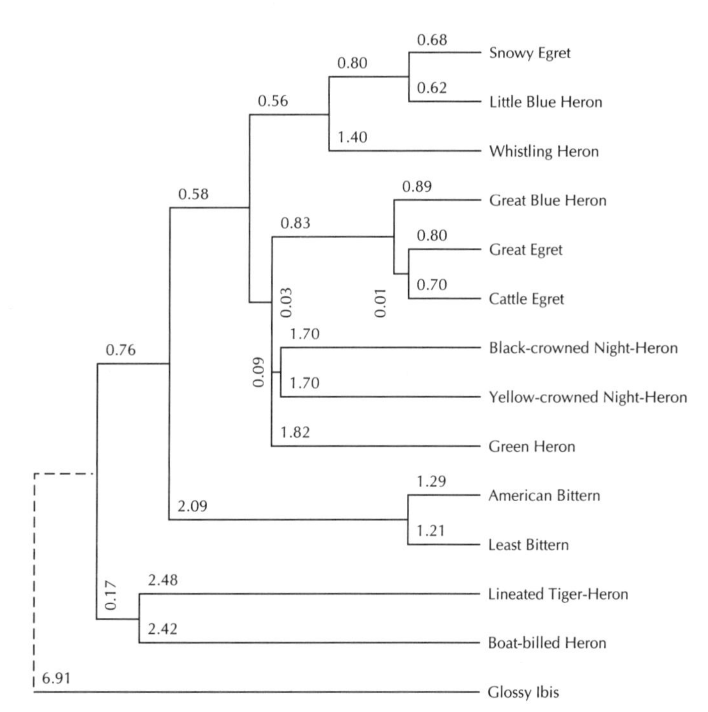

Branching diagram or evolutionary tree of heron relationships based on DNA hybridization experiments conducted by Fred Sheldon (1987a, b). Species are clustered according to their similarity, defined by the hierarchy of branch lengths.

The genetic distance between Great Blue Heron and Great Egret measured by ΔT_{mode} from the preceding figure is approximately the same as the branching distance in this tree, which is the sum of three intervening branch lengths (0.89 + 0.01 + 0.80 = 1.7). Notice that most branch tips on the trees line up with each other, because their DNAs diverged at approximately the same rate over time. The American Bittern and Least Bittern, however, exhibit relatively long total branch lengths, because their DNA diverged (accumulated base-pair mutations) faster. Conversely, the DNAs of the Lineated Tiger-Heron and Boat-billed Heron changed more slowly than the other species in this study.

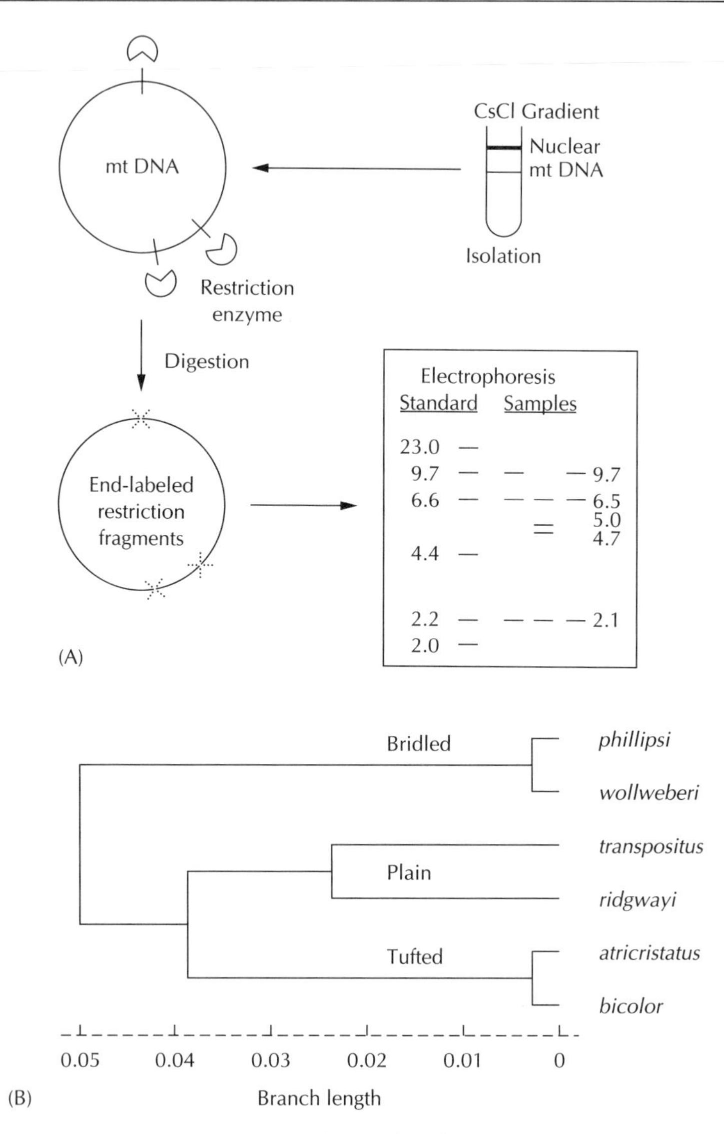

Figure 3–9 Restriction site analysis of mitochondrial DNA. (A) Comparison of the mtDNAs of different species starts with isolation of DNA samples by centrifugation. The isolated mtDNA is then cut into fragments of different sizes by specific restriction enzymes that recognize and digest particular DNA sites (called restriction sites) 4 to 6 base pairs long. The radiolabeled fragments (sizes given in kilobases) of different species are then separated on the basis of their different mobilities in an electrical current (electrophoresis). Fragment patterns are analyzed to identify the presence or absence of a particular restriction site. (From Shields and Helm-Bychowski 1988). (B) Tallies of shared restriction sites are then used to produce a tree of genetic relationships among taxa; the branch lengths are proportional to the differences in mtDNA sequence composition. In this example, the Tufted Titmouse and Plain Titmouse are more closely related to each other than either is to the Bridled Titmouse (Gill and Slikas 1992). The two subspecies of the Plain Titmouse—*transpositus* and *ridgwayi*—are more different from each other than are the subspecies of either the Bridled Titmouse or the Tufted Titmouse.

The affinities between flamingos and their possible relatives—ducks, storks, or stilts—still remain to be fully explored with biochemical technologies. Preliminary DNA hybridization experiments (Sibley and Ahlquist 1985) support the original hypothesis that flamingos are related to the storks, but these experiments require confirmation and a more comprehensive set of comparisons. Still to be attempted is a thorough analysis using newer biochemical techniques such as DNA sequencing.

The variety of methods for comparing DNA genetic codes of birds increases with each new advance in biochemical genetics (Barrowclough 1992; Sheldon and Bledsoe 1993). Especially fruitful are comparisons of the compositions of a small circular DNA molecule (mtDNA) found in the mitochondria of the cytoplasm. Short nucleotide sequences of mtDNA can be compared indirectly (Figure 3–9) or more directly by sequencing the nucleotides themselves. The mtDNA cytochrome *b* gene is currently the most popular target for sequence studies. Scott Lanyon and John Hall's (1994) analysis of 888 nucleotides of this gene confirmed one of the interesting results of Sibley and Ahlquist's DNA hybridization experiments regarding the relationships of barbets (Family Capitonidae), which are brightly colored, tropical, fruit-eating relatives of the woodpeckers. Both biochemical studies indicated the New World barbets were more closely related to the toucans (Family Ramphastidae) of Central and South America than to similar barbet species in the Old World (Figure 3–10). The large-sized and big-billed toucans evolved from New World barbets after the barbets had distributed themselves throughout the tropical regions of both hemispheres.

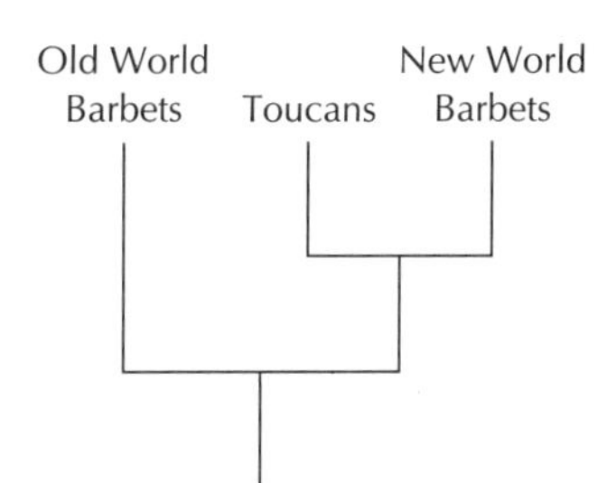

Figure 3–10 Evolutionary relationships among Old World barbets, toucans, and New World barbets based on a parsimony analysis of mitochondrial DNA, cytochrome *b* sequence data. Despite their different morphologies, the toucans and New World barbets are sister taxa. The Old World barbets and New World barbets are similar by virtue of sharing older ancestral traits.

Construction of a phylogeny based on biochemical or morphological characters, or both, is only the first step toward understanding the evolution of birds. The next step is to map other information onto the phylogeny to explore evolutionary trends in behavior, ecology, and biogeography. Examples in later chapters include the evolution of courtship displays of manakins (Chapter 9), nest building by swallows (Chapter 16), and brood parasitism by cowbirds (Chapter 19).

Summary

The species is the primary unit of biological classification. Theoretically each taxon is monophyletic, consisting of species more closely related to one another than to species in other taxa; and, in theory, the hierarchy of the classification and the evolutionary history of birds are the same. The formal naming of species and their orderly arrangement in a hierarchical classification facilitate scientific inquiry and communication.

The current ordinal classification of the birds of the world has changed little from the landmark work of Hans Friedrich Gadow in 1893, although the number of species recognized by ornithologists changes as a result of new discoveries and conceptual revisions.

The diversity of modern birds reflects historical patterns of speciation, extinction, and phyletic evolution. Conservative characters—attributes that do not change easily in the course of adaptation—enable ornithologists to decipher which groups of species share a common ancestor. Recognition of ancestral (primitive) versus changed (derived) character states aids in the reconstruction of the sequences of past evolutionary events.

Convergence—the independent evolution of similar adaptations by unrelated species—can cause unrelated species to appear to be related. Cases of convergence can be revealed by detailed study of complex characters, such as the internal anatomy of the toes, and by biochemical evidence.

Two recent revolutions, cladistic methodology and biochemical technology, have infused a new vitality into the study of the phylogenetic relationships among species and among major groups of birds. Biochemical studies, which permit comparisons of DNA compositions of species, tend to confirm conclusions based on morphology but sometimes suggest unsuspected affinities and new patterns of adaptive radiation. The evolutionary relationships of certain birds, such as the flamingos, continue to puzzle ornithologists.

Further Readings

Eldredge, N., and J. Cracraft. 1980. Phylogenetic Patterns and the Evolutionary Process. New York: Columbia University Press. *A provocative exposition of cladistic philosophy.*

Hillis, D.M., and C. Moritz. 1990. Molecular Systematics. Sunderland, Mass.: Sinauer Associates. *Excellent reviews of laboratory methods and data analysis.*

Mayr, E., and P.D. Ashlock. 1991. Principles of Systematic Zoology, 2nd ed. New York: McGraw-Hill. *The basics of classic systematics.*

Raikow, R.J. 1985. Problems in avian classification. Current Ornithology 2: 187–212. *A thoughtful review of the challenges currently facing systematic ornithology.*

Sheldon, F.H., and A.H. Bledsoe. 1993. Avian molecular systematics, 1970s to 1990s. Annu. Rev. Ecol. Syst. 24: 243–278. *A candid review of the successes and limitations of molecular systematics in ornithology.*

Sibley, C.G., and J.E. Ahlquist. 1990. Phylogeny and Classification of Birds. New Haven, Conn.: Yale University Press. *A comprehensive review of the historical literature on bird systematics, plus a controversial new classification based on the results of DNA hybridization experiments.*

Sibley, C.G., and B.L. Monroe, Jr. 1990. Distribution and Taxonomy of Birds of the World. New Haven, Conn.: Yale University Press. *A comprehensive list of species with their distributions and taxonomic comments, embedded in Sibley and Ahlquist's proposed new classification.*

PART II

Form and Function

CHAPTER 4

Feathers

FEATHERS, THE MOST DISTINCTIVE FEATURE of avian anatomy, are an extraordinary evolutionary innovation. Collectively referred to as plumage, feathers are unique structures of the skin that provide insulation for controlling body temperature, aerodynamic power for flight, and colors for communication and camouflage. Feathers also perform secondary roles. Modified feathers are important in swimming, sound production, hearing, protection, cleanliness, water repellency, water transport, tactile sensation, and support.

This chapter covers both feather structure and function. First we consider basic feather structure and functional modifications of this structure, followed by the major kinds of feathers in a bird's plumage, and the care of feathers by application of oily secretions of the preen gland. Feathers are replaced in seasonal molts. A discussion of the relationships of molts and plumages and the details of molt sequences in some feather tracts is followed by an evaluation of feather colors, both the basis of the colors themselves and the functions of plumage color pattern.

Feather Structure

Feathers consist mainly of keratin, an inert substance of insoluble microscopic filaments embedded in an amorphous protein matrix. Keratins are long-lasting biological materials resistant to attack by protein-digesting enzymes of microbes or fungi. Although keratins are a standard constituent of other hard epidermal structures, such as claws, hair, nails, and scales, the keratin in bird feathers and in the scales on bird legs is unique (Brush 1993). It is not, as was once believed, the same as that found in modern reptile scales. This finding suggests that the reptilian scales of the ancestors of birds changed their chemistry as well as their form and function.

The detailed structure of bird feathers has fascinated biologists for centuries; it is an enormous topic. We begin by reviewing the structure of a typical body feather—called a contour feather. The primary features of a typical contour feather are a long central shaft and a broad flat vane on either side of this shaft (Figure 4–1). The hollow base of the shaft—the calamus, or quill—anchors the feather in a follicle below the surface of the skin. The rest of the shaft—the rachis—supports the vanes. Lateral branches off the rachis—called barbs—are the primary elements of vane architecture. Each barb consists of a tapered central axis, the ramus (pl., *rami*), with rows of smaller barbules projecting from both sides. A barbule consists of a series of single cells linked end to end; the cells may be simple or may bear projections called barbicels, some of which are elaborate and hooklike. Barbs and barbules form an interlocking but flexible surface.

The vanes of a typical body feather grade from a hidden, fluffy basal portion, which provides insulation, to an exposed, cohesive outer portion, which has a variety of functions. The barbules on the barbs at the base of the body feather are long, thin, and flexible and do not have barbicels. With their similarly thin, flexible parent barbs, they create a downy—or plumulaceous—feather texture. In contrast, the outer part of the vane is a firmly textured, tightly interlocking—or pennaceous—structure. Well-developed hooklet barbicels are present on the pennaceous barbules that form this part of the vane. The cohesive structure

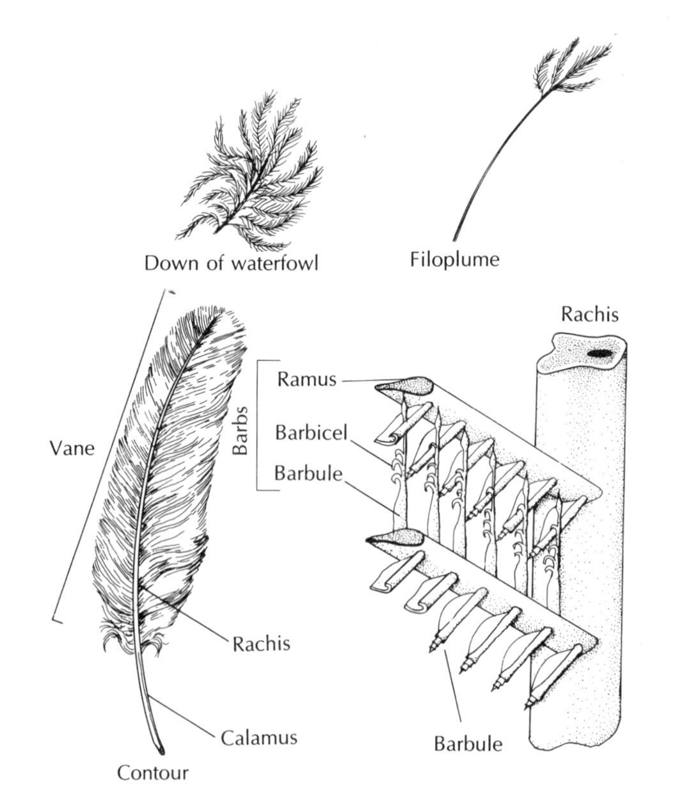

Figure 4–1 Structure of three kinds of feathers, with detailed structure of a typical contour feather.

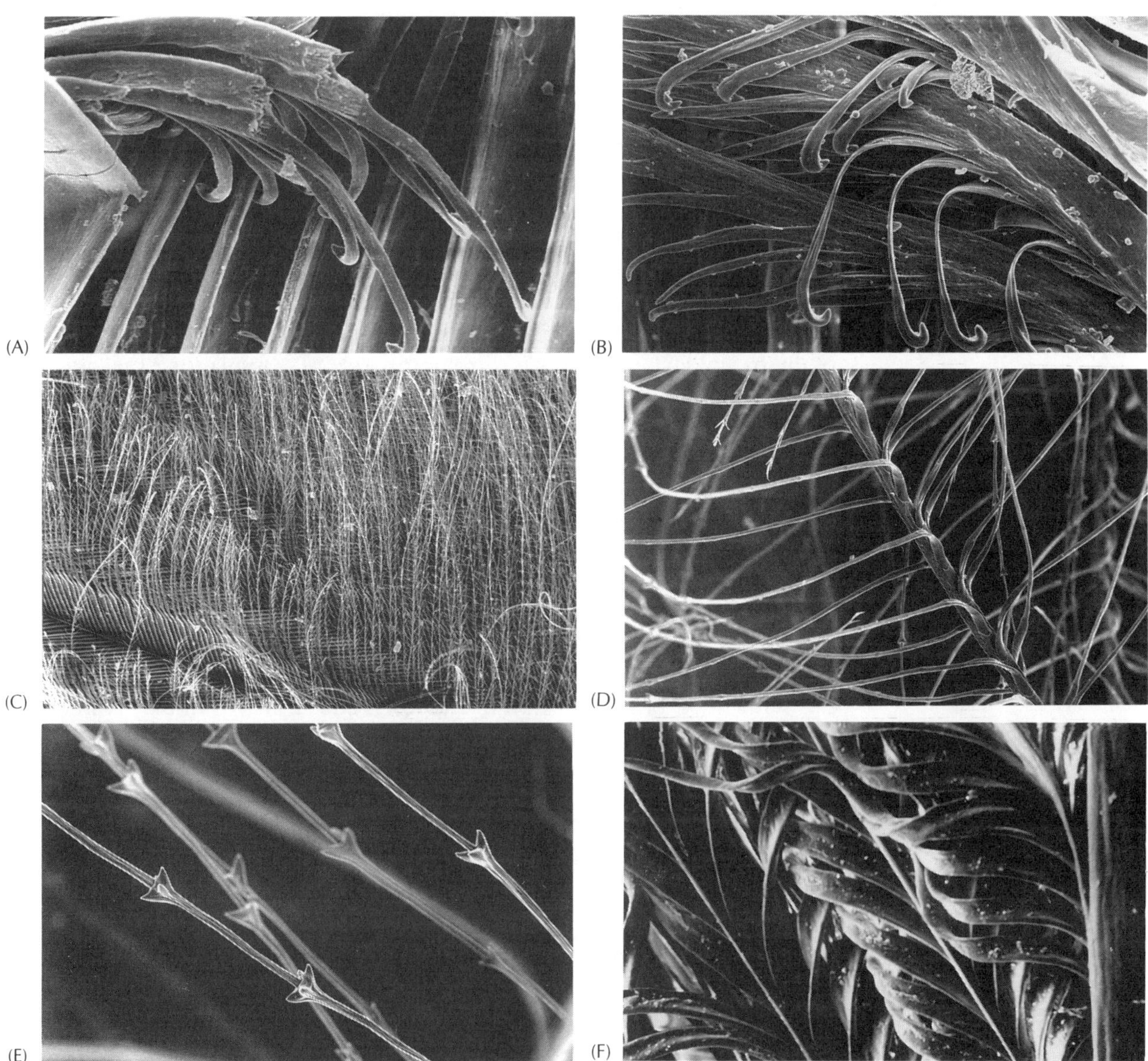

Figure 4–2 Scanning electron micrographs (SEMs) of feather structures: (A) Wild Turkey, tail feather. Oblique view of distal barbules with hooklets interlocking with proximal barbules. 358×. (B) Common Crow, wing feather. Distal barbules, displaced to show hooklets. Behind them are more distal barbules showing other, unhooked projections. 406×. (C) Barred Owl, upper wing covert. Dorsal oblique view. The vertical filaments are the tips of the distal barbules, which are unusually long and project upward from the plane of the feather. This creates the velvety nap that quiets the airflow over the wings, producing the silent flight of most owls. 215×. (D) Domestic Goose, body down feather. Downy barb. The oblique thicker element is the ramus of the bar and the thinner ones are the barbules. Although the down appears grossly to be a bunch of fluff, magnification shows that the barbules are arranged in a regular manner. 130×. (E) Domestic Goose, body down feather. Barbules on a downy barb, showing projections at each node, called nodal prongs. These prongs are homologous with the hooklets and other projections on pennaceous barbules. They are thought to serve in keeping the downy barbs from becoming entangled, thereby creating the fluffy texture, but how they do this is not known. 325×. (F) Namaqua Sandgrouse, abdominal feather. The vertical element on the right is the rachis and the oblique elements are the rami of the barbs, bearing the coiled barbules which serve for holding water. 153×. (Photographs courtesy of P. Stettenheim)

of the outer vane is based on the interlocking arrangement of pennaceous barbules. Those barbules on the distal side—the side away from the base of the feather—of the barb have barbicels that grasp the next higher, inner barbules of the adjacent barb. The barbicels, which can slide laterally along the next barb, are responsible for both cohesion and flexibility of the pennaceous vane (Figure 4–2).

The body feathers of birds typically include a secondary structure—or afterfeather, or aftershaft—that emerges from the underside of the rachis where the first basal barbs of the vane branch off. With rare exception, the barb and barbule structure of afterfeathers is plumulaceous. The afterfeather's primary function is to enhance insulation. Ptarmigans are grouse of high, cold alpine habitats. Providing essential insulation, the afterfeathers of ptarmigan winter plumage are three-fourths as long as the main feather. The afterfeathers of their summer plumage are much shorter.

Feathers are subject to striking modifications. Fusion of the developing barbs sometime produces feathers that look like strips of plastic, as, for example, do the crown feathers of the Curl-crested Araçari, a small Brazilian toucan, and the central tail feathers of the Red Bird-of-Paradise. The "plastic" feathers of the bird-of-paradise function in courtship display, but why the araçari has such feathers is not known. The familiar Cedar Waxwing of North America is named for its waxlike wing feather tips with fused bright red terminal barbs. Vane shapes of display feathers range from long and pointed, like those on a rooster's neck (called hackles), to short and round, like those of the head feathers of small birds. Rachises vary from thin and flexible, like those in the display tail feathers of some tropical hummingbirds, to stiff rods, like those in the bracing tail feathers of woodpeckers. The close spacing of large barbs with extra-long, curved barbicels produces water-repellent feathers in petrels, rails, and ducks. Coiled barbules on the belly feathers of sandgrouse help to transport water (Box 4–1).

Box 4–1
Desert sandgrouse carry water in modified feathers

The sandgrouse of African deserts are pigeonlike shorebirds that commute at dawn and dusk to the nearest water hole to quench their own thirst and also to get water for their young, which remain flightless at nests up to 30 kilometers away (Cade and Maclean 1967). At the water hole, male sandgrouse soak their belly feathers, which are modified to hold water in flight. Flattened and coiled barbules on the inside surface of the feathers (see Figure 4–2F) have hairlike extensions that absorb and hold water for transport back to the young. Upon return, the male adopts an upright watering posture, which attracts the young from their hiding places near the nest. The young drink by squeezing the wet elongate feathers in their bills. The chicks glean only 10 to 18 milliliters of the 25 to 40 milliliters of water soaked up each time the parent visits the water hole.

Vaned Feathers

The most conspicuous feathers are vaned feathers that include the smaller contour feathers covering the body surface and the larger flight feathers of the wings and tail. The smooth overlapping arrangement of vaned feathers reduces air turbulence in flight. The tiny, flat contour feathers that cover a penguin's body create a smooth, almost scaly, surface that reduces friction during swimming. As the feathers wear, increased friction reduces a penguin's swimming speed.

The flight feathers of the wing—called remiges (sing., *remex*)—are large, stiff, almost completely pennaceous feathers (Figure 4–3). They primarily serve aerodynamic functions and have very little importance in insulation. Rows of smaller feathers—called coverts—overlap the bases

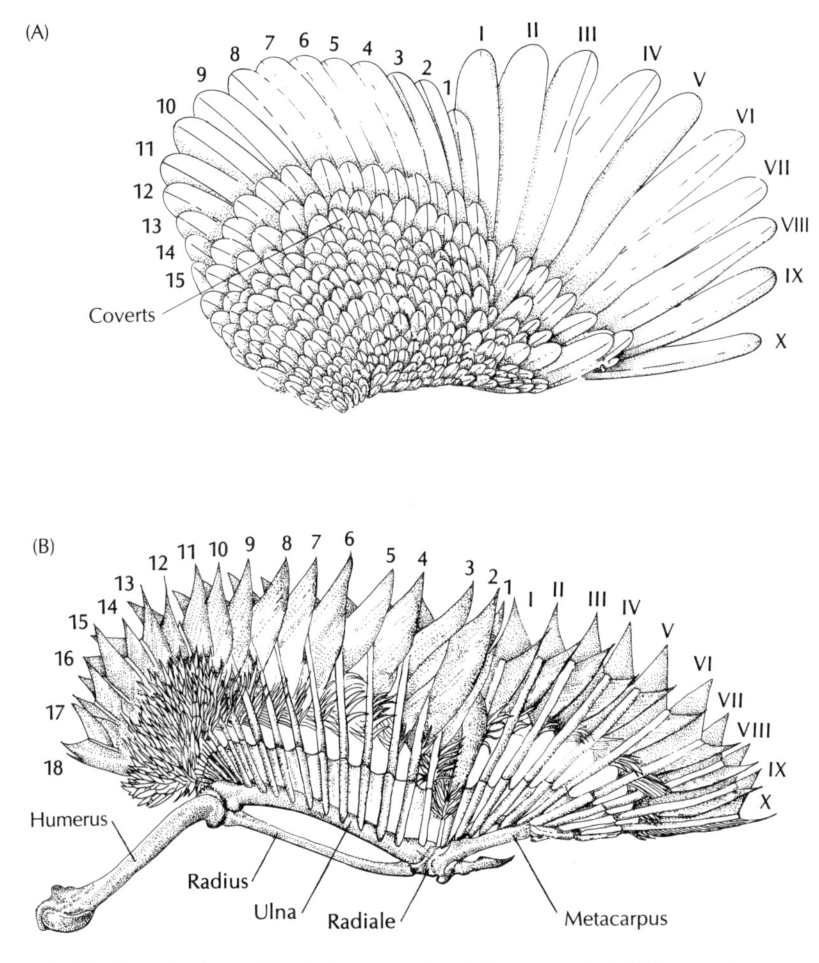

Figure 4–3 Dorsal view (A) of the extended left wing of a White Leghorn Chicken and (B) of the skeletal attachments of the primaries and secondaries of the same wing. Primary remiges are numbered I to X; secondary remiges are numbered 1 to 15 in A and 1 to 18 in B. The ends of the feathers have been omitted in B. (After Lucas and Stettenheim 1972)

of the remiges and cover the gaps between them. The long shafts of the outer (distal) remiges—the primaries—attach to the hand bones. These feathers provide forward thrust on the downstroke of the wing during flight. Flight feather vanes are asymmetrical, presenting a narrow outer vane that cuts the air. Most birds have 10 primaries; however, storks, flamingos, grebes, and rheas have 11, Ostriches have 16, and some songbirds have 9. The flightless kiwis have only three or four primaries.

Specialized barbules—called friction barbules—are found on the inner vanes of the outer (primary) wing feathers. Friction barbules have broad, lobed barbicels that rub against the barbs of overlying feathers, thereby reducing slippage and separation of feathers during flight. The longest friction barbules are found in the central part of the inner vane.

The silent flight of an owl, which enables it to surprise prey, results in part from two special structural features that muffle feather sounds. The barbs on the leading (forward) edges of the owl's primaries are long, curved, well-separated structures that reduce air turbulence. Unusually long barbules also help minimize the rubbing of overlapping feathers and create a soft, slightly fuzzy feather texture. Nightjars have a similar soft feather texture.

Because flight efficiency is directly linked to the structure of the primaries, major structural modifications of these feathers are uncommon. The narrow outer primaries of the male American Woodcock, which produce trilling noises during courtship flights, are an exception. The modified primaries of flightless cassowaries consist only of strong spinelike extensions of the calamus. These 28-centimeter-long spines protect a cassowary's flanks from abrasive vegetation. During the breeding season, long extensions of the second primary of male Standard-winged Nightjars grow out and are used in courtship (Figure 4–4). It is said that the nightjars discard the extensions by biting them off, but this notion is unverified.

The inner (proximal) flight feathers of the wing—the secondaries—attach to the ulna (Figure 4–3). Numbering from 6 in hummingbirds to 19 in some owls, and 40 in albatrosses, the secondaries form much of the inner wing surface. They have also been modified for display purposes, for example, in the broad flaglike inner secondary that is essential for courtship in the Mandarin Duck. Quite a different kind of modification are the thickened, clublike feather shafts of the central secondaries of the Club-winged Manakin, a tiny denizen of the thick undergrowth of Amazonian rain forests. The shafts make loud, castanetlike, snapping noises when the manakin claps its wings together in a courtship display flight.

The flight feathers of the tail—called rectrices (sing., *rectrix*)—attach to the fused caudal vertebrae—or pygostyle. There are usually 12 rectrices, which function primarily in steering and braking during flight (Figure 4–5). Among the exceptions are anis and grouse with 18 tail feathers and snipe with 24. The elaborate tails of peacocks, lyrebirds, birds-of-paradise, and some hummingbirds serve primarily in display and can be a handicap in flight. Some motmots, drongos, kingfishers, and hummingbirds have racket-shaped rectrices with bare shafts and terminal vaned

Figure 4–4 The "standards" of the Standard-winged Nightjar are highly modified primaries, which are dropped shortly after courtship has been completed.

sections. The circular tail tips of a male King Bird-of-Paradise are tight whorls of rachises and inner vanes (Stettenheim 1976). Tail feathers are also modified for sound production in some snipes and the Lyre-tailed Honeyguide, a West African relative of woodpeckers, and for bracing support in creepers, woodpeckers, woodcreepers, swifts, and penguins.

Down, Bristles, and Other Kinds of Feathers

Unlike firm-vaned feathers, down feathers are soft and fluffy (see Figure 4–1). The down feathers of chicks and also those of adult birds—called definitive down—vary from thick continuous distributions to restricted distributions. Down feathers typically lack a rachis, but as always there are exceptions, including the down feathers of waterfowl. Rather flexible plumulaceous barbs and barbules extend directly and loosely from the basal calamus. Downy barbules entangle loosely, trapping air in an insulating layer next to the skin. Down feathers provide excellent natural, lightweight thermal insulation.

Semiplumes are intermediate in structure between down and contour feathers. They have a large rachis with loose plumulaceous vanes. Some are close to down in structure, whereas others more closely resemble contour feathers. Semiplumes are distinguished from down feathers by the length of their rachises, which is always longer than the longest barb. Semiplumes are found at the edges of the contour feather tracts (see

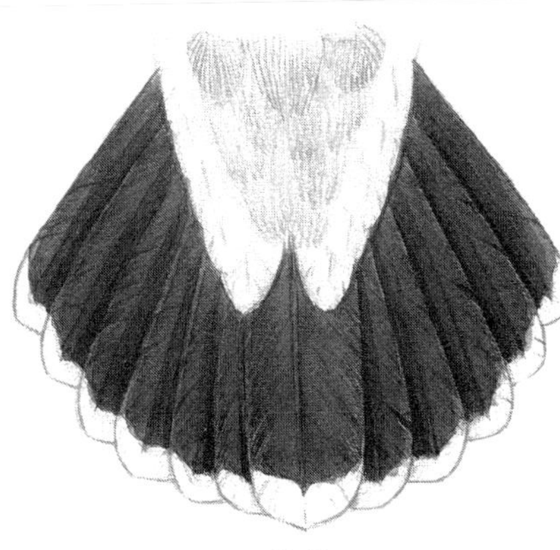

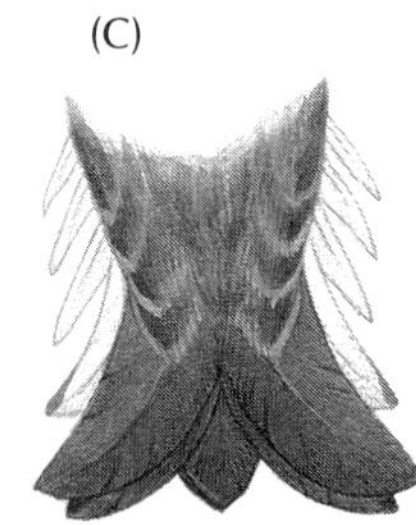

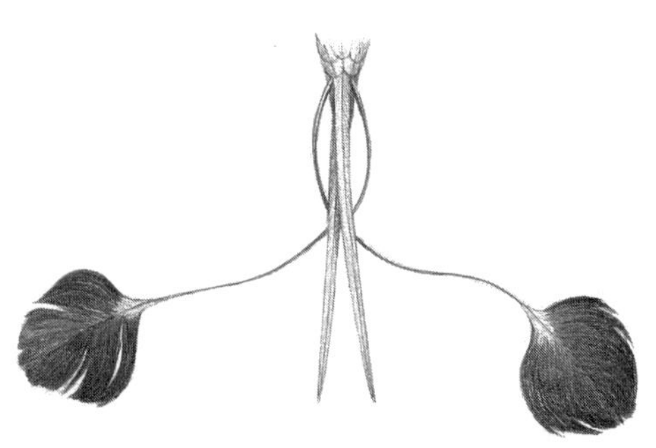

Figure 4–5 Tail feathers and their modifications: (A) unmodified tail of gull; (B) racket-shaped tail feathers of a motmot, a drongo, and a hummingbird (Marvelous Spatuletail); (C) sound-producing tails of a honeyguide (Lyre-tailed Honeyguide) and a snipe; and (D) supporting tail of a woodpecker.

Box 4–2
The display plumes of egrets nearly caused their extinction

The breeding display plumes of Great and Snowy Egrets (see Figure 1–1) were coveted for human display as well. These large semiplumes once commanded a high price for use in ladies' hats of the highest fashion. The resulting slaughter of egrets in their breeding colonies nearly caused their extinction early in the twentieth century. Concerned citizens rallied to the cause of bird conservation with an unprecedented outcry and founded the National Audubon Society, which adopted the egret as its emblem. Laws were passed to protect the egrets, which recovered first in guarded Audubon sanctuaries.

Figure 4–7) but are usually hidden from view. They enhance thermal insulation, fill out the aerodynamic contours of body plumage, and serve as courtship ornaments (Box 4–2).

Filoplumes are hairlike feathers which monitor the movement and position of adjacent vaned feathers (see Figure 4–1). Distributed inconspicuously throughout the plumage, they are most numerous near mechanically active or movable feathers; each flight feather may have 8 to 12 filoplumes. They extend beyond the contour feathers of songbirds, particularly on the back of their necks, a region called the nape (Clark and Cruz 1989). Filoplumes consist of a fine shaft—or rachis—that thickens distally, ending in a terminal tuft of one to six short barbs with barbules. Disturbance of a filoplume's enlarged tip is magnified and transmitted by the long, thin shaft to sensory corpuscles at its base, which then signal the muscles at the base to adjust the feather's position. Filoplumes associated with the flight feathers aid aerodynamic adjustments. Those in association with contour feathers may also help to monitor airspeed; filoplumes are absent in penguins and in Ostriches and other flightless ratites.

Bristles (Figure 4–6) are specialized feathers with both sensory and protective functions. Corresponding to their sensory functions, bristles, like filoplumes, have sensory corpuscles at their bases. Bristles are simplified feathers that consist only of a stiff, tapered rachis with a few basal barbs. Semibristles are similar but have more side branches. Except for those on the knees of the Bristle-thighed Curlew and on the toes of some owls, bristles are found almost exclusively on the heads of birds. The facial feathers of raptors tend to be simplified to bristles and semibristles, which are easier to keep clean than fully vaned feathers. This condition reaches an extreme in the carrion-eating vultures, which have bare heads with scattered bristles. The eyelashes of such birds as Ostriches, hornbills, rheas, and cuckoos consist of protective bristles, as do the nostril coverings of woodpeckers, jays, and crows. Most aerial insect-eating birds have bristles and semibristles around their mouths. The semibristles around the mouths of nightjars and owlet-nightjars are especially well developed, acting not only as insect nets, but possibly also as sensors of tactile information, in much the same way that a cat's whiskers do.

Figure 4–6 Bristles. (A) Whip-poor-will has well-developed bristles about the mouth. (B) Australian Owlet-Nightjar has elaborate bristles and semibristles around its bill. (C) An exception to the usual head locations of bristles are those on the knees of the Bristle-thighed Curlew.

The Feather Coat

The feather coat of most birds consists of thousands of feathers. A Tundra Swan has roughly 25,000 feathers, of which 20,000 (80 percent) are on its head and neck (Wetmore 1936). Songbirds typically have 2000 to 4000 feathers, of which 30 to 40 percent are on the head and neck. The lightness of a single feather belies the total weight of a bird's feather coat. In general, the feather coats of birds weigh two to three times as much as their bones. For example, the plumage of a Bald Eagle weighs about 700 grams, or 17 percent of its total mass (4082 grams), whereas its skeleton weighs only 272 grams, or 7 percent of its body mass—less than half that of the plumage (Brodkorb 1955).

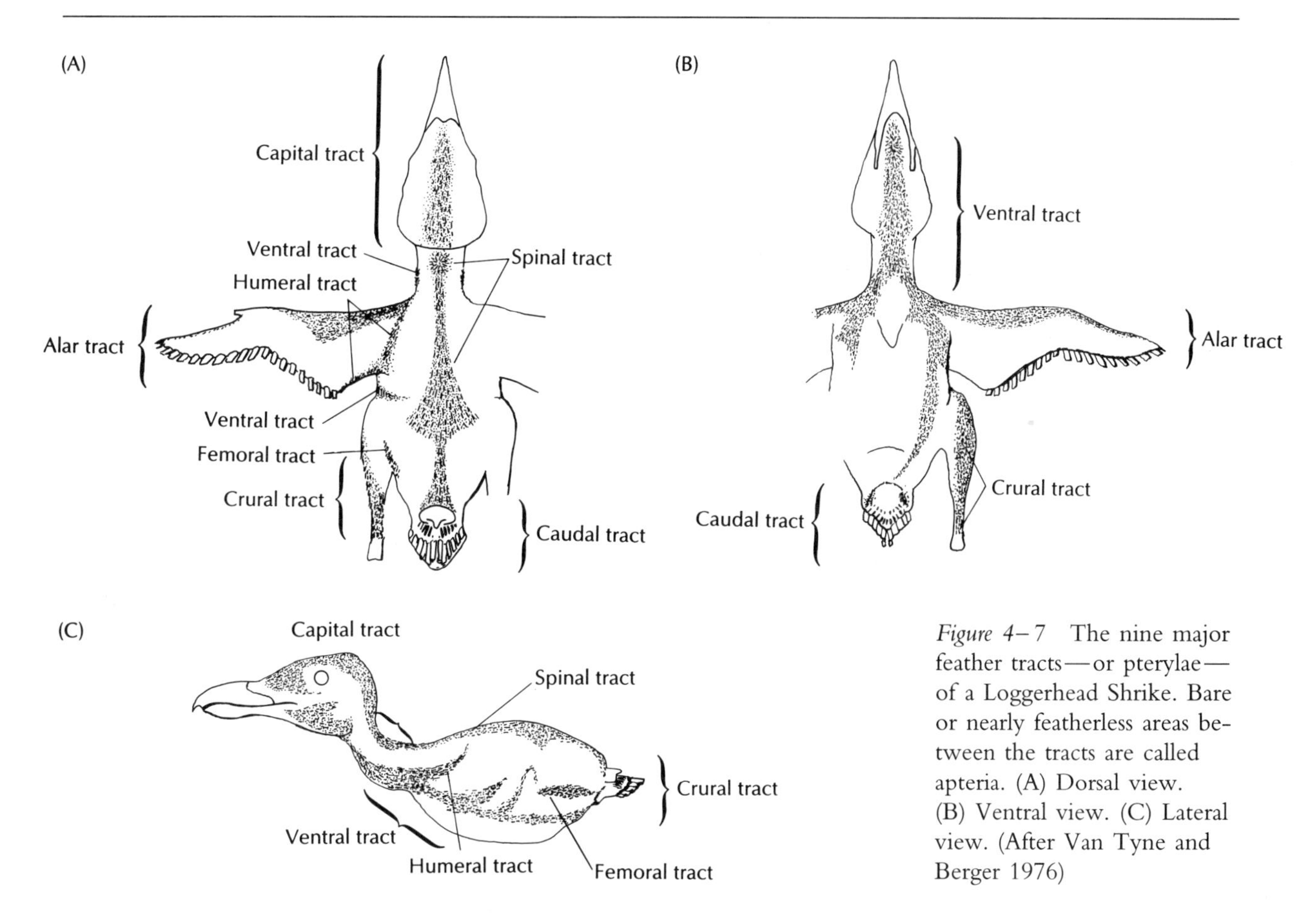

Figure 4–7 The nine major feather tracts—or pterylae—of a Loggerhead Shrike. Bare or nearly featherless areas between the tracts are called apteria. (A) Dorsal view. (B) Ventral view. (C) Lateral view. (After Van Tyne and Berger 1976)

Although feathers cover the entire body of a bird, they are not attached to the skin evenly or uniformly in most birds. Rather, feather attachments are grouped in dense concentrations called feather tracts—or pterylae—which are separated by regions of skin with few or no feathers—called apteria. The feather tracts are not evident without close examination because the feathers spread out from them to cover the entire body. There are eight major feather tracts (Figure 4–7); these are subdivided into as many as 100 separate groupings, which can be used to distinguish avian taxa. The study of these arrangements is called pterylosis. The functional significance of feather tracts and apteria has not yet been established. The bases of adjacent feathers are linked by an elaborate network of tiny muscles, which elevate and depress feathers for courtship displays or heat regulation. If functional sets of feathers were not grouped closely, but were spaced evenly over the skin, these controlling muscles would have to span greater distances. This arrangement would make them less effective unless their mass increased accordingly. The apteria probably facilitate wing and leg movements and provide spaces for tucking these appendages beneath the feather coat. Apteria themselves may also facilitate heat loss (see Chapter 6); penguins lack them. Screamers and mousebirds are the only flying birds that lack apteria.

Feather Care

Daily care of the feathers is essential. Feathers are inert and do not have an internal system of nourishment and maintenance. They would become brittle with age and exposure were it not for regular applications of the waxy secretions of the uropygial gland—or preen gland—located on the rump at the base of the tail. This gland, which is found in most birds,

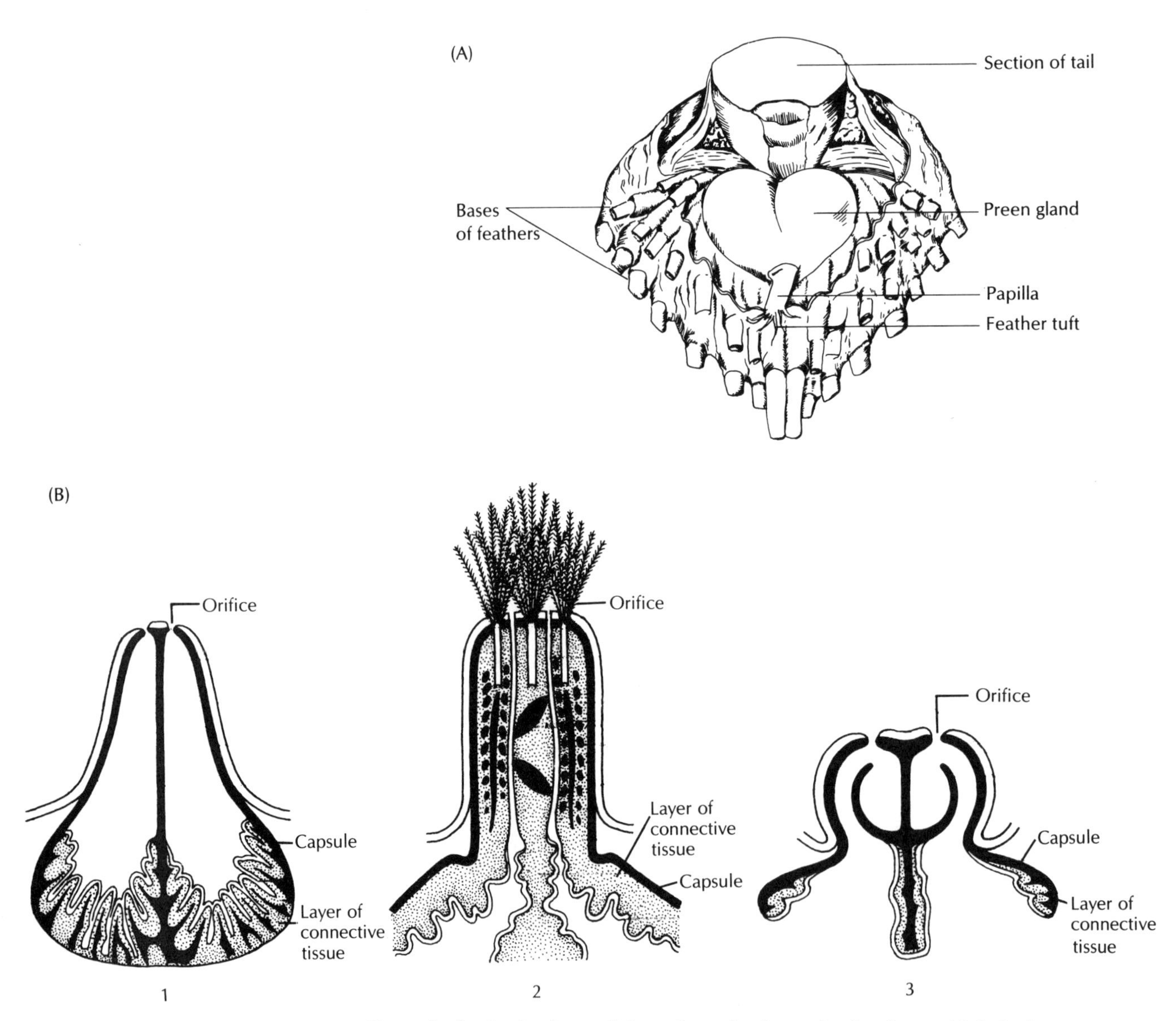

Figure 4–8 At the base of the tail on the lower back of most birds is the preen gland, which produces oily secretions that are essential for feather care. (A) Dorsal view of the gland and its environment on a White Leghorn Chicken. (B) Details of papilla: (1) delicate type; (2) compact type; (3) unique passerine type. (A after Lucas and Stettenheim 1972; B adapted from Jacob and Ziswiler 1982)

appears to have evolved as an essential accessory to feathers. Most preen glands are bilobed structures with a small tuft of downlike feathers encircling the glandular orifices of a well-differentiated papilla (Johnston 1988) (Figure 4–8).

The preen gland secretes a rich oil of waxes, fatty acids, fat, and water, which, when applied externally with the bill, cleans feathers and preserves feather moistness and flexibility (Jacob 1978; Jacob and Ziswiler 1982). Regular applications of the secretion to the plumage sustain its functions as an insulating and waterproofing layer. Water birds typically have large preen glands, but whether the secretions of this organ are essential for keeping feathers dry and maintaining buoyancy remains to be verified.

The waxy secretions of the preen gland also help to regulate the bacterial and fungal floras of feathers. Certain preen gland lipids protect feathers against fungi and bacteria that digest keratin (Baxter and Trotter 1969; Pugh and Evans 1970). Others may promote the growth of nonpathogenic fungi and discourage feather lice. Such chemical hygiene, researchers believe, is one of the most important functions of preen gland secretions. The foul-smelling preen gland secretions of hoopoes and woodhoopoes may also repel mammalian predators.

Birds may preen their feathers as often as once an hour while resting. They systematically rearrange their plumage with their bills and reposition out-of-place feathers. They also draw the long flight feathers individually and firmly through the bill to restore the vane's integrity and to remove parasites (Box 4–3). Feather-chewing parasites damage the structural integrity of feathers, reducing both winter survival and attractiveness of male pigeons to females (Clayton 1990, 1991) (Figure 4–9). Feather damage reduces the insulating quality of feathers and thereby causes metabolic heat production to increase by 8.5 percent (Booth et al. 1994).

Birds groom and delouse head and neck feathers by vigorous scratching. Herons, nightjars, and barn-owls have miniature combs on their middle toe claws that are used in grooming. Most birds scratch their heads directly, reaching up under the wing with a foot, although some scratch indirectly, over the wing (Simmons 1957, 1964; Burtt and Hailman 1978) (Figure 4–10). The advantage of one method over the other is not apparent but may reflect phylogenetic relationships. Crippled and one-legged birds cannot scratch their heads properly and often accumulate large, uncontrolled populations of lice on their heads.

Until recently it was thought that birds lacked poisonous chemical defenses like those of some brightly colored frogs and insects. However, certain New Guinea forest birds—the three species of shrike-thrushes called pitohuis—are now known to be toxic. John Dumbacher and his colleagues (1992) discovered that the skin feathers of shrike-thrushes contain a deadly neurotoxin. The Hooded Pitohui, in particular, carries large amounts of poison. Indigenous New Guinea peoples knew that shrike-thrushes made them sick if eaten without special preparations. Still to be learned are the answers to several key questions. Where do these birds produce this poison? Is it produced in the oil gland? How do they

(A)

(B)

Figure 4–9 (A) Scanning electron micrograph of a chewing louse (*Philopterus* sp.) on a host's feather. (B) Damage to abdominal contour feathers done by feeding lice: (left to right) no damage, average damage, and severe damage. Only the basal downy region and the barbules of the basal and medial regions of a feather are consumed, never the distal region. The barbs and shaft are not damaged, apparently because they are too large to ingest (Clayton 1990). (Courtesy of D. Clayton and K. Hamann)

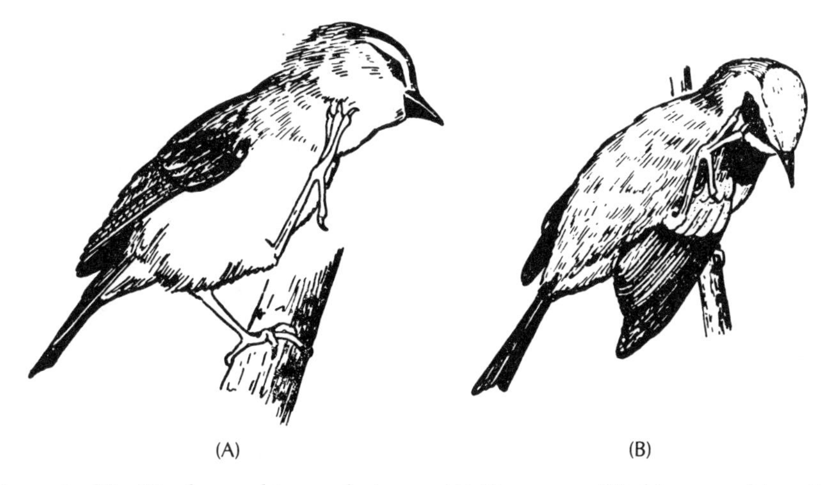

Figure 4–10 Head-scratching techniques. (A) Tennessee Warbler scratching directly, with foot under the wing. (B) Golden-winged Warbler scratching indirectly, with foot over the wing. (From Burtt and Hailman 1978)

keep from poisoning themselves when they preen? The feather poisons of pitohuis, some of which are bright rusty orange and black in color, are similar in chemical structure to alkaloids produced by the deadly arrow poison frogs of South America.

A dustlike substance resembling talcum powder is present on the contour feathers of many birds. Special feathers called powderdowns, which are dispersed throughout the feather coat, slough continuously from the surface of their barbs this waterproofing powder of keratin particles 1 micrometer in diameter. Powderdown feathers grow in dense, distinctly arranged patches on birds such as herons and the unique Courol (formerly called the Cuckoo Roller) of Madagascar and Kagu of New Caledonia.

Box 4–3
Feather ectoparasites are specialists

Living among the feathers themselves are unique bird parasites, which include chewing lice, louse-flies, and feather mites (see Figure 4–9). Chewing lice—or mallophaga—feed on the feathers themselves as well as on blood or tissue fluids. Up to 12 species may inhabit the plumage of one bird, with each species specializing on different kinds of feathers or parts of the body. Louse-flies are flat, tough, clawed, bloodsucking flies specialized for living in the feathers of birds and the fur of mammals. More than 150 species are known to parasitize birds. Louse-flies are the principal vectors of trypanosome and *Haemoproteus* blood parasites and also aid the transport of chewing lice and feather mites from one host to another. Feather mites live their entire life cycle on their avian host and include many species specialized for particular feather microhabitats. Some live on the feather surfaces, others live inside the feather shaft.

Feather Growth

Although a bird can change the position, visibility, and function of its plumage, it cannot change the structure of an individual feather. Feathers are dead structures. Once fully grown, they cannot change color or form except through fading or abrasion. No nerves, muscles, or blood vessels lie beneath the outer surface of the exposed feather. The only mechanism for repair of damage is replacement of the whole feather. Except for cases of accidental loss of feathers and their immediate regrowth, feather replacement takes place regularly with age and with season, and is termed molt.

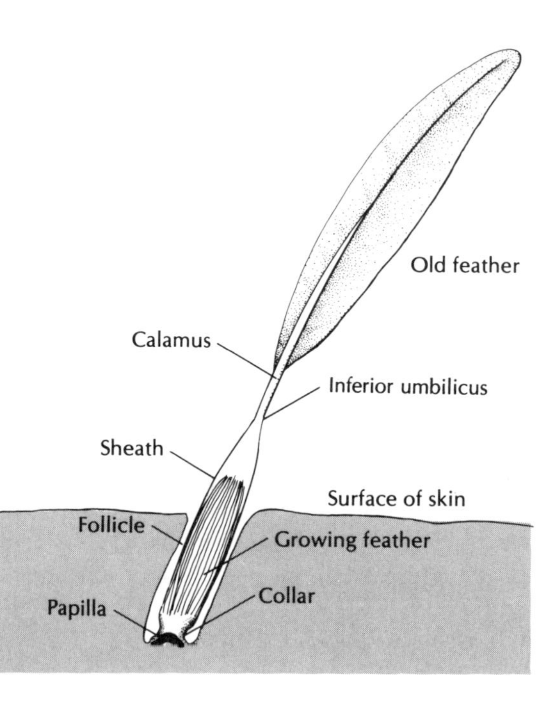

Figure 4–11 A new feather, growing from a papilla and collar in the follicle, pushes out the old feather. (After Watson 1963)

New feathers grow from specialized pockets of epidermal and dermal cells called follicles, which periodically produce new feathers (Figure 4–11). The growth of a new contour feather starts in the follicle with the formation of a thickened dermal papilla. A thin layer of epidermal cells covering the papilla gradually thickens into the collar that develops into the feather. The dermal papilla itself develops into the feather pulp that supports the delicate epidermal cylinder. It also supplies nutrients to, and removes wastes from, the cells of the growing feather. The cells of the epidermal cylinder divide rapidly to form a tubelike structure, the main axis of which will form the rachis and the secondary barb ridges. Cells then differentiate into barbs and barbules.

The new feather grows rapidly, and toward the end of its growth the basal cells form a simple cylindrical calamus that anchors the mature feather in the follicle. The emerging new feather then pushes its predecessor out of the follicle.

The transformation of soft epidermal tissue into a hard durable structure is the last phase of feather growth. A major shift in cell function causes the new feather to fill with keratin, which constitutes 90 percent of the mature feather. As the process of keratinization ceases in the oldest, outermost feather cells, the sheath encasing the young feather cracks, thereby allowing the keratinized feather tips to emerge and the barbs to unfold. The pulp, the core of living cells and blood vessels, is then resorbed by the follicle. The only evidence remaining of this early life-support system is a small hole at the end of the shaft, known as the inferior umbilicus.

The follicle grips the feather at the calamus by a combination of muscular tightening and friction. Substantial force—500 to 1000 grams for a single body feather of the average chicken—is required to pull a feather from this grip: The tight grip of follicular muscles, which are controlled by the autonomic nervous system, may relax when a bird becomes frightened. The resultant loss of feathers is known as fright molt. Nightjars, for example, easily drop their feathers when disturbed.

Molts and Plumages

Every bird goes through a series of plumages—or feather coats—in its lifetime. The first natal down plumage may consist of a few scattered down feathers, the feather coat found on most hatchling land birds; or it may be a dense, fuzzy covering like that of ducklings and chicks. Birds such as ducks, whose natal down covers their entire bodies, are said to be ptilopaedic; by contrast, psilopaedic chicks, such as a hatchling thrush, have only a few scattered down feathers. Such fragile feathers rarely last more than a week or two. They are soon replaced by a more substantial set of downy or vaned feathers. In loons and penguins, a second generation of down grows from the same follicles, pushing the old ones out. In hawks and waterfowl, a second coat of down grows from different follicles.

Most birds have only one coat of natal down, which is pushed out of its follicles by incoming juvenal pennaceous feathers in the first weeks of life. Wisps of down may remain attached for a time to the new feathers. The baby bird's first set of wing and tail feathers appear at this time and grow rapidly in preparation for flight. As the young bird—now called a juvenile—approaches independence, it exchanges parts of its juvenal feathers for new plumage.

The replacement of juvenal plumage by immature or adult plumage involves most of the feather coat, although not always those of the wings or tail. The young American Robin, for example, begins in midsummer to replace its spotted juvenal plumage with unspotted adult plumage (Figure 4–12). The first wing feathers remain. A few months later, its original flight feathers will propel the young robin on its first migratory flight. The bird will not molt again until it is just over 1 year old.

An adult bird typically molts after breeding, replacing its entire plumage. It may keep its new set of feathers for 12 months, or it may replace

Figure 4–12 The spotted plumage of a juvenile American Robin, with residual tufts of down still attached to incoming head feathers.

some plumage before nesting the following year, converting somber camouflage plumage to a brightly colored plumage for territorial display. Feathers of species that retain their plumage a full year may change in appearance because of wear. The European Starling, which is spotted in the winter, loses its spots as the feather tips wear off; by spring it is sleek and glossy. Meadowlarks also wear off the buffy feather tips of their winter plumage, exposing bold black and yellow underparts in the spring.

Molt Patterns and Terminology

Phillip Humphrey and Kenneth Parkes (1959) recognized the need for a terminology of molts and plumages that is independent of seasonal aspects, which vary among species and regions. They proposed that molts be related to the incoming generation of feathers, because that feather loss is a passive result of the growth of new feathers. They also proposed (1) that the plumage that is renewed after breeding be considered the main component—or the "Basic" plumage—of the annual cycle of plumages, and (2) that breeding adornments be considered temporary additions—or "Alternate" plumages.

Consider the plumages of a male Scarlet Tanager, which wears an olive-green Basic plumage in fall and winter and a bright red Alternate plumage in spring and summer. The plumage of a male Scarlet Tanager at any given time actually comprises a series of feather generations that bear testimony to the bird's age and the time of its last molt (Table 4–1). In its first month, a juvenile male Scarlet Tanager is olive-green with olive-brown wing feathers and streaked underparts. The first Prebasic molt in July and August produces the Basic 1 plumage, which resembles its unstreaked, olive-green mother, except for black wing coverts that identify it as a male. In less than a year, this male tanager will undergo the Prealternate 1 molt, which replaces most of its olive-green body plumage with red-orange. Molting males appear with peculiar mosaics of differently colored old and new feathers: green, yellow, and orange. Even when fully red-orange in May, the young tanager still has the olive-brown wings that signal its first-year status; adult males have black wing feathers.

In its second fall, the male tanager undergoes the Prebasic molt of its entire plumage, replacing red-orange body feathers with olive-green winter or Basic 2 plumage and replacing its olive-brown wing feathers with jet black feathers. The following spring the Prealternate molt replaces winter plumage with a bright red breeding plumage. The maturing tanager now resembles other adult males and proceeds through regular cycles of Prebasic molt into camouflaged Basic plumage after breeding and Prealternate molts into brightly colored Alternate display plumage before breeding.

The comparative study of molts and plumages reveals that some species, such as the American Robin and European Starling, undergo only a single annual molt, whereas others, such as the Scarlet Tanager, have extra seasonal molts (Dwight 1907; Stresemann and Stresemann 1966). One complete molt a year was probably the primitive pattern from which more complex molt patterns evolved, and it continues to be the typical

TABLE 4–1
Plumages and molts of a male Scarlet Tanager

Parameter	Year	Winter	Spring	Summer	Fall
Molt	1			Prejuvenal	First Prebasic
Plumage				Juvenal	Basic 1
Color[a]				—	Green/brown
Molt	2		Prealternate 1		Prebasic
Plumage		Basic 1	Alternate 1		Basic
Color		Green/brown	Red/brown		Green/black
Molt	3+		Prealternate		Prebasic
Plumage		Basic	Alternate		Basic
Color		Green/black	Red/black		Green/black

a. Body/wings.

pattern. Gradual feather replacement imposes the least metabolic stress on an individual, and a yearly molt is sufficient to offset normal rates of feather wear. Multiple molts have proved advantageous for some birds as aids to seasonal display or as adaptations to severe feather wear or parasites. For example, in deserts, where wind and sand rapidly destroy feathers, some African larks molt completely twice a year. European larks, which suffer less abrasion, molt only once a year. Species, such as the Bobolink (Figure 4–13) and Sharp-tailed Sparrow, that live in coarse grass habitats may also molt twice a year (Stresemann 1967). Shedding parasites is one apparent result of the double molt in this sparrow. It has fewer feather parasites than the Seaside Sparrow, which lives in the same marshes but molts only once a year (Post and Enders 1970).

A few birds molt three or four times a year, but the extra molts are only partial ones. The Ruff, a large shorebird with an unusual mating system, undergoes the Prebasic molt in the fall; the Prealternate molt in the spring, which produces most of its breeding plumage; and then a third supplemental molt, which produces the "ruff" (see Chapter 9). To match their camouflage to the seasonal changes in the tundra, ptarmigan have three partial molts a year, and some populations of the Willow Ptarmigan have four.

Whereas geese have simple annual molt and plumage cycles, many ducks of the north temperate regions have evolved more unusual se-

Figure 4–13 Bobolinks molt completely twice a year. The male changes from a brown streaky (left) plumage like the female's in the winter to a bold black-and-white plumage (right) in the spring.

quences, in which the Prealternate molt starts before the fall Prebasic molt finishes. After they breed, drakes undergo a rapid Prebasic molt that often includes simultaneous loss of all flight feathers, rendering them flightless and vulnerable for several weeks. The Basic plumage that follows is a dull, hen-colored "eclipse" plumage that does not last long. An early Prealternate molt produces the drake's handsome breeding (Alternate) plumage by early winter, the season of courtship and pair formation.

Sequences of Feather Replacement

Molting in most birds follows a regular sequence within each feather tract. The usual sequence for the primary flight feathers, for example, is from the innermost primary outward to the last feather of the wing tip. In contrast with groups such as the ducks that become flightless because they molt all their flight feathers at the same time, regular and symmetrical sequences of flight feather replacement help to maintain flight ability. The staggered replacement of the primaries and secondaries of the wings produces only small, temporary gaps in the wing surface and only a small reduction of flight power.

Like the flight feathers of the wing, tail feathers typically molt centrifugally from the innermost pair to the outermost pair, with some exceptions. Large Asian partridges called snowcocks use their enormous tails with 20 rectrices for additional aerodynamic lift in sailing across steep ravines. Snowcocks have a pattern of tail molt that differs from that of other partridges; it starts in the middle of each side of the tail and proceeds slowly in both directions, a process that maintains needed aerodynamic lift for flight. The extended display primaries of male Standard-winged Nightjars mentioned above emerge last, out of normal sequence, apparently because they are a liability in flight.

We do not know what triggers the serial replacement of feathers (Payne 1972). Possibly, each follicle has a different sensitivity to the hormones that trigger molt. If so, the feather with the greatest sensitivity would drop first. Adjacent follicles may interact, directly or indirectly, in association with local expansion of blood vessels. Much remains to be learned about the precise control of molt sequences.

Feather Colors

The intricacies of feather microstructure and pigments combine at times to produce stunning effects in plumage. Brilliant reds, greens, and blues are combined into bold plumage color patterns, as in the Painted Bunting. The tiny Many-colored Rush-Tyrant of South America is red, orange, blue, green, yellow, black, and white. Locally it is called *Ciete Colores*, meaning "seven colors." At the other extreme, however, are drab gray-olive birds such as the Northern Beardless-Tyrannulet of Central

America and the leaf-warblers (Family Sylviidae, Genus *Phylloscopus*) of Europe and Asia. Bold or subdued, plumage color patterns evolve in concert with behavior. The resulting conspicuous, distinctive, or cryptic elements of plumage color play a key role in visual signaling and communication, discussed in Chapter 9.

Feather colors come in all shades, hues, and tints, due either to biochrome pigments deposited in the feather microstructure or to special features of feather surfaces. Biochrome pigments are naturally occurring chemical compounds that absorb the energy of certain wavelengths of light and reflect the energy of other wavelengths to produce the observed colors. Structural colors result from physical alteration of the components of incident light on the feather surface.

Biochrome Pigments

The three major categories of feather pigments are melanins, carotenoids, and porphyrins. Melanins, which produce earth tones—grays and blacks, browns, and buff colors—are found in all birds. Carotenoids, which produce bright yellows, oranges, reds, and certain blues and greens, are present in a broad variety of birds. Porphyrins are more restricted or unstable, ephemeral pigments responsible for particular bright brown and green feather colors, and also a unique magenta.

With the exception of albinos, virtually all birds have some melanin pigment in their feathers. Melanin pigment is synthesized from the amino acid tyrosine by mobile pigment cells called melanoblasts, which creep about in the inner (dermis) layer of the skin. Melanoblasts manufacture and insert melanin granules into specific cells that are destined to become particular barbs and barbules. Periodic deposition into the embryonic feather structures during development produces subtle color patterns such as barring or speckling. Shades of brown or gray depend on the density of melanin deposition.

Two kinds of melanin prevail in the barbs and barbules of bird feathers. Eumelanins are large, blackish, regularly shaped granules that produce dark brown, gray, and black. Phaeomelanins are smaller, irregularly shaped, reddish or light brown granules that produce tans, reddish browns, and some yellows. Color patterns often result from having mostly eumelanins in some places and mostly phaeomelanins elsewhere. In the plumage of the Gray Catbird, for example, the lead-gray color of most of the plumage results from eumelanin, and the rusty color of the undertail coverts comes from phaeomelanin.

Melanins perform many functions. The extra keratin associated with melanin makes the feather more resistant to wear (Burtt 1979). Dense melanin concentrations in the black wingtips of high-speed aerial species, such as gulls and gannets, reduce fraying of those feathers. Melanins help protect the feathers of desert species from sand abrasion. Melanins also absorb radiant energy, which aids thermoregulation (see Chapter 6). There is speculation that melanin granules promote drying of damp feathers by absorbing and concentrating radiant heat in the feather microstructure (Gill 1973; Wunderle 1981). If true, this could help to explain why birds of wet climates tend to be dark colored, a phenomenon known as Gloger's rule, in honor of Constantin Gloger, who studied the relationship between climate and color variation in birds.

Most carotenoid pigments, which are responsible for bright red, orange, and yellow colors, are derived from the diet (Box 4–4). Once assimilated, however, they can be modified and used in many ways that are under genetic or physiological control. They dissolve easily in lipids or organic solvents and are often stored in egg yolk, body fat, and the secretions of oil glands. The solubilities of carotenoid pigments enable chemists to extract them easily and to identify those responsible for particular plumage colors. Carotenoid pigments accumulate in droplets of lipid in the cells of growing feathers and are then left embedded in the barbs and barbules when the natural fat solvents disappear during the last stages of keratinization.

A comparison of the placement of carotenoid pigments in the feathers of male Red Crossbills and male White-winged Crossbills shows how different effects have been achieved (Figure 4–14). The feathers of the Red Crossbill are deep brick red, whereas those of the White-winged Crossbill are pink. Red pigment is present in both barbs and barbules of Red Crossbill feathers but only in the barbs of White-winged Crossbill feathers. The mixture of clear barbules and red barbs in the feathers of the White-winged Crossbill produces pink, but the bird's plumage reddens as the clear barbules abrade.

Porphyrins, which are related chemically to iron-containing hemoglobin and liver bile pigments, show intense red fluorescence under ultraviolet illumination. Porphyrins are fairly common in the reddish or brown feathers of at least 13 orders of birds, notably owls and bustards. These pigments, however, are found primarily in new feathers, because they are chemically unstable and easily destroyed by sunlight. The best-known avian porphyrin pigment contains copper instead of iron. This pigment—called turacin or uroporphyrin III—produces the bright magenta in the wings of turacos, which are spectacular crow-sized birds of African forests.

Chemically related to turacin is another unusual copper porphyrin pigment called turacoverdin, which produces the bright green colors of a few birds (Dyck 1992). Most of the green colors of birds result from combinations of yellow pigments and structural blue—or iridescence (see next section). Olive-green colors usually result from combinations of melanins and carotenoids, one in the barbs, the other in the barbules. Turacoverdin, which is close in chemical composition to turacin, creates the true green colors of turacos. It is also responsible for the green back and wing feathers of the Crested Partridge and the Blood Pheasant, and the green wings of the Wattled Jacana. The pigment responsible for the green head feathers of the Pygmy Goose (Family Anatidae) and some eiders is different and essentially unstudied.

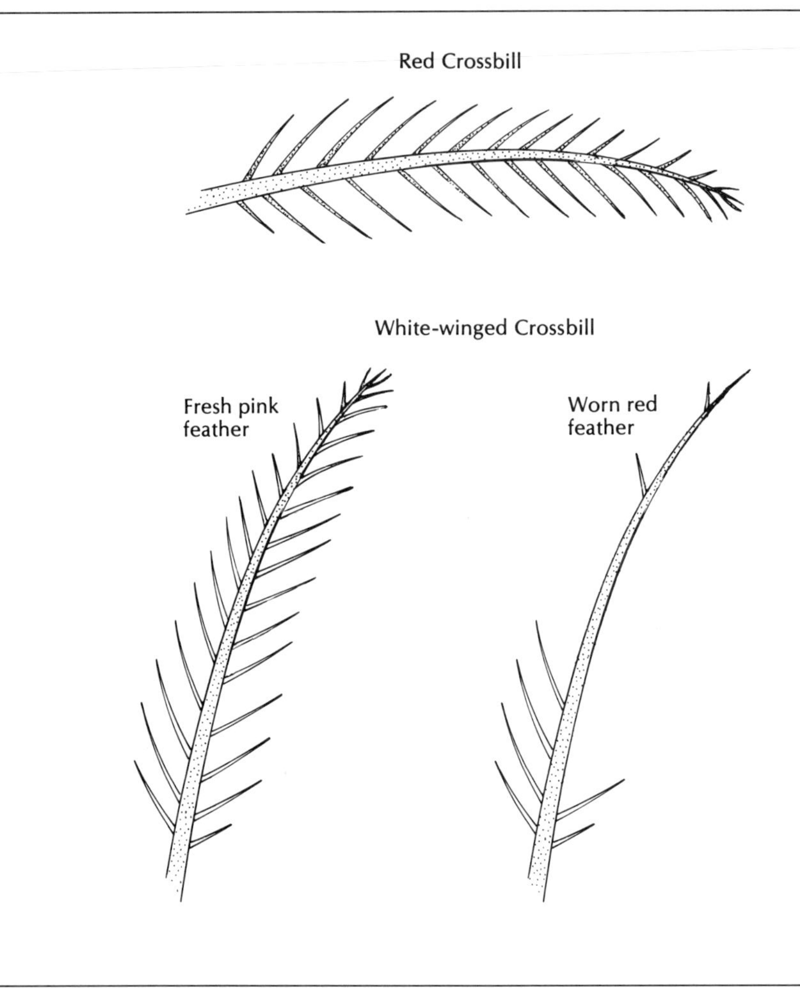

Figure 4–14 Red carotenoid pigment (stippling) is deposited in both the barbs and barbules of feathers of the Red Crossbill but only in the barbs of the similar White-winged Crossbill, leaving the barbules clear. New plumage of the White-winged Crossbill is therefore pink, not red; but the plumage reddens as the unpigmented barbules wear off.

Box 4–4
Bright red marks the best male House Finches

Male House Finches vary in plumage color from bright to pale red or even yellow-orange. With a series of elegant experiments, Geoffrey Hill (1990, 1991, 1992, 1993) demonstrated not only that plumage color of male House Finches reflects their diet but also that females prefer brightly colored males over dull-colored males.

High-quality males apparently have intrinsically superior foraging ability and better access to carotenoid-rich foods that brighten their colored badges. Both the intensity and extent of red carotenoid pigmentation in the plumage reflect the carotenoid pigments they eat while they are replacing their feathers during the annual molt; these finches cannot use carotenoids stored in advance of the molt to achieve a brighter plumage.

Because bright red pigmentation serves as an accurate badge of male quality, female House Finches use the red color badge to pair preferentially with the best males. Brightly colored males are better providers for their families—they bring more food to their females during incubation and to their nestlings. Females paired with dull colored males more often abandon the breeding effort, apparently because of inadequate provisioning by their mates. Bright red males also survive the winter better than dull males and thus are available for more than one breeding season.

Structural Colors

Many of the brightest feather colors such as parrot greens, bluebird blues, and hummingbird iridescences are structural colors that result from the physical alteration of incident light at the feather surface. In general, structural blues and greens result from the scattering of short (blue) wavelengths of incident light by tiny melanin particles in the surface cells on the feather barbs. The remaining longer (red and yellow) wavelengths pass through the surface layer to an absorbent melanin layer below. Blue is left as the apparent hue. Differently sized particles scatter different wavelengths of light; those responsible for parrot greens are larger than those responsible for bluebird blues. Different mechanisms based on interference among wavelengths of incident light are responsible for the structural blues and greens of small African parrots, called lovebirds (Dyck 1971; Brush 1972). A completely novel mechanism of structural color production is responsible for the brilliant green skin color of the Velvet Asity, a species found only on Madagascar. Rick Prum and his colleagues (1994) discovered that green color is produced by the reflection of light from precisely ordered hexagonal arrays of parallel protein (collagen) fibers in the fleshy growths over the eyes of the males in breeding condition.

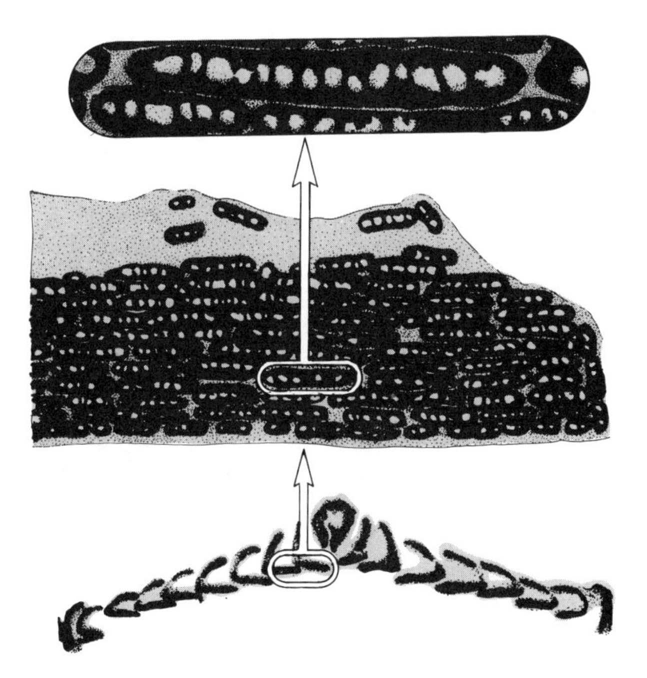

Figure 4–15 The bright, directional iridescent colors of hummingbirds result from interference and reinforcement of light components reflected by layers of hollow platelets on the surfaces of broadened barbules. Shown here are three electron photomicrographs of increasing magnification. (Bottom) The cross section of a barbule. (Middle) A tiny section is magnified 16,000 times to show the layers of platelets. (Top) The cross section of one platelet is magnified 45,000 times to show the internal air spaces. (From Greenewalt 1960a)

Carotenoid pigments can convert structural blues to green or violet. Wild Budgerigars, for example, are green because of an association of yellow pigment with structural blue. Mutant parakeets are blue rather than green because a single recessive gene blocks carotenoid pigment deposition. If the carotenoid pigment is red rather than yellow, violet or purple results, as in the Pompadour Cotinga (Brush 1969). Structural blue from the barbs plus red pigment in the barbules is responsible for the purple head feathers of the Blossom-headed Parakeet.

Iridescence projects glistening colors such as the "eyes" on a peacock's tail and the brilliant throat of a hummingbird. These colors depend on the angle of incidence of the light striking the feathers; iridescent feathers appear black from certain viewing angles because their colors result from interference of light waves reflected from the outer and inner surfaces of hollow granules or other structures (Figure 4–15). The brilliant iridescences of hummingbird feathers come from 7 to 15 closely stacked layers of tiny melanin granules, located on barbules. Each granule is a flat, hollow platelet with two reflecting layers that create particular colors by light interference and reinforcement (Greenewalt 1960a; Greenewalt et al. 1960). The intensity of the iridescence increases with the number of granule layers. The iridescent colors of some African starlings are caused by reflections from the interfaces between melanin granules and keratin layers. The iridescent colors of trogons are produced by air-filled melanin plates and hollow melanin tubes, uniform in some places to an accuracy of less than 0.01 micrometer and arranged in precise layers (Dürrer and Villiger 1966).

Genetic Control of Feather Color

Except for a few domesticated birds, especially chickens, little is known about the genetic control and inheritance of feather colors (Buckley 1987). The presence, absence, or pattern of deposition of particular pigments often is controlled by simple Mendelian genes that segregate and recombine in predictable combinations, just as the genes that control blue eyes versus brown eyes in humans do. Analogous to human eye colors are alternative plumage colorations—called color phases—in birds, including dark-colored versus white phases of herons, seabirds, and geese; rusty versus gray phases of owls and nightjars; black versus pied phases of oystercatchers and passerine songbirds; buffy versus grayish downy chick colors of swans, geese, and terns. The Gouldian Finch, a brightly colored Australian species commonly kept as a cagebird, comes in red-faced, black-faced, and yellow-faced color phases. White-throated Sparrows come in two color types that differ in the color of their head stripes, white versus tan. This difference corresponds to the presence or absence of a particular chromosome. Such color differences can serve as genetic markers for studies of population structure and mating preferences (Chapter 22). The details of genetic control of feather structure and color and how these relate to the evolution of elaborate display plumages, seasonal changes in plumage, or species differences remain major topics for future research.

Summary

The avian feather is a unique structure that is versatile in form and function. Feathers provide insulation, which enables birds to maintain a high body temperature; they are essential for flight and serve in visual communication and camouflage. Modified feathers aid in swimming, sound production, protection, cleanliness, water repellency, water transport, tactile sensation, hearing, and support of the bird's body.

The basic structure of a contour feather consists of a stiff, central rachis with side branches called barbs and secondary side branches called barbules. The interlocking system of barbs and barbules forms a flexible but cohesive flat surface called the vane. Loose barbs and barbules at the base of the feather enhance insulation. Other major kinds of feathers include the flight feathers, down feathers, semiplumes, filoplumes, bristles, and powderdown. The tough, inert molecules that form the feather are a unique form of keratin.

The entire feather coat consists of thousands of individual feathers, which are arranged in groups called tracts. Linking the bases of adjacent feathers is a system of tiny muscles that control feather position. The feather coat of a bird typically weighs two to three times more than its skeleton. The entire feather coat is replaced at least once a year in regular molts. Partial molts may supplement the main annual molt to produce composite plumages.

Feather coloration is controlled by carotenoid and melanin pigments, which are deposited in the barbs and barbules, and by structural alteration of light at the feather surface, which is responsible for most of the blue and green colors of bird feathers. Iridescent colors result from interference patterns of light as it is reflected by special layers of pigment granules.

Further Readings

Brush, A.H. 1993. The evolution of feathers: A novel approach. Avian Biology 9: 121–162. *A new perspective on the chemical structure and evolution of feathers.*

Butcher, G.S., and S. Rohwer. 1989. The evolution of conspicuous and distinctive coloration for communication in birds. Current Ornithology 6: 51–108. *A provocative application of the comparative method to the analysis of plumage color patterns in birds, including an extensive historical bibliography.*

Jacob, J., and V. Ziswiler. 1982. The uropygial gland. Avian Biology 6: 199–324. *A review of the structures of preen glands and the chemistry of their secretions.*

Lucas, A.M., and P.R. Stettenheim. 1972. Avian Anatomy: Integument. Washington, D.C.: U.S. Government Printing Office. *The standard reference on feathers.*

Stettenheim, P. 1973. The bristles of birds. The Living Bird 12: 201–234. *A well-illustrated review of these highly modified feathers.*

Stettenheim, P. 1976. Structural adaptations in feathers. Proc. Int. Ornithol. Congr. 16: 385–401. *An elegant summary of the adaptive modifications of feather structures.*

Stresemann, E., and V. Stresemann. 1966. Die Mauser der Vögel. J. Ornithol. 107: 1–445. *The encyclopedia of molt patterns of birds of the world.*

CHAPTER 5

Flight

FLIGHT IS THE CENTRAL avian adaptation. Yet birds do not merely fly. They are masters of the fluid that is air, just as fishes are masters of the fluid that is water. Birds can hover in one place, dive at breathtaking speeds, fly upside down and backward, and soar for days on end. In their evolution, birds have exploited an extraordinary range of specialized modes of flight.

This chapter first considers the anatomical bases of avian flight, particularly the skeleton and flight muscles. It then proceeds to the basic aerodynamic principles of flight, including the role of wings as airfoils, the phenomenon of lift, and the countering forces of drag. The gliding flight of soaring birds helps to illustrate these principles. An entirely different mode of flight is that of hummingbirds, which use propellerlike wings for precision, helicopterlike control of their movements. The more complex forms of flapping flight have recently been clarified through the experimental study of airflow in the wakes of flying birds. At the opposite extreme are flightless birds. Some diving birds (such as penguins) have traded aerial flight for underwater flight using highly modified flipperlike wings.

Avian Skeleton

The skeleton of a bird is uniquely structured for flight (Figure 5–1). Fusions and reinforcements of lightweight bones make the avian skeleton both powerful and delicate. Unusual joints not only make flight motions possible but also brace the body against the attendant stresses. The skeleton strategically supports the large muscles that provide the power for flight. Cross sections of bird bones reveal light, air-filled structures unlike the relatively dense, solid bones of many terrestrial animals. The hollow long bones of the wings are particularly strong (given their mass) and in many cases are strengthened further by internal struts. Instead of a heavy, bony jaw filled with dense teeth, birds have a lightweight, toothless bill. The huge, hollow bills of toucans, for example, are not the burden they seem. The sternum—or breastbone—typically has a large keel—or carina—which anchors the major flight muscles. A bird's flying ability is correlated with the size of its keel; some flightless birds lack the keel completely.

The skeleton of birds is able to withstand the strains imposed by flight. The thorax is more rigid and better reinforced than that of a reptile. The fully ossified dorsal and ventral ribs provide a strong connection between the backbone and the breastbone. Horizontal bony flaps—called uncinate processes—extend posteriorly from the vertical upper ribs to overlap the adjacent rib and reinforce the rib cage. In addition, the partially fused thoracic vertebrae have limited flexibility. The pectoral girdle includes, on each side, the coracoid, scapula, and furcula. Dorsally situated on top of the rib cage are the long, saberlike scapulae, each of which joins anteriorly to the coracoid and furcula—or clavicle—to complete a triangular system of struts that resist the chest-crushing pressures created by the wing strokes during flight. An acute angle between the attachment of the scapula to the coracoid reduces the stretch and increases the exertion force of the dorsal elevator muscles, which help pull the humerus—or upper wing bone—upward. This angle is oblique in flightless birds.

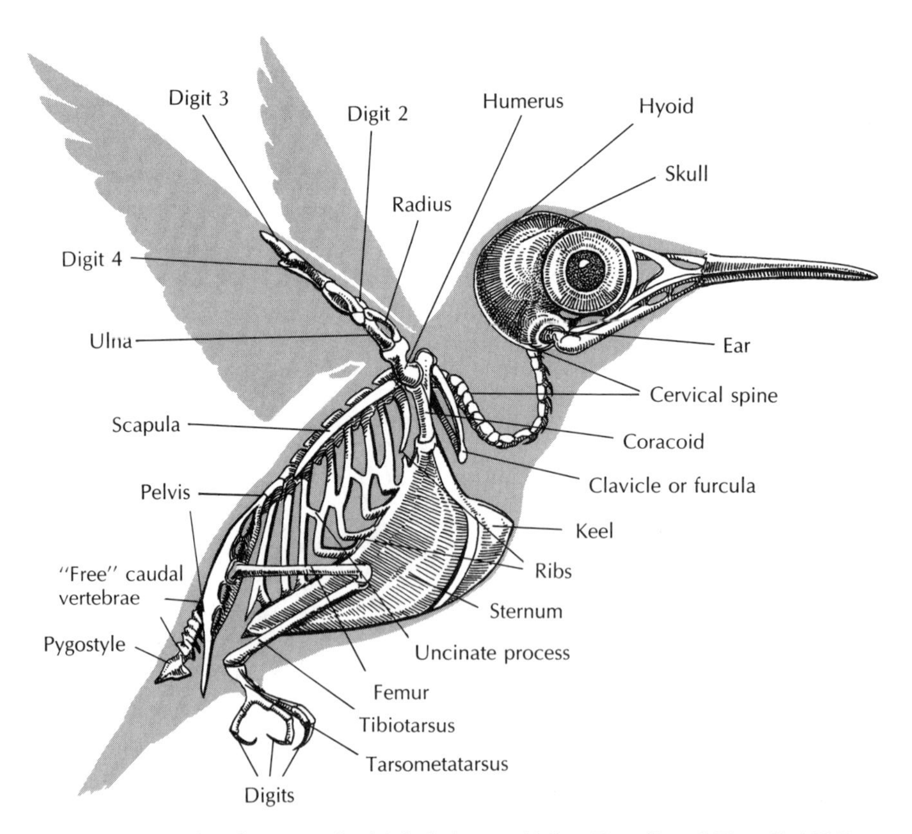

Figure 5–1 Major features of a bird skeleton. (After Tyrrell and Tyrrell 1985)

Box 5-1
The furcula is a flexible, elastic spring

In most birds the furcula—or wishbone of the holiday dinner turkey—is a fused pair of clavicles—or "collar bones"—and serves as a strut or spacer between a bird's shoulders. X-ray movies of flying European Starlings reveal that in flight the furcula can act as an elastic spring (Jenkins et al. 1988; Pool 1988). With each beat of the wing, the upper ends of the furcula spread widely, up to 50 percent wider than the normal resting width, and then contract. The furcula repeats this cycle of wide elastic expansion and contraction 14 to 16 times a second in synchrony with the starling's wing beats. Exactly how the spring action of the furcula aids flight is unclear, but it may enhance respiratory performance by pumping air through the air sacs (Chapter 6).

The avian wing is a modified forelimb. The humerus, ulna, and radius are homologous to the limb bones of tetrapods. Large surfaces at the joints between the limb bones allow the resting wing to fold neatly against the body. These elaborate joints also permit the wing to change positions and angles during takeoff, flight, and landing. When outstretched, these joints are strong enough to withstand the powerful wrenching forces created during wing strokes.

The fused hand and finger bones help provide strength and rigidity in the outer wing skeleton. Most of the wrist bones—the carpals and metacarpals—are fused into a single skeletal element called the carpometacarpus. There are only 2 free carpals in the avian wrist, far fewer than the 10 or more in most other vertebrate wrists. The hand itself includes three digits, rather than the five found in most tetrapods. A similar condition is found in some dinosaur fossils. The alula—or bastard wing—originates from the first digit—the thumb—and moves independently of the rest of the wing tip. Within the wing itself are powerful tendons and compact packages of tiny muscles that control the subtle details of wing position.

Flight Muscles

The two great flight muscles—pectoralis and supracoracoideus—originate on the pectoral girdle and insert onto the expanded base of the humerus. Their ventral positions help to lower the bird's center of gravity in flight.

The pectoralis muscle complex is the largest and accounts for about 15 percent of the total mass of a flying bird. Contraction of this muscle pulls the wing down in the power stroke. The pectoralis muscle attaches to the furcula and to the strong membrane between the coracoids and the furcula. It attaches also to the peripheral portions of the sternum, including the outer part of the keel. In tree-trunk climbing birds with shallow keels, such as woodcreepers, the pectoralis muscle spreads thinly over the rib cage for attachment.

The two supracoracoideus muscles housed on opposite sides of the sternum lift the wings on the recovery stroke (Figure 5–2). From each of these ventrally positioned muscles, a strong tendon passes upward and forward through the triosseal canal (formed by the junction of the coracoid, scapula, and furcula) and inserts on the dorsal side of the base of the humerus. The dorsal insertion of this pulleylike tendon enables the ventrally located supracoracoideus muscles to raise the wing. The supracoracoideus muscles are required for the rapid initial wing beats that are essential for clearing the ground quickly upon takeoff and achieving a minimal airspeed. A pigeon is unable to take off from the ground if its supracoracoideus tendons are cut experimentally (Sy 1936). Once

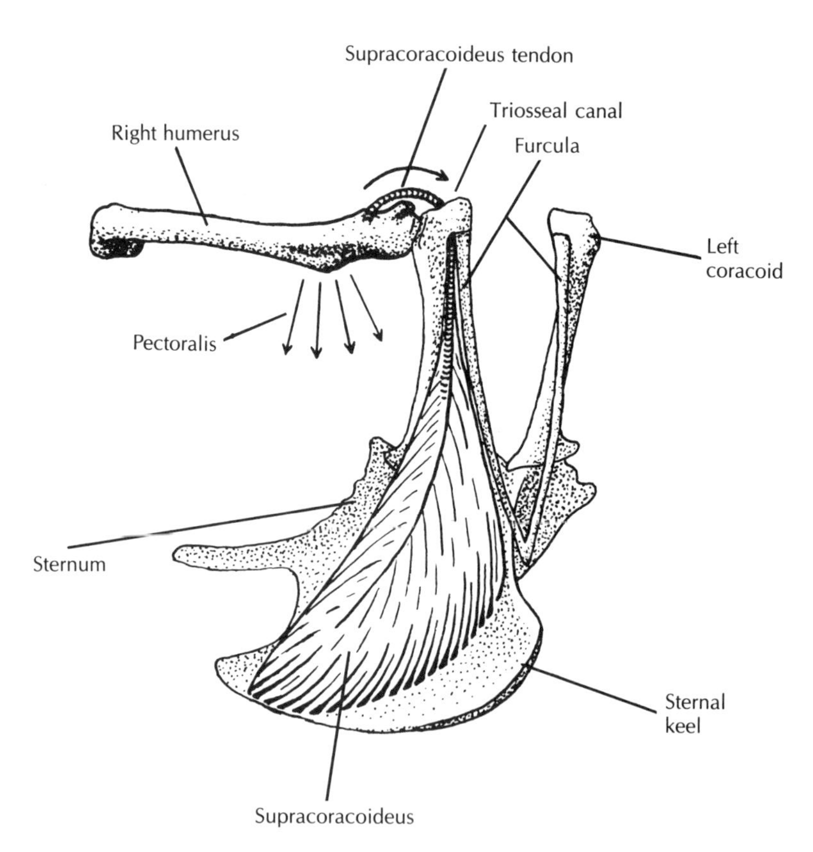

Figure 5–2 Right front view of the pectoral girdle and sternum of a pigeon. The ventrally located supracoracoideus muscle raises the wing by means of a pulleylike tendon that passes to the dorsal surface of the humerus through the triosseal canal between the furcula and coracoid bones, plus the scapula (not illustrated here). The curved arrow indicates the action of this tendon. The pectoralis muscle, which has been removed in this drawing, inserts onto the lower side of the humerus and pulls the humerus downward, as indicated. (From George and Berger 1966)

launched and airborne, however, pigeons can fly quite well without functional supracoracoideus muscles because the smaller dorsal elevators can handle the less-demanding recovery strokes of the wings during sustained flight.

The supracoracoideus muscles typically are smaller than the pectoralis muscles. They are large relative to body size in hummingbirds, which use the upstroke of the wing as a propelling power stroke rather than as a recovery stroke. The supracoracoideus of hummingbirds is five times larger relative to body size than in most other birds. It is half the size of the pectoralis muscle and constitutes 11.5 percent of total body mass, more than in any other bird. The supracoracoideus muscle is also unusually large in penguins, whose flippers propel them forward with a powered upstroke as well as downstroke.

The power for flight derives from the metabolic activity in the cellular fibers of flight muscles, some of which have an extraordinary capacity for aerobic metabolism. Certain muscle fibers are suited to specific modes of flight. The extremes of the variation are the red and the white fibers, but intermediate fiber types exist.

The sustained contraction power of red muscle fibers results from oxidative metabolism of fat and sugar. These narrow fibers have high surface-to-volume ratios and short diffusion distances, which aid the uptake of oxygen. They also contain abundant myoglobin, mitochondria, fat, and enzymes that catalyze the chain of metabolic reactions known as the Krebs cycle. Experimental studies of extracts from pigeon breast muscle, which is rich in red fibers and the associated enzymes, have contributed to our present knowledge of aerobic metabolism.

Sustained flight power derives from a high concentration of red muscle fibers in the flight muscles. The aerobic—or free-oxygen—capacity of the flight muscles of small songbirds and small bats is at the highest level known for vertebrates. Few birds have muscle that consists entirely of red fibers. Rather, blends of different fibers that combine short-term power with long-term endurance in flight are typical of most birds. The dark red breast muscle of pigeons consists mostly, but not exclusively, of red fibers. Conversely, the white breast muscle of fowl has a low proportion of red fibers. Extreme cases include sparrows, which have only red fibers in the pectoralis muscles, and hummingbirds, which have only red fibers in both pectoralis and supracoracoideus muscles.

White muscle fibers are powered by products of anaerobic—without free oxygen—metabolism. Unlike red fibers, they contain little myoglobin, few mitochondria, and a different set of enzymes. The white fibers are capable of a few rapid and powerful contractions, but they fatigue quickly as lactic acid—a product of anaerobic metabolism—accumulates. The light meat of the breast muscles of domestic fowl and grouse consists primarily of narrow, white muscle fibers, which enable these birds to take off with explosive power. The short-term power of white muscle fibers is useful as well for fast turns and evasive actions in flight, but the birds tire easily and cannot fly long or far.

Elementary Aerodynamics

To stay aloft, birds must overcome the force of gravity with forces that are equal to it and opposite. Four forces—weight, lift, drag, and thrust—must be in balance to maintain level flight at a constant speed. Upward lift counters the downward effect of a bird's weight. Forward thrust counters the slowing influence of the friction forces collectively called drag.

We can begin to understand how the wings of birds produce these forces by considering the nature of lift, an aerodynamic force produced by the flow of air past the surfaces of an airfoil. An airfoil is an asymmetrically curved structure that tapers posteriorly. The wings of birds are airfoils, as are the wings of airplanes. Correct orientation of the airfoil with respect to passing air produces the net upward force called lift, which keeps a bird or an airplane airborne. Lift results in part from the different speeds at which air flows past the upper and lower surfaces of the airfoil. Owing to the faster air speeds on the curved upper surface of the wing, pressure on that surface is less than the pressure on the undersurface of the wing, generating a net upward force (Figure 5–5A).

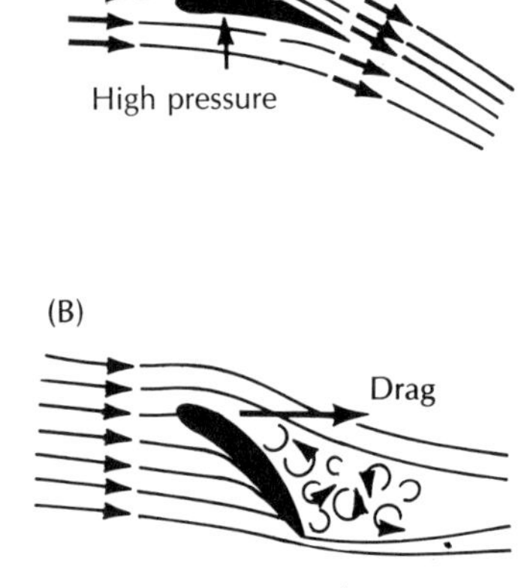

Figure 5–3 (A) The streamlined, asymmetrical shape of an airfoil produces lift by reducing pressure on the upper curved surface relative to that on the lower surface, due to different airspeeds (Bernoulli's principle) and by deflecting air downward, thereby producing an opposite, upward-directed physical force. (B) Lowering the rear edge of the airfoil to a more vertical position relative to the airstream increases its so-called angle of attack. Large angles of attack can cause the airstream to separate from the upper surface of the airfoil, thereby increasing turbulent air-flow—or drag—and reducing lift to the point of stalling. (After Rüppell 1977)

Lift due to differential air pressures is an expression of the Bernoulli principle, which relates air pressure to air velocity. The movement of gaseous molecules in the atmosphere creates pressure on any surface they strike. The random Brownian motion of molecules in still air is equal in all directions, creating omnidirectional pressure called static pressure. Wind, on the other hand, imparts directional, or dynamic, pressure, and the force increases with air velocity. As the energy in dynamic pressure increases, that present in static pressure decreases. The sum of the two pressures remains constant. Thus, fast-moving air imparts less pressure against an adjacent surface than slower-moving air, causing a net force upward in the case of the paired surfaces of an airfoil. Proof of this can be seen if one blows gently over the upper surface of a piece of tissue paper. The tissue rises or straightens out because of the net upward pressure on the lower side. The unequal pressures that develop as air flows over the surfaces of a bird's wing have the same effect.

The shape of an airfoil also deflects passing air downward (Figure 5–3). Elementary physics (Newton's third law) tells us that for every action there is an equal but opposite reaction. The downward deflection of air by the airfoil produces upward forces called lift, which counteract the pull of gravity. Correct orientation of an airfoil with respect to passing air keeps a bird, or an airplane, airborne. For level flight, the amount of lift generated must balance the bird's weight.

The potential for lift increases with the rate of downward deflection of air—a function of airspeed—and the volume of air deflected—a function of wing area. When a large bird, such as a gull, stands on the edge of a cliff facing into the wind, the flow of air across its outstretched wings generates lift, the amount of which increases by the square of the velocity of the airstream. If the wind is strong enough, the bird rises effortlessly into the air. Alternatively, in still air, the gull may have to jump off the cliff with wings outstretched; as it drops downward, its airspeed increases and lift also increases to the point of real flight. Birds that do not launch themselves from cliffs or trees may generate the initial forward thrust by running as they take off. Loons and some ducks run over the water until they become airborne.

The orientation—or angle of attack—of a wing in an air current affects the amount of lift that is generated. More lift is generated as the rear edge of the wing tilts downward, increasing the angle of attack and the downward deflection of air. If the wing is tilted too far downward, however, the airstream no longer follows the streamlined surfaces of the airfoil that direct it downward. Instead, the air separates from the surface and swirls up and forward from the rear edge of the wing, thereby increasing drag, blocking the backward flow of air over the upper surface, and causing a loss of lift—or a stall (Figure 5–3B). When landing, a pilot purposely stalls an airplane by increasing the angle of attack of the wings just before the wheels touch the runway. Birds also adjust the angle of their wings to stall just before landing.

Slots between adjacent flight feathers aid fine control of the air moving over the wing surface and thereby aid in the extraction of lift-producing energy. Slots are cracks or holes through which air squeezes, producing two kinds of results. First, some slots control the flow of air over the airfoil to maintain some lift at slow speeds or at high attack angles of the wing when a bird is stalling. The extended alula, or bastard wing, creates a slot that controls airflow, especially during landing and takeoff, when forward thrust is minimal and extra lift is essential. Second, air forced from beneath the wing through a slot expands on the upper side, reducing the pressure there and increasing lift. The slots in the wingtips of many soaring birds, such as the California Condor, function this way.

Figure 5–4 Slots aid fine control of airflow over the wing surface and prevent stalling at slow airspeeds. Slots constitute 40 percent of the wing area of a California Condor. Slight adjustment of the primaries and their associated slots control a condor's speed, lift, and aerial position as it searches the terrain for carcasses. (Courtesy of Santa Barbara Museum of Natural History)

Perfectly smooth, frictionless flow of air over an airfoil is only an ideal. Any slight air turbulence—or slowing of air molecules on contact with real surfaces—reduces lift. Negative forces that oppose a bird's movement through the air are called drag. Flight requires power to overcome drag to keep the airfoil moving forward at the speed required to generate adequate lift. The two primary categories of drag are induced drag and profile drag.

Induced drag is produced by turbulence of air about the airfoil that disrupts the lift-producing airstream. Eddies on the rear edge of the wing and especially at the wingtip reduce flight efficiency. Pointed wings generate less induced drag than rounded ones because the small area of the pointed wingtip creates less turbulence. Induced drag decreases with increasing airspeed, and therefore the component of flight power required to overcome it—called induced power—also decreases with flight speed.

Profile drag is caused by friction between the air and a bird's body and wing surfaces. Thin leading edges of the wings, for example, minimize this form of drag, which is the reason it is easier to throw a Frisbee than a soccer ball. Profile drag and the flight power required to overcome it—called profile power—increase with airspeed. Thus, it is harder to throw a Frisbee, or to fly, into a strong wind with high airspeeds than into a light wind. Because they reduce profile drag, long narrow wings are best for fast-flying birds.

Forward thrust provides the power to overcome the slowing effects of drag on airspeed. Propellers or jet engines provide thrust for airplanes. The downbeat of the wing provides thrust for most birds. The total thrust required to overcome the effects of drag is the sum of induced power and profile power. Because induced power decreases as profile power increases, total flight power requirements vary in a parabolic relationship to flight speed (Figure 5–5): The energy cost of flying is least at intermediate speeds and greatest at low and high speeds. Hovering in one place with no airspeed is the most energetically expensive mode of flight.

Minimum power speed is the speed at which a bird uses fuel most slowly and can stay airborne longest. Estimates of flight speeds (airspeeds) indicate that most birds fly at speeds of 30 to 60 kilometers per hour (Rayner 1985). Common Eiders are among the fastest fliers (80 kilometers per hour) clocked in steady flight. In general, birds fly at speeds with low power requirements. Yet birds do not always adhere strictly to flight speeds that minimize power costs (Pyke 1981; Norberg 1981; McLaughlin and Montgomerie 1990). Hummingbirds, for example, hover expensively in front of flowers to extract nectar, fly slowly and economically between adjacent flowers, or fly fast to beat competitors to nectar-filled flowers (Gill 1985). Peregrine Falcons dive on prey at extraordinary speeds. They may reach speeds of 180 kilometers per hour, although accurate measurements have never been made.

To achieve the maximum flight range with a given amount of fuel, a bird should fly faster than its minimum power speed, because the momentum added with increased speed carries it farther for the same total power investment. Flight at maximum range velocity is most characteristic of long-distance migrants, such as geese. Migrating Common Swifts travel at about 40 kilometers per hour, close to their predicted maximum range velocity. In contrast, while feeding, Common Swifts cruise slowly at only 23 kilometers per hour, close to their predicted minimum power speed.

Wing Sizes and Shapes

Flight speeds, aerial agility, and energy consumption all depend on the size and shape of a bird's wings. Lift and drag forces vary with wing di-

Box 5–2
Geese fly in formation

Flying in formation helps to save energy, especially in large or heavy birds with small wings relative to their mass (Badgerow 1988; Alexander 1992). By flying just off the wingtip of the preceding individual, each goose cancels some of the air turbulence at its own wingtips, which reduces induced drag and saves energy. In the familiar "Vee" formations of migrating geese, each individual flies off and behind the wingtip of the individual in front of it (see figure). The energetic advantage of formation flight could be as high as 50 percent.

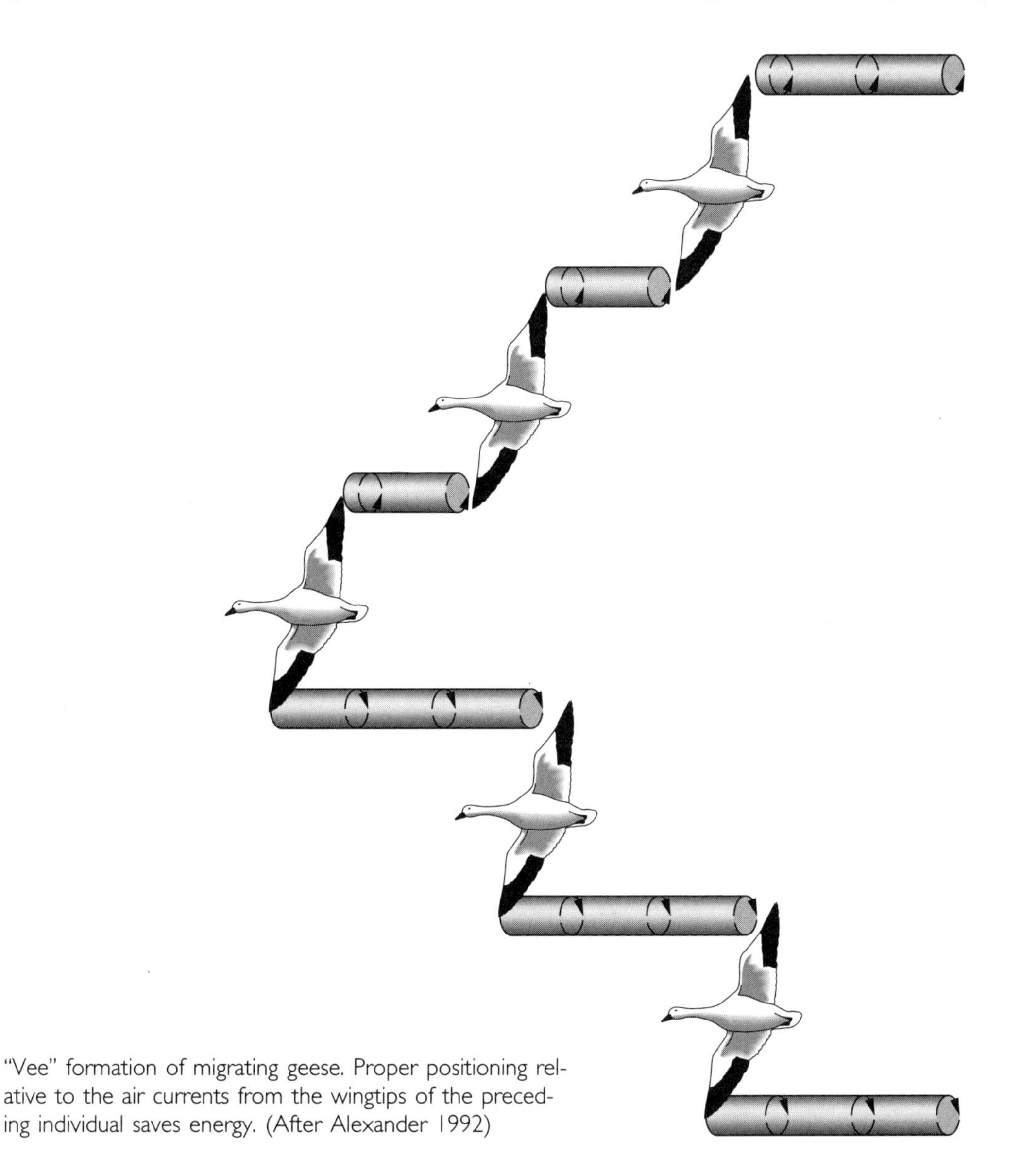

"Vee" formation of migrating geese. Proper positioning relative to the air currents from the wingtips of the preceding individual saves energy. (After Alexander 1992)

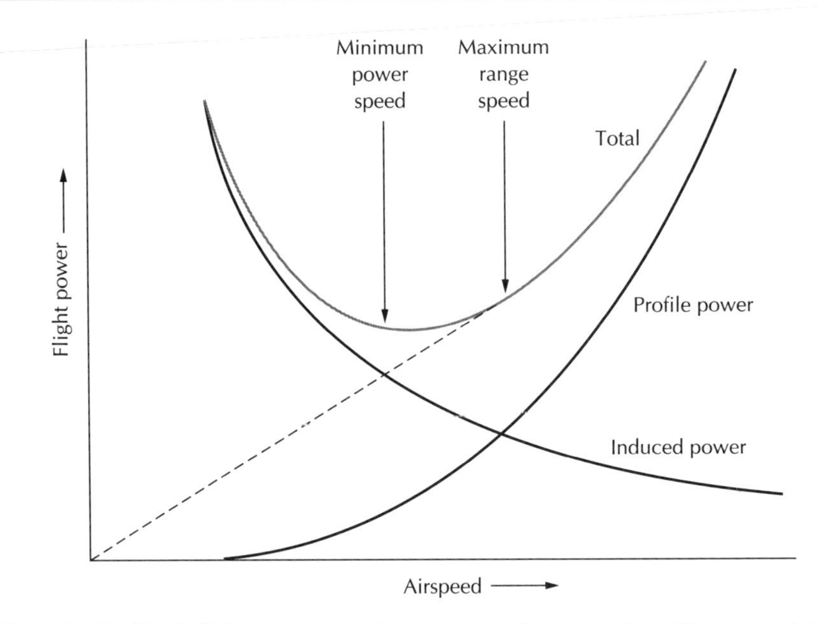

Figure 5–5 Total flight power requirements are the sum of profile power (which increases with speed) and induced power (which decreases with speed). This sum bears a parabolic or U-shaped relationship to airspeed. The total is least at minimum power speed, or the bottom of the parabola. The power required to fly a unit distance is least at the maximum range speed, which is defined by the lowest value intersection of the (broken) line drawn from the origin (zero speed, zero power). (After Alexander 1992)

mensions as well as with the patterns and speeds of airflow over the wing surfaces. Studies of variations in bird wings reveal the pervasive strength of evolutionary adaptation. Over the millennia, birds have exploited the advantages of different wing designs.

Aerial and open-country birds, such as shorebirds, swallows, and terns, typically have long, pointed wings, whereas species living in dense vegetation often have short, rounded wings (Figure 5–6). The wing shapes of migrant birds also tend to be longer and more pointed than those of related nonmigrant species. A falcon's pointed wings serve it well in high-speed chases in open country, whereas the short, rounded wings of a Sharp-shinned Hawk enable it to maneuver while chasing small birds in dense vegetation. Wrens and other passerines have short, rounded wings. Their rapid wing beats enable them to maneuver without collision amid seemingly impenetrable networks of branches and vines. The short, rounded wings of quail and pheasants permit short bursts of rapid acceleration, enhancing their chances of escaping predators.

The costs of flight are influenced by the relationship between total wing area and body mass, that is, by how many grams of mass each unit area of wing surface must carry. The relationship between wing area and body mass—called wing loading—is given in grams per square centimeter of wing surface area. Some birds have small wings relative to their body mass, and others have proportionately large wing areas. Small passerines have rather large wings and, consequently, low wing loadings of about 0.1 to 0.2 gram per square centimeter. At the other extreme are large birds with small wings such as albatrosses (1.7 grams per square centimeter) and the Thick-billed Murre (2.6 grams per square centimeter). Loons, grebes, auks, diving ducks, and even flamingos also have high wing loadings. To take off, they must run over the water, flapping their wings to gain enough lift for flight.

Although wing lengths and areas increase with mass among birds, the increase is less than that required to maintain a constant wing loading. The wing area increases as the product (square) of two dimensions, whereas body mass increases with volume, or the product (cube) of three dimensions. Therefore, to maintain parity with mass, the wing area must increase 1.5 times for each unit increase in mass. Yet the slope of the relationship averages less than that for most birds, as little as 1.275 for birds such as passerines, herons, and raptors (Greenewalt 1975). Only in hummingbirds does the wing area increase in proportion to the mass increase, keeping the per-gram cost of hovering the same for hummingbirds of all sizes.

Gliding Flight

The soaring flight of vultures and many other birds illustrates how the forces of weight, lift, and drag work in this simplest form of flight. Soaring vultures rely primarily on wing lift to balance the pull of gravity. Without flapping their wings to apply forward thrust, they gradually lose altitude—that is, they "sink"—because of drag. Glider airplanes also sink

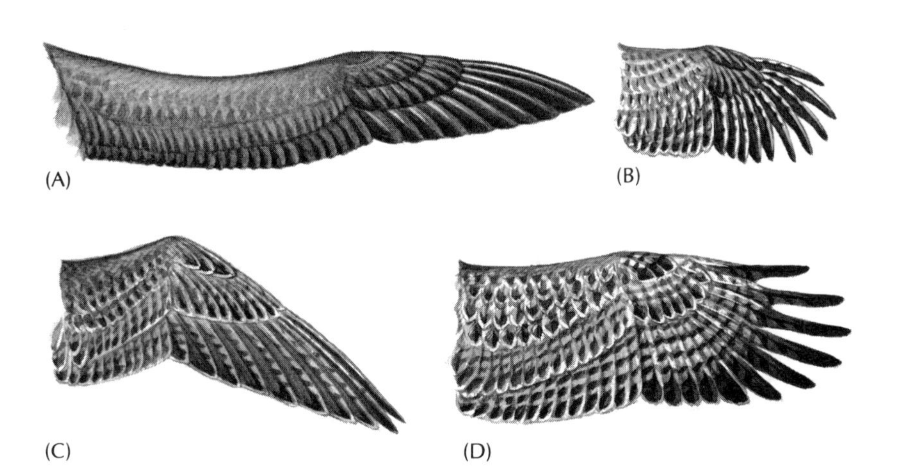

Figure 5–6 Flight abilities vary with the shape of bird wings. (A) Long, narrow wings, such as those of an albatross, are best for high-speed gliding in high winds. (B) Short, rounded wings, such as those of a grouse, permit fast takeoffs and rapid maneuvers. (C) The slim, unslotted wings of falcons permit fast, efficient flight in open habitat. (D) Slots in wings of intermediate dimensions increase lift and gliding ability of buteos.

at predictable rates that reflect their airspeed and wing dimensions. Sink rates are lowest when the combination of induced drag and profile drag is lowest at intermediate flight speeds. Even at their best, birds sink faster (1 to 2.5 meters per second) than a well-designed glider airplane (0.5 meter per second) (Alexander 1992).

Soaring birds and gliders both can counter their inevitable descent by taking advantage of upward air movements. The two principal ways of doing this are called thermal soaring and slope soaring. Thermal soaring exploits columns of warm air that rise when the ground is heated by the sun (Figure 5–7). The soaring bird circles upward within the column of rising air and then glides to the base of an adjacent thermal. Air rises in thermals at the rate of approximately 4 meters per second, which easily offsets a sink rate of 1 to 2 meters per second. Colin Pennycuick (1972) pioneered the study of gliding flight of birds by following vultures that commuted from their roosts to feeding grounds out on the Serengeti Plain of East Africa. He did so from his own plane, a motorized glider that could simulate the flight of the vultures but could also generate thrust when necessary. The vultures could travel 75 kilometers by using only six thermals that rose to heights of 1500 meters. Migrating hawks, such as the Broad-winged Hawk in eastern North America, also use thermals, a practice that allows them to cover long distances with maximum economy. Flocks of Broad-winged Hawks rising in a thermal of air are a special attraction at famous hawk-watching locations, such as Hawk Mountain in eastern Pennsylvania.

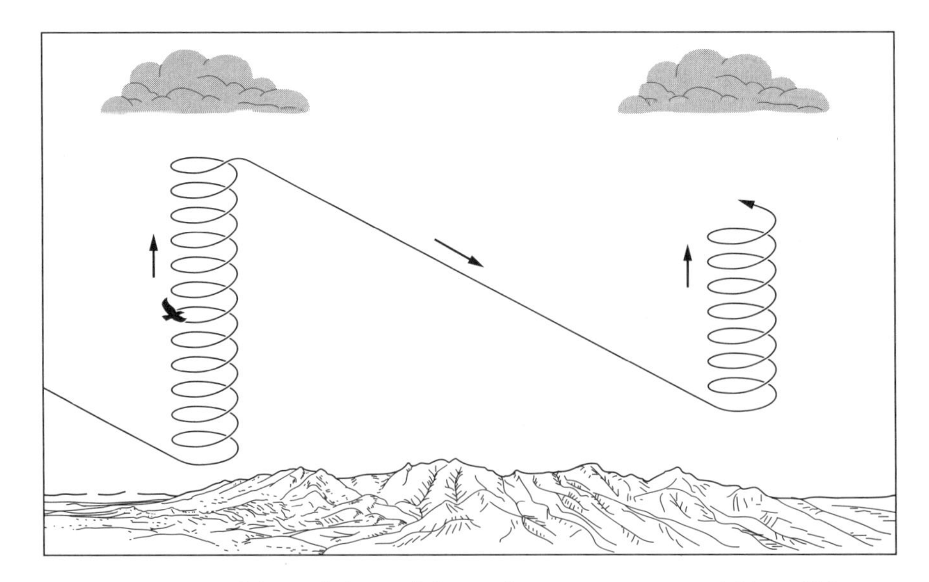

Figure 5–7 Use of thermals by a gliding vulture to counteract sinking. (After Pennycuick 1973)

Slope soaring takes advantage of a different kind of rising air, namely, the air that is deflected upward when it hits a terrestrial ridge or ocean wave. Migrating hawks soar along ridges and gulls hang effortlessly behind boats or above the ocean beach by riding the deflected air currents. Seabirds, such as the large, long-winged albatrosses, can cruise expertly close to large wave crests on the windward side. If the line of the wave crest is not in the intended direction of travel, the albatross can fly into a headwind by alternately rising off the crest of a wave and gliding at an angle to an adjacent wave. This so-called dynamic soaring takes advantage of the layers of different wind speeds above the ocean. The albatross accelerates downward from the fast-moving, upper air layers into the slower moving, lower air layers; then, using their momentum plus gravity and surface wind eddies, they swing upwind. As they lose speed and lift, they bank again into the fast-moving, upper air layers.

The wing shapes of birds affect their gliding abilities. The long, narrow wings of albatrosses, for example, produce more lift than do shorter, broad wings of equal total area because the leading edge of a wing produces the most lift; the rear half of a wing produces the least. In addition, induced drag becomes less important as wing length increases. Long, narrow, pointed wings, such as those of swallows, falcons, and also albatrosses, have a high lift-to-drag ratio. Consequently, swallows can glide better than, say, sparrows, which have short, rounded wings of low lift-to-drag ratio. Whereas long, high-lift wings enable Turkey Vultures to begin soaring early in the day, short, rounded wings force Black Vultures to wait for the assistance of rising warm-air currents until they can soar without expensive flapping. Conversely, long, narrow wings work best at high speeds. At low speeds, eddies from the rear edge interfere with the flow of air at the leading edge. Therefore, narrow-winged birds must fly fast to avoid stalling.

Flapping Flight

Gliding flight minimizes the use of powered thrust to overcome the negative effects of drag. Flapping flight, on the other hand, adds thrust to the balance of controlling forces. Each primary flight feather functions as an airfoil, as can the wing itself. When these airfoils change their orientations downward from the horizontal, a portion of the upward lift they generate changes to forward thrust. This principle is perhaps best illustrated by the performance of the rotating blades of a helicopter. Each blade is an airfoil positioned at the best angle of attack relative to the sweep of the rotor. When the blades rotate in a horizontal plane, air is deflected directly downward to generate lift that offsets the weight of a hovering helicopter. By tilting the rotor forward, the blades drive air backward as well as downward, thereby imparting forward thrust and causing the helicopter to move forward (Figure 5–8).

The same principles apply to the wing action of hummingbirds. Crawford Greenewalt (1960a) took high-speed movies of hummingbird flight and then studied them at slow speeds to discover how hummingbirds achieve their remarkable control. He concluded that hummingbird flight resembles that of a helicopter or, more precisely, a novel combination of airplane and helicopter in which the propellers rotate about a horizontal axis to produce various combinations of lift and forward thrust. Greenewalt describes the action thus:

> In hovering flight the wings move backward and forward in a horizontal plane. On the down (or forward) stroke the wing moves with the long leading edge forward, the feathers trailing upward to produce a small, positive angle of attack. On the back stroke the leading edge rotates nearly a hundred and eighty degrees and moves backward, the underside of the feathers now uppermost and trailing the leading edge in such a way that the angle of attack varies from wing tip to shoulder, producing substantial twist in the profile of the wing. (Greenewalt 1960a, p. 233)

A hummingbird can move forward from stationary hovering, or even fly backward, just by changing the direction of the wing beat, because every angle produces a different combination of lift and thrust. Forward velocities increase as the wings beat in an increasingly vertical plane. This rotation of the wing is made possible by the unusual structure of the humerus and its articulation with the pectoral girdle. The secondaries of a hummingbird's inner wing are short, and the outer primaries are elongated to form a single, specialized propeller. The complete stroke of the wingtip describes a figure-eight pattern (Figure 5–9).

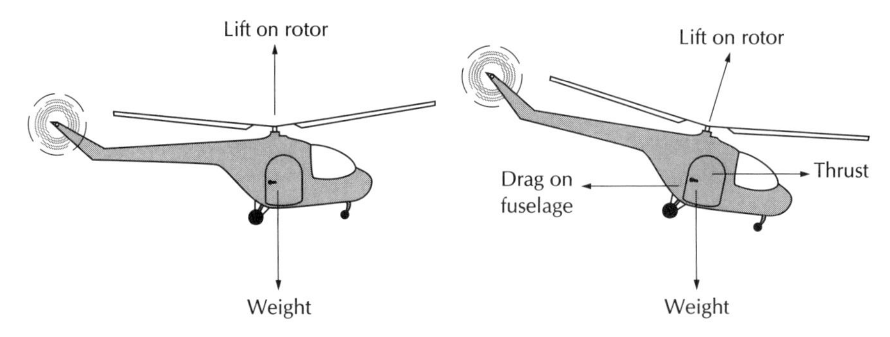

Figure 5–8 Hovering (left) and forward (right) flight of a helicopter. When the blades of the rotor rotate in a position that is horizontal to the ground, the lift they generate balances the downward pull of gravitation—called weight. Tilting the rotor directs some of the lift in a forward direction—called thrust—and also generates negative forces—called drag.

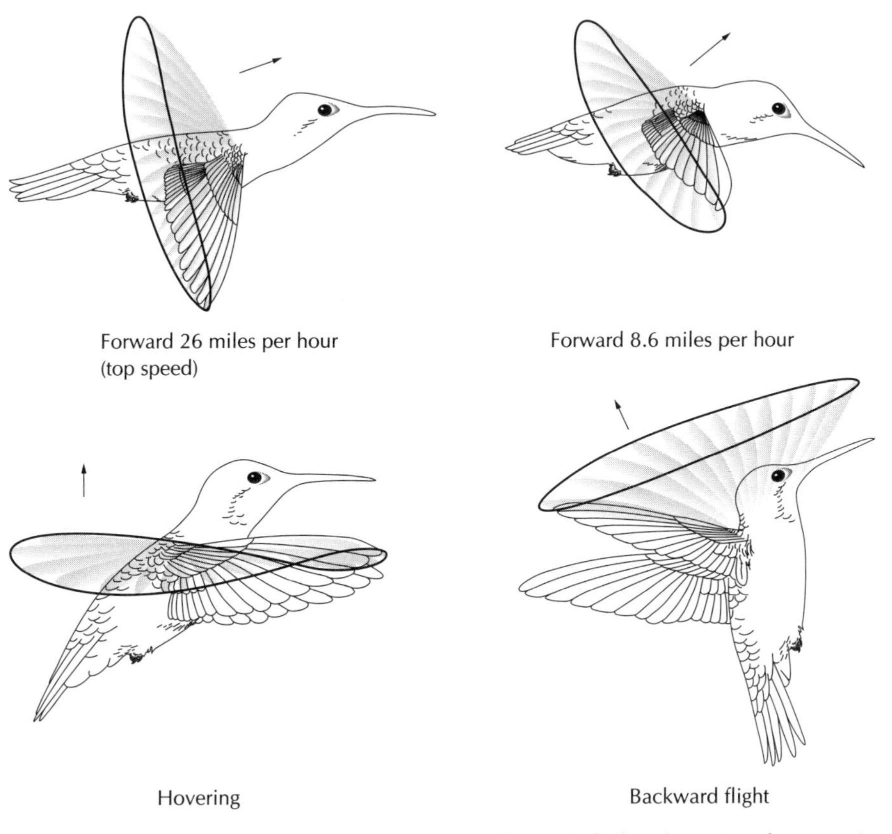

Figure 5–9 Hummingbird wing motions. In forward flight, the wings beat vertically to generate forward thrust. In hovering flight, the wings beat horizontally in the pattern of a flattened figure eight. To fly backward, the hummingbird tilts the angle of wing action to create rear-directed thrust. (After Greenewalt 1960a)

The wings of other birds also act like large propellers, with some accessory propellers in the wingtip and with some distinction between the contributions of the inner and outer wing. In the outer half of the wing, each primary functions as a smaller, separate airfoil; together they can produce forward thrust as does the propeller of an airplane. To produce forward thrust, the airfoils of propellers and of primaries move vertically rather than horizontally through the air. As the leading edge of the primary slices the air column during the downstroke, the net pressure on the back surface pushes the feather forward. Control of the angle of attack of each primary by tendons and muscles and by the natural responses of the flexible vanes to air pressure results in a continuously integrated system of feather positions as the wing flaps through the air. The forward forces of thrust produced by the propellerlike primaries are transferred to the inner wing, the horizontal movement of which creates the upward forces of lift. The result is forward flight.

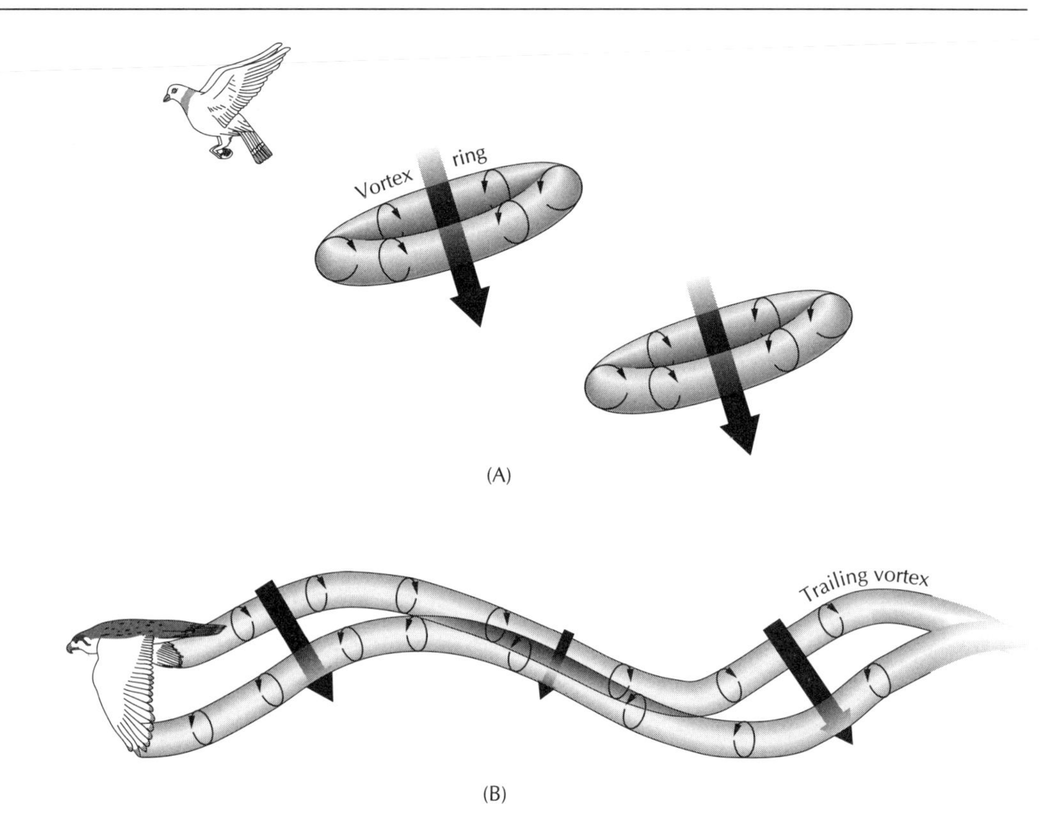

Figure 5–10 Wing beats leaving trailing currents of swirling air. (A) The powered downstroke of slow-flying pigeons produces doughnut-shaped patterns of air currents called vortex rings. (B) The pattern of vortices trailing behind a fast-flying kestrel with continuously integrated powered downstrokes and recovery upstrokes. (After Alexander 1992)

Like the wings of insects, the wings of birds and their controlling musculature oscillate mechanically with intrinsic elasticity (Greenewalt 1960b). The rate of wing beats of a Ruby-throated Hummingbird, for example, is essentially constant at about 53 strokes per second, and the durations of the upstroke and downstroke are equal, as in mechanical devices. Also, the wing beat rates of various species of hummingbirds and most other birds decrease predictably with increasing wing length, as oscillation theory predicts. These observations have important implications for the neuromuscular basis of avian flight. Once the wing beat rate reaches its natural oscillating frequency, the nerves and muscle fibers responsible for sustaining the rhythm need to fire perhaps only once every four beats, like a child on a swing with only an occasional push by the parent.

Supplementing the wings are the tail's contributions to lift, which may be more important in young birds that are learning to fly than in skilled adults. Immature raptors, in particular, tend to have longer tails than do adults. The size difference (up to 15 percent) is most pronounced in short-tailed species such as the Bateleur and sea-eagles, but also in the familiar Red-tailed Hawk of North America. Reflecting the increased lift, immatures of these raptors have a more buoyant flight than do adults. Extra lift apparently reduces the chance of injury when they strike prey and also facilitates their mastery of flight and hunting skills (Amadon 1980).

Birds in flight control the patterns of lift and thrust. No aircraft yet approaches the average bird's acrobatic maneuverability. About 50 different muscles control the wing movements. Some muscles fold the wing, others unfold it. Some pull the wing upward, others pull it down, and still others adjust its orientation. In most small birds only the downstroke is the power stroke. Little lift is achieved on the recovery stroke, during which the primaries are separated to minimize air resistance. Powered downstrokes followed by simple recovery strokes produce doughnutlike rings of swirling air—called vortex rings—in the wake of the flying bird (Rayner 1988) (Figure 5–10A). The forces of lift and thrust on various parts of the wing are continuously integrated during the normal wing beat, and in fast flight will convert the trailing vortices from doughnutlike rings to continuous streams that reflect the integration of forces from the wings' downstrokes and upstrokes (Figure 5–10B).

Even more important than the integration of lift and thrust is the bird's control of each wing independently. Asymmetrical wing actions enable the bird to steer, turn, and twist. By flapping with one wing oriented for forward flight and the other wing oriented for backward flight, the bird can execute an abrupt turn. Setting the wings in a partially folded position reduces the amount of lift, enabling the bird to lose altitude gradually while gliding. By setting one wing back further than the other, the bird adds curvature to its glide path. Slow-motion photographs of birds during aerial maneuvers, chases, and landings reveal the precise changes in wing position that control body orientation and airspeed. Birds rarely crash.

Box 5–3
It's not easy to land on elevated perches

Elaborate use of the wings in landing is a feature of avian flight. Landing on elevated or arboreal perches, for example, requires exceptional control of flight trajectory. Birds are unique among flying vertebrates in the way they land (Caple et al. 1983, 1984). Aerial species such as bats, flying squirrels, and certain lizards make contact with their forelimbs and then rotate their bodies downward until the hind feet touch the landing surface. Variations exist, but only birds rotate their center of mass upward to stall directly over the landing site.

Flightless Birds

Not all birds fly. Besides the ratites (Ostriches and cassowaries, for example), there are flightless grebes, pigeons, parrots, penguins, waterfowl, cormorants, auks, and rails (Livezey 1988, 1989a, b, 1993). The original birds of remote predator-free islands, such as the Hawaiian islands in the Pacific Ocean and the Mascarene islands in the Indian Ocean, included a host of flightless birds: geese, ibises, rails, parrots, and the now-extinct Dodo. If flight and mobility are so clearly advantageous to the majority of birds, why are some birds flightless? The answer lies largely in the costly development and maintenance of the anatomical apparatus required for flight. An enlarged, keeled, calcified sternum and large pectoralis muscles, for example, are expensive to produce, and their maintenance requires much energy. In the absence of advantageous uses, such as the need to fly from predators, natural selection favors reduced investment in the material and energy for flight. In rails, conversion of a cartilaginous sternum to a bony sternum made of calcium is delayed until the bird is nearly full grown, predisposing them to the evolution of flightless forms, especially on islands where predators are absent (Olson 1973) (Figure 5–11). Evolutionary reduction of the sternal keel and the mass of flight muscles is, in fact, a first sign of reduced flight ability. The angle between the scapula

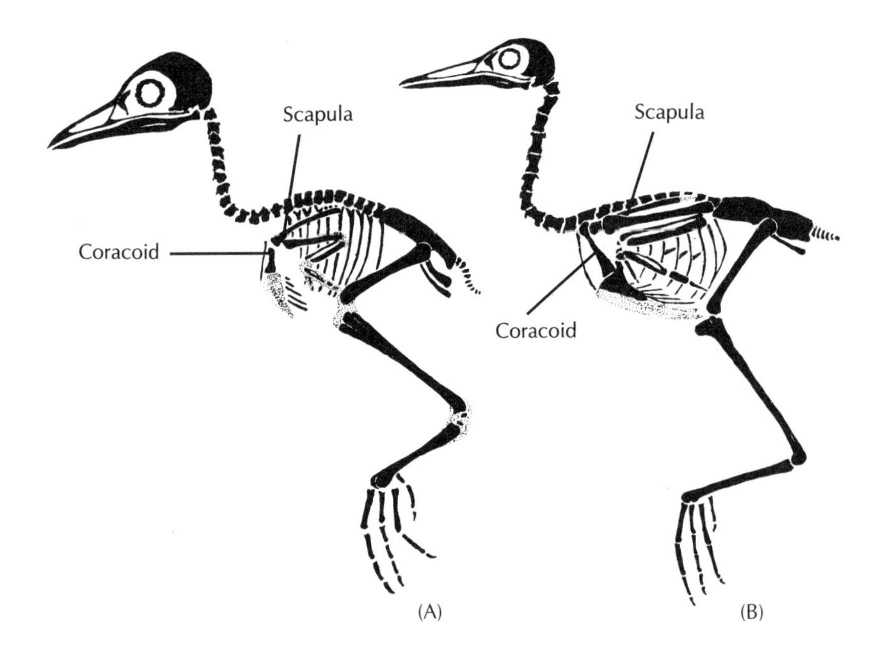

Figure 5–11 Skeletons of the King Rail, a flying rail, at (A) 17 days and (B) 47 days after hatching (size reduced so that femur lengths in the two drawings are equal). Stippled areas represent cartilage. Note the obtuse angle formed by the articulation of the scapula and coracoid in the younger form and the acute angle in the older form. (From Olson 1973)

Figure 5–12 Flightless Cormorants preening and drying wings. (Courtesy of E. and D. Hosking)

and coracoid also becomes more obtuse, and ultimately the wing bones become smaller. Kiwis, for example, have only vestigial wings.

Other routes to the evolution of flightlessness are seen in specialized diving birds. Foot-propelled divers such as loons, grebes, and cormorants evolve powerful legs and feet that function as paddles. If evolution favors hindlimbs for locomotion, wings and associated pectoral development may regress and render a diving bird nearly or completely flightless. Extreme cases are those of the flightless Short-winged Grebe of Lake Titicaca, Peru, and the Flightless Cormorant of the Galápagos Islands (Figure 5–12). Reduced-sized wings trap less air, thereby reducing buoyancy that interferes with diving.

Penguins, which are wing-propelled divers, represent another route to flightlessness in specialized diving birds. Their wings propel them through water rather than through the air; their feet act as rudders rather than as paddles. The evolution of such forms has occurred not only in penguins but also among the auks in the Northern Hemisphere.

The evolution of wing-propelled divers from flying birds proceeds through an intermediate state in which wings are used for both underwater propulsion and aerial flight. Diving-petrels represent the intermediate stage in the evolution from flying petrels to flightless penguins. Auks, such as the Razorbill, with dual-purpose wings represent the intermediate stage in the evolution of specialized divers from flying gull-like ancestors to the flightless Great Auk of the North Atlantic. The progressive specialization of wing skeletal structure is evident in the changes from the slim wing bones of a gull through shorter and heavier bone structures to the broad, flat wing skeleton of a penguin's flipper (Figure 5–13).

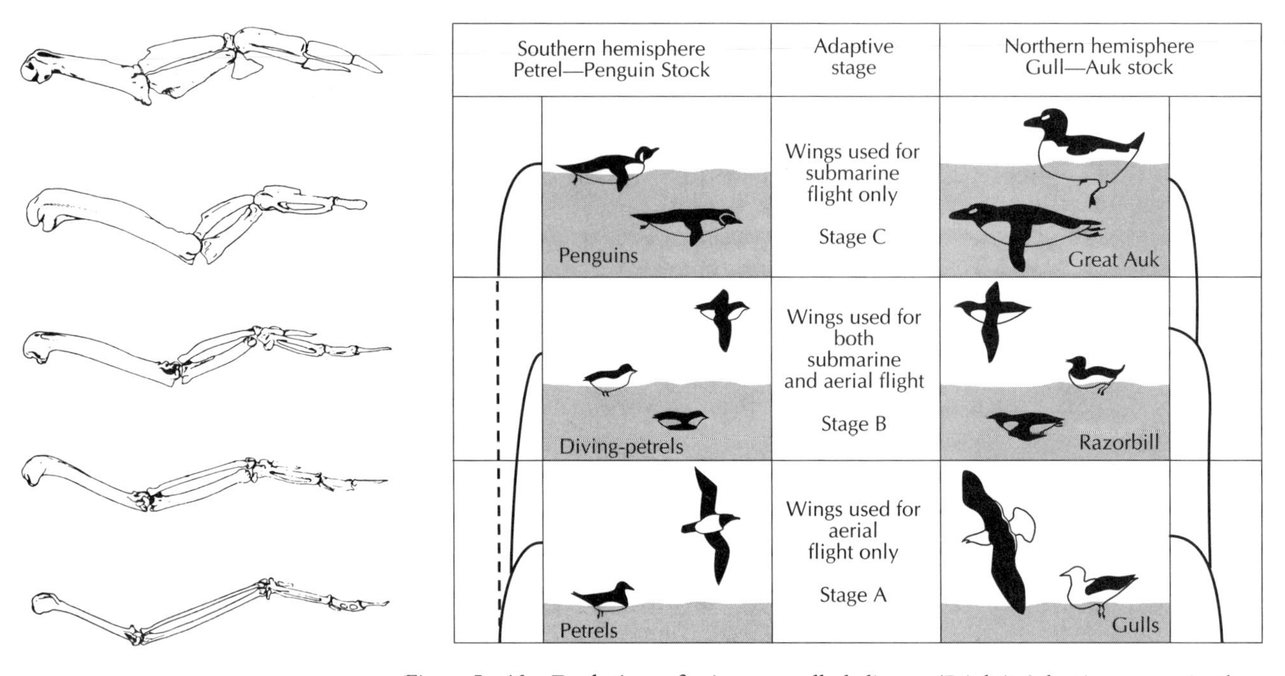

Figure 5–13 Evolution of wing-propelled divers. (Right) Adaptive stages in the parallel evolution of two stocks of wing-propelled diving birds. (Left) Modifications of the wing skeleton in wing-propelled diving birds: (bottom to top) an aerial gull, an auk, the flightless Great Auk, an extinct penguinlike auk, and a penguin. (From Storer 1960)

Summary

Structural adaptations for flight dominate avian anatomy. Fusions and reinforcements of lightweight bones are among the adaptations of the avian skeleton for flight. Of particular importance are the keeled sternum, which supports the powerful pectoralis and supracoracoideus flight muscles, and the strutlike arrangement of the pectoral girdle. The tendons of the ventrally located supracoracoideus muscles pass through the triosseal canal to dorsal insertions on the humerus. The red fibers of avian flight muscles have an extraordinary capacity for aerobic metabolism and sustained work.

The form of the wing and of the individual flight feathers is that of an airfoil, which generates a force called lift as air passes over and is deflected downward by the asymmetrical surfaces. Control of flight is achieved through changes in wing and wing feather positions and through the use of slots between feathers. Gliding birds exploit rising air currents, both heated thermals and slope-deflected air, to gain altitude without the exertion of flapping. Hummingbirds achieve extraordinary maneuverability in flight by beating their wings at different angles in a figure-eight pattern that includes a powered upstroke as well as a powered downstroke. Flight power requirements are least at intermediate flight speeds, but birds often fly faster or slower than this speed to maximize distances traveled or to feed. Particular wing shapes adapt birds to specific modes of flight because they influence the penalties of induced and profile drag relative to the wing's ability to generate lift and thrust. Long, narrow wings sacrifice maneuverability for high-speed flight with low drag. Birds are the only vertebrates that can land with precision on elevated or arboreal perches.

Some birds have become flightless, particularly on remote islands that lack mammalian predators. Delayed ossification of the sternum in rails predisposes them to the evolution of flightlessness. Specialized diving birds rely either on hindlimb locomotion or wing-propelled underwater locomotion. Extremes of both kinds of diving birds have lost the power of flight. Penguins, for example, have flipperlike wings.

Further Readings

Alexander, R.M. 1992. Exploring Biomechanics. New York: Scientific American Library. *An elegant, nontechnical summary of the principles of flight and other forms of animal locomotion.*

Pennycuick, C.J. 1989. Bird Flight Performance. Oxford: Oxford University Press. *A practical calculation manual that includes computer software plus background discussions of relevant aerodynamics.*

Rayner, J.M.V. 1988. Form and function in avian flight. Current Ornithology 5: 1–66. *A comprehensive review of the aerodynamic principles of flapping flight, especially recent theoretical and experimental advances, plus application of these principles to the wing shape dimensions of birds.*

Rüppell, G. 1977. Bird Flight. New York: Van Nostrand Reinhold. *A readable, nontechnical introduction to avian flight.*

CHAPTER 6

Physiology

FEATHERS AND FLIGHT are conspicuous features of birds. Less conspicuous, but no less fundamental, are the internal systems of metabolism and excretion—collectively called physiology—which sustain daily activities and adapt individual birds to their particular environments, hot or cold, wet or dry. Activity that is unconstrained by low ambient temperatures is one of the conspicuous advantages of maintenance of high body temperatures. Birds are fully active in the early morning cold, in midwinter, and in the high mountains. The advanced metabolic physiology of birds provides both power and endurance. Power and endurance derive from the maintenance of high, constant body temperatures by means of metabolic heat production, which in turn demands much energy and water, two resources that often are in short supply. Adaptive systems of heat conservation, heat loss, and water economy enable birds to live in extreme or seasoned environments.

The high body temperatures of birds require active control of heat exchange with the environment to conserve metabolic energy in cold environments and to lose heat in hot environments or in flight. Supporting the demands of sustained aerobic metabolism are a highly efficient respiratory system coupled to a powerful heart and circulatory system. Water reserves, essential for evaporative cooling and the excretion of electrolytes, are easily depleted; therefore, birds maintain a delicate physiological balance of the conflicting needs for temperature regulation, activity, and water economy. Other features of avian physiology, such as digestion, the senses, reproduction, and hormonal control of the annual cycle are discussed in later chapters.

The High Body Temperatures of Birds

The physiology of birds, specifically their metabolism, relates directly to the maintenance of a high body temperature through the production of metabolic heat—or endothermy. Most birds, large and small, in the frigid

Arctic and in the hottest deserts, keep their core body temperature high, at about 40°C. High body temperatures enhance intrinsic reflexes and powers, and thus enable birds to be active, fast-moving creatures. In general, the rates of physiological processes increase with temperature. Transmission speed of nerve impulses increases 1.8 times with every 10°C increase in temperature. The speed and strength of muscle fiber contractions triple with each 10°C rise in temperature.

The maintenance of high body temperatures through endothermy, however, is energetically expensive; birds consume 20 to 30 times more energy than do similar-sized reptiles. The maintenance of high body temperatures also risks lethal overheating. Above 46°C, most proteins in living cells are destroyed more rapidly than they are replaced (Figure 6–1), and changes in chemistry of the brain cause death.

More important than speed or strength is endurance (Bennett 1980). Warm amphibians and reptiles can escape or strike with lightning speed, but they are quickly exhausted. Birds fly for hours or days. Increased aerobic metabolism and insulation were among the major changes that we believe accompanied the evolution of reptiles into birds. These changes made possible regulated high body temperatures and the many advantages of constant and dependable rates of muscle function (see, for example, the discussion of red versus white muscle fibers in Chapter 5). Higher activity levels coupled with greater endurance opened a new range of ecological opportunities for birds. However, the high metabolic demands of temperature regulation and of the daily activities of birds require extraordinary delivery rates of energy and oxygen to the body's cells as well as

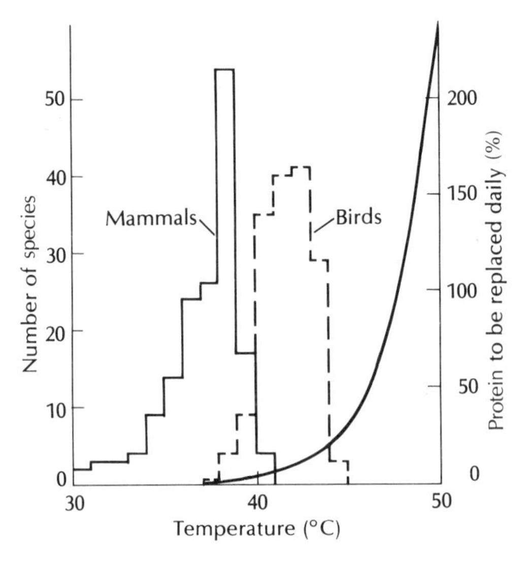

Figure 6–1 Birds and mammals regulate their body temperatures to be just below temperatures that destroy body proteins. Shown here are the body temperatures of many bird and mammal species. The curved line represents the rate of protein replacement as a function of body temperature.

rapid removal of poisonous metabolic waste products. Powerful respiratory and circulatory systems help to meet these demands and to keep a bird's body chemistry in balance.

Avian Respiratory System

The avian respiratory system is different in both structure and function from the mammalian respiratory system. Four anatomical features—the nostrils, tracheal system, lungs, and air sacs—transport air between the atmosphere and the circulatory system of a bird. With each breath, a bird replaces nearly all the air in its lungs. Because no residual air is left in the lungs during the ventilation cycle of birds, as it is in mammals, birds transfer more oxygen during each breath. Thus, overall, birds have a more efficient rate of gas exchange than do mammals.

Birds lack a diaphragm, which promotes inhalation and exhalation in mammals by increasing or decreasing the volume of the chest cavity, thereby causing the lungs to inflate or deflate accordingly. Instead, birds inhale by lowering the sternum, which enlarges the chest cavity and expands the air sacs. On exhalation, contraction of the sternum and ribs compresses the air sacs and pushes fresh air from them through the lungs. During flight, the movements of the sternum are enhanced and are complemented by springlike expansions and contractions of the furcula (Jenkins et al. 1988). Early beliefs that wing and breathing movements were synchronous have proved wrong. Both respiration and wing beats are controlled by the central nervous system; wing action does not directly drive the breathing motions of the thorax.

Rates of breathing decrease with increasing size of a bird. When not flying, a 2-gram hummingbird breathes about 143 times a minute, whereas a 10-kilogram turkey breathes only 7 times a minute. In flight, birds meet the increased oxygen demand by increasing their ventilation rates to 12 to 20 times their normal resting rates.

Most birds inhale through nostrils—or nares—at the base of the bill. A flap—or operculum—covers and protects the nostrils in some birds, such as diving birds, which must keep water from entering their nostrils, and flower-feeding birds, whose nostrils might become clogged with pollen. Air passes through the external nares into complex, paired nasal chambers separated by a nasal septum (Figure 6–2). Each chamber has elaborate folds called concha that increase the epithelial surface area over which air flows. The conchae cleanse and heat the air before it enters the respiratory tract. Olfactory tubercles sample (or smell) its chemistry. The conchae also are well supplied with nerves and a network of blood vessels—*rete mirabile*—that help to control the rate of water and heat loss from the body.

In contrast to human and other mammalian lungs, which are large, inflatable, baglike structures that hang in the chest cavity, avian lungs are small, compact, spongy structures molded among the ribs on either side

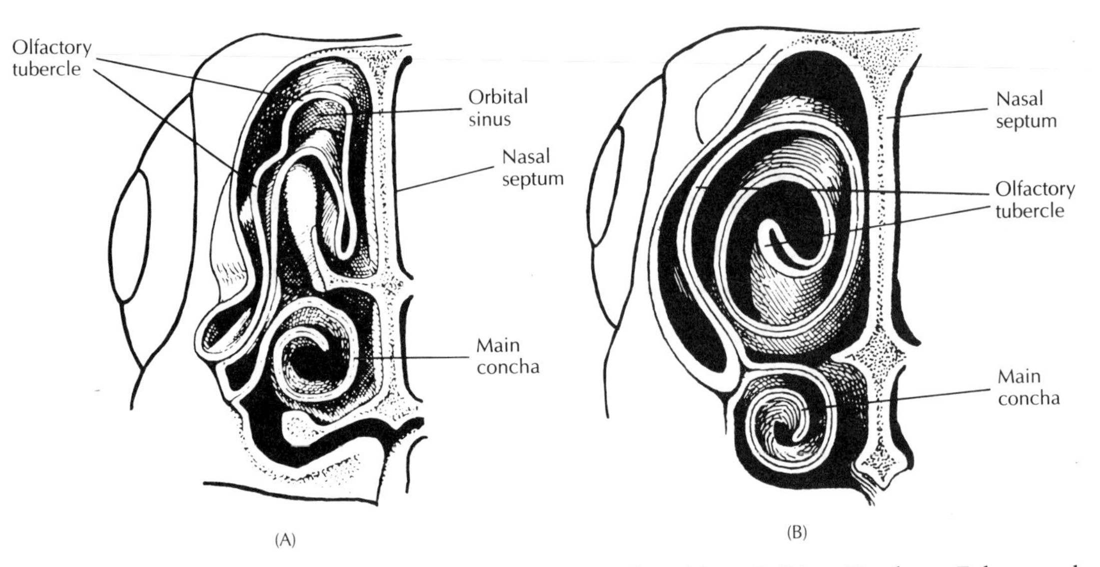

Figure 6–2 Cross section of the nasal cavities of (A) a Northern Fulmar and (B) a Turkey Vulture, showing the elaborate folds—called concha—that cleanse and heat inhaled air and remove water from exhaled air. (Adapted from Portmann 1961)

of the spine in the chest cavity. Avian lungs weigh as much as the lungs of mammals of equal body weight, but because avian lungs have much greater tissue density they occupy only about half the volume. Healthy bird lungs are well vascularized and light pink in color.

The internal structure of the mammalian lung resembles a bush, with many subdividing bronchial stems and branches. The bronchial branching patterns and connections in bird lungs more closely resemble the plumbing of a steam engine (Figure 6–3). Branching from each of the two pri-

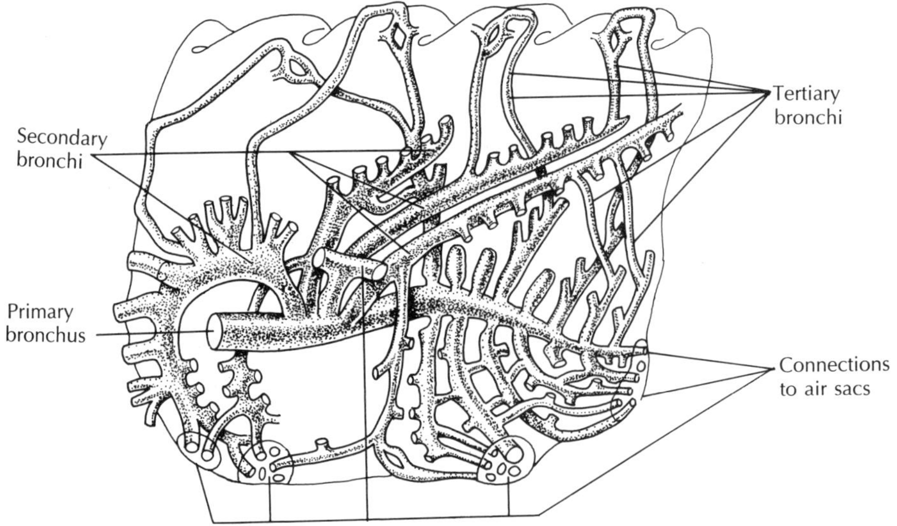

Figure 6–3 Interconnecting bronchial tubules form the internal structure of the bird's lung. Tertiary bronchi and fine air capillaries constitute most of the lung tissue. (After Lasiewski 1972)

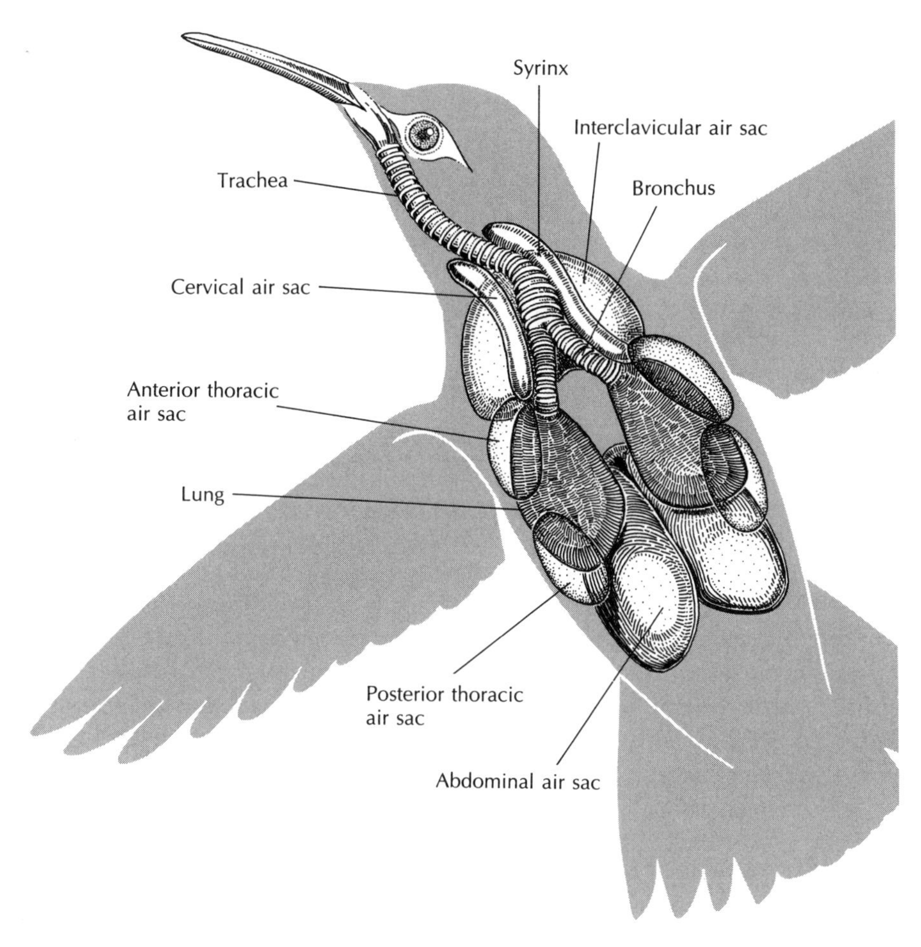

Figure 6–4 Positions of the air sacs and lung in a bird's body. (After Tyrrell and Tyrrell 1985)

mary bronchi that traverse the entire lung are about eleven secondary bronchi, four of which—the craniomedial bronchi—service the anterior and lower parts of the lung. Most of the lung tissue comprises roughly 1800 smaller interconnecting tertiary bronchi; these lead into tiny air capillaries that intertwine with blood capillaries, where gases are exchanged.

The air sac system is an inconspicuous but integral part of the avian respiratory system (Figure 6–4). Air sacs are thin-walled (only one to two cell layers thick) structures that extend throughout the body cavity and into the wing and leg bones. The air sacs connect directly to the primary and secondary bronchi and in some species connect indirectly to some tertiary bronchi. As detailed below, the air sacs make possible a continuous, unidirectional, efficient flow of air through the lungs. They not only help deliver the huge quantities of oxygen needed, but also help remove the potentially lethal body heat produced during flight. Inflated air sacs also help protect the delicate internal organs during flight.

The number of air sacs varies from 6 in weavers and 7 in loons and turkeys to at least 12 in shorebirds and storks. Most birds, however, have 9 air sacs:

1. The paired cervical sacs located in the neck, which are apparent in displaying male frigatebirds in the form of great, inflatable red sacs
2. A pair of anterior thoracic sacs, which fill the forepart of the body cavity
3. The large, paired posterior thoracic sacs, which fill the upper chest
4. The large, paired abdominal sacs, which cushion the abdominal organs and carry air to leg and pelvic bones
5. A single interclavicular sac, branches of which penetrate the wing bones, sternum, and syrinx; pressure in the syrinx from the interclavicular sac is essential for vocal sound production (see Chapter 10).

Avian respiration maximizes contact of fresh air with the respiratory surfaces of the lung. Air flows to the avian lung through one set of bronchi to enter the gas-exchange areas and exits through another set. Most of the air breathed in during the first inhalation in the cycle passes via the primary bronchi to the posterior air sacs (Figure 6–5). During the exhalation phase of the first breath, the inhaled air moves from the posterior air sacs into the lungs where it flows through the air-capillary system, the primary site of oxygen and carbon dioxide (CO_2) exchange. The next time the bird inhales, this oxygen-depleted air moves into the anterior air sacs. During the second exhalation, the CO_2-rich air is then expelled from the anterior air sacs, bronchi, and trachea back into the atmosphere. The continuous airflow through the avian lung allows more efficient extraction of oxygen than occurs in the mammalian lung. The flow of air seems mostly passive in response to differences in local air pressures.

BOX 6–1

Birds hyperventilate without severe penalty

Rapid breathing during exercise or at high, oxygen-poor altitudes expels large amounts of carbon dioxide. Loss of carbon dioxide increases the alkalinity of the blood (normally the pH lies between 7.3 and 7.4), which causes blood vessels to constrict, severely reducing the flow of oxygen-rich blood to the brain. In mammals, blood flow to the brain may drop 50 to 75 percent during such hyperventilation, an effect that causes fainting and, sometimes, death. Remarkably, for reasons that remain unknown, this does not happen in birds even at pH 8, which would kill a mammal (Grubb et al. 1978, 1979). Without this safeguard, birds would be unable to fly at high altitudes.

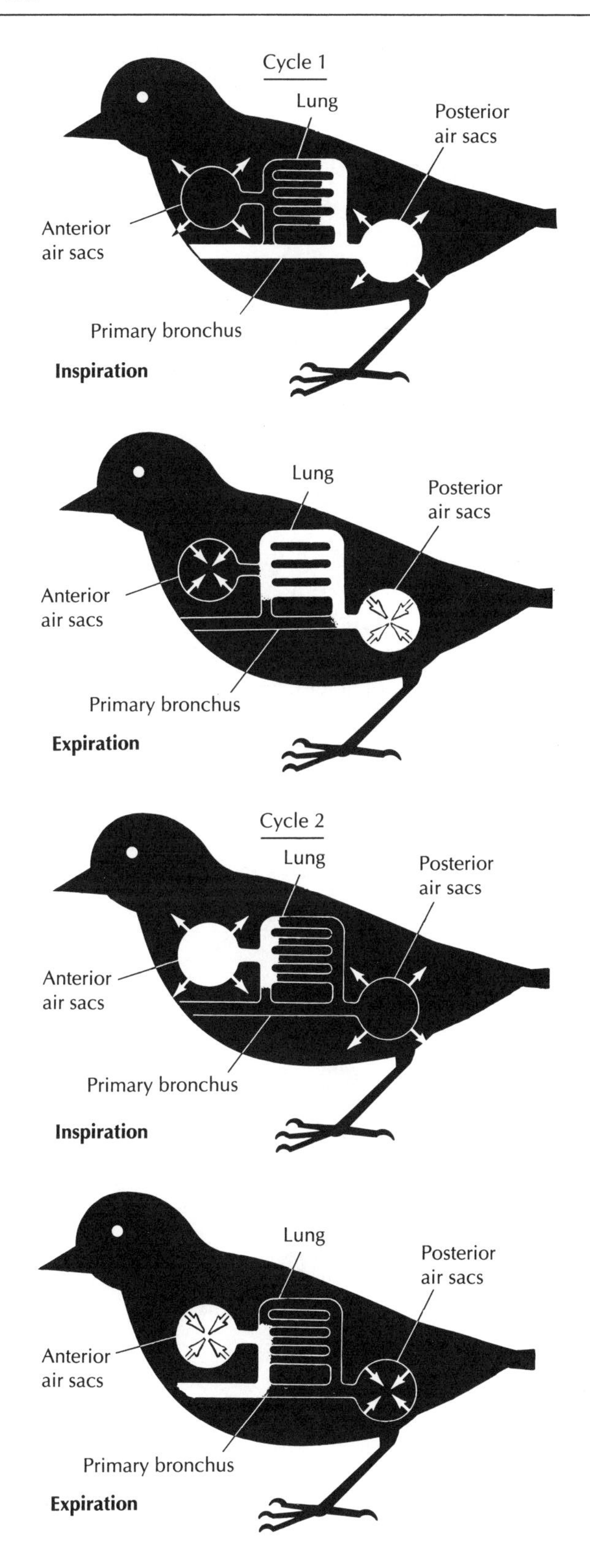

Figure 6–5 The unidirectional movement of a single inhaled volume of air (shown in white) through the avian respiratory system. Two full respiratory cycles—inspiration, expiration, inspiration, and expiration—are required to move the gas through its complete path. This diagram does not show the tracheal connection from the main bronchus to the mouth. (After Schmidt-Nielsen 1983)

Avian Circulatory System

The circulatory system of birds is matched to the demands of their metabolism (Jones and Johansen 1972). The high metabolism of birds requires rapid circulation of high volumes of blood between sites of pickup and delivery of metabolic materials. The circulatory system must deliver oxygen to the body tissues at rates that match usage and must simultaneously remove carbon dioxide for exhalation. It must also meet the demands for fuel, delivered in the form of glucose and elementary fatty acids, and remove toxic waste products for excretion. Birds have high peak metabolic demands, and they sustain high levels of activity metabolism for long periods. The demands on the avian circulatory system are far greater than on those of reptiles, and even exceed those in most mammals.

Like mammals, birds have a double circulatory system and a four-chambered heart (Figure 6–6). The hearts of birds must be powerful to meet the demands of their metabolism. A large structure in most birds, the heart accounts for 2 to 4 percent of the total weight of a hummingbird. Avian hearts are 50 to 100 percent larger and are more powerful than those of mammals of the corresponding body size. The evolution of the avian heart with two ventricles from the reptilian, three-chambered heart (only one ventricle) relates to the increased oxygen requirements of endothermy. Circulation of blood to the lungs for gas exchange—

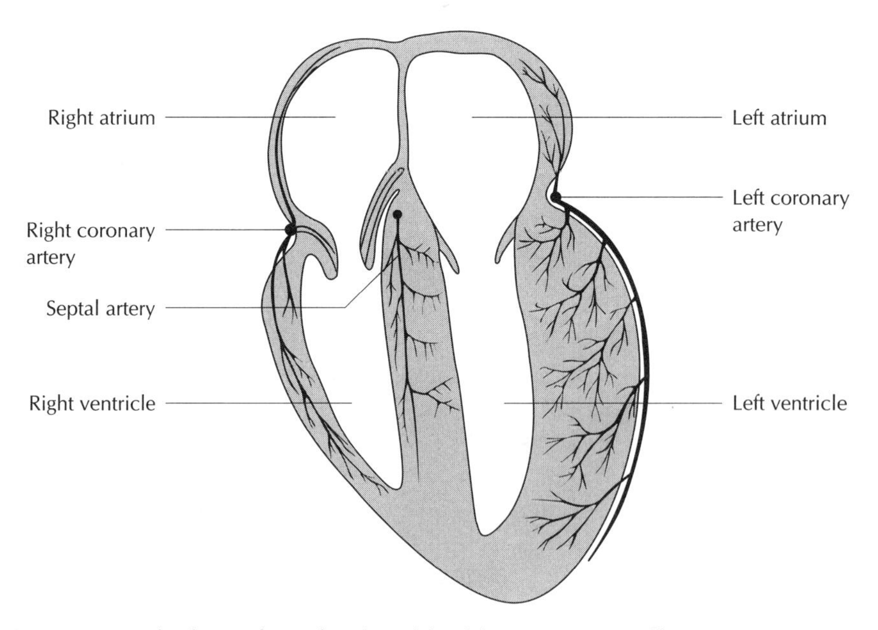

Figure 6–6 The large, four-chambered bird heart supports efficient oxygenation of blood. (From Jones and Johansen 1972)

pulmonary circulation—is separated from general body—systemic—circulation. Blood returning from the tissues is pumped from the left ventricle directly to the lungs. Newly oxygenated blood returns from the lungs to the right ventricle and then is pumped from it directly to the demanding tissues. In the three-chambered, reptilian heart, newly oxygenated blood mixes with blood returning from the tissues in the single ventricle.

The performance of the heart is measured in terms of cardiac output—or the rate at which the heart pumps blood into the arterial system. Defined as heart rate times stroke volume from one ventricle, it averages 100 to 200 milliliters of blood per kilogram of mass per minute in birds. A large proportion of the oxygenated cardiac output from the left ventricle goes directly to the legs for the purpose of heat loss (see page 136). The legs get three times as much blood per heartbeat as the pectoral muscles and twice as much as the head, the next most important target. Together, the legs and brain receive 10 to 20 percent of the total cardiac output.

Birds and mammals achieve high cardiac outputs in different ways. The resting heart rates of birds are about half those of similar-sized mammals. Normal resting heart rates in medium-sized birds range from 150 to 350 beats per minute and average about 220. Heart rates of small birds are much higher and exceed 1200 beats per minute in small hummingbirds (Lasiewski et al. 1967). The resting cardiac output of ducks is apparently higher than that of other birds and averages 200 to 600 milliliters of blood per kilogram per minute. Although bird hearts beat more slowly at rest than mammalian hearts, their cardiac outputs are similar because of large stroke volumes. Not only is the avian heart large, but the ventricles empty more completely than do those of mammals on each contraction. In addition, at high heartbeat rates, the ventricle fills more completely between contractions.

The avian ventricle—or outgoing pump—is made up of more muscle fibers than the mammalian ventricle. Reflecting a greater capacity for aerobic work and endurance at high activity levels, each fiber is thinner and its cells contain more mitochondria—energy-producing organelles that depend on the supply of oxygen—than do mammalian heart muscle fibers. The thinness of avian fibers speeds the transfer of oxygen.

The high-performance features of the avian heart have their costs. The high tension of avian heart muscles and the strength of the ventricular contractions lead to high arterial blood pressures, up to 300 to 400 millimeters of mercury in some strains of domestic turkeys—the maximum known for any vertebrate. (A blood pressure of 150 millimeters of mercury is high for a human.) Not surprisingly, aortic rupture is a common cause of death in these turkeys, which are raised on high-fat diets for weight gain. Small, wild birds also die easily of heart failure, aortic rupture, or hemorrhage when frightened or otherwise stressed. A male Prairie Warbler, for example, is reported to have died of a heart attack during courtship (Ketterson 1977). Also contributing to such problems is the stiff structure of the avian arteries, which improves smooth, peripheral blood flow but increases susceptibility to atherosclerosis, especially in those fat domestic turkeys.

Specialized diving birds have an unusual tolerance to asphyxia—or lack of air. Mallard ducks, which dive only modestly, can reduce oxygen consumption by as much as 90 percent during prolonged dives. They ration the available oxygen sparingly to some sensitive tissues, especially the central nervous system, sensory organs, and endocrine glands. Blood flow to most other organs and skeletal muscles stops during a dive; they rely on anaerobic metabolism during this period. Selective vasoconstriction and a profound slowdown of heart rate—bradycardia—start within 6 seconds of immersion of the head, and the heart rate drops to half that of predive levels within 8 seconds. The diving reflex is triggered when water touches special receptors in the nares.

Levels of Metabolism

An orderly approach to understanding the energy requirements of birds as they relate to environmental physiology begins with the examination of aerobic metabolism and, in turn, associated rates of oxygen consumption. Metabolic rates change rapidly with different levels of activity, dropping to a minimum when a bird rests at night and rising to a maximum when it flies. Intermediate rates of metabolism support the regulation of body temperatures during periods of cold or heat stress. The discussions below proceed from an introduction to minimum—or basal—metabolism, which provides a comparative index to physiological adaptations, to the ways birds regulate their body temperatures during cold and heat stress.

Basal Metabolism

Cellular metabolism never stops completely. Even resting birds use energy. Carefully controlled measurements of the minimal metabolic requirements of resting birds fasting at nonstressful—thermoneutral—temperatures give estimates of basal metabolism. All birds have high basal metabolic rates (BMRs); and, for their various sizes, passerine birds have the highest rates of any group of vertebrate animals. The average basal metabolic rate of a passerine bird is 50 to 60 percent higher than those of nonpasserines of the same body size (Walsberg 1983). There are exceptional passerine birds, for example, some Australian honeyeaters with low metabolic rates (MacMillen 1986), and exceptional nonpasserine birds with high metabolic rates (Prinzinger and Hänssler 1980).

Basal metabolism relates directly to mass, although not in a 1:1 relationship. Large birds expend less energy per gram of mass than do small birds. Although an 8-kilogram bustard is 100 times larger than an 80-

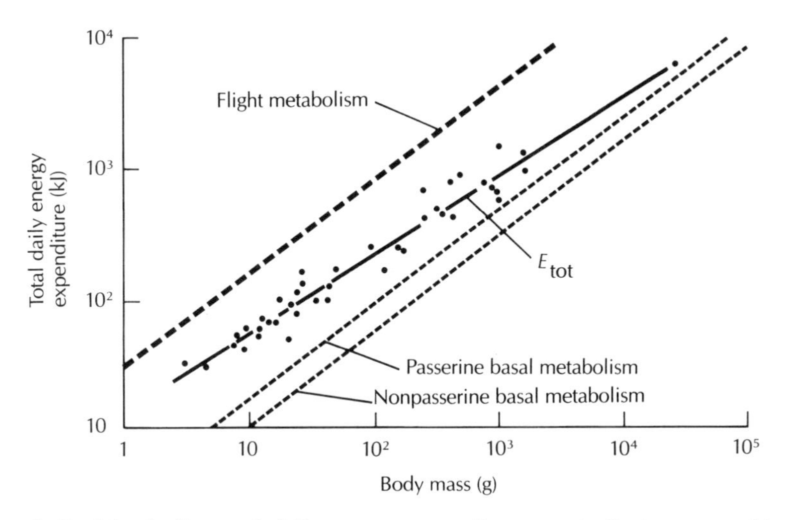

Figure 6–7 Metabolism and daily energy expenditures typically increase with body size. Passerine birds tend to have significantly higher basal metabolism than nonpasserines. Total daily energy expenditures (E_{tot}, in kilojoules) do not increase as fast with increasing body size as does basal metabolism, possibly because small birds are more active than large birds. Energy expenditures of birds during flight—flight metabolism—vary from 2 to 25 times higher than basal metabolic rate. (Adapted from Walsberg 1983)

gram falcon, it expends only 30 times as much energy per unit of time. The slope of this fundamental physiological relationship—that is, the increase of basal metabolism with mass—is predictably 0.72 to 0.73 for different-sized birds (Figure 6–7) as well as for different-sized mammals.

The surface area from which heat is lost partly explains the relationship between metabolism and mass. The surface area of a sphere increases as the 0.67 power of mass in accord with the basic two-thirds ratio of a two-dimensional area to the contained three-dimensional volume. A large bird cannot lose heat as fast as a small bird because it has less surface area per gram of heat-generating tissue. If an Ostrich's tissues produced heat at the same rate as a sparrow's tissues, the Ostrich would soon reach boiling temperatures internally because it could not dissipate heat from its body surfaces fast enough.

Activity Metabolism

Basal metabolism represents the expenditure of minimal energy. A bird usually spends only a fraction of its day at this low metabolic level, most of its time being spent instead in activities that require more energy and oxygen. The simple digestion of a meal, the slight muscle actions associated with awareness and attention, a strenuous sprint, or a vertical takeoff all increase energy expenditures. Birds are capable of extraordinary levels of aerobic metabolism. Small birds in flight can operate at 10 to 25 times their basal metabolic rate for many hours, whereas small mammals can sustain an activity level of metabolism of only 5 to 6 times their BMR (Bartholomew 1982).

Just being awake or resting increases metabolic rate by 25 to 80 percent above the basal rate. The increased metabolism characteristic of birds that are resting quietly in small cages without temperature stress apparently reflects only increased muscle tension and mental activity.

Metabolic costs increase with exertion. The oxygen consumption of a Bronze Sunbird increased directly with activity in a small cage (Wolf et al. 1975). A Greater Rhea's metabolism is 3.5 times BMR when it strolls along at 1 kilometer per hour but is 14 times BMR when it trots at 10 kilometers per hour. The metabolism of swimming Mallards is 3.2 times their BMR at their most efficient swimming speed and 5.7 times their BMR when they swim as fast as they can; Mallards prefer to swim at the more economical speed (Prange and Schmidt-Nielsen 1970).

Although it costs more per unit time, flight is generally a more efficient form of locomotion than running. To fly 1 kilometer, a 10-gram bird uses less than 1 percent of the energy that a 10-gram mouse uses to run the same distance. The energy expended in power flight per unit time generally exceeds that of other modes of locomotion, but estimates of flight metabolism range from 2 to 25 times the BMR, with variations that reflect flight mode, flight speeds, wing shape, and/or laboratory constraints (Farner 1970; Berger and Hart 1974). Low values of flight metabolism come from swallows and swifts in forward (partly soaring) flight, and high values come from finches and hovering hummingbirds. Hovering in one place is extremely expensive, costing hummingbirds, for example, an average of 0.286 watt per gram, or, depending on body size, 7 to 17 times their BMR.

Insulation, Heat Loss, and the Bird's Climate Space

A bird's thermal relations with its environment are central to its survival. Endothermy is part of a dynamic relationship between internal heat production and external heat loss to the environment. Heat, a direct product of metabolism, is an inevitable result of the inefficiency of biochemical reactions. Rates of heat production or loss are expressed in watts or joules per hour. For example, the average student at rest produces heat at the same rate as does a 100-watt lamp.

The rate of heat loss from the body is proportional to (1) the absolute difference between body temperature (or more exactly, surface temperature) T_b and the environmental temperature T_a (or T_b - T_a) and (2) the rate of heat transfer across the surface layers. The rate of metabolic heat production H must balance the rate of this loss, as represented in the equation

$$H = (T_b - T_a)/I$$

where I is the insulation coefficient—or resistance to heat flow. In special situations (for example, in a nest hole or a burrow free of wind in which wall temperature equals air temperature), ambient air temperature provides an accurate index to the rate of heat loss or heat gain; but in more complex environments, in which the sun shines and the wind blows, a bird's thermal relationships with its environment become a complex function of the intensity of radiation and convection, which are incorporated into a physiological measurement called standard operative temperature (Weathers et al. 1984).

Bird feather coats are among the best natural, lightweight insulations. The resting metabolism—and presumably the rate of heat loss—of frizzled chickens (Figure 6–8), whose abnormal feathers provide little insulation, are twice those of normal chickens at 17°C (Benedict et al. 1932). Contour feathers in the plumage contribute to a bird's insulation, but the true down feathers underneath the contour feathers are of primary importance. Thus, Arctic finches have dense down, and tropical finches do not.

Insulation increases with the amount of plumage. Some birds enhance their insulation during cold seasons by molting into fresh, thick plumage (Calder and King 1974). Nonmigratory House Sparrows increase plumage weight 70 percent, from 0.9 gram of worn plumage per bird in August to 1.5 grams of fresh plumage in September. Seasonal adjustments in insulation are not to be expected in tropical birds or in migratory species that escape major seasonal changes in environmental temperatures.

Birds adjust the positions of their feathers to enhance either heat loss or heat conservation. Fluffing the feathers in response to cold creates more air pockets and increases the insulation value of the plumage. For example, a Ringed Turtle-Dove—the domesticated variety of the African Collared-Dove—continuously adjusts the position of its feathers in relation to air temperature (Figure 6–9). Holding the wings out from the body, and extreme elevation of the back feathers—scapular feathers—enhance heat loss by exposing the bare apterial skin to convection. Tropical seabirds that nest in the open sun often elevate their plumage to avoid overheating.

Dark pigmentation aids temperature regulation by absorbing the energy-rich short wavelengths of the solar spectrum. When rewarming in the sun from mild overnight hypothermia, as many birds do, the Greater Roadrunner erects its scapular feathers and orients its body so that the morning sun shines on strips of black-pigmented skin on its dorsal apteria (Ohmart and Lasiewski 1971). Metabolism, and heat production, of gray-plumaged Zebra Finches decrease as a result of exposure to strong radiant energy, but that of white-plumaged finches does not (Hamilton and Heppner 1967; Heppner 1970). Light-colored plumage reflects, rather than absorbs, more of the impinging radiant energy than does dark plumage.

Figure 6–8 "Frizzled" chickens have high metabolism rates because their abnormal plumage does not provide as much insulation as that of normal chickens. The following description appeared in *Ornamental and Domestic Poultry* (Edmund Saul Dixon 1848, p. 344): "It is difficult to say whether this be an aboriginal variety, or merely a peculiar instance of the morphology of feathers; the circumstance that there are also Frizzled Bantams would seem to indicate the latter case to be the fact. School-boys used to account for the up-curled feathers of the Frizzled Fowl, by supposing that they had *come the wrong way out of the shell.* They are to be met with of various colours, but are disliked and shunned, and crossly treated by other Poultry. Old-fashioned people sometimes call them French Hens. The reversion of the feathers rendering them of little use as clothing to the birds, makes this variety to be peculiarly susceptible of cold and wet. They have thus the demerit of being tender as well as ugly. In good specimens every feather looks as if it had been curled the wrong way with a pair of hot curling-irons. The stock is retained in existence in this country more by importation than by rearing. The small Frizzled Bantams at the Zoological Gardens, Regent's Park, are found to be excellent sitters and nurses."

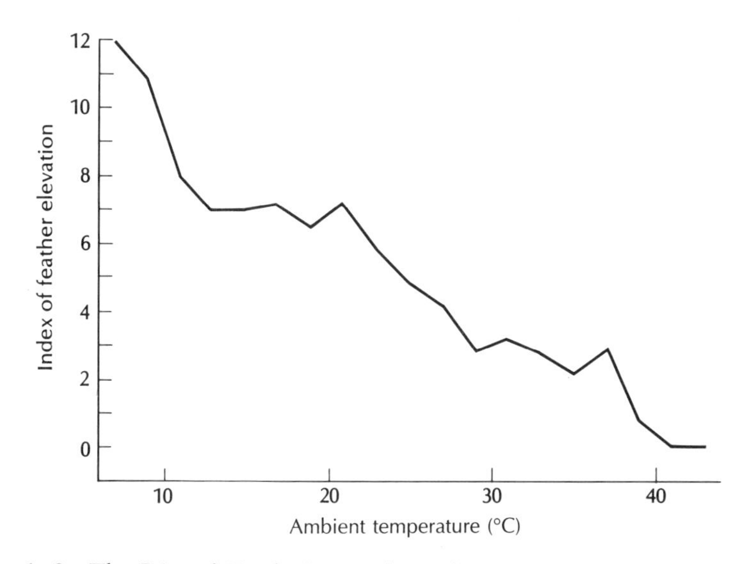

Figure 6–9 The Ringed Turtle-Dove adjusts the rate of heat loss from its body by subtly raising and lowering its feathers in response to ambient temperature. The higher the index, the greater the elevation of the feathers. (After McFarland and Baher 1968)

The net thermal effect of plumage, however, is influenced by the wind. Wind increases the rate of heat loss by decreasing I, the insulation coefficient. Metabolism must increase to compensate for such heat loss due to convection. A Snowy Owl's metabolic rate is directly proportional to the square root of wind velocity (Gessaman 1972). The thick plumage of this species provides excellent insulation, but the rate of heat loss in winds of only 27 kilometers per hour is three times that in still air. The use of wind-sheltered sites, including holes or burrows for roosting and nesting, protects birds from this heat loss. Small birds are particularly vulnerable to convective heat loss because they have more surface area relative to mass than do large birds.

The cooling effects of wind are most pronounced on black feathers, which concentrate solar heat near the surface of the plumage and can increase the amount of heat a bird's body absorbs from the environment when there is no breeze. A light breeze, however, removes the accumulated surface heat and reduces further penetration of the radiant heat. The black plumage of desert ravens increases convective heat loss, providing the same effect as the black robes and tents of Saharan desert tribes.

The body sizes of nonmigratory birds reflect geographical gradients in temperature and humidity (James 1970; Aldrich and James 1991). Climatic rules, such as Bergmann's Rule—the increase of body size with cooler temperatures—refer to these correlations in a simplistic and sometimes misleading way (Zink and Remsen 1986). Widespread North American birds, such as the American Robin and the Downy Woodpecker, tend to be smallest in hot, humid climates and largest in cold dry climates (Figure 6–10). The potential for heat loss by evaporative cooling

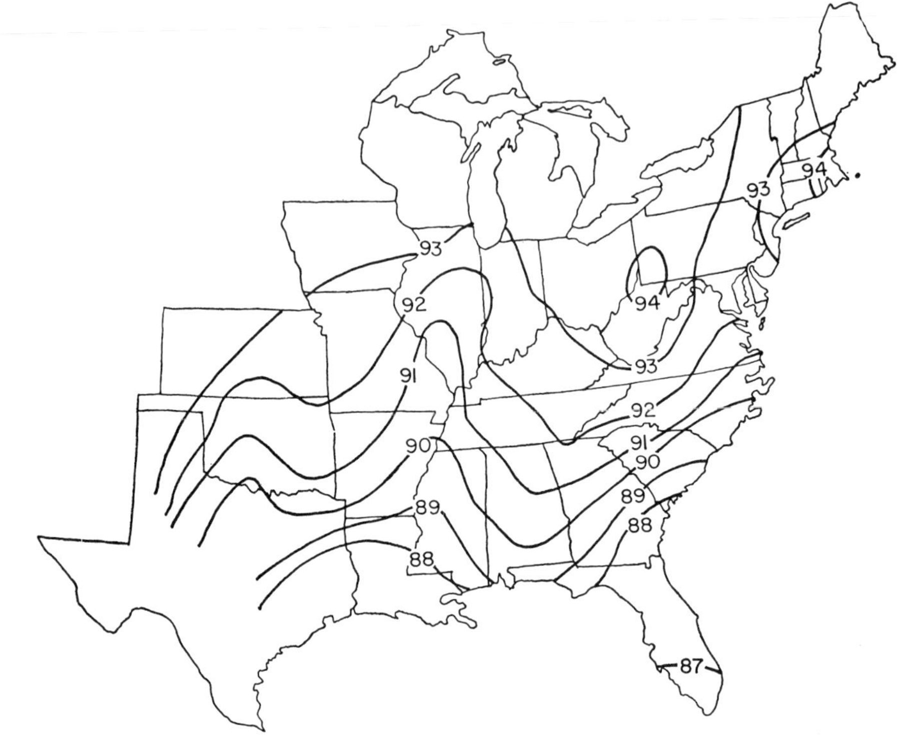

Figure 6–10 Size variation in Downy Woodpeckers. Body size (which is directly proportional to wing length) increases to the north, but individuals in the warm, humid Mississippi Valley and coastal areas are small compared with those at other localities at similar latitudes. Numbers indicate average wing lengths in millimeters. (From James 1970)

Box 6–2
Climate affects body size in House Sparrows

Geographical differences in body size as well as plumage color can evolve rapidly among populations—within 100 years in the case of House Sparrows introduced to both North America and New Zealand (Johnston and Selander 1971; Baker 1980; Lowther and Cink 1992). Body size of House Sparrows is positively correlated with seasonality and annual temperature range in both North America and Europe (Murphy 1985). Increased fasting ability appears to be the primary advantage of large size in seasonal environments. Conversely, small size minimizes individual maintenance costs in equable and more predictable or aseasonal environments.

Studies of the effects of severe winter weather have documented the survival advantages of size for House Sparrows. The classic study by Herman Bumpus compared the measurements of survivors and victims of a severe winter storm in Rhode Island in 1898 (see Grant 1972; Johnston et al. 1972). Body size also affected the survival of sparrows in Kansas during the devastating winter of 1978–79 (Fleischer and Johnston 1982). Large-sized males were favored in both cases, apparently because of their superior thermoregulation efficiencies or fasting abilities—or perhaps because of their greater access to well-protected roost sites (Buttemer 1992).

is lowest in hot humid climates, which favor small individuals with more heat losing surface area relative to mass. Conversely, cool, dry air favors larger bodies with reduced surface areas that conserve heat. Large body sizes also increase potential energy stores and ability to fast. Nevertheless, energy constraints eventually limit the distributions and abundances of species (Root 1988).

Temperature Regulation

The model of endothermy, developed by Per Fredrik Scholander and his colleagues (1950), is one of the foundations of avian thermobiology (Figure 6–11). In the thermoneutral zone, the amount of oxygen consumed by resting birds does not change with temperature. At lower and higher temperatures outside the thermoneutral zone, temperature regulation requires increases in metabolism. The Scholander model of endothermy is oversimplified, because it assumes absolutely constant body temperatures over a wide range of ambient temperatures and an abrupt transition at the lower critical temperature between so-called physical thermoregulation—adjustment of heat loss—and chemical thermoregulation—adjustment of heat production. Many birds violate these conditions. However, the model usefully provides the basis for an orderly consideration of the processes that are involved in the control of body temperature.

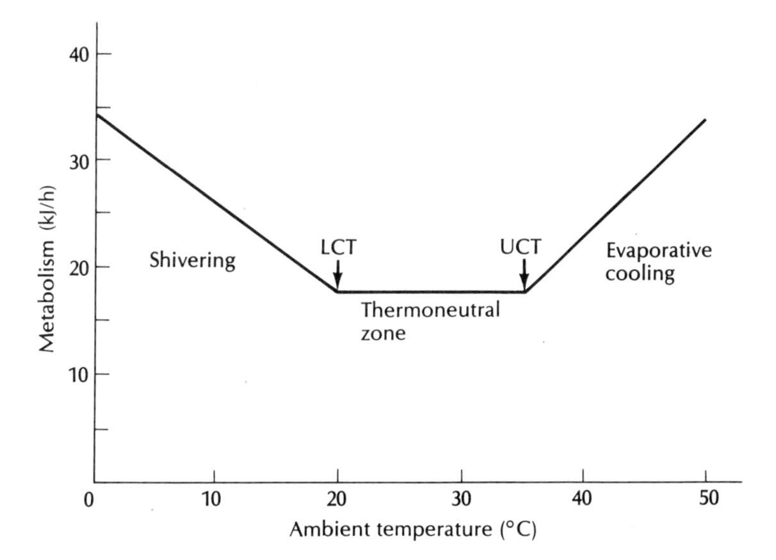

Figure 6–11 Scholander's model of endothermy. Metabolism increases below the lower critical temperature (LCT), primarily as a result of shivering heat production. Metabolism increases above the upper critical temperature (UCT) due to active loss of heat through panting and evaporative cooling, as well as the direct effects of higher temperatures on cellular functions. Metabolism is relatively insensitive to changing ambient temperature in the zone of thermoneutrality. (After Calder and King 1974)

Most birds do not have to change their rates of heat production to maintain an average body temperature of 40°C in the thermoneutral zone. Birds can control the rates of heat loss by changing feather positions, by varying rates of return of venous blood flow from the skin, by manipulating blood circulation in their feet, and by changing the exposure of their extremities, all of which require little direct energy expenditure. For example, fluffing feathers increases insulation, and constriction of veins decreases the flow of blood near the body's surface, thereby further reducing heat loss.

Responses to Cold Stress

A cooling bird first tenses its muscles and begins to shiver, a response that increases oxygen consumption. The temperature at which shivering begins is called the lower critical temperature (LCT). The pectoralis muscles are the major source of heat produced by shivering, supplemented by the leg muscles in some species. Mammals can produce heat by nonshivering thermogenesis (heat production) in a particular kind of fat—called brown adipose tissue (BAT)—but birds appear to lack BAT and associated capabilities for nonshivering thermogenesis (Olson et al. 1988; Saarela et al. 1989).

The lower critical temperatures of large birds are lower than those of small birds, a pattern seen also in mammals. In the absence of special adaptations, therefore, small birds are more sensitive to cold than are large birds; they start to shiver at a higher temperature. The temperatures included in the thermoneutral zone of bird species partly reflect adaptations to the average environmental temperatures in which they live (Weathers 1979). For example, birds living in colder northern climates do not shiver until air temperature drops below 9°C (Snow Bunting) or 7°C (Gray Jay). By contrast, similar-sized southern species, such as the Northern Cardinal and the Blue Jay, start shivering at 18°C (Calder and King 1974).

Birds also select microclimates—localized climates—that reduce their rate of heat loss. Roosting in holes or protected sites such as evergreen trees greatly reduces heat loss, which is especially important during cold winter nights for small passerine birds (Mayer et al. 1982; Dawson et al. 1983). Grouse and ptarmigan burrow into the snow to insulate themselves from cold air temperatures; so do Willow Tits, Siberian Tits, and Common Redpolls (Cade 1953; Sulkava 1969; Korhonen 1981). Huddling together reduces heat loss. Sometimes, however, birds go to extremes: About 100 Pygmy Nuthatches roosted together in one pine tree cavity, so densely huddled that some suffocated (Knorr 1957). On cold days, Inca Doves sit on top of one another between flock feeding forays, forming two-or three-row "pyramids" with up to 12 birds (Mueller 1992). Pyramiding doves face downwind with feathers fluffed in a sheltered sunny place. In large pyramids, doves exposed on outside positions

BOX 6–3
Goldfinches acclimatize to winter cold

Natural physical adjustments to seasonal changes in temperature are called acclimatization. This process may go on over a period of weeks or months, but on a daily basis it reduces the costs of thermoregulation. Winter-acclimatized American Goldfinches, for example, can maintain normal body temperature for 6 to 8 hours when subjected to extremely cold temperatures of −70°C (Dawson and Carey 1976; Carey et al. 1983). Summer-acclimatized goldfinches, however, cannot maintain normal body temperature for more than 1 hour when exposed to such frigid temperatures. The ability of winter-acclimatized goldfinches to withstand cold stress—called thermogenic endurance—is based on restructuring the metabolic pathways that mobilize and use energy substrates, especially fatty acids; small mammals simply increase their aerobic metabolism (Marsh et al. 1990; Dawson et al. 1992). The seasonal changes in goldfinches include both daily metabolic replenishment of fat reserves plus increased activity of the enzymes that catabolize fatty acids and inhibit the use of carbohydrate (glycogen) stores.

frequently try for better positions in the top row and cause the whole pyramid to readjust.

A simple corollary to thermoregulation is that daily energy requirements vary geographically and seasonally in relation to prevailing climates. Ambient temperatures decrease and metabolic costs increase with distance from the equator. For example, House Sparrows that live in the tropical climates of Panama at 9 degrees north latitude spend 58 to 67 kilojoules per day on all activities in summer and winter, respectively (Kendeigh 1976). House Sparrows in Churchill, Manitoba, at 50 degrees north latitude, spend 117 kilojoules per day in summer and 151 kilojoules per day in winter, twice as much as their Panamanian counterparts.

Hypothermia and Torpor

As an energy-saving measure, avian body temperatures fluctuate a few degrees during the day and may drop 2° to 3°C at night. The physiological condition in which the body temperature drops below normal is called hypothermia. Some birds, such as the Turkey Vulture, lower their body temperatures about 6°C at night until T_b equals 34°C, that is, they become mildly hypothermic. Black-capped Chickadees and other northern titmice lower their body temperatures 8° to 12°C at night during extreme winter cold (Moreno et al. 1988). Some small birds, including chickadees but especially hummingbirds, can lower their body temperature even further and enter a state of torpor—or profound hypothermia—in which they are unresponsive to most stimuli and incapable of normal activity. The oxygen consumption of a hummingbird drops by 75 percent when it allows its body temperature to drop 10°C (Calder and King 1974). Hummingbirds potentially save up to 27 percent of

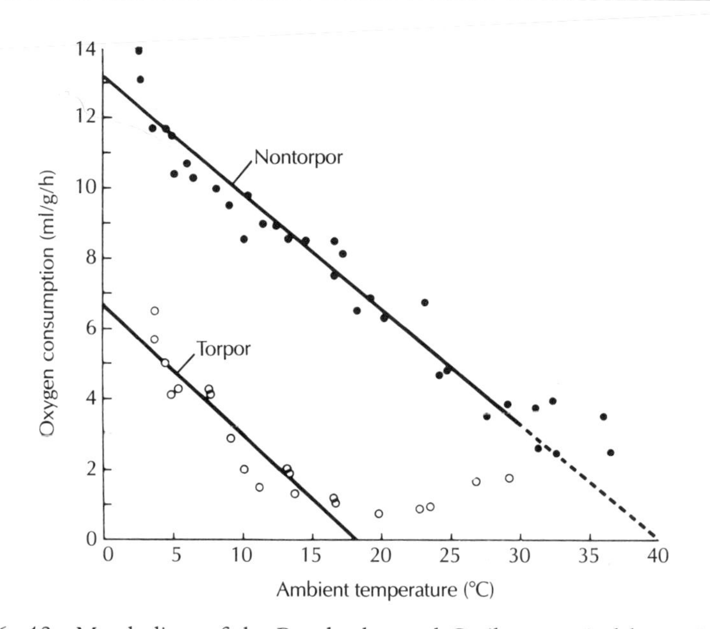

Figure 6–12 Metabolism of the Purple-throated Carib, a tropical hummingbird, during torpor and nontorpor. Nontorpid birds increase their metabolism (measured here in terms of oxygen consumption) as temperature decreases below the lower critical temperature of about 30°C. Torpid birds regulate their body temperature to about 17.5°C. (Adapted from Hainsworth and Wolf 1970)

their total daily energy expenditures by allowing their nighttime body temperatures to drop 20° to 32°C below normal. However, a torpid hummingbird does not abandon control of its body temperature by letting it come to equilibrium with air temperature (Hainsworth and Wolf 1970). Instead, it regulates a lower body temperature, increasing oxygen consumption as needed at low air temperatures (Figure 6–12).

Warming up is the main difficulty with hypothermia and especially deep torpor. Birds waking from torpor begin to show good muscular coordination at 26° to 27°C, but require body temperatures of at least 34° to 35°C for normal activity. A small hummingbird requires about an hour to arouse from torpor at 20°C, but a medium-sized bird such as an American Kestrel would require 12 hours to warm up from a hypothermic body temperature of only 20°C. The extra costs of reheating a large, cool body are prohibitive. Thus, small birds such as hummingbirds can become torpid overnight, whereas kestrels cannot.

Although full torpor usually is not practical or economical for short periods in larger birds, Common Poorwills (55 grams), a small nightjar that lives in the southwestern United States and Mexico, "hibernate" at body temperatures as low as 5° to 6°C for 2 to 3 months during the winter (Ligon 1970; Csada and Brigham 1992). This habit was long known to native Americans; the Hopi people refer to the poorwill as Hölchoko, "the sleeping one." Torpor reduces a poorwill's oxygen consumption by over 90 percent. Poorwills are capable of spontaneous arousal at low ambient temperatures but require about 7 hours to warm up fully (Ligon 1970). Poorwills save energy by using torpor during the summer and spring, especially when faced with cold, wet weather; they even occasionally go torpid during incubation.

Responses to Heat Stress

Birds generally respond to externally imposed heat loads through avoidance behaviors, through physiological defenses, and through controlled elevation of body temperature—called hyperthermia (Figure 6–13). Simple heat avoidance behaviors include reduced activity at midday, seeking shade, bathing, or thermal soaring to cooler air. Domestic pigeons can be trained to turn on cooling fans, especially when thirsty (Schmidt and Rautenberg 1975). Desert-adapted birds also tend to have low metabolic rates, which foster lower rates of water turnover and lower food requirements (Dawson 1984). Poorwills, for example, endure severe heat stress when baked by the desert sun. At T_a = 47°C, poorwills can dissipate up to five times their metabolic heat production; their low metabolic rates and highly efficient evaporative cooling systems allow them to tolerate high temperatures.

The air temperature above which birds expend energy to actively lose heat by evaporative cooling and other means is called the upper critical temperature (UCT). Metabolism increases also because of panting and other efforts that facilitate heat loss. Panting increases evaporative cooling from the upper respiratory tract. Birds typically ventilate faster during heat stress, when body temperatures rise to 41° to 44°C and above. Different modes of shallow panting may enhance evaporative water loss without inducing a shift in blood acidity (Dawson 1984). Evaporative cooling is a highly effective method of heat loss—it can dissipate 100 to 200 percent of heat production—but it presents additional risk in the form of water loss.

Figure 6–13 The Sooty Tern, a bird that is subject to great heat stress at the nest. On a hot day, the bird uses a variety of heat-dissipating mechanisms: (1) exposing the bend of the wing; (2) panting; (3) ruffling crown feathers; (4) ruffling back feathers; (5) wetting abdomen periodically; and (6) exposing the legs. (From Drent 1972)

To supplement panting when they are hot, some birds rapidly vibrate the hyoid muscles and bones in their throats (Lasiewski 1972). This action—called gular fluttering—increases the rate of heat loss through evaporation of water from the mouth lining and upper throat. Many seabirds, both adults and young, which are exposed to the hot sun shining on their exposed nests, regulate body temperature by means of gular fluttering. Desert species, such as Common Poorwills, achieve over half of their evaporative cooling this way.

Birds do not have sweat glands but evaporate water directly through the skin. The amount of water lost increases during heat stress and undoubtedly involves dilation of the blood vessels in the skin, as it does in humans. Evaporative water loss through the skin is especially well developed in certain pigeons and doves (Marder et al. 1989). The precise mechanisms that control this form of heat loss are still obscure. As our understanding of them improves, so may our appreciation of the role of the apteria and the evolution of feather tracts.

When necessary, birds, especially large-footed water birds such as herons and gulls, can lose most of their metabolic heat through their feet (Figure 6–14). Alternatively, when heat conservation is important, they can control blood flow to reduce this loss by more than 90 percent. Control of heat loss from the feet is made possible by special blood vessels in the avian leg, which act to conserve or dissipate heat, as needed. The arteries and veins serving the leg intertwine at the base of the legs in such

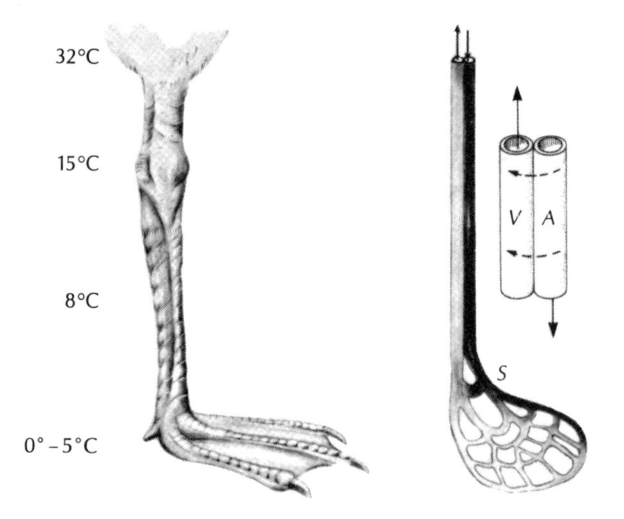

Figure 6–14 Gulls regulate the rate of heat loss from their feet by varying the amount of blood shunted from the base of the leg, where the temperature is roughly 32°C, to veins at the base of the foot, where the temperature may be close to 0°C, They can decrease circulation through the foot, where the rate of heat loss is high, by opening a shunt (*S*) and constricting the blood vessels in the feet, thereby providing a more direct return of the blood. Also heat from outgoing arterial blood can be transferred directly to incoming venous blood. Arrows indicate the direction of arterial (*A*) and venous (*V*) blood flow and dashed arrows the direction of heat transfer. (From Ricklefs 1990)

a way that heat carried by arterial blood from the body core can be transferred directly to returning cool blood in the veins of an exposed extremity. This so-called countercurrent exchange conserves body heat at low air temperatures. For cooling, the blood can completely bypass the network and go directly into the extremities. An overheated Antarctic Giant-Petrel can increase by 20-fold the rate of blood flow through its feet. Storks and New World vultures increase heat loss through evaporative cooling from the legs by excreting directly onto their own legs (Kahl 1963).

By using countercurrent heat exchange in the blood vessels of the head, most birds maintain the temperature of their brains about 1°C cooler than that of their bodies. Helmeted Guineafowl have colorful, naked heads with large protrusions—helmets—and wattles that enhance convective heat loss, as do the wattles of chickens and other fowl (Whittow et al. 1964). Heat loss from these wattles may be so great that a guineafowl's head cools faster than its body, beyond the ability of increased blood flow from the body core to replace lost heat (Crowe and Withers 1979). Unlike most birds, guineafowl brain temperatures vary as much as 6.5°C without serious consequence.

The heat produced during flight could cause lethal increases in body temperature (Berger and Hart 1974). Rock Doves, for example, produce seven times more heat in flight than at rest, and their body temperatures quickly rise 1° to 2°C. The body temperatures of Budgerigars flying at 35 kilometers per hour (in 37°C air) rise to 44°C; they store 13 percent of the heat they produce. Some birds apparently will not fly at temperatures above 35°C because of their inability to control hyperthermia. White-necked Ravens, for example, fly only short distances in the heat or fly in the cooler air at high altitudes (Hudson and Bernstein 1981). However, flight itself increases convective heat loss. The airstream compresses the plumage to the skin, and extension of the wings exposes the thinly

Box 6–4
Controlled hyperthermia has both advantages and risks

In heat-stressed birds, especially dehydrated individuals, body temperatures may rise 4° to 6°C above normal, approaching the near-lethal threshold of 46°C. Such controlled hyperthermia reduces the rate of heat gain from the environment by bringing body temperature closer to air temperature. If body temperatures exceed air temperatures, the hyperthermic bird can lose heat without evaporative cooling and save water. Ostrich body temperatures increase 4.2°C during the daily cycle, a response that saves liters of water per day that otherwise would be lost in evaporative cooling. Controlled hyperthermia during the warm daylight hours also allows for storage of extra heat needed to save fuel at cooler nighttime temperatures, especially in large birds. Elevated body temperatures routinely accompany sustained, heat-generating activity and may enhance physical capabilities (see the discussion at the beginning of the chapter).

feathered ventral base of the wing. As a result, the rate of heat loss by flying parakeets increases to 3.1 times the resting value at 20°C and that of Laughing Gulls increases to 5.8 times the resting value. Nonevaporative heat loss directly from these bare skin areas accounts for 80 percent of a bird's total heat lost at low and medium temperatures.

Water Economy

Satisfying daily energy requirements is only one side of the physiological coin. Water economy is equally important in arid environments. The potential for debilitating water loss is a corollary of the high body temperatures and activity levels of birds, especially during exposure to midday heat. Enhanced evaporative heat loss is essential to avoid heat stress during strenuous activity. For example, evaporative water loss in a desert finch—the Brown Towhee—quadruples when ambient temperature increases from 30° to 40°C, whereas oxygen use only doubles (Bartholomew and Cade 1963). Water is used and replaced at high rates as a result of high evaporative water losses and the limited capacity of birds for concentrating electrolytes in the urine (see "Excretory Systems" below).

Birds replace lost water from several sources. Water present in food satisfies the fluid needs of many birds, particularly nectar-eating or fruit-eating birds and meat-eating raptors such as Peregrine Falcons and Sooty Falcons, which can nest in arid parts of the Sahara where midday shade temperatures exceed 49°C. Likewise, insect-eating birds get most of the water they need from the body fluids of consumed insects; unlike seed-eating birds, they rarely visit water holes. Swallows, which drink surface water often to replace that lost from sustained flight metabolism, are an exception. Because they specialize on dry foods, seed-eating birds experience the greatest need for free-standing water and visit natural water sources in large numbers. California Quail and Rock Wrens, however, obtain adequate water by supplementing their diet of seeds with insects.

Metabolic water, which is produced as a by-product of the oxidation of organic compounds containing hydrogen molecules, supplements ingested water. Because of their high metabolism, birds produce more metabolic water in relation to body size than do most vertebrates. Metabolism of 1 gram of fat yields 38.5 kilojoules of energy *plus* 1.07 grams of water. Metabolism of a gram of carbohydrate or protein produces about 0.56 and 0.40 gram of water, respectively. Metabolic water production increases directly with oxygen consumption and thus with increased metabolism at both colder and higher temperatures. Small birds can replace less of their evaporative water loss with metabolic water than can large birds (Figure 6–15) (Bartholomew 1982), but certain exceptional seed-eating birds, primarily small desert passerines such as Zebra Finches, can survive, drinking not a drop, on a diet of air-dried seeds containing less than 10 percent water, which they supplement with metabolic water.

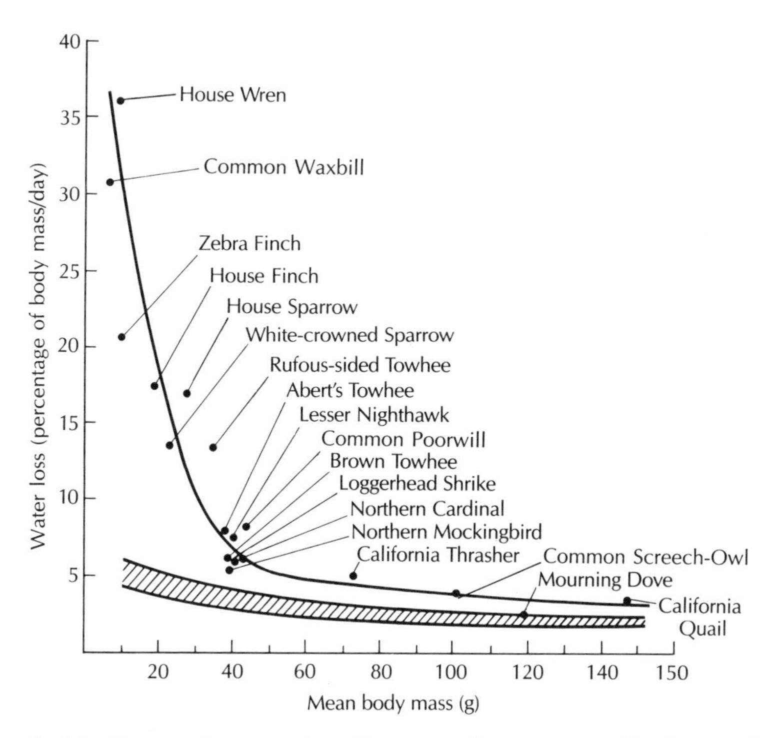

Figure 6–15 Evaporative water loss (in terms of percentage of body mass lost per day) at nonstressful ambient temperatures (near 25°C) decreases sharply with increasing size (and therefore reduced surface area relative to mass) of small birds. Metabolic water production, the projected range of which is indicated by the cross-hatching, partially offsets evaporative loss. (After Bartholomew and Cade 1963)

Much water potentially lost by evaporation is conserved by countercurrent cooling in the nasal and respiratory passages, a general function of the nares and pharynx of small birds. The respiratory passages are cooled by the evaporation of water from the nasal passages during inhalation. During exhalation, the warm, humid air from the lungs comes in contact with the cool respiratory passages, causing water vapor to condense on the nasal passages before the air is finally exhaled. The condensed water then evaporates and is recovered by the cool, dry air that is inhaled when the bird takes its next breath. The amount of water recoverable by these means increases with decreasing ambient temperature. This recovery system reduces evaporative water losses, especially at colder ambient temperatures, to levels closer to that replaced by metabolic water production.

Drinking free water from streams, water holes, dew, raindrops, and even snow is a casual, incidental activity in most mesic habitats—those having a moderate amount of moisture; in deserts, however, daily visits to isolated springs or water holes, where predators wait, may be necessary. Dean Fisher and his colleagues (1972) conducted dawn-to-dusk watches at water holes in the arid regions of western and central Australia. Over half of the 118 species of birds in the area appeared to be independent of surface water. Other species came in spectacular numbers to drink (Figure

6–16). In a 2-hour period one hot day with a shade temperature of 32°C, 1500 to 2000 Spiny-cheeked Honeyeaters came from all directions into the mulga trees that surrounded a water hole. The drinking frequencies of parrots also correlated closely with maximum daily temperatures. One day, during an unusually dry period, Dean Fisher recorded 67,000 visits to one water hole.

Figure 6–16 (Top) Huge flocks of birds regularly visit water holes in arid Australia. (Bottom) Budgerigars at a water hole. (Courtesy of C.D. Fisher)

Excretory Systems

Excretion of water and nitrogenous wastes by birds combines processing by the kidneys and the intestines (Figure 6–17) and, in some species, the action of salt-secreting glands. Avian kidneys—flat structures sited against the fused vertebrae on the dorsal wall of the abdominal cavity—differ in structure and function from those of reptiles or mammals (Braun 1982). Urine produced by the kidneys mixes with fecal components in the lower intestine, where additional water can be resorbed as needed (Thomas 1982).

The most conspicuous physiological adaptation for promoting water economy in birds is the excretion of nitrogenous wastes in the form of uric acid, white crystals that give bird droppings their usual color (mammals excrete urea). The turnover of proteins in the maintenance of body structures produces nitrogenous products, which become toxic if allowed to accumulate. Excretion of nitrogen as urea in aqueous solution requires flushing by large quantities of water, but uric acid can be excreted as a semisolid suspension in which each molecule of uric acid contains twice as much nitrogen as a molecule of urea. Therefore, birds require only 0.5

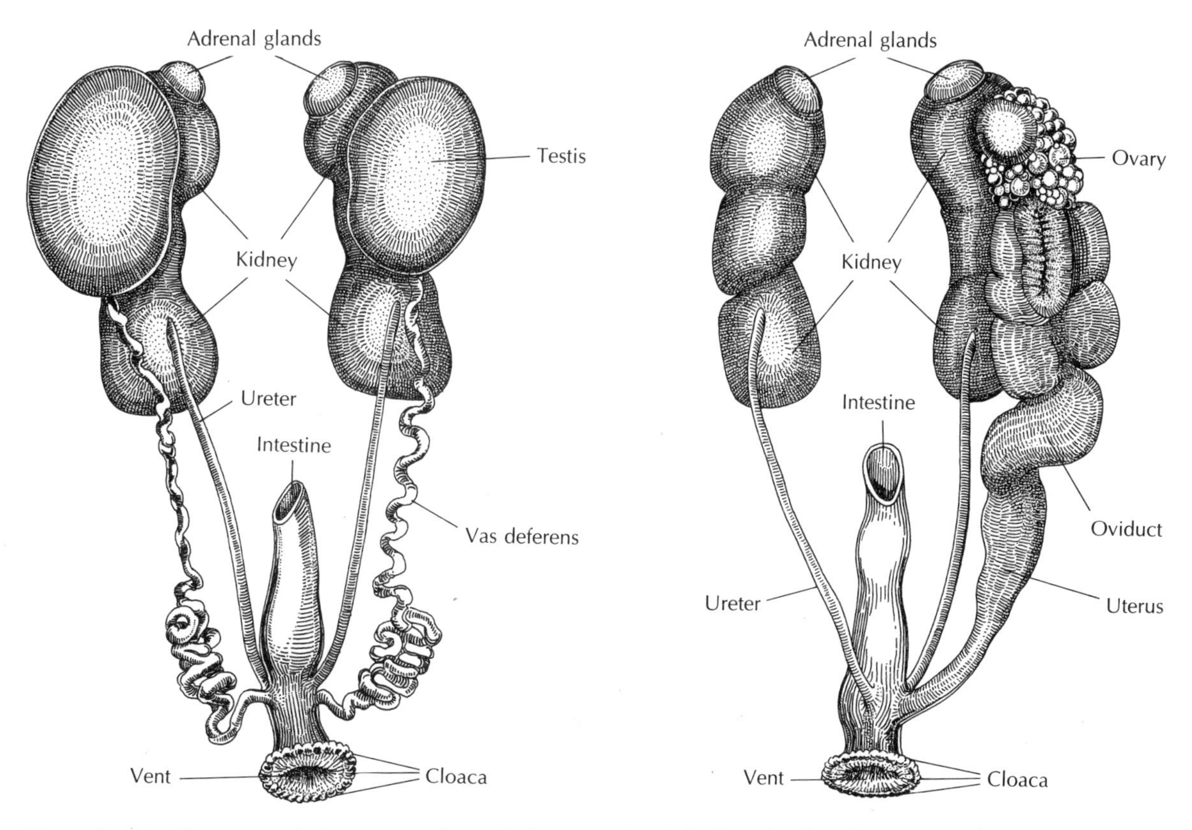

Figure 6–17 The urogenital systems of a male hummingbird (left) and a female hummingbird (right). These organs are located inside the body on the dorsal wall at the front of the abdominal cavity. Only the left ovary is functional in most female birds.

to 1.0 milliliter of water to excrete 370 milliliters of nitrogen as uric acid, whereas mammals require 20 milliliters of water to excrete the same amount of nitrogen as urea. Birds can concentrate uric acid in the cloaca, just prior to defecation, to amazing levels—up to 3000 times the acid level in their blood. Kangaroo Rats, one of the most efficient mammalian water conservationists, can concentrate urea to levels only 20 to 30 times those in the blood.

Water typically constitutes 75 to 90 percent of the excrement of birds with access to plenty of water, but can drop to 55 percent in desert birds. Although avian kidneys can concentrate nitrogenous wastes, they usually cannot concentrate salt or electrolytes much above normal blood levels. Mammalian kidneys, especially those of the Kangaroo Rat, excel at this because of their long loops of Henle—structures that withdraw water molecules from solution. In contrast, the loops of Henle in the avian kidney are short, presumably associated with the excretion of uric acid instead of urea. This anatomic feature presents a problem, particularly for oceanic birds, which must drink seawater containing about 3 percent salt. The body fluids of birds contain 1 percent salt. The high salt content of their marine foods further increases their need to excrete electrolytes. For this reason, seabirds, as well as other birds with water-conservation problems, rely on extrarenal structures called nasal salt glands (Figure 6–18).

Salt Glands

Salt glands, which are often large, conspicuous structures located in special depressions in the skull just above the eyes, enable seabirds to drink

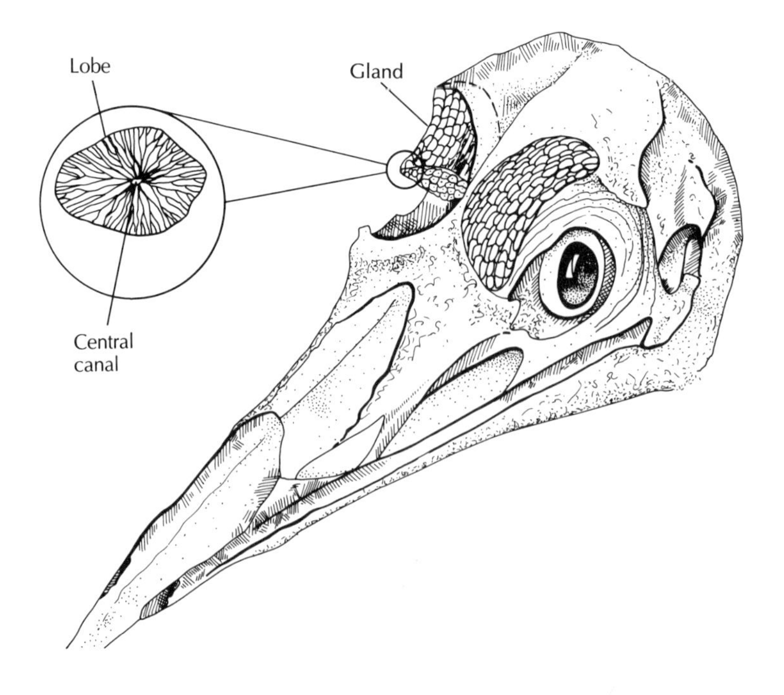

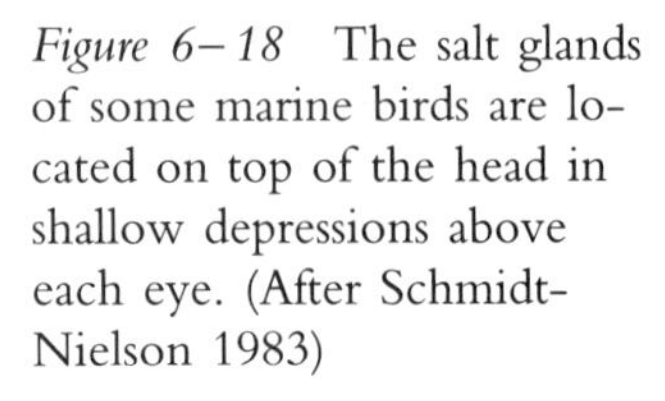
Figure 6–18 The salt glands of some marine birds are located on top of the head in shallow depressions above each eye. (After Schmidt-Nielson 1983)

seawater and to unload the newly ingested salt rapidly via concentrated salt solutions. For example, if a gull drank one-tenth of its body weight in seawater, it would excrete 90 percent of the new salt load within 3 hours (Schmidt-Nielsen 1983). These amazing glands produce and excrete salt solutions containing up to 5 percent salt, more concentrated than seawater.

Salt glands are special infoldings of the cellular lining of the nares. Inside the salt gland are many secretory tubules arranged in lobes. The tubules, each of which is composed of a modified epithelial cell, extract salt from blood in associated capillaries of the ophthalamic arteries (which also service the eyes) and empty directly into a central canal leading to the main duct. Salt concentration and removal by these tubules involve transport of the ions mediated by sodium-activated and potassium-activated enzymes (ATPase). Each of the pairs of glands has a main duct that leads to the anterior nasal cavity. The salt concentrate runs out of the nostril and down grooves to the bill tip before dripping off. Some birds, such as storm-petrels, eject the fluid forcibly. The activity of the salt gland is stimulated directly by the intake of salt, or sometimes just by overload of salt in the blood. These energy demanding, active transport processes in the salt gland may increase the resting metabolic rate by up to 7 percent (Peaker and Linzell 1975).

Salt glands are widespread among nonpasserine birds subject to potential electrolyte imbalances resulting from salty diets. Salt glands are largest and best developed in oceanic birds such as albatrosses, which must drink seawater. When individual birds, such as Mallards, drink saltwater instead of fresh water, their salt glands increase in size. The size of the gland reflects the number of lobes in it and varies in different birds from 0.1 to 1 gram per kilogram of body mass. Auks and gulls have particularly large glands, with as many as 20 lobes. Surprisingly, and for no apparent reason other than evolutionary chance, no passerines have salt glands, not even those that live in salt marshes or feed on intertidal invertebrates on the seacoast.

Daily Time and Energy Budgets

The daily activities of birds integrate the combined challenges of metabolic rates, thermoregulation, and water economy. How much time each day a bird spends resting, feeding and drinking, or flying affects its total daily energy expenditures and therefore its food and water requirements. Although time and energy resources can be separately defined, they are nearly inseparable considerations in the problem of daily energy balance and survival. The amount of energy a bird spends each day is roughly the sum of the hourly costs of sleeping, walking or hopping, flying, and other activities, but the figures are only approximations (see Weathers et al. 1984; Nagy 1989). A more direct way of measuring daily energy expenditures of free-living birds is to inject birds with radioactive water, release and then recapture them, and measure the rate of loss of the labeled water that normal metabolism converts into carbon dioxide during respiration (Williams and Nagy 1984). This technique incorporates microclimate variables, such as heat loss due to convection, not reflected in the conversion of time budgets to energy budgets and thus avoids crude estimates of the metabolic costs of different kinds of activities.

Studies of the daily activities of Gambel's Quail illustrate the working components of the environmental physiology of one species (Goldstein 1984; Goldstein and Nagy 1985). A medium-sized quail with a dangling topknot, the Gambel's Quail is one of the few ground-dwelling vertebrate species that is active during the summer day in the seasonally hot and dry Colorado Desert. They divide their time each day between foraging (6.2 hours per day) and resting (6.7 hours per day); they sleep or rest overnight (Figure 6–19). They run only to move quickly across hot, sunlit areas to another bush, and fly short distances only two or three times a day; the time and energy costs of these activities are minor. The costs of thermoregulation also are minor because nighttime temperatures remain within the zone of thermoneutrality and because the quail can avoid heat stress by resting in the shade during the heat of the day.

Goldstein and Nagy (1985) estimated the total daily energy costs of these quail both by conversion of the time budgets into energy budgets, using laboratory estimates of metabolism for each activity, and directly using the standardized radioactive water technique. The two procedures produced similar results, 81.8 and 90.8 kilojoules per day, respectively, both well below simple predictions. It turns out that the quail depress their resting metabolism well below rates predicted for their size, appar-

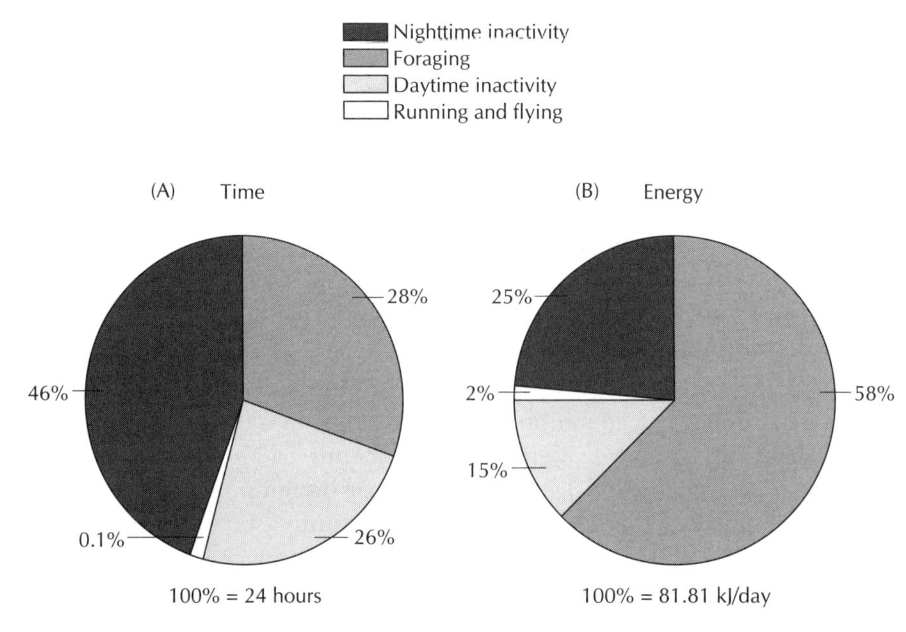

Figure 6–19 Time and energy budgets of Gambel's Quail in summer. (A) Fractions (%) of the 24-hour day spent foraging, resting, and running and flying. (B) Time budgets are multiplied by metabolic rates appropriate to each activity to estimate total daytime energy expenditures. Note that feeding, a costly activity, makes up a larger part of the energy budget than of the time budget. (From Goldstein and Nagy 1985)

ently as a way to reduce metabolic heat production during the heat of the day.

Gambel's Quail cannot exist on dry seeds alone but must supplement metabolic water production with preformed water ingested in foods or by drinking. Provision of free water can increase the size of local quail populations. The estimates of daily metabolic rate suggest that these quail produce metabolic water at a rate of 2.7 milliliters per gram per day, which they must supplement with at least 5.1 milliliters per gram per day from water in their food or elsewhere. In the absence of free drinking water, the quail eat insects, particularly ants, and succulent fruits of cacti to get this water. Curiously, by eating ants, which contain 62 percent water, the quail also reduce the numbers of one of their competitors for food seeds in the desert sands.

Summary

Birds have high metabolism. Flight and the maintenance of high body temperatures use large amounts of energy. Both the circulatory and respiratory systems have evolved exceptional capacities for the delivery of fuel and the removal of metabolic products. Water loss linked to high metabolic rates also poses difficulties for some birds.

Birds regulate their body temperatures at 40° to 42°C by adjusting plumage insulation, by increasing heat production through shivering when cold, and by evaporative water loss through panting and gular fluttering when hot. Regulation of blood flow through the feet aids heat loss or retention. To save energy, some birds, notably hummingbirds, swifts, and nightjars, can lower body temperature and become torpid. Birds can also elevate body temperature a few degrees to reduce the need for evaporative water loss and to store body heat for cold nights. Heat produced during flight can be lost quickly with little loss of water. However, birds have little latitude for higher body temperatures: 46°C is lethal.

Birds depend on metabolic water as well as on that ingested in their food or drunk as free water. Desert water holes attract huge aggregations of thirsty birds, although the excretion of nitrogenous wastes as uric acid rather than as urea promotes water economy in birds. Seabirds have well-developed salt glands embedded in their skulls over their eyes. These glands void concentrated salt solutions and thereby enable the birds to drink seawater and to eat prey with high salt contents.

Further Readings

Calder, W.A. 1984. Size, Function, and Life History. Cambridge, Mass.: Harvard University Press. *An analysis of the functional relationships of changes in overall size and the dimensions of body components.*

Calder, W.A., and J.R. King. 1974. Thermal and caloric relations of birds. Avian Biology 4: 259-413. *A comprehensive review of energetic aspects of avian physiology.*

Dawson, W.R. 1984. Physiological studies of desert birds: Present and future considerations. J. Arid Environ. 7: 133–155. *A comprehensive review of how birds cope with heat stress and water loss.*

Jones, D.R., and K. Johansen. 1972. The blood vascular system. Avian Biology 2: 157–285. *A detailed review of the avian circulatory system.*

King, A.S., and J. McLelland, Eds. 1980–1988. Form and Function. *In* Birds, Vols. 1–4. London: Academic Press. *An encyclopedic reference on functional anatomy and physiology of birds.*

Scheid, P. 1982. Respiration and control of breathing. Avian Biology 6: 406–453. *A detailed review of the avian respiratory system.*

Skadhauge, E. 1981. Osmoregulation in Birds. Berlin: Springer. *A comprehensive review of water turnover and economics.*

Sturkie, P.D. 1986. Avian Physiology, 4th ed. New York: Springer-Verlag. *A general review with an emphasis on poultry.*

CHAPTER 7

Feeding

BECAUSE BIRDS EXPEND ENERGY at a tremendous rate, they must feed frequently to refuel themselves. Birds sit, walk, hop, fly, and dive in search of food. Sitting shrikes simply wait for prey, whereas crows walk methodically across fields, and warblers hop from twig to twig. Vultures soar, scanning for carrion below, and falcons swoop down on fast-flying quarries. Ducks, grebes, and loons dive in deep waters to catch fish or to pluck invertebrates from rocky moorings.

Adaptations for feeding are a conspicuous feature of avian evolution. These adaptations include not only the modes of locomotion birds use while feeding but also anatomical specializations of the bill and tongue, legs and feet, crop, stomach, and intestines. Following a review of avian bills and digestive systems, this chapter considers feeding as a highly refined behavior. The study of the search for food and the choices inherent in feeding behaviors constitutes a major field of exploration. Of special interest is the balance between feeding effort and energy requirements and the role of learning and practice to increase foraging efficiency. Finally, birds store both internal and external food reserves for later use. The recovery of cached seeds draws on the extraordinary spatial memories of some seed-eating birds.

Bill Structure

A bird's bill is its key adaptation for feeding. The size, shape, and strength of the bill prescribe the potential diet (Figure 7–1). The land carnivores—eagles, hawks, falcons, and owls—have strong, hooked beaks with which they tear flesh and sinew. Other bill types tear meat, spear fish, crack seeds, probe crevices, or strain microscopic food from the mud. The broad, flat bill of a duck is suited for straining mud, whereas the chisellike bill of a woodpecker is suited to dig into trees to reach ants. Marine predators, such as penguins and cormorants, have bills with curved projections that direct fish toward the esophagus. The varied lengths and curvatures of shorebird bills determine which prey they can

Figure 7–1 The bills of birds reflect their feeding specialties. (A) Red Crossbills extract seeds from pine cones. (B) Northern Cardinals crack large, hard seeds. (C) Northern Shoveler ducks strain food from the mud. (D) Reddish Egrets spear small fish. (E) Golden Eagles tear apart the flesh of their prey.

reach by probing into the mud (Figure 7–2). Even slight differences in bill dimensions influence the rate at which food can be consumed.

The bills of birds may seem specialized to one particular food or way of feeding, but more often than not they are multipurpose organs. Most birds feed on a variety of foods or change diets with the season. The heavy bill of a Song Sparrow, for example, is well suited to a winter diet of hard seeds but also serves for catching and eating soft-bodied insects in the summer. Long-billed shorebirds can pick up insects from the surface in the breeding season and probe deeply into the mud while migrating and in winter.

Four major features make up the general morphology of bird bills. The upper mandible—or maxilla—attaches to the braincase by a thin, flexible sheet of bone called the nasofrontal hinge. The lower mandible articulates with the quadrate, a large, complex bone at the lower mandible's posterior end. The large jaw muscles, which enable a bird to bite, attach to the posterior surfaces of the mandible. Covering both jaws is a horny sheath—or rhamphotheca—which may have sharp cutting edges (as in boobies), numerous toothlike serrations (as in mergansers), or well-developed notches (as in falcons and toucans).

The avian bill is not rigid; birds can flex or bend the upper half of the bill, an ability called cranial kinesis (see Zusi 1984). The maxilla of most birds flexes only at the nasofrontal hinge. In some birds, the dorsal ridge of the bill itself bends—rhynchokinesis—at the base of the bill, near the tip of the bill, or at both sites. A woodcock, a long-billed, snipelike bird, can open just the tip of its bill to grasp an earthworm deep in the mud.

The bone configurations that constitute the bill, jaws, and palatal region are an engineer's delight; their structure relates directly to the amount of force caused by using the bill (Bock 1966). The maxilla is a flattened, hollow, bony cone reinforced internally by a complex system of bony struts called trabeculae. The maxilla is reinforced where the greatest forces are manifest. The trabeculae located near the nasofrontal hinge help distribute the stress on the hinge that is caused by biting. The curvature of the continuous upper jawbone surface also adjusts for stress (Figure 7–3).

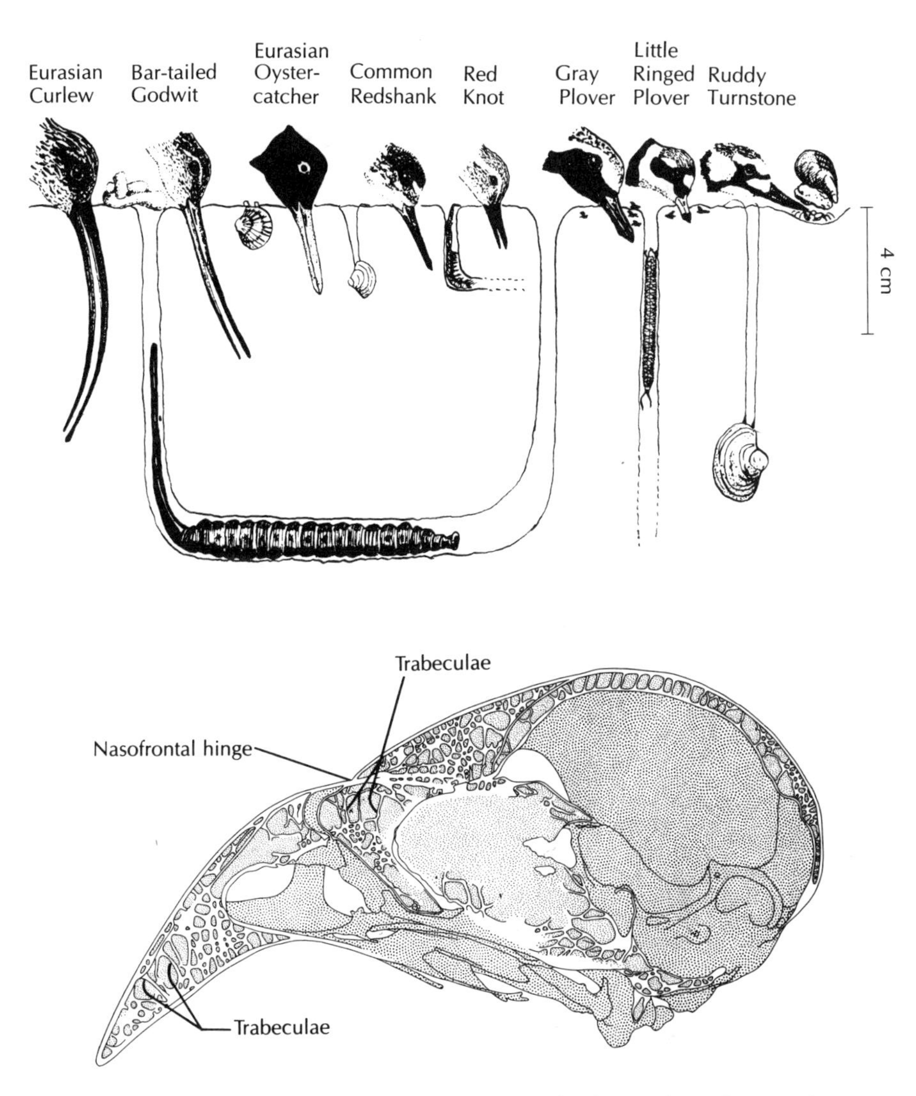

Figure 7–2 Varied bill lengths enable shorebirds to probe to various depths in the mud and sand for food. Plovers feed on small invertebrates, mainly by surface pecking with their short bills. Common Redshanks and other species of waders with moderate bill lengths probe the top 4 centimeters of the substratum, which contains many worms, bivalves, and crustacea. Only the long-billed birds such as curlews and godwits can reach deep-burrowing prey such as lugworms. (Adapted from Goss-Custard 1975)

Figure 7–3 The form of their large bills enables finches such as the Northern Cardinal to bite hard seeds without straining the nasofrontal hinge (located between bill and skull) with excessive shear forces. Shown here is a cross section of a cardinal skull, revealing the bony struts (trabeculae) in the upper jaw and forehead. The deeper, nontrabecular areas of the upper jaw are shown in fine stippling; other nontrabecular bone is shown in heavier stippling. Lower jaw is not shown. (From Bock 1966)

The functional refinements of bill structures can be seen in the details of the filter-feeding bills of flamingos, the seed-cracking bills of finches, and the probing bills of hummingbirds and other nectar-feeding birds.

The Filter-feeding Bills of Flamingos

A flamingo's filter-feeding bill is one of the most distinctive and specialized avian bills (Jenkin 1957). It consists of a large, troughlike lower mandible housing a powerful, fleshy tongue that creates suction as it pumps back and forth. Sets of projections called lamellae, which vary in structure from hard ridges and large hooks to velvet-textured fringed platelets, sort tiny food items from debris in the water. The solid, narrow, upper mandible contains sets of lamellae, each set opposing another set on the inner surfaces of the lower mandible. With head and bill upside down, flamingos strain fine food particles from water or mud pumped by the tongue through the lamellae. The size of the filtering apparatus determines a flamingo's diet. The relatively coarse filters on a Greater Flamingo's bill strain out small invertebrates, whereas the Lesser Flamingo's fine filters strain out tiny blue-green algae. By virtue of this difference, the two species can feed side by side and eat different foods (Figure 7–4).

The Seed-cracking Bills of Finches

Seeds sustain a variety of birds. Doves and seedsnipes swallow seeds whole and grind them in the gizzard. Jays and titmice hammer them until they split open. Most passerines that specialize in seed-eating crack and shuck the seed husks with powerful bills. Evening Grosbeaks and Hawfinches can crack the hardest seeds, such as olive and cherry pits. Many finches have elaborate, hard ridges and grooves inside the maxilla and the anterior palatal region, which enable them to hold the seed in place while it is cut by the sharp edges of the jaws. Finches extract seed kernels by either crushing or cutting the seed hull. In the crushing method, the seed, held by one or both margins of the bill, is pressed against a central ridge in the horny roof of the mouth to pop the kernel from the shell. In the cutting method, a finch uses its tongue to lodge the seed in furrows of the hard palate and then cuts the husks with rapid forward and backward movements of the sharp edges of the jaws. In both cases, the cut husks fall out of the mouth, and the clean kernel is swallowed.

The husking speed of a finch depends on the relationship of bill size and strength to seed size and hardness (Willson and Harmeson 1973). In general, large-billed finches can husk seeds of a wider range of size and hardness than can small-billed finches. The Northern Cardinal has a much larger bill than the White-throated Sparrow, which in turn has a bigger bill than the American Tree Sparrow. Large, hard hemp seeds are husked in 13.5, 13.9, and 19 seconds, respectively, by these birds. However, small-billed birds are more efficient at exploiting small seeds; American Tree Sparrows can husk small millet seeds in 1.6 seconds, about

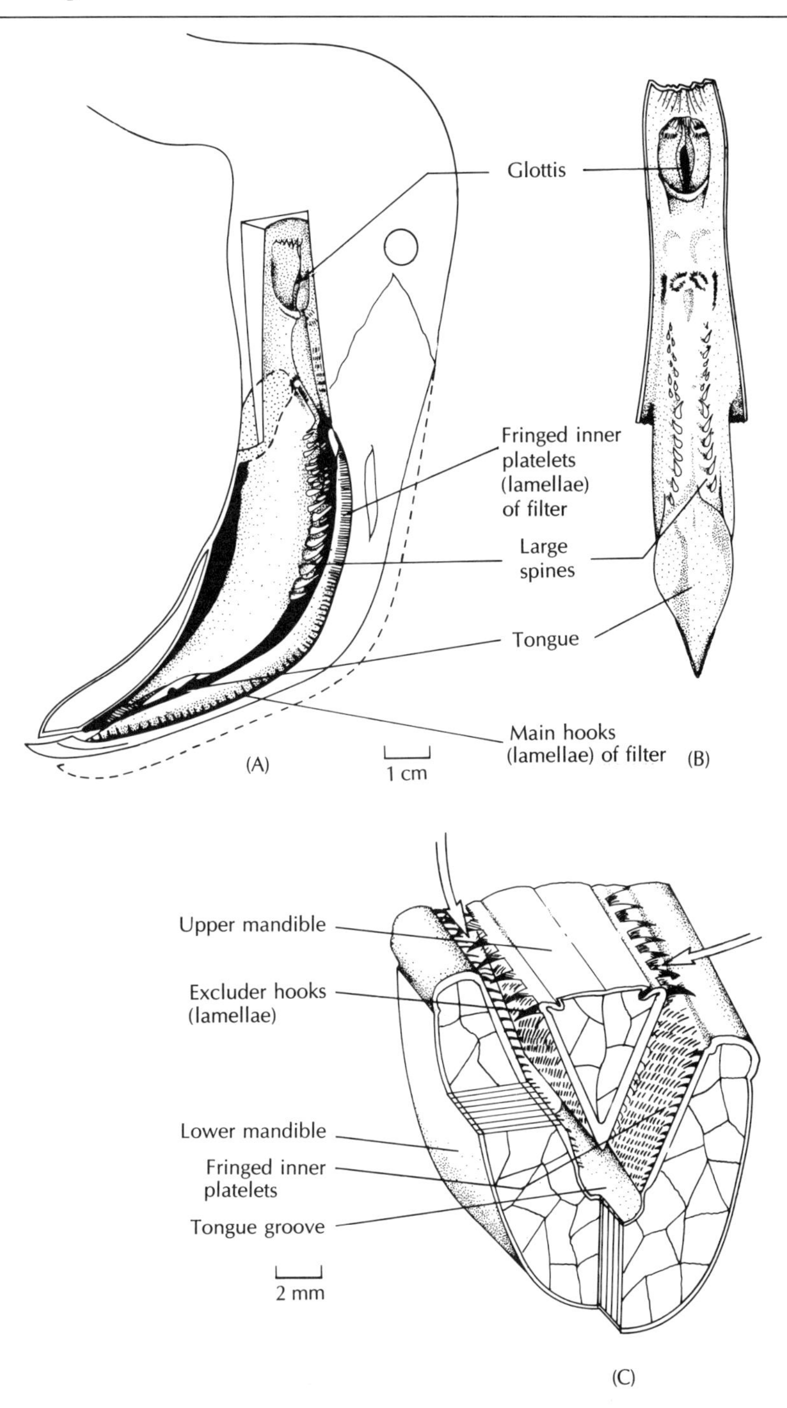

Figure 7–4 (A) Head of a flamingo in normal "upside-down" feeding position. (B) The tongue. (C) Cross section of the bill of a Lesser Flamingo, showing flow of water currents caused by the pumping action of the tongue. When the tongue (omitted from diagram) is depressed in the tongue groove, inflowing currents (arrows) are drawn through the large, hooklike lamellae of the upper mandible, which exclude large objects. When the tongue is elevated, outflowing currents pass through the succession of filters on both jaws formed by the smaller, velvet-textured lamellae—called fringed inner platelets. (After Jenkin 1957)

three times faster than the 4.9 seconds needed by the larger White-throated Sparrows.

A large finch bill can be so advantageous in times of food shortage that the average bill size in a population increases from one year to the next (Boag and Grant 1981). In 1976 and 1977, a severe drought gripped Daphne Island in the Galápagos archipelago. Plants failed to produce new crops of seeds, and seed densities dropped sharply, especially the densities of small seeds. Many finches starved. Medium Ground-Finches with large, deep bills survived in greater numbers than small-billed individuals because those with large bills could crack the remaining larger, harder seeds. The result was a dramatic increase in average bill size over only one year's time, due to natural selection (Figure 7–5). This intense natural selection was later reversed by the improved survival of small-billed birds during wet years, when small seeds were again plentiful.

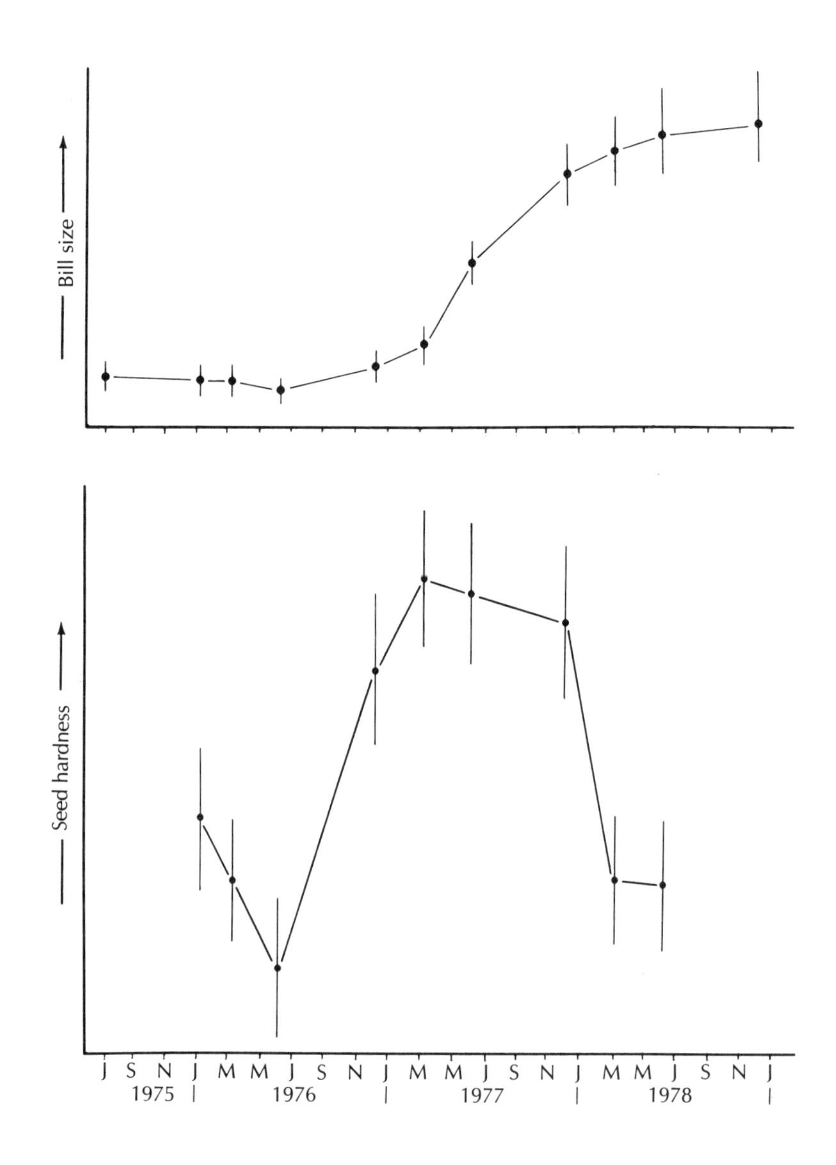

Figure 7–5 Increase in bill size (top) in the Medium Ground-Finch during a period of drought that resulted in intense natural selection. Failure of the usual seed crop on Daphne Island favored individuals with large bills able to crack the more abundant, large, hard seeds (bottom). (Adapted from Boag and Grant 1981)

BOX 7–1

One finch species has three bill sizes

The Black-bellied Seedcracker of West Africa is a unique bird. Some individuals have small bills, others have large bills, and still others have truly massive bills; they pair randomly with each other (Smith 1990a, b). The differences in bill size are not related to age, sex, or locality. Instead the differences represent a rare case of trimorphism in a complex anatomical trait. The bill sizes match a trichotomy in the seedcracker's main food, the seeds of grasslike sedges *(Scleria)*. Large-billed individuals specialize on one kind of very hard sedge seeds; small-billed individuals specialize on soft seeds of another species of sedge. The large-billed form probably evolved from small-billed forms when there was a shortage of and intense intraspecific competition for soft sedge seeds during the dry season, a time of low seed availability. Individuals with intermediate bill sizes are disadvantaged, particularly juveniles. Selection against intermediate-sized individuals, which produces two size classes, is called disruptive selection. Sexual dimorphism in size or color is a more typical result of this kind of natural selection.

The Probing Bills of Nectar-feeding Birds

Nectar feeders, such as hummingbirds and sunbirds, probe their long, thin bills into floral nectar chambers and draw up nectar through tubed tongue tips. Bill forms tend to match the lengths and curvatures of preferred flowers, which, in turn, depend on the birds for pollination (Figure 7–6). Specialized nectar-feeding birds feed from their own well-matched flowers when these are available, but sometimes they must feed from flowers of other sizes and shapes. When this occurs, the effects on foraging of slight differences in bill dimensions are most apparent.

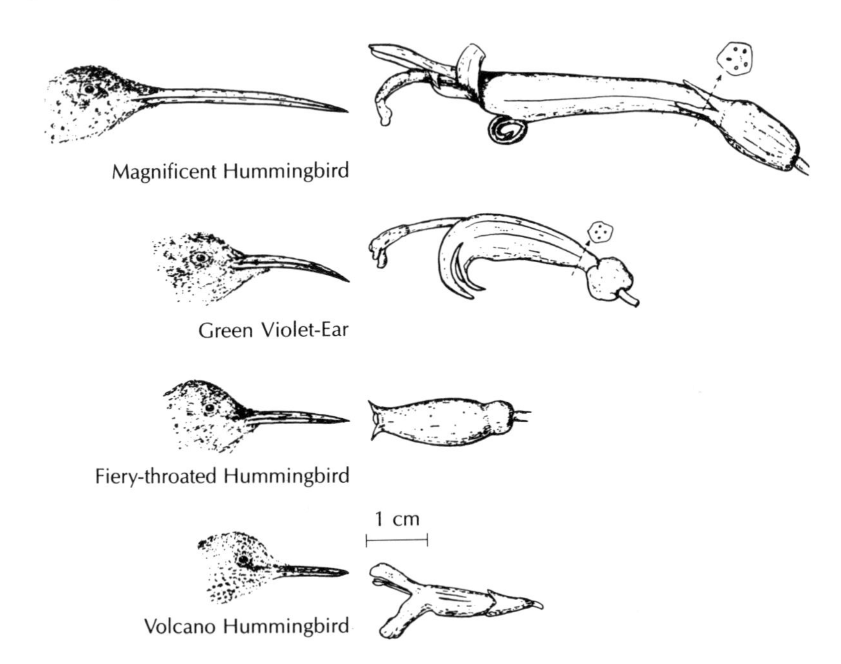

Figure 7–6 The lengths and curvatures of hummingbird bills match those of their preferred flowers. (Adapted from Wolf et al. 1976)

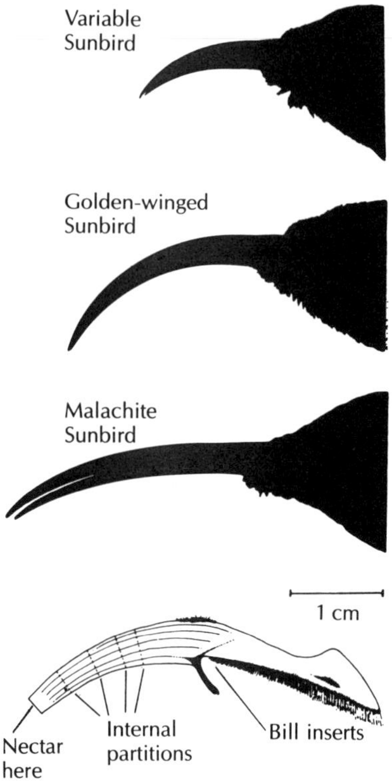

Variable Sunbird

Golden-winged Sunbird

Malachite Sunbird

1 cm

Nectar here

Internal partitions

Bill inserts

Figure 7–7 These three sunbirds all feed on flowers of a mint plant. The Golden-winged Sunbird can extract nectar faster than the other two because its bill closely matches the length and curvature of the flower, enabling it to reach nectar at the base of the flower more easily. (Adapted from Gill and Wolf 1978)

TABLE 7–1
Effect of sunbird bill dimensions on nectar extraction

Sunbird species	Bill length (mm)	Time per flower(s)	Nectar removal (%)	Rate of extraction (μL/s)
Golden-winged	30	1.3	90	3.6
Malachite	33	1.8	82	2.4
Variable	20	2.8	62	1.1

From Gill and Wolf 1978.

Differences in the bills of three species of East African sunbirds, for example, affect not only the time they take to probe a flower of the mint Leonotis but also how much of the available nectar they obtain from a chamber at the base of the corolla (Figure 7–7, Table 7–1). The long, strongly curved bill of the Golden-winged Sunbird matches the curvature of the flower and thereby enables it to visit flowers faster and extract more of the available nectar than do the other species. Pollen transfer is effected by pressing forward against the pollen laden anthers and the stigma at the entrance to the flower. The Malachite Sunbird forces its straighter bill down the curved corolla with several jabbing motions and often misses the opening through the diaphragm that protects the nectar chamber. The Variable Sunbird's bill and tongue are too short to reach down the full length of the flower. Instead, it stabs the base of the flower to gain access to the nectar chamber. This approach is slow (2.8 seconds per flower), removes only 62 percent of the available nectar, and fails to effect pollen transfer.

Avian Digestive System

Because birds lack teeth that chew food before swallowing, the avian digestive system is specialized to process unmasticated food. The major parts of this system—the oral cavity, esophagus, crop, two-chambered stomach (proventriculus and gizzard), liver, pancreas, intestine, and cloaca—are further specialized to accommodate particular types of diets and feeding practices (Figure 7–8). Linked to the development of flight and high metabolic rates, the digestive systems of birds extract nutrients and energy with high efficiencies from small volumes of rapidly processed food (Place 1991).

The lack of true teeth in the avian bill usually precludes the elaborate chewing and maceration of bulk that is typical of mammals. Instead, most birds swallow food whole. Raptors rip their food apart. The plantcutters (Phytotomidae) of South America chew vegetation with toothlike serrations on the edges of their bills, and some fruit-eating birds, such as the

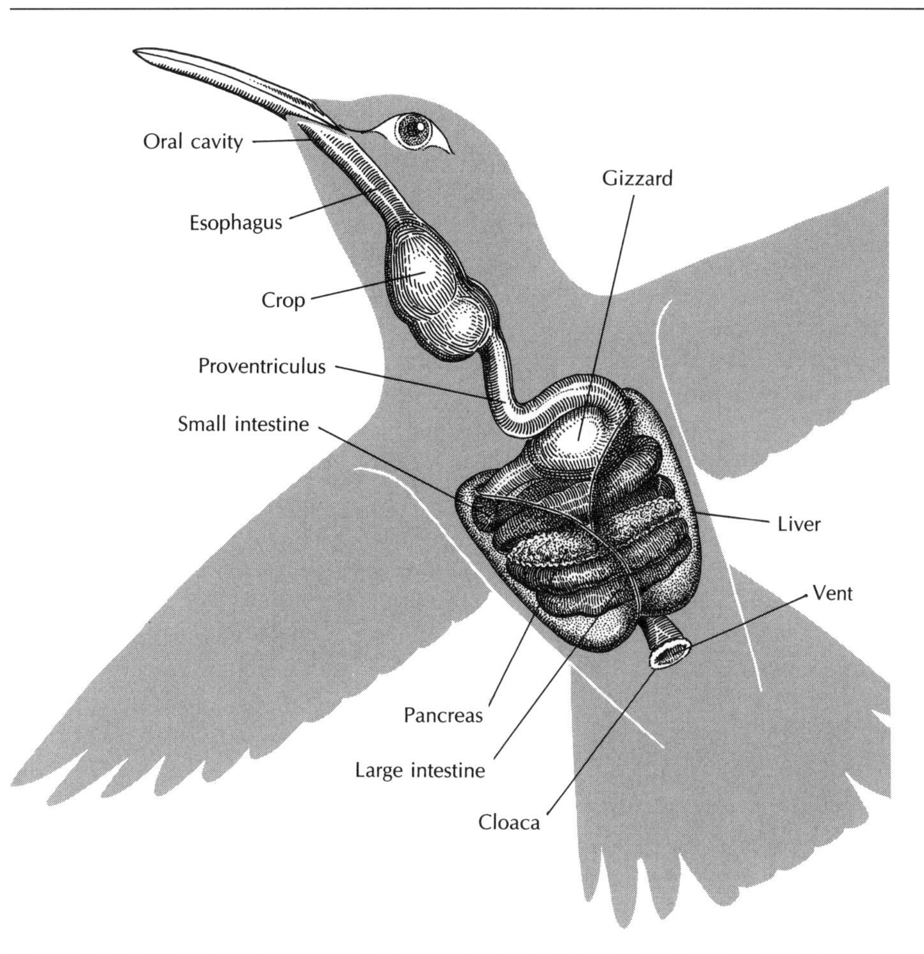

Figure 7–8 The digestive tract of a hummingbird. (After Tyrrell and Tyrrell 1985)

Eurasian Bullfinch crush berries before swallowing them. Most other fruit eaters gulp down fruits whole. The lack of teeth in birds appears to be a weight-reducing adaptation for flight because teeth require a heavy jawbone for support.

The oral cavity houses taste buds, salivary glands, and a tongue that is often specialized. Taste buds in the soft palate aid in food selection. Three major sets of salivary glands and a variety of smaller ones provide lubrication, essential for the passage of food toward the esophagus.

Although the primary purpose of the salivary gland is to provide lubrication, some birds produce salivary secretions for other uses as well. The salivary secretions of woodpeckers are sticky, which helps them extract insects from wood crevices and ants from nests. Gray Jays use sticky salivary secretions to form lumps of food—called boluses—that they store for future use (Bock 1961; Dow 1965). Swifts use salivary fluids to glue together and attach their nests to cave walls. Edible-nest Swiftlets of Southeast Asia make their nests almost entirely of hardened saliva (Figure 7–9), which is the primary ingredient of the gastronomic delicacy called bird's nest soup.

The tongue aids in the gathering and swallowing of food. Most bird tongues have rear-directed papillae that aid in swallowing. Extremely sensitive structures, bird tongues are filled with tactile sensory corpuscles, especially at the tip. These corpuscles are best developed in the spoon-tipped tongues of seed-eating songbirds, which manipulate tiny seeds, and in the strong, club-shaped tongues of parrots.

Bird tongues usually are not muscular structures but operate by means of sheathed, rear-directed bony extensions—called the hyoid apparatus. Hummingbirds and woodpeckers have extremely long hyoidal extensions of the tongue that extend back to the rear of the skull and then curl around to lie on top of the skull. These hyoids, which double the reach of the tongue and bill, permit hummingbirds to extract nectar from long, curved flowers and woodpeckers to reach deep into tree trunks for insects.

The woodpecker tongue is also fitted with barbs. Similarly, the tongues of penguins and other fish-eating birds often have rear-directed hooks that top slippery fish from wriggling back up to freedom. Convergent with flamingos, the tongues of filter-feeding waterfowl and small, southern ocean petrels—called prions or whalebirds—have fringes and peripheral grooves that strain tiny food particles from the mud. Ducks such as the Northern Shoveler draw mud and water into their mouth and then force it out through the filtering system (Figure 7–10). Like flamingos, different species of the filter-feeding prions have different bill sizes and filter straining characteristics.

In most birds, food passes from the oral cavity to the stomach via the esophagus, a muscular structure lined with lubricating mucous glands. In birds that swallow large prey whole—fish-eating birds, for example—the esophagus expands as needed. No mere passageway, the esophagus is a versatile organ. The esophagus of pigeons produces nutritious fluid—called pigeon milk—for their young. The esophagus of pigeons also can

Figure 7–9 An Edible-nest Swiftlet by its nest, which is made almost entirely of hardened saliva, the main ingredient of bird's nest soup. (Courtesy of the Director, Sarawak Museum)

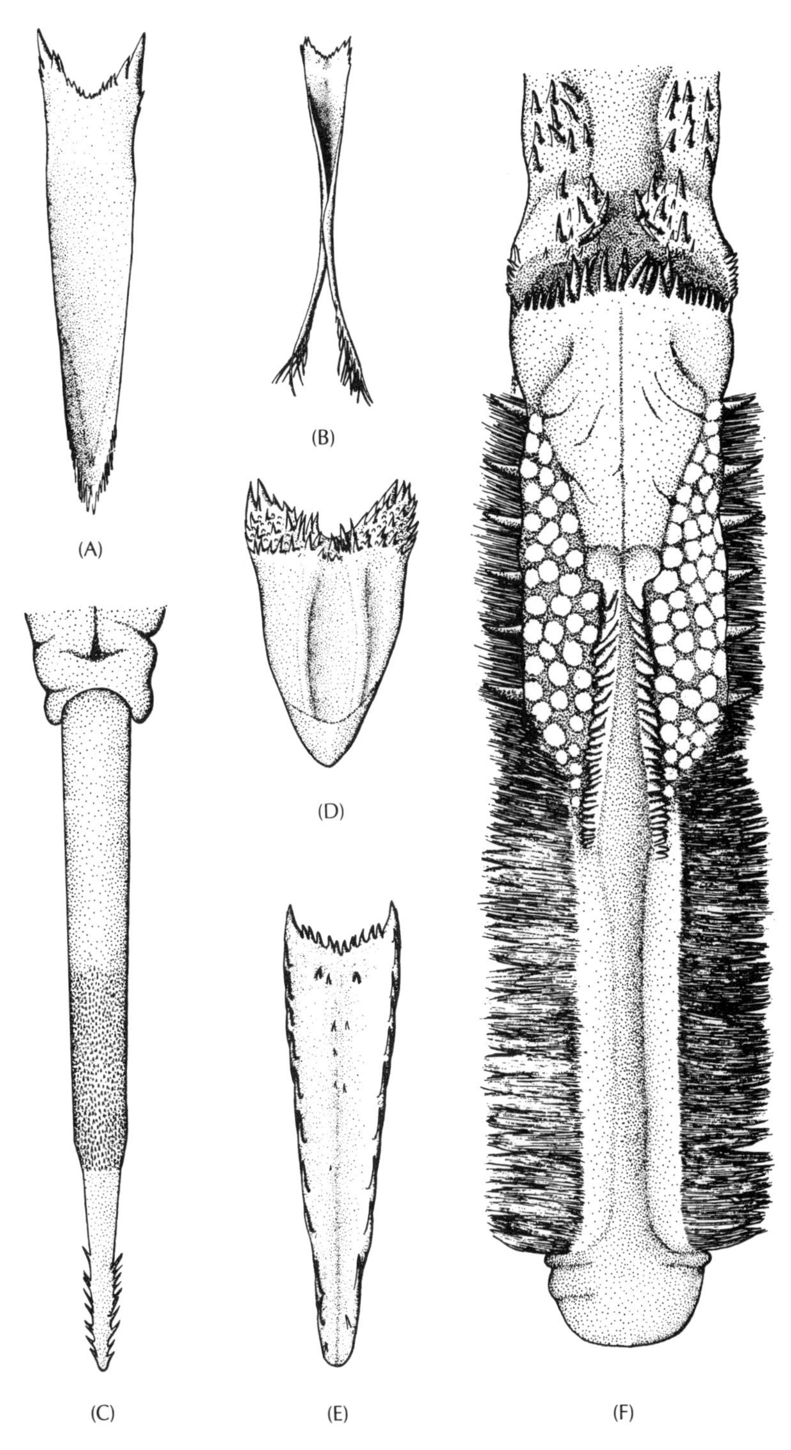

Figure 7–10 Bird tongues (dorsal view): (A) generalized passerine tongue with terminal fringes (American Robin); (B) tubular, fringed nectar-feeding tongue (Bananaquit); (C) probing and spearing woodpecker tongue (White-headed Woodpecker); (D) short, broad tongue of a fruit eater (Diard's Trogon); (E) fish-eater tongue (Sooty Shearwater); (F) food-straining tongue (Northern Shoveler). (Adapted from Gardner 1925)

be inflated for display and sound resonance, as in Ostriches, bustards, prairie-chickens, and female painted-snipes.

The crop—an expanded esophageal section found in many birds—stores and softens food and regulates its flow through the digestive tract. In this respect, the crop serves the same function as the stomach of a monogastric (nonruminant) mammal. The crop varies greatly in shape from a simple, expanded section of the esophagus, as found in cormorants, ducks, and shorebirds, to a lobed, saclike diverticulum, as found in vultures, fowl, pigeons, parakeets, and certain northern finches, the redpolls (Figure 7–11). In two herbivorous birds, the Hoatzin of South America and the flightless Kakapo, a nocturnal parrot of New Zealand, the crop has evolved into a glandular muscular stomach, wherein the tough leaves that make up their spartan diets are digested. A concave section of sternal keel accommodates the Hoatzin's well-developed crop-stomach, which is 50 times the weight of its poorly developed true stomach (Sick 1964).

Birds have two-chambered stomachs composed of an anterior glandular portion—or proventriculus—and a posterior muscular portion—or gizzard (Figure 7–12). Shapes and morphological structures of the stomach differ more than any other internal organ, reflecting the dietary habits of different species (Place 1991). The proventriculus, a structure not present in reptiles, is most developed in fish-eating birds and raptors. It secretes acidic gastric juices (pH 0.2 to 1.2) from its glandular walls, thereby creating a favorable chemical environment for digestion. Peptic enzymes in the proventriculus dissolve bones rapidly. The Lammergeir, a huge vulture, can digest a cow vertebra in 2 days. A shrike can digest a mouse in 3 hours. In addition to the usual functions, petrels use their well-developed proventriculus to store oil by-products of digestion, which they regurgitate as food for their young—and sometimes spew at predators and ornithologists.

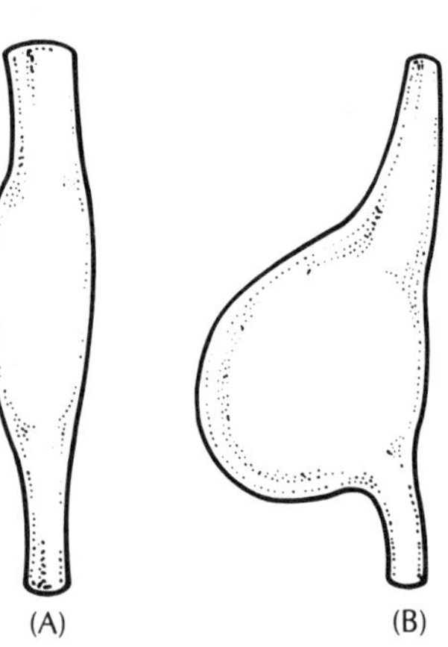

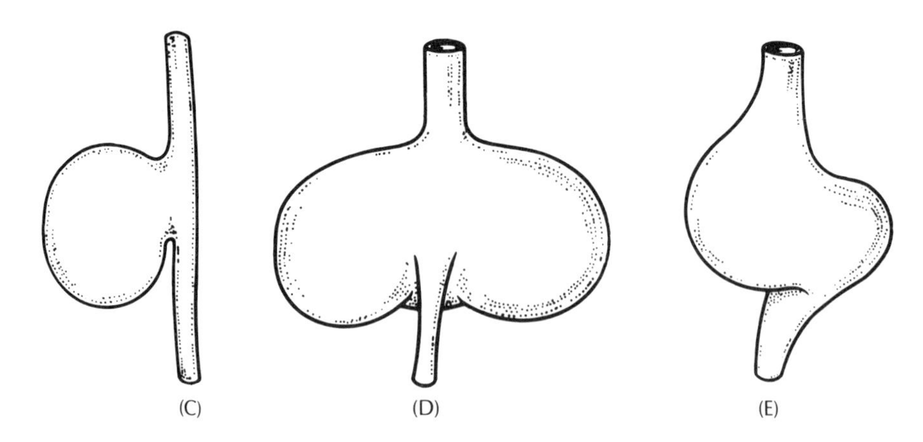

Figure 7–11 Avian crops: (A) cormorant; (B) vulture; (C) fowl; (D) pigeon; (E) parakeet. (From Pernkopf and Lehner 1937)

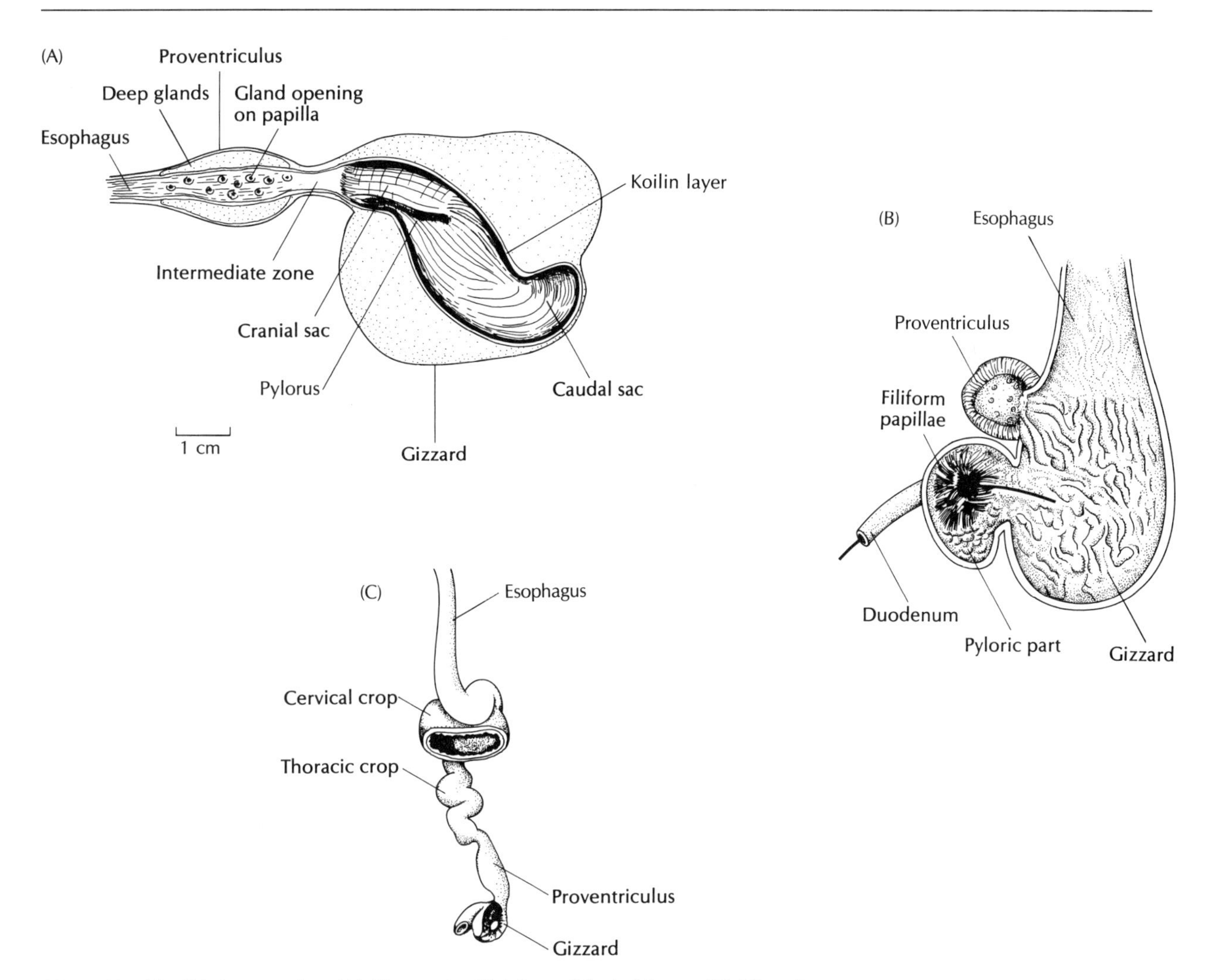

Figure 7–12 Bird stomachs: (A) Domestic Chicken; (B) Anhinga; (C) Hoatzin (all 66 percent of natural size). (After McLelland 1975; Garrod 1876; Pernkopf and Lehner 1937)

The avian gizzard—the functional analogue of mammalian molars—is a large, strong, muscular structure used primarily for grinding and digesting tough food. The gizzards of grain eaters and seed eaters, such as turkeys, pigeons, and finches, are especially large and have powerful layers of striated muscles. Turkey gizzards can pulverize English walnuts, steel needles, and surgical lancets. Grebes swallow their own feathers, which accumulate between the gizzard and intestine. Acting as a filter, the feathers stop the further passage of sharp fish bones and indigestible remains, which form boluses for regurgitation. In other birds, indigestible items, such as the beaks of squids or pieces of plastic, may remain in the gizzard for months.

The internal grinding surfaces of the gizzard are covered with a strongly keratinized koilin layer, a rough pleated or folded surface with many grooves and ridges. In some pigeons it has strong, tooth-shaped projections. The gizzard can also contain large quantities of grit and stone, which grind food. Quartz particles, the size of which corresponds to the coarseness of the bird's diet, are a common form of grit. The gizzards of moas, extinct ostrichlike birds of New Zealand, have been found to contain as much as 2.3 kilograms of grit. The gizzard is not so muscular in birds that eat softer foods such as meat, insects, or fruit; and in raptors and herons it may take the form of a large thin-walled sac. In some birds, gizzard structure changes seasonally from large and hard to small and soft in relation to dietary changes. For example, in winter, when seeds are the main food of the Bearded Parrotbill, its gizzard is a large, muscular, keratinized structure containing grit. In summer, when insects are the main food, its gizzard is smaller and less muscular (Spitzer 1972).

The length of a bird's intestinal tract averages 8.6 times its body length but varies from 3 times body length in the Common Swift to 20 times body length in the Ostrich. The intestine tends to be short in species that feed on fruit, meat, and insects and long in species that feed on seeds, plants, and fish. The detailed histology and patterns of relief of the absorption surfaces also vary in accord with diet.

Passage time of food through the digestive tract—from the esophagus through the glandular stomach and gizzard into the intestine and finally out the cloaca as feces—varies from less than half an hour in the case of berries ingested by thrushes and the Phainopepla to half a day or more for less easily digested food. Each stage in this passage may include special processing. The sections of the digestive tract are adapted to the specific requirements of a bird's normal diet. Consider, for example, birds that feed principally on mistletoe berries, which include some flowerpeckers in the Old World tropics and euphonias (a kind of tanager) and the Phainopepla in the New World (Docters van Leeuwen 1954; Walsberg 1975). The stomachs of these flowerpeckers and tanagers act as diverticula for digesting insects only; the easily digested berries bypass the stomach after being shucked and go straight to the intestine. A Phainopepla's stomach shucks 8 to 16 mistletoe berries in succession, popping the seed and pulp directly into the intestine and retaining a stack of outer layers (exocarps), which then are passed as a group into the intestine and defecated between sets of undigested seed and pulp (Figure 7–13).

Feeding adaptations among nectar-feeding birds include a tubular tongue for nectar extraction, a distensible esophageal pouch (crop) for nectar storage, and juxtaposition of the entrance to the digestive area (proventriculus) and the opening into the intestine (pylorus). This anatomical arrangement allows nectar to bypass the stomach—quite like the arrangement in euphonia tanagers—while diverting insect food into the stomach for longer digestion. The gizzards of nectar-feeding birds are thin-walled structures, in contrast to the hard, muscular structures that pulverize ingested seeds before digestion begins in finches.

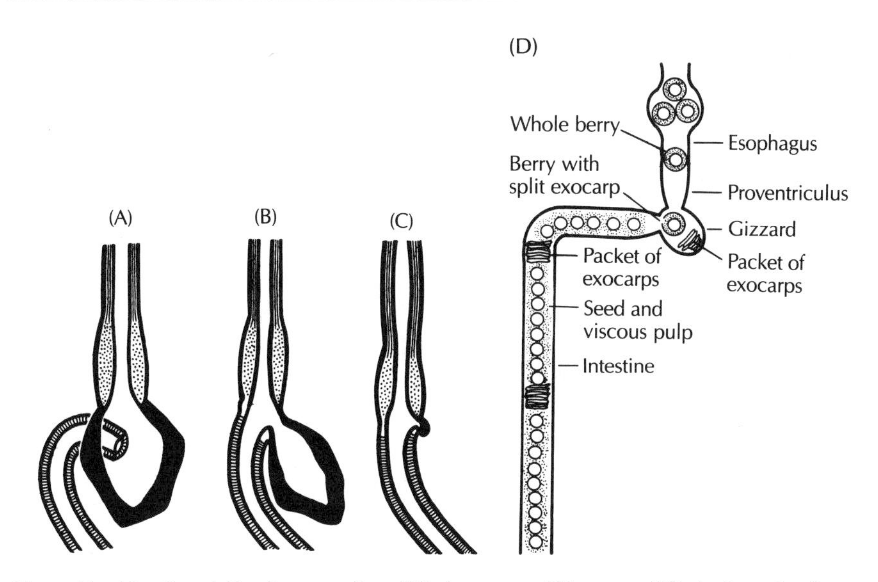

Figure 7–13 Specialized stomachs of fruit eaters: (A) unmodified gizzard of a primitive flowerpecker; (B) more specialized stomach of the Black-sided Flowerpecker, which allows fruit to bypass the gizzard and insects to enter the gizzard for grinding; (C) rudimentary gizzard of the Violaceous Euphonia; (D) gizzard of the Phainopepla, which can shuck the outer layer skin (exocarp) from mistletoe berries and then defecate a pack of skins at intervals between the undigested parts of the berries. (Adapted from Desselberger 1931; Walsberg 1975)

Assimilation of digested food through the intestinal walls depends on the nature of the food ingested (Ricklefs 1974). Hummingbirds assimilate 97 to 99 percent of the energy in nectar, which consists primarily of simple sugars and water. Hummingbirds quickly absorb glucose from their fluid meals by means of unusually high densities of sites that actively bind sugar and transport it across cell membranes (Karasov et al. 1986). Other foods are less easily assimilated. Raptors assimilate 66 to 88 percent of the energy contained in ingested meat and fish. Herbivores assimilate as much as 60 to 70 percent of the energy contained in the young plants they ingest but only 30 to 40 percent of the energy in ingested mature foliage. Spruce Grouse assimilate only 30 percent of the energy contained in the spruce leaves they eat (Pendergast and Boag 1971).

Many passerine songbirds cannot digest sucrose—a complex sugar—because they lack the enzyme sucrase, which breaks sucrose into smaller components amenable to assimilation (Martinez del Rio and Stevens 1989). Ingestion of sucrose can cause sickness due to malabsorption. As a result, European Starlings learn to avoid sucrose in laboratory tests. Curious whether fruits with simple sugars (glucose and fructose) could be protected from predation by increasing their sucrose content, Larry Clark and Russell Mason (1993) found that only impractically high concentrations of sucrose would serve as a repellent.

Small side sacs—called ceca—aid digestion of plant foods. The ceca (sing., *cecum*) are attached to the posterior end of the large intestine of some birds. They are most prominent in fowl and Ostriches, in which they functionally resemble the rumen of cattle. They also tend to be larger in young birds than in adults. The precise role of ceca in digestion remains unclear, but it appears that bacteria in the ceca further digest and ferment partially digested foods into usable biochemical compounds that are absorbed through the cecal walls (Gasaway 1976a, b). Ceca may also function to separate the nutrient-rich fluid in partially digested food from the fibrous portion, which is eventually eliminated (Fenna and Boag 1974). Ceca are poorly developed or nonexistent in most arboreal birds, perhaps because of the unacceptable weight of watery, partially digested food in the intestine and the large structures required to handle them. Indeed, well-developed cecal fermentation is restricted to ground-dwelling and flightless birds and is much more common in mammals than in birds (Figure 7–14).

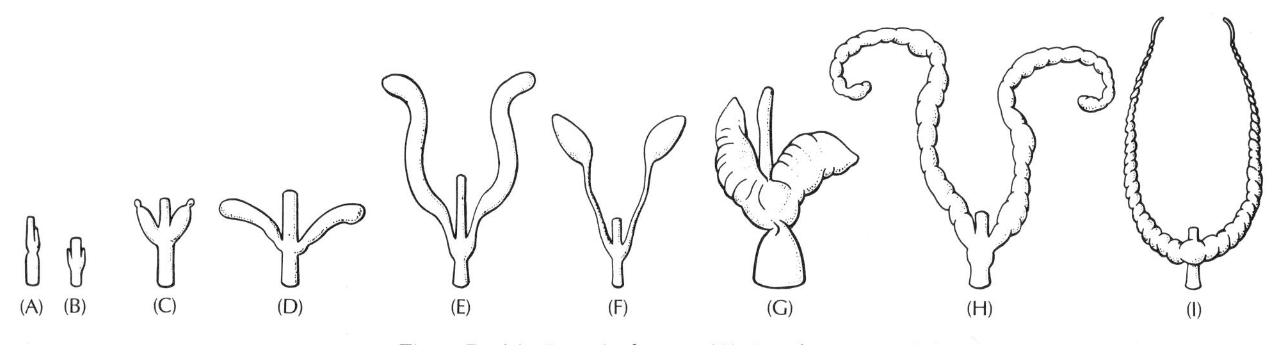

Figure 7–14 Intestinal ceca: (A) Purple Heron; (B) Eurasian Sparrowhawk; (C) Marabou Stork; (D) a rail; (E) Helmeted Guineafowl; (F) Barn Owl; (G) Northern Screamer; (H) Great Bustard; (I) Ostrich. (From McLelland 1979)

Wax as Food

Waxes, which consist of saturated, long-chain fatty acids, are among the least digestible of all foods, but several birds have special capacities for using wax as a source of metabolic energy. Seabirds, including petrels and auklets, metabolize the wax compounds found abundantly in the marine crustacea they eat (Roby et al. 1986; Place 1991). Indeed, wax is a major source of energy for both adults and chicks of these seabirds. Waxy foods, together with bile and pancreatic digestive juices, recycle several times from the small intestine back to the churning actions of the gizzard and proventriculus to break the complex fatty acids into smaller, usable elements. Allan Place (1991) suggests that birds generally and these seabirds

BOX 7–2

Certain birds live off the wax of the land

Yellow-rumped Warblers and Tree Swallows consume large quantities of wax-coated bayberries along the Atlantic coast. Laboratory experiments by Allen Place and Edmund Stiles (1992) revealed that both warbler and swallow are capable of high assimilation efficiencies (80 percent) of bayberry wax. Their special gastrointestinal traits include elevated gallbladder and intestinal bile-salt concentrations, slow gastrointestinal transit of dietary lipids, and probably also the return of the partially digested food to the gizzard from the small intestine. The ability to use an unusual food source like bayberry wax allows them to occupy northern regions during periods when insects are not available.

in particular have evolved digestive systems that optimize the churning cycles so as to extract high yields from food processed in small volumes.

Honeyguides have long been famous for their ability to eat and assimilate pure wax, usually from the honeycombs of bees but occasionally from candles on the altars of Christian missionaries (Friedmann and Kern 1956; Diamond and Place 1988). Still not known is whether wax digestion by honeyguides is facilitated by special intestinal flora, as once believed, or by unknown biochemical means similar to those used by Yellow-rumped Warblers (Box 7–2). The Greater Honeyguide of Africa recruits help to gain access to beehives in hollow trees. The honeyguide leads animals with a sweet tooth, such as the ratel (honey badger), as well as people to beehives it has found. First the honeyguide solicits attention by approaching closely and giving distinctive churring calls. If it gets the helper's attention, the honeyguide flies a short distance in the direction of the beehive, returning frequently to ensure progress. In this manner the honeyguide leads its recruited assistant a kilometer or more to the beehive, which it announces with a new set of excited vocalizations. Honey is prized by many African peoples, who open up the hive and leave the wax for the honeyguide. Some Kenyan tribes believe that it is best not to reward the honeyguide, which will be more inclined to lead them to another nest if it is hungry (Isack and Reyer 1989).

Feeding Behavior and Energy Balance

The feeding behavior of birds is influenced not only by their anatomical equipment but also by the availability of food. Availability reflects ease of capture as well as abundance. Some conspicuous foods, such as red berries, are easily located and eaten, but many live prey are scarce, cryptic, skilled at avoiding capture, or distasteful. Food gathering, therefore, requires sophisticated hunting skills and even creative practices.

Innovation and Use of Tools

Some birds use tools in feeding (Boswall 1977, 1983), and some, when faced with dramatic changes in resources, have developed innovative behaviors. The Woodpecker Finch of the Galápagos pries grubs from crevices with a stick or a cactus spine held in its bill (Millikan and Bowman 1967) (Figure 7–15). The nuthatchlike Orange-winged Sittella of Australia pries out grubs with sticks (Green 1972). Egyptian Vultures crack Ostrich eggs with stones (Lawick-Goodall 1968). Green Herons sometimes use pieces of bread as fishing bait (Lovell 1958). For example, a Green Heron was seen dropping bait (bread and pieces of tissue paper) into pools of a stream and waiting for fish to gather. When currents began to carry off the bait, the heron retrieved it and used it again.

Forty years ago, a few Great Tits in the British Isles learned to rip open milk bottle caps to drink the cream (Fisher and Hinde 1949). Apparently this was a novel application of normal bark-tearing behavior (Morse 1980). The skill passed rapidly to other titmice, forcing milk companies to replace the cardboard caps with sturdier aluminum ones. The tits learned to open these too. Darwin's finches of the Galápagos Islands are renowned for their novel feeding efforts (Bowman and Billeb 1965). On one island, the Sharp-beaked Ground-Finch, for example, drinks the nutritious blood of nestling seabirds by pecking open their growing pin feathers. (Galápagos Mockingbirds drink blood from iguanas as well as seabirds.) To uncover food, the Small Ground-Finch and Large Cactus-Finch have learned to push aside sizable stones with their feet by first bracing their heads against a large rock for leverage; Paul DeBenedictis (1966) observed a 27-gram finch move a 378-gram stone in this manner.

Figure 7–15 The Woodpecker Finch of the Galápagos uses twigs to pry insect larvae out of small holes. (Courtesy of I. Eibl-Eibesfeldt)

Search Behaviors

In general, birds prefer familiar foods. Common Wood-Pigeons, for example, search for familiar types of grain, and titmice choose familiar caterpillars (Murton 1971a; Boer 1971). The preference for familiar foods lessens the number of unpleasant surprises—poisonous, or otherwise dangerous prey. Familiar food also can be found more readily than unfamiliar food if the foraging bird uses a specific "search image," as we do when we look for a friend in a crowd or for a jigsaw puzzle piece with a particular shape. For example, when captive Blue Jays are shown color slides of tree trunks with and without cryptic moths, they learn to search quickly for a particular kind of moth and to peck a key 10 times when they spot one (Pietrewicz and Kamil 1979). A Blue Jay's skill at spotting moths increases rapidly with experience.

Food supplies vary with site and time, and birds respond to variations in several ways. Area-restricted search is one such response. When food sites are concentrated, a bird improves its success by staying in or near sites of high food density and by moving rapidly past sites of low food density. By turning sharply and moving short distances after food is found, the bird concentrates its search near concentrations of food. Eurasian Blackbirds, for example, adopt area-restricted searching when the low-density distribution of artificial prey (pastry caterpillars) is changed from uniform to random or clumped (Smith 1974). Caged Ovenbirds capture housefly pupae by moving shorter distances and turning more sharply, especially when the pupae are clustered (Zach and Falls 1977). Experienced Ovenbirds apply area-restricted searching to known productive sites even before they find prey (Figure 7–16).

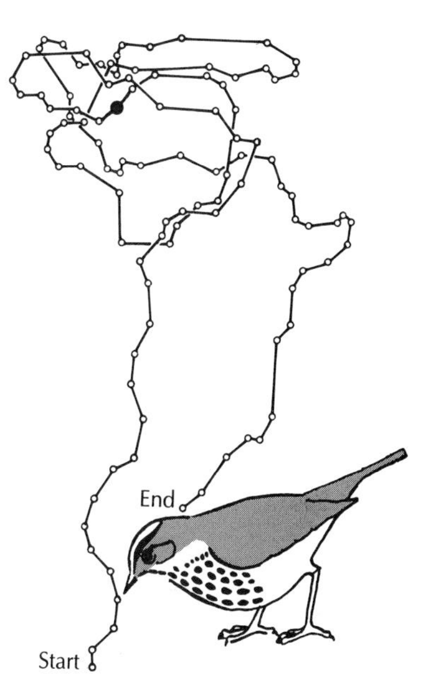

Figure 7–16 When prey distribution is locally dense rather than uniform or random, captive Ovenbirds search primarily near a site where they have been successful (solid circle). (From Zach and Falls 1977)

Feeding success affects the exact location in the forest where a bird hunts for food or which flower it probes for nectar. Combined with morphological predisposition and experience, the choices of profitable feeding sites can lead to routine use of feeding stations—or ecological niches—which differ according to species. In a New Hampshire forest, for example, there are ground feeders such as thrushes, Ovenbirds, and Dark-eyed Juncos; tree-trunk feeders such as woodpeckers and nuthatches; general canopy feeders, such as Scarlet Tanagers, some vireos, and one species of flycatcher, which search widely in both deciduous and coniferous trees; and specialized canopy feeders, such as Blackburnian Warblers and Black-capped Chickadees, which tend to restrict their searches for food to the outer twigs of conifers (Holmes et al. 1979).

Similar species also segregate themselves by feeding station. Small species of titmice tend to feed at small, high branches and on the outermost twiglets in British woods; large species concentrate their efforts on the ground and on large branches that offer more support (see Chapter 23). In his classic study of the foraging niches of wood warblers that coexist in northern spruce forests, Robert MacArthur showed that Bay-breasted Warblers and Blackburnian Warblers concentrated their time in the top half of the tree, Bay-breasted near the trunks and Blackburnians in the outer branches, while Cape May Warblers and Black-throated Green

Warblers concentrated below in the middle of spruce trees, Cape Mays taking insects from the trunk bark and Black-throated Greens from the foliage. A fifth species, the Yellow-rumped Warbler, stayed below heights of 3 meters. The vertical segregation of niches reduced overlap of foraging effort and therefore competition among these similar species for food (MacArthur and MacArthur 1961).

Increasing Energy Profit

Birds are sensitive to the net energy profits of their feeding efforts. We can define the energetic profit of foraging as the rate of net energy gain, or

$$\text{Profit} = \frac{\text{energy gain} - \text{energy cost}}{\text{foraging time}}$$

where energy gain is the assimilated caloric value of a bird's food; energy cost is the caloric value of energy expended in finding, capturing, and eating the food; and foraging time includes the time required to locate food and to consume it. The ways birds adjust their profit margins—called optimal foraging theory—was a major research topic in the 1970s and 1980s.

Birds tend to choose food of high energetic profit. For example, White Wagtails prefer medium-sized flies, even though large flies with greater energy content are more common (Davies 1977). Medium-sized flies yield comparatively more energy per second of foraging time because large flies take too long to subdue and swallow relative to their higher energy content (Figure 7–17). Northwestern Crows on the coast of British Columbia, Canada, selectively choose large whelks, a kind of marine snails (Zach 1979). On the basis of size only, the crows select whelks that supply about 8.5 kilojoules of energy, a yield that exceeds the high foraging costs of 2.3 kilojoules per whelk. To eat whelks, the crow must work hard to drop them repeatedly from the air onto rocks until they break; 20 drops per whelk is not unusual.

When a hungry bird encounters an assortment of prey that differ slightly in energy yield, it must decide whether to pause long enough to eat low-yield prey or to continue seeking better prey. Captive Great Tits, allowed to select large and small mealworm pieces from a continuous supply on a little conveyor belt, ate large and small pieces when both were scarce but selected only large pieces when they were so common that the investment of time in the small pieces would lower their foraging efficiency (Krebs et al. 1977).

If a bird flies to distant feeding areas, commuting costs affect its foraging performance. The costs of flying far from a nest or a favorite perch to flowers in a field or to a nectar feeder are important energy considerations affecting the foraging habits of hummingbirds and sunbirds. Rather than fill their crop with nectar, hummingbirds stop feeding when the

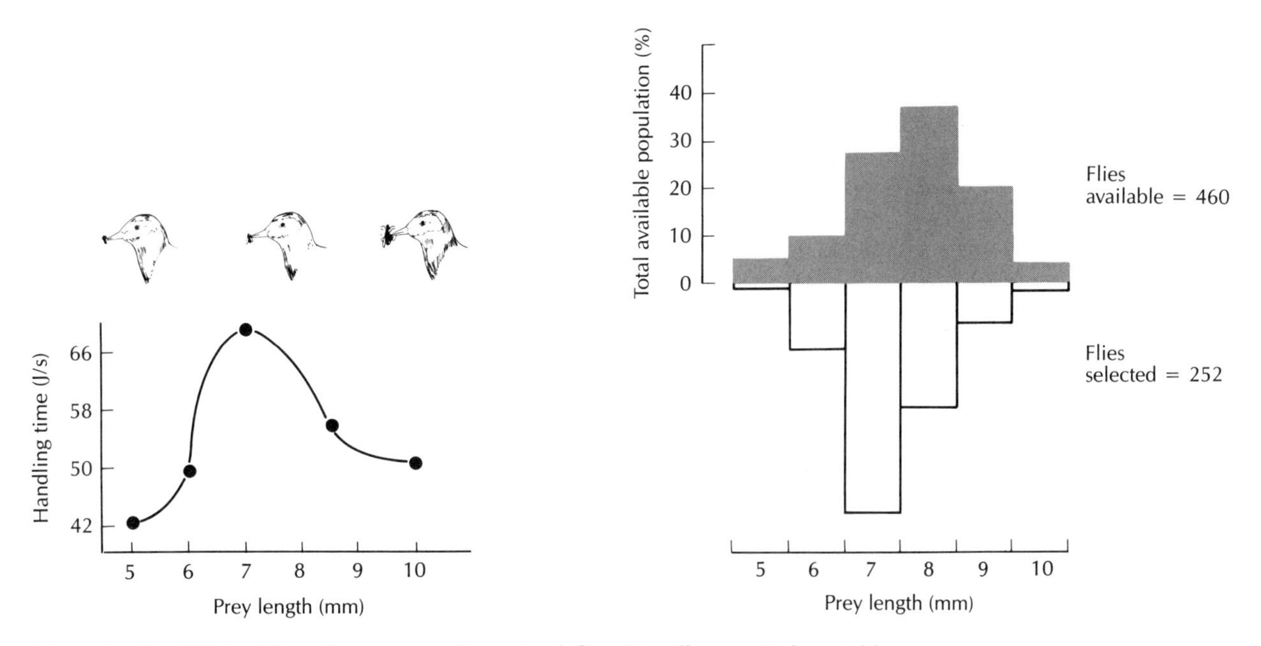

Figure 7–17 White Wagtails select medium-sized flies (7 millimeters) that yield the most energy per second of handling time. (Adapted from Krebs 1978)

increasing flight costs imposed by the added weight of more nectar would reduce their net gain (DeBenedictis et al. 1978). Most birds face similar costs when they commute between nests and food sources to feed nestlings. Northern Wheatears, for example, compensate for increased travel to distant food cups containing maggots by consuming greater quantities at greater distances (Carlson and Moreno 1982) (Figure 7–18).

Birds do not always follow theoretical foraging patterns or fit them simply. Recall from the discussion of these studies at the beginning of the chapter that seed husking times correlate with bill dimensions and seed size or hardness. The finches in those studies preferred seeds they could husk quickly, regardless of energy yield, rather than seeds that yield the highest rate of energy intake (Willson 1971). Prolonged exposure to predators is of paramount concern to such birds, which eat quickly lest they be eaten. Birds also occasionally visit low-yield sites or eat inferior food, despite the availability of better options. Some degree of error may be inevitable in food selection. Possibly, they select inferior food as a way of monitoring future foraging possibilities in case there is a change in food availability. Great Tits, for example, continue to explore low-yield patches while concentrating their search for mealworms in the site with the greatest prey density (Smith and Sweatman 1974). When the density of food of the highest yielding patch drops, the tits switch to the best available alternative patch.

Only rarely does a bird exhaust localized food resources. Foraging therefore becomes a process of regular harvesting of renewable resources by returning regularly to productive foraging sites, such as backyard bird

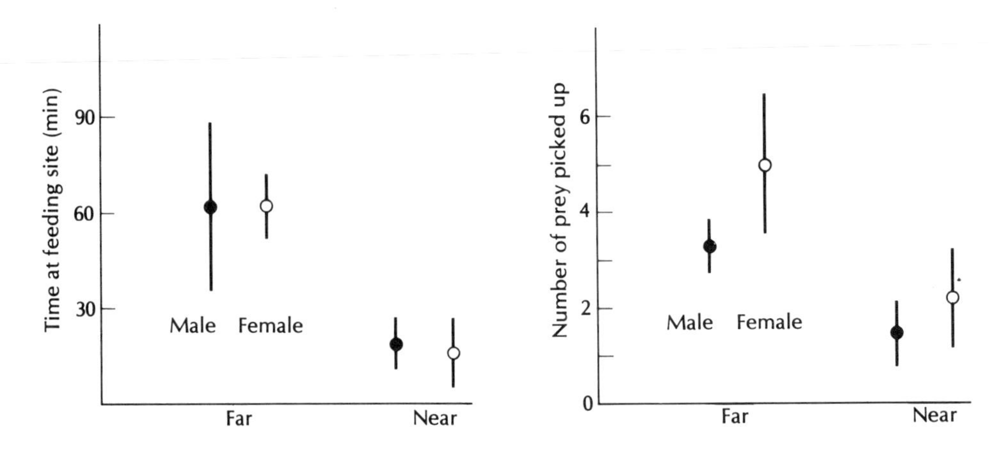

Figure 7–18 (A) Northern Wheatears spend more time at feeding sites far from the nest than at feeding sites near the nest. (B) They also pick up more prey (maggots) far from the nest, which compensates for the increased travel effort. The average time at feeding sites and average number of prey picked up are indicated by the circles, the variation (standard deviations) by the vertical bars. Differences between males and females were minor when compared with differences between far and near sites. (Drawn from data in Carlson and Moreno 1982)

feeders. Pileated Woodpeckers cycle regularly among partially excavated trees on their large territories. Hawks move repeatedly between favored perches. How often a bird should return to productive site depends on the rate of food renewal, but this is a little studied aspect of avian foraging behavior. Hummingbirds repeatedly return to flowers to harvest nectar produced since their last visit. Some hummingbirds are specialized "trap-liners" that cycle among widely scattered flowers, just as fur trappers check their circuit of widely scattered traps—called a trapline. In one field study, color-marked Long-tailed Hermits, traplining hummingbirds of the rain forest understory, learned to time their returns to coincide with the interval of experimental refill (Gill 1988). Marked birds also learned to wait longer if they could get more nectar by doing so. Counteracting the advantages of waiting, however, was the prospect of losing the nectar to a competitor. When that happened, the hermit hummingbirds returned frequently to harvest small amounts of nectar, keeping flowers almost empty and unattractive to competing individuals.

Learning to Feed

Foraging profits and efficiencies reflect experience and ability. Young birds typically improve their foraging skills through learning and practice (Wunderle 1991). First, the chick must develop adequate motor abilities for pecking, refined flight, and, in some species, the ability to stalk prey. Young terns, pelicans, and herons, for example, miss the fish they dive at more often than adults. The young bird must also learn what is food and what is not, and which prey are potentially harmful. For example, young European Reed-Warblers and Spotted Flycatchers progressively improve their foraging efficiencies with age; they peck less often at inedible objects, increase the complexity of capture techniques, and fly longer distances to catch prey (Davies and Green 1976; Davies 1976a).

The detailed patterns of adult foraging behavior may be set by early experiences. Hand-raised nestlings of the Chestnut-sided Warbler are reluctant to take food in novel situations, but not in the situations they were exposed to while very young (Greenberg 1983, 1984). Adult Chestnut-sided Warblers are also reluctant to feed in novel situations. They are neophobic, that is, they exhibit little opportunism in the way they look for food. Their conservative ways are manifest on the breeding grounds, on the tropical wintering grounds, and in the laboratory.

The period of parental care after young have left the nest can be an important training period that relates to the success of a young bird's foraging techniques, among other survival skills. Young raptors need parental training and much practice to become skilled in prey capture and killing. The training period relates directly to diet in Eurasian Oystercatchers, which by family tradition are either polychaete worm specialists or crab and bivalve mollusk specialists. Feeding on marine worms is relatively easy; they simply have to be caught peeking out of their burrows. Successful spearing of slightly ajar bivalve mollusks, however, is difficult. Young oystercatchers that are learning to feed on worms require only 6 to 7 weeks to become independent of parental feeding. Learning to feed on mollusks takes months, and adult oystercatchers that specialize on them help feed their young for up to a year (Norton-Griffiths 1969). Similarly, the long parental care periods of tropical insect-eating birds provide time to train young to find cryptic insect prey. To develop the skills of feeding on the wing, swifts and swallows practice taking meals from parents in flight. Juvenile terns practice picking up pieces of seaweed or trash from the ocean surface and may repeatedly release and retrieve their trial objects, or "toys," as if they are playing (Ashmole and Tovar 1968).

Nutrition

Little is known about the degree to which the diets and foraging behavior of wild birds are directed toward nutrition. In one exception, Willow Ptarmigan prefer particular heather leaves that are rich in nitrogen and phosphorus (Moss 1972; Moss et al. 1972). It is usually assumed that birds passively obtain adequate nutrition in the course of their daily foraging to meet their energy needs and rarely suffer malnutrition or nutritional stress (King and Murphy 1985). Recent laboratory experiments, however, suggest that White-crowned Sparrows are sensitive to concentrations of certain amino acids, namely, valine and lysine, in synthetic diets; they were adept at selecting diets that satisfied their amino acid requirements (Murphy and King 1989). Nutrition should become a new focus of future studies of foraging behavior.

Energy Balance and Reserves

Whether hungry or temporarily sated, all birds face the challenge of maintaining their energy balance. Energy balance is the dynamic relationship between energy intake and energy expenditure. Ideally, intake and expenditure are roughly equal so that the bird neither gains nor loses much weight. Preceding migration or winter, a bird may eat more than it metabolizes each day so that the excess can be stored as fat, which provides reserves needed to compensate for periods of inadequate intake.

Foraging Time

The amount of time a bird must feed each day depends on its total energy requirements and achieved rate of energy intake. As requirements increase, so must foraging times or, alternatively, the rates of net energy gain. Roughly speaking, a bird's foraging time must double when its rate of net energy gain is reduced by half. If a short foraging time is sufficient for self-maintenance, individuals can afford to build up energy reserves or undertake energy-expensive activities such as migration, molting, and breeding (see Chapter 12). Low foraging time also allows birds more time to remain hidden from predators, select favorable microclimates, establish dominance and property rights over other individuals, court mates, and rear young.

Birds vary their foraging time and effort in relation to their energy requirements and foraging success. Sunbird foraging efforts, yielding mainly short term energy profits, for example, decline with an increase in floral nectar content (Figure 7–19). Foraging times also vary with seasonal changes in food availability. Anna's Hummingbirds forage 14 percent of the day in the nonbreeding season but only 8 percent of the day while breeding in the vicinity of nectar-rich flowers. Small titmice and goldcrests in England feed 90 percent of the day in winter when food is scarce, their metabolism is high, and days are short (Gibb 1960). They must find an insect about every 2 seconds just to survive. Many fail to meet this requirement. Goldcrests in arctic Finland and Norway sometimes suffer high mortality rates in harsh winters because they are unable to balance their energy budgets (Österlöf 1966; Hogstad 1967).

Fat Reserves

Most birds maintain minimal fat (lipid) reserves, probably because the survival benefits of large energy reserves do not offset the costs of carrying excess weight (Blem 1990). Small Temperate Zone passerines typically

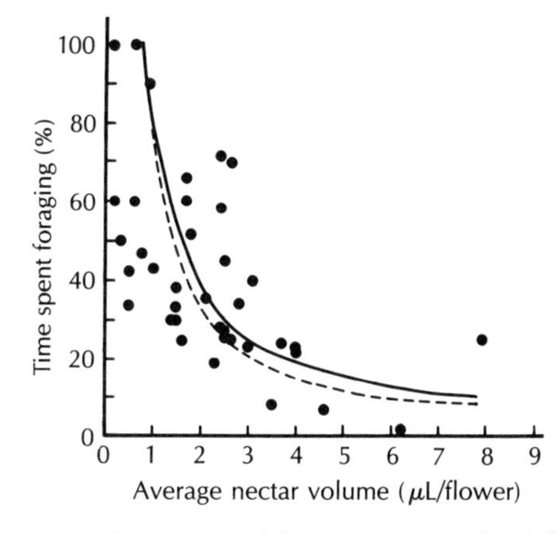

Figure 7–19 The amount of time a Golden-winged Sunbird feeds depends on the average amount of nectar it gets from a flower. The solid line is the predicted relationship, assuming that the sunbird visits only as many flowers as it needs to replace total daily expenditures. The dashed line is fitted to the actual field measurements of foraging efforts. (Adapted from Gill and Wolf 1979)

have fat reserves of no more than 10 percent of their body mass to cover their fasting needs during midwinter. Yellow-vented Bulbuls in tropical Singapore maintain fat reserves throughout the year of only 5 percent of body weight, little more than is needed to survive overnight and to begin feeding the next morning (Ward 1969). In general, large birds can store more fat and can fast longer than smaller birds, a pattern also seen in mammals. At moderately low temperatures (1° to 9°C), a 10-gram warbler, for example, may not survive a day without food, whereas a 200-gram American Kestrel can survive for 5 days (Kendeigh 1949; Calder 1974). Male Emperor Penguins fast for 90 to 120 frigid days during their incubation vigils of the Antarctic winter and may lose 45 percent of their mass during this period (del Hoyo et al. 1992). Nestling Common Swifts can survive 10 days of fasting by using torpor but lose about half of their mass (Koskimies 1950). Adult swifts can survive for 4.5 days of fasting with a 38 percent weight loss.

Fasting birds draw first on their glycogen deposits, then lipids. Only the lipid reserves of cardiac muscle are exempt from normal use, and, as a last resort, body tissues can be metabolized after fat deposits are exhausted. Pectoral muscles begin to atrophy during periods of food stress; even gonadal tissues may be sacrificed. Metabolism of these tissues causes free amino acid levels in the blood to increase rapidly, 500 percent in the case of lysine during a 36-hour fast (Fisher 1972).

Food Caches

Hoarding food for future use is one way of preparing for food shortages (Vander Wall 1990; Källander and Smith 1990). Groups of Acorn Woodpeckers, for example, build large granaries of acorns for the winter.

Figure 7–20 A Common Yellowthroat impaled on barbwire by a Loggerhead Shrike. (Courtesy of S. Grimes)

Meat eaters such as hawks, owls, and shrikes routinely set aside a fraction of their prey for future use. Shrikes are notorious for impaling their prey on thorns for later consumption (Figure 7–20).

Titmice depend on autumn caches for winter feeding; Crested Tits of Europe obtain up to 60 percent of their winter food from provisions built up earlier in the year. These seed caches are more difficult to relocate than other stored foods, such as the acorn granaries of woodpeckers and the impaled prey of shrikes. Recovery of widely dispersed, concealed seed caches requires extraordinary spatial memory, which is processed by an enlarged hippocampal complex of the forebrain (see Chapter 8). With two known exceptions, the Great Tit and the Blue Tit, all Temperate Zone species of titmice regularly cache seeds (Chapter 8). The seed-caching titmice form small discrete flocks (fewer than 10 individuals) with a membership hierarchy. Flock members defend a winter territory, a behavior that reduces loss of cached seeds to "scrounging" by other tits.

Summary

Because they have high energy demands, birds have a constant need for food. Specialized adaptations of their locomotory morphology, bill structures, digestive systems, and foraging behavior highlight the evolutionary responses of birds to this urgent need.

The avian bill consists of bony extensions of the jawbones, which are covered by a horny sheath, the rhamphotheca. Modern birds lack teeth. The internal structure of the bill permits substantial bending and directs stress away from weak points. Subtle differences in bill structures affect the abilities of finches to husk seeds and hummingbirds to extract nectar from flowers.

The avian digestive system is specialized to process unmasticated food. Salivary glands lubricate food before it is swallowed. The gizzard pulverizes ingested food for digestion. The gizzard is particularly well developed in birds that eat hard or tough foods. Ceca, sacs attached to the posterior end of the intestine, are found primarily in terrestrial birds and also aid the digestion of foods.

Whereas the study of the functional feeding morphology of birds has a long history, the study of their foraging behavior and, particularly, their sensitivity to the costs and gains of alternative prey or feeding sites is a recent but popular topic. In subtle ways, birds bias foraging time to favor the most profitable feeding sites, select prey of the size that enhances their energetic profit, and innovate unusual foraging techniques, sometimes using tools. When they are very hungry, birds are more likely to risk starvation by spending time searching for an area with a high concentration of food. The amount of daily foraging time reflects a bird's energy needs. Birds usually maintain small to moderate fat reserves and so must balance their expenditures with new intakes on a daily basis. When food is not available, birds draw first on fat and glycogen reserves and then on other body tissues. Some birds, particularly nutcrackers and their relatives and titmice, hoard food in caches for use during future periods of food stress.

Further Readings

Källander, H., and H.G. Smith. 1990. Food storage in birds: An evolutionary perspective. Current Ornithology 7: 147–208. *A timely review of the literature on this increasingly popular topic.*

Krebs, J.R., and N.B. Davies. 1984. Behavioral Ecology, 2nd ed. Sunderland, Mass.: Sinauer Associates. *Includes important reviews of research on the foraging behavior of birds.*

McLelland, J. 1979. Digestive system. *In* Form and Function in Birds, A.S. King and J. McLelland, Eds., pp. 69–181. New York: Academic Press. *A summary of the classical literature on the anatomy and function of the digestive tract.*

Morse, D.H. 1980. Behavioral Mechanisms in Ecology. Cambridge, Mass.: Harvard University Press. *A fine review of the early literature on foraging behavior of birds.*

Wunderle, J.M., Jr. 1991. Age-specific foraging proficiency in birds. Current Ornithology 8: 273–324. *An up-to-date review of the rapidly growing literature.*

PART III

Behavior and Communication

CHAPTER 8

Brains and Senses

ORNITHOLOGISTS ONCE ASSUMED that birds perceive the world the same way as people do. The sensory experience of birds, however, extends far beyond the sensory experience of humans. The highly developed color vision of birds reaches into the near-ultraviolet range of the spectrum. The broad hearing range of birds encompasses sounds of very low frequencies—called infrasounds—and some owls can track prey in complete darkness by their hearing alone. Birds can navigate by means of patterns of the Earth's magnetism and can orient themselves in flight "automatically" because of their extreme sensitivity to minuscule shifts in gravity and barometric pressure.

Recent research reveals that birds have well-developed brains and are more intelligent than is implied by the familiar and inappropriate slur "bird brain." Contrary to public "wisdom," birds in general and songbirds in particular have large brains relative to body mass. Indeed, birds have large, well-developed brains that are 6 to 11 times larger than those of like-sized reptiles. The brains of most birds and most mammals account for 2 to 9 percent of their total body mass. Bird brains and primate brains both exhibit functional lateralization, with left hemispheric dominance associated with learning and innovation in vocal repertoires. Substantial learning by birds guides the mastery of complex motor tasks, social behavior, and vocalizations.

This chapter first reviews avian intelligence and then the major features of the avian brain. Included are two examples of the functional relationships between brain evolution and behavior, the motor control of song production by particular nuclei in the forebrain, and the role of the hippocampus in processing spatial memory used by nutcrackers and chickadees to relocate cached seeds. Then follow sections on the sensory abilities of birds, with particular emphasis on their two primary senses, vision and hearing.

Avian Intelligence

Birds master complex problems in the laboratory, outperforming many mammals in advanced learning experiments (Kamil 1985, 1988). Crows and magpies do especially well in laboratory experiments that test higher faculties. In one of these—the Krushinsky problem—the bird looks through a slit in a wall at two food dishes, one empty and one full, that move out of sight in opposite directions (Figure 8–1). The bird must then decide which way to go around the intervening wall to get to the dish that contains food. Cats, rabbits, chickens, and pigeons do poorly in this test, but crows and dogs solve the problem immediately. Blue Jays, which are related to crows, quickly learn general problem-solving strategies, that is, stay with a stimulus that produces food (win–stay) and avoid one that does not (lose–shift). Neither cats nor squirrel monkeys are able to do this in laboratory experiments (Kamil et al. 1977).

For most mammals, learning to count is a formidable problem. Monkeys require a training ordeal of 21,000 trials to learn to distinguish between sound series with two or three different tones; rats never learn to make this distinction. Birds, however, easily master complex counting problems. Ravens and parakeets, for example, can learn to count to seven and can learn to identify the box containing food by counting the number of small objects in front of it (Koehler 1950).

One of the most advanced forms of learning, insight learning—or learning by observation and imitation of others—may be routine among birds. Blue Jays, for example, learn the difference between edible and inedible butterflies by watching the feeding behavior of jays in another cage (Brower et al. 1970). The spread of the milk bottle feeding habit among English titmice (see Chapter 7) is attributed to learning by imitation. Novel behavior involving tool use, such as throwing stones at Ostrich

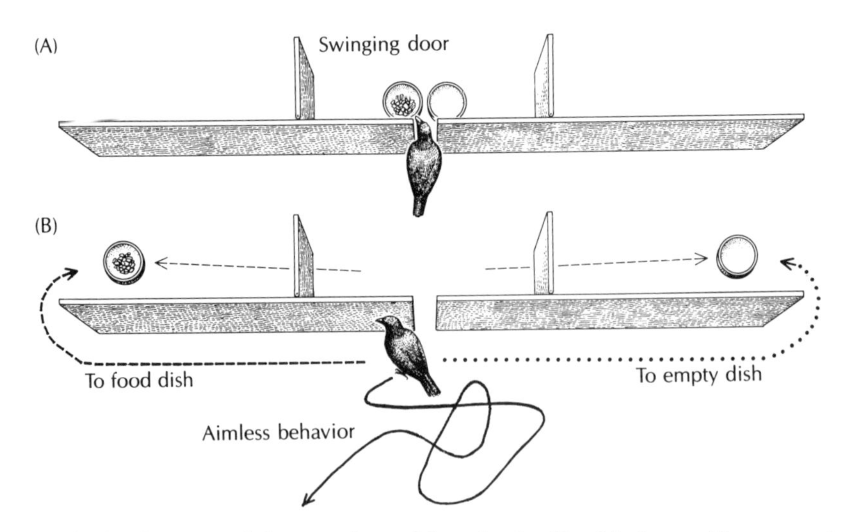

Figure 8–1 Crows and dogs performed best in the Krushinsky problem experiment, in which food dishes, (A) viewed by the subject through a slit in a wall, move out of sight behind swinging doors. (B) The subject must then choose to proceed left or right to find the food dish. (From Stettner and Matyniak 1968, with permission of Scientific American)

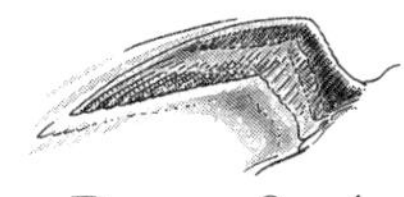

BOX 8–1

Conversations with a parrot explore avian intelligence

A Gray Parrot named Alex has provided deep insights into the intellectual abilities of some birds (see figure). Irene Pepperberg (1981, 1987, 1988) first taught Alex a vocabulary of English vocalizations to identify, request, refuse, or comment on over 80 objects of different colors, shapes, and materials. Alex's comprehension of categories and labels was illustrated by his performance in a series of trials. When presented with an array of objects—purple truck, yellow key, green wood, orange paper, gray peg wood, and red box—and then asked, "What object is green?" Alex replied, "Wood." He responded with an accuracy of 81 percent over 48 such trials. The ability for two-way communication between Irene and Alex increased, enabling increasingly complex tests of Alex's abilities. He could provide additional information about an object that was uniquely defined by the conjunction of two other categories (Pepperberg 1992). For example, to the question "What color is the three-corner (shape) key (object)?" Alex would answer, "Yellow."

Alex, an African Gray Parrot that has changed our understanding of avian intelligence, and his companion, Irene Pepperberg, discuss the different objects between them. (Courtesy of I. Pepperberg)

In general, birds quickly learn to recognize the odd object, not only in a set of familiar objects but also in sets of unfamiliar objects; monkeys master this task with difficulty. Alex went a step further; he learned to report on the absence or presence of similarity and difference between two objects (Pepperberg 1988). When asked either "What's same?" or "What's different?" he responded, "None," if the two objects were, respectively, totally dissimilar or identical. The required concepts of nonexistence or absence reflect advanced cognitive (and linguistic) abilities to deal with discrepancies between the expected and actual state of affairs.

eggs by Egyptian Vultures and stick-probing by the Galápagos Woodpecker Finch, also probably spreads by imitation of innovative individuals. Cultural transmission of novel behavioral traits can thus be important elements in the evolution of behavior in birds.

The Avian Brain

The brain analyzes incoming signals, integrates them with past experience, channels them through genetically programmed neural switches, and activates a series of motor instructions throughout the body. Classic studies of brain function stressed the results of direct electrical stimulation of the brain, which, for example, can cause doves to adopt particular display behaviors (Åkerman 1966a, b). As described below, recent studies of how the avian forebrain controls song provide a deeper understanding of the functional relationship between brain cells and behavior.

The main divisions of the avian brain are typical of all vertebrates. The forebrain is responsible for complex behavioral instincts and instructions, sensory integration, and learned intelligence. It includes the olfactory bulbs and cerebral hemispheres. The midbrain regulates vision, muscular coordination and balance, physiological controls, and secretion of neurohormones that control seasonal reproduction. It includes the optic lobes and chiasma and the cerebellum. The hindbrain—or medulla—links the spinal cord and peripheral nervous system to the major control centers of the brain. Cranial nerves, except those controlling vision and smell, enter the brain through the medulla. The forebrain and midbrain in both birds and mammals are conspicuously more highly developed than those of reptiles (Figure 8–2). Otherwise, avian and mammalian brains are quite different.

A cross section of the avian forebrain reveals three major layers: the cerebral cortex on the dorsal surface, the hyperstriatum below it, and the neostriatum in the deep interior (Figure 8–3). Together, the hyperstriatum and the neostriatum make up the corpus striatum, which accounts

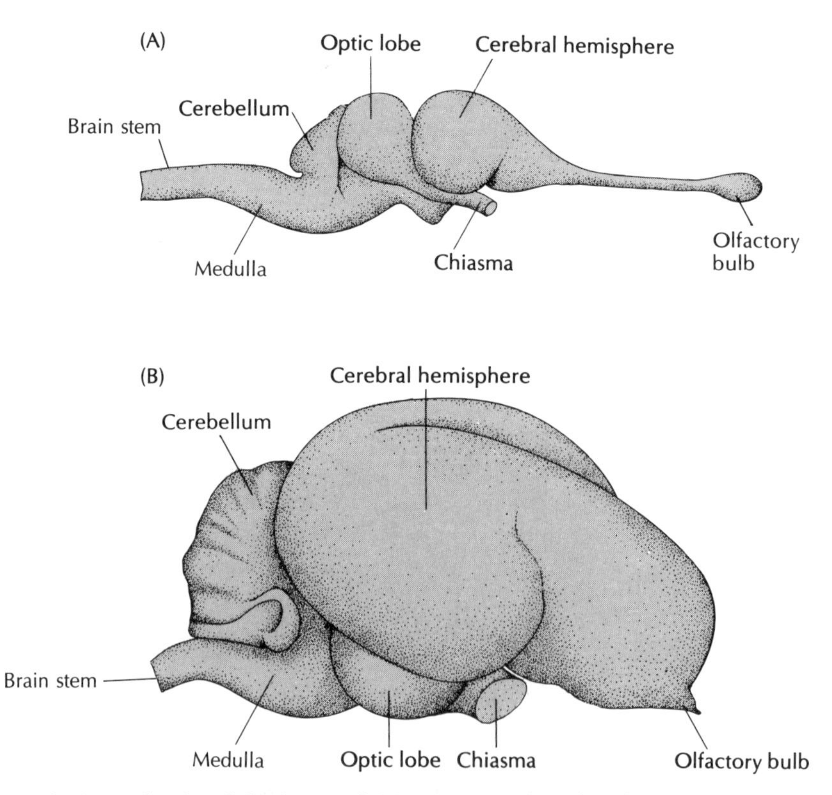

Figure 8–2 Left side of the brain of (A) a monitor lizard and (B) a macaw, drawn to the same scale. Note the well-developed cerebral hemisphere and cerebellum in the avian brain. (After Portmann and Stingelin 1961)

for the bulk of the forebrain. Small and deeper sections include the archistriatum and the paleostriatum. Parts of the corpus striatum control eating, eye movements, locomotion, and the complex behavioral instincts central to reproduction: copulation, nest building, incubation, and care of young. In general, the left cerebral hemisphere controls complex

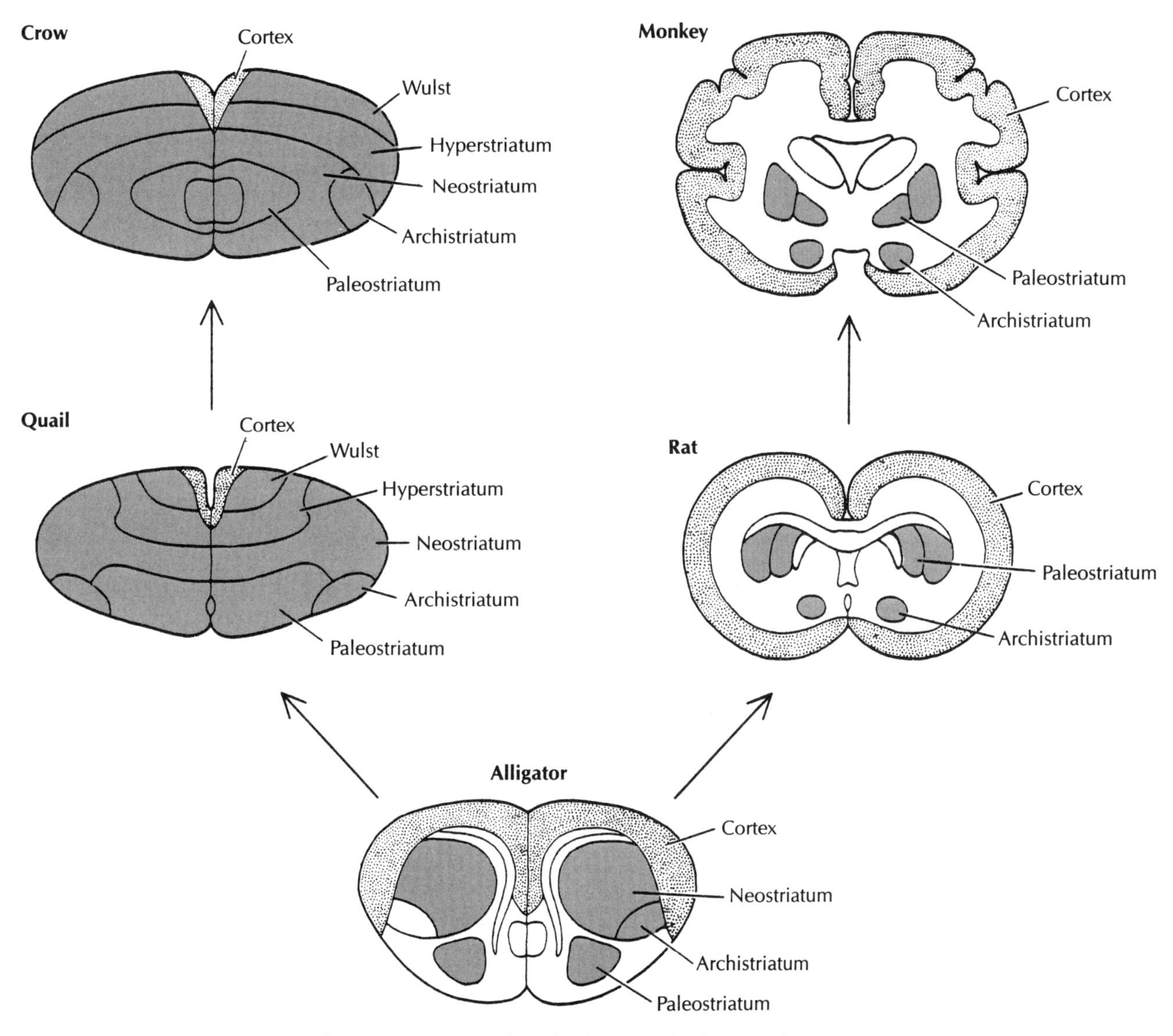

Figure 8–3 Cross sections of avian and mammalian forebrains. The brains of birds and mammals evolved in different directions after diverging from those of their reptilian ancestors (represented here by an alligator). The mammalian brain has been restructured by the enlargement of the cortex, which serves as the seat of intelligence. In birds, an alternative and unique feature—the hyperstriatum and associated Wulst—has become the seat of intelligence. (From Stettner and Matyniak 1968, with permission of Scientific American)

integration and learning processes and suppresses sexual and attack behavior. The right cerebral hemisphere monitors the environment and selects novel stimuli for further processing, which may entail memorization by the left side.

Intelligence in birds has evolved differently from that in mammals, in which the cerebral cortex is the principal feature of the forebrain (Figure 8–3). It overgrows the small corpus striatum and reaches its largest and most deeply fissured state in higher primates such as chimpanzees and humans. The thin cortex of the avian brain, on the other hand, has little to do with intelligence. Removal of the cortex has scarcely any effect on a bird's performance. A pigeon whose cortex has been removed still performs normally in visual discrimination tests and can mate and rear young successfully.

The hyperstriatum, unique in birds, is the center of avian learning and intelligence. It is best developed in intelligent birds such as crows, parrots, and passerines. Domestic Chickens, Japanese Quail, and Rock Doves, which do not perform as well in laboratory intelligence tests, have smaller hyperstriata. Damage to the hyperstriatum severely impairs a bird's behavior. The anterior hyperstriatum, particularly a bump called the Wulst, may be the seat of higher learning processes in birds. Removal of the Wulst does not impair a bird's normal motor functions nor its ability to make simple choices; however it does destroy the bird's ability to learn complex tasks. For example, quail whose Wulst has been removed cannot master multiple reversal learning tests, where they must learn to switch from one rewarding symbol to another that was previously nonrewarding.

Peripheral organs receive signals from all the senses—vision, hearing, touch, taste, and smell—and feed them to the brain for processing, integration, and response. Before reaching the main integration centers of the forebrain, sensory signals pass through their respective control centers. The sensory control centers in a bird's brain are compartmentalized as they are in the brains of fish and amphibians. Visual information goes to the optic lobes of the midbrain; information on body orientation and localized pressure goes to the cerebellum in the midbrain; acoustical information goes to its related processing centers in the hindbrain; olfactory information goes to the olfactory bulbs and then to the olfactory lobe in the forebrain.

The optic lobes and the cerebellum dominate the avian midbrain. The two optic lobes of birds are huge in relation to the rest of the brain. Together with large eyes, this visual apparatus displaces the rest of the brain from the ventral and lateral portions of the skull, the usual positions in other vertebrates. Balance and coordination during flight require extensive input from sensory receptors throughout the body and in the middle ear; the large size of the cerebellum reflects the importance of this input.

Studies of the avian brain have helped to understand how the central nervous system controls complex behavior. Two examples—the control of song and the caching of seeds—are presented here.

Control of Song by the Central Nervous System

Bird song is controlled through a distinct pathway from the brain to the sound production apparatus—the syrinx—located at the base of the tracheal windpipe (Figure 8–4). Nerve impulses start in the forebrain (in the High Vocal center of the corpus striatum), proceed to the robust archistriatal and intercollicular nuclei of the midbrain, then to the tracheosyringeal motor (hypoglossal) nucleus in the brain stem, and finally through the tracheosyringeal motor nerves to the syringeal muscles that control the syrinx. A variety of calls, some recognizable and some abnormal, can be evoked by electrical stimulation of the midbrain nuclei. Destruction of the nuclei in the High Vocal center renders a songbird mute.

Bird song is normally controlled by the left hemisphere. The left hemisphere of a Common Canary's forebrain is dominant, as is that of a human (Nottebohm 1980). The right cerebral hemisphere assumes control of the functions of the left hemisphere only if the left hemisphere is damaged. Impairment of a young canary's song-control centers in the left hemisphere leads to formation of an alternative set in the right hemisphere and the acquisition of a new song repertoire. Such functional

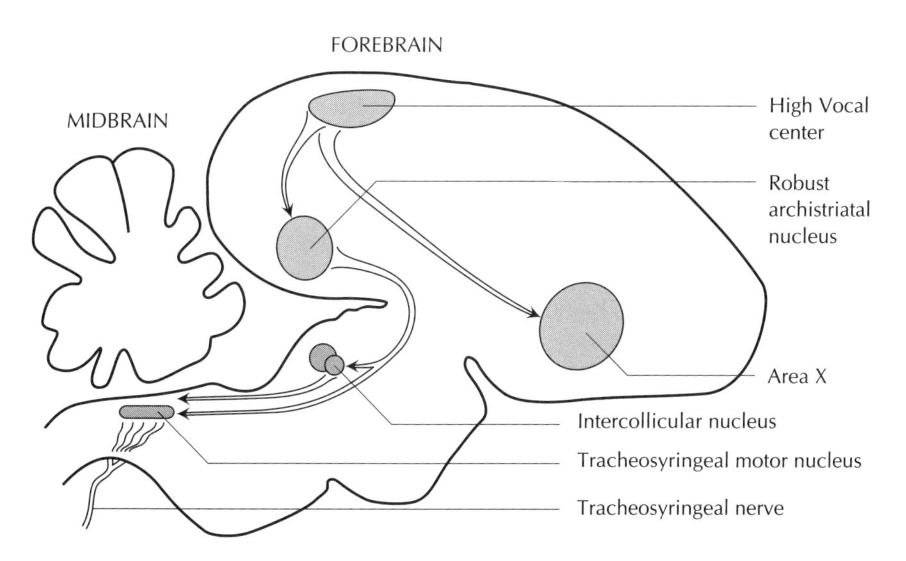

Figure 8–4 A sagittal section through the brain of a Zebra Finch, showing the descending portion of the song system. Signals that originate in the High Vocal center (formerly called the hyperstriatal nuclei) of the forebrain proceed to the intercollicular nucleus and to the tracheosyringeal (hypoglossal) nerves that control the muscles of the syrinx. Signals from the High Vocal center also transmit to the part of the brain called simply Area X, which controls other information feedback loops not shown here. (After Williams 1990)

Box 8–2

The amount of brain space that controls song is flexible

The development of brain tissue controlling song increases with the size of individual song repertoires (Nottebohm et al. 1981). Male canaries with large repertoires have larger song-control nuclei than do male canaries with small song repertoires. Also, song-control nuclei are larger in male Zebra Finches, which sing, than in female Zebra Finches, which do not sing (Arnold 1980). Populations of Marsh Wrens that differ in song repertoire size also differ in the amount of brain space allocated to the song-control (High Vocal) centers (Canady et al. 1984; Kroodsma and Canady 1985). Marsh Wrens from California learn three times more songs than do Marsh Wrens from New York and have 40 percent larger volumes of the song-control nuclei. This difference in brain space and song learning ability appears to be genetically controlled and related to the competition among males for mates, competition that is more intense in the West than in the East.

lateralization of the brain was once thought to be an exclusively human attribute, associated with extraordinary language abilities.

The song-control centers of Common Canaries and Zebra Finches are controlled by sex hormones and also grow new or replace old synapses in relation to learning, even by adult birds (Alvarez-Buyilla et al. 1990). Experimental exposure to the sex hormone estradiol at an early age enlarges the sizes of the nuclei as well as their sensitivity in the adult birds to sex hormones that stimulate singing during the breeding season (Arnold and Saltiel 1979; Gurney and Konishi 1980). The male hormone testosterone increases the length and branching complexity of dendrites—receptor branches—of some neurons in the song centers.

In canaries, new dendritic connections—called synapses—form in spring when the birds are learning new songs; these connections disintegrate in the fall when the birds stop singing. Adult songbirds can thus form new neurons and replace old ones. Research on such neurons in songbird brains has direct application to medical research concerned with treating brain and spinal cord damage and AIDS. When he tried to isolate the neural growth factor in Zebra Finches, Mark Gurney (1988) and his colleagues found a large protein molecule, which they named neuroleukin. Analysis of the functional structure of neuroleukin led them to the part of the AIDS retrovirus that destroys neurons in the human brain, thereby causing dementia. Once again basic research on birds has led to a discovery of medical importance.

Spatial Memory

The hippocampal complex of the avian brain is another target of analysis of the relation between brain evolution and social behavior in birds

(Sherry 1992; Sherry et al. 1992). Homologous in structure and function with the same structures in mammals, the avian hippocampal complex includes the hippocampus and associated parahippocampus of the forebrain. The hippocampal complex is a well-delineated, paired anatomical structure that lies adjacent to the midline of the dorsal forebrain. The hippocampi of birds and mammals are functionally equivalent with respect to controlling certain memory tasks, including spatial orientation and cognitive memory. Spatial memory processed in the hippocampus controls the daily behavior of highly mobile animals such as birds that accurately revisit feeding places, nests, and remote wintering grounds. For example, lesions in the hippocampus of homing pigeons—a specialized, domesticated breed of Rock Doves—disrupt their ability to learn a navigational map (Bingman 1988).

Members of three families of passerine birds—Corvidae (crows, jays, and nutcrackers), Sittidae (nuthatches), and Paridae (titmice and chickadees)—cache thousands of seeds as a means of exploiting temporary food surpluses and providing reserves for future use. The spatial memory requirements of seed-caching birds seem prodigious. Each autumn, individual titmice may stock over 50,000 caches of one spruce seed each (Haftorn 1959). They recover (and sometimes recache) seeds up to 28 days later (Sherry 1989; Hitchcock and Sherry 1990). Crows, jays, and nutcrackers are especially diligent hoarders. The development of spatial memory varies among species in relation to their dependence on cached seeds. Probably the most able is the Clark's Nutcracker (Figure 8–5), which hides an average of two pine seeds in each of 1400 to 2000 caches in order to survive the winter and early spring (Balda and Kamil 1989). The ability to recall the precise locations of about 2000 caches for spans of as long as 8 or 9 months reflects a phenomenal spatial memory. Stephen Vander Wall (1982) concluded that the nutcrackers memorize the locations of thousands of caches by making spatial references to surrounding large objects. When he moved the large objects that surrounded hidden caches, the nutcrackers adjusted their search in the direction of the moved objects. Experiments conducted by Al Kamil and Russ Balda (1985) showed that nutcrackers remembered the obscure locations preselected by the researchers and thus were not limited to a simple response strategy, such as to dig near a big rock or a big tree.

The extraordinary spatial memory of seed-caching birds is processed by an enlarged hippocampal complex (Krebs et al. 1989; Sherry et al. 1989). The three families of passerine birds that cache seeds for later recovery have significantly larger hippocampal volumes than do other passerine birds. Experimental studies of chickadees and nutcrackers have demonstrated that spatial memory for seed recovery is indeed based in the hippocampus and includes both cognitive and working components (Sherry 1989, 1990). Chickadees with experimental lesions to the hippocampus continue to hide seeds in normal fashion but cannot find them again, except by chance.

Figure 8–5 Clark's Nutcracker, a seed-caching bird with extraordinary spatial memory.

Avian Vision

Birds are highly visual animals. They have large eyes, with which they search visually for food and detect predators at great distances. They also engage in complex, colorful courtship displays, matched it seems by an exceptional system of color vision. The true nature of avian vision, however, remains to be determined. Experiments have not yet confirmed historical beliefs of extraordinary visual acuity—the ability to resolve fine detail at distance. Passerines and raptors, believed to have the keenest sight of all birds, can resolve details at 2.5 to 3 times the distance humans can, not 8 times as was once thought (Pearson 1972). One of the distinctions of avian vision may lie in the ability to capture at a glance a whole picture rather than to piece a scene together after a laborious scan, as humans do.

Avian eyes are large, prominent structures. The European Starling's eyes account for 15 percent of its head mass. The eyes of eagles and owls are as big as human eyes. Unlike the uniformly round, rotating eyes of mammals, the eyes of birds vary in shape from round to flat to tubular. They fill the orbits fully but are capable of only limited rotation, mostly toward the bill tip.

Because birds' eyes are generally set on the sides of their heads, birds see better to the side than to the front. Penguins and passerines, for example, examine nearby objects with one eye at a time. The resulting image is relatively flat because monocular vision does not achieve depth perception with the same accuracy as binocular vision. To compensate, birds bob their heads quickly, viewing an object with one eye from two different angles in rapid succession. Some birds, such as swallows, nightjars, hawks, and owls, restrict lateral monocular vision to close objects and use forward binocular vision for distant viewing. Generally, binocular vision is atypical. Among ducks, only the Blue Duck of New Zealand can stare forward; other ducks use one eye at a time. Parrots have a binocular field of only 6 to 10 degrees (Walls 1942). Bitterns stare forward with binocular vision while pointing their bills skyward. Quite the opposite are woodcocks, whose huge eyes are set far back on the head allowing broad rearview binocular vision.

Eye Anatomy

A cross section of the avian eye reveals a small anterior component that houses the cornea and lens and a larger posterior component that is the main body of the eye (Figure 8–6). The two sections are separated by a scleral ring composed of 12 to 15 small bones—called ossicles. Two striated muscles—Crampton's muscle and Brucke's muscle—attach to these ossicles and are responsible for focusing on objects. The lens is large and conspicuous. The pecten, a distinctive and intriguing feature of the avian eye whose function remains unclear, projects from the rear surface of the eye near the optic nerve into the large cavity filled with vitreous humor—the clear substance that fills the eye behind the lens.

Cornea and Lens

In birds, both the cornea and the lens change their curvature while focusing; only the lens does this in mammals. Contraction of Crampton's

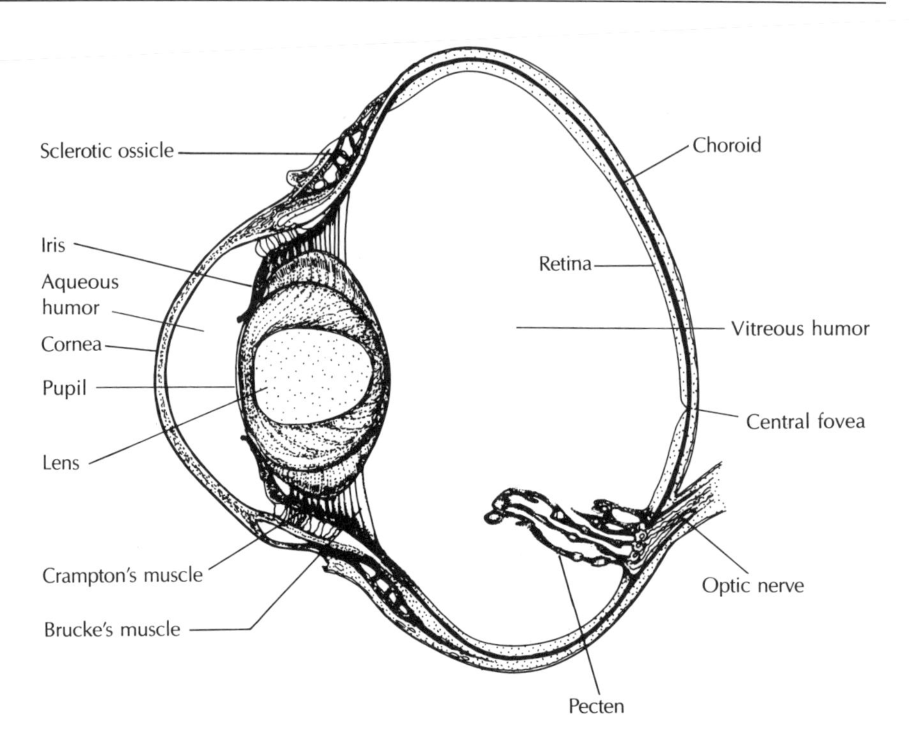

Figure 8–6 Cross section of the avian eye. (After Grassé 1950)

muscle increases the corneal curvature and thus the cornea's refractive power. A change in the cornea has little effect underwater, however, because the refractive index of the cornea is nearly the same as that of water. As we might expect, diving birds, such as cormorants, have weakly developed Crampton's muscles; to compensate, they have strong Brucke's muscles, which control the shape of the soft, flexible lens. Plunge divers such as kingfishers can even keep a target fish in focus as they dive.

Lens shape varies more among bird species than in other vertebrates. The lenses of parrots, storm-petrels, and hoopoes have flat anterior surfaces but strongly convex posterior surfaces. Ducks, owls, and nightjars have lenses that are strongly convex in both front and back, whereas in passerines and raptors the convex posterior curvature is noticeably greater than the anterior curvature. The reasons for and benefits of these differences are not yet known.

The pupil opening is round in all birds except the gull-like skimmers (Rhyncopidae), whose opening forms a catlike vertical slit in bright light, returning to the rounded shape in dim light (Zusi and Bridge 1981).

Iris colors of birds vary from the common deep brown to bright red, white or bright yellow, green (cormorants), or pale blue (gannets) and may aid species recognition. At night, some birds' eyes shine bright red in the beam of a flashlight or automobile headlights. This "eyeshine" is not the iris color but that of the vascular membrane—the tapetum—showing through the translucent pigment layer on the surface of the retina. Kiwis, thick-knees, the Boat-billed Heron, the flightless Kakapo, many nightjars, owls, and other night birds share this distinctive trait.

Retina and Fovea

The highly developed anatomy of the avian retina and its light receptor cells suggest excellent vision. The large number of cones—the daylight (color) receptors of the retina—enables birds to form sharp images, no matter where light strikes their retina. The number of cones can be as high as 400,000 per square millimeter in House Sparrows and 1 million per square millimeter in the Common Buzzard. By way of comparison, the human eye has at most 200,000 cones per square millimeter (Walls 1942). Away from the densest concentrations in the foveae, cone concentrations in the human retina drop sharply to only one-tenth of those of birds.

Foveae—concave depressions of high cone density (Figure 8–7)—are known to be the sites of greatest visual sharpness in humans. Most birds have one fovea in each eye, located in the center of the retina near the

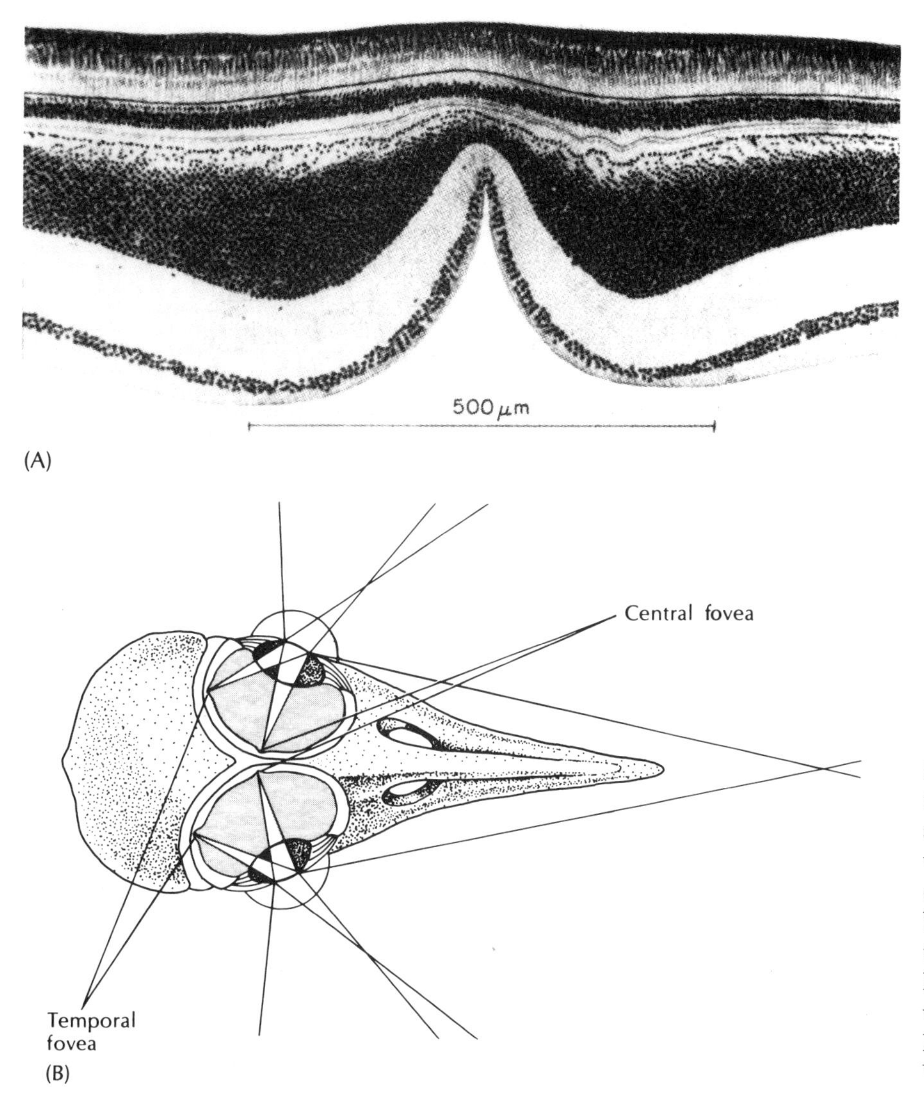

Figure 8–7 (A) Cross section of a Least Tern retina, showing the visual-cell layer with rods and cones and the deep central fovea. (B) Some birds, such as raptors, have temporal foveae, which enhance forward binocular vision. (A from Rochon-Duvigneaud 1950; B after Wilson 1980)

optic nerve. This central fovea is deeper and more complex in its cell structure in visually acute passerines, woodpeckers, and raptors than it is in pigeons and Domestic Chickens. Whether deep foveae enhance avian visual acuity is not clear. They may, however, aid in detection of the movements of small images. Fast-flying birds that must judge distances and speeds accurately, such as hawks, eagles, terns, hummingbirds, kingfishers, and swallows, have temporal as well as central foveae. These birds also have forward-directed eyes and, therefore, good binocular vision, which projects images onto the temporal foveae. Images of their peripheral or lateral monocular vision fall on the central foveae. Although cones are most abundant in the foveae, high cone densities also occur in horizontal, ribbonlike strips around the retina in albatrosses, grebes, plovers, and other birds. These ribbons apparently increase a bird's ability to perceive the horizon and work in concert with the semicircular canals of the inner ear to achieve proper body orientation.

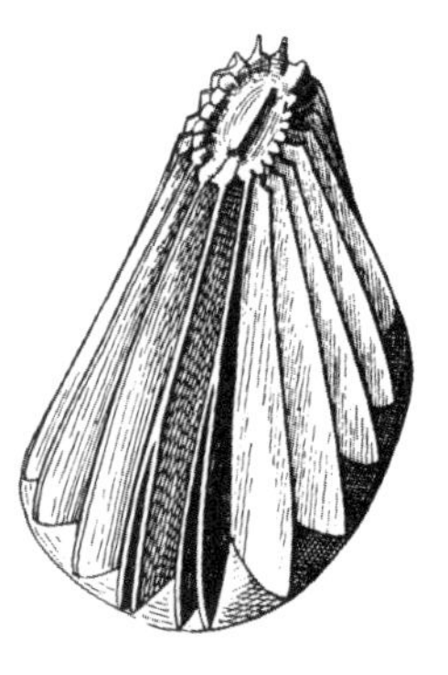

(A)

(B)

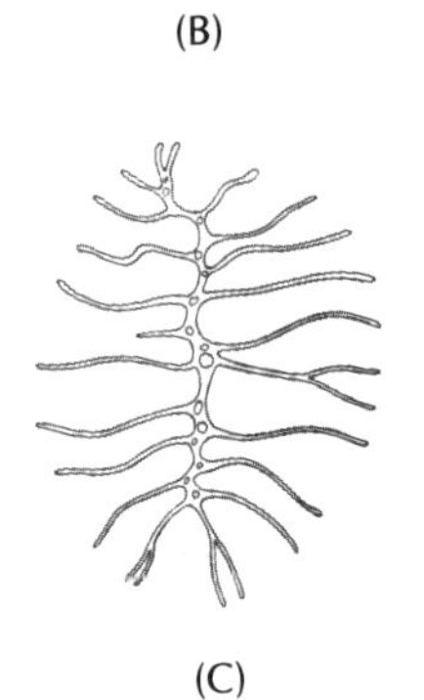

(C)

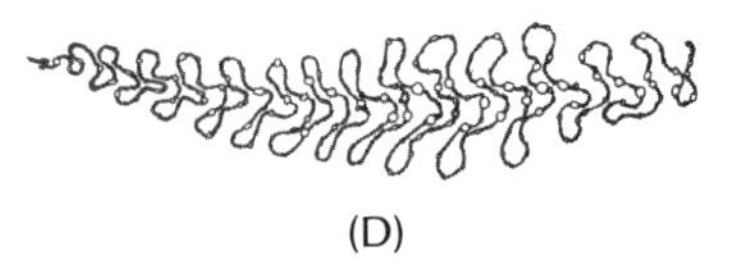

(D)

Figure 8–8 Structure of the pecten of (A) an Ostrich and (B) most modern birds. (C) Basal cross section of the structure of A, including central web and lateral vanes. (D) Dorsal view of the typical pleated structure of avian pectens. (Adapted from Walls 1942)

The Pecten

The pecten, a remarkable feature of the avian eye, is a large, pleated, vascularized structure attached to the retina near the optic nerve. Protruding conspicuously into the vitreous humor and, in some birds, almost touching the lens, the large, elaborate avian pecten is unique among vertebrates. In most birds the pecten has 20 or more accordion-pleated fins, giving it a superficial resemblance to an old-fashioned steam radiator (Walls 1942) (Figure 8–8). Nocturnal birds have fewer folds. The pectens of owls, nightjars, and the Kakapo have only four to eight folds, and the simple, reptilelike pecten of kiwis has no folds at all; it probably represents an evolutionarily degenerate condition (Sillman 1973).

The avian pecten has fascinated scientists for centuries. At least 30 theories have been proposed to explain its existence. Some researchers believe the pecten is involved in the regulation of internal eye temperatures and hydrostatic pressures; some suggest it reduces glare; others hypothesize that it might be a sextant for navigation or a dark mirror for indirectly viewing objects near the sun. The majority opinion, however, holds that the avian pecten functions primarily as a source of nutrition and oxygen for the retina. Unlike its mammalian counterpart, the avian retina has no embedded blood vessels. The assumption is that, instead, the vascular supply system is concentrated in a single structure, the pecten, which interferes less with visual functions than would a complex network of blood vessels.

Color Vision

The richness of avian color perception is probably beyond that of human experience (Goldsmith 1980). We speculate that primitive mammals, including the ancestors of primates, were night creatures that lost the retinal oil droplets associated with sensitive color vision. Once lost, these droplets did not evolve again in placental mammals. Instead, the color vision of humans and other primates reevolved on a different basis, without pigmented oil droplets. Very likely, the avian retina—with its high cone densities, deep foveae, near-ultraviolet receptors, and colored oil droplets that interact with several cone pigments—is the most capable daylight retina of any animal.

The presence of large numbers of cone receptors, which contain the visual pigments, suggests that diurnal birds have well-developed color vision. By contrast, the retinas of nocturnal owls contain mostly rods, which are simple light receptors important in black-and-white (and hence night) vision. Color vision is based on visual pigments, which convert the electromagnetic energy of light into neural energy. In addition to visual pigments, the cones of diurnal birds often contain colored oil droplets. Carotenoid pigments (Chapter 4) in the oil act as red–yellow filters, but their contribution to color vision is not understood. Perhaps the yellow oil droplets enhance the contrast of objects seen against the sky by filtering out much of the blue background. Similarly, red oil droplets may enhance the contrast of objects against green backgrounds, such as fields and trees, by filtering out the prevailing green background. The yellow oil droplets are concentrated in the central and lower retina, where distant images such as those in the sky usually fall. Red oil droplets are concentrated in cones of the peripheral and upper retina, where nearby images such as those on land usually fall.

Unlike humans, birds are sensitive to light in the near-ultraviolet spectrum. The lenses of the human eye absorb ultraviolet light; in birds, the lenses transmit ultraviolet light to the retina, where some cones have peak sensitivity in the near-ultraviolet spectrum (Chen et al. 1984). Melvin Kreithen and Thomas Eisner (1978) demonstrated that homing pigeons, in addition to having the normal vertebrate sensitivity to blues and greens (at 500 to 600 nanometers), are sensitive to the near-ultraviolet spectrum (325 to 360 nanometers). Black-chinned Hummingbirds, Belted Kingfishers, Mallards, and several passerines also are sensitive to ultraviolet light (Goldsmith 1980; Parrish et al. 1984). Given the taxonomic diversity of species tested, the majority of birds probably possess this trait.

Detection of Natural Magnetic Fields

Birds use magnetic information for navigation. (See Chapter 13 for a detailed discussion of this topic.) Tiny crystals of magnetite near the olfactory nerves between the eyes of pigeons discovered by Charles Walcott and his colleagues (1979) were initially thought to be the basis of a directional sensory system for the detection of magnetism. Similar crystals exist in bacteria and honeybees that respond to magnetism.

More recent research, however, points to an entirely different reception mechanism (Semm and Demaine 1986; Wiltschko and Wiltschko 1988;

Phillips and Borland 1992). The photopigment rhodopsin is theoretically capable of converting both light (electromagnetic energy) and magnetic fields to nerve impulses that are then processed by the central nervous system (Leask 1977). Hence, a hypothesized circular array of specialized photoreceptors within the eye organizes a directional compass that can be calibrated through experience. The relative roles of magnetite and these photoreceptors are still unknown.

Avian Hearing

Sounds provide birds with essential information. From territorial defense to mate choice and recognition of individuals and from song learning to prey location, predator avoidance, and navigation, birds depend on their hearing for a wide range of activities. Although the anatomy of the avian ear has been fully described, ornithologists are just beginning to appreciate the full range of sounds that birds hear.

Ear Structure

The three sections of the avian ear are the external ear, the middle ear, and the inner ear. The first two sections funnel sound waves from the environment into the cochlea—the fluid-filled, coiled section of the inner ear that is the base of the hearing organ. Hair cells in the cochlea monitor vibrations transmitted by the fluid and encode them into a temporal sequence of nerve impulses that register in the acoustical centers of the brain. The avian ear is structurally simpler than that of mammals. Its acoustical efficiency, however, is the same as that of the mammalian ear.

The external ears of birds are inconspicuous structures located behind and slightly below the eye. They lack the elaborate pinnae—or projecting parts—of mammalian ears. In contrast to the three bones in the middle ear of a mammal, the middle ear of a bird has only one bone—the columella, or stapes—which connects the eardrum—or tympanic membrane—to the pressure-sensitive fluid system of the inner ear (Figure 8–9). Located next to the attachment of the columella to the bony cochlea is the flexible round window, which protects the inner ear from pressure damage. The shape of the columella varies with taxon, but most birds have a simple columella similar to that of reptiles (Feduccia 1977). Compared with those of mammals, avian inner ears have a short basilar membrane, no division between inner and outer hair cells, and a simple system of cochlear nerves.

Specialized auricular feathers on the external ear protect the hearing organ from air turbulence during flight while permitting sound waves to pass inside. Diving birds, such as auks and penguins, have strong, protective feathers covering the external ear openings. These birds can protect their middle and inner ears from pressure damage in deep water by

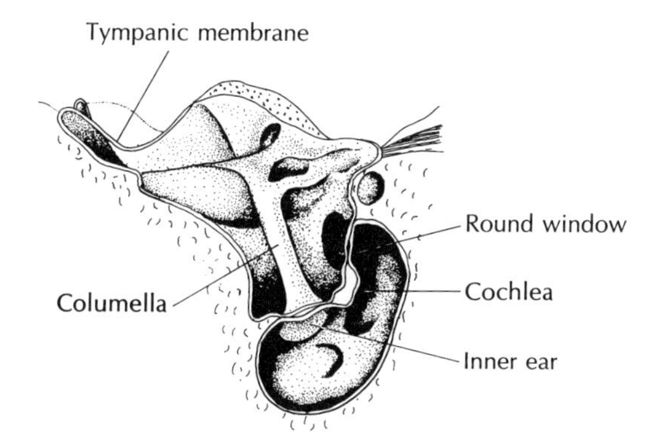

Figure 8–9 Middle ear region of a chicken. A single bone—the columella, or stapes—transmits sound from the eardrum to the fluid-filled cochlea of the inner ear. (After Pohlman 1921)

closing the enlarged rear rim of the external ear. The entire muscular rim to which the auricular feathers are attached forms an enlarged, though inconspicuous, ear funnel in some birds, especially passerines, parrots, and raptors. Among raptors, the shape varies from no funnel at all in the fish-eating Osprey to well-developed funnels in harriers (Accipitridae), which locate field mice acoustically (Rice 1982). The superb hearing of nocturnal owls is related to their exceptional ear funnels. Large anterior and posterior ear flaps regulate the size of the ear opening and enhance acoustical acuity more than fivefold (Schwartzkopff 1973). In many owls, the external ears and in some cases the skull are bilaterally asymmetrical, a condition that aids precise location of prey.

Acoustical information is processed primarily by auditory nuclei in the hindbrain. The basic plan of the avian auditory central nervous system is the same as that of reptiles, with some derived specializations (Carr 1992). Specialized dark-hunting owls that rely on sound have an extraordinary number of ganglionic cells in the medulla for processing sound and spatial information. The Barn Owl, for example, has about 47,600 ganglionic cells in one half of the medulla; the Carrion Crow has about 13,600; and the Little Owl, which hunts in the early morning light, has about 11,200 (Winter 1963). Oilbirds, which use sound to navigate in the dark, also have highly developed auditory centers.

Hearing Ability

Still uncertain is what birds actually hear and how their hearing compares with ours. Although still in its infancy, research on avian hearing abilities already is yielding some surprising results. Contrary to past impressions, most birds may not have extraordinary acoustical acuity (Dooling 1982; Fay 1988); humans can hear fainter sounds than most birds at most frequencies. Furthermore, the frequency range of good hearing is narrower in birds than in mammals. The frequencies of sound are measured in terms of cycles per second (hertz) or, for high-frequency sounds, thousands of cycles per second (kilohertz). Maximum sensitivity is confined to frequencies between 1 and 5 kilohertz. Sensitivity decreases rapidly at both lower and higher frequencies. Owls are an exception: Great Horned Owls hear low-frequency sounds and Barn Owls hear high-frequency sounds (up to 12 kilohertz) better than do humans. Oscine songbirds tend to hear high-frequency sounds better and low-frequency sounds less well than do other birds (Figure 8–10).

Surprisingly, small songbirds are not particularly sensitive to high-frequency sounds. Unlike bats and some other mammals, birds do not hear ultrasonic sounds—sounds with frequencies higher than those audible to humans. Pigeons, chickens, and guineafowl, however, hear very low frequencies (infrasound below 20 hertz) extremely well; pigeons can hear much fainter sounds (50 decibels lower) in the 1 to 10 hertz range than humans can hear (Kreithen and Quine 1979; Schermuly and Klinke 1990). The significance of this ability is not yet understood.

Birds are sensitive to small changes in the frequency and intensity of sound signals, but not unusually so, and are not as sensitive as humans. Birds can discriminate temporal variations in sound, such as duration of notes, gaps, and rate of amplitude modulation, as well as other vertebrates can, including humans. Laboratory tests do not support the idea that birds have exceptional powers of temporal resolution, a result that bears di-

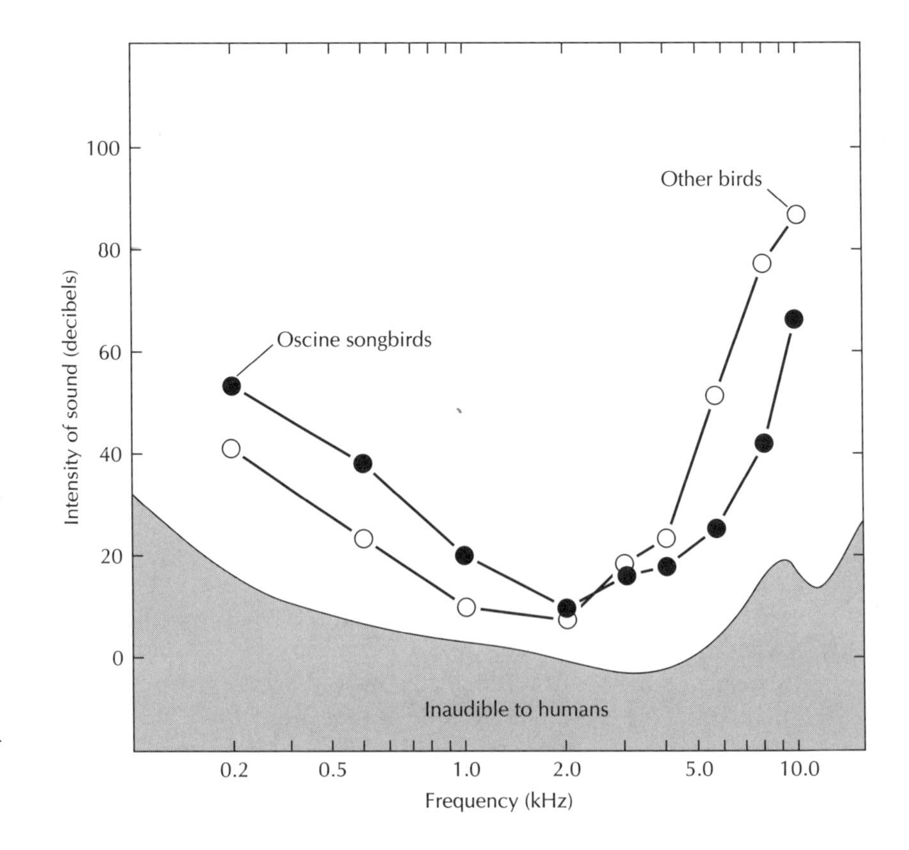

Figure 8–10 Hearing thresholds of nine species of oscine songbirds (solid circles) and nine other bird species (open circles). The lower curve shows how loud a sound must be at a particular frequency for humans to hear it. For both birds and humans, the higher the required intensity, the poorer is an individual's hearing. Birds hear well (required intensity less than 40 decibels) over a narrower range of frequencies than do humans. (From Dooling 1982)

rectly on the abilities of birds to recognize subtle call variations and on the evolution of vocalizations (see Chapter 10).

Budgerigars enhance reception of their own signals by filtering out environmental noises (Dooling 1982). Their greatest signal-to-noise sensitivity is at 2 to 4 kilohertz, the principal frequencies of their own vocalizations. Parakeet ears are also buffered against acoustical trauma; they are not temporarily deafened by loud noises as are most vertebrates.

Use of Echolocation for Navigation and Prey Detection

A few birds use echolocation—reflected vocalizations—for navigation. Some cave swiftlets of Southeast Asia find their way through dark cave corridors by emitting short, probing clicks of 1-millisecond duration at normal frequencies (2 to 10 kilohertz); they do not employ ultrasound as bats do (Medway and Pye 1977). Echolocation at these normal hearing frequencies is at best only one-tenth as functional as the ultrasound sonar system of bats. For example, the cave-nesting Oilbird (Family Steatornithidae), a fruit-eating nightjar of South America, echolocates with sharp clicks 15 to 20 milliseconds long over a broad frequency spectrum (1 to 15 kilohertz) (Konishi and Knudsen 1979). They can avoid disks that are 20 millimeters or more in diameter but collide with smaller objects.

Owls can locate prey by sound in complete darkness (Payne 1971). The Barn Owl can catch a running mouse in total darkness because it can pinpoint sounds to within 1 degree in both the vertical and the horizontal plane. The Barn Owl can also determine the direction and speed of a mouse's movement. Humans can locate sounds in the horizontal plane about as well as a Barn Owl but only one-third as well in the vertical plane.

Both owls and humans locate the sources of sounds by means of differences in the intensity and time of arrival of sounds at the two ears. Looking directly at the source equalizes these stimuli. The asymmetrical arrangement of the ears of some owls enhances reception differences and thus the ability to locate prey quickly and accurately. This ability is well developed in the Barn Owl, which locates sounds in the vertical plane by means of its asymmetrical ear openings and the troughs formed by the feathered facial ruff (Figure 8–11). The left ruff faces downward, thereby increasing sensitivity to sounds below the horizontal, and the right ruff faces upward, increasing sensitivity above the horizontal. The owl need only tilt its head up or down to equalize input to the two sides and thus to pinpoint the location of a mouse.

The process of sound localization by songbirds is still poorly understood but may be based on mechanisms other than those used by either owls or humans. The small heads of songbirds and their poor hearing at high frequencies reduce the possible differences between the two ears in timing and intensity of arriving sound signals (Klump et al. 1986). The effective separation of the ears of small birds, however, is increased by passage of sounds through a canal that connects the inner ears. Sounds entering one ear reach the tympanic membrane and inner ear of the other ear, thereby increasing interaural time differences used to localize the source of the sound (Calford and Piddington 1988; Carr 1992).

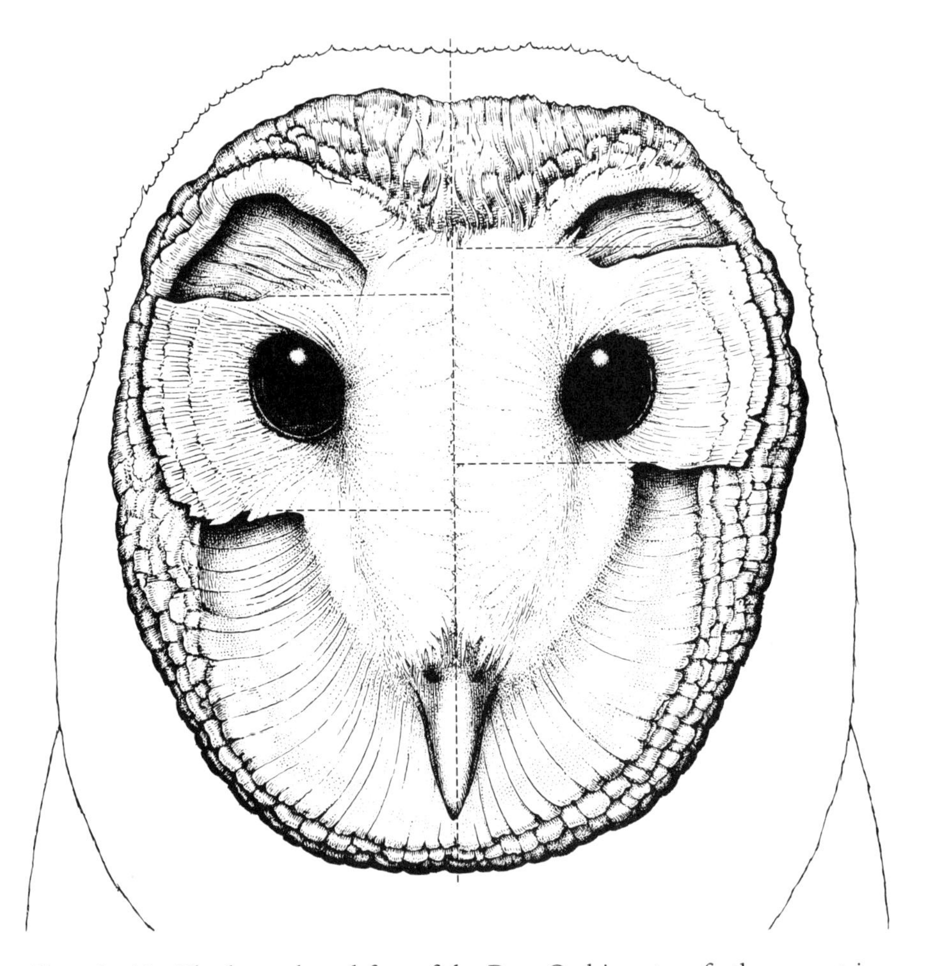

Figure 8–11 The heart-shaped face of the Barn Owl is not perfectly symmetrical. The left ear, which is higher than the right ear, is most sensitive to sounds from below the horizontal (an imaginary horizontal plane that is parallel to the ground and passes through the owl's head). Conversely, the lower right ear is most sensitive to sounds from above the horizontal. Enhancing the difference in ear positions are the downward-oriented, feathered ruff on the left side of the face and the upward-oriented ruff on the right side. This asymmetry causes a sound to arrive at each ear at slightly different times, thereby enabling the owl to pinpoint the source of the sound. (From Knudsen 1981, with permission of Scientific American)

Mechanoreception

Birds are extremely sensitive to mechanical stimulation (Schwartzkopff 1973). Included among the senses of mechanoreception are tactile reception, equilibrium (or balance), and detection of barometric pressure. Mechanoreception reaches its highest level in the hearing organ.

Tactile corpuscles, the primary source of skin sensitivity, also monitor changes in muscle tension—proprioception. These are cells specialized for tactile response that are found at the ends of sheathed nerve fibers. The ellipsoidal Herbst corpuscles are the largest and most elaborate of the different kinds of tactile corpuscles, which consist of an outer multilayered sheath and an inner core. The onionlike layers of the outer sheath are adapted for elastic reception and transfer of rapid pressure changes; the inner, cylindrical core is an elaborate sensory nerve fiber (Figure 8–12). Herbst corpuscles are abundant in the sensitive bill tips of sandpipers and snipes, which use tactile foraging to find small prey in the mud (Clara 1925; Bolze 1968), and in the tips of woodpecker tongues. They are also concentrated in feather follicles that have sensory functions, especially those of filoplumes and bristles (see Chapter 4) and are numerous in the wing joints of birds, where they help govern wing positions in flight.

The organs of equilibrium—the semicircular canals located in the ears and the associated sets of specialized sensory cells—are among a bird's most important sensory organs because they regulate the balance and spatial orientation so essential to skilled flight. They give birds an excellent sense of balance and body position, enabling them to reorient automatically with respect to gravity, even when blindfolded.

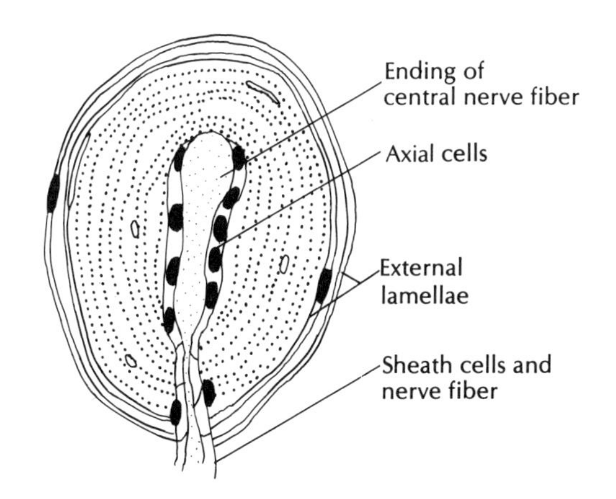

Figure 8–12 Herbst corpuscle from the bill of a duck. The most elaborate of avian tactile sensors, it consists of up to 12 onionlike layers of external lamellae that transfer slight pressure changes to the elaborate nerve ending of the receptor axon in the center. (After Portmann 1961)

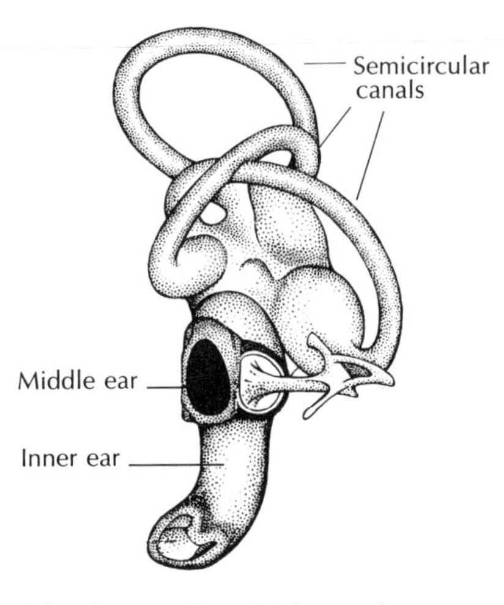

Figure 8–13 The semicircular canals, which are the organs of equilibrium, are located next to the apparatus of the middle ear. (Adapted from Pumphrey 1961)

Three semicircular canals, one oriented horizontally and two oriented vertically, connect directly to the cerebrospinal fluid system (Figure 8–13). When the position of the bird's head changes, fluid moves through the canals (Werner 1958). At the bases of the semicircular canals are delicate sets of membranes equipped with sensory hair cells, which detect the movements of small crystals of calcium carbonate—statoliths—floating in the fluid. Spatial variations in the pressure of the crystals on the hair cells cause different patterns of excitation, which enable the bird to sense the direction of gravity and of linear and circular acceleration.

The size of the semicircular canals is related to flight performance: pigeons, owls, thrushes, ravens, and falconiform birds have relatively larger canals than do galliform birds and ducks. Among galliform birds, the size of the semicircular canals increases with the mobility of a species (Sagitov 1964). So does the size of the cerebellum of the avian midbrain, which is responsible for balanced muscular coordination.

Responses to Barometric Pressure

For years, ornithologists have appreciated that birds are aware of an approaching winter storm and feed actively to build their energy reserves. Birds also know how to choose altitudes for migration. These abilities suggest sensitivity to differences in barometric pressure. Homing pigeons, the only birds that have been studied in this regard, are, in fact, extremely sensitive to small changes in air pressure, comparable to differences of only 5 to 10 meters in altitude (Kreithen and Keeton 1974a).

Avian Taste and Smell

Birds can taste and smell, although how well birds taste is still unclear (Wenzel 1973). The few studies of taste acuity in birds suggest only that they may be equally or less sensitive than mammals with respect to some ingredients. A few taste buds are located on the rear of the avian tongue and on the floor of the pharynx: about 24 in the chicken, 37 in the pigeon, and 62 in Japanese Quail. Avian taste buds are similar in structure to mammalian taste buds but negligible in number by comparison. Humans, for example, have roughly 10,000 taste buds.

The avian sense of smell, which is based in the surface epithelium of the posterior concha of the olfactory cavities (see Chapter 6; see also Bang and Wenzel [1986] for comparative descriptions of avian olfactory cavities) has traditionally been underestimated. The small size in most birds of the olfactory bulbs (relative to brain size) fostered the belief that only a few exceptional birds—those with large olfactory bulbs, namely, vultures, kiwis, and petrels—used olfaction in their daily activities. Now that view is changing—most birds probably can smell and use odors in their daily routines (Waldvogel 1989; Clark et al. 1992). Northern Bobwhites, Common Canaries, Mallards, Domestic Chickens, Manx Shearwaters, Turkey Vultures, Brown Kiwis, and Humboldt Penguins all detect odors in the laboratory. Even passerine songbirds, which have minimal olfactory bulb sizes (1.5 millimeters) and were assumed to lack a sense of smell, can detect certain odors with the same acuities as rats and rabbits (Clark et al. 1992). Simple and critical olfactory functions can be accommodated by very small amounts of olfactory tissue. Among the orders of birds, however, odor detection thresholds are correlated with the size of the olfactory bulb relative to the size of the cerebrum (Figure 8–14). Olfactory bulbs also tend to be larger in nocturnal birds, which

---

BOX 8–3
Birds like chili peppers

Recent studies by Donald Norman and his colleagues (1992) revealed an ecological link between birds and chili peppers. The active chemical ingredients in chili peppers—called capsaicins—have a familiar, flaming effect on the oral epithelia and taste buds of mammals. The normal concentration of these chemicals (1000 ppm) in wild chilies repels rodents but does not make food distasteful to birds. Indeed, birds are attracted to wild capsicum fruits—called bird peppers—because they are high in vitamins, protein, and lipids. A reasonable hypothesis is that capsaicins protect the pepper seeds from consumption by rodents while allowing and even attracting birds to eat the fruits and disperse the seeds.

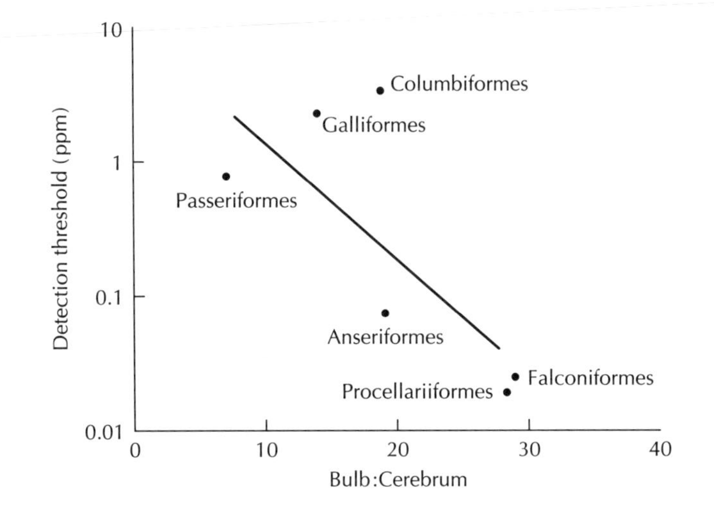

Figure 8–14 The relationship between olfactory acuity, expressed by the detection threshold for a chemical in parts per million (ppm), and the proportion of brain tissue allocated to the olfactory bulb for six orders of birds. (From Clark et al. 1992)

need expanded nonvisual sensory abilities, than in related diurnal taxa (Healy and Guilford 1990).

Odors can affect daily sexual and foraging behaviors of birds. The sexual prowess of male Mallards depends on smelling a female's breeding odors. Experimental cutting of the olfactory nerves inhibits courtship and sexual behavior (Balthazart and Schoffeniels 1979). Female odors apparently come from oil gland secretions, which change in composition during the breeding season. Black-billed Magpies and Turkey Vultures can smell putrid meat. Wild flying Turkey Vultures, which find carcasses by smell, are attracted to the source of ethyl mercaptan fumes released into the air to simulate the smell of rotting meat (Stager 1964, 1967; Smith and Paselk 1986). Engineers have used the remarkable olfactory abilities of Turkey Vultures to locate leaks in a pipeline 42 miles long by pumping the same chemical through it and then spotting where the vultures gathered.

Tube-nosed seabirds also locate food by odors, particularly carboxylic acids, which are odorless to humans. Leach's Storm-Petrels potentially can detect the volatile compounds released by krill, small marine crustacea, from distances of 2.5 to 25 kilometers from the source and then locate the food by following the odor upwind (Hutchinson and Wenzel 1980; Clark and Shah 1992). Leach's Storm-Petrels also use their well-developed sense of smell to locate their nesting burrows in the dark, conifer forests on islands in the Bay of Fundy (Grubb 1972, 1974).

Honeyguides, which lead animals and people to beehives (Chapter 7), can find the beehives by their pungent smells. In experiments, honeyguides can find concealed beeswax candles and are attracted by the odor plume of a burning candle. Goslings learn to choose and reject food plants by smell at an early age (Würdinger 1979) and European Starlings select appropriate nest construction materials by smell (see Chapter 16).

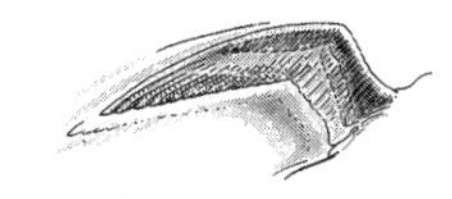

BOX 8-4

New Zealand kiwis sniff for their food

Kiwis are flightless, chicken-sized forest birds found only in New Zealand (see Figure A-1 in the Appendix). Active only at night, kiwis probe their long bills into wet soil to find earthworms, which they locate by sniffing through nostrils located at the bill tip. All other living birds have nostrils at the base of the bill.

A series of classic experiments demonstrated that kiwis rely on their highly developed sense of smell to find food (Wenzel 1968, 1971). Screened tubes containing either fragrant food or just dirt (the control) were buried 3 centimeters deep in a large cage. The captive kiwis quickly found the baited tubes and punctured them to extract the food, but ignored the control tubes with only dirt. Parallel laboratory experiments demonstrated increased respiration and brain neural activity with exposure to food odors and conditioned aversion to food containing noxious chemicals.

Senses and Behavior: A Projection

The senses of a bird are an integral part of its feeding behaviors, affecting food detection, innovative feeding, and effective relocation of caches. These senses reach full expression in the process of communication via visual and vocal displays, the topics of the next two chapters. The outstanding navigational abilities of birds also are derived from high-level integration of their wide-ranging sensory capabilities (see Chapter 13). Hormones set the physiological environment for behavior, including sensitivity to particular stimuli. For example, seasonal activities are controlled by photoreceptors in the midbrain that trigger the release of pituitary hormones; these hormones, in turn, govern details of reproductive behavior, molt, and migration (see Chapter 11). Also mediating the interaction between heritage and environment is a continuum of sensory experiences ranging from brief imprinting to prolonged intelligent learning. The sensory reactions of birds, therefore, only open the door to the full expression of the adaptive behaviors of birds presented in the next chapters.

Summary

Birds exhibit greater intelligence than implied by the popular slur "bird brain." They outperform mammals in many laboratory problem solving experiments. Two-way verbal communication experiments with a Gray Parrot have revealed advanced abstracting and conceptual abilities.

The evolution of the avian brain has taken a different course from that of mammals. The basis for avian intelligence lies in the hyperstriatum layer of the forebrain, primarily in tissue called the Wulst. Particularly exciting have been the identification of the brain centers for control of song in passerine birds and the definition of the neural pathways that control this complex motor skill. Another example of the relationship between brain development and behavior in birds is the control of spatial memory by the enlarged hippocampus of the forebrain in seed-caching birds.

Birds have a full repertoire of well-developed senses. Large eyes and well-developed optic lobes of the brain provide excellent vision, including an ability to follow small moving objects. Birds may also have the most highly developed color vision of any vertebrate. The hearing of birds as a group is good but not extraordinary, except perhaps for the ability of Rock Doves to hear extremely low frequencies (infrasound) and the ability of Barn Owls to pinpoint sounds made by potential prey. Birds are sensitive to slight differences in barometric pressure and to magnetism. The senses of smell, taste, and touch are also better developed in birds than once was thought.

Further Readings

Carr, C.E. 1992. Evolution of the central auditory system in reptiles and birds. In Evolutionary Biology of Hearing, D.R. Webster, R.R. Fay, and A.N. Popper, Eds., pp. 511–544. New York: Springer-Verlag. *A thorough comparative perspective.*

Dooling, R.J. 1982. Auditory perception in birds. In Acoustic Communication in Birds, Vol. 1, D.E. Kroodsma and E.H. Miller, Eds., pp. 95–130. New York: Academic Press. *An excellent review of avian hearing abilities.*

Hartwig, H.-G. 1993. The central nervous system of birds: A study of functional morphology. Avian Biology 9: 1–119. *A long-awaited review of the functional anatomy of the avian brain.*

Kamil, A.C. 1988. A synthetic approach to the study of animal intelligence. In Nebraska Symposium on Motivation, Vol. 35. Comparative Perspectives in Modern Psychology, D.W. Leger, Ed., pp. 257–308. Lincoln: University of Nebraska Press. *A modern perspective of a classic topic.*

Walls, G.L. 1942. The Vertebrate Eye and Its Adaptive Radiation, Cranbrook Institute of Science Bulletin No. 19. Bloomfield Hills, Mich.: Cranbrook Institute of Science. *The classic reference.*

Wenzel, B.M. 1973. Chemoreception. Avian Biology 3: 389–415. *A detailed review of the subject.*

Ziegler, H.P., and H.-J. Bischof, Eds. 1993. Vision, Brain, and Behavior in Birds. Cambridge, Mass.: MIT Press. *A much-needed, new reference on avian vision, integrating the latest information on anatomy, neurobiology, and behavior.*

CHAPTER 9

Visual Communication

BIRDS COMMUNICATE WITH one another by means of displays—specialized acts that transmit information between a sender and a receiver. The exchange of information by visual or vocal signals helps birds to resolve conflicting individual purposes as well as to foster cooperation. Territorial defense, attraction and courtship of mates, and maintenance of winter flock structures all require birds to communicate. Courtship displays combine plumage ornaments and colors, postures, and vocalizations, which have evolved together. Each accentuates the other, thereby enhancing the impact of the display itself and enriching its informational content regarding identity, status, intentions, and potential as a mate. The displays and communication behaviors of birds are the blended results of evolutionary inheritance and innately channeled experience (Gould and Marler 1987).

This chapter concerns visual aspects of communication by birds; vocal communication is the theme of Chapter 10. Many birds have conspicuous and distinctive plumage color patterns with associated feather ornaments. Sometimes, as in the King Bird-of-Paradise (Figure 9–1), their feather colors and ornaments are extravagant. Complementing feather colors may be bright colors of the bill, facial skin, eyes, or feet. Some birds are brightly colored, but others are not. Some species exhibit striking color differences between the sexes (sexual dimorphism), ages, or seasons, whereas others do not. When birds are sexually dimorphic, males usually are more colorful than females. When the age classes differ, adults usually are more colorful than young birds. Also, birds in the breeding season tend to be more colorful than nonbreeding birds. On a case by case basis, bird colors and associated displays serve to identify an individual, to signal attack, escape, or neutral locomotive intentions, and to communicate location or the desire to play, mate, or take over a territory.

This chapter reviews first the general categories of plumage coloration, and the recognition of species and individuals by visual cues, sometimes

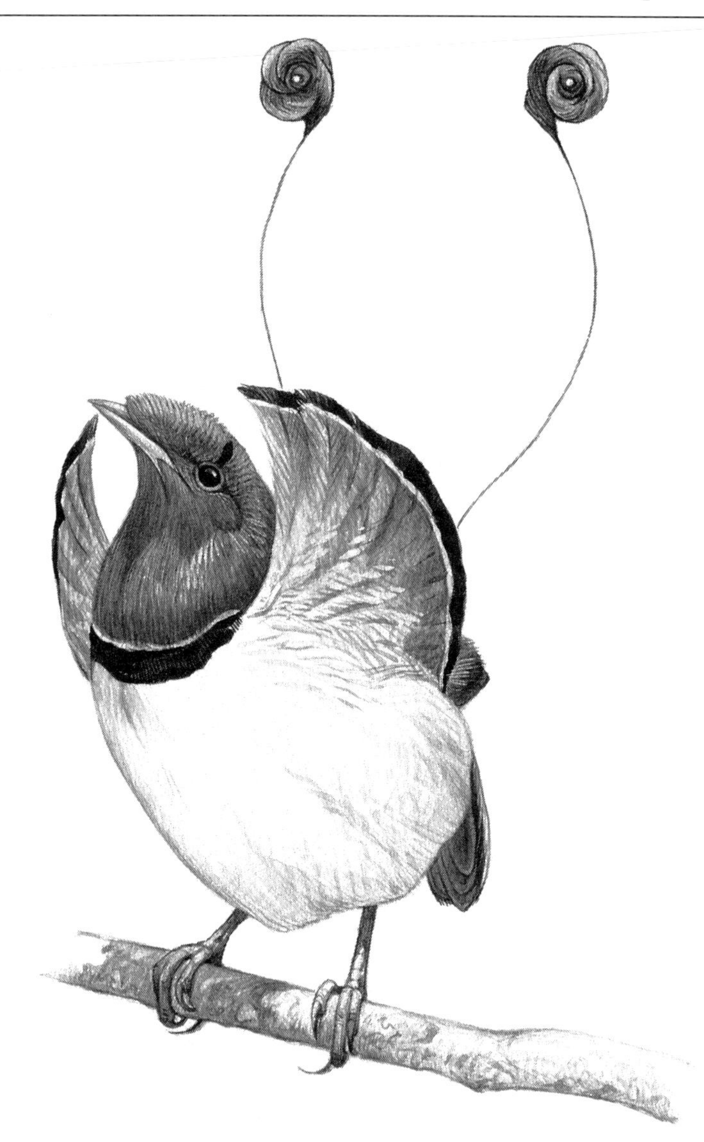

Figure 9–1 The courtship display of the King Bird-of-Paradise, as described by Ingram (1907, pp. 227–228): "He always commences his display by giving forth several short separate notes and squeaks, sometimes resembling the call of a Quail, sometimes the whine of a pet dog. Next he spreads out his wings, occasionally quite hiding his head; at times, stretched upright, he flaps them, as if he intended to take flight, and then, with a sudden movement, gives himself a half turn so that he faces the spectators, puffing out his silky-white lower feathers. Now he bursts into his beautiful melodious warbling song, so enchanting to hear but so difficult to describe. . . . Then comes the finale, which lasts only for a few seconds. He suddenly turns right around and shews his back, the white fluffy feathers under the tail bristling in his excitement; he bends down on the perch in the attitude of a fighting cock, his widely opened bill shewing distinctly the extraordinary light apple-green colour of the gullet, and sings the same gurgling notes without once closing his bill, and with a slow, dying-away movement of his tail and body. A single drawn-out note is then uttered, the tail and wires are lowered, and the dance and song are over."

mediated by imprinting of young birds on their parents (Chapter 18). Then follows the evolutionary ritualization of elaborate visual displays, which can reflect the evolutionary history of groups of birds, such as pelicans and manakins. The well-studied pair formation displays of the Great Blue Heron illustrate the kinds of information transmitted and the ways in which herons enhance the process of communication. The attack and escape elements in the agonistic behavior of birds then introduce the game of communication and the current controversy over whether displays are "honest" or "devious." The chapter concludes with a section on the process of sexual selection, which is responsible for the elaboration of color patterns and display ornamentation of many birds.

Plumage Color Patterns

Plumage colors vary in hue from drab to bright and in pattern from cryptic (concealing) to bold. Concealment is the first role of bird color patterns, not just of those that are obviously cryptic, but also of many bright and bold ones that match the bird's usual environment. Ptarmigan are nearly pure white in winter, when they blend with the mountain snows. In spring, when patches of snow remain on the alpine meadows, the birds are white and brown. In summer, when herbs and lichen cover the rocks, ptarmigan are finely barred black and brown (Figure 9–2A). Woodcocks and Whip-poor-wills rest invisible to us on a forest floor of dead leaves. The American Bittern points its bill skyward, aligning its body contours and the stripes on its breast with the surrounding vertical marsh grasses. The wood-colored Common Potoo of tropical America conceals itself by assuming the posture of a dead stump (Figure 9–2B).

(A)

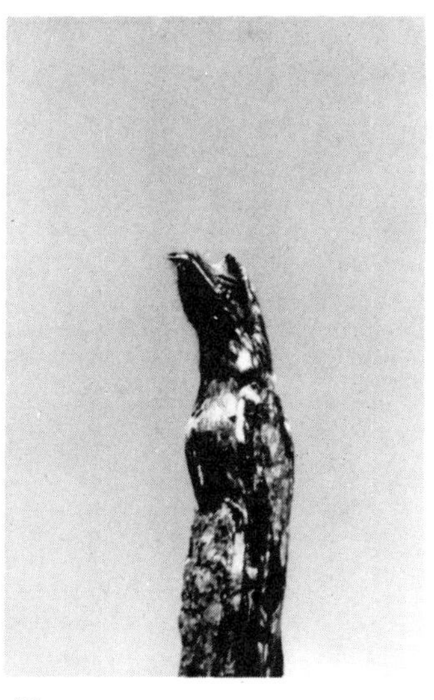

(B)

Figure 9–2 Plumage coloration provides excellent camouflage. (A) The White-tailed Ptarmigan blends into an alpine meadow. (B) The Common Potoo looks like a dead stump. (A courtesy of A. Cruickshank/VIREO; B courtesy of J. Remsen/VIREO)

Some bold color patterns reduce the contrast between a bird's shape or outline and its background. The breast bands of the small plover, a classic example of a disruptive pattern, visually separate the outline of its head from that of its body. To be most effective, the contrast between disruptive patches on a bird's body should be as great as that between the bird and its background. In other words, the color patches on a bird provide the most effective concealment when their sizes match those of light and dark elements in the background. The finely patterned summer plumage of a ptarmigan blends with the finely patterned alpine grasses and lichens, and the boldly patterned plumages of the wood warblers of North America blend with the small leaves, branches, and lighting particular to their arboreal niches.

Abbott Thayer and his son Gerald (1909) were the first to identify the principle of countershading in concealment (Figure 9–3). Lower reflectivity of the dorsal surface of a bird interacts visually with contrasting light undersides to disguise its outline, helping it to match its background. The value of contrast increases with the intensity of illumination from above. Open-country birds, such as plovers, have strongly contrasting colors on their upper and lower surfaces. White underparts work particularly well in this regard as a neutral (achromatic) reflector that takes on the hue of the nearest surface. Because they are closer to the sand or mud, white breasts on small plovers function more effectively for countershading than do the white breasts on large, long-legged shorebirds. Large shorebirds with long legs often have dark underparts rather than white.

The advantages of bold color patterns in visual display can supplement or take precedence to the need for concealment. Whereas countershading (dark upperparts and white underparts) enhances concealment, reverse countershading (white upperparts and dark underparts) makes birds conspicuous. Breeding male Spectacled Eiders, Bobolinks, and Gray Plovers have striking reverse countershading (Figure 9–3B). The triangular, black throat patches of Hooded Orioles and Golden-winged Warblers are more restricted forms of reverse countershading. The triangle points to the bill and thus may help focus attention on bill movements.

There are many other ways of increasing the signal values of plumage patterns (Figure 9–4). The uniform coloration of the all-red Northern Cardinal enhances its outline and renders it more conspicuous than would a mixed color pattern, and the crest probably enhances this effect. Contrasting edgings enhance striking signal patches, such as the white crest of a Hooded Merganser, the orange crown stripe of a Golden-crowned Kinglet, or a Mallard's blue wing patch. Unusual shapes, especially those that are geometrically regular, such as the triangular white wing patches of an adult Sabine's Gull or the rectangular wing patches of ducks, are highly visible because they do not normally match the elements in a natural background. Regular repetition, such as in the tail spots of a Yellow-billed Cuckoo or the head stripes of a White-crowned Sparrow, achieve similar conspicuous results.

(A)

(B)

Figure 9–3 (A) The plumage pattern of a Killdeer combines countershading, achromatic reflectance of substrate by white underparts, disruptive head and breast markings, and breast bands that help match horizontal breaks in the shoreline or horizon. (B) The breeding plumage of a Gray Plover is an example of reverse countershading. (A courtesy of A. Cruickshank/VIREO; B courtesy of C.H. Greenewalt/VIREO)

Figure 9–4 Conspicuous plumage signal patterns: (A) triangular wing pattern of an adult Sabine's Gull; (B) repeated white tail spots of a Yellow-billed Cuckoo; (C) outlined crest of a male Hooded Merganser.

Visual Identity

Species Recognition

Many bird species, such as breeding male North American wood warblers, have distinctive plumage color patterns. Face or head color patterns in particular tend to differ among related species, enabling rapid identification by birdwatchers as well, presumably, as by the birds themselves. Seabirds generally have drab black, white, gray, or brown plumages, but differ in color of the bill and the feet and also of the eye (Pierotti 1987). Species that coexist almost always differ in the colors of their facial skin, bills, and feet; if they do not, then they hybridize occasionally or commonly. Apparently, these soft part, skin color differences serve as isolating mechanisms. However, proof that birds use distinctive color patterns to recognize members of their own species remains scarce.

A chick's early visual experience with its parents may affect its reaction to alternative species-recognition color patterns and its eventual choice of a mate. One case concerns the white and dark "blue" phases of the Snow

Goose of the Canadian Arctic (Cooke 1978; Geramita et al. 1982; Rockwell et al. 1985). The different colors of this species have a simple genetic basis, analogous to that of blue eyes and brown eyes in humans. The dark phase results from a single dominant allele; the white phase, from recessive alleles in homozygous condition. As a rule, white geese pair with other white geese, and dark geese also pair with each other.

Early visual imprinting on family color is the force behind these mating preferences. Young Snow Geese choose mates of the same color as their families, principally that of their parents. Regardless of their own color phase, geese raised by white parents later choose white mates; geese raised by dark parents choose dark mates, and geese raised by mixed pairs

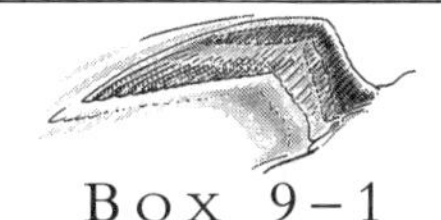

Box 9–1

Endangered species projects accommodate the sexual preferences of hand-reared birds

When hand-raised by humans, captive baby birds tend to imprint on their human keepers and then to orient their adult sexual interests toward them. Disguises and models of parent birds are essential proxies for rearing California Condor chicks that will later exhibit proper species-recognition behavior (see figure). Improper recognition behavior, however, sometimes has scientific advantages. Captive birds that have imprinted on their human keepers will ejaculate onto the keeper's hand, providing sperm for artificial insemination. This technique has been used for captive propagation of endangered species, such as the Peregrine Falcon (see Chapter 24).

A model condor head was created as the surrogate parent to prevent hand-raised California Condor chicks from imprinting on their human keepers. (Courtesy of the San Diego Zoological Society)

choose either white or dark mates. The color of siblings has a secondary effect, especially in mixed families with white-phase offspring produced by heterozygous dark-phase parents. Young geese from such families occasionally choose mates unlike their parents.

The effects of early imprinting on a bird's sexual response to species-specific color patterns are further illustrated by the results of cross-fostering experiments, in which young are raised by parents of another species (Immelmann 1972a, b). Cross-fostering causes the sexual interests of ducks, doves, finches, and gulls to shift to the foster species. For example, male Zebra Finches raised by Bengalese Finches, the oldest domesticated cagebird and whose natural species origins are unknown, prefer to court Bengalese Finch females instead of Zebra Finches. When Zebra Finches are doubly imprinted on Zebra and Bengalese Finches, they prefer to court hybrids with visual features of both species (ten Cate 1987).

Individual Recognition

In addition to species differences, birds can distinguish among individuals by means of variations in plumage patterns, size, voice, and behavior. The extent of yellow on the bills of Bewick's Swans and the variable, harlequin color patterns on the heads of Ruddy Turnstones provide a simple basis for individual recognition (Figure 9–5). Field ornithologists learn to recognize individuals by these and more subtle differences—extent of plumage wear or a missing feather in combination with eye colors or plumage colors typical of certain age and sex classes. Budgerigar parakeets can learn to discriminate among individuals of their own species based on the slide-projected photographs (Trillmich 1976).

Outstanding in the world of birds with respect to individual variability are the large male Eurasian sandpipers call Ruffs (Figure 9–6). The genetically controlled colors of their prominent neck feathers and ear tufts vary from black to brown to buff to white and various combinations of these; no other bird species is so individually variable. Ruffs, which are openly promiscuous, assemble and display on communal courtship grounds—called leks (see Chapter 17). Each male defends its own territory and tries to attract as many females as possible and to mate with them (one at a time). Before mating, the female nibbles at the male's feathered ruff, which is flared during a courtship dance.

Figure 9–5 The harlequin face and neck patterns of Ruddy Turnstones vary among individuals. (From Ferns 1978)

In correspondence with their varied colors, Ruffs comprise two castes—or social classes. Resident males, which have dark ruffs (black, brown, or variously patterned), share their lek territories with "satellite" males, which have white ruffs. The more conspicuous, subordinate satellite males help to attract females to the territory of the more aggressive resident males and then steal copulations with visiting females when their "partner" is busy defending his territory (Hogan-Warburg 1966; Rhijn 1973; Shepard 1975). Inheritance studies of captive Ruffs by David Lank (personal communication) reveal that the caste behavior passes genetically from father to son even though sons never know their father; females nest and raise their young alone, away from the lek. The lek behaviors of both resident and satellite males are "hard-wired," not learned. Ruffs raised from the egg in captivity developed the whole lek system with both strategies and without any association with experienced birds.

Figure 9–6 The Ruff is an unusual species with individually variable male plumage. Two social classes of males act as partners on a display territory of the lek. The white-ruffed satellite males are subordinate to the variably colored dark-ruffed resident males. (Courtesy of E. and D. Hosking)

Evolution of Displays

Communication behavior reflects evolutionary heritage. In fact, we can infer the phylogeny of behavioral traits in much the same way we infer the phylogeny of morphological traits. Indeed, the two are often correlated: Closely related birds tend to have similar behaviors. Behavioral traits, for example, distinguish the main groups of pelecaniform seabirds (Van Tets 1965). Members of the Suborder Pelecani—pelicans, boobies, anhingas, and cormorants—have a similar "bowing" courtship display that is absent from the courtship behavior of tropicbirds and frigatebirds. The closely related boobies and gannets have "head-wagging" displays, whereas the closely related anhingas and cormorants have "kink-throating" and "pointing" displays. Pelicans, however, lack these displays as well as "sky pointing," "wing waving," and the "hop" used by boobies, anhingas, and cormorants (Figure 9–7).

Figure 9–7 Displays of pelicaniform birds. (A) Bowing displays: 1, quiver-bowing of Brown Booby; 2 and 3, front-bowing of Great Cormorant; 4 and 5, wing-bowing of Northern Gannet; 6 and 7, front-bowing of Red-footed Booby. (B) Pointing (left) and kink-throating (right) displays of male Anhinga. (C) Sky-pointing displays (left to right): Brown Booby, Masked Booby, Red-footed Booby, and Blue-footed Booby. (D) Wing-waving displays (left to right): Great Cormorant, Neotropic Cormorant and Pelagic Cormorant. (Adapted from Van Tets 1965)

The Neotropical manakins (Pipridae) are tiny, promiscuous forest birds which, like Ruffs, perform elaborate courtship rituals on communal lek display grounds. David Snow describes the "grunt jump" and other dances of one species, the White-bearded Manakin (Figure 9–8):

> Landing transversely on one of the uprights within a few inches of the ground, it becomes momentarily tense, with beard extended—a slowed down film shows the bird quivering as if bracing itself for the effort—then . . . it projects itself at lightning speed headfirst down to the ground, turns in the air to land on its feet for a split second, and with a peculiar grunting noise rockets up to land in a higher position than the one it has just left. The whole evolution—I called it the "grunt jump"—lasts about a third of a second. It may then do what I called its "slide-down-the-pole": with fanning wings and taking such short rapid steps that it seems to slide, it moves down the perch for a foot or so and remains near the bottom of the upright for a moment, usually to resume its to-and-fro leaping and snapping. (1976, p. 42)

Other manakin species perform circuslike, whirling cooperative displays (see Chapter 17), have highly modified wing feathers that make mechanical noise (Chapter 4), and strut up and down fallen logs. Manakin displays are rich in identifiable components that are shared among species.

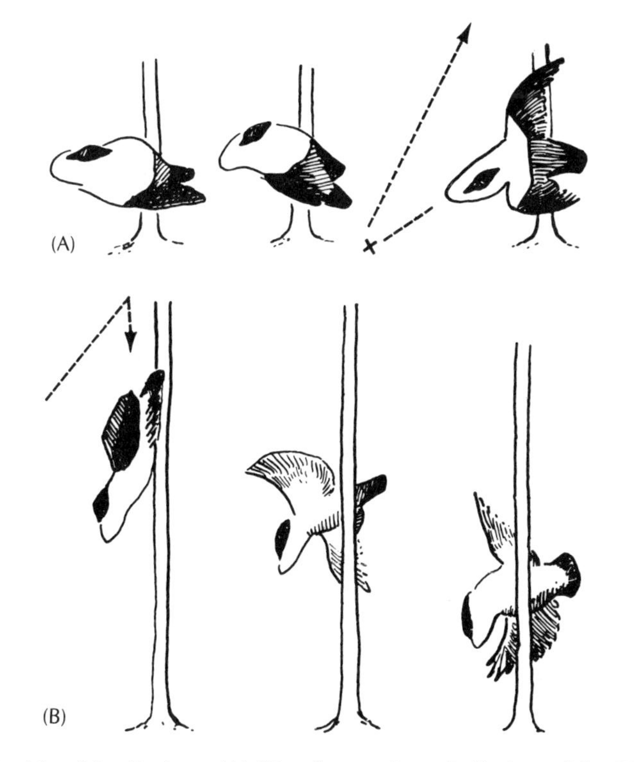

Figure 9–8 Manakin displays. (A) The "grunt-jump" display of the White-bearded Manakin. (B) The "slide-down-the-pole" display. (From Snow 1976)

Using the methods of parsimony analysis (Chapter 3), Rick Prum (1990) estimated the phylogenetic relationships among species of manakins on the basis of both the components of their courtship displays and the anatomical characters of the syrinx (see Chapter 10). The two different data sets produced trees with many points of agreement–or congruence (Figure 9–9). The basic components of manakin courtship displays were surprisingly conservative and reflected the genealogical history of the family. Such discoveries of highly conserved behavior reaffirm a premise of ethology, that is, that behavioral characters can carry phylogenetic information (Lorenz 1981).

Prum's analyses also suggest that changes in display behavior tend to precede and perhaps "drive" the evolution of enhancing plumage traits. In one example, three species—Crimson-hooded Manakin, Band-tailed Manakin, and Wire-tailed Manakin—share the primitive "wing-shivering"

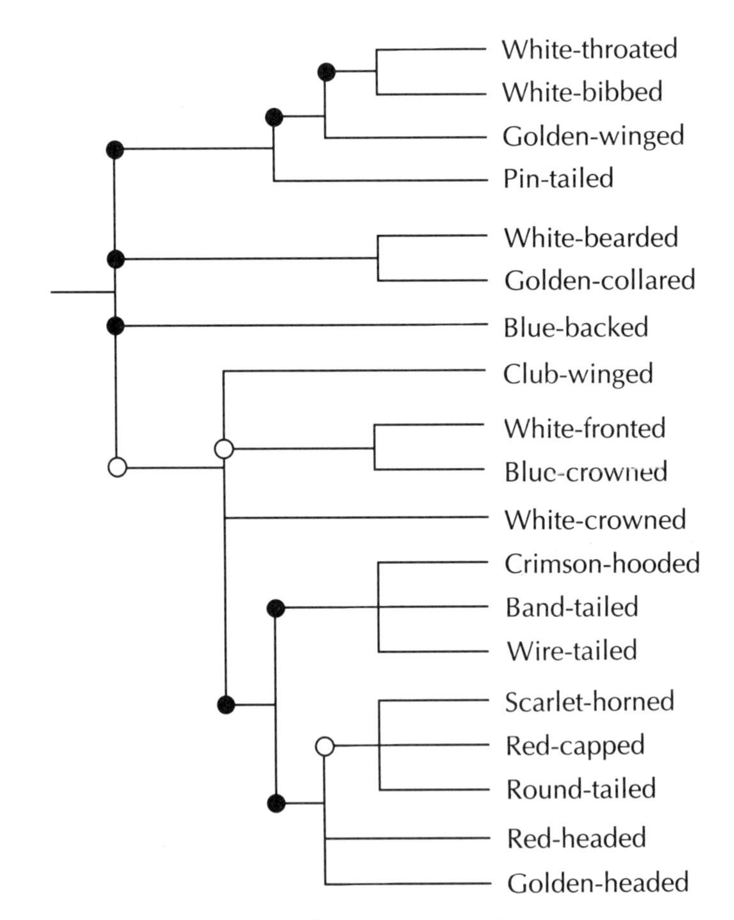

Figure 9–9 Phylogenetic relationships among manakins based on parsimony analysis of 44 display behavior characters for the 19 behaviorally best known species. The eight clades marked with a closed circle agreed with a tree based on anatomical characters. The three clades marked with an open circle did not agree with the other tree. (After Prum 1990)

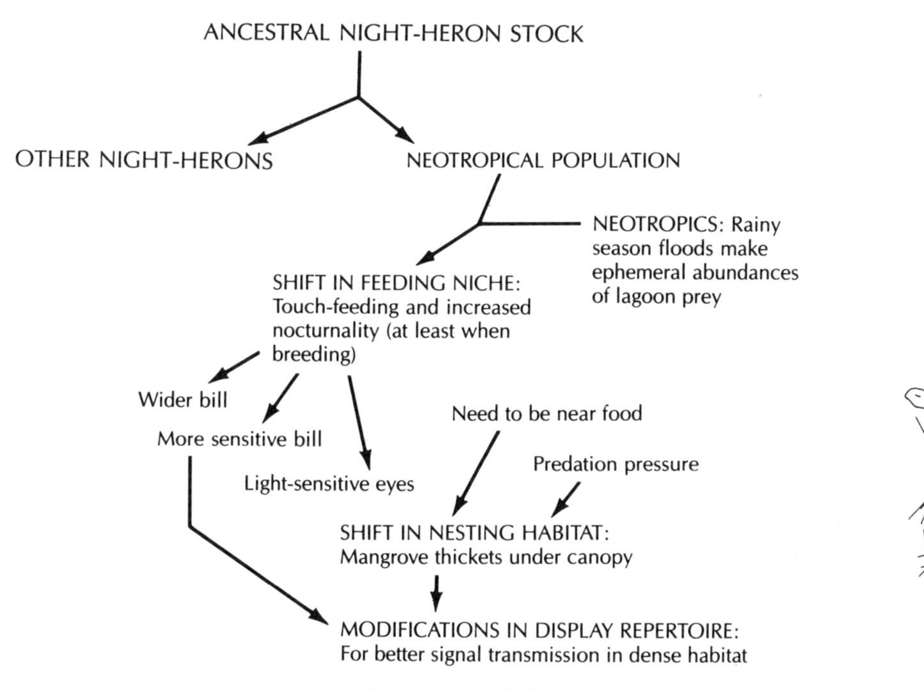

Figure 9–10 Tall-rocking display of the Boat-billed Heron (right), a species in which shifts in ecology (above) have led to major display modifications. (Adapted from Mock 1975)

and "about-face" displays. The Wire-tailed Manakin enhances these displays with lengthened, paintbrushlike, barbless extensions of its tail feathers, which swish back and forth across the face of the visiting female before mating.

Some displays may be evolutionarily conservative; others are more likely to change with ecology. In the Boat-billed Heron, a shift in feeding ecology prompted the transformation of displays (Mock 1975). The Boat-billed Heron is a Neotropical night-heron with a wide bill evolved for touch-feeding in seasonally flooded lagoons. Its repertoire of social displays differs substantially from that of other herons. For example, the displays emphasize sounds made with the bill. The heron also uses dramatic visual displays such as the tall-rocking display (Figure 9–10). A shift in the Boat-billed Heron's feeding niche fostered the evolution of a wider, more sensitive bill and more sensitive eyes for nocturnal feeding. The ecological and accompanying morphological transformations then led to radical modifications of displays for better communication in the dense nesting habitat where visibility was poor.

We see, therefore, that avian displays involve various visual and vocal signals and coordinated behaviors that reflect the evolutionary history of a species. Symbolic displays often represent unrelated explicit behaviors transformed into symbolic ones by the process of evolutionary ritualization.

Ritualized Behavior

Ritualization is the process of the evolution of signals and displays from nonsignal movements. Ritualized feeding movements, for example, are incorporated into the courtship displays of quails, pheasants, and peacocks (Schenkel 1956; Williams et al. 1968). In these "tidbitting" displays, the male bows before the female, spreads its wings and tail to various degrees, and in some species, gives a food call (Figure 9–11A). Male Northern Bobwhites feed their mates as part of the tidbitting display just before copulation. Ring-necked Pheasants and Domestic Chickens display by manipulation of food and mock pecking, but do not feed their mates. The Himalayan Monal, a pheasant, bows low before the female and pecks vigorously at the ground amidst full sexual display of its wings, tail, and head feathers. In the most simplified and ritualized form of the tidbitting display, the Common Peafowl merely points its bill at the ground. The origin of the display movement would not be apparent without comparisons among related species.

Virtually any nonsignal behavioral pattern can evolve into a ritualized display element with particular functions. Feather positions initially used to control heat loss from the body are a common source of display features (Morris 1956). Most displays, however, evolved from incomplete locomotor movements—called intention movements—such as the initial postures associated with leaping into the air to fly or flexing the head and bill to peck. The head-throw display of courting male Common Goldeneye ducks seems to have evolved from the locomotory movement of leaping out of the water (Figure 9–11B).

Seemingly inappropriate behaviors, such as beak wiping, feather preening, and drinking, often appear in aggressive situations (Tinbergen 1952; Ziegler 1964). In the middle of a fight, a Blue Jay may suddenly wipe its bill several times as if it had a compelling itch. Alternatively, a bird may redirect its actions: Instead of attacking a mate, it may attack another individual. Herring Gulls that are facing combat redirect their pecking at the ground. Such redirected behaviors may be incorporated into the formal display repertoire of a species. Preening movements, for example, have become a ritualized element of the courtship displays of terns and ducks (Lorenz 1941; Van Iersel and Bol 1958).

The traditional view of ritualized displays is that ritualization increases the efficiency and clarity of information transfer (Cullen 1966; Zahavi 1977). To this end, ritualization also increases the repeated precision of a display. The 1.3-second head-throw display of the Common Goldeneye, for example, varies with a standard deviation of only 0.08 second (Dane et al. 1959). (Standard deviation is a measure of how much variation there is above and below an average.) Clarity rather than ambiguity would seem to be the overriding value of stereotyped displays. Clarity may be enhanced further by the increased discreteness or contrast of the display. The elaborate cartwheel display of several male manakins is a

(A)

(B)

Figure 9–11 Two ritualized displays. (A) Tidbitting display of a peacock (male Common Peafowl). (B) Head-throw display of a Common Goldeneye. (Courtesy of the San Diego Zoological Society)

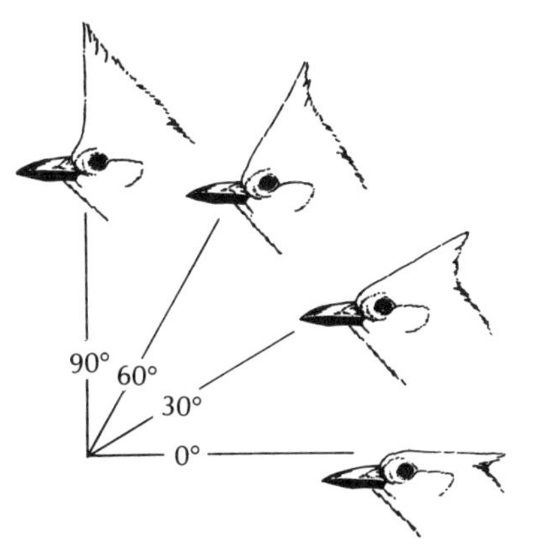

Figure 9–12 The positions of the crest of a Steller's Jay signal the likelihood of attack (high crest) or retreat (low crest). (From Brown 1964a)

well-delineated, all-or-nothing courtship display with clear intention (Chapter 17).

Graded or variable displays convey information about intensity of motivation and the probability of a sender's subsequent actions. The high-crest positions assumed by a defensive Steller's Jay indicate that it will probably attack rather than flee its opponent (Figure 9–12). Variations in stereotyped displays do not necessarily make for confusion; they may enhance communication once a reliable standard has been established (Zahavi 1980).

Deciphering Displays

Deciphering the information transmitted by a display remains one of the greatest challenges in the study of bird behavior. Ornithologists can only guess at the message of a display from correlations between preceding and succeeding actions of sender and receiver. We now recognize not only that a single display may contain several messages but also that the messages themselves may vary with the context of the display (W.J. Smith 1969, 1977).

Douglas Mock's studies of the displays of Great Blue Herons and Great Egrets illustrate the ways ritualized displays are used to communicate with

TABLE 9–1
Contexts and messages of Great Blue Heron displays

	Displays														
	Stretch	Snap	Wing preen	Circle flight	Landing call	Twig shake	Crest raising	Fluffed neck	Arched neck	Forward	Supplanting	Bill duel	Bill clappering	Tall alert	Static-optic
I. Uses/contexts															
External disturbances						X	X	X						X	
Nest defense						X	X	X	X	X				X	X
Male advertisement	X	X	X	X		X	X								X
Female–female encounters							X		X	X	X			X	
Greetings at nest[a]	X				X		X	X	X				X		
Intrapair appeasement	X				X								X		
Intrapair aggression							X					X			
II. Messages (Smith 1969)															
Identification	X	X	X	X	X	X	X	X	X	X	X	X	X	X	X
Probability	X	X	X	X	X	X	X	X	X	X	X	X	X	X	X
General set							X								
Locomotion					X	X	X				X				
Attack							X	X	X	X	X	X			
Escape							X								
Nonagonistic	X												X		
Association	X	X	X	X									X		
Bond-limited	X		X		X								X		
Play															
Copulation															
Frustration															

a. Includes greeting ceremony, nest relief ceremony, and stick transfer ceremony.

From Mock 1976.

mates or potential mates (Mock 1976, 1978, 1980). The two heron species nest side by side in open habitats and communicate primarily by means of visual displays. Each of the 15 displays in the repertoire of the Great Blue Heron has its particular set of contexts and conveys particular messages (Table 9–1 and Figure 9–13). Great Blue Herons increase the complexity of a signal by adding vocalizations, by erecting particular plumes, and by coding information into the variability of the display itself, in the same way that the Steller's Jay reveals the probability of attack with the position of its crest. Plume positions, body positions, duration of the display, and repetition of movements are some of the display variables that provide information.

The highly stereotyped stretch display is a long, conspicuous display that exhibits a heron's bright bill colors and chestnut wing linings to the fullest. The erected neck plumes enhance the visual impact of the neck motions during this display. The stretch display, which signals withdrawal or submission, occurs in several contexts: in male advertisement; as part of the nest relief ceremony, when mates relieve each other of incubation duties; and when the female sends her mate to collect another stick for the nest, a form of pair-bonding cooperation that Mock refers to as intrapair appeasement.

The forward display of the Great Blue Heron (see Figure 9–13B) is a variable display that projects the probability of attack, particularly in nest defense and female-to-female encounters. Males refrain from using the forward display in the early stages of pair formation, when attraction of a potential mate is the goal, but use it more often as courtship proceeds through phases of critical assessment and possible rejection. The use of this display subsides once the pair bond is firmly established.

The Great Egret achieves similar ends by different means. The 16 displays in its repertoire are each less variable in form than those of the Great

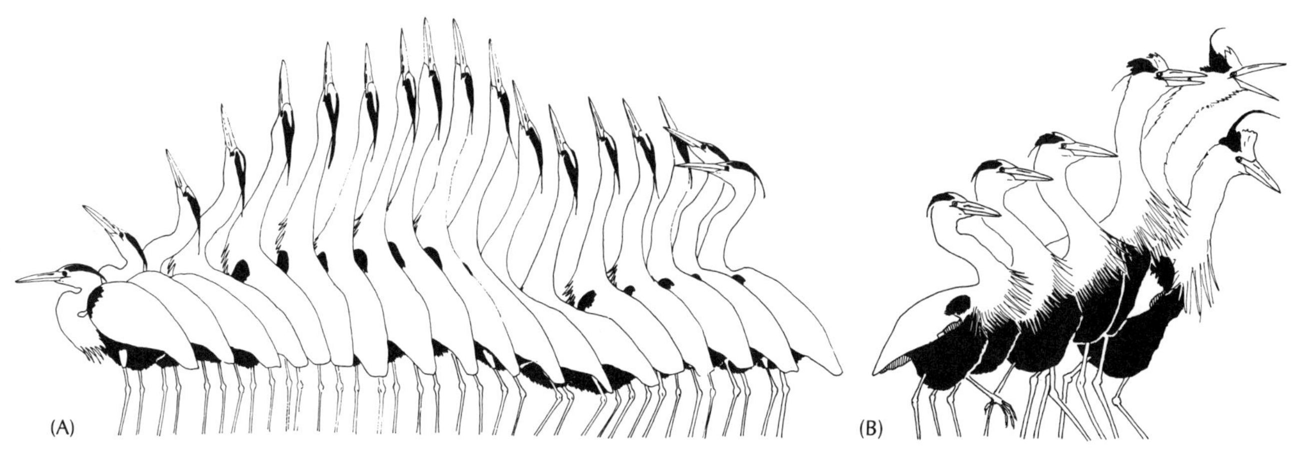

Figure 9–13 (A) Stretch display and (B) forward display of the Great Blue Heron. These sequences are drawn directly from movie frames of a filmed display. (From Mock 1976)

Blue Heron and are less likely to be accompanied by acoustical signals. Whereas the Great Blue Heron uses the stretch display in a variety of contexts, the Great Egret uses it in only one—spontaneous advertisement by unpaired males. Instead of varying the presentation of one display, the Great Egret increases its power of communication by varying the sequence in which different displays are presented.

Agonistic Behavior

When two birds interact, each has selfish purposes that can foster either hostility or cooperation. Birds can manipulate one another to individual or sometimes mutual benefit. Inherent in all social interactions governed by rules is the threat of cheating by those that would take advantage of the existing system. For many years, students of bird behavior have tended to assume the morality of truthfulness in their interpretations. Now it appears that avian social communication may not be as straightforward and honest as we once supposed. Individual birds serve their own interests in many ways. The nature of communication between rivals as well as between partners, therefore, invites our attention. We begin this discussion with a traditional view and finish with the provocative idea of avian deceit.

The competitive encounters between rivals—complex mixtures of aggression (attack, threaten) and escape (submit, flee)—are called agonistic behavior. When birds fight over something—mates, food, or territory—they usually avoid direct contact and risk of injury by using threat and appeasement displays.

Threat displays, which emphasize the bill and wings as weapons, herald a real attack if the issue is not resolved quickly. Appeasement or submission displays signal the opposite intent, a willingness to yield on the point at issue, a signal that defuses the conflict and thereby protects the yielding individual from direct attack (Figure 9–14). Often the submissive bird turns its head and bill away from a threatening rival, a movement that reduces the level of provocation and avoids a physical attack. An appeasing avocet, for example, hides its long bill beneath its back feathers and adopts a sleeping posture. Other species fluff their feathers, in contrast to the sleeked postures associated with threat displays.

Communication of aggressive intent or submission is a central function of the social displays of birds. Even courtship usually starts with aggression by the male toward the female. If the female stays, the actions of the male shift from hostility to appeasement, subordination, solicitation, and ultimately establishment of a pair bond with regular contact and copulation. The courtship of Common Black-headed Gulls illustrates this process (Moynihan 1955). Prior to the arrival of the females, male Common Black-headed Gulls gather in large areas near the nesting colony called "clubs," where each bird establishes a small, temporary pairing territory. Rival males avoid physical battle and risk of injury by ritualized aggression. As they await the arrival of a potential mate, they threaten each other with the upright-threat display, with the combination long-call and oblique display, and with the forward display attack posture (Figure 9–15). The contrast between the upright-threat display and normal posture defines unambiguously whether the recipient is welcome. If stylized threats do not succeed, physical attacks may ensue.

Figure 9–14 Threat and submissive postures. (A) Threat display of Great Tit. (B) Great Tit (right) about to attack a Blue Tit (left), which is in submissive posture. (Courtesy of E. Hosking)

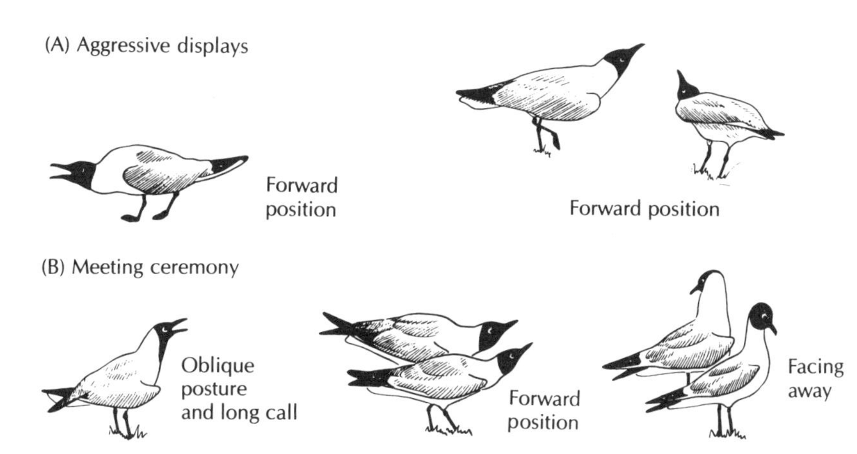

Figure 9–15 Displays of the Common Black-headed Gull. (A) Aggressive displays using forward postures. (B) Greeting ceremony display, including oblique display with long-call display (top), forward display (center), and facing-away display (bottom). (From Tinbergen 1959)

Females visit pairing territories in the club to find a mate. The male greets an approaching female as a potential rival with the aggressive oblique display and long calls. Instead of fleeing or challenging him as would a rival male, the female Common Black-headed Gull stretches her neck upward and faces away (facing-away display), revealing her sex and her potential as a mate. In response, the aggressive male reduces the severity of his threat by redirecting the oblique display to the side. The female keeps returning to a selected male and stays longer each time. Ritualized appeasement replaces ritualized aggression. The potential mates engage in mutual displays, such as facing away. Gradually, the female moves closer and begs for food, which the male regurgitates onto the ground in front of her. Copulation follows. The male then deserts his pairing territory and assists the female in nesting and rearing of young.

What information do Common Black-headed Gulls actually communicate by such displays? The traditional view is that threat and appeasement postures signal the probability that an individual will attack or escape. Reduced ambiguity through ritualization of agonistic displays should lessen the possibility that a bird will misread its opponent's intentions (Cullen 1966). Agonistic displays, therefore, should accurately communicate the probability of attack or escape and thus reliably inform a receiver of what will follow.

In fact, varied threat displays, such as the upright- and the forward-threat displays of the Common Black-headed Gull, convey different probabilities of subsequent attack, but these probabilities may not be very high. Critical studies reveal that attacks do not reliably follow even the highest ranked threat displays (Caryl 1979). For example, the most intense threat displays of the Blue Tit and the Great Skua are followed by attacks only about half of the time (Stokes 1960; Andersson 1976) (Figure 9–16). Furthermore, the degree of the reactions that threatened individ-

Figure 9–16 Threat displays of the Great Skua. (A) Bend posture with long call and wing raising, which indicates that the skua will stay and will probably attack. (B) Bend posture, which indicates that the skua will stay but is less likely to attack than in A. (C) Relaxed posture with neck withdrawn. (Adapted from Andersson 1976)

uals show is not directly related to the degree of threat in the displays, as we would expect them to be if "probability of attack" were the message being conveyed.

Information about the probability of escape, however, is a more trustworthy aspect of agonistic displays. Submissive or appeasement displays predict escape more reliably than threat displays predict attack. Escapes almost always follow certain postures, such as the crest-erect-facing-rival display of the Blue Tit. Such instances suggest that we might view complex sequences of agonistic displays more productively as contests between individuals, that is, as games of bluff. It appears that escape signaling is straightforward and that threat displays tend to be bluffs. It follows that the submissive bird determines the outcome in many ritualized contests.

The Game of Communication

Is avian communication fundamentally different from human communication? To what degree do birds use their repertoires of visual and vocal signals to manipulate other birds in conscious or unconscious ways? In later chapters, we shall present cases in which birds apparently have evolved convergent plumage color patterns, manipulating other birds by mimicking them. South Pacific orioles thereby gain access to fruit trees that are controlled by aggressive friarbirds; several species of tanagers in the Andes flock together with reduced interspecific strife (see Chapter 14). Cases of social mimicry illustrate the evolutionary games that guide the development of visual signals. Still more striking examples of trickery in avian communication are found in brood parasites that mimic their hosts (see Chapter 19).

The view that avian communication may have elements of deceit or ambiguity challenges the traditional constructs that the evolution of displays is toward a reliable transfer of truthful information (Krebs and Dawkins 1984). In fact, aggressive displays do not carry a reliable message, as we have just seen. Rather, they signal some possibility of continuing the encounter and may be only a bluff. In the game of communication, senders may try to manipulate the attention and muscle power of their observers to some selfish advantage; other individuals are considered to be pawns to be used if possible. For example, the unmated male not only announces his availability with displays and song but also does his best to get prospective females interested enough to move into his territory. Listeners try to decipher the true intentions of a sender and to respond accordingly in their own best interests. If the female Common Black-headed Gull concludes that, despite her submissive response to a male on a club territory, she will be attacked rather than tolerated, then she will leave and perhaps try again, or she may move to an alternative prospect.

Sexual Selection

Striking sexual differences in plumage (and size) are typical of many birds, including the Ruffs and manakins discussed earlier. Darwin concluded that exaggerated sexual differences like the tail of a peacock evolve as a result of what he called sexual selection, namely, contests among males for mates and female preferences for particular males. Sexual selection usually promotes elaborations of male displays. Males rather than females tend to compete for mates, and their reproductive success varies more than that of females, especially in nonmonogamous mating systems (see Chapter 17) (Payne 1979). As potential male reproductive success increases, so does the value of the characteristics—such as large size, fancy plumage, intricate songs, and striking displays—that are responsible for the success. Darwin's insights into the evolutionary role of sexual selection are now largely confirmed, but the precise nature of the process invites more study. The effects of competition among males, female choice, and resources other than mates intertwine in ways that only well-designed field experiments can tease apart.

The Red-winged Blackbird, a North American species that is marked by large variation in male sexual success and by striking sexual dimorphism, has become a focus of these studies (Searcy and Yasukawa 1983). Male Red-winged Blackbirds are jet black with bright red and yellow shoulder patches—or "epaulettes"; females are smaller and plainer, streaked brown.

Males establish and defend large territories in marshes throughout North America. Those with the best territories attract harems of up to 15 females. Female Red-winged Blackbirds consistently choose high-quality territories rather than particular males. Water level, nest cover, and food abundance determine territory quality and, therefore, a female's nesting success. Male age is a secondary criterion for females in some parts of the country. Females prefer older, more experienced males in Indiana, where males help feed the young, but not in Washington, where males do not help feed the young. Thus, sexual selection among male Red-winged Blackbirds operates through competition for the best territories (Figure 9–17).

Figure 9–17 A territorial male Red-winged Blackbird in aggressive display posture. (From Orians and Christman 1968)

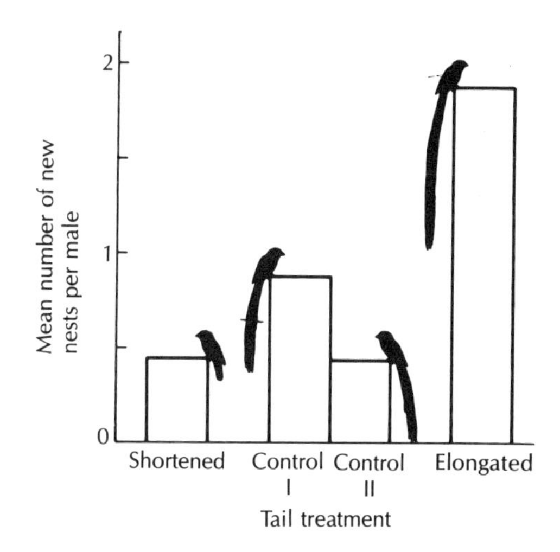

Figure 9–18 Female Long-tailed Widowbirds prefer males with long tails. In this experiment, the tails of some males were shortened and the tails of others were extended. Control I males had their tails cut off and then restored; control II males had unaltered tails. The ability of males to attract females to their territories directly reflected their tail length. (Adapted from Andersson 1982)

Experiments have demonstrated that the male's red epaulettes are essential in the male–male competition. Males on which the red is dyed black suffer more frequent challenges and usually lose their territories, although those that are not challenged still attract mates. We can conclude that the epaulettes have evolved in relation to male competition, not to female choice.

An extraordinary convergent counterpart of the Red-winged Blackbird lives in East Africa. The male Long-tailed Widowbird, another polygynous species that is jet black with bright red epaulettes, defends marshland territories in the highlands of Kenya. Sexual selection seems to have gone a step or two further in the enhancement of the male display that is characteristic of this species. True to its name, the Long-tailed Widowbird has an enormous tail, up to half a meter long. Like female Red-winged Blackbirds, female widowbirds are small, brown-streaked birds. Sexual selection favors the long tail of the male because it enables females to spot them from afar (Andersson 1982). Humans can spot a displaying Long-tailed Widowbird from over a kilometer away.

In a particularly elegant experiment, Malte Andersson increased the tail lengths of some male widowbirds by 25 centimeters and decreased the tail lengths of others by that same amount (Figure 9–18). Males with "super" tails attracted more females to nest on their territories than males with shorter tails or tails of normal length. These experimental manipulations did not, however, affect a male's ability to hold his territory. Female preference, rather than male competition, was the source of sexual selection.

Sexual Selection and the Extrinsic Displays of Bowerbirds

Bowerbirds are large songbirds found only in New Guinea and Australia. They construct and decorate architecturally elaborate stick or grass structures—called bowers—that provide platforms for both courtship and copulation (Marshall 1954; Diamond 1986; Borgia 1986) (Figure 9–19A). Male bowerbirds build bowers of two general kinds: maypole bowers and avenue bowers. Five species of bowerbirds build maypole bowers, and eight species build avenue bowers. Phylogenetic analyses of DNA base pair sequences indicate that maypole bower builders are nearest relatives of one another; the same is true for the avenue bower builders (Kusmierski et al. 1993).

Maypole bowers consist of sticks built around a central sapling—or maypole; avenue bowers are walled structures placed on the south side of a display court. Bowers of both kinds are decorated with brightly colored objects. The decorations are as extraordinary as the bower structures themselves. Some species paint the walls of their bowers with fruit pulp, charcoal, or shredded dry grass mixed with saliva. Other species decorate their bowers with mosses, living orchids, fresh leaves turned upside down, or colorful fruits. They restore the original arrangements after external disturbance, discard unsuitable items placed near their bowers by experimenters, and replace wilted flowers or leaves with fresh ones daily. Males often steal each other's decorations. The Spotted Bowerbird of Australia is notorious for household and camp pilferage—of scissors, knives, silverware, coins, jewelry, car keys, and even a glass eye, snatched from a man's bedside.

Ornithologists initially considered bowers to be ritualized courtship nests, but the origin of bowers probably had little to do with nests (Borgia et al. 1985). Rather, bowers have been added to male display courts as markers of male social status and ability.

The visual stimuli of bowerbird courtship displays have somehow been transferred from plumage color patterns to the bowers. Thomas Gilliard

Figure 9–19 (A) Bowers of bowerbirds include simple forest clearings with ornaments on the ground (1); a mat of lichens decorated with snail shells (2); a maypole built of sticks about a central sapling or fern and surrounded by a raised, ornamented court (3–5); and a decorated avenue built with varying complexity of stick walls opening onto a platform (6–9). (B) Breeding behavior of the Satin Bowerbird. The male builds an avenue bower of sticks (top left), at which it courts visiting females (top center). Females judge bower quality and then may solicit copulation by crouching in the bower (top right). Other males destroy the bower in the absence of the owner (bottom left) and may try to interrupt copulation (bottom center). Mated females lay their eggs and rear their young without male help at nest sites away from the bower (bottom right). (From Borgia 1986, with permission of Scientific American)

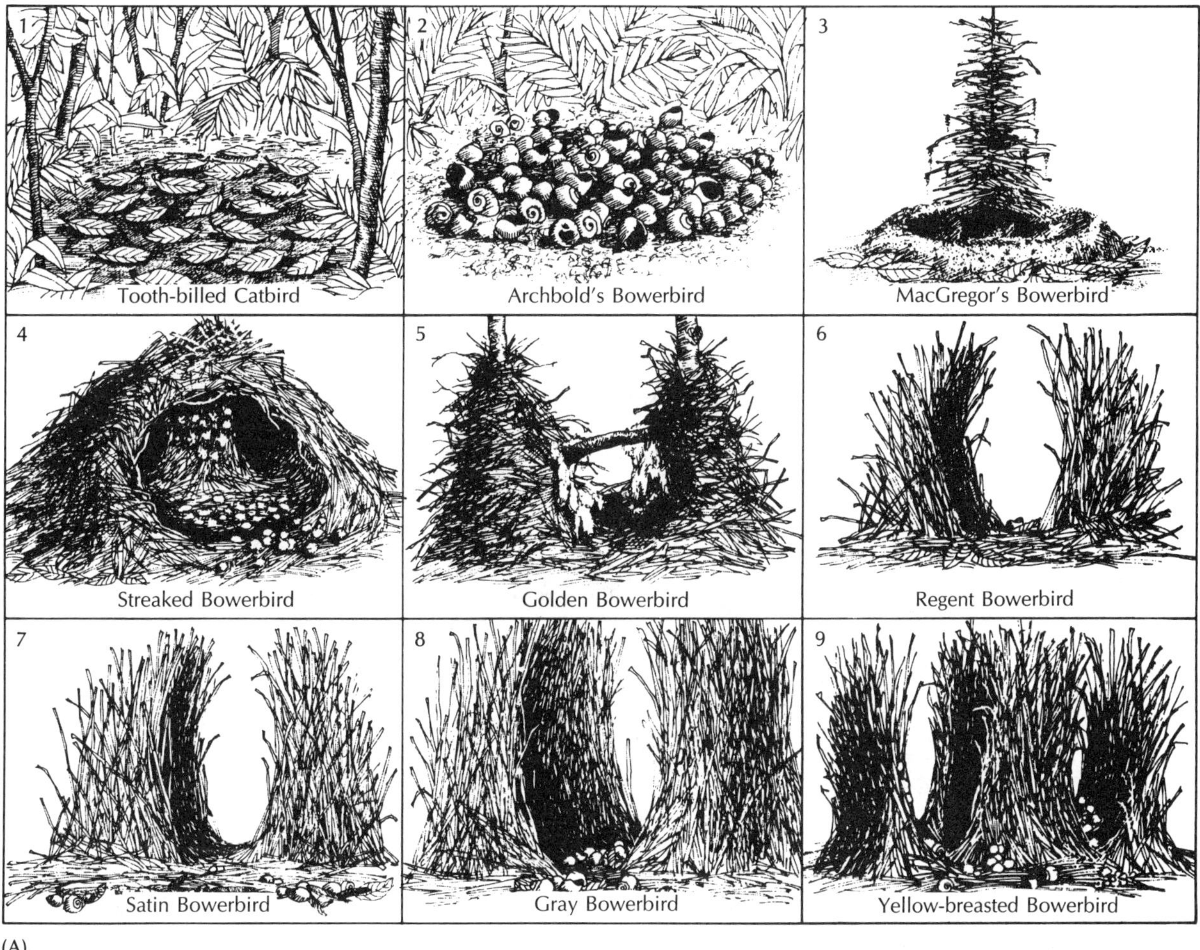
1
Tooth-billed Catbird
2
Archbold's Bowerbird
3
MacGregor's Bowerbird
4
Streaked Bowerbird
5
Golden Bowerbird
6
Regent Bowerbird
7
Satin Bowerbird
8
Gray Bowerbird
9
Yellow-breasted Bowerbird

(A)

(B)

(1956, 1969), an adventurous ornithologist who pioneered the study of the birds of New Guinea, noticed an apparent trade-off, or transfer of function, between plumage elaborations and bower displays. Modestly colored bowerbirds tend to have more elaborate bowers than do brightly colored species. The Streaked Bowerbird, which has only a short orange crest, builds an elaborate, well-decorated hut around the maypole, whereas the related MacGregor's Bowerbird, which has a long and conspicuous orange crest, builds only a simple column of sticks without much decoration.

The Satin Bowerbird, which has brilliant blue eyes, builds a large avenue bower and decorates the structure with anything blue it can find (Figure 9–19B). Originally confined to natural objects including parrot feathers and flowers, this species now exploits human trash. One bower was decorated with glass fragments, patterned crockery, rags, rubber, paper, bus tickets, candy wrappers, fragments of a blue piano caster, a child's blue mug, a toothbrush, hair ribbons, a blue-bordered handkerchief, and blue bags from domestic laundries (Marshall 1954).

The construction of a bower and provisioning it with fresh decorations require experience and considerable effort. By monitoring the activity at bowers with video cameras triggered by infrared sensors, Gerald Borgia and Mauvis Gore (1986) discovered that male Satin Bowerbirds tear apart one another's bowers, if they can, and steal prized decorations of rival males. Decorations that are rare in the environment, such as blue parrot feathers in northern Queensland, are particularly prized and subject to theft. Dominant males, better able to protect their bowers, have more time to visit and degrade the bowers of nearby competing males, which must constantly rebuild and struggle to keep up a minimally acceptable bower. The ability and status of a male are directly reflected in the quality of its bower.

Borgia's video cameras also recorded the bower visits and preferences of female Satin Bowerbirds for well-made and well-decorated bowers (Borgia 1985a, b; Borgia et al. 1985). A female Satin Bowerbird visits an average of 3.6 bowers in a local area before mating with a particular male. The females clearly prefer well-made bowers with special decorations. Five of 22 males accounted for 56 percent of the 212 copulations recorded in 1981. These males had the most blue parrot feathers, snail shells, and leaves as decorations, as well as the best bower structures, judged in terms of symmetry, stick size, stick density, and quality of construction. Males whose leaf decorations were experimentally removed from their bowers obtained fewer matings than did control males.

Good Genes versus Fashion Icons

Exactly why females prefer larger males or fancier displays has not been obvious, but there are two sets of working hypotheses: the "good genes" hypotheses and the "arbitrary choice" or fashion icon hypotheses.

Good genes hypotheses propose that exaggerated male plumage and displays truthfully signal genetic or physiological superiority. Females should recognize superior males and select them to sire offspring. What aspects of genetic or physiological superiority might exaggerated courtship displays serve to index? Amotz Zahavi (1975) suggested that the enormous tail of a peacock might actually be a handicap and similarly that the bright colors announcing a male's presence to potential rivals or mates would also attract predators. Males that survived to display such handicaps had superior stamina or abilities to escape predators. Evolution would tend to favor bigger and bolder badges of superiority, if females preferred to mate with the males that bore them. In the case of House Finches (Box 4–4), females prefer brightly colored males, which have better survival rates and are better family providers.

Another version of a good genes hypothesis asserts that ornamented plumage provides an index to a male's state of health, particularly its resistance to pathogens and parasites (Hamilton and Zuk 1982). Females could diagnose disease-prone males by the lower quality of their display plumage or reduced display stamina. Consistent with this hypothesis, species with brightly colored males tend to carry more parasites than species with dull-colored males, especially when they reside year-round in the disease-rich Tropics (Zuk 1991; Pruett-Jones et al. 1991).

More convincing are the results of experiments in which parasites actually affect the quality of male ornaments that serve as the basis for female choice in Barn Swallows (Box 9–2) and in Red Junglefowl, which are the ancestors of Domestic Chickens (Zuk et al. 1990a, b; Ligon et al. 1990). Marlene Zuk and her colleagues first established that hens of the

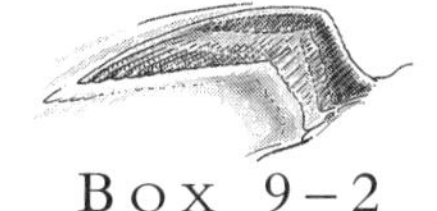

Box 9–2
Sexual selection has connected parasites and long swallow tails

The long, forked tail of the familiar Barn Swallow may be a male's most important ornament (Møller 1990, 1991; Smith and Montgomerie 1991). Male swallows attract females by singing and displaying their outermost tail feathers, which in Denmark are 16 percent larger than in females. Tail length serves as an index to a male's load of ectoparasites, particularly bloodsucking mites, which reduce weight, tail feather length, survival of nestlings, and reuse of nests. Unmated males have more parasites than mated males; and mated pairs exhibit similar parasite loads—that is, males and females with the lowest parasite loads tend to pair with each other. Female swallows mate more readily with males that have longer tails. Often these are older males because tail length increases with age, but females still prefer same-aged males with the longest tails. Once paired, females also prefer longer-tailed, unmated males as partners for extra-pair copulations. Balancing enhanced attractiveness, however, is a reduced ability of males with long tail streamers to guard their mates from other males (Smith et al. 1991). Thus, long-tailed male Barn Swallows are cuckolded more often than their less attractive neighbors.

Red Junglefowl mated quickly with roosters bearing large, fleshy, red combs on their heads. Comb size is strongly affected by the level of blood testosterone, which in turn reflects the individual's physical condition. Zuk and her colleagues then infected some of the junglefowl with an intestinal nematode worm, which significantly reduced comb size. Hens preferred roosters without worms over those that were infected, and thus seem to use comb size as an index to the state of health of a potential mate.

Paralleling the good genes hypotheses for female selection of exaggerated male ornamentation is the concept of arbitrary choice and runaway selection (Fisher 1930; Lande 1981), which stresses a more arbitrary, sometimes uncontrolled, process of ornament and display elaboration based on intrinsic female preferences for fancier males—dubbed "fashion icons" (Ridley 1992). Assessment of male quality via a good genes hypothesis was not part of the original concept, but the two hypotheses potentially overlap and reinforce each other. Once the process of favoring slightly more elaborate displays or plumages begins, it may go to extremes, as in the case of the bizarre plumage displays of birds-of-paradise (Figure 9–20). Genetic models of runaway selection require the genes

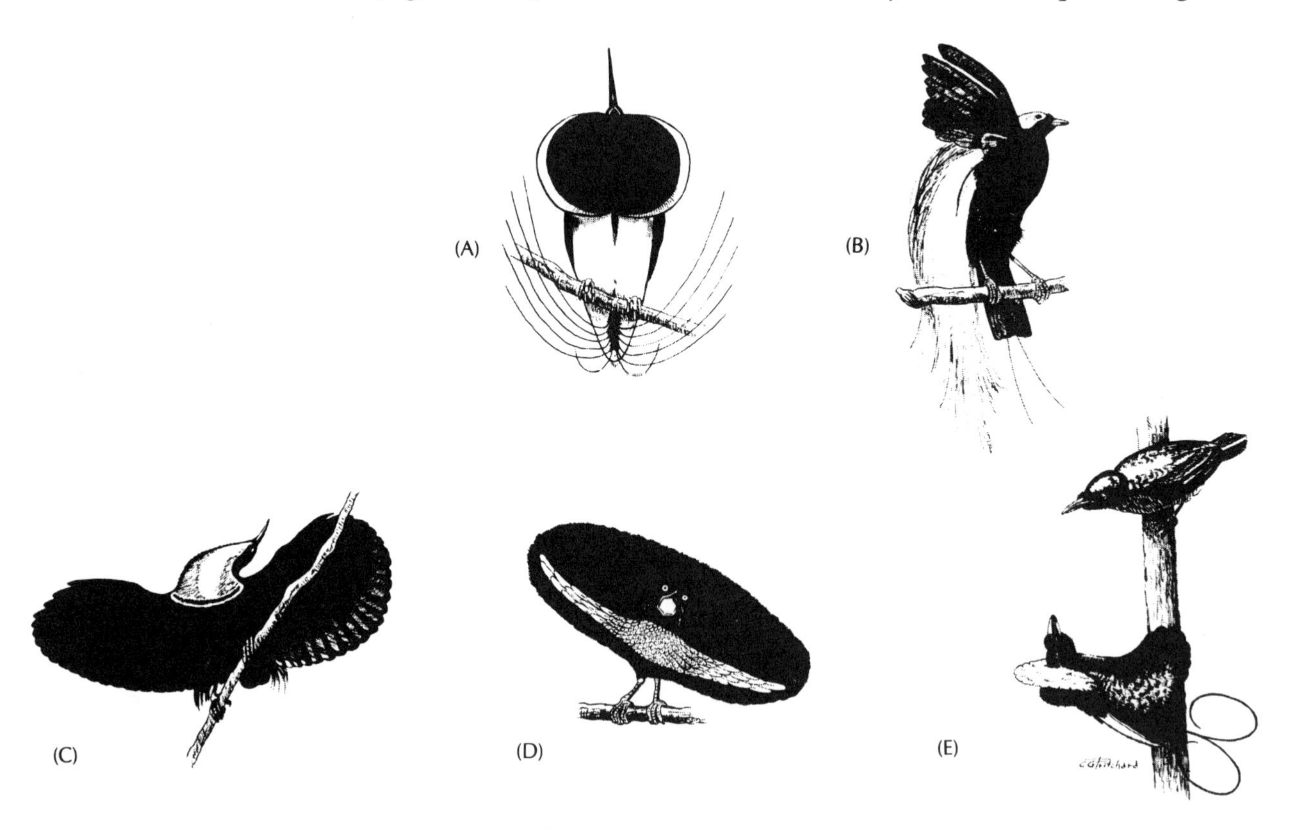

Figure 9–20 Elaborate plumages and displays of male birds-of-paradise: (A) Twelve-wired Bird-of-Paradise; (B) Lesser Bird-of-Paradise; (C) Magnificent Riflebird; (D) Superb Bird-of-Paradise; (E) Magnificent Bird-of-Paradise. (Adapted from Johnsgard 1967)

controlling male traits to be linked with the genes controlling female preference for those traits. It is more likely, of course, that a process of imprinting innately fosters new and fancier ornamentation.

Recall that Zebra Finches imprint on the color patterns of their parents and that this affects their later selection of courtship targets (see beginning of chapter). Two subtle aspects of this process suggest how imprinting could aid runaway sexual selection (ten Cate and Bateson 1988). First, males with artificially brightened or contrasting color patterns prove to be more attractive models to the young finches than are normally colored males. Second, young finches show a slight imprinting preference for novel or unfamiliar stimuli, including foster parents of a different species. Other studies show that imprinting biases for the unfamiliar reduce the probability of inbreeding with siblings. If they prove to be general phenomena, imprinting biases for bright, unusual male color patterns would give those novelties an advantage and help to catalyze the process of sexual selection leading initially to increasingly different ornamentation and ultimately to the evolution of new species from small, isolated populations.

Summary

Birds communicate with one another via displays, which are acts specialized to exchange information. Avian displays serve primarily as a means of identification and to communicate locomotory intentions (attack, escape, move, or stay still) or other intentions such as a desire to mate, play, or claim ownership. Visual displays function in concert with plumage color and plumage elaborations. Differences in color patterns serve as the basis for species recognition, which is mediated in many species by visual imprinting of young birds on their parents. Variations in plumage color also allow the recognition of individuals. In certain cases, such as Ruffs, color variations are correlated with inherited differences in social behavior.

Displays evolve from nonsignal behavior patterns through a process called ritualization, which leads to increased uniformity of performance as well as modification of behavior patterns. Avian display repertoires are used in a variety of contexts or only in specific contexts. The conservation of basic display elements among related species indicates that display behaviors reflect evolutionary relationships, as illustrated by two groups of birds, pelican allies and manakins. Agonistic displays inform opponents about probabilities of attack or escape, but birds may also bluff, that is, they may not always be truthful with one another.

The elaboration of display ornamentation reflects a process of sexual selection based on the advantages of ornaments in direct competition among males for mates, or as they relate to female choice. In bowerbirds, which build external displays of elaborate accumulations of colorful objects, females prefer bowers that are well built and well decorated. Female preferences for exaggerated ornamentation may reflect the ways in which ornaments signal male genetic or physiological superiority. Experiments with Barn Swallows and Red Junglefowl support the hypothesis that parasites reduce ornament quality and that females prefer the healthiest parasite-resistant males with the best ornaments. The evolution of exaggerated ornamentation may be mediated by imprinting biases for novel, brighter visual stimuli.

Further Readings

Alcock, J. 1993. Animal Behavior, 5th ed. Sunderland, Mass.: Sinauer Associates. *An excellent text with many bird examples.*

Butcher, G.S., and S. Rohwer. 1989. The evolution of conspicuous and distinctive coloration for communication in birds. Current Ornithology 6: 51–108. *A detailed review of the alternative hypotheses.*

Lorenz, K. 1965. Evolution and Modification of Behavior. Chicago: University of Chicago Press. *A classic.*

Loye, J.E., and M. Zuk, Eds. 1991. Bird–Parasite Interactions. New York: Oxford University Press. *The starting point of a rapidly growing new field, including important papers about the relationship between parasites and sexual selection.*

Miller, E.H. 1988. Description of bird behavior for comparative purposes. Current Ornithology 5: 347–394. *A forward-looking review of ways to improve the methods of comparative behavior analysis.*

Pruett-Jones, S.G., M.A. Pruett-Jones, and H.I. Jones. 1991. Parasites and sexual selection in a New Guinea avifauna. Current Ornithology 8: 213–245. *A detailed comparative analysis and discussion of the limitations of the comparative approach.*

C H A P T E R 1 0

Vocal Communication

Animals in general communicate by means of visual, acoustical, tactile, chemical, and even electrical signals. Birds typically use only two of these modes, visual and acoustical—or vocal—signals. Complementing the use of visual displays to mediate social interactions are rich vocabularies of sounds. As noted in Chapter 1, birds have the greatest sound-producing capabilities of all vertebrates. Vocalizations serve birds especially well for communication over long distances, at night, and in dense cover. One key message—the species identity of the sender—pervades both visual and vocal displays of birds. Then follows individual identity, which mediates social status, pair bonds, and family relationships.

The scientific literature on bird vocalizations began almost 400 years ago with the observation by Ulyssis Aldrovandus that ducks and chickens called even after their heads were chopped off; the source of the vocalizations was apparently sited in the body and not the head. The source of avian vocal abilities is, in fact, a unique organ—the syrinx—that operates with nearly 100 percent physical efficiency to create loud, complex sounds and can even produce two independent songs simultaneously. In other respects, we now know that bird song has much in common with human music and speech, sharing similar sounds, tones, and tempos. Furthermore, bird song is produced by a series of rapid and complex motor activities, such as those controlling the tongue during speech or the fingers of a skilled violinist playing an intricate passage (Marler 1981).

This chapter presents the physical characteristics of bird vocalizations and examines how the syrinx produces these sounds. Then follows a discussion of the functional aspects of vocal communication by birds: What kinds of information do particular songs and calls convey? What information enables birds to recognize their own species or to discriminate among individuals? Subject to sexual selection just as plumage ornaments and visual displays are, song repertoires or vocal displays vary in size

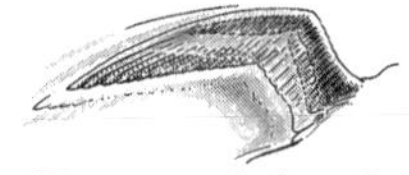

BOX 10–1

There is a technical vocabulary describing vocal communication

Amplitude Loudness or maximum energy content of a sound.

Fundamental tone See **Harmonic.**

Frequency Number of complete cycles per unit time completed by an oscillating sound waveform; usually expressed in hertz or kilohertz.

Glissando A blending of one tone into the next in a scalelike passage.

Harmonic A tone in the series of overtones produced by a fundamental tone. The frequencies of the tones in a harmonic series are consecutive multiples of the frequency of the fundamental (see Figures 10–2 and 10–3A).

Hertz (Hz) Unit of frequency equal to one cycle per second.

Modulation Defining the form of a sound (technically the carrier wave) by variation of either frequency or amplitude.

Oscillograph Device that records oscillations as a continuous graph—called an oscillogram—of corresponding variations in an electric current, as would be generated by a tape recording of a sound (see Figure 10–1).

Overtone See **Harmonic.**

Pitch Relative position of a tone in a scale, as determined by its frequency.

Resonance The intensification and prolongation of sound, especially of a musical tone, produced by sympathetic vibration.

Sinusoidal waveform Simple, pulsed cycles of energy that describe a regularly rising and falling sine curve, defined by the equation $y = \sin x$.

Sonogram Visual display of the frequency content of a sound distributed in relation to time (see Figure 10–1).

Tone A sound of distinct pitch and quality; in music, the interval of a major second.

among species and even among males of the same species. The advantages of song variety and the role of vocal mimicry are considered before the final topic—how young birds learn their songs—which is an active area of modern ornithological research. Discussions of bird vocalizations require a small, specialized working vocabulary of terms from music and from the scientific study of sounds (acoustical physics). Included in Box 10–1 are some of these terms.

Physical Attributes

Bird vocalizations are not easily classified. They range from the short clicks of swifts, to the quavering whistles of the tropical, partridgelike tinamous, to the long, tinkling melodies of wrens, to the seemingly endless imitations of other birds by mockingbirds. The variety of sounds emitted by birds reflects adjustments to enhance information content and to enhance the physical transmission of information to listeners. A traditional distinction exists between "songs" and "calls." The term *song* connotes long vocal displays with specific, repeated patterns often pleasing to the human ear (Pettingill 1984). Song refers primarily to the vocal displays of territorial male birds. The term *call* connotes a short, simple vocalization, usually given by either sex. Various calls include distress calls, flight calls, warning calls, feeding calls, nest calls, and flock calls. There is, however, no real dichotomy between songs and calls in either their acoustical structure or their function. Yet the term *song* is so entrenched and alternatives so lacking that continued use seems certain.

A fundamental dichotomy, unlinked to the perception of songs versus calls, does exist in the acoustical structure of bird vocalizations: whistled songs versus harmonic songs (Greenewalt 1968). Whistled songs lack, or appear to lack, harmonic content. Whistled songs consist of nearly pure (lacking harmonics) sinusoidal waveforms, which result from the varying compression of escaping air. The higher the pitch, the more frequently the sound waves oscillate. Both the basso profundo (80 to 90 hertz) of a Spruce Grouse and the high, thin notes (9000 hertz) of Blackpoll Warblers are, technically speaking, whistled songs (Figure 10–1).

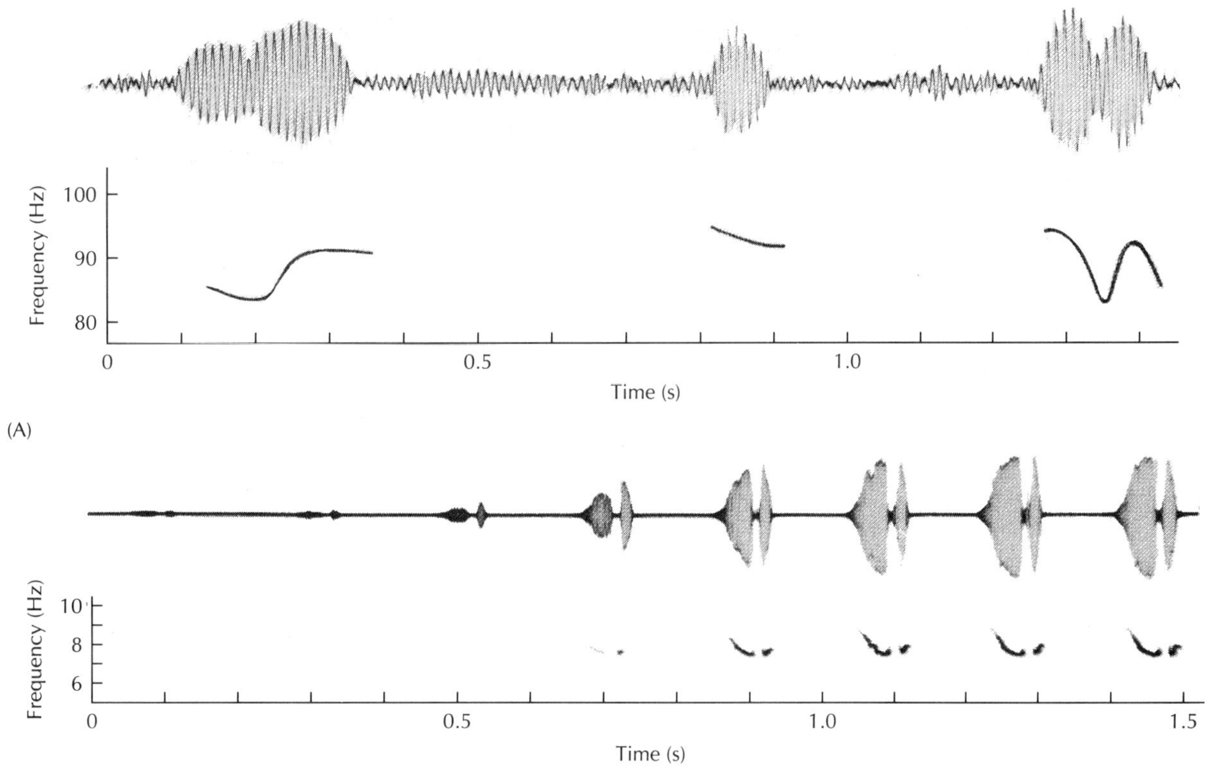

Figure 10–1 Paired oscillograms and sonograms of (A) the bass notes (90 cycles per second = 90 hertz) of the whistled song of a Spruce Grouse and (B) the high, thin notes (9 kilocycles per second = 9 kilohertz) of the whistled song of a Blackpoll Warbler. The oscillograms (upper records) display patterns of amplitude modulation as the vertical deflection (above and below the midpoint) of the sinusoidal waveform; frequency is calculated from the number of complete cycles per second. The sonograms (lower graphical records) display the distribution of energy (1 kilocycle per second = 1 kilohertz) in a song with respect to time. (From Greenewalt 1968)

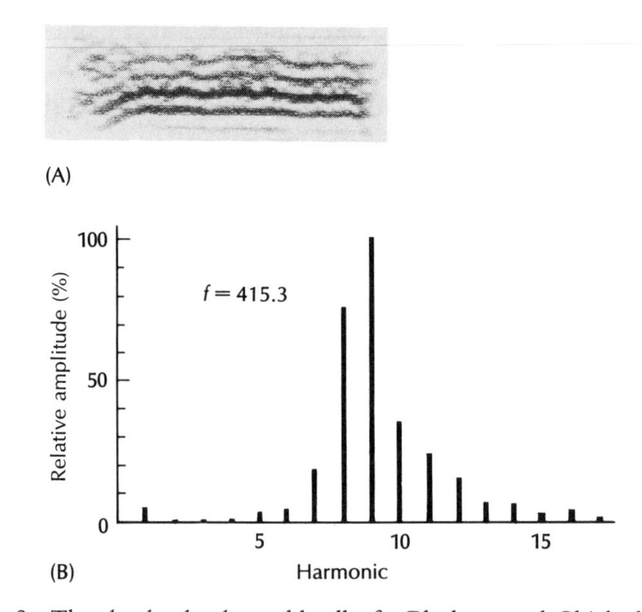

Figure 10–2 The *dee-dee-dee-dee* scold call of a Black-capped Chickadee consists of a series of harmonic phrases. (A) Sonogram of one *dee* phrase. (B) Relative amplitudes of the tones of the harmonic series; the fundamental frequency *f* is 415.3 hertz. Numbers on the abscissa represent the frequencies of the harmonics as multiples of the fundamental (1 ≡ *f* at 415.3 hertz). The loudest tone is assigned an amplitude of 100 percent, and the amplitude of the other tones are calculated relative to that value. (From Greenewalt 1968)

Harmonic songs employ harmonics—or overtones—tones with frequencies that are multiples of the fundamental frequency. One dominant harmonic has more energy—greater amplitude—than do the others in the spectrum. The number of harmonics and their relative amplitudes determine the timbre—or general tonal quality—of the notes of bird songs (and musical instruments). Qualities such as clarity, brilliance, and shrillness, as well as nasal and hornlike tones, reflect various combinations and emphases of harmonics (Figure 10–2). For example, the distinctive sounds of a clarinet and a Hermit Thrush result from an emphasis on the odd-numbered (3, 5, 7, etc.) harmonics (Marler 1967, 1969).

The physical structure of a sound affects the ease with which a listener—predator or neighbor—can locate its source (Marler 1955). The calls that birds use to locate or attract one another, for example, are made up of short notes with a broad frequency range. The assortment of frequencies in such notes enriches the information about direction and distance. In contrast, alarm calls are faint, thin (narrow frequency range), high-pitched calls of long duration that conceal the sender's whereabouts (Figure 10–3). The physical structure of a particular sound also determines the distance it will travel and how much physical distortion it will sustain before reaching the listener. Interference, absorption, and scattering of the sound waves by vegetation, the ground, and air progressively distort a sound (Morton 1975; Wiley and Richards 1982).

Low-frequency sounds, such as the calls of grouse, bustards, cuckoos, doves, and large owls, are the most effective for long-distance communication; they are less subject to attenuation and interference than are high-frequency sounds. The calls of tropical forest birds, which depend on long-distance communication through dense vegetation, are usually lower in frequency than those of tropical species living in open habitats (Chappuis 1971; Morton 1975). Birds of the forest floor, such as antbirds and curassows, have low-pitched calls that suffer minimal distortion from ground reflections. The complex buzzlike songs of open field birds, such as Clay-colored Sparrows, contrast with the simpler whistles of forest birds. Reverberations in forests mask the fine temporal structure of bird songs. Forest-dwelling birds, therefore, tend to produce simple sounds (Wiley and Richards 1982). Conversely, broadband songs, rich in temporal structure (with complex frequency modulations), are advantageous in open habitats because simple, sustained tones tend to be distorted by strong temperature gradients and air turbulence.

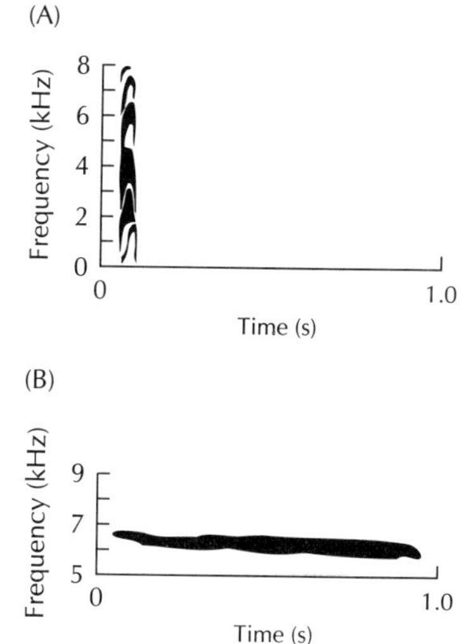

Figure 10–3 Sonograms of two vocalizations of a Eurasian Blackbird. (A) The call used when mobbing an owl is of short duration and has a broad frequency range; its source is easy to locate, and it attracts other birds to the site. (B) The alarm call, used when a hawk flies over, is of long duration and has a narrow frequency range; its source is difficult to locate, and thus this call does not reveal the blackbird's location to the hawk. (Adapted from Marler 1969)

Sound Production by the Syrinx

Birds range from virtually silent to garrulous. At one extreme, Mute Swans, Turkey Vultures, and Greater Rheas only hiss and grunt occasionally. At the other extreme, mynas, parrots, mockingbirds, and skylarks possess seemingly unlimited vocabularies.

The vocal virtuosity of birds stems from the structure of their unusual and powerful vocal apparatus. All songs and calls come from the syrinx, a unique avian organ located in the body cavity at the junction of the trachea and the two primary bronchi (Figure 10–4). The syrinx may form from tracheal tissues, as it does in Neotropical woodcreepers, antbirds, and their relatives; from bronchial tissues, as in most cuckoos, nightjars, and some owls; or from tissues of both structures, as in most birds. The avian larynx, located at the top of the trachea, at the back of the oral cavity, does not include vocal cords but serves only to open and close the glottis and thereby keep food and water out of the respiratory tract.

Crawford Greenewalt (1968, 1969; see also Gaunt and Wells 1973) defined the main principles by which the syrinx produces sound. Contraction of thoracic and abdominal muscles forces air from the main air sacs through the bronchi to the syrinx. On each side of the syrinx is a thin, glass-clear membrane—the internal tympaniform membrane *(membrana tympaniformis interna)*. Sound is caused by the vibration of the air column as air passes through the narrow (syringeal) passageways, which are bounded on opposite sides by corresponding projections called the internal labium and external labium (Figure 10–5). Vibrations of the internal tympaniform membrane—regulated by its mass, internal tension, and

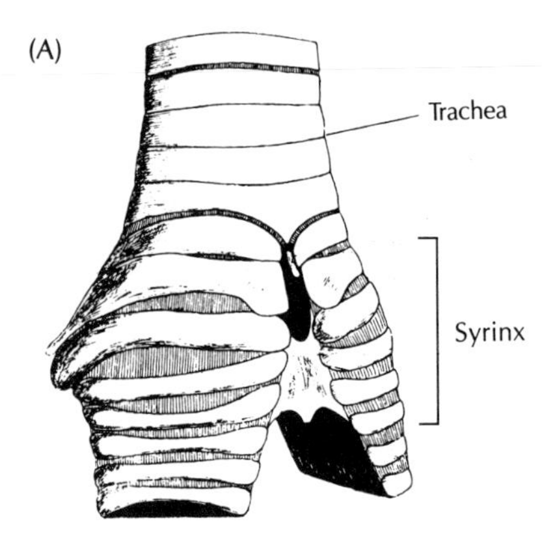

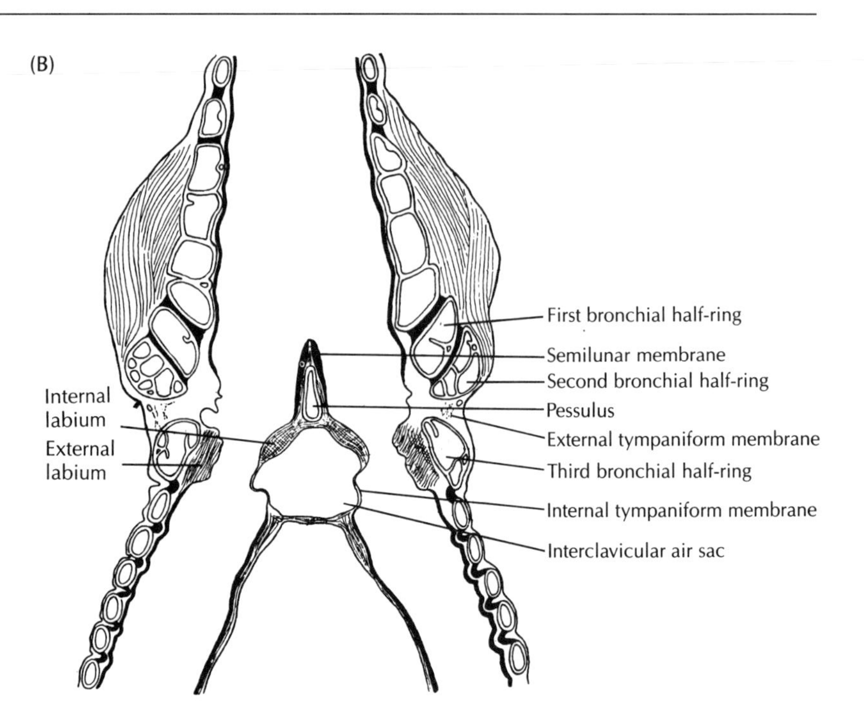

Figure 10–4 (A) Bird vocalizations originate from the syrinx, an elaboration of the junction of the base of the trachea and the two bronchi. (B) The main elements of the syrinx are its vibrating tympaniform membranes, the muscles that control tension in these membranes, and the supporting cartilage. (Adapted from Häcker 1900)

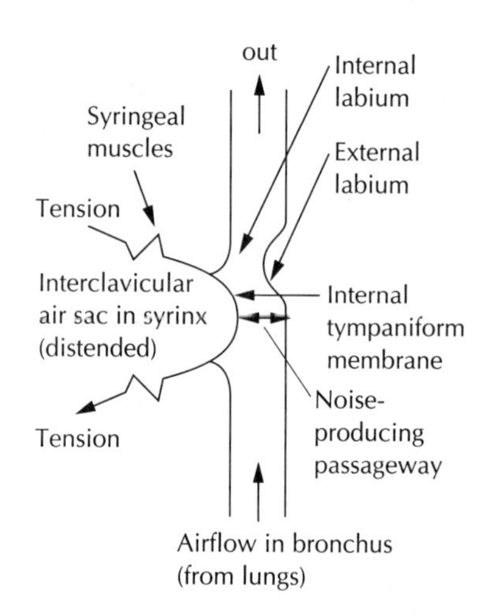

Figure 10–5 Sound production in the syrinx depends on the tension of the internal tympaniform membrane, which is controlled by pressure in the interclavicular sac, contraction of the syringeal muscles, and the diameter of the air passageway. (After Greenewalt 1968)

protrusion into the adjacent air column—determine the sound characteristics. The efficiency of sound production is extraordinary; nearly 100 percent of the air passing through the syrinx is used to make sound, compared with only 2 percent in human sound production.

Surrounding the bronchial junction—the syrinx—and within the pleural cavities is a single interclavicular air sac. Pressure in the interclavicular air sac pushes the thin membranes into the bronchial air space, into position for vibration and creation of sounds. A needle puncture of the interclavicular air sac prevents buildup of the pressures needed to move the tympaniform membranes, thereby rendering a bird voiceless. Sound tone depends on the precise tension of the membrane. Syringeal muscles change the tension of the tympaniform membrane as a bird sings. When the membrane vibrates without constraint, it produces a whistled song without harmonics. Rippling distortions of the membrane's free movement result in the production of harmonics, but the exact origin of these distortions is still under investigation.

Syringeal muscles control the details of syrinx action during song production. Species that lack functional syringeal muscles, such as ratites, storks, and New World vultures, can only grunt, hiss, or make similar noises. Most nonpasserine birds have two pairs of narrow muscles on the sides of the trachea above the syrinx; these are called extrinsic muscles because they originate outside of the syrinx. More elaborate musculature is characteristic of oscine songbirds, which have up to six pairs of intrinsic syringeal muscles in addition to the extrinsic muscles (Figure 10–6). The intrinsic syringeal muscles originate within the syrinx and insert onto the bronchial rings, the internal and external tympaniform membranes, and the syringeal cartilage. Despite such well-developed syringeal muscles, the songs of oscines are not acoustically much more complex than those of species with simpler syringeal muscle arrangements. Possibly the elaborate syringeal musculature of oscines enhances the frequency range of their songs (Brackenbury 1982).

Song complexity is due to modulation of the frequency or amplitude or both of a sound signal over time and is achieved through changes in the diameter of the passageway and the tension of the tympaniform membrane. Modulation in bird song is defined as a change of the constant amplitude or frequency of a phrase. The rapidity of the modulations, their continuous orchestration, and the variable coupling of frequency and amplitude changes enable the rather simple syringeal anatomical system to produce a variety of modulated sounds.

The frequency and amplitude modulations of bird songs are unique among animal sounds. Simple tones, such as the notes of a White-throated Sparrow, contain little modulation, whereas the variable songs of

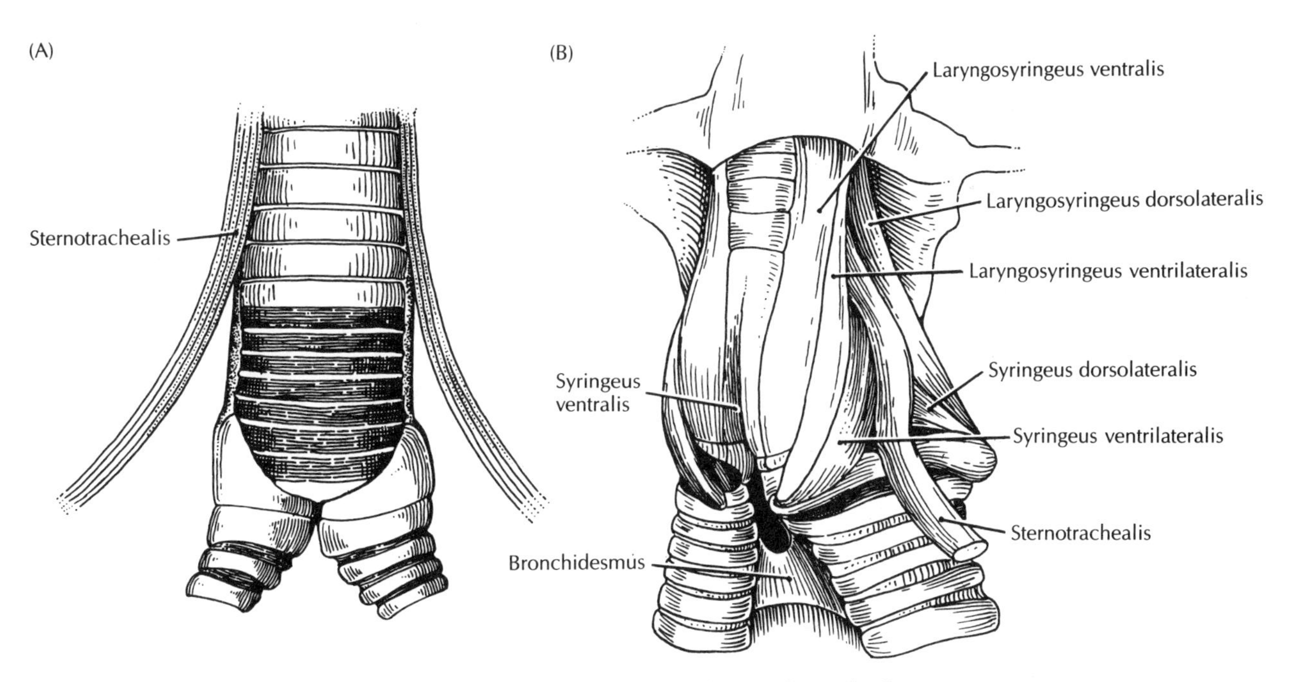

Figure 10–6 Simple and complex syringeal musculature. (A) The simple tracheal syrinx of the Chestnut-belted Gnateater with its single pair of extrinsic syringeal muscles. (B) The elaborate tracheobronchial syrinx of the Little Spiderhunter, with six pairs of intrinsic syringeal muscles. The bronchidesmus is a wide band of tissue that ties the two bronchi together. (Adapted from Van Tyne and Berger 1976)

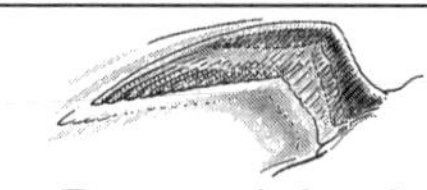

Box 10–2
Birds have two independent voices

The syrinx consists of two independent halves that can produce different, complex songs simultaneously (see figure). In addition to having different frequency content, the notes produced by the dual voices can be modulated independently of one another. First discovered in an analysis of the song of a Brown Thrasher, the phenomenon of two independent voices has since been reported for grebes, bitterns, ducks, sandpipers, bellbirds (Cotingidae), and many songbirds (Greenewalt 1968; Miller 1977). The two sources also can be coupled to produce a single, complex sound (Nowicki and Capranica 1986).

Initial attempts to apply musical instrument and human voice sound production theories to bird song were discouraged by the discovery of two separate sound sources in birds. It seemed physically impossible to modulate two sounds separately in a single trachea, oral cavity, or instrumental sound chamber. Critical experiments were then conducted using a helium atmosphere in which sound travels faster than in a normal nitrogen-based atmosphere, with predictable effects on frequency and harmonic structure (Nowicki 1987). (Recall your own high-pitched voice after you inhaled helium from a balloon!) The helium experiments revealed that bird song, like human speech, is the result of rapid, coordinated output of two or more motor systems acting in concert. The bird's vocal tract filters the harmonic spectrum produced by the syrinx and concentrates the energy at single frequencies. A bird controls the filtering process by varying tracheal length, by constricting the larynx, or by flaring its throat and beak. The rapid beak and throat movements seen in singing birds thus help to produce their complex songs.

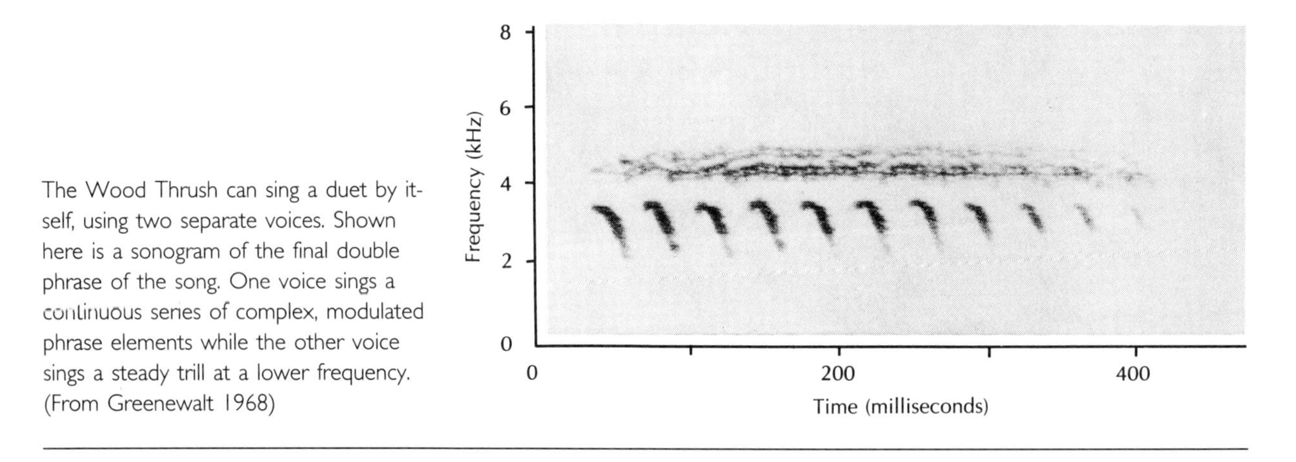

The Wood Thrush can sing a duet by itself, using two separate voices. Shown here is a sonogram of the final double phrase of the song. One voice sings a continuous series of complex, modulated phrase elements while the other voice sings a steady trill at a lower frequency. (From Greenewalt 1968)

a Song Sparrow and the brief notes of a Tree Swallow contain complex, rapid modulations. Even short phrases within songs may include rapid modulations. The brief *glug glug glee* song of the Brown-headed Cowbird encompasses a four-octave interval from 700 to 11,000 hertz, the greatest frequency range in a single bird song. In one 4-millisecond fraction of the *glee,* the signal rises continuously from 5 to 8 kilohertz, an amazingly rapid glissando (Greenewalt 1968) (Figure 10–7). Females apparently select males on the basis of their abilities to perform such vocal gymnastics.

(A)

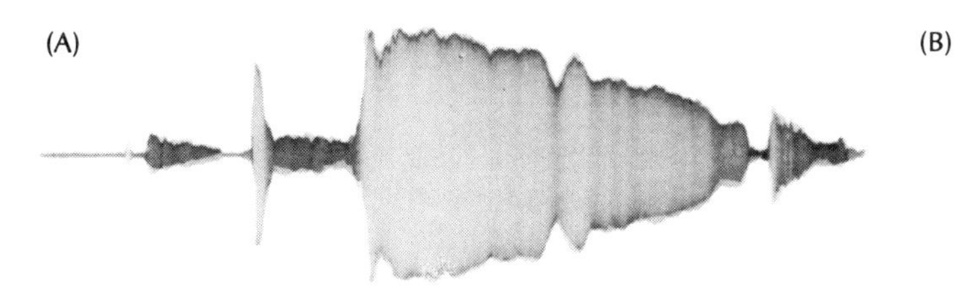

(B)

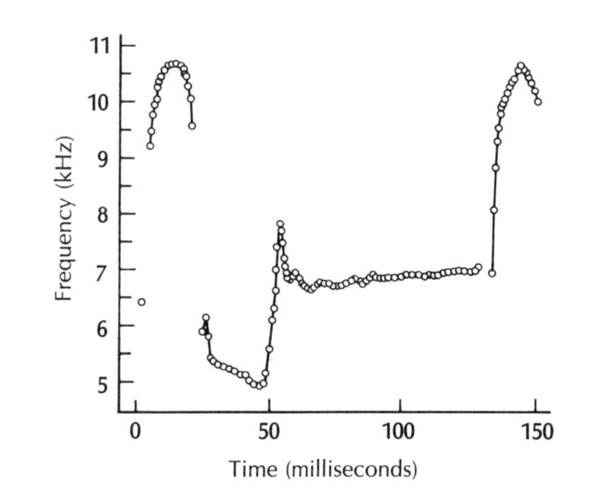

Figure 10–7 Complex modulations of frequency and amplitude characterize the song of the Brown-headed Cowbird. (A) Oscillogram of the *glee* phrase, in which the rapid cycles of the sinusoidal waveform cannot be individually distinguished. (B) A summary of the succession of frequencies composing the phrase. Note the rapid frequency modulations at 50 and 130 milliseconds. (From Greenewalt 1968)

Rapid, pulsed bursts of air, when expired, cause amplitude modulations in the songs of some species. The trilled whistles of young chicks, for example, are produced by rapid vibrations of the abdominal muscles (up to 50 cycles per second) and the resulting pressure pulses of air forced into the syrinx (Phillips and Youngren 1981). Birds such as the Grasshopper Warbler, which have sustained songs, apparently breathe and sing simultaneously by using shallow "minibreaths" (Brackenbury 1982).

Sounds produced by the syrinx can be filtered and modified by changes in the length of the trachea (Nowicki 1987). The loud, resonating, trumpetlike calls of swans, cranes, some curassows, and guineafowl are due in part to an unusually long trachea that is coiled in the body cavity or in the bony sternum itself (Figure 10–8). Two genera of birds-of-paradise also have elongated tracheae coiled between the skin and breast musculature (Clench 1978).

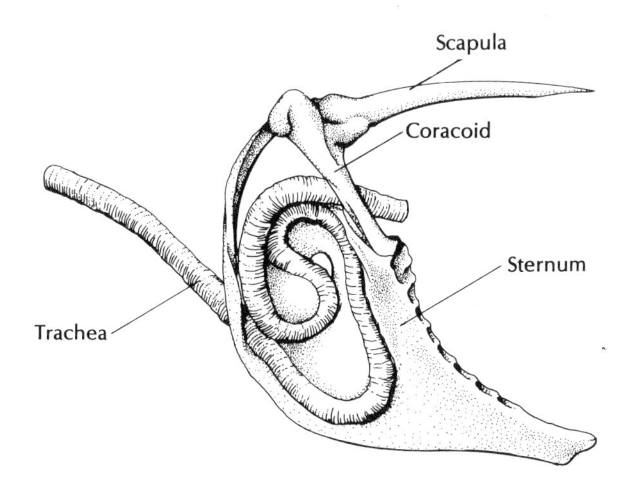

Figure 10–8 A crane's elongated trachea is coiled inside the sternum. (After Grassé 1950)

Avian Vocabularies

Just as a bird can have a repertoire of visual displays, it also can have a repertoire of calls and songs. The Blue Jay gives the familiar *jay jay* alarm call but also has a large repertoire of distinct calls that include a gurgling "pump handle" call and excellent imitations of some hawks. Birds generally have 5 to 14 distinct calls with a variety of overlapping functions (Thorpe 1961; Armstrong 1963). The Chaffinch has 12 adult sounds, 7 of which are used only in the breeding season—6 by the male and 1 by the female (Table 10–1). These functions include proclamation of territorial ownership, attraction of mates, broadcast of personal characteristics (species, age, sex, competence), warning of potential dangers, and maintenance of social contact. Most birds have some calls that are used only occasionally for special purposes. Contact or association calls, for instance, help birds to keep track of one another while flocking or when in dense vegetation. Alarm calls signal danger and advise escape flight. Precopulatory trills and postcopulatory grunts integral to mating ceremonies are heard at no other time.

TABLE 10–1
Repertoire of the Chaffinch

Vocalization	Transcription	Context
Flight call	*tupe* or *tsup*	Flight or flight preparation
Social call	*chink* or *spink*	Seeking companion of unknown whereabouts
Injury call	*seeee*	Injured in flight
Aggressive call	*zzzzzz* or *zh-zh-zh*	Fighting (captive males only)
Alarm calls	*tew*	Danger, used especially by young birds
	seee	Escaping a real threat, just after copulation (breeding males only)
	huit	Moderate danger or after real danger (breeding males only)
Courtship calls	*kseep*	Active courtship (breeding males only)
	tchirp	Ambivalence toward approach and copulation with female (breeding males only)
	seep	Ready for copulation (females only)
Subsong		Practice of real song
Song		Territoriality, identification, and courtship; average is 2 to 3 per male, up to 6

From Marler 1956.

The loud territorial songs of birds are among their most conspicuous and familiar vocal displays. Usually these long-distance signals, which carry 50 to 200 meters or more, convey information about the identity, location, and motivation of the singer (Morton 1986). Territorial songs serve as signals to potential rivals that the territory is occupied by a resident male prepared to protect his exclusive use of that space and any associated females. When a territorial male Great Tit, for example, is removed from its territory, another male will take over within 10 daylight hours unless territorial song is broadcast from loudspeakers on the territory (Krebs 1977). When a song is broadcast, rival males take three times as long (30 daylight hours) to exploit the vacancy.

Inseparably coupled to the warning message is advertisement to unmated females. Female attraction to territorial male song is the first step toward courtship and pair formation. Strong but indirect evidence lies in observations that males sing less frequently after they acquire a female than before. Experimental removal of females from paired, male White-throated Sparrows results in a resurgence of song, which then subsides after a second mate is obtained (Wasserman 1977).

Territorial song also includes information about a singing male's location in the territory. The spectrum of frequencies in a call note reveals or conceals the singer's location, as will be explained. So do variant song forms in a male's repertoire. Male Chestnut-sided Warblers, for example, reveal their location by using accented song forms in the centers of their territories, an unaccented form exclusively at the boundaries of their territories, and two other song forms at intermediate sites (Lein 1978) (Figure 10–9). Territorial Yellow-throated Vireos also select particular elements from their repertoire to prescribe their location and to indicate whether they are about to change location (Smith et al. 1978)

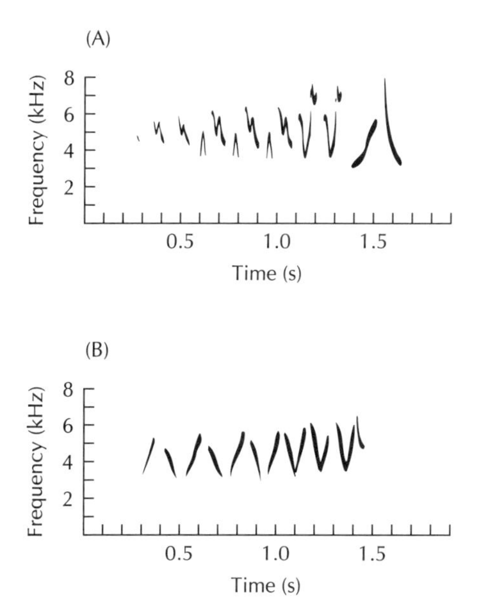

Figure 10–9 Chestnut-sided Warblers use (A) mostly accented forms of their territorial song (phoneticized as *I wish to see Miss BEECHER*) in the centers of their territories and (B) unaccented forms near the edges of their territories. Accented territorial songs of wood warblers attract females; unaccented songs help to repel rival males.

Vocal Identity

With few exceptions, each species of bird has distinct vocalizations that facilitate the attraction of mates, the maintenance of the pair bond, and social cohesion. Vocalizations mediate the process of species recognition, that is, interactions among members of the same species as opposed to members of other species. The vocalizations of many birds, such as nightjars or tyrant-flycatchers, are more distinct than are their cryptic plumage colors. The Willow Flycatcher and the Alder Flycatcher, which have different songs but are otherwise indistinguishable in size or color, were discovered to be different species only recently (Stein 1963) (Figure 10–10). Previously they were regarded as a single species—the "Traill's Flycatcher"—with two song forms, only one of which was used by the individuals that lived side by side in the same swamps. Critical field studies revealed a difference in the structure of their nests and thus in the behavior of the associated females. In addition, laboratory studies demonstrated that the vocal difference between these two sibling species—species that are nearly identical in appearance—of flycatchers was genetically determined (Kroodsma 1984). This is just one example of the many cases in which vocal differences led to the separation of sibling species or to confirmation of species status in taxonomically difficult groups of birds (Payne 1986).

Birds respond readily to playback of tape-recorded songs and discriminate in field experiments between songs of their own and those of other species. The two kinglets of Europe—Common Goldcrest and Firecrest—have similar, high-pitched, warbling songs. Yet males predictably approach the taped broadcast of the song of their own species but only occasionally show interest in the song of the other species (Becker 1976, 1982). The use of song playback from tape recorders not only enables birdwatchers to draw out and view secretive species but also allows ornithologists to census populations of cryptic birds such as owls and rails and assess the species-recognition behavior of geographically separated populations. Wesley Lanyon (1967, 1978) concluded that particular island populations of *Myiarchus* tyrant-flycatchers in the West Indies belonged to the same species, the Lesser Antillean Flycatcher (see Figure 2–14) because they reacted aggressively to each other's songs. Their behavior in playback experiments suggested that they would compete for territories and would interbreed if they met in natural conditions.

Vocal displays, especially if they are innate rather than learned, can also reflect the evolutionary history—or phylogeny—of species, as can visual displays (Payne 1986). The wax-eating honeyguides, for example, have advertising calls based on the repetition of a single, simple note. Three characteristics of the advertising call help to define the relationships among four species of the honeyguide genus *Indicator* (Figure 10–11).

Features that birds use for species recognition are embedded in the acoustical structures of their songs. The Common Goldcrests mentioned earlier rely on frequency range, frequency changes between notes, and

Figure 10–10 The Alder Flycatcher, shown here at its nest, can be distinguished reliably from the virtually identical Willow Flycatcher only by its song. The Alder Flycatcher calls *fee-beeo*, whereas the Willow Flycatcher calls *fitz-bew*. Both songs are innate, not learned. (Courtesy of A. Cruickshank/VIREO)

alternation of long and short elements to discriminate their songs from those of Firecrests (Becker 1982). White-throated Sparrows rely on the regularity of the pattern of their typical song—pure, sustained tones without harmonics—and pitch itself for species recognition (Falls 1982). The song of this common North American sparrow consists of spaced, clear whistles that differ in pitch and fit the words *Old Sam Peabody*. Male sparrows do not respond strongly to songs with wavering notes, notes with multiple tones or harmonics, frequent changes in pitch between notes, or long intervals between notes.

Syntax—the sequence of particular notes or syllables—is sometimes an important recognition feature in songs. Short, discrete vocal elements are often called syllables, several of which combine to form a phrase. Brown

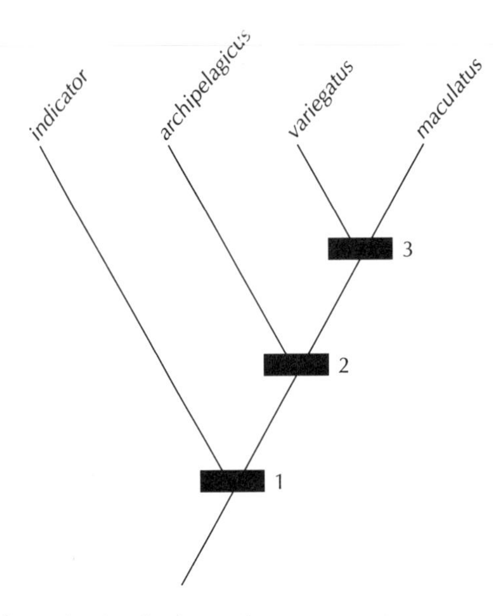

Figure 10–11 A hypothesis of relationships among four species of *Indicator* honeyguides based on their advertising calls. This hypothesis is consistent with one based on morphological characters. The defining characters—or synapomorphies—are indicated by three, numbered, black bars: (1) All four species have a *churr* phrase that is unknown in the other honeyguides. (2) A prolonged *purr* phrase of at least 4 seconds' duration is present in the advertising calls of three species, excluding *Indicator indicator.* (3) Only two species, *variegatus* and *maculatus*, have low-pitched calls at a frequency below 2 kilohertz. (From Payne 1986)

Thrashers distinguish their songs from those of Gray Catbirds by the number of repetitions of each syllable (Boughey and Thompson 1976). Syntax, however, is not an essential recognition feature of the songs of either White-throated Sparrows or Indigo Buntings. Rearrangement of the notes and intervals of the sparrow song does not diminish a male's interest in the playback. Male Indigo Buntings use the rhythmic timing and patterns of frequency change in individual notes, but not the distinctive couplet arrangement of notes, to recognize other buntings (Emlen 1972).

Most experimental studies of species recognition in songbirds test the aggressive responses of territorial males to male song. Because females do not usually respond aggressively to song playback, it is difficult to study the vocal cues they use to identify others of their own species. Yet female responses to male vocalizations may be critical to pair formation. Females sometimes respond directly to male song with either precopulatory trills or copulatory postures (King and West 1977; Searcy and Marler 1981). Female Song Sparrows and Swamp Sparrows whose sex drive has been experimentally enhanced with the hormone estradiol will respond more strongly to songs of their own species than to alien songs. They discriminate between the two by recognizing distinctive syllable structures and patterns of syllable delivery (Searcy et al. 1981).

The songs of male Brown-headed Cowbirds vary in their potency, defined in terms of how readily estradiol-treated females solicit copulation (West and King 1980; West et al. 1981). Male cowbirds, even those hand-raised in isolation, are capable of singing high-potency songs, but only the top-ranked dominant individuals in a group actually do so. If a subordinate dares to use potent vocalizations while displaying, it invites attack by the dominant male. As a result, subordinate males deliberately downgrade their vocalizations and apparently wait for an opportunity to sing their best songs without risk.

Individual Recognition

Details of song pitch, phrase structure, syntax, and composition serve as individual signatures that enable birds to identify offspring, parents, mates, and neighbors (Beer 1970; Falls 1982). White-throated Sparrows, for example, use variations in pitch to this end. Ovenbirds use variations in the structure of the phrase that can be verbalized as *tea-cher,* and Indigo Buntings use groups of repeated syllables as individual signatures.

Recognition of parents by the young or of young by the parents provides an essential bond for feeding and protection, especially in dense nesting colonies, where confusion is likely. Mutual voice recognition between parents and young has been documented in penguins, auks, swallows, and the Pinyon Jay. Although young gulls recognize their parents by voice, parents recognize their young by behavioral responses to parental calls rather than by the youngster's calls (Beer 1979).

Discrimination of individual vocalizations enables mates to recognize each other. Colonial seabirds, in particular, must somehow distinguish their partners from the hordes of potentially antagonistic neighbors. Black-legged Kittiwakes, among others, use frequent individually distinctive calls to maintain productive perennial pair bonds and to coordinate sharing of a narrow nest ledge (Wooller 1978). Emperor Penguins, Northern Gannets, Least Terns, and various species of gulls respond quickly to playback of their mates' calls but not to those of other individuals (Jouventin et al. 1979; White 1971) (Figure 10–12).

Individual vocal differences enable birds to distinguish neighbors from strangers and to respond accordingly. Playback experiments have shown that birds recognize the specific calls of their neighbors (Falls 1982). Territorial males use this information to concentrate their defense efforts against strangers who are attempting to locate a mate or a territorial vacancy. Territorial males do not react aggressively to a neighbor as long as the neighbor is where it belongs, in its own territory, but they will attack a singing neighbor if it happens to trespass. Territorial males also will

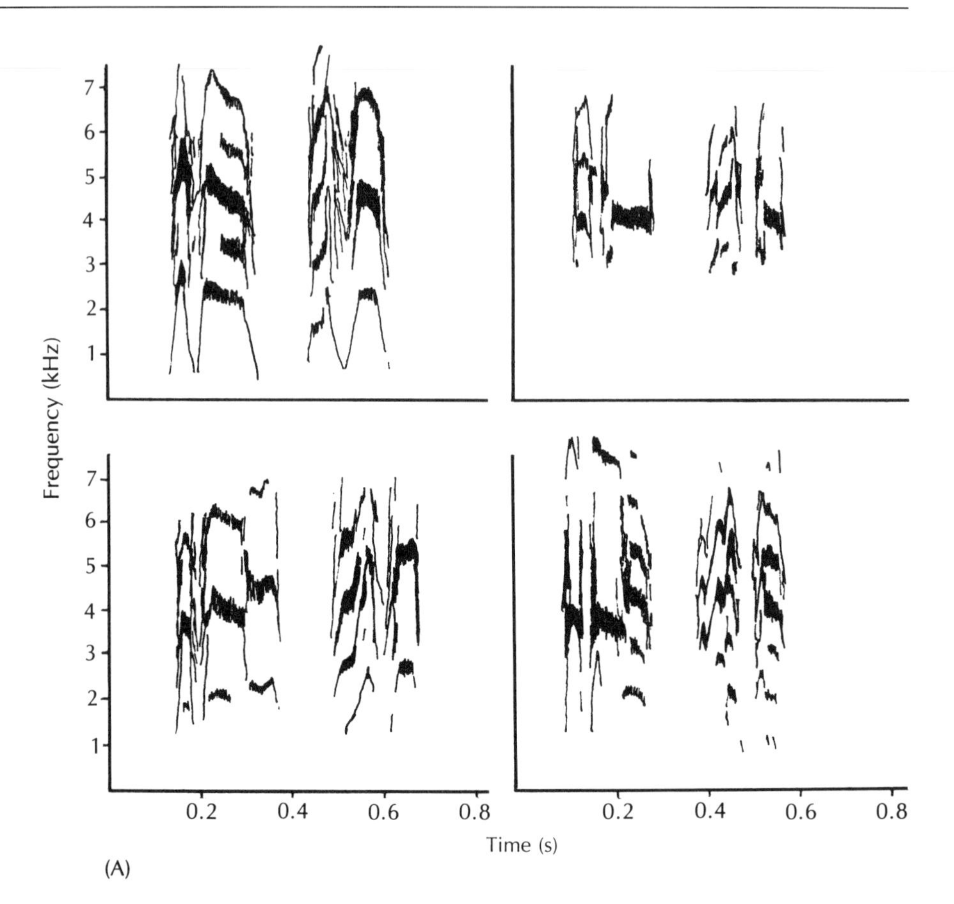

(B)

Figure 10–12 Individual differences in the *purrit-tit-tit* call enable Least Terns to recognize each other. (A) Sonograms of the calls of four individuals. (B) Two Least Terns at a nest with one egg and one chick. (A from Moseley 1979; B courtesy of A. Cruickshank/VIREO)

respond aggressively if the neighbor's songs are experimentally broadcast from the wrong side of the territory, which may indicate a greater threat of trespass.

Some birds use distinctive vocal duets for individual recognition. Vocal duets are coordinated, overlapping bouts of sounds by members of a mated pair or extended family group (Farabaugh 1982). At least 222 species in 44 families are known to sing duets. Most of these are monogamous, tropical birds that defend year-round territories. The duets function both in joint defense of territorial space against encroaching neighbors and in maintenance of the pair bond.

Box 10–3
Bush-shrikes duet with precision

Each pair of Tropical Boubous, a kind of African bush-shrike, develops a unique set of duetting patterns, which they use to keep track of each other in dense vegetation, to synchronize their reproductive cycles, and to maintain their territorial integrity (Thorpe and North 1966). Either member of the pair can initiate the duet. The respective note contributions are so well synchronized that few people realize that two birds, not one, are singing. A pair of Tropical Boubous increases the complexity of its duet patterns as the density of shrikes and, perhaps, the need for distinction increase (see figure).

Duetting bush-shrikes respond to cues—preceding notes—in only a fraction of a second and with astonishing precision (Thorpe 1963). These reaction times can be measured quite accurately in the duets of the Black-headed Gonolek, a bush-shrike with a simpler duet than that of the Tropical Boubou. The female gonolek responds to the male's lead *youck* with a sneezelike hiss. The average response time of one female was only 144 milliseconds, with a standard deviation of 12.6 milliseconds. Another female responded in 425 milliseconds, with a standard deviation of 4.9 milliseconds. These values (12.6 and 4.9 milliseconds) are exceedingly low. Human auditory reaction times, not nearly as precise, have a standard deviation of 20 milliseconds.

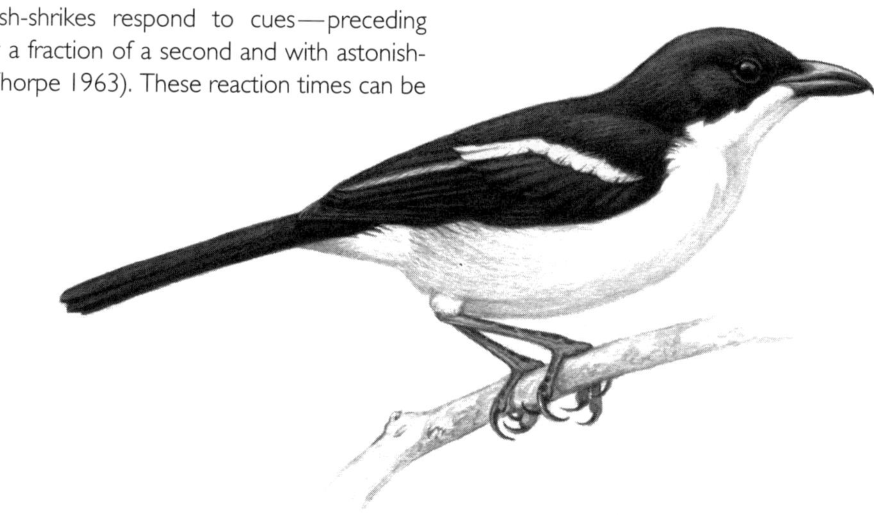

Tropical Boubou, an African bird famous for its duets.

Song Repertoires

The vocal repertoires of birds are among the richest and most varied in the animal kingdom and are comparable to those of nonhuman primates (Marler and Hamilton 1966). The repertoires of territorial songs vary in size from the single song type of a White-throated Sparrow and the two distinct territorial songs of many species of North American wood warblers to hundreds of songs used by some wrens and mockingbirds. Large repertoires constructed from assorted syllables are typical of some wrens, several thrushes, and accomplished mimics such as the Northern Mockingbird and the Superb Lyrebird of Australia. Among wrens, the Canyon Wren has but three simple songs per individual, whereas individual Sedge Wrens and Marsh Wrens have over 100 songs (Kroodsma 1977). Even though Winter Wrens in Oregon have a relatively small repertoire of roughly 30 songs per individual, the songs are extraordinary; lasting a full 8 seconds, they are composed of organized sets of syllables, each consisting of 50 notes selected from a pool of 100 (Kroodsma 1980).

Rules govern the formal sequences of bird song delivery (Lemon and Chatfield 1971; Lemon 1977). Swainson's Thrushes sing one song type after another (Dobson and Lemon 1977). Song Sparrows sing one of their 10 to 20 patterns a dozen times before switching to another pattern (Mulligan 1966). Northern Cardinals and Marsh Wrens string different songs in rigid sequences, in which each song (except the first) is predictable from the preceding one.

Subject to sexual selection, large repertoires may enhance a male's attractiveness to females and his ability to compete with neighbors, discourage would-be territorial males, or stimulate continued interest by listeners. A male's reproductive success may increase with repertoire size, which is as subject to sexual selection as plumage ornamentation (Chapter 9). Female Swamp Sparrows and Song Sparrows that are treated with the sex hormone estradiol adopt copulation postures more readily in response to a male repertoire that varies than to an invariant repertoire (Searcy and Marler 1981). In experiments with taped song sequences, female Common Canaries respond to large repertoires by building nests faster, laying the first egg sooner, and laying larger clutches than females exposed to taped song sequences with only five different syllables. Females may also use large repertoires to identify older, more experienced mates (Nottebohm and Nottebohm 1978; Yasukawa 1981).

Male Great Tits with large repertoires achieve greater success in reproduction—in terms of heavier young—than those with small repertoires. This may indicate that they were able to stimulate females more, as just mentioned, or to acquire and protect better territories (Krebs and Kroodsma 1980). The broadcast of large tape-recorded repertoires on territories that were experimentally emptied of Great Tits in one study and of Red-winged Blackbirds in another study resulted in a slower reoccupancy rate than repeated broadcasts of simple song types (Krebs

1977; Yasukawa 1981). However, it is not yet clear whether large repertoires discourage potential settlers because they mimic the presence of several males rather than one or because they signal the high social status of the singer.

Eugene Morton (1982) suggests that varied repertoires thwart the ability of neighbors to assess accurately one another's location. The patterns of signal degradation due to interference and distortion are so predictable that males of at least one species, the Carolina Wren, can use that information to judge the distance to a singing rival (Richards 1981). This ability requires detailed familiarity with a neighbor's distinctive song phrases. Variable repertoires might make critical evaluation of each phrase much more difficult, thereby forcing neighbors to be more attentive to their territorial boundaries. The cost to rivals of time and energy, which translates directly into survival and reproductive output, may eventually work to the versatile singer's advantage.

Large repertoires potentially prevent declining interest—or habituation—by the listener (Hartshorne 1973; Kroodsma 1978). Great Tits, Red-winged Blackbirds, and Song Sparrows show renewed interest in playbacks of taped songs after they hear new song types (Dobson and Lemon 1975). If habituation by listeners helps to promote a diversity of song types, continuous singers might project a greater variety of songs than those that sing periodically. In fact, continuous singers do tend to be most versatile in their use of distinctive song forms. North American wrens, in particular, are likely to present new songs after short intervals of silence (Kroodsma 1982a). More research is needed to clarify the adaptive significance, if any, of complex variations in avian song.

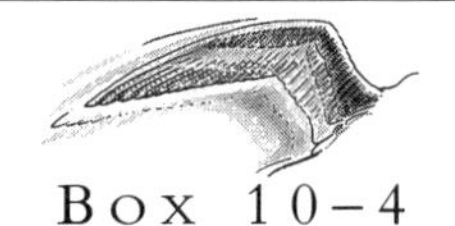

Box 10–4
Both bird vocalists and human fiddlers have technical duels

Large song repertoires may provide ammunition for vocal duels among competing males. Marsh Wrens, for example, duel vocally for control of quality territories, which attract multiple females. Increased competition for limited marsh habitat in the western United States has selected for larger repertoires and related brain space (Chapter 8). Using their large repertoires of songs that can be arranged in complex, varied sequences, neighboring males try to match each other's sequences or take the initiative in a duel (Kroodsma 1979). Leadership in the duels, which draws on both singing skill and repertoire size, promotes social dominance and increased reproductive success. Such avian vocal duels parallel guitar or banjo duels (see the movie *Deliverance*) or dominance contests among human fiddlers to demonstrate technical mastery of their musical instrument. How well a fiddler plays and embellishes the traditional tune "Orange Blossom Special" quickly establishes his or her rank among the masters.

Vocal Mimicry

Imitating the calls of other species is one way that some birds enlarge their vocal repertoires (Baylis 1982). The most renowned vocal mimics include the Northern Mockingbird, European Starling, Marsh Warbler, Australian lyrebirds, bowerbirds, scrub-birds, and the African robin-chats. Fifteen to twenty percent of the passerine birds in most regions of the world practice vocal mimicry (Vernon 1973).

Male Northern Mockingbirds (Figure 10–13) can have a repertoire of more than 150 songs, which both change from year to year and increase in number with age (Derrickson and Breitwisch 1992). Mockingbirds enhance their repertoire by imitating other birds, the calls of frogs and insects, and mechanical sounds. One Northern Mockingbird can imitate dozens of different species, broadcasting in sequence the songs of the American Robin, Blue Jay, Northern Cardinal, and a variety of other common species of the eastern United States. In Texas, mockingbirds broadcast the calls of Bell's Vireos, Great-tailed Grackles, and Dickcissels, among others. Some mockingbirds can imitate species found hundreds of kilometers away. For example, Jim Tucker of Austin, Texas, was surprised one morning to hear a mockingbird imitate a Green Jay, a species that is found only in the Rio Grande valley 500 kilometers to the south. Was this song learned directly from a Green Jay in the Rio Grande valley, or was it passed northward through a series of mockingbird generations?

Ornithologists remain fascinated by the possible functions of vocal mimicry such as the songs of the mockingbird (Derrickson and Breitwisch 1992). One possibility is that Northern Mockingbirds challenge and exclude the species they imitate (Baylis 1982). More likely, vocal

Figure 10–13 Northern Mockingbird singing. (Courtesy of A. Cruickshank/VIREO)

mimicry increases the potential size of the song repertoire, which is a measure of the quality of both a male and his territory. Mating success and subsequent reproductive success both increase with increased repertoire size. Male mockingbirds duel vocally with each other but direct most of their singing toward females inside their territory (Breitwisch and Whitesides 1987). Mounting evidence suggests that mockingbirds sing primarily to their mates or potential mates. For example, males sing during copulation, sing more when their mate is removed, and sing more, especially at night, when they are unmated.

Migratory species may have international repertoires. Marsh Warblers, among Europe's most versatile vocal mimics, spend much of the year in Africa (Dowsett-Lemaire 1979). Although it imitates some European species, most of the songs a Marsh Warbler broadcasts are those of African birds, which it hears during migration and on the wintering grounds. Territorial male Marsh Warblers thus may inform potential mates where they spend the winter. It could be to a female's advantage to pair with males adapted for wintering in the same part of Africa as she does, and thus to produce young with similar tendencies.

Some birds use vocal mimicry to attract help in the mobbing of predators. Thick-billed Euphonias, a tropical tanager, imitate the mobbing calls of other tropical birds when a nest is threatened (Morton 1976). Neighbors of the species being imitated then gather to scold and help discourage the predator.

The ability to imitate new sounds is important to the development of a young bird's vocal repertoire. Copying the vocalizations of neighbors also leads naturally to the formation of regional dialects, quite like the local accents of humans. These two important topics of current research—song development and local song dialects—are addressed in the concluding sections of this chapter.

Learning to Sing

Avian vocalizations can be inherited, learned, or invented. The calls of chickens and doves, as well as the songs of certain flycatchers, are inherited. So are the calls of brood parasites, such as cuckoos, honeyguides, and cowbirds, which lay their eggs in the nests of other birds. When these birds are raised in acoustical isolation or are deafened before they hear their fellows sing or call, they nonetheless sing normal songs as adults (Konishi 1963; Nottebohm and Nottebohm 1971; Kroodsma 1984). The Eastern Meadowlark inherits its call notes but learns its songs from other meadowlarks (Lanyon 1957). Inherited mechanisms guide the learning process even in species that learn songs after much practice. The hearts of young Song Sparrows actually beat faster the first time they hear the song of their species. There is no change in their heartbeat when they hear the songs of another kind of sparrow. Inherited mechanisms screen out irrelevant sounds, such as those made by insects, frogs, waterfalls, and trains, and respond to appropriate song models.

Most groups of birds have innate vocalizations. Learning, however, guides the vocal development of parrots, hummingbirds, and oscine songbirds. Although we do not yet fully understand the complex physiological dynamics and neural mechanisms involved, the stages of song development in birds with fixed repertoires are well documented (Box 10–5). Our current knowledge of the early development of bird song may provide one of the best pictures to date of how a complex, learned motor skill develops (Marler 1981, 1983).

Although birds such as Gray Parrots and Northern Mockingbirds add new vocalizations to their repertoires throughout their lives, vocal learning is most intense during, and is often restricted to, early age. Most Marsh Wrens, for example, learn their songs before they are 2 months old (Kroodsma and Pickert 1980). An experimentally isolated White-crowned Sparrow, exposed by a loudspeaker to its species song, has a critical learning period lasting from 10 to 50 days of age. In its natural habitat, however, this sparrow has a learning period that may last several more months.

Some other songbirds, represented by the well-studied Chaffinch (Thorpe 1958), are typically receptive to song models for 10 to 12 months. The critical learning period in such birds lasts about a year—that is, into the first breeding season—at which time first-year males have a chance to learn songs from more experienced males. Termination of the critical learning period of the Chaffinch corresponds to the rise of testos-

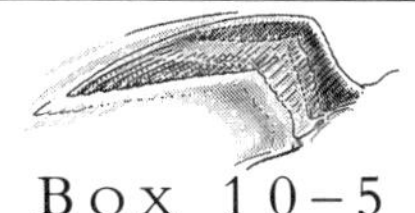

Box 10–5
Birds with fixed repertoires learn songs in four stages

Observations of and experiments on the development of singing behavior of hand-reared baby birds have revealed four key periods that influence adult songs.

1. Critical learning period The early period during which information is stored for use in later stages of learning. In most species, the critical learning stage lasts less than a year—sometimes much less.

2. Silent period The long period (up to 8 months) in which syllables learned during the early critical learning period are stored without practice or rehearsal.

3. Subsong period This practice period is analogous to infant babbling. It apparently bridges the gap between the perceptual and sensorimotor stages of vocal learning. The subsong period is a period of practice without communication, perhaps subsong is a form of vocal play. (See text for a discussion of subsong.)

4. Song crystallization The next practice period during which the young bird transforms plastic song into real song by selecting a few syllables from its unstructured repertoire, perfecting them, and then organizing them into correct patterns and timing. (See text for a discussion of plastic song.)

terone levels in the spring. Castrated Chaffinches can learn new songs for up to 2 years (Nottebohm 1967). Sarah Bottjer and her colleagues (Bottjer et al. 1984) discovered a discrete forebrain nucleus that connects to the main song production control nuclei in the High Vocal center and that appears to control the learning of songs by young Zebra Finches (see Chapter 8). Lesions made in this nucleus—named MAN—during the early critical period permanently impair a finch's song-learning abilities. However, lesions made in the nucleus did not affect the maintenance of song patterns of adult birds. This nucleus accumulates the sex hormone testosterone, which mediates both the timing and sexual basis of song learning.

Isolation experiments demonstrate the importance of experience during the early critical learning periods (Nottebohm 1975). Isolation from the model songs of adults at this stage permanently handicaps a bird's singing ability; it will never develop a normal song. Although individuals isolated at an early age still sing, their songs are less complex, have fewer notes per syllable, and have less frequency modulation than normal songs. Nevertheless, the innate songs of isolated birds resemble those of their species. The syllables are similar in form and are repeated with approximately the same timing; the tonal quality also resembles that of normal song.

The second stage of song development—the silent period—can be characterized as a time during which syllables that have been memorized during the critical learning period are stored in the brain. Swamp Sparrows store memorized song syllables for 240 days (Marler and Peters 1981). When this period has elapsed, young sparrows start practicing by listening to themselves and matching some of their vocalizations to previously memorized syllables. The initial, sensitive perceptual phase of song learning thus is well separated from the later sensorimotor phase by a period of silence.

The practice stages begin with subsong—a long, soft, unstructured series of syllables and ill-formed sounds. Distinctly formed sounds begin to emerge, some of them recognizable as syllables heard during the sensitive period. Within a month or so, depending on the species, subsong develops into the first attempts at producing mature song. This so-called plastic song contains only rudiments of the final structure. In a matter of weeks, during song crystallization, the young bird transforms plastic song into real song. Even birds that have been isolated as nestlings can develop the timing typical of their species. In their final songs, young male Swamp Sparrows use only one-fourth of the syllables they learned and practiced in the earlier phases of song development (Marler and Peters 1982b).

Auditory feedback is essential for song development. No oscine songbird produces a normal song if it has been deafened before song crystallization begins. In the deaf bird, recognizable structural entities seldom appear, and when they do, they deteriorate quickly. Frequency modulation of syllables is extremely poor in deaf birds, and they do not repeat sounds accurately (Konishi and Nottebohm 1969). Experimental deafening of male White-crowned Sparrows during their silent period (70 to 100 days of age) erases their original song memory or interferes with a necessary matching process. Songs of such males do not differ from those of males that have been deafened so early that they never heard model songs. Deafening after song is crystallized, however, has little effect. Apparently, auditory feedback becomes less important after the correct motor patterns are developed (Marler and Mundinger 1971).

Learning and imitation are not the only elements of song acquisition. Individuality is important, too. Young birds transform and improvise as they develop individual signatures in their songs. They systematically transform memorized themes or mix syllables from several models into unique themes. A single song of the Swamp Sparrow, for example, may contain invented, improvised, and imitated elements (Figure 10–14).

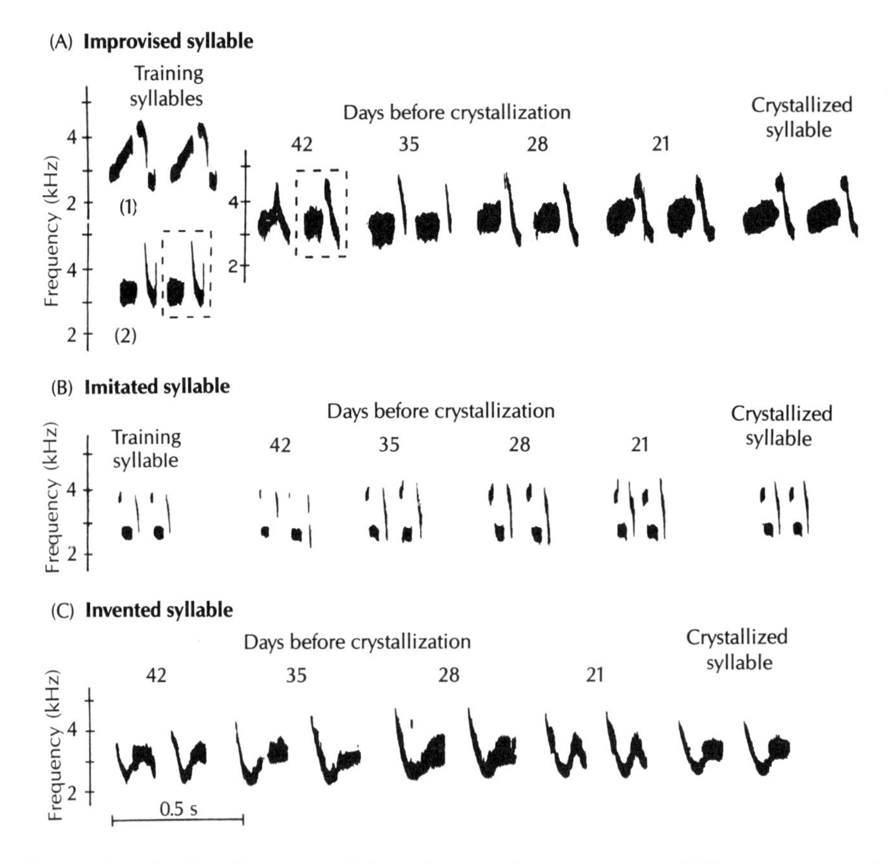

Figure 10–14 Development of three Swamp Sparrow song syllables by means of (A) improvisation, (B) imitation, and (C) invention. Improvisation involved changing training syllable A2 to a new syllable by combining the second note of training syllable A1 with the first note of syllable A2. The invented syllable was unlike any training syllable but proceeded through the same developmental stages as the improvised and imitated syllables to a final crystallized form. (From Marler and Peters 1982a)

However, in this effort the sparrow rarely breaks up series of notes that constitute a syllable. In fact, the syllable may be a natural perceptual unit, designed to map readily onto patterns of sound production (Marler 1981).

Song learning is mediated and constrained by inherited sensory templates—neural filters that pass only particular sounds. A young bird must select appropriate song models with precision from a rich sound environment. Studies of Swamp Sparrows and Song Sparrows illustrate this aspect of song learning (Marler and Peters 1977, 1981, 1982a). A Swamp Sparrow's song is a repetitious trill of a single syllable, whereas a Song Sparrow's song uses a pattern of several complex syllables. To discover how the young of these species learn their own songs despite the fact that they grow up hearing both songs, Susan Peters and Peter Marler isolated nestling sparrows and then exposed them to taped songs during the critical learning period. Syllable structure is the key to song learning for young Swamp Sparrows, whereas temporal pattern is the key for young Song Sparrows. Swamp Sparrows do not learn the Song Sparrow song because they cannot learn its syllables. Song Sparrows do not learn the Swamp Sparrow song because they cannot learn its temporal pattern.

Dialects

Bird songs can vary within a species from coast to coast, or from one hilltop to the next. The importance of imitative learning in the song acquisition of young birds leads naturally to local dialects similar to those of human societies.

Regional variation in song characteristics, such as syllable structure or delivery pattern, is typical of oscine birds. Carolina Wrens in Ohio, for example, sing faster than those in Florida, and Bewick's Wrens in California, Arizona, and Colorado have regionally distinctive song patterns (Kroodsma 1982a) (Figure 10–15). The persistent song dialects of White-crowned Sparrows on the central California coast are restricted to areas of only a few square kilometers (Kroodsma et al. 1985; Trainer 1985). One dialect, the *Berkeley* dialect, has persisted for decades. The average life span of local song neighborhoods of Indigo Buntings in southern Michigan was three times that of the individual buntings (or roughly 15 years) (Payne 1983). The song traditions of birds may provide excellent models for the analysis of rates of culturally related change in human language and in bird behavior (Payne 1981a, b; Payne et al. 1981).

To the degree that vocal dialects result from the process of song learning, patterns of geographical song variation may simply reflect recent history. New song traditions arise when young birds colonize new areas. The distinct songs of Chaffinches in each valley of northern Scotland presumably arose this way, as did the mosaic distribution of local song types of Eurasian Tree-Creepers (Thielcke 1961, 1969). This theory

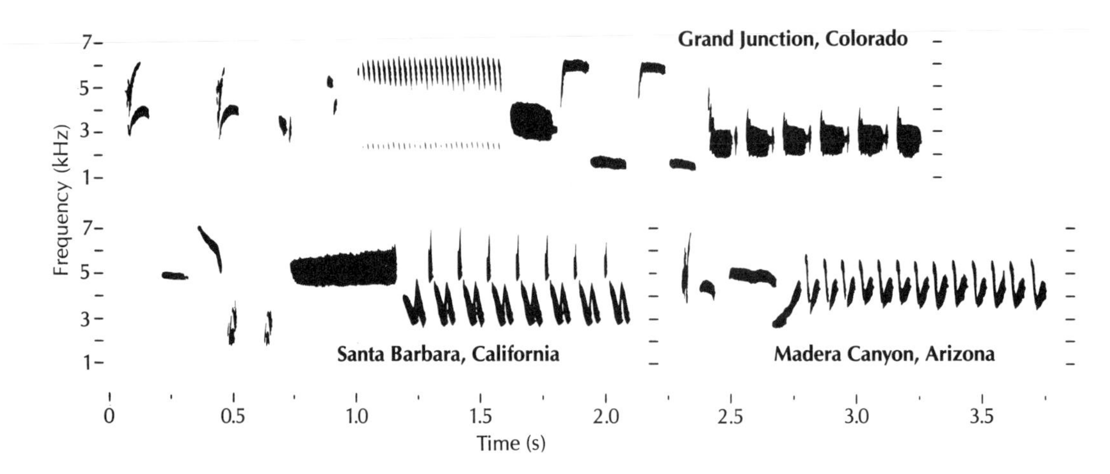

Figure 10–15 Song dialects: Bewick's Wrens sing strikingly different songs in Colorado, California, and Arizona. (From Kroodsma 1982a)

states that the historical model of song dialect formation parallels that of geographical speciation, in which small, isolated populations accidentally evolve differences (see Chapter 22).

Alternatively, social forces may mold the patterns of local song variation. Local song dialects arise when young males increase their reproductive success by imitating their older neighbors. First-year male Indigo Buntings copy the song of an established neighbor and thereby increase their chances of holding a territory, pairing with a female, and fledging young of their own (Payne 1982b). In Africa, young males of an unrelated, promiscuous finch, the Village Indigobird, increase their chances of attracting females by mimicking the song details of the dominant males that do most of the mating in a local area (Payne and Payne 1977).

A third theory—the ecological hypothesis—holds that song dialects influence the genetics of local populations (Baker 1982). To the degree that local song dialects mark an environment, such as the one in which a young bird was raised successfully, they could potentially guide an individual's choice of territory or mate and lead to genetic differences between dialects or local populations. This hypothesis was not supported by Robert Payne and David Westneat's (1988) studies of Indigo Buntings in Michigan. In this species, local song neighborhoods are not genetically distinct and provide no apparent information for selection of genetically similar mates. The controversial possibility of such an effect in Pacific coast populations of White-crowned Sparrows is under continuing study to tease apart and control the many variables involved (Kroodsma et al. 1985). The functional relationship among song dialects, the genetic structures of local populations, and speciation will be an important topic of future research.

Summary

Birds use vocalizations to mediate social interactions, particularly over long distances, at night, and in dense cover. The physical characteristics of vocalizations affect their information content and their transmission effectiveness through the environment.

The vocal virtuosity of birds springs from the structure of the syrinx, a sound-producing organ located at the junction of the two bronchi at the base of the trachea. Sound results from the vibration of a thin membrane, the tension and position of which are controlled by syringeal muscles and air pressure in the interclavicular air sac. Many birds can stimulate the two sides of the syrinx independently and thus can sing two songs simultaneously. The vocal tract, particularly the trachea, filters the sounds produced by the syrinx and can add resonant qualities to the calls of cranes, swans, and some birds-of-paradise, which have long, coiled tracheas.

The vocal repertoires of birds are among the richest in the animal kingdom. The loud broadcasts of territorial birds, which are among the most familiar vocal displays, convey information about the identity, location, and motivation of the singer, including ownership of territorial space. More varied song repertoires help to attract females and foster superiority in vocal duels between competing males. Included in the acoustical structure of songs are features that birds use for both species and individual recognition. Precise duets used by mated pairs also serve as distinctive vocal signatures. Vocal mimicry is one way some species increase the size of their vocal repertoires.

Avian vocalizations may be inherited, learned, or invented. Learning guides vocalization development in songbirds, parrots, and hummingbirds. Four stages of song learning are evident: an early critical learning period, a long silent period, and two practice periods, which involve subsong production and song crystallization. Guiding this process are inherited templates of song characteristics that screen out irrelevant sounds.

The formation of song dialects in a local culture is one possible consequence of the process of song learning. Dialects may reflect accidents of history and cultural change, may be used to enhance the reproductive success of young males, and may foster the evolution of local genetic differences among bird populations.

Further Readings

Gaunt, A.S., and S.L.L. Gaunt. 1985. Syringeal structure and avian phonation. Current Ornithology 2: 213–245. *A review of recent work on the mechanics of syrinx function.*

Greenewalt, C.H. 1968. Bird Song: Acoustics and Physiology. Washington, D.C.: Smithsonian Institution Press. *A pioneering acoustical analysis of bird song.*

Kroodsma, D.E., and E.H. Miller, Eds. 1982. Acoustic Communication in Birds. New York: Academic Press. *A rich collection of papers on all aspects of avian vocalizations.*

Payne, R.B. 1986. Bird songs and avian systematics. Current Ornithology 3: 87–126. *A current review of the use of song differences as taxonomic characters.*

Pepperberg, I.M. 1991. Learning to communicate: The effects of social interaction. Perspectives in Ethology 9: 119–164. *An enlightened perspective of vocal learning in birds with reference to human social models.*

Spector, D.A. 1992. Wood-warbler song systems: A review of paruline singing behaviors. Current Ornithology 9: 199–238. *An interesting taxonomic perspective of the functions of different song types in the wood warblers of North America.*

Thorpe, W.H. 1961. Bird Song. London: Cambridge University Press. *A classic.*

PART IV

Behavior and the Environment

CHAPTER 11

The Annual Cycles of Birds

BIRDS FACE SEASONS of stress and seasons of opportunity. Month-to-month changes in day length, climate, and resources are inherent in the Earth's daily spin and annual solar orbit. Whereas seasons in temperate climates are regulated by temperature, tropical seasons are due to rainfall. Each year an adult bird invests time and energy above and beyond that required for daily survival into three main efforts—reproduction, molt, and, in some cases, migration. Tight scheduling of these added costs during the most favorable seasons often is required. The conflicting demands of these efforts, combined with seasonal resources and opportunities, define a bird's annual cycle (Figure 11–1).

This chapter first describes the basic components of avian annual cycles and then proceeds to the physiological clocks—called circadian rhythms and circannual cycles—that control the annual calendars of birds by synchronizing a bird's internal state with its seasonal environments. Photoperiod—the length of daylight—is an essential environmental cue for the clocks. The chapter concludes with the timing and costs of breeding and molting. Energy for these activities must be above and beyond that required for self-maintenance, as discussed in Chapter 6.

Annual Cycles

The typical year of some permanent residents has only three main, sequential tasks: breed, molt, and survive until the following breeding season. In equatorial Borneo (Sarawak), for example, where the climate is unusually stable and day length is unchanging, small birds normally start to nest when the heavy rains begin in December (Fogden 1972). Adults begin to molt shortly after the young have left the nest in May and continue molting until the beginning of the 2-month "dry" season, when

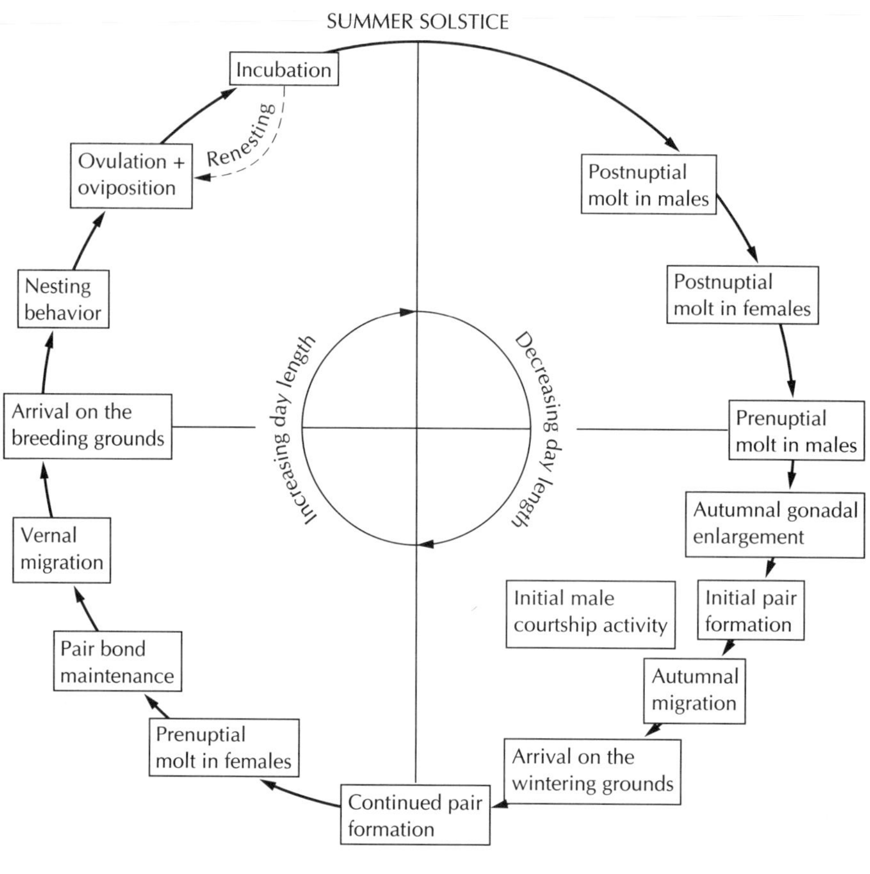

Figure 11–1 Annual cycle of the Mallard. Most individuals attempt to breed when they are 1 year old. The terms *prenuptial molt* and *postnuptial molt* from Dwight's (1900) system of molt and plumage nomenclature correspond to *prealternate molt* and *prebasic molt*, respectively (see Chapter 4). (From Bluhm 1988)

food starts to become scarce. When heavy rains resume and food supplies increase, gonads increase in size and the cycle repeats itself (Figure 11–2).

Similar sequences of reproduction and molt are typical of permanent residents of northern temperate localities, including Song Sparrows in Ohio, Black-capped Chickadees in Wisconsin, and Chaffinches in Britain. After the quiescent winter months, sex hormones flow, gonads increase in size, and males proclaim their territories with conspicuous songs and, if necessary, brutal fights. Pair bonds are established or reaffirmed and mating occurs. Young are hatched in May and June and generally reach independence by late July. Molt follows in August and September. At this time, young birds leave their natal territories, and some residents, among them chickadees (Paridae) and Chaffinches (Fringillidae), aggregate into well-organized flocks for the winter. Social competition for territories, food, and mates—all resources essential for reproduction—

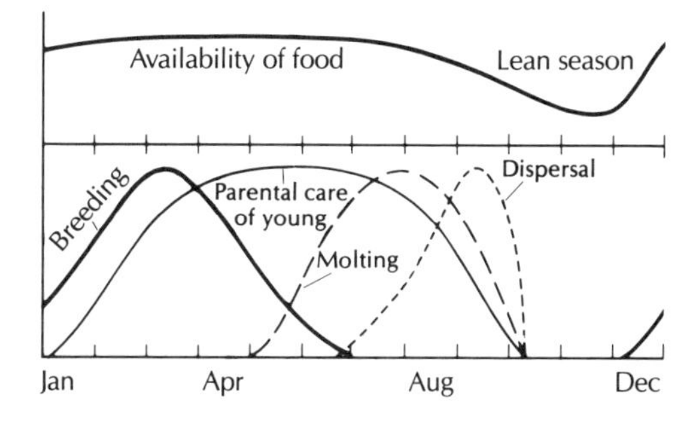

Figure 11–2 Birds have well-defined breeding and molting seasons, which coincide with the months of greatest food availability, even in the equatorial rain forests of Borneo. (After Fogden 1972)

may start in the autumn and may directly influence gonadal development and later reproduction (Hegner and Wingfield 1986)

Migration complicates the annual cycle. After they breed and often molt, migratory birds generally gather in flocks and eat tremendous amounts of food, fueling themselves for their trips. As the date for departure to the south approaches, they become restless after dark. After migration, when they have reached their wintering grounds, their physiology returns to "normal." Migratory preparations are repeated the following spring for the return north, where the cycle of reproduction, molt, and preparation for migration repeats. Many Temperate Zone birds, especially those that migrate, molt twice a year, once after breeding and again in late winter or early spring. The spring molt usually does not include the flight feathers (see Chapter 4).

The annual cycle varies among species and among populations of the same species. White-crowned Sparrows on the Pacific coast of North America illustrate such variation (Cortopassi and Mewaldt 1965; Mewaldt and King 1978). The White-crowned Sparrow breeds throughout northern Canada and from southern Alaska to central California (Figure 11–3). Populations on the Pacific coast differ in the extent of their annual migrations and in other aspects of their annual cycles. Those that breed in Alaska and in northwestern Canada *(Zonotrichia leucophrys gambelii)* are long-distance migrants that winter primarily in California, where they mix with winter flocks of the local nonmigratory sparrows *(Z. l. nuttalli).* Members of another population *(Z. l. pugetensis),* which breed on the coasts of Washington, Oregon, and British Columbia, also mix with *nuttalli* flocks during the winter.

White-crowned Sparrows from northern localities nest later in the spring than those from southern localities. The southern resident *nuttalli* come into breeding condition first, then *pugetensis,* and finally the *gambelii* of the far north. Differences in the timing of gonadal enlargement and breeding activities characterize not only the three subspecies but also the geographical gradients of populations within each subspecies.

Some, but not all, of these White-crowned Sparrows molt in the spring before breeding. This extra "prenuptial" molt is known as the prealternate molt (Chapter 4). Most of the migratory northern coastal sparrows *(pugetensis)* undergo a complete prealternate body molt in the spring before migration and nesting, whereas southern resident *nuttalli* merely molt some head feathers in the spring, if any. All *gambelii* molt before migrating northward, and northern populations of this race molt faster than southern populations. Their molt is not faster because their individual feathers grow faster but proportionately more plumage is replaced at any one time. Northern populations also start this spring molt later than do southern populations, about 3.4 days for each latitudinal degree north (Mewaldt and King 1978).

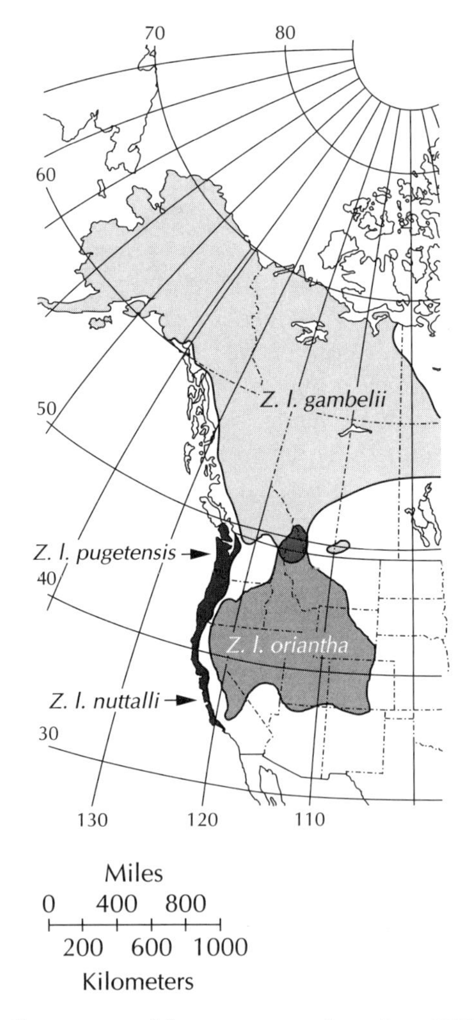

Figure 11–3 Breeding ranges of four western subspecies of White-crowned Sparrows, *Zonotrichia leucophrys*. The most northern races, *Z. l. gambelii* and *Z. l. pugetensis*, migrate to central California, where they winter with resident *Z. l. nuttalli*. The Rocky Mountain race, *Z. l. oriantha*, migrates south to Arizona and Mexico. (From Cortopassi and Mewaldt 1965)

Circadian and Circannual Rhythms

The annual cycle involves an orderly sequence of complex integrated behavioral and physiological conditions. A network of physiological controls regulates the schedules of reproduction, molt, sleep, feeding, and migration. Biological clocks in the cells of all plants and animals control the release of hormones and other chemicals that regulate metabolism, reproduction, and behavior.

Birds possess an elaborate system of biological clocks (Gwinner 1975; Meier and Russo 1985). Neuroendocrine systems synchronize imprecise cellular rhythms so that the entire bird is organized internally and appropriately synchronized with its periodic environment. In addition to regulating the basic daily functions of general activity and cycles of body temperature, these internal clocks are essential to the proper functioning of the sun compass by which birds navigate (see Chapter 13) and are used for the measurement of day length itself. They also govern migratory restlessness, premigratory fattening, and egg laying. Some biological clocks—called circadian rhythms—match the daily 24-hour cycle of the Earth's rotation on its axis. Others—called circannual cycles—are synchronized to the annual cycle of the Earth's revolution around the sun.

Circadian Rhythms

Circadian rhythms are the basic units of the higher order of organization of the complexes of hormonal and neural activities that directly control seasonal activities. Daily cycles of daylight and darkness have a profound effect on the biology of all animals. Twilight triggers a switch in animal physiology from diurnal to nocturnal systems. Every individual has an intrinsic rhythm approximately 24 hours in length in which the rate of metabolism, body temperature, and level of alertness fluctuate in predictable ways. Because they are not exactly 24 hours in length, these internal cycles tend to depart gradually from real time, starting slightly earlier or later each day, unless they are somehow synchronized or entrained by external cues—or *Zeitgebers,* literally, "time givers."

When Chaffinches are kept in constant dim light, their endogenous rhythms—for example, activity and metabolic rate—function in a period of about 23 hours and therefore drift about 1 hour per day (Figure 11–4). White-crowned Sparrows have a regular cycle of activity and sleep that is just under 24 hours when they are kept in a dimly lit experimental cage. Natural, external light–dark cycles, however, adjust (technically, entrain) the endogenous rhythm, keeping it synchronized with the 24-hour cycle.

Circannual Cycles

Endogenous rhythms control the annual cycles as well as the daily cycles of some birds. Self-sustaining circannual rhythms have a period of approximately 1 year. When captive European Starlings, Garden Warblers,

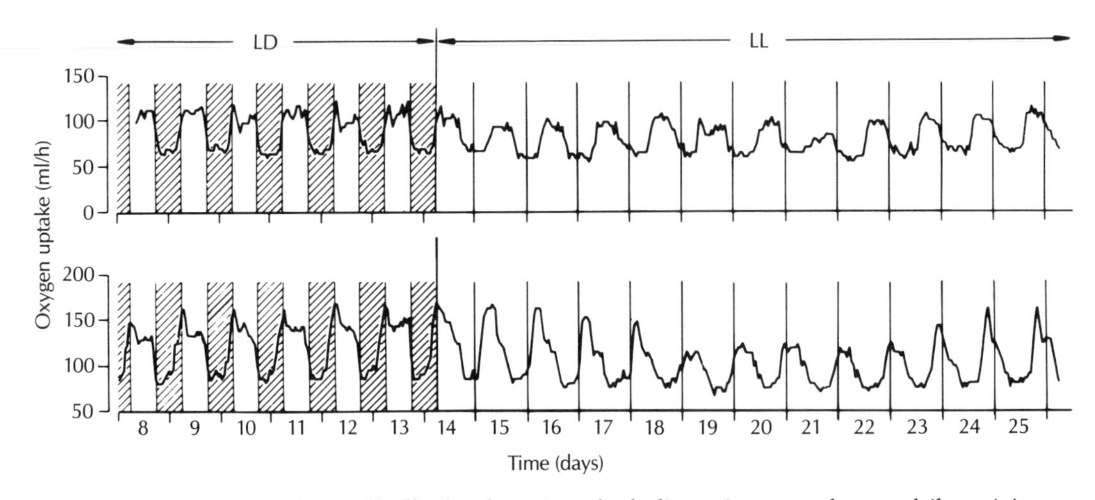

Figure 11–4 Chaffinches kept in a dimly lit environment have a daily activity cycle (measured here in terms of oxygen uptake, milliliters per hour) of just under 24 hours. This experiment demonstrates that under constant dim illumination (LL) the cycle drifts 1 hour of clock time unless it is synchronized by an external stimulus such as regular 24-hour light–dark cycles (LD). (Adapted from Aschoff 1980)

and Blackcaps are kept in a constant daily environment of 12 hours of light and 12 hours of dark, they come into breeding condition and molt in a predictable annual cycle (Figure 11–5).

Drift characterizes circannual cycles in constant environments, just as it does in circadian cycles. Seasonal changes in day length probably entrain the endogenous circannual rhythms, just as daily light–dark cycles entrain the circadian rhythms, but this hypothesis requires more study (Farner 1980b). In certain equatorial birds, such as the Sooty Tern, which functions on a 9.6-month internal cycle rather than a 12-month cycle, natural selection has favored the uncoupling of the internal annual cycle from seasonal change (King 1974).

Role of Photoperiod

Seasonal reproduction by birds has favored the evolution of a control system that synchronizes the physiologies of individuals with the environment. Day length—or photoperiod—plays a key role in this control system. Specifically, the avian photoperiodic control system uses two kinds of information (Farner 1980c): environmental light, which stimulates neural receptors; and clock information from an internal circadian cycle, which enables the bird to measure day length. This system allows birds to respond at the mean optimal time for reproduction, to synchronize reproductive function in mating pairs, and to terminate reproductive function—three fundamental requirements for control of the annual reproductive cycle.

In recent years the pineal gland has been regarded as the probable location of the biological clock and the mechanisms of photosensitivity in birds, partly because the pineal glands of many reptiles are photosensitive organs. The pineal glands are essential to normal circadian rhythms in at least some species of birds. Experimental removal of the pineal gland causes normal 24-hour cycles to disappear. Unlike the glands in reptiles, however, avian pineal glands do not house the primary light receptors.

William Rowan (1929) pioneered in research on photoperiodic control of avian gonadal cycles. He showed that increases in photoperiod of only 5 to 10 minutes per day caused the testes of Dark-eyed Juncos to increase

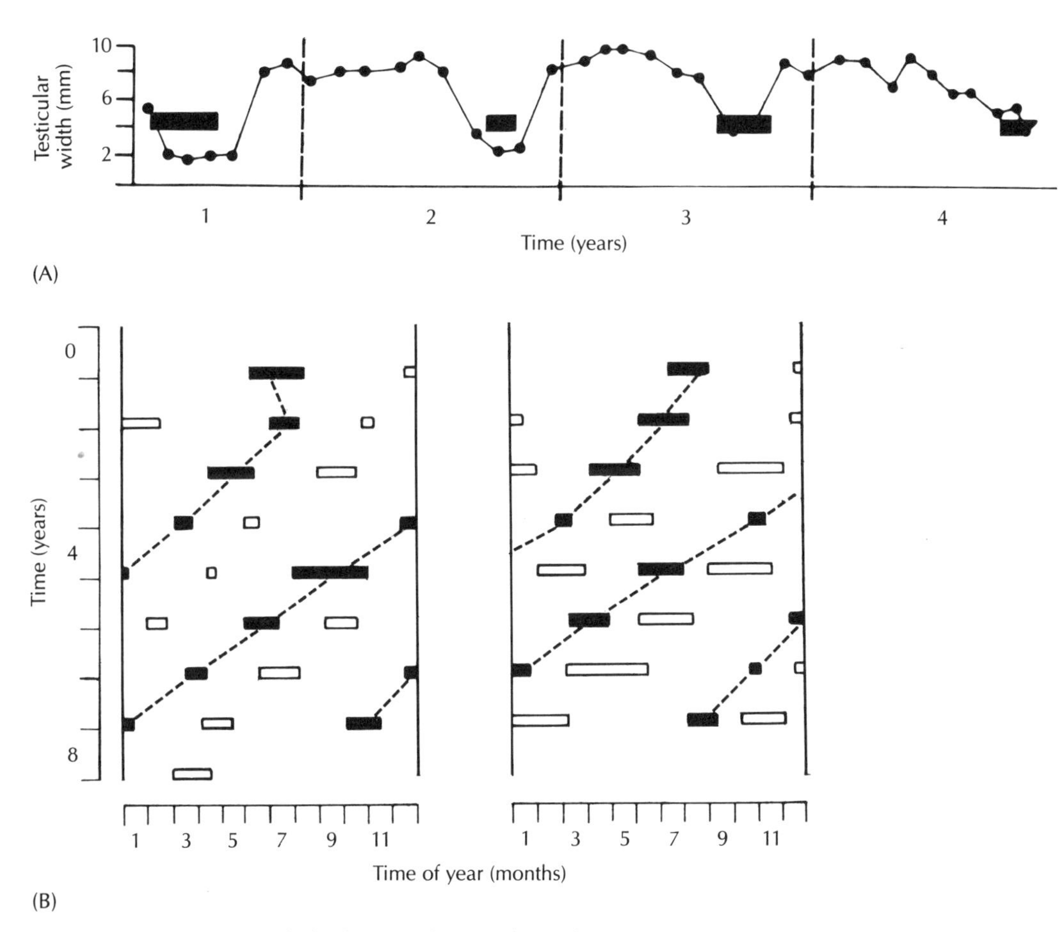

Figure 11–5 Circannual rhythms under constant photoperiodic conditions. (A) Rhythms of testicular width (curves) and molt (bars) in a European Starling. The undamped oscillations in testes size and the intervals between successive molts deviate irregularly from a 12-month cycle. (B) Rhythms of summer molt (solid bars) and winter molt (open bars) in a Garden Warbler (left) and in a Blackcap (right), both maintained in captivity for 8 years. Both molts occur progressively earlier each year because the birds have an internal rhythm with a mean period of about 10 months. (Adapted from Gwinner 1977 and Berthold 1978)

in size, an effect that was reversible and repeatable up to three times between autumn and spring (Figure 11–6). The phenomenon of photoperiodic control of gonad cycles has since been recognized in over 60 north temperate bird species in various orders and families.

The control of the White-crowned Sparrow's annual cycle is one of the best examples of this physiological phenomenon (Farner 1980c). Increasing photoperiods during late winter and early spring stimulate events in the annual cycle. The longer days of early spring stimulate gonad development, the prealternate molt, and spring migration. Supplemental information such as warmer temperatures, rainfall, and the springtime display behavior of other sparrows provides fine tuning of physiological events later in the year; they stimulate the final stages of gonad development on the breeding ground and, as a result, the increased secretion of sexual hormones. After the birds breed, the shortened days of late summer

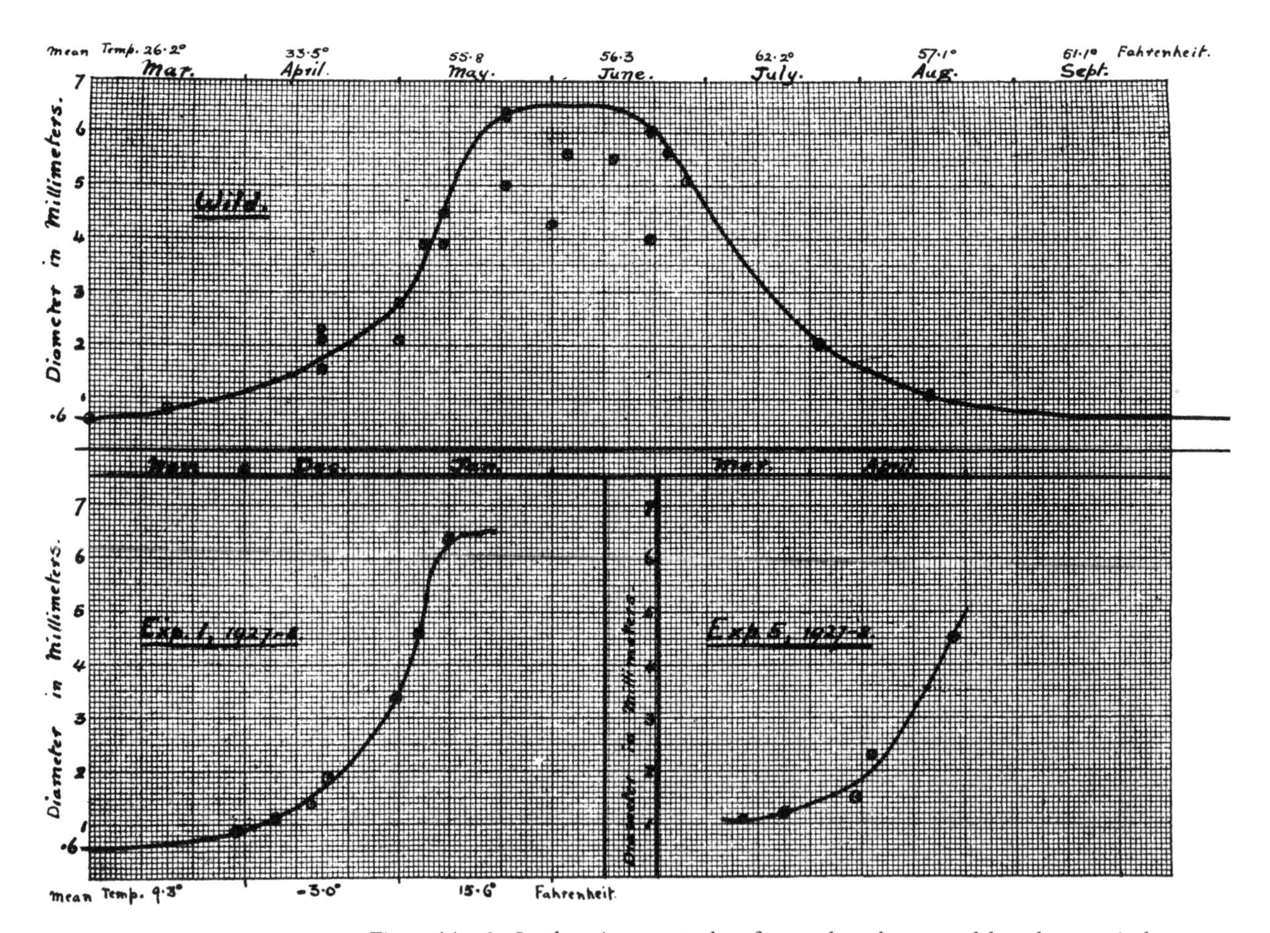

Figure 11–6 In the pioneer study of annual cycle control by photoperiod, William K. Rowan demonstrated that longer day lengths caused the testes of captive Dark-eyed Juncos to increase prematurely to full size in January (lower left) and again in April (lower right), instead of in May and June as in wild juncos (upper). Mean temperature is the average air temperature in that month. (From Rowan 1929)

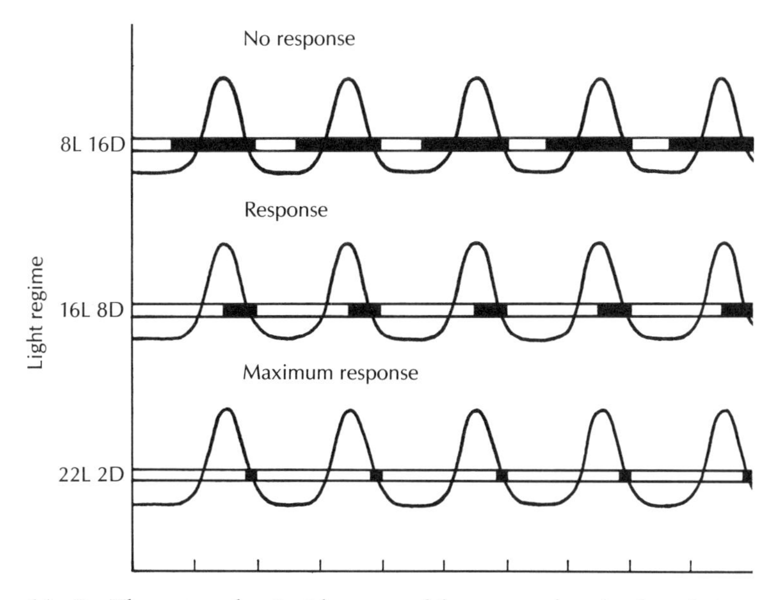

Figure 11–7 The external coincidence model suggests that day length is measured via the increased amount of time that daylight periods (open bars) coincide with the photosensitive phase of the circadian rhythm (oscillation peaks). L, number of hours of light; D, number of hours of dark. Response was measured in terms of gonadal enlargement, which was greatest for a 22-hour light–2-hour dark cycle. (From Farner 1980a)

stimulate the main (prebasic) molt. The timing of this molt, as well as of a light-insensitive—or photorefractory—period of the testis and changes in body fat in preparation for fall migration, is actually triggered the preceding spring by increasing day lengths. Finally, the very short days of early winter, which inhibit gonad growth, terminate the refractory period, and thereby restore sensitivity to the stimulus of long photoperiods —and the cycle begins anew. Short winter days are essential to the control of the annual cycle: The testes will not grow in response to the long days of spring unless the bird has experienced a prior period of short day lengths. Thus, White-crowned Sparrows stay in nonbreeding condition for several years if exposed only to long photoperiods.

Circadian rhythms of photosensitivity are probably a basic adaptation of cellular organisms to the 24-hour light–dark cycle of the planet (Farner 1980a). Circadian rhythms include a limited photosensitivity period, during which external light stimulates receptors in the brain, which in turn trigger a series of physiological reactions. As day length increases, so does the chance that there will be daylight during the photosensitive period (Figure 11–7). Not only does the chance of overlap—or coincidence—increase with increasing day length, but the duration of the period of overlap also increases. The amount of overlap enables birds to measure day length. Originally developed by Erwin Bünning for plants, we now have evidence of this "external coincidence" model for at least 10 species of birds. The external coincidence model, however, is the simplest working model of the mechanisms of circadian rhythm. Other, more complicated models based on the interactions of two or more internal rhythms exist, and their refinement is a high priority in the study of the control of biological rhythms.

The Receptor-Control System

Birds do not monitor day length visually, as do mammals, but do so by means of special receptors in the hypothalamus of the brain. Longer day lengths induce gonad development and migratory behavior even in eyeless birds. The light receptors of the White-crowned Sparrow, for example, lie in the ventromedial hypothalamus of the lower midbrain. They are structurally unspecialized elements that are sensitive to extremely low light intensities such as those that directly penetrate brain tissues. Pinpoint illumination of the hypothalamic receptors by means of a single, thin light-conducting optical fiber induces both testicular growth and migratory behavior (Yokoyama and Farner 1978). This technique enables precise determination of the locations of the receptor cells, which are most affected by light wavelengths of 500 nanometers (probably absorbed by the pigment rhodopsin).

After stimulation of the photoreceptors, neurosecretory cells in the hypothalamus induce the release of neurohormones from the ends of axons in the median eminence, the neural portion of the pituitary that links this organ to the midbrain (Figure 11–8). The released neurohormones are then carried in the blood to the anterior pituitary gland, where they induce the synthesis and release of hormones that directly affect the activity of the gonads themselves. Thus, a series of neural and physiological events translates increasing day length into sexual activity.

The avian control system differs from those of other tetrapod vertebrates, not only in the use of hypothalamic photoreceptors and a

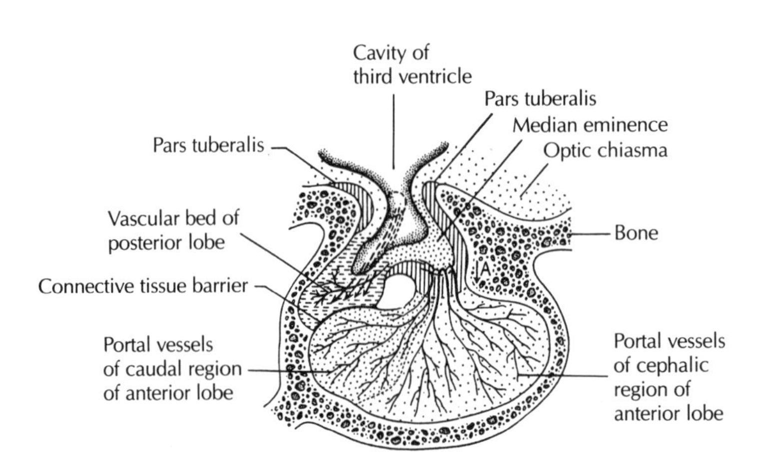

Figure 11–8 Avian pituitary gland and adjacent structures. Daylight stimulates special photoreceptors in the tuberal region (pars tuberalis) of the lower hypothalamus of the midbrain. Neurohormones are released in the median eminence and carried to the anterior pituitary gland via the hypophyseal portal blood vessels. They stimulate gonadal hormone production and, as a result, gonadal activity. (From Höhn 1961)

circadian system for measuring day length but also by morphological separation of function within the median eminence of the hypothalamus, within the system of portal blood vessels, and within the pars distalis—or main part of the pituitary. The divisions of these organs may ultimately be related to the evolution of separate controls of follicle-stimulating hormone (FSH) and luteinizing hormone (LH), the two hormones that are most important in controlling gonadal development and function (see Chapter 15). Research on male birds has shown that photostimulation during the photosensitive phase of the circadian rhythm causes an increase in the pulsed release of LH. After release, pulses of plasma LH travel throughout the bird's body and stimulate gonadal activity. Release of FSH also increases with photostimulation, although the speed of its release lags behind that of LH. In combination with LH, FSH stimulates the ovaries and testes to make gametes.

It appears that after photoperiodic regulation of the annual cycle evolved, some additional safeguards and corrections were essential. Photorefractory physiology is one of these. The gonadal cycle normally concludes with a rapid collapse and reabsorption of gonadal tissue. Then follows the photorefractory period, during which long days do not induce gonadal regrowth. The mechanisms of photorefractory physiology are still a mystery, but the phase relationships of two different circadian rhythms may be responsible (Meier and Russo 1985).

The photorefractory period is best developed in migratory Temperate Zone species. Some nonmigratory temperate birds also have a photorefractory period; but it is weak or absent in most photosensitive tropical species examined to date. The photorefractory physiology of adults, however it works, seems to be an adaptation for scheduling molt and migratory preparations during the favorable conditions of late summer by discontinuing reproductive activity while days are still long (Miller 1959; Farner 1980c). Reproductive activity by yearling birds in response to late summer photoperiods would likewise be disadvantageous.

Physiological Control of Molt

Molt and preparations for migration are also triggered by changes in day length and can be experimentally manipulated. Stephen Emlen (1969), for example, accelerated the annual cycle of Indigo Buntings, inducing an extra molt into the year by suddenly increasing the photoperiods to which captive birds were exposed. The endocrine pathways that tie molt directly to photoperiod, or those that tie it indirectly to photoperiod via the gonadal cycles, are not as well defined as the links between the gonads and the hypothalamus through the pituitary gland. Yet the endocrine hormones clearly affect the timing and course of molt. Historically, thyroid hormone was thought to regulate molt in birds, but the relationship is not a simple or direct one; rather, indirect interactions between thyroid activity and the gonadal cycle probably are involved.

The gonadal hormones—androgens and estrogens—appear to inhibit molt, because molt begins as the influence of the hormones on the bird's breeding physiology wanes (Hahn et al. 1992). Nonbreeding and reproductively unsuccessful individuals begin to molt earlier than successful breeders. Experimental injections of gonadal hormones into molting birds slow or even stop molt. Molt cycles, however, are not directly coupled to gonadal cycles because castrated birds continue to molt on schedule. Thus, the inhibition of molt by gonadal hormones may control the timing of molt to some degree, but they cannot be responsible for the initiation of the process itself. Furthermore, Rock Doves, Anna's Hummingbirds, and Great-tailed Grackles, among others, start to molt in the spring while their gonads are enlarging and their sex hormone levels are high.

Breeding Seasons

Guiding the evolution of the timing of seasonal physiological controls have been such factors as the timing of adequate food supplies for both parents and their young, the availability of nest sites, the locations of favorable climates, and areas or times of low predation risk, all of which ornithologists call ultimate factors. Over many generations, the control systems are tuned to the best average time for reproduction. However, they provide no guarantee against the vagaries of particular years. Drought or parasites may cause widespread nesting failure in some years. The birds cannot predict such disasters before starting to nest, but they can, of course, make last-minute adjustments.

Proximate factors are the external conditions that actually induce reproduction. The correct habitat, new vegetation or abundant food, the ritualized displays and aggression among neighbors, and social stimulation in general are all proximate factors that help to bring on the final stages of gonad enlargement and ovarian development. For desert species, good rains can trigger breeding, irrespective of the season. Temperature is probably the most important modifier of annual gonadal cycles (Farner and Mewaldt 1952). For example, a 10-degree increase in average spring air temperatures stimulates Eurasian Skylarks to lay their eggs (Delius 1965).

Pinyon Jays begin their breeding activities when they encounter young, green pine cones. Their annual cycle is closely tied to the availability of the seeds of the pinyon pine, one of their primary foods. In southwestern New Mexico, Pinyon Jays breed sporadically, sometimes in the autumn if there is a bumper crop of pine seeds. To determine whether green pine cones, the first visual evidence of future food abundance, actually trigger breeding by these jays, David Ligon (1974) isolated two groups of 10 male jays for a year. He gave one group fresh green pine cones daily in the summer, when their testes decreased in size after the normal spring enlargement. This offering caused a dramatic reversal of their

Box 11–1
Precisely when do American Robins nest?

The American Robin is one of the most widespread and familiar species of North American birds. Like many other species that must await the arrival of warm spring climates, robins nest progressively later at more northern and western (mountain) locations. Frances James and Hank Shugart (1974) developed a model that used climatic variables to predict when robins would nest in a particular region. Using dates for the nestling period in 8544 nests on file with the Cornell Laboratory of Ornithology's Nest-record Card Program, they showed that in the East an average robin nested 3 days later for each degree of increasing latitude and at progressively cooler temperatures as spring progressed northward.

Combinations of temperature and humidity were the best predictors of the nestling period. Robins typically had nests with young in late April and early May when the relative humidity was about 50 percent and the temperatures were between 45° and 65° F. These environments define either directly or indirectly the environments that allow successful nesting. The robins nested later at localities with higher or lower relative humidities. The model also showed that certain localities where robins do not breed, such as San Diego, California, and El Paso, Texas, fall outside the species-defined climate space.

reproductive state. The testes of the control group of jays (no pine cones) continued to shrink. Thus, the abundance of food is an important modifier of the annual gonadal cycles of this jay.

The joint action of proximate and ultimate factors delimits the characteristic breeding seasons for the majority of species in a particular region. April, May, and June, the spring months when temperatures rise and insects emerge, make up the primary breeding season in the north temperate region; September to December are the primary breeding months in the south temperate region. A few species nest before the weather seems to be suitable. Red Crossbills in the Rocky Mountains, for example, will nest in January and February, surrounded by snow, if conifer seeds, their primary food, are abundant. Omnivorous Common Ravens and Rooks, as well as Gray Jays in Canada and their counterparts in Siberia, also nest early in the spring if they are able to find adequate food. Many large raptors can find food more easily in the open winter woods than after new leaves emerge; Great Horned Owls, for example, incubate their eggs in January and February.

Rainfall usually defines the seasons in the tropical lowlands, especially in arid environments (Bourne 1955; Marchant 1960). Although some individuals can be found breeding in most months in the Tropics, nesting activity in lowland Costa Rica reaches a peak for most birds at the end of the dry season and early in the rainy season. Kingfishers are an exception, preferring to breed during the dry season when streams run shallow and clear, making fish easier to capture. Hummingbirds, too, nest at the beginning of the dry season when flowers begin to bloom.

Tropical nesting seasons last longer than those in the Temperate Zone. Favorable tropical climates permit nesting for 6 to 10 months, or even, in some cases, throughout the year. On Trinidad, for example, nests of the Ruddy Ground-Dove, the Barred Antshrike, and the Palm Tanager are among those found throughout the year (Snow and Snow 1964). Nesting seasons at temperate latitudes usually last 3 to 4 months. In the high Arctic, where only a month or so is suitable for breeding, birds must start nesting immediately after migration, and sometimes they gain a few days head start by using old nests.

Long-term studies have defined the annual cycles of the 18 species of seabirds that nest at various times of the year on Christmas Island, an atoll in the tropical Pacific Ocean (Figure 11–9A). The independent responses of each species to seasonal changes in feeding opportunities and to the specific needs of its young are tempered by internal physiological rhythms (Schreiber and Ashmole 1970). Gray-backed Terns and Wedge-tailed Shearwaters have well-defined spring and late summer nesting seasons, respectively, whereas Christmas Shearwaters and Common White-Terns nest throughout the year, with ill-defined peaks at opposite seasons. Sooty Terns and Brown Boobies have two distinct nesting seasons each year: the terns in winter and summer, the boobies in spring and fall. In the case of the terns, the second peak reflects another breeding attempt by individuals that were unsuccessful in the first season.

The El Niño Southern Oscillation in 1982 and 1983 severely disrupted nesting by Christmas Island seabirds (Schreiber and Schreiber 1984). El Niño was known historically as the periodic warm water disruption of cold upwelling off the coasts of Ecuador and Peru; it destroys the anchovy fishing industry and causes severe crashes in the local seabird populations. Not just a local phenomenon, the entire equatorial Pacific Ocean changes in response to atmospheric changes that influence global climates. The sudden changes in ocean currents and temperatures and associated flooding rains from August 1982 to July 1983 caused wholesale reproductive failure, severe adult mortality, and the disappearance of the entire Christmas Island seabird community. With the return of normal oceanic and atmospheric conditions, representatives of all seabird species returned to nest again. This event revealed to ornithologists for the first time the sensitivity of tropical bird populations to unpredictable, anomalous global climate changes.

Local populations of a species respond to local conditions. Nesting by Brown Pelicans, for example, is strongly seasonal at northern sites but is prolonged at tropical sites (Schreiber 1980a). Cold water temperatures, which depress food supplies, appear to delay the onset of nesting at all sites. After food availability, the hurricane season is the second most important factor controlling the onset of nesting in these pelicans (this observation holds true for tropical seabirds in general). Pelicans nest irregularly throughout the year in the Caribbean and northern South America, more predictably after the hurricane season during the winter and spring in Florida, and from March to June in Louisiana and the Carolinas (Figure 11–9B).

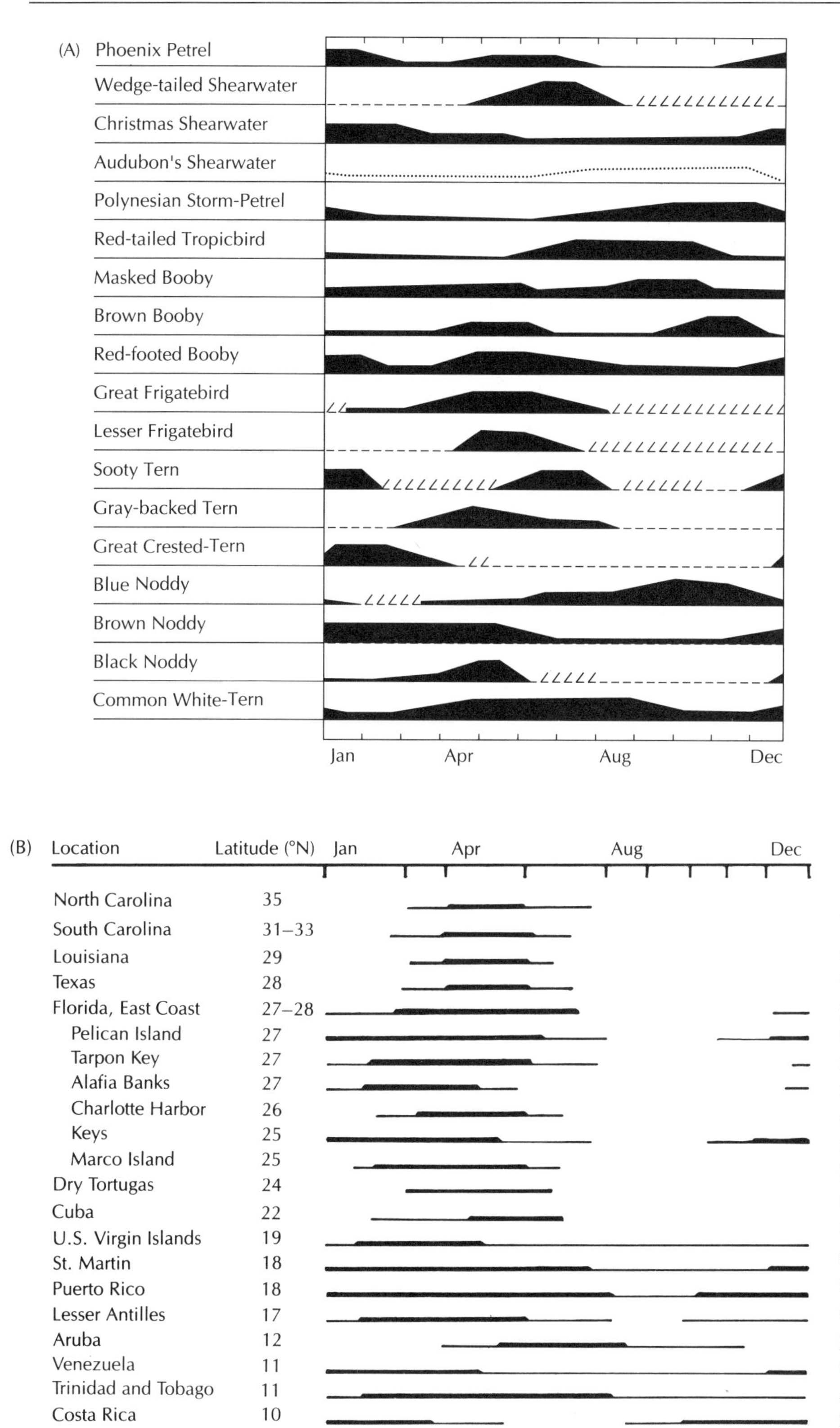

Figure 11–9 (A) Projected breeding seasons of seabirds on Christmas Island, Pacific Ocean. Black indicates the presence of eggs, slanted lines the presence of chicks but not eggs, and dashed lines the presence of adults without eggs or chicks. (B) The time and length of the breeding season (line) of the eastern race of Brown Pelican vary geographically as shown by the date that eggs are laid. The thicker portion of the lines indicates the probable presence of eggs. (A adapted from Schreiber and Ashmole 1970; B adapted from Schreiber 1980a)

Energetic Costs of Reproduction

The general correspondence between breeding season and food availability reflects the central issue of the annual energy budgets of birds. The energetic costs of reproduction and molt favor segregation of these stages during the annual cycle. Birds can assume the costs of migration, reproduction, or molt only after they have first met the costs of self-maintenance, their highest priority, and such basic social interactions as may be necessary to obtain food or a roost site, their second-highest priorities (King 1974). Some seasons, such as a northern winter, permit only self-maintenance for most species, whereas others accommodate additional activities. Reproduction and molt must be scheduled during the months when a bird's requirements for self-maintenance are lowest or when food availability is great. Usually, the costs of only one extra activity can be accommodated.

Peak reproductive activities increase total daily energy expenditures by as much as 50 percent (Ricklefs 1974; Walsberg 1983). Daytime activity costs may actually double or even triple, but overnight costs remain relatively constant. At the beginning of the breeding season, courtship, territoriality, and nest building demand significant effort. Only minor amounts of productive energy are channeled into growth of the gonadal tissues, but subsequent egg formation and egg laying by females impose new demands on energy and nutrition (see Chapter 15). The production of large clutches of big, richly provisioned eggs by waterfowl is especially expensive and may temporarily double a female's total daily energy requirement. Incubation can also create an energy shortage because it limits the amount of time a bird can forage for its own maintenance. The parents then face another surge of demands on their time and energy when the chicks hatch and require food and brooding.

Birds draw on their reserves to get through periods of peak energy demand. In some species, a female's protein reserve may control when she breeds. The protein reserves of female Gray-backed Camaropteras and Red-billed Queleas fall substantially during egg laying, apparently because the females metabolize proteins to extract essential amino acids not available in adequate amounts in their diets. Protein reserves can be estimated from the lean dry mass of the pectoral muscles. If food is not plentiful enough to permit birds to build a reserve, they will not breed. Large waterfowl, such as Snow and Canada geese, rely almost exclusively on body stores of nutrients and energy to produce their large eggs; smaller ducks must feed to supplement their endogenous reserves (Bluhm 1988).

Costs and Timing of Molt

The complete molt is a major undertaking. The bird sheds and then regenerates thousands of feathers, roughly 25 to 40 percent of its lean dry mass (that is, excluding fat and water content). Molt draws significantly on protein and energy reserves to synthesize feather structure and also to offset the effects of poorer insulation and flight efficiency. Thomas Bancroft and Glen Woolfenden (1982) estimated that adult Scrub Jays and Blue Jays must increase daily metabolism 15 to 16 percent during peak periods of feather production. Sheldon Lustick (1970) found that at moderate air temperatures, molting Brown-headed Cowbirds consume 13 percent more oxygen than do nonmolting cowbirds. At lower air temperatures, however, when increased heat production was needed to maintain a constant body temperature, molting cowbirds consumed 24 percent more oxygen than did nonmolting birds. The difference of 11 percent reflected the cost of poorer insulation during molt. Molting during the warm summer months thus can be advantageous.

Molt is a period of intense physiological change (Dolnik and Gavrilov 1979; Murphy and King 1991, 1992). Accompanying the replacement of worn feathers is the synthesis of keratin by the skin, increased amino acid metabolism, increased cardiovascular activity to supply blood to the growing feathers, shunting of water to the developing feathers, changes in bone metabolism and calcium distribution, and a daily cycling of the body protein content, plus increased need for iron for red blood cell production and for calcium for bone formation. Together these metabolic changes impose substantial hidden costs beyond those required simply to convert amino acids into feather proteins. Only about 7 percent of the energy used by molting birds is actually incorporated into the feathers themselves.

Molt typically follows breeding and often precedes migration. To ensure that they finish molting before the end of the summer, birds may not nest a second time. Molt is so important for the Sooty Terns of Ascension Island that it may prevent continuous year-round nesting (Ashmole 1965). The pace of molt and its costs vary in relation to the time available. In Denmark, Common Ringed Plovers molt their entire plumages in about 1.5 months, after breeding and before migrating a short distance to southwestern Europe. Their relatives that breed in the Arctic first migrate to southern Africa and then molt at a leisurely pace for nearly 4 months. The northernmost *(pugetensis)* populations of the White-crowned Sparrow complete their molt in 47 days, whereas the slow-molting southern *(nuttalli)* populations typically molt in 83 days. Thrush Nightingales, Lapland Longspurs, and renesting White-crowned Sparrows molt so fast at high latitudes that, temporarily, they become virtually flightless. Arctic Peregrine Falcons and American Golden-Plovers as well as many other shorebirds that nest in the Arctic begin their molt on the breeding ground but are unable to complete it in time to leave for the south. Molt of the flight feathers stops just before migration and then resumes for several more months on the wintering grounds.

Few species breed and molt at the same time, but the exceptions to the rule are instructive. Some female hornbills molt while imprisoned in sealed nest cavities, incubating eggs and brooding young. The energy requirements for self-maintenance are minimal for these domestics, and as a

Box 11–2
Molt by Gambel's White-Crowned Sparrow requires energy and special nutrition

Mary Murphy and Jim King at Washington State University are deciphering the costs—both energy and nutrition—of the rapid fall (prebasic) molt in Gambel's White-crowned Sparrow. The complete molt of this sparrow lasts about 54 days, with peak feather production and energy costs from day 18 to day 36 (see figure). The actual energy costs of molt total 605 kilojoules to 876 kilojoules, with daily investments that are proportional to the molt intensity. The daily energy costs of peak molt (58 percent of BMR) are higher than those associated with reproduction.

Obtaining adequate nutrition for the molt is probably not a major problem for sparrows in the wild. Muscle tissues can be broken down as needed to provide most of the amino acids required. Keratin synthesis, however, requires disproportionately high proportions of sulfur-containing amino acids, especially cysteine. To have cysteine available in amounts sufficient to continue feather growth overnight when the sparrows fast, they store extra reserves in the liver during the day, feeding selectively on foods containing such amino acids if needed; the stored cysteine is liberated for use at night.

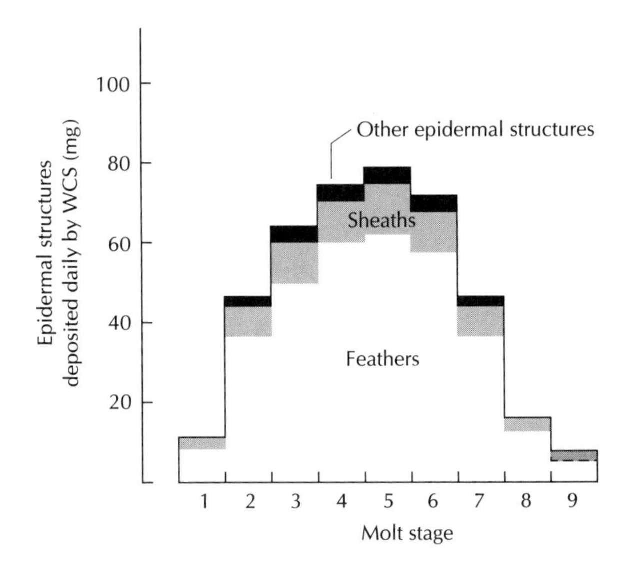

Plumage, sheaths, and other epidermal structures deposited daily during the 54-day prebasic molt period of Gambel's White-crowned Sparrows (WCS). Each of the nine molt stages lasts 6 days. (From Murphy and King 1992)

result, the added costs of molt can be accommodated. Also the flight feathers may not be essential during this sedentary period. Loss of feathers and reduced insulation may, in fact, be advantageous at this time because of the high temperatures that build up inside the nest cavity. Male hornbills, which feed the incubating females, wait to molt until their families leave the nest. Similarly, some female raptors, such as the Eurasian Sparrowhawk, molt their primaries while incubating and being fed by their mate. Gulls and sandpipers that breed in the high Arctic, where the reproductive season is short, must start molting before they finish breeding to be ready for migration. The Dunlin, for example, begins to molt its primaries just before incubation and then finishes 4 to 5 five weeks later. It leaves the breeding grounds later than other sandpipers that winter farther south. Resident petrels of Antarctica molt and breed at the same time during the brief summer (Maher 1962).

Cases of synchronous molt and breeding occasionally occur in birds that live in productive, tropical environments with minimal seasonal variation. Three to four percent of the African birds examined by Robert Payne (1969a) were molting while breeding. Eight to ten percent of the

Costa Rican birds examined by Mercedes Foster (1975) bore signs of both molt and reproductive activity. The prolonged molts of tropical birds apparently minimize daily costs in the absence of strong seasonal constraints. Tropical birds molt more predictably than they breed, because reproduction may be tied to irregular periods of rain or may involve several renesting attempts due to high rates of nest failure because of predation. Because occasional overlap between molting and nesting is likely, it is sometimes advantageous to have a flexible system of physiological controls (Foster 1975).

Desert birds such as Darwin's finches of the Galápagos and the Zebra Finch of Australia molt on a regular schedule but nest whenever the unpredictable rains begin, stopping the molt temporarily to do so. Molt resumes after nesting is completed. Molt interruption also occurs in the Gray-backed Camaroptera in Uganda (Fogden and Fogden 1979), which breeds in the rainy season, and in Pinyon Jays of New Mexico, which breed when food is abundant. Tropical terns such as the Common White-Terns on Christmas Island turn the molt on and off to breed whenever possible (Figure 11–10) (Ashmole 1968). This delicate seabird has no pigment in its flight feathers, which consequently wear easily and must be replaced more often than in most other terns. Wave after wave of molt is initiated in the flight feathers. The innermost primaries often begin to molt again before the outermost primaries are replaced in the preceding molt. As many as three successive molts may be in progress

Figure 11–10 Common White-Terns molt almost continuously to replace their worn, unpigmented feathers, but they interrupt the molt upon laying an egg. (Courtesy of Ralph W. Shreiber)

simultaneously. When a Common White-Tern starts to nest (it simply lays an egg precariously on a bare branch!), the molt stops suddenly, no matter which feathers may be missing—the molting equivalent of musical chairs. After the tern has finished nesting, molt resumes as if there had been no interruption in the complicated pattern of feather replacement.

The Annual Cycles of Waterfowl

Whereas geese have simple, annual molt and plumage cycles, many ducks of the northern latitudes have evolved more unusual sequences, in which the prealternate molt (also called the prenuptial molt) starts before the fall, prebasic molt (also called postnuptial molt) finishes (Figure 11-1). Drake Mallards drop all their flight feathers in a rapid postnuptial molt and become flightless for several weeks. The postnuptial molt coincides with the gonadal refractory period. The resulting, hen-colored, basic plumage is called an "eclipse" plumage because it does not last long. Before the eclipse plumage is complete, the drake begins the prenuptial or prealternate molt, which produces the handsome, familiar breeding plumage by late October. Courtship and initial pair formation may precede fall migration in this and other duck species. Pair formation continues and is completed on the wintering grounds. Hen Mallards undergo their prealternate, or prenuptial molt, in normal fashion as day length increases prior to spring migration.

Nonannual Cycles

Not all birds follow a 12-month cycle. The Rufous-collared Sparrow ranges from Mexico to Chile. Near the equator in Colombia, this sparrow breeds and undergoes a complete molt twice a year (Miller 1962). These cycles correspond to the two dry seasons each year. To the north in Costa Rica, Rufous-collared Sparrows also breed twice a year, but they only have one complete molt and one partial molt each year, like their relative, White-crowned Sparrows in Washington (Wolf 1969).

Year-round availability of adequate food fosters double breeding seasons among tropical bird species. The Sooty Terns of Christmas Island breed every 6 months, although the individuals that breed twice in the same year are those that failed during the first breeding season. Successful individuals wait 8 to 9 months before breeding the following year (Ashmole 1963a). In another case of double breeding seasons, two, temporally separated populations of Band-rumped Storm-Petrels of the Galápagos Islands alternate use of nesting burrows (Harris 1969). Good nest sites may be limited for this species.

In only a few cases is the breeding cycle independent of calendar year. Unlike Sooty Terns on Christmas Island, the Sooty Terns on Ascension

Island in the tropical Atlantic nest every 9.6 months, in different months in successive years. Successful nesting is possible at any time of the year, so ample food must be available every month (Ashmole 1965) (Figure 11–11). Brown Boobies on Ascension Island nest at 8-month intervals, and White-tailed Tropicbirds nest at 10-month intervals if they are successful and renest in 5 months if they are not. Successful Audubon's Shearwaters and Swallow-tailed Gulls on the Galápagos Islands nest at 9-month intervals.

A few very large birds cannot fit their extended reproductive efforts into a single year and hence may skip a year between nestings. Frigatebirds, Crowned Eagles, Eurasian Griffons, and Wandering Albatrosses nest once every 2 years. King Penguins take 2 months to incubate their eggs, 10 to 13 months to raise their nestlings, and then molt. As a result, they breed only twice every 3 years (del Hoyo et al. 1992; see also Weimerskirch et al. 1992).

Birds adapt to local opportunities whether they exist as predictable seasons or irregular occasions, probably by means of a variety of physiological mechanisms. Many tropical and desert birds cannot breed on a regular schedule but do so opportunistically whenever unpredictable rains permit. Zebra Finches, which begin nesting as soon as the rainy season starts, maintain partially developed gonads in the nonbreeding season, thus minimizing the time required to develop functional organs. Increased water intake and, possibly, the consequent effects of changes in the salt

Figure 11–11 Sooty Terns on Ascension Island do not have a regular 12-month breeding cycle, but instead breed approximately every 9.6 months and, consequently, in different months in successive years. (Courtesy of Cruickshank/VIREO)

concentration of body fluids on the hypothalamic receptors, rather than photoperiod, induce the final stages of gonadal development. Termination of reproductive gonad activity can be induced by the unavailability of drinking water after the rains end. Donald Farner (1967) proposed that in such species the hypothalamus stimulates the release of gonadotropins from the pituitary on a steady, predictable basis unless inhibited by unfavorable external information such as cold weather or drought.

Summary

Birds face seasonal cycles of stress and opportunity. Physiological cycles, guided by internal cellular clocks, prepare the bird for each season. In general, seasonal changes in day length, or photoperiod, control gonadal activity and therefore reproductive efforts by directly stimulating receptors in the midbrain and, in turn, the secretion of gonadal hormones by the pituitary gland.

Ultimate factors such as food supplies, nest sites, climate, and predator risk determine the evolution of breeding seasons in birds. Proximate factors such as temperature, rainfall, and green vegetation adjust the actual onset of reproduction to local conditions. Warm spring and summer months constitute the main breeding season in the Temperate Zone. Rainfall usually defines tropical breeding seasons.

Birds generally do not breed and molt at the same time but undertake these efforts, which require substantial energy, in different months. In some exceptional cases, molt and breeding do take place simultaneously; for example, female hornbills, confined to the nest and fed by the males, can afford to molt, and some sandpipers must molt and nest to accommodate the short Arctic summer. Opportunistic breeders such as the Common White-Tern interrupt molt while they nest.

The simplest annual cycles proceed from breeding to molting to surviving seasons of reduced food availability to breeding again. Seasonal migrations and extra molts complicate the annual cycles of many birds. A few, mostly tropical, birds have 6-month cycles, breeding twice a year. Others have 9- or 10-month cycles, thus breeding in different months each year.

Further Readings

Bluhm, C.K. 1988. Temporal patterns of pair formation and reproduction in annual cycles and associated endocrinology in waterfowl. Current Ornithology 5: 123–185. *A fine integration of current literature on hormonal control of annual cycles as they apply specifically to ducks and geese.*

Farner, D.S. 1980. The regulation of the annual cycle of the White-crowned Sparrow, *Zonotrichia leucophrys gambelii.* Acta XVII Congressus Internationalis Ornithologicus 1: 71–82. *An excellent review of the physiological controls of the annual cycle of one of the most intensely studied North American bird species.*

Gwinner, E. 1977. Circannual rhythms in bird migration. Annual Review of Ecology and Systematics 8: 381–405. *A summary by one of the field's pioneers.*

King, J.R. 1974. Seasonal allocation of time and energy resources in birds. In Avian Energetics, pp. 4–85. Publication of the Nuttall Ornithological Club No. 15. *A basic reference for the topics discussed in this chapter.*

Meier, A.H., and A.C. Russo. 1985. Circadian organization of the avian annual cycle. Current Ornithology 2: 303–343. *A technical review of current research.*

Murphy, M.E., and J.R. King. 1992. Energy and nutrient use during moult by White-crowned Sparrows *Zonotrichia leucophrys gambelii.* Ornis Scandinavica 23: 304–313. *Reviews pioneering work on the physiological costs of molt.*

Payne, R.B. 1972. Mechanisms and control of molt. Avian Biology 2: 103–155. *A detailed review of the subject.*

Silver, R., and G.F. Ball. 1989. Brain, hormone and behavior interactions in avian reproductive status and prospectus. Condor 91: 966–978. *A visionary review of recent progress.*

Walsberg, G.E. 1983. Avian ecological energetics. Avian Biology 7: 161–220. *A comprehensive review of avian energy budgets, with an emphasis on reproductive activities.*

CHAPTER 12

Migration

ANCIENT RECORDS OF THE SEASONAL appearances and disappearances of birds perplexed early naturalists, who were not certain whether birds migrated or hibernated. Even though Aristotle understood that cranes moved seasonally from the steppes of Asia Minor (then Scythia) to the marshes of the Nile, he believed that small birds—swallows, larks and turtle-doves—hibernated. Later anecdotes about swallows that were found frozen in marshes and that flew off after being thawed fueled this misconception. Another legend, which persisted for five centuries until the 1600s concerned the Barnacle Geese of northern Europe. Their high Arctic breeding grounds were unknown in medieval times, and they appeared mysteriously each winter, arising it was said directly from the curiously shaped goose barnacles *(Lepas)* that rode ashore on driftwood (Lockwood 1984).

We now know that every fall an estimated 5 billion land birds of 187 species leave Europe and Asia for Africa (Moreau 1972), and that a similar number of over 200 species leaves North America for Central and South America. At a newly discovered migration hotspot in Veracruz, Mexico, a team from Hawks Aloft Worldwide counted during the fall of 1992 more than 2.5 million raptors, including more than 900,000 Broad-winged Hawks, 500,000 Swainson's Hawks, and 12,000 Mississippi Kites (Bildstein et al. 1993).

Migration allows year-round activity, unlike dormancy and hibernation, the means by which many animals live through severe seasons. The advantage of migration is that birds can exploit seasonal feeding opportunities, while living in favorable climates throughout the year. The costs of migration also are potentially great. It takes radical physiological adjustments and sustained fine tuning to survive such extended travel.

The decline of some North American populations of Neotropical migrants since 1950 is a major conservation concern (Terborgh 1989; Askins et al. 1990; Hagan and Johnston 1992). Chapters 21 and 24 discuss different aspects of the precarious population ecology of Neotropical migrants. The conservation of declining migrant species presents special challenges because they face widespread loss of suitable habitats on their wintering grounds in Central America and the Caribbean, on their breeding grounds in North America, and at critical stopover or refueling sites along their migration corridors. Compounding the loss from fragmentation of forests are increased losses of nests to predators and to brood parasitic Brown-headed Cowbirds. Increased mortality and decreased reproductive success could cause some of the most familiar migrants to become rare and endangered species in the next few decades.

This chapter first presents the main patterns of bird migrations and the leading theories on the origin of migration. Then follow the dimensions of long-distance migration, which requires extraordinary physiological endurance and large fuel supplies. Direct extensions of the physiological and ecological controls that manage other aspects of the annual cycles of birds determine the timing of migration, which relates particularly to favorable weather conditions.

Patterns of Migration

Migration routes and patterns are almost as varied as the migrants themselves, reflecting the histories of populations, their abilities to cross large barriers, the positions of topographical barriers, and the relative locations of summering and wintering grounds. Details are available for hundreds of species as a result of the extensive marking and recovery programs of the past 50 years (Dorst 1962; Moreau 1972; Lövei 1989; Lane and Parish 1991). The main migration routes of North American land birds are oriented north–south, partly because wintering ranges of most species lie south of breeding ranges and partly because the coasts, major mountain ranges (Appalachian Mountains, Rocky Mountains, and the Sierra Nevada), and major river valleys (Mississippi) trend north–south. In the Old World, birds initially migrate east–west in accordance with the longitudinal displacement of seasonal ranges and the east–west orientation of the Alps, the Mediterranean Sea, the North Sea coasts, and the great deserts of North Africa and the Middle East. In general, birds of the Southern Hemisphere do not migrate as far north as Northern Hemisphere birds migrate south. In South America, Kelp Goose and Buff-necked Ibis, both of which nest at the southern reaches of the continent, migrate north only as far as central Chile and Argentina. Some swallows and a few flycatchers, such as the Crowned Slaty Flycatcher, move north into tropical South America, but they are exceptions to the rule. Similarly, only about 20 southern African species winter as far north as equatorial Africa, in contrast to 183 Palearctic species that move to sub-Saharan Africa to spend the winter months.

Migration routes sometimes reflect the recent distributional histories of birds. Those individuals that inhabit newly colonized areas tend to retrace the population's historical expansion routes. Northern Wheatears that colonized Greenland from the British Isles return there and then head south on their way to Africa for the winter. Pectoral Sandpipers from

Alaska recently established a new breeding population in Siberia. Instead of migrating south through the Orient, as do most Siberian shorebirds, these "Siberian" Pectoral Sandpipers fly back to Alaska and then south with the rest of their species to South America. Conversely, Arctic Warblers, Yellow Wagtails, and Northern Wheatears, species that have spread recently into Alaska from Siberia, return to Siberia before migrating south.

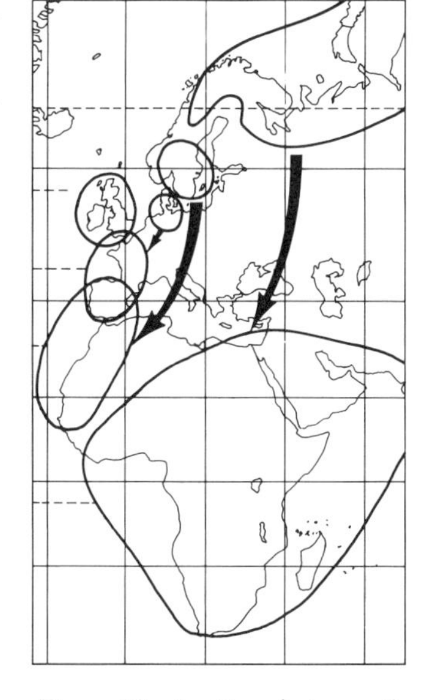

Figure 12–1 Populations of the Common Ringed Plover maintain distinct wintering and breeding ranges. The populations that breed farthest north winter farthest south. The Common Ringed Plovers of the British Isles do not migrate at all. (After Dorst 1962)

Why Do Birds Migrate?

Why birds migrate is still one of the most challenging questions in ornithology, despite a century of effort to frame a satisfactory answer (Wallace 1874; Dorst 1962; Gauthreaux 1982). The goal is to formulate a convincing theory of the evolutionary steps a sedentary species might have taken in becoming a migratory species. Why, for example, do some populations of a species, such as the Common Ringed Plover, migrate whereas other populations of the same species do not (Figure 12–1)? Why do some birds migrate farther than others? Sanderlings, the sandpipers that scurry back and forth with the waves on sandy beaches, fly from their breeding sites in the high Arctic to wintering sites as near as the state of Washington and as far as southern Chile (Myers et al. 1985) (Figure 12–2). The energetic price of the 230-hour, 7500-kilometer flight to Chile matches the cost of living one midwinter month in California. Why should a Sanderling invest so much to go so far when apparently suitable beaches line the Pacific coast from Washington to Chile?

Migration, which is tied to predictable, seasonal opportunities, is different from nomadic wandering, which is tied to unpredictable, aseasonal opportunities. Sporadic, scattered pine seed crops or insect infestations

Figure 12–2 A flock of migrating Sanderlings. The lead birds are Black Turnstones. (Courtesy of J.P. Myers/VIREO)

attract opportunistic feeding by nomadic species such as Red Crossbills, which wander great distances in search of pine seeds and which may breed wherever food is abundant. In the Tropics, fruit-eating and nectar-feeding birds wander locally in search of their unpredictable sources of food. In contrast, seasonal cycles of climate or insect abundance attract corresponding cycles of breeding, flocking, and migratory relocation. To take advantage of predictably favorable conditions, birds undertake both local and long-distance movements between seasonal residencies. On a local scale, tropical hummingbirds simply migrate up and down mountain slopes. On a global scale, Arctic Terns leave their nesting colonies in the far northern Atlantic and Arctic oceans (70°N latitude) for the waters of Antarctica over 12,000 kilometers away. Buff-breasted Sandpipers fly similar distances from breeding grounds in the vast high Arctic to winter on the pampas of Argentina. More common are migrations to less distant wintering grounds. Many species of wood warblers that breed in the northern United States and southern Canada spend the winter in Central America and in the West Indies.

The benefits of migration must be substantial, because migration is costly. Long-distance travel is a hazardous undertaking because of its energy costs and risk of exposure, exhaustion, and other physical calamities. More than half the small land birds of the Northern Hemisphere never return from their southbound migration. In North America alone, roughly 100 million waterfowl migrate south to their wintering grounds each fall, but only 40 million return to their breeding grounds in the spring. Ocean and desert crossings, aside from posing extreme physical demands, carry the threat of devastating hurricanes and sandstorms. Weakened migrants are also vulnerable to diurnal predators such as Eleonora's Falcon, which breeds in the fall so that it can feed its nestlings on migrants trying to cross the Mediterranean (Figure 12–3). Adding to

Figure 12–3 Eleonora's Falcon breeds in the fall, when small birds that migrate past its Mediterranean breeding areas provide an abundant source of food for its nestlings. (Courtesy of Harmut Walter)

losses due to natural predation are the large numbers of songbirds and raptors shot or trapped in southern Europe as they attempt to migrate to Africa for the winter (Woldhek 1980). Losses on the wintering grounds and on the return trip also raise the mortality figures.

The potential benefits of migration are species- or population-specific. Discussions in the past have emphasized the need to escape from inhospitable climates, probable starvation, social dominance, shortage of nest or roost sites, or competition for food. A more positive view of the same ecological forces is that migrants aggressively exploit temporarily favorable opportunities. Following this view, many northern latitude migrants are tropical birds that temporarily exploit the long days and abundant insects of high-latitude summers, rather than temperate birds that tolerate the Tropics to escape the northern winter (Stiles 1980; Levey and Stiles 1992). Hummingbirds, flycatchers, tanagers, and many wood warblers may be tropical American birds that have expanded their seasonal activities northward. In Europe and western Asia, barriers such as the Mediterranean Sea and the Sahara desert discourage northward travel. In eastern Asia, the absence of such barriers has permitted a variety of tropical birds—kingfishers, orioles, pittas, drongos, white-eyes, and minivets—to migrate north. Attractive nesting opportunities also invite migration to temperate latitudes. The large expanses of northern Temperate Zone habitats facilitate dispersed, low-density breeding. Reduced predation of nests may be one result of low densities, breeding opportunities for yearlings another. Several years' wait for a breeding space is often the case in the Tropics. Such factors are often incentives to migrate.

Populations can acquire or lose the migratory habit. The European Serin, for example, has spread throughout Europe from the Mediterranean in the last 100 years. Whereas the ancestral Mediterranean populations are resident, the new northern populations are now migratory (Mayr 1926). Conversely, the resident Fieldfares of Greenland are recent colonists from the migratory populations of Europe. Several Palearctic species that winter in southern Africa, including the European Bee-eater and Black Stork, have established resident breeding populations in South Africa (Moreau 1966). Similarly, Barn Swallows wintering in Argentina during the austral (southern) spring and summer have lately started to nest there (Martinez 1983). The migratory habit thus changes in relation to new geographical and ecological circumstances.

Generally speaking, birds that migrate to the Tropics survive the winter better than do those that stay in the Temperate Zone (Table 12–1), but Temperate Zone residents achieve higher per capita reproductive success than do returning migrants. Tropical residents trade low productivity for high survivorship. Few nests succeed, clutch sizes are small, and each pair attempts several nests a year, but adults are long-lived.

Partial migration evolves when local conditions promote opportunistic movement of birds to better conditions in nearby regions. In the case of the European Robin, unpredictable winter conditions favor both migrants and nonmigrants. Climatic conditions swing from mild winters

TABLE 12–1
Life history traits of residents and migrants

Trait	Temperate resident	Migrant	Tropical resident
Productivity	High	Moderate	Low
Adult survival	Low	Moderate	High
Juvenile survival	Low	Moderate	Moderate to high

that favor residents to severe winters that favor migrants. Migratory behavior in the population is maintained as a behavioral polymorphism (Figure 12–4). Resident individuals, which make up about one-fifth of the robin population in southwestern Germany, remain within 5 kilometers of their breeding territories, do not put on large reserves of premigratory fat, and do not exhibit sustained migratory restlessness (see below) in the laboratory. In contrast, migrant individuals fatten in the fall, exhibit intense migratory restlessness, and travel an average of 1000 kilometers to their winter habitats. Parents pass their heritable migratory behavior to their offspring.

The evolution of a migratory species presumably has three stages: partial migration, the division of a species into migratory and resident

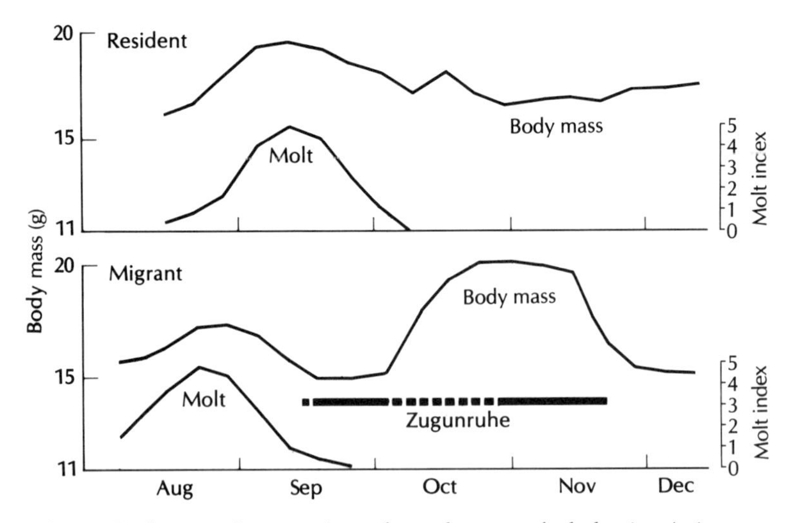

Figure 12–4 Body mass (in grams), molt, and *zugunruhe* behavior (migratory restlessness) of (top) a young resident and (bottom) a young migratory European Robin in the laboratory. Breeding experiments revealed a genetically based polymorphism for migratory behavior, including early molt, premigratory fattening, and migratory restlessness in the two forms of this species. A molt index of 1 indicates the beginning or the end of the molt. A molt index of 5 indicates a heavy molt that includes most of the feather coat. (After Biebach 1983)

populations, as seen in the White-crowned Sparrow (Chapter 11), and fully disjunct migration due to the gradual elimination of resident populations (Cox 1968, 1985). Periodic droughts probably foster partial migration by some species that were once residents of the arid Mexican Plateau and southwestern United States but now migrate annually to the north. George Cox hypothesized that increased severity of the dry, summer season during the Pleistocene epoch eliminated the resident populations of these species, which were replaced by other species better adapted to the new desert conditions, leaving only the migratory populations. At first, such populations might have migrated short distances of a few hundred miles, but the distance between summer and winter ranges gradually increased to thousands of miles as a species expanded its latitudinal range or as required habitats become more separated by changes in the Earth's climates.

Individuals of fully migratory species vary in the distances they migrate. From their breeding range throughout the northern United States and southern Canada, Dark-eyed Juncos migrate south to wintering grounds throughout the eastern United States to the Gulf Coast, but the migration distances of individual juncos translate into different average wintering distributions of males and females, adults and young (Ketterson and Nolan 1982, 1983). Adult females migrate farthest south to the southernmost states, young males stay farthest north in Indiana and Ohio, and adult males and young females settle at intermediate latitudes. Ellen Ketterson and Val Nolan hypothesize that three factors—mortality during migration, overwinter survival, and reproductive success as a function of time of arrival on the breeding grounds—differently affect young and old, male and female juncos. Greater mortality among the young selects for shorter migrations by the young of both sexes. Territory establishment by the earliest arrivals on the breeding grounds favors shorter migrations by males than by females, and especially by young males, which are at a disadvantage in the competition for breeding territories. Adult females migrate farther south to regions of lower junco densities, where overwinter survival is greatest (Figure 12–5).

Philip Hockey and his colleagues (1992) asked similar questions about the migration of long-distance shorebird migrants, such as the Sanderling mentioned at the beginning of this chapter. They concentrated their analyses on the species that winter on estuaries distributed from northern Europe to southern Africa. The Sanderling was one of these. Contrary to previous assumptions, survival rate was not affected by migration distance; models based on differences in survival in benign versus temperate climates did not apply. Instead, the shorebirds distributed themselves widely in relation to how much winter food was available in each estuary. The densities of shorebirds were strongly correlated with local food resources. Many individuals migrated farther to exploit the seasonally rich feeding opportunities south of the equator. Thus, the wintering distributions of shorebirds track the availability of food in coastal wetlands across a broad range of latitudes.

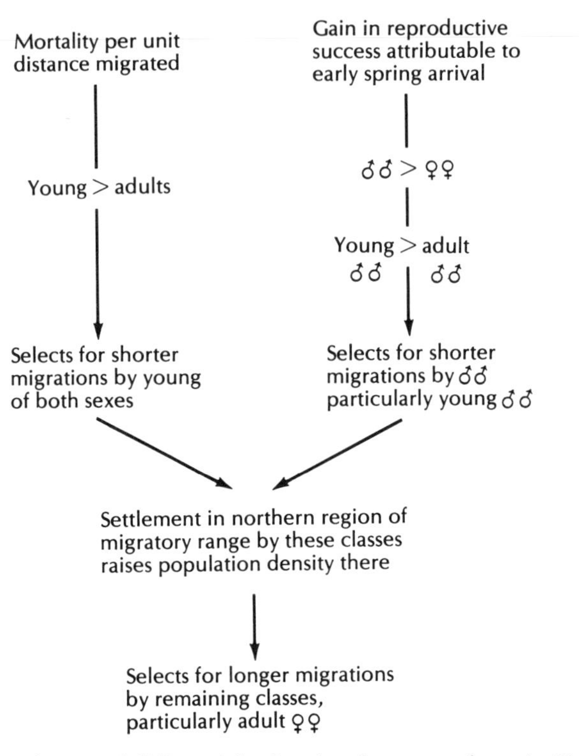

Figure 12–5 Evolution of differential migration by age and sex in Dark-eyed Juncos. (After Ketterson and Nolan 1983)

Migratory Feats

Long-distance migratory flights are extraordinary feats of physiological endurance. Arctic Terns commute 25,000 kilometers round trip each year. The migrations of Arctic shorebirds regularly exceed 13,000 kilometers one way from the high Arctic to distant South America. Red Knots, for example, fly from Baffin Island to Tierra del Fuego. A banded Lesser Yellowlegs flew 3220 kilometers from Massachusetts to Martinique, West Indies, in 6 days or less. These and other migrants cross thousands of kilometers of open ocean or inhospitable terrain without stopping, stretching their fuel reserves and physical abilities to the limit. Dangerous as nonstop crossings may be, they are often the only way to reach a destination or they may be preferable to longer, safer routes because of shorter flight time.

Every fall, vast numbers of migrants leave the coasts of New England and the maritime provinces of Canada, heading southeast over the ocean (Figure 12–6). The capacity and predilection of larger, faster shorebirds such as the American Golden-Plover for such flights have been known for many years, but radar studies now reveal similar efforts by millions of small land birds (Williams and Williams 1978; also see opposing views of Murray 1989). Up to 12 million birds have passed over Cape Cod in one night, embarking on a nonstop journey of 80 to 90 hours. Wave after wave of the migrants, such as the Blackpoll Warbler, depart at intervals of several days, heading past Bermuda and from there continuing on to the Lesser Antilles. Radar stations on Bermuda and Antigua pick up the approaching and passing waves of migrants. As these migrants reach the latitudes of Florida, they encounter strong trade winds from the northeast. The migrants then fly with the wind to the southwest toward the north coast of South America. The strong tailwinds enable the tired travelers to make the last half of the journey somewhat more easily.

Evidence of the strenuous nature of the trip and of the way that the migrants stretch their physical capabilities to the limit can be seen in the exhausted condition of birds that stop at Curaçao, short of their

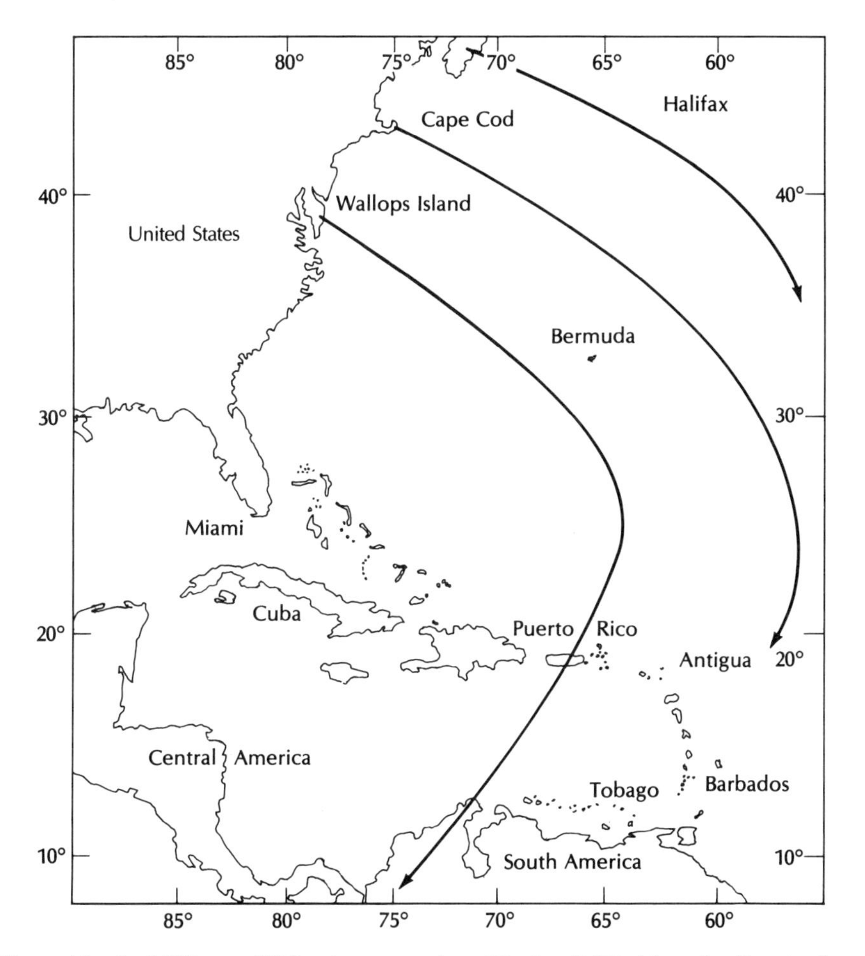

Figure 12–6 Millions of fall migrants such as Blackpoll Warblers fly directly from northeastern North America to northeastern South America. This 86-hour marathon flight takes them southeast past Bermuda to the trade winds, which then assist them on a southwesterly course to the Lesser Antilles and the coast of South America. (After Williams and Williams 1978, with permission of Scientific American)

Box 12–1
Radar records document the decline of migrant birds

Radar is a powerful tool for tracking birds in flight. Military radars can track and identify (using flap rates, for example) single birds and assess their flight speeds, altitudes, and compass orientation. Weather surveillance radar stations also provide continuous monitoring of migration activity (Gauthreaux 1992). Migrating birds show up on the radar screen as small targets that move at predictable speeds. Flocks of migrants can be distinguished from single individuals, and the intensity of migration each night can be quantified. Records from a battery of weather radar stations along the Gulf Coast monitor the arrival of trans-Gulf migrants from Central America. Comparison of these records for the spring seasons of 1965 through 1967 and 1987 through 1989 indicates a 50 percent decline in migratory activity on days with favorable weather.

destination, when flight conditions have been poor. Little more than feathered skeletons, they have depleted their fat reserves, metabolized much of their protein, and drained the remnants of their precious body water (Voous 1957). As Tim and Janet Williams (1978) point out, "The trip does, however, require a degree of exertion not matched by any other vertebrate. For a man the metabolic equivalent would be to run 4-minute miles for 80 hours. . . . If a Blackpoll Warbler were burning gasoline instead of its reserves of body fat, it could boast of getting 720,000 miles to the gallon!"

Eurasian migrants also face herculean challenges (Moreau 1972; Lövei 1989). Northern Wheatears from Greenland start their journey to the British Isles by crossing 2000 to 3000 kilometers of open ocean with no assurance of favorable winds. Many European migrants fly 1100 kilometers directly across the Mediterranean and then, almost immediately thereafter, 1600 formidable kilometers nonstop across the Sahara desert (Moreau 1961). In the spring, they return across the Sahara, proceed 400 kilometers across the eastern Mediterranean, fly 600 kilometers over bleak, foodless Anatolia (in Turkey), and finally travel another 650 to 1100 kilometers across the open water of the Black Sea. Still another route between Asia and Africa, used by birds that breed in northern Russia, includes traversing 1600 kilometers of Caspian desert plus 1700 kilometers of Saudi Arabian deserts and the intervening water passages. Some migrants such as falcons and bee-eaters cross from India directly to East Africa over 4000 kilometers of the Indian Ocean.

Migratory passages across deserts and major bodies of water may be followed by a local grounding, or spectacular "fallouts," of thousands of exhausted birds, especially when they encounter bad weather or headwinds at their landfall. The fallouts on the coasts of Louisiana and Texas of Neotropical migrants after they have flown across the Gulf of Mexico are legendary. These take place in April when a cold front passes the coastline, catching the migrants in midflight. The barrier of bad weather and opposing winds forces the northbound migrants to land on the first available land. Victor Emmanuel, who grew up in Houston, Texas, and who has probably witnessed as many fallouts as anyone, describes his experience when the entire trans-Gulf migration was grounded in late April 1960:

> There were trees decorated with tanagers, orioles, and grosbeaks. Trees dripped with warblers of many species—ten or more varieties in one tree. *Birds were everywhere.* In the trees, in the bushes, on fenceposts, on fence wires, around houses, and most remarkably, in the grass. Sometimes a hundred orioles and buntings would fly up from the grass and perch in dead stalks. What impressed and delighted me most was seeing warblers in the grass, and even hopping on the ground! Here were these tiny birds, the "butterflies of the bird world," not hidden amid the foliage of tall trees but literally at my feet. I've seen twenty or more Bay-breasteds, a dozen Blackburnians, and many others on the ground. In such a situation, you can approach warblers quite closely and enjoy every detail of their brilliant plumage. (Emmanuel 1993, p. 1)

Fatty Fuels for Migration

Migrants develop stores of fat especially for migration. Fat yields two times more energy and water per gram metabolized than does either carbohydrate or protein (Table 12–2). Fat is stored in adipose tissues under the skin, in the muscles, and in the parietal cavity. For example, in White-crowned Sparrows, subcutaneous fat is deposited initially at 15 separate sites; and with continued deposition, the fat stores spread laterally and coalesce into a continuous layer between the skin and muscles (King and Farner 1965) (Figure 12–7). Some fat is also stored in most muscles and in internal organs. Unlike the human heart, the avian heart does not accumulate much fat, even when the migrant reaches peak obesity (Odum and Connell 1956; King and Farner 1965).

Adipose tissue does not consist simply of large, inert globs of fat but supports a dynamic system for synthesis, storage, and release of lipids (George and Berger 1966). The enzyme lipase breaks down fat into free

TABLE 12–2
Fuels for migration

Fuel	Energy yield (kj)	Metabolic water (g)
Fat	38.9	1.07
Carbohydrate	17.6	0.55
Protein	17.2	0.41

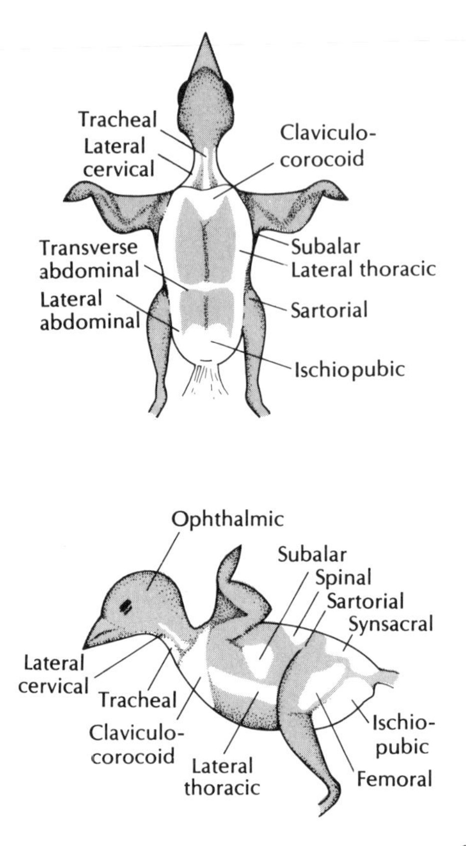

Figure 12–7 Principal sites of subcutaneous fat deposition in the White-crowned Sparrow. (After King and Farner 1965)

fatty acids and glycerol for transport to sites of use. Fatty acids are transported by the blood to mitochondria in the muscle cells for oxidation. In the mitochondria, fatty acids are progressively chopped into 2-carbon (acetyl) fragments, which are oxidized further in the Krebs cycle, eventually producing adenosine triphosphate (ATP) molecules that power the contractions of muscle fibers. In the muscles, lipase activity, which is a good index of the capacity of muscles for fat metabolism, increases in relation to migratory activity. Specialists in the physiology of endurance exercise are exploring the metabolic capacities of the pectoralis muscles of migrant birds (Dawson et al. 1983).

Migrants fatten rapidly just before migration by consuming enormous quantities of energy-rich food. Blackpoll Warblers nearly double their weight, from an average of 11 grams to an average of 21 grams. Ruby-throated Hummingbirds, which cross 500 to 600 miles of open water in the Gulf of Mexico, also nearly double their normal weight of 3 grams to make this trip.

How much fat migrants store reflects their requirements. Fat makes up 3 to 5 percent of the normal mass of small nonmigrating birds. Short-range migrants, which can refuel regularly, carry low to moderate fat reserves of 13 to 25 percent (Berthold 1975). In one study, the average reserves of White-throated Sparrows increased to 17 percent of total body weight just before migration, dropped to 6 percent when they reached their destination, and then increased to 12 percent as the winter progressed (Odum and Perkinson 1951). Dunlins wintering at Teesmouth, England, developed moderate midwinter fat reserves of 15 percent to survive unpredictable periods of inclement weather, and then fattened rapidly prior to spring migration to Arctic nesting grounds (Davidson 1983). Such long-range and intercontinental migrants as the Dunlin may build up fat deposits that account for 30 to 47 percent of their total weight, mainly in preparation for long, nonstop flights (Berthold 1975).

Regular refueling usually accompanies long-distance migrations. Songbirds typically fly only several hundred kilometers and then pause for 1 to 3 days of rest and refueling (Winker et al. 1992a, b). Some songbirds, however, press on several nights in succession until their reserves are nearly exhausted. Three to four refueling stopovers are a strategic aspect of the extraordinary migrations of Arctic shorebirds, which fly up to 30,000 kilometers round trip from the Arctic tundra of North America to the southern tip of South America, and back. Migrating shorebirds congregate by the millions at key staging areas such as the Copper River Delta in Alaska, the Vendee in France, the Bay of Fundy in eastern Canada, and the Delaware Bay coastlines of New Jersey and Delaware. Five to twenty million shorebirds pass through the Copper River Delta every spring, including almost the entire Pacific Coast populations of two species, Western Sandpipers and Dunlins (Figure 12–8). Their migratory movements are timed to coincide with the appearance of abundant food at these sites, where they build up fat reserves required for the next leg of their journey. Conservation programs for shorebirds currently are directed toward the protection of these critical staging areas.

Figure 12–8 Millions of shorebirds gather at key staging areas such as the Copper River Delta in Alaska to refuel for the next (in this case, final) leg of their migration to northern breeding grounds. (Courtesy of D. Norton)

Box 12–2
International cooperation will conserve migratory species

The Western Hemisphere Shorebird Reserve Network (WHSRN) was formed in 1985 to address shorebird conservation issues arising from research by ornithologists at the Academy of Natural Sciences of Philadelphia, the Manomet Bird Observatory, the Canadian Wildlife Service, and others. This research revealed that many species of shorebirds were declining in numbers, apparently as a result of habitat loss. WHSRN, a voluntary collaboration of private and government organizations, gives international recognition to critical shorebird habitats and promotes their cooperative management and protection. The shorebirds serve as a symbol for uniting countries in a global effort to maintain the Earth's biodiversity. Using data from private and government sources, in less than a decade the network comprised and protected 21 of the most important stopover sites in North and South America (see figure). These reserves contain 4 million acres of wetlands on which the continued existence of 30 million shorebirds depends.

Sites included in the Western Hemisphere Shorebird Reserve Network. (Courtesy of J.M. Sibbing, WHSRN)

Flight Ranges of Migrants

How far migrants can fly nonstop depends both on their fat reserves and on how quickly they use their fuel. David Hussell and his associates at Long Point Observatory on the north shore of Lake Ontario captured and weighed nocturnal migrants arriving at various times of the night after flying north across Lake Erie. If one assumes that all the birds that were weighed had taken off at the same time and that those arriving later at Long Point had flown longer, then weight loss is seen to relate directly to time in the air. These data suggest an average weight loss of 0.9 percent of body weight per hour of flight (Hussell and Lambert 1980). This loss can be extrapolated to rate of fuel use if the loss is attributed to fat metabolism only, and if other components of body weight, especially water, are assumed to be in balance (Berger and Hart 1974). Thus, weight losses of about 1 percent, typical of the small migrant passerines weighed, project to expenditures of about 418 joules of energy per gram of body weight per hour of flight (Figure 12–9). The Blackpoll Warbler appears to be more fuel efficient than most other migrants; Hussell's studies indicate weight losses of only 0.6 percent per hour of flight for them, or energy expenditures of 250 joules per gram per hour of migration.

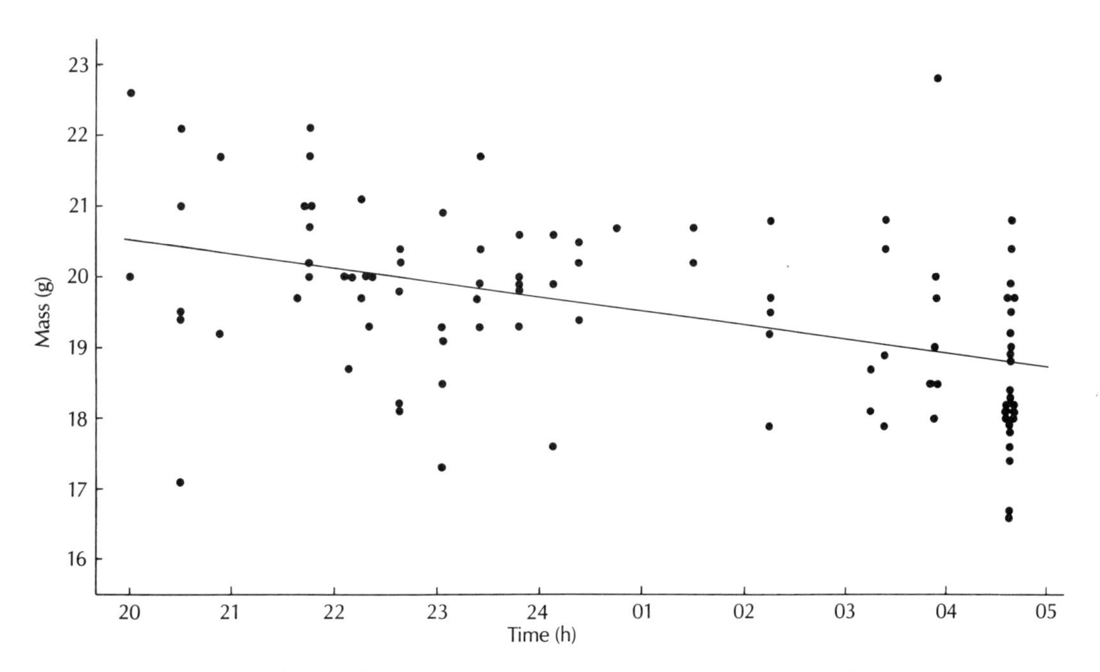

Figure 12–9 Ovenbirds, weighed on arrival at Long Point Observatory on the north shore of Lake Ontario, decreased in mass by an average of 0.2 gram per hour as the night proceeded. Assuming that those that arrived later had flown longer than those that arrived earlier, one can use data like these to estimate the energy costs of migratory flights. (Adapted from Hussell 1969)

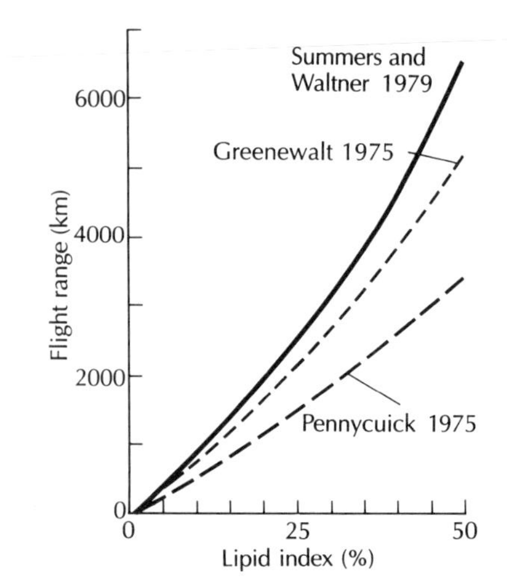

Figure 12–10 The potential flight range of birds, such as that of the Dunlin shown here, increases with fat load (lipid index); but estimates of flight range vary greatly, depending on authors' assumptions about physiology and aerodynamics. (Adapted from Davidson 1983)

The potential flight range of a migrant is a function of its fat load (lipid index), moderated by tail- or headwinds and, in some cases, water loss. The various equations available for projecting flight ranges of migrants are based on different assumptions about physiology and aerodynamics and therefore produce varied results (Berger and Hart 1974; Pennycuick 1989; Greenewalt 1975; Summers and Waltner 1979; Davidson 1983). Peter Berthold (1975) concluded that small birds that expend 418 joules per gram per hour during migratory flight and that have fat reserves of 40 percent of total live weight can fly about 100 hours and cover about 2500 kilometers. At that rate, they should be able to cross the most extensive barriers with energy to spare, unless they encounter strong headwinds. Migrant shorebirds such as the Dunlin have estimated flight range potentials of 3000 to 4000 kilometers (Davidson 1983) (Figure 12–10).

Efforts to understand the migration flights of Ruby-throated Hummingbirds illustrate how knowledge of flight physiology has yielded increasingly realistic projections of flight range. Ornithologists had long wondered how such a tiny bird could carry enough fuel to cross the Gulf of Mexico. Some doubted that hummingbirds crossed at all, suggesting that they took a less direct route overland to Central America. Others proposed that hummingbirds hitched rides on the backs of larger migrants. The problem intensified when Oliver Pearson (1950) and Eugene Odum and C.E. Connell (1956) projected an inadequate maximum flight range of only 640 kilometers for this hummingbird. Subsequent studies of hummingbird flight metabolism suggested much lower values (roughly half) for the caloric costs of flight; so, if a hummingbird consumed fat at the rate of 9.18 watts, carried 2 grams of fat, and flew at a velocity of 40 kilometers per hour, it should be capable of flying over 1000 kilometers nonstop in about 26 hours, more than enough to cross the Gulf of Mexico (Lasiewski 1962).

Some migrants save energy by utilizing tailwinds. The land birds that fly to South America orient their trip so that they pick up the trade winds as they enter the tropical Caribbean region. They backtrack to land during their first night out at sea if wind conditions seem unfavorable for intercontinental flight (Richardson 1978). The observation that the ground speed of migrants does not increase in proportion to the known strength of tailwinds suggests that some, perhaps many, birds throttle back and coast with tailwinds, thereby saving energy and potentially increasing their flight range (Bellrose 1967).

Timing of Migration

Precise arrival and departure dates are an impressive feature of migration in some species. Every year, after their transequatorial migration, Short-tailed Shearwaters arrive at their breeding colonies off southern Australia within a week of the same date. The traditional return of Cliff Swallows the week of March 19 to the San Juan Capistrano mission in California has become a symbol of the arrival of spring itself.

Internal rhythms that are linked to other aspects of the annual cycle guide the timing of migration. Caged migratory passerines predictably become restless just before the time they would migrate in the wild. This phenomenon—called migratory restlessness, or *zugunruhe*—has been familiar to bird fanciers for at least 200 years. Typically, a captive bird wakes shortly after dark and then jumps or flutters in the cage until at least midnight. Because the amount of activity is easily measured, it lends itself to experimental study of both the physiology of migration and orientation behavior (see Chapter 13). Nonmigratory birds do not exhibit *zugunruhe* behavior (see Figure 12–4). Adrenocortical hormones are known to act in concert with prolactin in stimulating this behavior in White-crowned Sparrows. More generally, however, our knowledge of the endocrine controls of the many different facets of migratory behavior of birds is still poor (Wingfield et al. 1990).

We now know that increasing day length in winter stimulates early spring restlessness, hyperphagia (eating to excess), fat deposition, and weight increases in many migratory birds. Extending Rowan's findings about the photoperiodic control of the annual cycle (see Chapter 11), Albert Wolfson showed that Dark-eyed Juncos from migratory populations respond to increasing day length by adding fat stores, whereas sedentary juncos do not (Wolfson 1942). Spring fat deposition and migratory activity of White-crowned Sparrows are under the direct control of increasing day length, mediated precisely by an internal clock. The average date of onset of springtime premigratory fat deposits in captive White-crowned Sparrows has been shown to remain virtually constant over the course of 8 years (King 1972).

The timing of preparations for fall migration is indirectly set by the spring activities. The normal fall sequence of photorefractory testes, prebasic molt, and preparations for migration in White-crowned Sparrows, for example, depends on prior exposure to long photoperiods, but the pace is proximately influenced by shortening days (Farner and Lewis 1971). Rowan suggested some causal relations between gonadal cycles and migration, but the available evidence now indicates that sex hormones do not directly regulate migration (Wingfield et al. 1990). In one set of pioneering experiments, for example, castration did not prevent male Golden-crowned Sparrows from becoming restless and putting on their premigratory fat deposits at the appropriate time of the year (Morton and Mewaldt 1962).

The timing of migration relates first to internal physiological rhythms, but extrinsic weather factors also play a role, primarily one of fine tuning (Saunders 1959). Northward movements of migrants in the spring correlate with the warming of the higher latitudes. Both American Robins and Canada Geese move north in the eastern United States, just behind the main spring thaw, along a front of regions that have a mean daily temperature of 2°C. A line connecting these points is called the 2°C isotherm. Willow Warblers in Europe move north with the 9°C isotherm.

Daily weather conditions and favorable winds, in particular, also influence departure times (Raynor 1956; Richardson 1978). In spring, major northward movements in the United States coincide with a depression (lowering of barometric pressure) toward the southwest, followed by a strong flow of warm southern winds from the Gulf of Mexico toward the northeast. The sizes of migration waves relate directly to the intensity of the depression and strength of the favorable winds (Bagg et al. 1950). The value of favorable winds is clearly seen in records of arrivals of northbound migrants at Baton Rouge, Louisiana (Gauthreaux 1971). Migrants from Central America usually reach Louisiana in midafternoon after crossing the Gulf of Mexico, but when they have strong southern tailwinds, they arrive several hours earlier, in the late morning. On rainy days with adverse winds, they arrive later in the evening and do not arrive at all on days when there are cold fronts or east winds.

Fall migration departures are also stimulated by favorable weather conditions. Good flights of raptors at Hawk Mountain, Pennsylvania, and of land birds at the tips of peninsulas like Cape May, New Jersey, are the result of strong northwest winds due to a barometric depression moving east from the Great Lakes region. Departures from the New England coast are related to favorable tailwinds (Richardson 1978), and peak flights south across the Gulf of Mexico in early October coincide with improved flight conditions to the north (Buskirk 1980).

Exactly how migrants forecast weather conditions is a mystery, but birds are sensitive to changes in barometric pressure and feed more intensely as storms approach and barometers fall. Wind directions aloft, however, are not easily judged from the ground. Meteorologists track weather fronts by monitoring infrasound with a special system of microphones. Pigeons, too, seem to be sensitive to infrasound and may use this source of information in some way (Kreithen and Quine 1979).

Some variations in the timing of migration relate to age and sex. Males generally migrate north before females to compete for breeding territories. Male Red-winged Blackbirds, for example, arrive on the breeding grounds 1 to 5 weeks before their potential mates. In Europe, male Eurasian Skylarks spend a month alone on their territories. In the fall, young Least Flycatchers, Common Swifts, and Chaffinches migrate south before their parents, using unrefined navigation systems. The females of these species tend to move south ahead of the males. Adult shorebirds generally leave before their young, a whole month earlier in the case of Hudsonian Godwits.

Migrants should fly at times of the day and at heights where travel is least costly, safest, and most rapid. Some birds, therefore, migrate by day and others by night, and still others, such as waterfowl and shorebirds, at both times (Figure 12–11). Diurnal and nocturnal flights offer different advantages (Kerlinger and Moore 1989). Hawks migrate during daylight hours when they can take advantage of warm rising air currents. Swifts

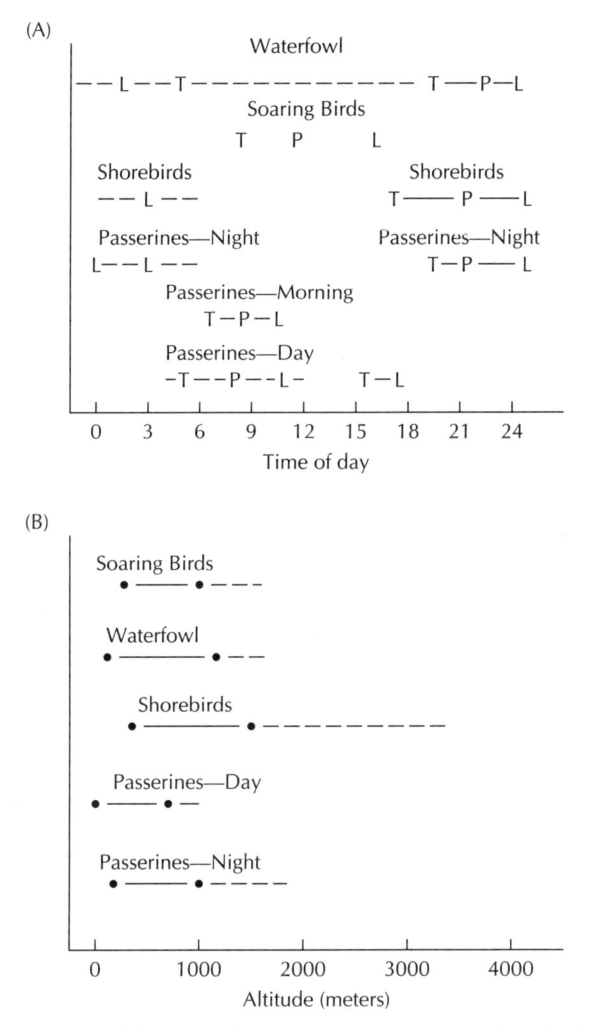

Figure 12–11 Times of day and altitudes of migration. Each solid line represents the normal time of migration including takeoff (T), peak migration (P), and landing (L). Dashed lines indicate variations in the data. Behaviors of diurnal (day) and nocturnal (night) passerine migrants are shown separately. Some nocturnal passerines fly again shortly after dawn (morning). (From Kerlinger and Moore 1989)

and swallows, which feed on the wing, also migrate by day. However, many small land birds, including most flycatchers, thrushes, and wood warblers, as well as rails and woodcocks, depart shortly after sunset and migrate by night. Nocturnal migration usually stops by 2 A.M. These nocturnal migrants can sometimes be seen through a telescope as silhouettes against a full moon. Predation by hawks and gulls is less likely at night, and the migrants can refuel by day. Most important, more stable night atmospheres with weaker horizontal winds and less turbulent vertical motion create favorable flight conditions. Also, cooler and more humid night air favors heat loss and water retention.

Nocturnal passerines usually migrate at altitudes below 700 to 800 meters, although they climb to over 3000 meters, sometimes to 7000 meters to escape turbulent air in the boundary layer near the Earth's surface or to ride good tailwinds (Kerlinger and Moore 1989). Migrating shorebirds typically fly higher than do songbirds, often at altitudes of 2000 to 4000 meters, and sometimes much higher. With some exceptions, migration over water takes place at higher altitudes than over land, usually more than 1000 meters. Single migrants from the Yucatan approach Louisiana flying low over the Gulf of Mexico at 244 to 488 meters, but when they reach land, they climb to 1220 to 1524 meters and coalesce into small flocks of about 20 birds each (Gauthreaux 1972). All migrants tend to fly lower into opposing winds to increase ground speed and to reduce costs of transport.

In addition to the physical feats involved, migration via particular routes between precise breeding territories and wintering stations represents a navigational feat. What cues do birds use to maintain a steady southward course for thousands of miles? What kinds of information do young birds use on their first migratory flight? Such questions are the topic of the next chapter.

Summary

Billions of birds migrate every fall and spring to exploit seasonal feeding and nesting opportunities. The recapture of birds wearing numbered aluminum rings has enabled ornithologists to map the migration routes of many birds. Corresponding in part to the topography of the continents, major migration routes orient north–south in North America and east–west in Europe. The migratory routes of some birds retrace the history of the expansion of the range of their species.

Why some populations migrate and others of the same species are sedentary is an age-old and unresolved question. The migratory habit may appear in newly established populations of nonmigratory species or, in contrast, may be lost by colonizing populations of migratory species. In general among land birds, migrants achieve moderate levels of reproductive success and adult survivorship, whereas residents sacrifice productivity for high survivorship (tropical residents) or survivorship for high productivity (temperate residents). Trade-offs between the costs and benefits of migration determine how far individuals migrate. Wintering shorebirds distribute themselves widely in relation to the food availability in coastal wetlands, with apparently no extra cost to migrating long distances.

The flights of many long-distance migrants require extraordinary physical endurance. Nonstop 3- to 4-day journeys across the open ocean or desert regions are fueled by reserves of fat. Small land birds have a maximum flight range of about 2500 kilometers, and shorebirds can fly 3000 to 4000 kilometers. Regular refueling stops, however, are typical of most migrants. Shorebirds, for example, gather in vast numbers at critical enroute staging areas such as the Copper River Delta in Alaska. In addition to the major migrations between North and South America and between Eurasia and Africa, many species migrate short distances up and down mountain slopes or between southern Canada and the central United States.

Precise arrival and departure dates are an impressive feature of migration. Internal rhythms, linked to other aspects of the annual cycle, guide the timing of migration. Changes in day length stimulate preparations for migration such as accumulation of fat deposits and migratory restlessness. Weather conditions such as favorable winds and changing temperatures control the day-to-day efforts of migrants. Many birds migrate at night, when flight conditions are more favorable and predators are few.

Further Readings

Alerstam, T. 1990. Bird Migration. Cambridge: Cambridge University Press. *Translation of respected 1982 edition in Swedish.*

Blem, C.R. 1980. The energetics of migration. *In* Animal Migration, Orientation, and Navigation, S.A. Gauthreaux, Ed., pp. 175–224. Orlando, Fla.: Academic Press. *A review of energy requirements and fuels for migration.*

Dunne, P., Ed. 1989. New Jersey at the Crossroads of Migration. Franklin Lakes, N.J.: New Jersey Audubon Society. *A powerful, popular statement about the need to protect migrating birds through local conservation efforts.*

Gauthreaux, S.A., Jr. 1982. The ecology and evolution of avian migration systems. Avian Biology 6: 93–168. *A comprehensive review of the historical literature on the diversity and origins of migration systems.*

Gwinner, E., Ed. 1990. Bird Migration. Berlin: Springer-Verlag. *A compendium of timely review papers.*

Hagan, J.M., III, and D.W. Johnston, Eds. 1992. Ecology and Conservation of Neotropical Migrant Landbirds. Washington, D.C.: Smithsonian Institution Press. *Collected papers from a major, timely symposium.*

Kerlinger, P., and F.R. Moore. 1989. Atmospheric structure and avian migration. Current Ornithology 6: 109–142. *Reviews the literature on timing and altitudes of migration flights to test the hypothesis that migrants use favorable atmospheric conditions.*

Lane, B.A., and D. Parish. 1991. A review of the Asian–Australasian bird migration system. *In* ICBP Technical Publication No. 12. Cambridge: International Council for Bird Preservation. *An important overview of this poorly studied migration system and its conservation requirements.*

Lövei, G.L. 1989. Passerine migration between the Palearctic and Africa. Current Ornithology 6: 143–174. *A recent review of the best-studied global migration system across the Sahara desert.*

CHAPTER 13

Navigation

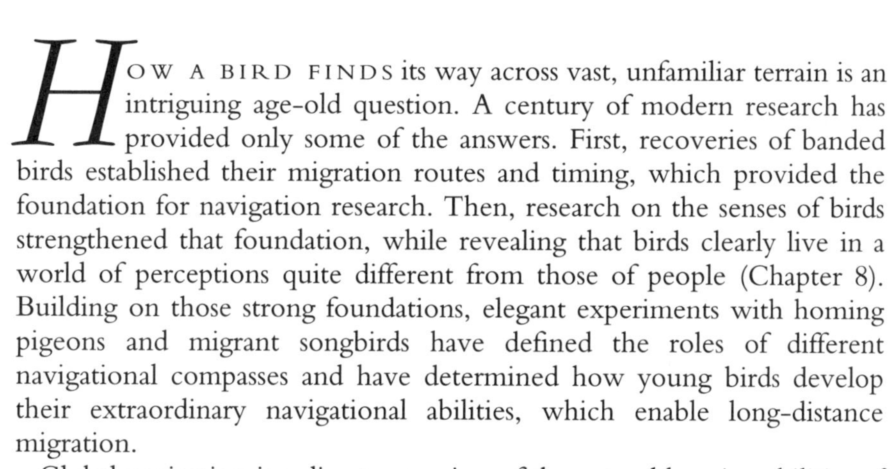

HOW A BIRD FINDS its way across vast, unfamiliar terrain is an intriguing age-old question. A century of modern research has provided only some of the answers. First, recoveries of banded birds established their migration routes and timing, which provided the foundation for navigation research. Then, research on the senses of birds strengthened that foundation, while revealing that birds clearly live in a world of perceptions quite different from those of people (Chapter 8). Building on those strong foundations, elegant experiments with homing pigeons and migrant songbirds have defined the roles of different navigational compasses and have determined how young birds develop their extraordinary navigational abilities, which enable long-distance migration.

Global navigation is a direct extension of the natural homing abilities of birds, which enables them to find their nests, their feeding grounds, or their breeding territories after a long winter absence. Individual birds return to a particular tree in Canada after wintering in South America or migrate annually between particular sites in Europe and Africa (Moreau 1972; Lövei 1989; Rappole and Warner 1980). In one of the earliest experiments, an Eastern Phoebe, wearing a silk thread placed on its leg by John James Audubon in 1803, returned the next spring to Audubon's house in Mill Grove, Pennsylvania, after wintering somewhere in the southern United States. Conversely, banded Northern Waterthrushes, which breed in northern bogs, returned predictably every year to the same wintering sites in Venezuela (Schwartz 1964).

Figure 13–1 The Black-chequered Homing Pigeon. Homing pigeons are special breeds of the Rock Dove that were developed for their powerful flight and ability to return to their home loft after being transported great distances. Homing pigeons were used originally to carry messages. Now they are used for sport, namely, races. (From Lumley 1895)

The homing feats of displaced birds also testify to their navigational abilities. Homing pigeons return to their lofts by flying as much as 800 kilometers per day from unfamiliar places (Figure 13–1). Ancient Egyptians and Romans developed these messengers by enhancing the natural orientation abilities of feral Rock Doves. Shearwaters and sparrows, as well as a variety of other birds, can return to a home site after being transported thousands of miles away (Table 13–1). A Manx Shearwater, for example, returned to its nest burrow in Wales only 12.5 days after being released in Boston (Mazzeo 1953). White-crowned Sparrows that

TABLE 13–1
Abilities of birds to return to the site of capture after transport to a distant, unfamiliar release site

Species	Number of birds	Distance (km)	Return (%)	Speed (km/day)
Leach's Storm-Petrel	61	250–870	67	56
Manx Shearwater	42	491–768	90	370
Laysan Albatross	11	3083–7630	82	370
Northern Gannet	18	394	63	185
Herring Gull	109	396–1615	90	112
Common Tern	44	422–748	43	231
Barn Swallow	21	444–574	52	278
European Starling	68	370–815	46	46

From Griffin 1974.

were shipped to Baton Rouge, Louisiana, returned the following winter to their wintering grounds in San Jose, California, where they were recaptured. They returned to California again after a second displacement to Laurel, Maryland (Mewaldt 1964) (Figure 13–2).

This chapter reviews the kinds of information that birds use to navigate while migrating, while commuting between nest sites and feeding grounds, and while flying home after being displaced by a curious ornithologist. Birds use several sources of information, often preferring one source if it is available and using the others when necessary. In addition to using visual landmarks such as landscapes and buildings, migrants use the sun by day and the stars by night. Birds also use olfactory cues, the Earth's magnetic field, and perhaps very low sound frequencies (infrasound). An innate magnetic compass that includes both direction and duration serves as the platform for the development of advanced navigation abilities.

Use of Visual Landmarks

First and foremost, birds often rely on visual landmarks for both local travel and long-distance migration. Both diurnal and nocturnal migrants, especially waterfowl, follow watercourses and coastlines but are often reluctant to cross large, open bodies of water unless the winds are favorable. They will gather en route in great numbers where restricted corridors function as funnels. The Strait of Gibraltar and the Bosporus at Istanbul are major funneling points for Eurasian migrants that detour around the Mediterranean Sea. The coasts of Central America funnel thousands of migrating raptors—Broad-winged Hawks, Swainson's Hawks, and

Turkey Vultures—over Panama City. Crowds of birdwatchers gather to view the spectacle of migrants funneled to the tips of peninsulas such as Point Pelee, Ontario, and Cape May, New Jersey.

Underlying the use of visual landmarks, however, are more sophisticated navigational compasses. Birds use cues other than landmarks and, perhaps, senses other than sight. In one study, homing pigeons ignored obvious landmarks until they came within sight of a tall building near Boston, which they used to make a final correction toward home

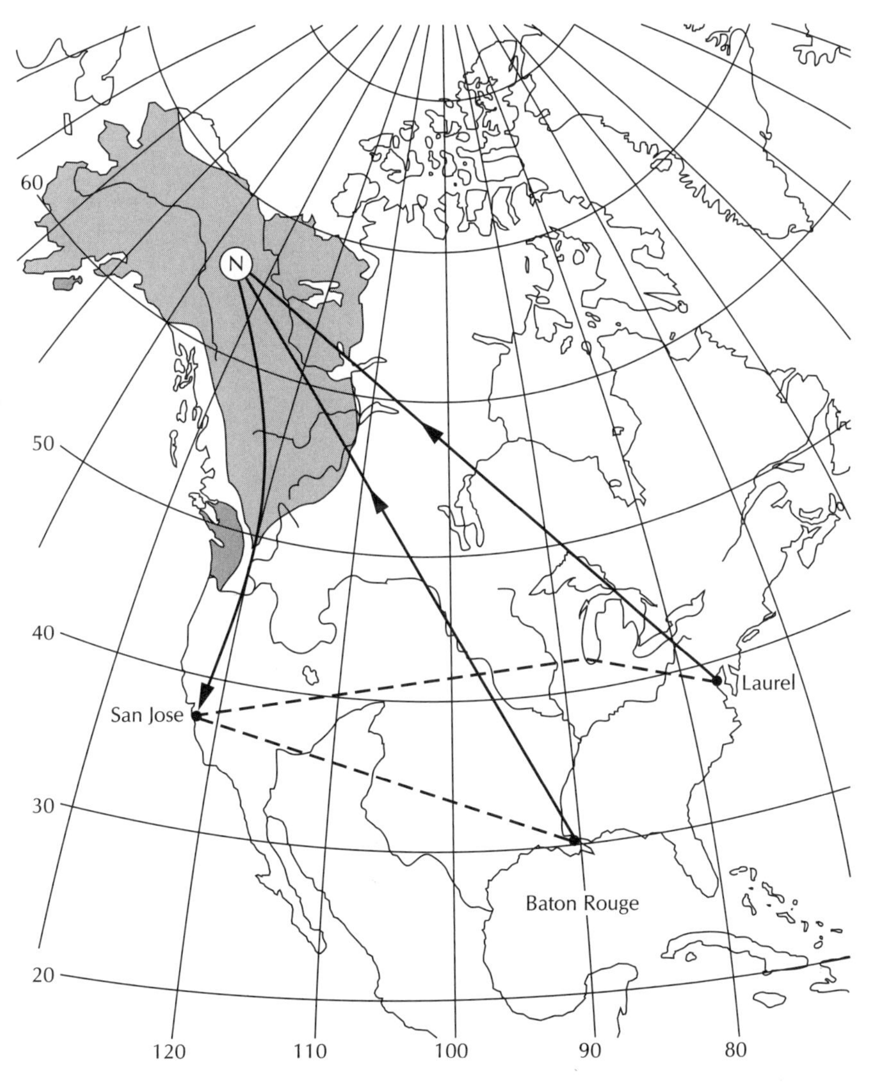

Figure 13–2 White-crowned Sparrows returned to their wintering grounds in San Jose, California, after being carried by aircraft (broken lines) to Baton Rouge, Louisiana, and to Laurel, Maryland. These marked sparrows apparently spent the intervening summers on their nesting grounds in Alaska. The solid lines show their probable flight paths. (Adapted from Mewaldt 1964)

(Michener and Walcott 1967). In another experiment, well-trained homing pigeons were fitted with frosted contact lenses that eliminated image formation beyond 3 meters (Schlichte and Schmidt-Koenig 1971; Schlichte 1973). These severely myopic birds flew "blind" for over 170 kilometers directly back to their lofts. When they reached the vicinity of their lofts, they hovered and then landed much like helicopters. Not all such pigeons performed perfectly, some crashed and some missed the loft altogether, but many oriented well without being able to see landmarks.

The Solar Compass

Scientists long suspected that birds navigate by the sun, but proof of this ability awaited experiments conducted with starlings and homing pigeons in the 1950s. In Germany, Gustav Kramer (1950, 1951) studied the orientation of migratory restlessness *(zugunruhe)* in European Starlings. The birds were housed in circular cages and placed in a large pavilion with windows through which they could see the sun, including its change of position as the day progressed. As long as they could see the sun, they focused their attention to the northeast, the correct direction for spring migration. On overcast days, however, the starlings showed no directional tendency (Figure 13–3).

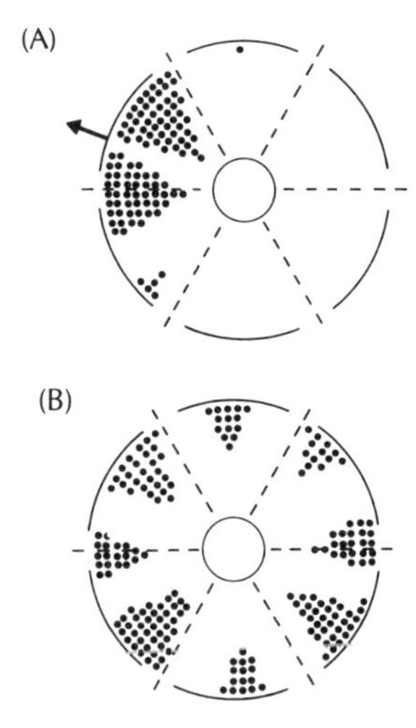

Figure 13–3 European Starlings use the sun to orient in a circular cage. (A) As long as they could see the position of the sun in the sky, they oriented their restless spring migratory behavior to the northeast. (B) On overcast days when they could not see the sun, they showed no directional orientation. Each dot represents 10 seconds of fluttering activity. (From Kramer 1951 and Emlen 1975a)

About the same time, in Great Britain, Geoffrey Matthews (1951, 1953, 1968) established that homing pigeons use the sun to guide them back to their lofts. He released them from unfamiliar sites away from the loft under a variety of weather conditions. The pigeons flew directly home when they could see the sun, but they fared poorly under overcast skies. As a result of releasing the pigeons at different times of the day, Matthews discovered a key feature of this orientation behavior: Not only could the pigeons use the sun for directional information, but they also compensated for its changing position as the day progressed, as if they could "tell the time."

The position of a point on the Earth relative to the sun changes continuously by 15 degrees per hour. To orient consistently in one direction, a bird must somehow understand the changing position of the sun relative to direction throughout the day; that is, the solar compass must be time compensated, as Matthews proposed. In an important experiment that extended his other pioneering experiments with the starlings, Kramer and his colleagues showed that birds compensated for the apparent motion of the sun. They trained starlings (and some other birds) to feed from the northwest cup of a series of cups placed around the perimeter of a circular cage. The birds reliably chose the correct cup when they could see the sun. However, when trained to accept a stationary light bulb as a substitute for the sun, they fed from cups increasingly farther to their left as they compensated for the assumed hourly change in the position of the "sun" (Kramer 1952).

BOX 13–1

A migrating thrush outperforms ornithologists

William Cochran and his coworkers (1967) captured a migrating Gray-cheeked Thrush in central Illinois one afternoon and attached a tiny radio transmitter to it. At dusk the thrush took off on the next leg of its journey, followed by the ornithologists in a small plane. A severe thunderstorm and shortage of fuel forced their plane down during the night, but the thrush flew on. After refueling, the Cochran group took off again and, remarkably, relocated the thrush in the vast night sky by dead reckoning. The thrush landed at dawn in Wisconsin after flying 650 kilometers on a firm compass bearing all night—without refueling.

Adelie Penguins, which swim and walk to their destinations rather than fly, also use a time-compensated solar compass. John Emlen and Richard Penney (1964) took penguins from their coastal breeding rookeries to the interior of Antarctica and released them. On cloudy days, the penguins wandered about randomly without significant orientation. When the sun was shining, however, the penguins headed north-northeast to the coast, compensating for the sun's apparent counterclockwise movement in the Southern Hemisphere by correcting their orientation 15 degrees per hour clockwise relative to the sun's position.

The next step in the study of a time-compensated solar compass was to change a bird's internal clock and thereby trick it into misreading the sun's position. Konrad Hoffmann (1954) did this by studying orientation behavior in "clock-shifted" European Starlings trained to find food in a particular compass direction. He kept these birds on a 12-hour-dark and 12-hour-light cycle that was 6 hours out of phase with natural daylight (the lights went on at 1200 instead of 0600). Accustomed to this schedule, the starlings predictably misread the sun's position in the sky. When they awakened, for example, it was actually noon, not 0600, and the sun was in its southern midday position, not its eastern dawn position. The clock-shifted starlings, however, interpreted this to be the dawn position of the sun; therefore, their "east" was really south. As a result they looked for food at a position 90 degrees clockwise from the correct bearing (Figure 13–4). This is a standard result: A 6-hour clock shift causes a 90-degree disorientation. Experiments with many other clock-shifted birds, including homing pigeons, have since confirmed the widespread use by birds of time-compensated solar cues.

(A)

(B)

Figure 13–4 When the internal clock of a European Starling is set 6 hours behind natural time (by changing the schedule of light and dark), it misreads the sun's position and looks for food 90 degrees (white arrow) from the correct location (black arrows). (A) Behavior during training, showing correct orientation; (B) behavior after the 6-hour clock shift in internal schedule. Each dot shows an attempt to find food. (Adapted from Hoffmann 1954 and Emlen 1975a)

The Stellar Compass

Land birds and waterfowl maintain their direction when they migrate at night by using the stars as a source of directional information. Franz and Eleanore Sauer (Sauer 1957, 1958) first demonstrated the ability of migrating passerine birds to use the stars for navigation in experiments with hand-reared Garden Warblers. The warblers were kept in circular experimental cages in a planetarium. When ready to migrate, they became hyperactive and restless; even when caged, they tried to fly or hop in their migratory direction. The Sauers watched the birds through the glass bottom of their cage. The warblers oriented north in the "spring" and south in the "fall" under the simulated night skies of the planetarium. When the Sauers turned off the "stars," the warblers became disoriented. When the Sauers rotated the north–south axis of the planetarium sky 180 degrees, the warblers also reversed their compass headings.

Stephen Emlen (1967a, 1975b) duplicated the Sauers's results with a North American migrant, the Indigo Bunting. The cage in which Emlen placed his buntings was a paper cone with steep sides sloping downward to an ink pad at the apex. A restless bird would hop up the side of the cone, leaving inky foot marks, and slide back down onto the pad. The intensity of ink on various sections of the cone indicated the relative degree of activity in each direction (Figure 13–5). Even in these

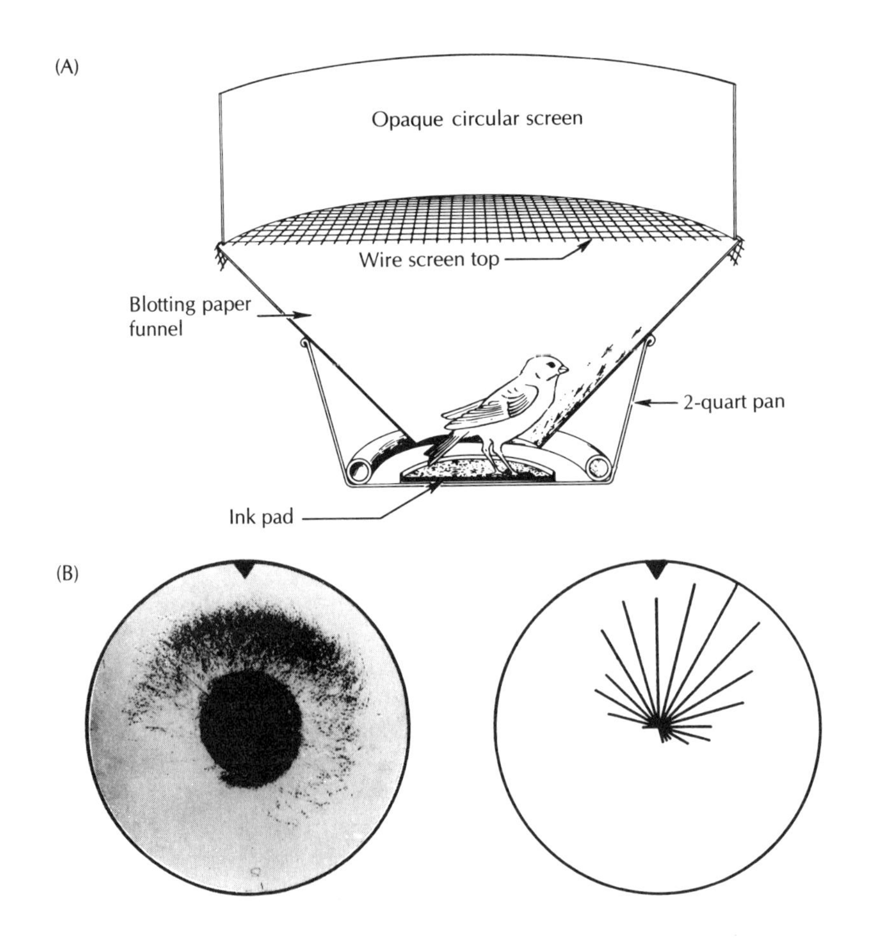

Figure 13–5 (A) The Indigo Bunting migrates at night between its summer range in the eastern United States and its winter range in Central America. Buntings in a state of migratory restlessness orient by the stars at night, even when confined to a funnellike cage placed under a planetarium sky. (B) Inky footprints record the orientation direction; the lengths of line vectors measure the intensity of ink left in each 15-degree sector. (Adapted from Emlen and Emlen 1966)

experimental cages, Indigo Buntings oriented north when a spring night sky was simulated in a planetarium and south when a winter night sky was simulated. Like the warblers, the buntings became disoriented when the planetarium sky was turned off and reversed their orientation when the axis of the sky was reversed (Figure 13–6).

Whether nocturnal migrants compensate for the movement of the stars as other birds do for the movement of the sun is not yet clear. The warblers studied by the Sauers changed their orientation when the planetarium sky was set ahead or behind the correct time, a behavior suggesting that they navigate by a time-compensated stellar compass. Emlen's Indigo Buntings, on the other hand, maintained their northward orientation under phase-shifted planetarium skies, a behavior suggesting that the buntings refer to star patterns or constellations rather than to a specific star. Studies of White-throated Sparrows and Mallard ducks also suggest that use of stellar cues does not include, and perhaps does not require, time compensation. Short-distance migrants such as the buntings may use the stars differently, perhaps with less sensitivity and less time compensation, than do long-distance migrants such as the warblers. Transequatorial migrants not only may need greater precision in their initial orientation because of the long distance they fly, but they also must orient under the different skies of the Southern and Northern Hemispheres.

Stephen Emlen (1967b) attempted to identify the stars that buntings use for orientation by systematically blocking out various constellations. He assumed, logically, that the buntings orient by the North Star, the one obvious, fixed point in the night sky, but they did not. Instead, they used the constellations that were within 35 degrees of the North Star. Moreover, the buntings were familiar with most of the major constellations in the Northern Hemisphere, including the Big Dipper, the Little Dipper, Draco, Cepheus, and Cassiopeia; if one of these constellations was blocked from view, the buntings used the others. Such redundancy is useful when sections of the sky are overcast; it also allows the birds to be flexible in their choice of guideposts in the complex ever-changing night sky. In addition, different birds appear to use different parts of the sky, their use apparently based on experience.

It is easy to change a bird's hormonal physiology by changing day length—or photoperiod (Chapter 11). Simulating the seasons by increasing or decreasing day lengths can bring caged birds into breeding condition, can cause them to molt more often than is natural, and can cause them to accumulate premigratory fat at the wrong time of the year. Using unnatural photoperiod regimes, Emlen (1969) manipulated the seasonal physiology of two groups of Indigo Buntings. He induced readiness for northward spring migration in one group and readiness for southward fall migration in the other group. Exposed to the same planetarium sky, buntings in the two groups oriented north and south, respectively. These results showed that migratory orientation is under physiological control, at least in some birds.

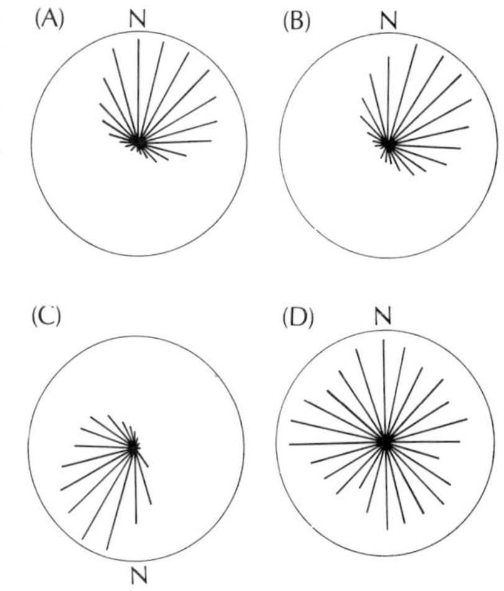

Figure 13–6 Line vectors, such as the ones described in Figure 13–5, show how Indigo Buntings use the stars to orient north in the spring. They do so under (A) natural night skies and (B) simulated night skies in a planetarium. (C) When the planetarium stars are shifted so that the North Star, *N*, is at true south, the birds reverse their orientation. (D) When the stars are turned off and the planetarium is diffusely illuminated, the buntings do not orient. (Adapted from Emlen 1975b)

Navigation by Olfaction

Petrels and pigeons can smell their way home. Leach's Storm-Petrels, for example, locate their nesting burrows on forested islands in the Bay of Fundy by using olfactory cues. On returning from the sea, they fly into the forest a short distance downwind from their nest burrows, which they then locate precisely by the smell of the nests (Grubb 1974). Storm-petrels with an experimentally impaired sense of smell did not return to their burrows after 1 week, whereas controls did. In the laboratory, these petrels found their dismantled nest material in an experimental maze.

The use of olfactory cues by homing pigeons for distance navigation remains uncertain because of conflicting experimental results (Waldvogel 1989). In one early experiment, pigeons with their nostrils plugged with cotton or with their olfactory nerves cut did not find their lofts as well as normal controls (Papi et al. 1971, 1972). However, other experiments with homing pigeons failed to show that impaired olfaction affects their navigation abilities (Keeton et al. 1977). Also, regional differences produced conflicting experimental results; Italian pigeons relied more on olfactory navigation than German pigeons. At best it appears that olfactory cues may supplement other navigation systems.

Responses to Geomagnetism

The geomagnetic fields of the Earth provide a map of horizontal space, just as gravity and barometric pressures give information about vertical space. The intensity and dip angle—or inclination of the magnetic field—change with latitude in ways that provide reliable, omnipresent information about geographical position.

Ornithologists were slow to accept the hypothesis that birds might use the Earth's magnetic field for orientation. An early report that magnets disrupted a pigeon's homing ability (Yeagley 1947) was discredited, largely because the results could not be repeated. Then Frederick Merkel and Wolfgang Wiltschko (1965) showed that captive European Robins could orient in experimental solid steel cages without celestial cues. In continued experiments that aroused general interest, these researchers showed that a robin reverses its orientation when the magnetic field imposed on the steel cage is reversed.

Several years later William Keeton (1971, 1972) showed that free-flying homing pigeons wearing bar magnets often did not orient properly on cloudy days, whereas control pigeons wearing brass bars usually did (Figure 13–7). Failures to repeat Yeagley's earlier experiments were due in part, Keeton revealed, to the use of the solar compass in preference to the magnetic compass on sunny days by both the experimental and control pigeons. Finally, in experiments that swayed the skeptical, Charles Walcott and Robert Green (1974) fitted homing pigeons with electric caps (containing Helmholtz coils) that produced a magnetic field through the bird's head. Under overcast skies, reversing the field's direction by

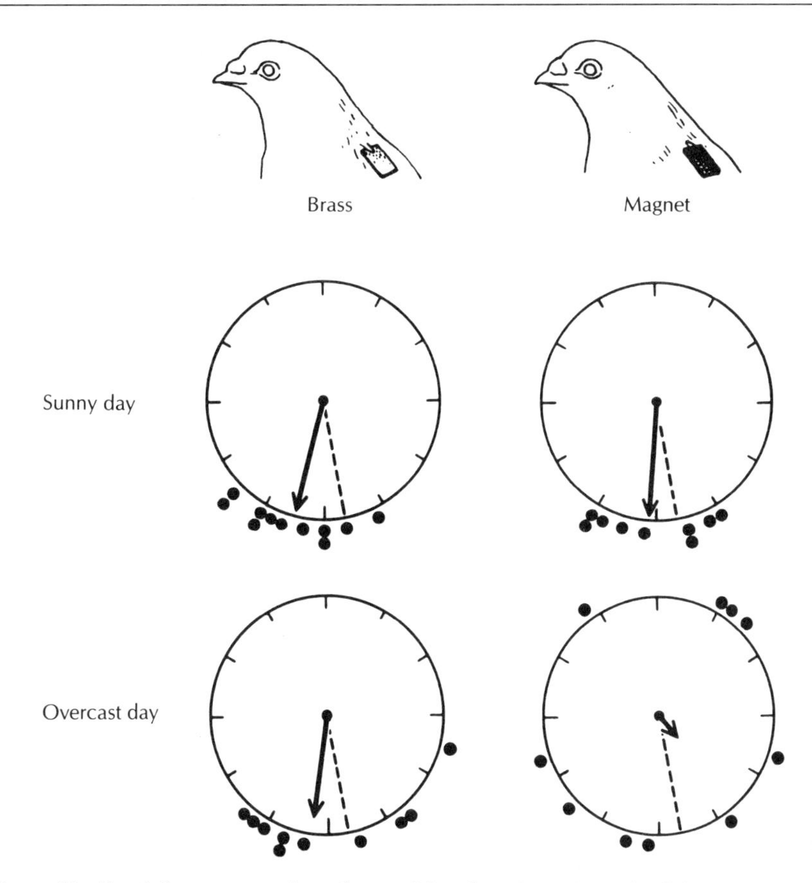

Figure 13–7 A bar magnet interferes with a homing pigeon's ability to return to its loft on overcast days. On sunny days, pigeons wearing magnets and control pigeons wearing brass bars both adopt accurate home bearings at unfamiliar release sites. On overcast days when they cannot orient by the sun (their preferred cue), the pigeons wearing magnets become disoriented. The control group, however, orients by means of the Earth's magnetic information. Vectors (arrows) show mean direction and consistency of orientation among individuals: Long vectors show consistent orientation, the short vector shows variable orientation. Dots represent bearings recorded for each pigeon tested. The dashed line represents the correct orientation. (From Keeton 1974)

reversing the current caused free-flying pigeons to fly in the direction opposite to their original course (Figure 13–8).

We now appreciate that many migrating bird species navigate by using the Earth's magnetic fields. Specialized photoreceptors in the visual system appear to be sensitive to a bird's orientation relative to these fields (Chapter 8).

The magnetic compass of birds differs from the mechanical compass people use to navigate (Wiltschko and Wiltschko 1988). First, the magnetic compass of birds is narrowly tuned to the total intensity of the ambient magnetic field. A slight change (10 percent) in a field's intensity disrupts orientation for 3 days, the time it takes for the bird to adjust to the new field. Second, birds are sensitive to extremely weak magnetic

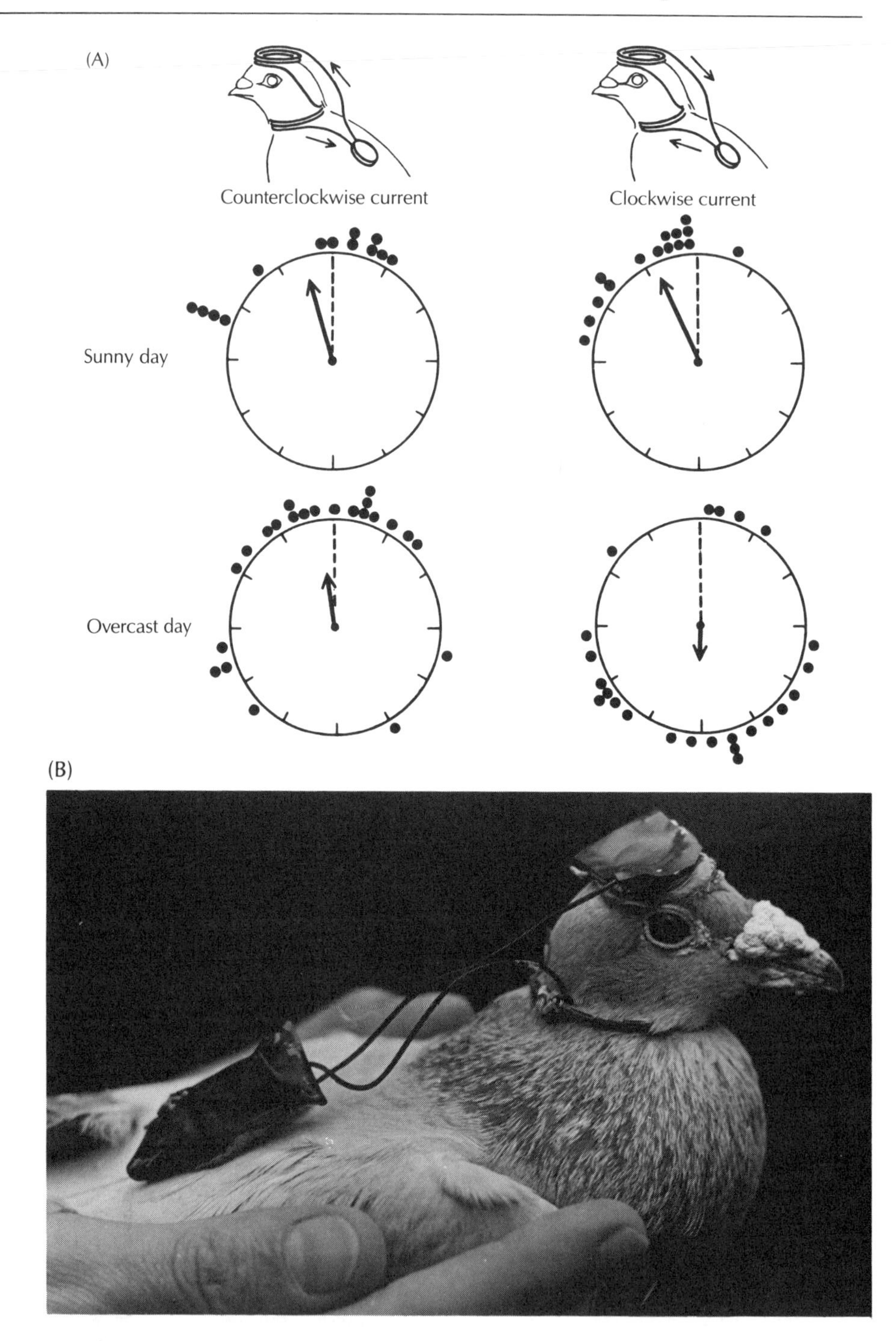

Figure 13–8 (A) By attaching Helmholtz coils to the heads of homing pigeons, Charles Walcott and Robert Green generated artificial magnetic fields by allowing an electric current to flow through the coils. The reversal of electric current, which reversed the magnetic field, caused the pigeons to reverse their orientation direction on overcast days. Vectors are portrayed as in Figure 13–7. (B) A homing pigeon equipped with Helmholtz coils. (A adapted from Walcott and Green 1974 and Keeton 1974; B courtesy of C. Walcott)

fields, as low as 10^{-7} to 10^{-9} gauss, which correspond to the natural fluctuations in the Earth's magnetic field caused by sunspots and hills of iron ore. These natural fluctuations disrupt the orientation of passerine birds migrating at night (Moore 1977). For example, solar storms, which change the Earth's magnetic field, disrupt the orientation in Ring-billed Gull chicks under cloudy skies (Southern 1971, 1972). Many early laboratory experiments failed because researchers used magnetic fields that varied in intensity and did not allow the birds enough time to adjust; or they used fields that were too strong, thereby exceeding a bird's range of sensitivity. Third, the magnetic compass of birds responds to the "poleward" or "equatorward" angles of inclination of a magnetic field, not to its "north–south" polarity. Reversal of a field's polarity per se does not always reverse a bird's orientation.

Learning to Navigate

The navigational abilities of birds are partly innate and partly dependent on early experience, with the result that inexperienced young migrant birds become lost more often than experienced adults. The rare visitors that excite birders, for example, are often lost immature birds (De Sante 1983).

Temporal patterns of migratory activity have some endogenous basis. Not only are migratory preparations and migration itself linked directly to endogenous circannual rhythms (Chapter 12), but the duration and pace of migration may be linked to these rhythms as well. The length of nocturnal activity in the laboratory relates directly to the distances these warblers migrate to their respective winter ranges (Gwinner 1977) (Figure 13–9). Long-distance migrants such as Garden Warblers show more total

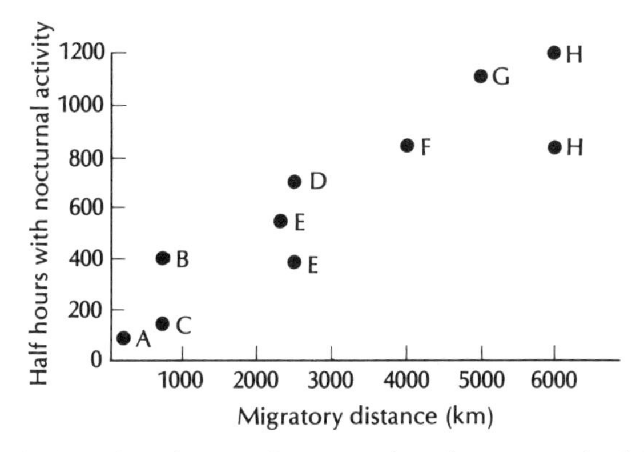

Figure 13–9 The lengths of time of nocturnal restlessness in the laboratory are well correlated with the migration distances covered by eight species of European warblers: (A) Marmora's Warbler; (B) Dartford Warbler; (C) Sardinian Warbler; (D) Blackcap; (E) Eurasian Chiffchaff; (F) Subalpine Warbler; (G) Garden Warbler; (H) Willow Warbler. Results for Willow Warblers and Eurasian Chiffchaffs tested under different conditions are shown separately. (After Gwinner 1977)

Box 13–2
Young starlings orient by general direction only

Young European Starlings can orient to a simple migratory direction but not to a particular goal. In a classic experiment, A.C. Perdeck (1967), a Dutch ornithologist, captured thousands of starlings that were migrating to their wintering grounds. He took them southeast from the capture site in Holland to Switzerland, where he released them. Normally these birds move west-southwest through The Hague from breeding grounds in Denmark and Poland to wintering grounds in southern England, Belgium, and northern France. Recoveries of the displaced adult starlings revealed that they reoriented and flew northwest toward their wintering grounds. The young starlings, however, failed to compensate for their geographical displacement and continued to migrate west-southwest, ending up in Spain and southern France. Adults recognized and corrected for the displacement relative to their goal, whereas young birds practiced simple directional orientation (see figure).

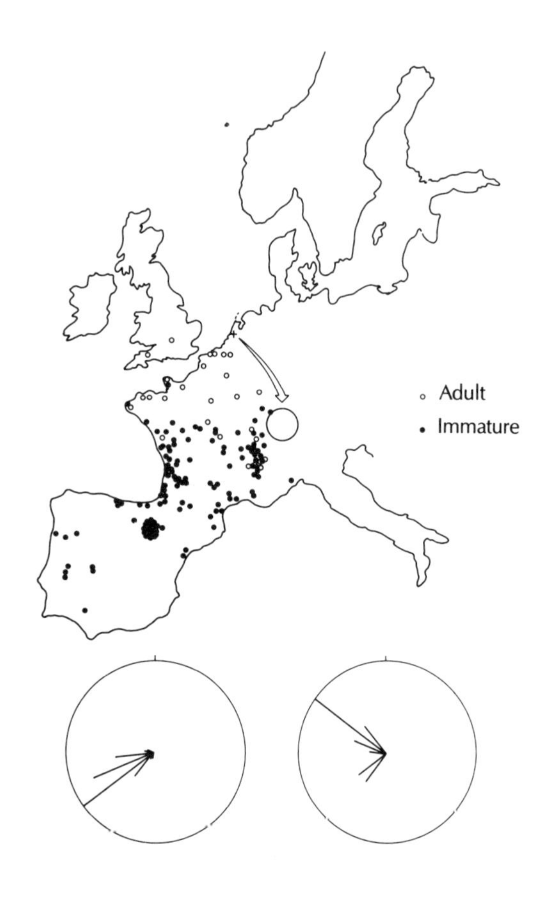

Recapture locations of adult and immature European Starlings displaced from Holland to Switzerland (arrow to large circle in center of map) during autumn migration. The immatures (black dots) continued their migration to the southwest (left vector diagram), whereas the adults (white dots) reoriented to their correct wintering grounds in southern England, Belgium, and northern France (right vector diagram). Vector diagrams summarize recoveries in 15-degree orientation sectors. (From Emlen 1975a, after Perdeck 1958)

nocturnal activity than do short-distance migrants such as the related Marmora's Warbler. The Willow Warbler normally takes 3 to 4 months to migrate from Europe to southern Africa; intense migratory restlessness of this warbler in the laboratory lasts over 4 months. The Eurasian Chiffchaff takes only 1 to 2 months to migrate from southern Europe to northern Africa; intense migratory restlessness in the laboratory lasts 60 days. These results suggest a basis for the simple directional navigation shown by Perdeck's young starlings (Box 13–2). Programmed for general orientation and the approximate amount of time they should migrate in that direction, a young starling should reach some part of its species' winter range. It can then develop the more refined abilities required for goal orientation on subsequent migrations.

Like the other European warblers, populations of the Blackcap differ from one another in the seasonal course and magnitude of *zugunruhe,* and the differences also correspond directly to the distance each population normally migrates. Initial evidence of direct genetic control of the programming for directional migration comes from the study of hybrids between the migratory German population and the nonmigratory African population. The hybrids exhibited intermediate *zugunruhe* activity (Figure 13–10). In separate laboratory experiments, Peter Berthold (1988) demonstrated that Blackcaps responded to selection for migratory behavior. He divided Blackcaps from a partially migratory population in southern France into two groups, one for the young of migratory parents and one for young of nonmigratory parents. A fully resident, nonmigratory population was established in four to six generations; a fully migratory population was established in only two generations. Recent studies of hybrids between southeast-migrating and southwest-migrating Blackcaps provide direct evidence for a genetic basis of migratory direction (Helbig 1991); the orientations of the first-generation hybrids were intermediate between and significantly different from those of the parent populations.

In the last 25 years, Blackcaps have added Great Britain to their list of wintering grounds. Instead of migrating southwest to the Mediterranean, increasing numbers of German Blackcaps migrate each year northwest to England and Ireland (Berthold et al. 1992). These new wintering grounds require a shorter migration and facilitate earlier returns to the breeding grounds. The offspring of Blackcaps that winter in England exhibit an innate orientation to the northwest, which suggests a rapid evolutionary change in the genetic program that controls their migratory behavior.

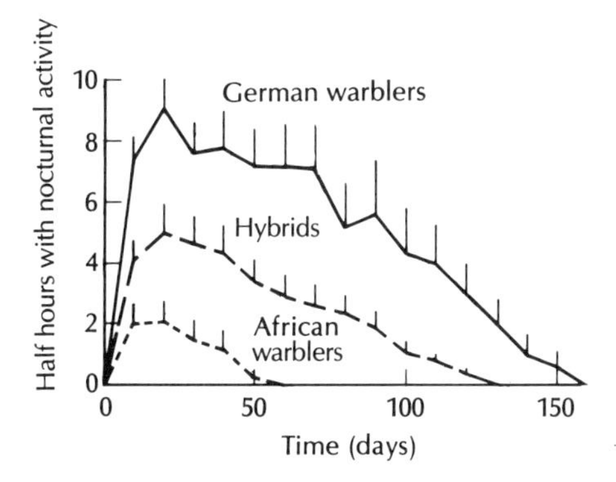

Figure 13–10 Blackcaps from migratory populations in Germany show intense and prolonged migratory restlessness, whereas individuals from a nonmigratory population in Africa show little migratory restlessness. Hand-raised hybrids of these forms have intermediate migratory behavior. (After Berthold and Querner 1981)

Box 13–3
Baby buntings learn the night sky

Baby Indigo Buntings, hand-reared without seeing the stars, cannot orient when they are first exposed to the night sky. In fact, they must see the sky regularly during the first month of life to be able to choose their migratory direction. The axis of rotation of the night sky, which centers on the North Star, establishes their north–south frame of reference (Emlen 1970) (see figure). They then learn the constellations associated with this axis. If the axis of rotation of the planetarium sky is switched from the North Star to Betelgeuse, the brightest star in the constellation Orion in the southern sky, the baby buntings orient south in line with the new axis of rotation.

The reference of the innate directional systems of nocturnal migrants to the axis of celestial rotation appears to be independent of their innate reference to geomagnetic fields (Wiltschko and Wiltschko 1988, Figure 15). In practice, the two navigational systems appear to complement each other, especially for transequatorial migrants; these birds must pass through the horizontal equatorial magnetic fields, which provide little directional information.

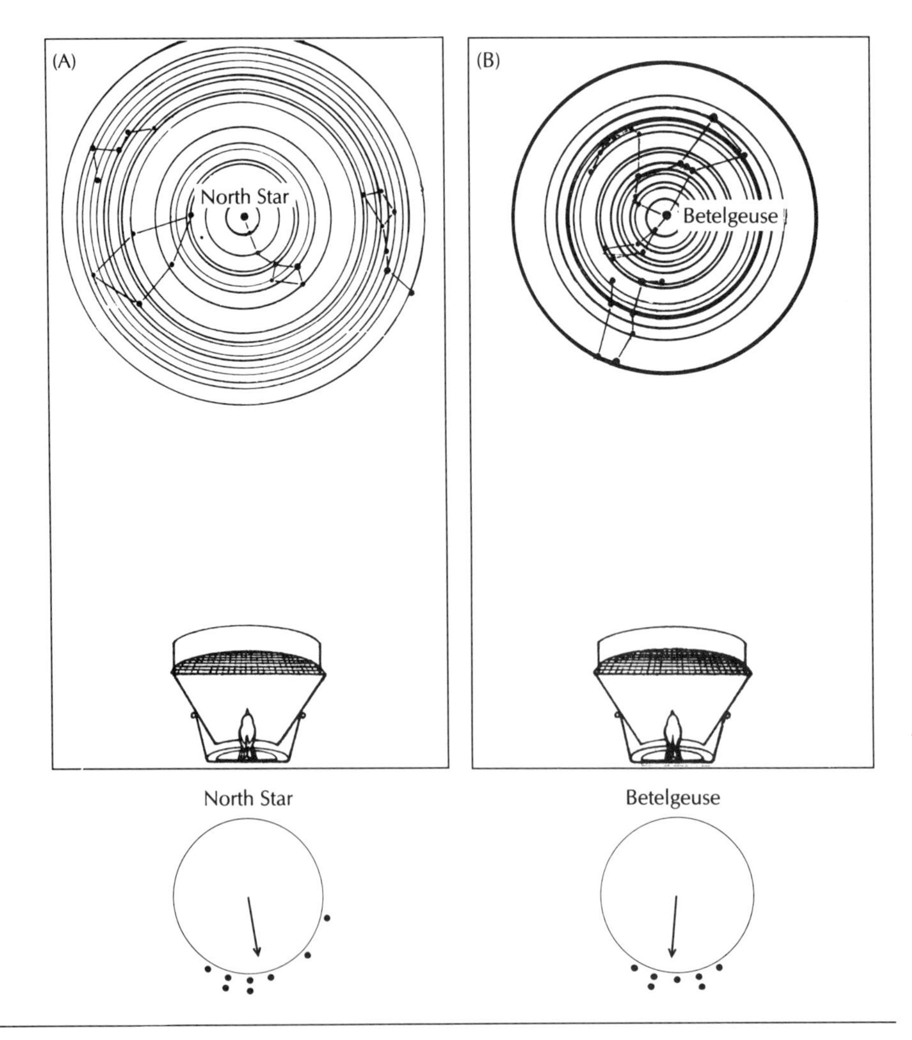

(A) Early visual experience of the natural night sky entrains an Indigo Bunting's use of the stars for orientation. (B) Buntings raised under a modified night sky that rotated around Betelgeuse instead of the North Star adopted Betelgeuse as the pole star and consistently oriented from it. Each dot represents the direction selected by one young bunting. The vectors (arrows) show the general direction of orientation. (Adapted from Emlen 1975b)

Hand-reared, caged migrants not only exhibit well-defined orientation behavior but also change their compass direction in ways that correspond to their natural migration routes. Garden Warblers change direction during their fall migration from southwest initially to south-southeast on their route from Spain to southern Africa. Devoid of cues other than magnetism, the orientation of migratory restlessness in the laboratory shows a corresponding shift. Restless, caged Garden Warblers orient southwest in August and September and then shift their heading to south-southeast from October to December (Gwinner 1977). Such internal programs, however, cannot guide migrants precisely to their final winter residences. External forces including food availability, climate, and competitive interactions come into play at various stages of the journey and may be the dominant factors, especially in short-distance migration.

Birds clearly possess an innate capacity for orienting in relation to the Earth's geomagnetic field. Wolfgang and Roswitha Wiltschko (1988) suggest that the magnetic compass serves as the initial basis for directional reference by young birds and then links other learned compasses, such as the solar compass, to form an integrated orientation system. A pigeon's ability to use magnetic compass information develops before solar compass abilities are manifest. On their first flight, juvenile homing pigeons record the general direction of their outbound journey, which is based on magnetic field information. Reversal of this direction establishes the "home direction" that guides the development, with continued experience, of full sensitivity to the polarity and declination lines of the Earth's magnetic field. Young pigeons do not establish a home direction if they are transported in a distorted magnetic field from their nest on their first trip or if they are made to carry magnets on their maiden flight. Once the home direction is established through route reversal and use of the magnetic compass, however, the young pigeons learn to use other clues including the solar compass.

Homing pigeons reared without seeing the sun and exercised only on overcast days do not, and probably cannot, use the solar compass (Keeton and Gobert 1970). Six-hour clock-shift experiments with such pigeons do not produce the 90-degree orientation error characteristic of young birds that have seen the sun while growing up. Exposure to the sun for less than 1 hour, however, is enough to activate the solar compass, which becomes increasingly refined with experience. Pigeons do not inherit a knowledge of solar compass positions but calibrate their solar compass in reference to their initial magnetic compass (Wiltschko et al. 1983).

Adding to the complexities of the interactions between compass systems is the potential role of the sunset point, which defines an east–west axis of directional information, not only by the position of the sun itself but also by the distinct pattern of sky light polarization that accompanies each sunset (Moore 1982; Able 1982). How the east–west sunset point compass is calibrated relative to the solar and magnetic compasses is unknown.

Maps and Bicoordinate Navigation

Choice and maintenance of a compass direction are only part of the challenge of navigation. If a bird is to reach a goal, such as a loft in the case of the homing pigeons, it must also know its own position relative to its goal. It must have a sense of location, or a map; a bird displaced to the north must fly south to the loft and a bird displaced to the west must fly east. To get oriented in an unfamiliar place, humans would look at a map or ask someone which way to go. It is not known how birds solve this dilemma. The Earth's magnetic field varies predictably with latitude and longitude in ways that potentially form a navigational grid. But whether birds can use this information has not yet been demonstrated.

Sun position potentially provides information on longitude as well as latitude. For example, at northern temperate latitudes, the sun is higher at noon in the south than in the north. The sun also rises progressively later as one travels west and, therefore, will be at a different position in its arc relative to any given absolute clock time (people adjust for this by having official times that compensate roughly for later sunrises in the west). A simple rule for westward navigation in North America would be: If the sun at your present location is higher than it would be at your goal at this time of day, fly away from it; if it is lower, fly toward it (Griffin 1974). The "sun-arc" hypotheses of Colin Pennycuick (1960) and Geoffrey Matthews (1968) embrace these possibilities, which require a bird to have both an accurate memory of sun positions and an acute sense of time. Clock-shift experiments, however, have failed to show that homing pigeons can use sun position for anything more than simple compass direction (Emlen 1975a). Clock shift should drastically alter a bird's interpretation of the sun's height above the horizon at a particular time of day. A bird whose internal clock had been advanced experimentally would view the sun as too low and therefore should fly east to compensate. But homing pigeons do not compensate in this way.

Summary

In their daily routines and on annual migrations, birds navigate great distances, sometimes across unfamiliar terrain. Enabling accurate orientation is a battery of diverse navigational tools that increase in sophistication with experience. Birds rely on acute visual memories for short-distance travel and local orientation. Birds also use the positions of the sun by day and the stars by night. The solar compass compensates for the ever-changing position of the sun in the course of the day. The stellar compass probably does not, but instead focuses on the constellations close to the North Star, the fixed axis of rotation of the night sky. Recent studies confirm that birds use the Earth's magnetic fields to define their initial compass directions and then add celestial compass information onto this foundation. Orientation by magnetic field information alone is practiced when clouds obscure celestial cues. Some, perhaps most, migratory songbirds inherit genetic programs that route them to traditional wintering grounds by controlling their orientations and flight distances.

Further Readings

Berthold, P., Ed. 1991. Orientation in Birds. Basel: Birkhauser Verlag. *A compendium of important review papers by leaders in the field.*

Emlen, S.T. 1975. Migration: Orientation and navigation. Avian Biology 5: 129–219. *An excellent, comprehensive, and critical review of the early literature.*

Keeton, W.T. 1980. Avian orientation and navigation: New developments in an old mystery. Acta XVII Congressus Internationalis Ornithologicus 1: 137–158. *An extraordinary summary of the topic.*

Papi, F., and H.G. Wallraff, Eds. 1982. Avian Navigation. Berlin: Springer-Verlag. *An important set of review papers.*

Wiltschko, W., and R. Wiltschko. 1988. Magnetic orientation in birds. Current Ornithology 5: 67–121. *A comprehensive review by two pioneers.*

CHAPTER 14

Social Behavior

BIRDS ARE BOTH PREDATORS and prey. Their needs for food and for protection—the most pressing requirements of any living creature—determine where and how they live and whether they are social or asocial, cooperative or competitive. Sometimes an individual should go it alone; at other times there is safety in numbers. Whether a bird lives alone or with others, the fact remains that, ultimately, birds must share limited space. Whether breeding or not, birds may space themselves at regular intervals over large territories, congregate in large numbers, or cluster in small groups. At one extreme, Solitary Eagles live in pairs on exclusive expanses of tropical forest. At the other extreme, dozens of pairs of Sociable Weavers cluster together in gigantic communal nests.

This chapter examines the spacing behaviors of birds and outlines the specific costs and benefits of territoriality, dominance, and flocking, with primary emphasis on nonbreeding associations. Discussion of the advantages and disadvantages of colonial nesting is deferred to Chapter 16.

Individual Spacing Behavior

Most birds maintain a small individual space around themselves wherever they go. Swallows, for example, space themselves at regular intervals on a telephone wire (Figure 14–1). Sparrows and sandpipers feeding in large flocks also maintain small distances between one another. This space reduces the frequency of hostile interactions and increases individual foraging efficiencies. Certain social species overcome the usual tendency to stay apart and actually touch one another while roosting, sometimes in large groups to stay warm (Chapter 6), and also preen each other's head feathers—called allopreening.

The usual tendency of individuals to space themselves promotes uniform dispersion patterns. If birds landed on a field at random, some sites in the field would remain empty and others would receive several birds in succession, resulting in random patterns of association. In all probability,

Figure 14–1 Cliff Swallows space themselves at regular intervals on a telephone wire. (Courtesy of A. Cruickshank/VIREO)

however, the birds would not sit quietly after landing. Individuals close to one another would move apart and fill the unoccupied spaces. Such regular, or uniform, dispersion patterns are typical of birds that occupy relatively uniform habitats. Killdeers residing in large fields, American Robins nesting in suburbia, and American Kestrels wintering along roadsides space themselves in a regular manner. Spacing patterns depend on the scale of one's perspective. When birds fly in a flock, the distances between individuals within the flock may be small but the distances between different flocks may be large. Flocking Snow Geese in winter fields clump together, but on a larger scale, the distributions of the flocks themselves may be random, uniform, or clumped.

Territorial Behavior

Birds establish, maintain, and protect their spatial relationships; aggressive individual assertions of status or rights to resources are normal parts of avian social life. Assertion of spatial rights is most apparent in territorial birds, which continually assert their exclusive rights to particular areas, food supplies, or mates. Ornithologists once thought that the territorial behavior of birds was genetically programmed and static. In fact, territorial behavior is flexible and dynamic, reflecting the balance between its costs and its benefits. Great Tits, for example, forgo defense of their winter territories on the coldest days to save essential energy (Hinde 1956).

Box 14-1
Three features define the territories of birds

Territorial behavior is a primary form of aggressive spacing behavior that has intrigued naturalists since Aristotle. H.E. Howard's *Territory in Bird Life* (1920) formally introduced scientific inquiry into the subject. Research has established three major aspects of territorial behavior (Brown and Orians 1970):

1. A territory is a fixed area defended continuously for some period of time, even if only hours, in either or both the breeding and nonbreeding seasons. It can move in location if centered on a mobile resource.

2. Acts of display or defense discourage rival birds that would otherwise enter or approach the territorial space.

3. Primary if not exclusive use of a territory is thereby limited to the defending individual and, perhaps, its mate and progeny.

The simplest territories are those with only one type of resource, such as the feeding territories of hummingbirds in fields of flowers or those of sandpipers on a beach at low tide. At the other extreme are the all-purpose nesting territories of land birds, which are used for male display, courtship, nest seclusion, and feeding. These territories enable individuals to space themselves to reserve essential resources, to reduce predation, and to control sexual interference by neighbors and vagrants. In suitable habitats, territories are usually contiguous areas separated by well-defined, though invisible boundaries. The dense nest territories of Royal Terns actually pack into a hexagonal configuration resembling the cells in a bee's honeycomb (Grant 1968; Buckley and Buckley 1977).

The average sizes of territories increase directly in relation to body size, energy requirements, and food habits of the various species of birds (Figure 14–2). This observation suggests a general importance of food resources to the territorial individual. Variations within species are even more revealing. Pomarine Jaegers, for example, defend small breeding territories of 19 hectares when lemmings, their principal food, are abundant and defend large territories of 45 hectares when lemmings are scarce (Pitelka et al. 1955; Maher 1970). The feeding territories of Golden-winged Sunbirds and Rufous Hummingbirds decrease in size as flower density, and thus the density of nectar, increase (Gill and Wolf 1975; Kodric-Brown and Brown 1978).

Simple relationships between food abundance and territory size, however, do not necessarily demonstrate that food and energy requirements alone control territory size. Territory size also depends on the density of competitors for the available space. When population density is low, nesting American Tree Sparrows regularly use only 15 to 18 percent of their large territories (Weeden 1965). They concentrate their activities in the core section but also defend a less frequently used buffer zone. In years of high population density and increased competition for breeding space, denser packing of smaller territories eliminates the buffer zones.

Territorial defense incurs costs as well as producing benefits (Figure 14–3). Conspicuous display can attract predators. The time and energy required to display, patrol territorial boundaries, and chase intruders can be a major investment. Territoriality is favored when the resulting benefits outweigh the incurred costs. The central requirement is that adequate resources are economically defensible (Brown 1964b). Two features of resource distribution—temporal variability and spatial variability—determine whether territories are economically defensible. Resources that change rapidly in time invite opportunistic use, not site-specific investment or long-term commitment. Aerial insects whose locations and densities shift frequently, for example, are usually not defensible food resources. Territorial sunbirds, which do not tolerate one another near choice flowers, will sit side by side in a bush while they catch passing insects.

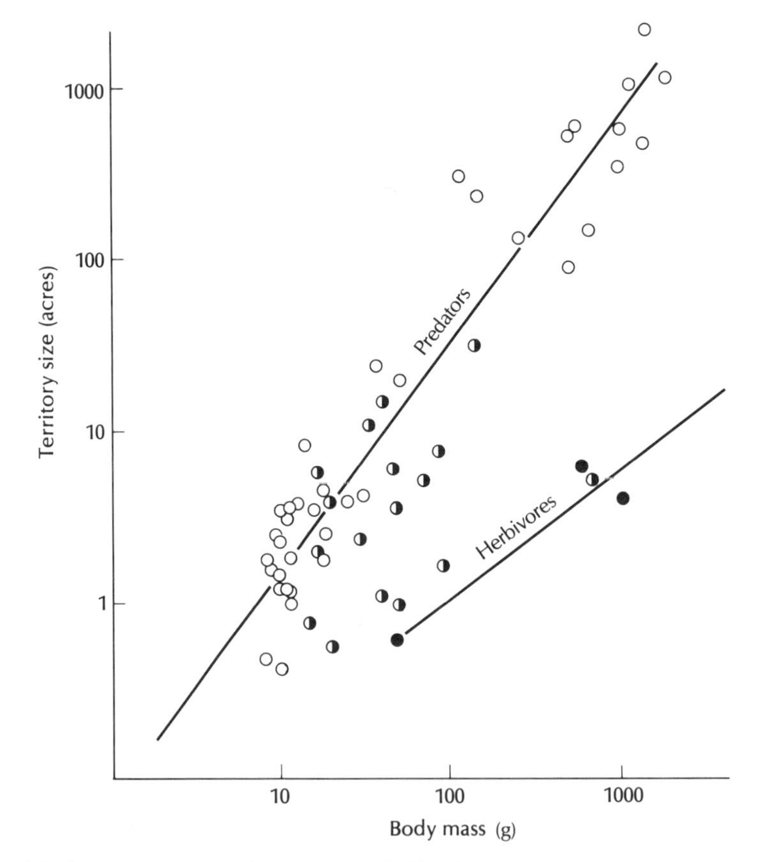

Figure 14–2 Territories or home ranges of birds increase directly in relation to body size, energy requirements, and selection of food types. The correlation suggests that territory size is geared to the food and energy requirements of the bird. Predators (open circles) have higher daily energy requirements than do herbivores (solid circles), which have correspondingly smaller territories. Half-shaded circles indicate species with mixed diets. (After Schoener 1968)

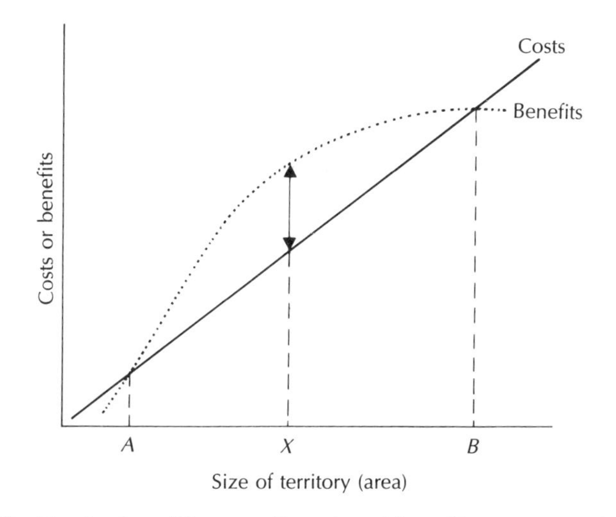

Figure 14–3 Territories of intermediate sizes (*A* to *B*) are economically defensible because the benefits exceed the costs. The costs of defense increase as territory size increases. The benefits relative to need (dotted line) increase rapidly at first but then reach a maximum value when needs are filled, as would be the case when food is in excess. Optimum territory size is at *X*, where the net benefit is greatest. (From Davies 1978a)

Sites rich in resources attract hordes of competitors and may be indefensible as a result. No gull would attempt to maintain a feeding territory on a garbage dump where thousands of other gulls vie for the same scraps. Similarly, Sanderlings do not defend their feeding territories on California beaches, when prey is either abundant or scarce (Myers et al. 1979). Beach space with dense concentrations of prey (isopods) is not defensible because no single Sanderling can keep the hordes of other Sanderlings away. Low prey densities also are not worth defending. Sanderlings, however, vigorously defend beach territories at intermediate prey concentrations. The size of the territories they defend then reflects the necessary defense effort; increased competition forces smaller territories. The territorial behavior of Sanderlings is affected by two additional factors, tide and predation risk. The territorial behavior of Sanderlings is manifest only at low tide; at high tide this beach sandpiper feeds or roosts in flocks (Figure 14–4). In years when Merlins, a small, predatory falcon, take up residence in their area, Sanderlings stay in flocks even when intermediate prey concentrations would favor territorial behavior, because isolated territorial Sanderlings would be vulnerable prey themselves (Myers et al. 1985).

Figure 14–4 Sanderlings may defend exclusive feeding territories or feed in large flocks. Note the leg color bands. This territorial individual returned each winter to defend its section of a Texas beach. (Courtesy of A. Amos/VIREO)

Birds typically defend territories against others of the same species. Sometimes they also expel other species. Golden-winged Sunbirds defend their territories against a variety of nectar-feeding birds, as do territorial hummingbirds. Wintering Northern Mockingbirds defend berry-rich feeding territories against other species, especially those that would eat some of the berries. The intensity of a mockingbird's defense increases

Box 14–2
Territory defense by sunbirds depends on economics

The costs and benefits of the feeding territories of nectar-feeding birds are unusually straightforward and easily defined. Hummingbirds and sunbirds defend clumps of flowers for several days to several weeks or longer. Golden-winged Sunbirds in Kenya, for example, defend about 1600 mint flowers, which produce enough nectar each day to satisfy an individual's energy requirements (see figure). Territorial sunbirds benefit by having an assured, adequate food supply. They defend these territories when the energetic benefits exceed the energetic costs of defense (Gill and Wolf 1975, 1979).

A territorial sunbird invests energy at a rate of approximately 12.5 kilojoules per hour chasing intruders. It recovers this investment and more by feeding at nectar-rich flowers on its territory, this feeding time being less than the feeding time required at nectar-poor, undefended flowers visited frequently by other sunbirds. Raising the average nectar volume from 1 to 2 microliters per flower cuts feeding time in half. The territorial sunbird, therefore, can spend more time sitting than a non-territorial sunbird and save energy. In this example, a defense investment of 20 minutes costing 3.7 kilojoules reduces the sunbird's total costs from 32 kilojoules per day to 26 kilojoules per day, a net savings of 6 kilojoules (Table 14–1). When the projected savings are less than the investment, sunbirds do not defend a feeding territory.

Golden-winged Sunbird, a species that often defends territories of nectar-rich flowers. (Courtesy of C.H. Greenewalt/VIREO)

Table 14–1
Energy costs of feeding on undefended and defended flowers, for the Golden-winged Sunbird

	Undefended flowers (1 μL nectar/flower)			Defended flowers (2 μL nectar/flower)		
Activity	Time spent (h)	Energy rate (kj/h)	Energy spent (kj)	Time spent (h)	Energy rate (kj/h)	Energy spent (kj)
Foraging	8	4.0	32.0	4	4.0	16.0
Sitting	—	—	—	3.7	1.7	6.3
Defense	—	—	—	0.3	12.5	3.7
Total energy spent			32.0			26.0

Energy saved by feeding on defended flowers: 6.0 kj.

From Gill and Wolf 1975.

with the potential threat to its food supplies (Moore 1978). Some birds defend nesting territories against other closely related species (Orians and Willson 1964; Murray 1971).

Territories may be occupied and defended by a single bird, a mated or cooperating pair of birds, an extended family, or even a group of unrelated individuals. Small groups of wintering tits and chickadees, for example, defend woodlot territories containing both food and roosting holes. Groups of unrelated Black-capped Chickadees establish common winter territories by late summer (Smith 1991). Group membership, which includes male and female pairs of both resident adults and newly settled first-year birds, is stable throughout the winter. In addition to protection of food stores for the winter, spring territorial breeding opportunities emerge from the communal winter effort.

Dominant Behavior

Birds assert themselves more effectively when they are on familiar ground or home territories than when they are strangers in a new place. Territory owners usually win encounters with intruders. For one thing, during high-speed attacks and chases, the owner can use familiar details of the territory to its own advantage. Because territorial owners have an investment to protect, they do not usually give up a fight as easily as a newcomer. Acorn Woodpeckers, for example, vigorously defend their tree granaries against squirrels, jays, and other Acorn Woodpeckers (MacRoberts and MacRoberts 1976; Koenig 1981a). The granaries hold valuable stores of winter food; in addition, each of the many holes (up to 11,000 per tree) represents an investment of 30 to 60 minutes of drilling time. These woodpeckers defend trees that are riddled with empty holes as well as those with holes that contain acorns (Figure 14–5).

Territoriality is one expression of dominance behavior. Dominance and aggressive reinforcement of status are a normal part of the social lives of birds. Individuals that win aggressive encounters achieve dominance, and consistent losers become subordinate. As social ranks are established in new groups of birds, losers cease challenging dominant individuals. Dominants use threat displays to assert their status and reserve their access to mates, space, and food. They move without hesitation to a feeder or desirable perch, supplanting subordinates and pecking those that do not yield at their approach. Subordinates are tentative in their actions and frequently adopt submissive display postures (see Chapter 9).

Dominance status is directly related to age and sex. Generally, large birds dominate small ones, males dominate females, and old birds dominate young ones. Within an age group or gender, physiology, genetics, and possibly parasite load affect dominance. Aggressive tendencies and dominance status are correlated with slight differences in adrenal gland activity and brain chemistry (Brown 1975). Aggressive, dominant strains

Figure 14–5 The granaries of Acorn Woodpeckers are valuable, defensible resources that contain essential supplies of acorns for the winter. (Courtesy of M.H. MacRoberts and W. Koenig)

of Domestic Chickens can be developed by artificial selection (Craig et al. 1965).

Rank has its advantages. High-ranking Dark-eyed Juncos and Field Sparrows survive longer than low-ranking ones (Baker and Fox 1978; Fretwell 1968). Subordinate Common Wood-Pigeons obtain less food per hour than dominants, a situation that increases their probability of starving (Murton 1967; Murton et al. 1971). Low-ranking individuals have less access to good feeding sites and are usually the first to emigrate. Weakened physical condition plus the extra costs and dangers of travel through unfamiliar situations all increase the risk of death. In one experiment, the feeding behavior of White-throated Sparrows in winter was affected by both their dominance status and the distance of food from protective cover (Schneider 1984). When food was placed at different distances from shelter, dominant individuals fed more often near shelter than did subordinates, even when this meant a sacrifice in their foraging efficiency.

The dominance status of individuals changes with location. The ability of territorial male Steller's Jays to win fights, for example, decreases with distance from their nesting areas rather than ceasing abruptly at a territorial boundary (Brown 1975). Although expressions of dominance and territoriality relate to specific resources such as food and may be initiated over rather large distances, the two behaviors differ with regard to the site

Box 14–3

Dominant Blue Tits are more cautious than their subordinates

Experiments with captive flocks of Blue Tits demonstrated that dominant individuals were more cautious during periods of danger than were subordinate individuals. Robert Hegner (1985) flew a model Eurasian Sparrowhawk over his aviary and watched which tits were the first ones to emerge from their hiding places to feed. Low-ranked individuals fed first, followed by high-ranked individuals. Hegner suggests that high-ranked individuals can afford to be cautious because they have the ability to control food sources and thus to ensure adequate foraging, whereas low-ranked individuals must take more chances to get to food ahead of their dominant flock mates.

defended, which is fixed in the case of territoriality and movable in the case of dominance. Dominance and territoriality, however, become indistinguishable in the site-dependent dominance systems of Steller's Jays and Bicolored Antbirds. Temporary residency also blurs the lines of definition between the two behaviors. Roving male Bronze Sunbirds, for example, shift from dominance behavior to territoriality through intermediate states of aggressive behavior. They often displace subordinate sunbirds to feed on certain flowers and then leave (Wolf 1978), but they also defend flowers for an hour or so of exclusive access and then leave, only to return later for another period of temporary residence. When conditions are poor and flowers scarce, they defend the territory constantly for several days to several weeks.

Sometimes territorial birds defend a nonstationary resource. Constant defense of a female and her immediate area, for example, borders on territorial defense of a well-defined resource. Such behavior is typical of the Cassin's Finch and related (carduelline) finches, particularly when an excess of males competes for mates (Newton 1972; Samson 1976). Glaucous Gulls and Glaucous-winged Gulls defend feeding eiders—a kind of sea duck that brings food to the surface—against other gulls (Ingolffson 1969; Prys-Jones 1973). Sanderlings will defend Willets from other Sanderlings when the Willet has a large sand crab, bits of which fall to the defending Sanderling (Myers et al. 1979).

Flocking Behavior

Birds often form flocks, a behavior that reduces the risk of predation and enables cooperative foraging. Like territoriality, flock formation reflects trade-offs between its costs and its benefits. Whereas stable food resources and defensible spaces promote territoriality, unstable food resources and indefensible areas promote flocking. White Wagtails in Britain join flocks when food on their territories becomes inadequate (Davies 1976b). They also stop defending territories at experimental piles of food and join feeding flocks when the same amount of food is evenly dispersed over a large area (Zahavi 1971b). Similarly, crows, jays, and magpies abandon territories and form flocks when feeding conditions deteriorate, mainly when food supplies become less stable and more patchy in distribution (Verbeek 1973).

Flocks range in composition from loose temporary aggregations to organized foraging associations of diverse species. At one extreme are the millions of blackbirds in the United States or the Bramblings in Europe that converge each evening at traditional roost sites. Temporary feeding aggregations of herons and seabirds also are open gatherings of individuals responding opportunistically to special situations. Multispecies flocks of tropical birds, which feed together as a group throughout the year and actively exclude new individuals from membership, are closed social systems, similar in many ways to much smaller family units.

A stable flock membership facilitates personal recognition among individuals and the development of a dominance hierarchy. Most dominance hierarchies in stable flocks are linear—or "peck right"—hierarchies, in which each individual clearly ranks above or below a set of others. In closed, stable social units, social rank increases gradually in relation to time, individual tenure, and occasional changes in group composition. Stable dominance relationships lower the frequency and intensity of overt hostility. Aggressive peck rates in stable groups of caged hens, for example, averaged 1 per minute when a bowl of food was placed in the cage. When group composition changed weekly, peck rates increased to 3 per minute (Guhl 1968). Social dominance relationships in flocks are based on mutual recognition among individuals. Hens easily recognized up to 10 other individuals, and one hen reliably recognized 27 individuals in four different flocks (Welty 1982). Physical details serve for individual recognition and, if altered, disrupt the social organization. Hens whose combs are dyed or covered with bonnets are regarded as strangers when returned to their flock (Schjeldrup-Ebbe 1935).

Dominance hierarchies are a conspicuous feature of the associations of birds that follow raiding parties of tropical army ants, which flush large numbers of insects and small reptiles that are usually camouflaged and hard to find. Tropical antbirds and woodcreepers habitually associate with ant swarms (Willis and Oniki 1978) and over 50 species of Neotropical birds are "professional" ant followers, that is, they obtain over half of their food from the vicinity of ant swarms. Large, dominant species, such as the large Ocellated Antbird, control the central zone of the ant swarm where prey are most likely to be flushed by the dense, leading columns of ants. Smaller, subordinate species, which are chased from this zone, take up stations in peripheral, less productive foraging zones but move toward the center when opportunity arises. In the presence of Ocellated Antbirds, Bicolored Antbirds occupy the intermediate zone, and the small Spotted Antbirds are shunted to the edge (Figure 14–6).

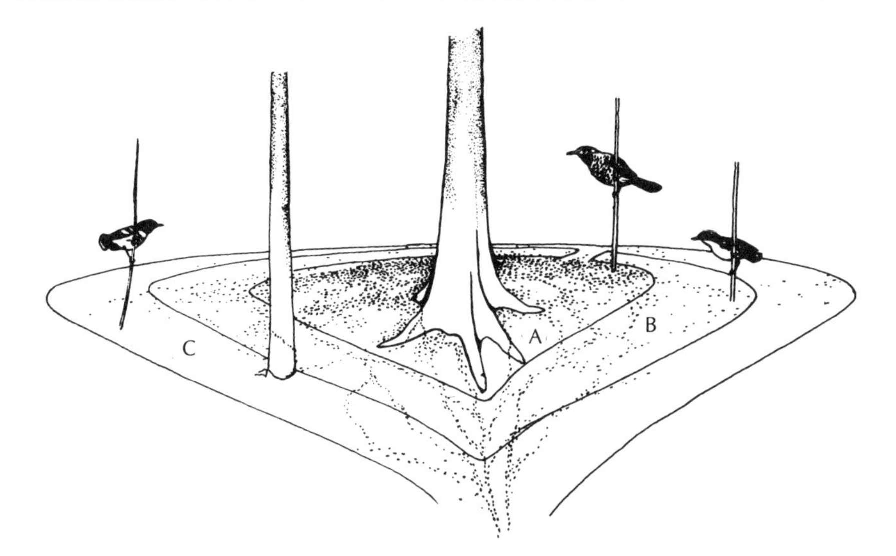

Figure 14–6 The hierarchy of interspecific dominance among birds that follow army ants. Large, dominant species such as the Ocellated Antbird control central sites (zone A), where foraging for flushed insects is best; they displace smaller species to outer zones, for example, Bicolored Antbirds to zone B. In turn, the Bicolored Antbirds displace Spotted Antbirds to zone C. Sometimes a subordinate species can infiltrate the central zone, but only such zone C antbirds as the White-plumed Antbird do this regularly. (From Willis and Oniki 1978)

Feeding in Flocks

Casual aggregations of individuals at rich feeding grounds are fortuitous, but why do unrelated individuals form stable foraging partnerships? Social tensions and the frequency of fights increase with group size. Subordinate individuals could avoid dominant "bullies" by feeding alone. Competition for conspicuous or rich food items also increases in groups. What, then, are the advantages of feeding together in organized flocks? The answers lie in the inescapable daily concerns of foraging efficiency and predation risk. Some of the advantages are straightforward, practical ones, including cooperative feeding. Flocks of pelicans encircle and trap schools of fish in shallow water; groups of cormorants and mergansers drive fish toward the shore where they are more vulnerable (Bartholomew 1942; Emlen and Ambrose 1970). Common Ravens steal Black-legged Kittiwake eggs more easily when hunting in groups than when alone, and subordinate birds profit by moving together onto defended food sources where they can overwhelm the territorial individual.

Flock members also benefit from the "beater effect"; prey that is flushed (and missed) by one bird can be grabbed by another. Groundhornbills in Africa, for example, walk in a line across fields to catch insects flushed by one another. Drongos and flycatchers participate in

Box 14–4
Plumage color signals social status

The plumage colors of Harris's Sparrows serve as badges of their social status (Rohwer 1975, 1977, 1982) (see figure). Top-ranked, dominant individuals have conspicuous, contrasting black markings on the plumage of the head and neck; low-ranked, subordinate individuals have few such markings. Many individuals are intermediate in appearance. Such variations facilitate individual recognition among the members of the large flocks that these sparrows typically form during winter. The evolution of the variability seems directly tied to the advantages of being dominant versus the advantages of being subordinate. Dominant individuals assert the prerogatives of their rank, including access to food. Conversely, subordinates of plain appearance benefit from flock membership, which they can maintain because they do not threaten the dominant individuals with visual badges of high status. When dyed with black to look like a dominant individual, subordinates suffer more frequent attacks but do not rise in status because they are not inherently aggressive.

Head color pattern also controls social status and access to communal feeding groups of the Jackass Penguin of South Africa (Ryan et al. 1987; Wilson et al. 1987). Adult penguins, which have bold black and white head patterns, feed communally on schools of fish. They aggressively exclude juveniles from the communal feeding groups, because juveniles, which are poor swimmers, interfere with the coordinated adult effort. Also, the conspicuous black and white color pattern of adult plumage causes loose fish schools to coalesce, making them easier to capture. Some older immatures, however, acquire an adult appearance by undergoing partial head molt, which reduces adult aggression and allows access to exclusive adult feeding clubs. One hypothesis is that the partial head molt signals honestly that these young birds have come of age, are of good health and able to assume additional energy expense, and perhaps are acceptable additions to the feeding group. An alternative explanation is that simply by changing their appearance, these immatures can illegally parasitize exclusive feeding groups.

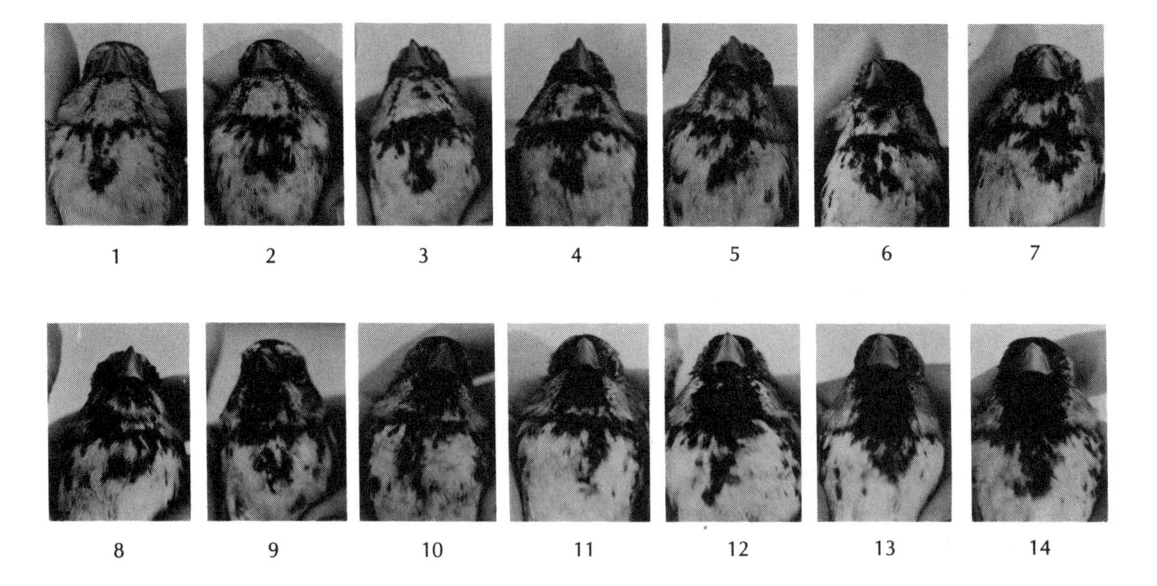

Plumage variations in male Harris's Sparrows reflect their social status. Dominance is correlated with an increasing extent of black markings on the head and neck: Numbers 1–3 are most subordinate; numbers 12–14 are most dominant. (Courtesy of S. Rohwer and *Evolution*)

Box 14–5
Harris's Hawks hunt in teams

Family hunting parties of two to six Harris's Hawks cooperate to catch rabbits (Bednarz 1988) (see figure). The hunting party assembles at dawn and then splits into small subgroups that search for prey by moving in a coordinated "leapfrog" fashion through the desert scrub. They then converge on a rabbit that is spotted and kill it with successive, relay strikes by different individuals. When a rabbit hides in thick cover, the group surrounds the area and waits for one or two of its members to deliberately flush the rabbit into the open. All members of the party then feed on the kill. Team hunting improves the probability of catching a rabbit and raises the average amount of energy available to each individual relative to that available when hunting alone. Team hunting also enables these hawks to kill larger prey than they could by hunting alone. Prior to this study, cooperative hunting and sharing of prey had been documented only for large social mammals such as lions.

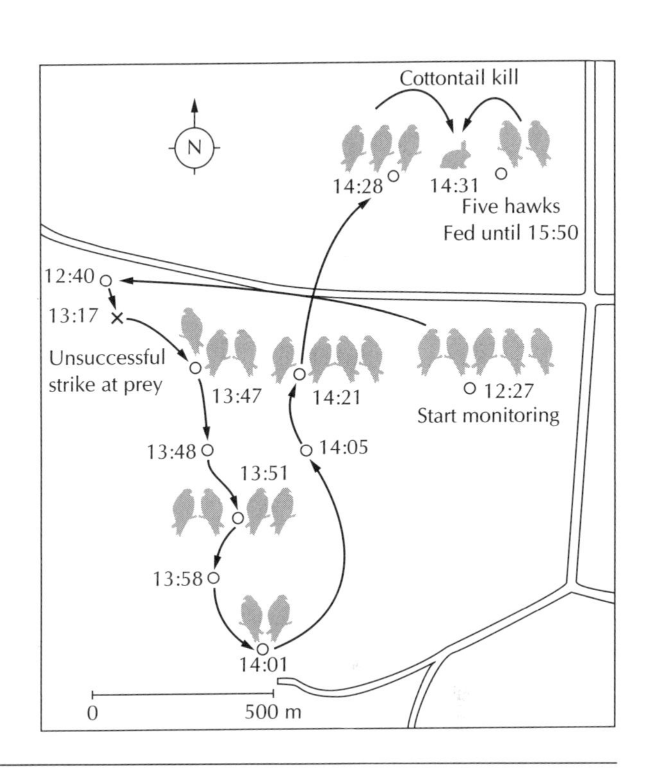

Sequence of movements of Harris's Hawks that culminated in the capture of a desert cottontail rabbit. Although all five hawks in the group remained in view, Jim Bednarz and his assistants specifically monitored a subunit that included hawk #995, which wore a radio. Perched hawks in this figure indicate the number of individuals that joined the subunit at each location. Subunit size remained unchanged from the previous location if no hawks are pictured (From Bednarz 1988, with permission from *Science*).

mixed foraging flocks and specialize on prey flushed by other birds. Flock membership also improves foraging in more subtle ways. Group foraging by pigeons and titmice helps them locate food because members can join successful individuals at rich clumps or concentrate their search efforts nearby (Murton 1971b; Krebs et al. 1972; Krebs 1973). Groups of four titmice in captivity found more hidden food together than alone. They watched one another's successes and modified the intensity and direction of their searches accordingly. Dominant individuals tend to benefit most because they can usurp the sites discovered by subordinate members of the flock. Instead of increasing or maximizing individual foraging efficiency, foraging in flocks may help ensure that an individual finds at least

BOX 14–6

Colonies and flocks may function as foraging information centers

Members of a flock or breeding colony can use one another to find food, particularly when the precise location of good feeding sites varies from hour to hour (Wittenberger and Hunt 1985). Observations that seem to support this "information center hypothesis" have been reported for birds as diverse as herons, blackbirds, and swallows. Seabirds track the locations of small schools of fish by following the line of individuals returning to the colony with food. Observations of marked Cliff Swallows, which feed on aerial insects that concentrate in the eddies of shifting breezes, suggest that they may derive a similar advantage (Brown 1986). Cliff Swallows that were unable to find food returned to their colony, located a neighbor that was successful and then followed that neighbor to its food source. All individuals in the colony were equally likely to follow or to be followed and thus contributed to the sharing of information that helped to ensure their reproductive success.

some food on a regular basis, before its reserves are exhausted. This effect would be most important during the stressful seasons, like winter, when food is scarce. The observation that small birds with limited fasting abilities tend to flock more than large birds also supports this idea (Morse 1970, 1978; Thompson et al. 1974).

Safety in Flocks

The security of a large group enables an individual to relax its personal vigil for predators and hence to feed more deliberately. Joining a flock theoretically decreases the risk of being caught and eaten, because there is safety in numbers. Predator confusion—the difficulty a predator has in focusing on one bird when many flush—is one advantage. A falcon, for example, risks injury from incorrect contact during a high-speed strike and is reluctant to swoop down on a fast-moving, swirling flock of birds. It cannot crash full speed into the center of a flock and hope to strike safely. When European Starlings sight a hawk, the flock packs more tightly together. An individual's chances of being a victim decrease as the number of potential victims in the flock increases and, as in nesting colonies, decrease even further for individuals near the center of the flock (Hamilton 1971) (Figure 14–7). The hunting success of a Merlin varied according to the size of sandpiper flock it attacked. It fared poorly with medium-sized sandpiper flocks, but did well with isolated individuals and with large flocks, which were less able to maintain a tight formation (Page and Whitacre 1975).

Predator detection also improves in flocks; greater individual security is the result. Spotted Antbirds seem less nervous and wary when in flocks than when alone (Willis 1972). Northern Goshawks cannot attack large flocks of Common Wood-Pigeons without being detected (Kenward 1978). Ostriches stick their heads up randomly to look for approaching lions; at any given time, at least one in the flock functions as a lookout (Bertram 1980). Flock members warn one another of danger and communicate so that they can flee at the same time. Ducks flush together at the approach of a predator because the individuals synchronize their takeoffs with a series of flight-intention movements that prime every duck's readiness for flight. Flight calls enable longspurs to flush as a group rather than singly.

Contact calls enable birds to associate and to maintain a cohesive flock structure, even in dense vegetation. Alarm calls, which are difficult to pinpoint with respect to source (Chapter 10), serve to alert other members of the social group to possible danger. When one member of a flock spots a predator, it gives an alarm call, and the rest of the flock either freezes or dives for cover. Giving an alarm call would seem advantageous to all but the one that thus revealed its position. Warning calls may seem to be heroic or altruistic acts, but they carry benefits for the caller as well if others in the flock are genetic relatives, such as siblings, parents, or offspring. Each flock member also can count on a certain degree of reciprocity. Most important, by calling loudly, the potential victim robs a predator of the element of surprise and thereby reduces the likelihood of attack. The intended victim reduces its own danger as it alerts kin and neighbors.

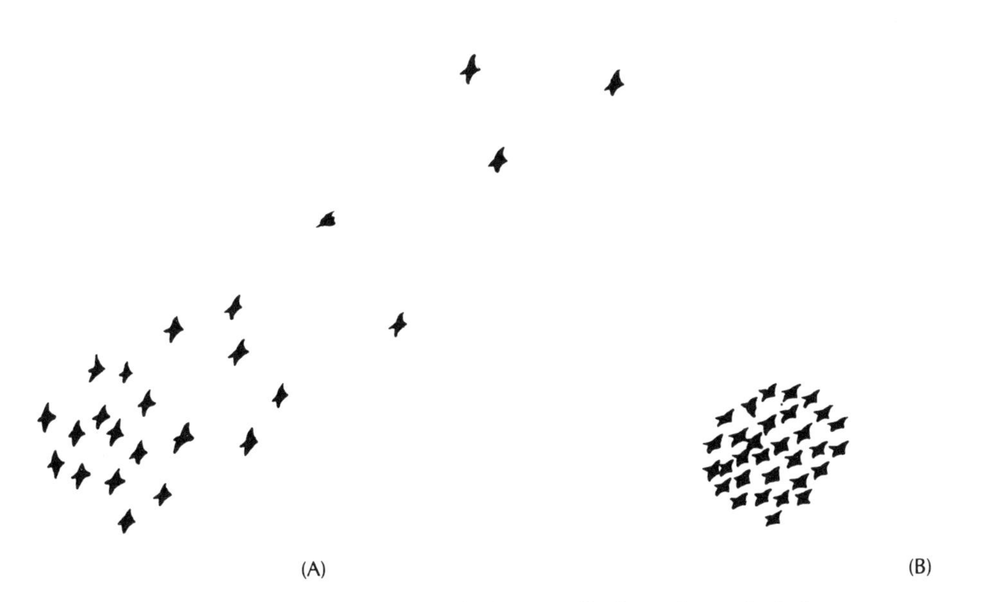

Figure 14–7 (A) European Starlings, which normally fly in loose flock formations, (B) form tight formations when threatened by a hawk. (From Tinbergen 1951)

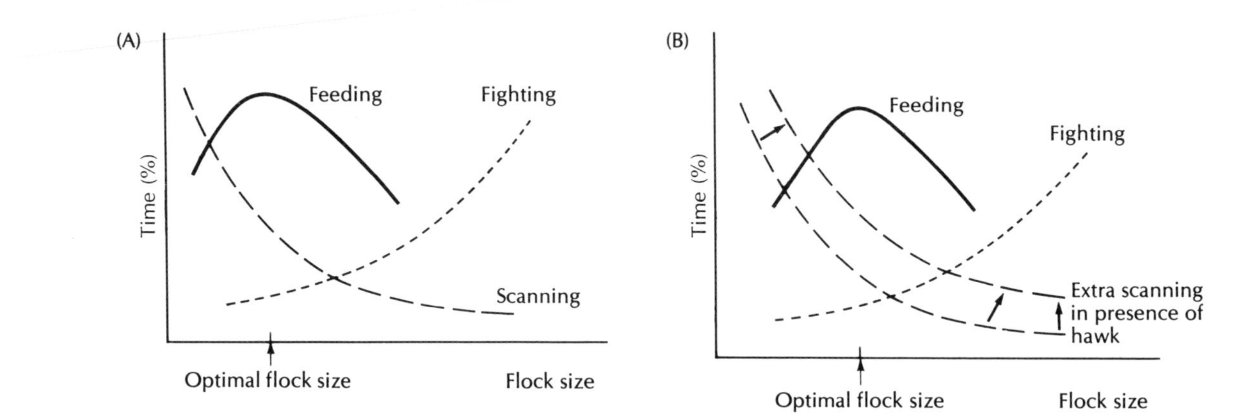

Figure 14–8 (A) The optimal flock size theoretically results from a balance between time spent fighting other members of the flock, time spent scanning for predators, and time devoted to feeding. An intermediate flock size permits the most feeding time. (B) When a predator hawk is present, more time must be spent scanning and the optimal flock size increases. (After Caraco et al. 1980b)

By relying in part on such mutual protection, each individual in a flock can spend less time looking for predators and more time feeding than when alone. However, the time an individual in a flock saves because of decreased surveillance is offset to some degree by the time it loses to involvement in aggressive interactions, which increase in frequency with group size (Caraco 1979). The amount of time available for feeding should, therefore, be greatest in flocks of intermediate size. Moreover, optimum group size should increase when predators are near and when each bird must spend more time in surveillance. Thomas Caraco and his colleagues (1980b) confirmed this in studies of Yellow-eyed Juncos in Arizona. In one experiment, average flock size increased from 3.9 to 7.3 juncos when a tame Harris's Hawk flew over the feeding grounds (Figure 14–8).

When they discover a predator, such as an owl or a snake, birds scold them vocally and sometimes attack them physically (Curio 1978; Montgomerie and Weatherhead 1988). The advantages of mobbing behavior include both discouraging or driving away an enemy plus the refinement of an individual's ability to recognize predators, which reduces future risk to self and family. Inexperienced birds quickly associate potential danger with the commotion of mobbing behavior. They then learn to recognize predators by observing the mobbing behavior of their parents or their flockmates. Eurasian Blackbirds will even learn to attack a detergent bottle if, in experiments, they are tricked into associating mobbing behavior of others with such an inanimate object (Curio et al. 1978). Species that join mixed-species foraging flocks tend to respond reciprocally to one another's alarm calls and to mob predators cooperatively, possibly sharing their knowledge of potential enemies (McLean and Rhodes 1991).

Box 14–7
Wanted: Experienced parrot flocks for conservation

The only two parrots native to North America disappeared 50 years ago. Subsistence hunting and habitat degradation exterminated their populations. One of these, the Carolina Parakeet of the eastern United States, is surely extinct, but declining numbers of the other species, the larger Thick-billed Parrot, persist in high mountain refuges in western Mexico. Noel Snyder and his colleagues have undertaken an ambitious conservation program to restore Thick-billed Parrots into the rugged Chiracahua Mountains of southeastern Arizona, where they once lived (Snyder et al. 1989). They used the captive-bred offspring of confiscated, illegal pet trade Thick-bills. Their initial efforts often were thwarted because the young parrots lacked predator-avoidance training by experienced flockmates.

Experienced wild Thick-billed Parrots protect themselves from hawks through their wary, vigilant, social behavior and through their ability to outfly a pursuer. Northern Goshawks, which are common in the Arizona mountains, found inexperienced, captive-bred parrots to be easy prey because they did not scan the sky for predators, did not freeze or flee when they saw one, and did not react quickly enough to alarm calls of experienced wild birds. The captive-bred birds seemed fearless, despite the fact that they had seen raptors in action from their cages.

Study of their behavior revealed that strong socialization with experienced flockmates is required to learn essential survival skills. The captive-bred birds also required lots of exercise to attain the condition required to keep up with wild flocks and to fly faster than a pursuing hawk. Further, flockmates teach one another to identify pine cones as food sources and how to extract the seeds from them. Finally, well-socialized parrots develop an essential sense of security. Without this sense of security that comes from joining other parrots to feed, pairs of adults seem unwilling to undertake the risks of breeding. The future success of Noel Snyder's parrot conservation program will depend on the training and gradual release of socially mature flocks of Thick-billed Parrots that work together to find food and avoid predators.

Mixed-Species Flocks and Social Signals

Flocks are not limited to members of the same species. Rich assemblages of species form foraging flocks. Flocks of chickadees, titmice, nuthatches, woodpeckers, creepers, and other associates are familiar both in the United States and in Europe, and several species of warblers may join them in the warmer months. Noisy gatherings of antbirds, antwrens, woodpeckers, flycatchers, and honeycreepers surge through the rain forests of South America. Tropical flocks may include 60 birds of 30 different species, whereas temperate flocks average 10 to 15 birds of 6 or 7 species. Curiously, flock size increases primarily as a result of the addition of new species, not more individuals of a few species (Powell 1979). Furthermore, flock composition changes regularly as the flock moves along, a result of new individuals joining and others leaving (Powell 1979; Munn and Terborgh 1979; Greenberg and Gradwohl 1983). Individuals join the flock as it moves through their territory, only to be replaced by neighbors as the flock moves from one territory to the next.

Nuclear species provide the main element of a flock structure. In Temperate Zone woodlands of North America, for example, titmice and chickadees are nuclear species. Large antbirds and greenlets take this role in lowland tropical forests. In eastern Peru, the Bluish-slate Antshrike and the Dusky-throated Antshrike assemble 30 other species with their loud rallying calls early every morning (Munn and Terborgh 1979). Plain-colored Tanagers and Blue-gray Tanagers are the usual nuclear species in canopy flocks in lowland Panama but are replaced by Sooty-capped Bush-Tanagers in highland habitats (Moynihan 1962). Other species, the "followers," join the flocks opportunistically and are subordinate to the nuclear species.

Why do birds of various species assemble to feed together; and, in particular, why do subordinate species join the nuclear species? Reduced predator vigilance and increased foraging efficiency are part of the answer. Kimberly Sullivan (1984a, b) showed that Downy Woodpeckers had high levels of vigilance and low levels of feeding when foraging alone. They stopped frequently to look for predators with a distinctive head-cocking behavior. However, when the woodpeckers associated with one to two birds of other species, they cocked their heads less often and fed more. When they fed with large mixed-species flocks, they cocked their heads infrequently and fed at high rates. The woodpeckers monitored the calls of their flockmates to assess their numbers of working associates and their tendency to be alarmed by possible predators.

In multispecies assemblages, the protection inherent to flocks can be achieved without the costs of increased conspecific competition for food (Krebs 1973; Buskirk 1976; Morse 1980). Territorial or rare species that are unable to put together a flock of their own kind can benefit by flocking with other species even though they are subordinate. Foraging success also increases in such flocks. Mixed flocks of several species, each with its

TABLE 14–2
Birds that flock together in the mountains of tropical America

Western Panama (black and yellow)	Northern Andes (blue, blue and yellow)	South Central Andes (blue, chestnut)
Yellow-thighed Finch	Blue-and-black Tanager	White-browed Conebill
Yellow-throated Brush-Finch	Blue-capped Tanager	Blue-backed Conebill
Sooty-capped Bush-Tanager	Santa Marta Mountain-Tanager	Chestnut-bellied Mountain-Tanager
Silver-throated Tanager	Hooded Mountain-Tanager	Black-eared Hemispingus
Slate-throated Redstart	Masked Mountain-Tanager	Golden-collared Tanager
Collared Redstart	Blue-winged Mountain-Tanager	Plushcap
Black-cheeked Warbler	Buff-breasted Mountain-Tanager	

From Moynihan 1968.

own searching skills, increase total scanning efforts for clumped, unpredictable prey. Social learning thus enables each participating individual to profit from the successes and failures of its associates. Individuals of different species can monitor one another's foraging success and modify their search efforts accordingly. Similar advantages pertain to mixed-species nesting colonies of herons, storks, and ibises. Their combined numbers maximize antipredator behavior, social interactions, and information transfer, but their ecological differences minimize nest competition and food competition (Burger 1981b).

The advantages of interspecific feeding associations are so marked that unrelated bird species have evolved similar plumage color patterns that promote flock cohesion. Subordinate species may gain acceptance by resembling dominant flock members. The color patterns of birds that flock together in the mountains of Central and South America offer striking examples of such social adaptations (Moynihan 1968) (Table 14–2). Whereas the neutral, nonthreatening plumage colors of nuclear species such as the Plain-colored Tanager may promote flock cohesion, species that habitually flock together tend to have similar brightly colored plumage patterns. Those species that participate regularly in the montane flocks of western Panama are typically black and yellow, sometimes variegated with brown and white, whereas bright blue or combinations of blue and yellow prevail in the humid temperate region of the northern Andes. Farther south, in Bolivia, the flock colors switch to blue or blue-gray above and chestnut below. Such distinctive color patterns possibly serve as flock "badges," which enhance the social integrity of multispecies flocks. The evolution of plumage colors of flocking birds, however, remains a controversial topic (Hamilton 1973; Powell 1985). Countering the potential value of cohesive "social mimicry" is the need for species distinctiveness. The bold color patterns of many tropical flocking birds may well promote recognition of conspecifics, a phenomenon known to occur in flocks (schools) of tropical reef fish.

Summary

The defensibility of a given space, the variability of food resources, and the probability of attack determine spatial relationships and social behavior. Territorial behavior is characterized by acts of display intended to discourage the presence of rivals and by the exclusive continued use of a defined area by the defending individual and perhaps its mate and progeny. The relative costs and benefits of territorial behavior govern its flexible expression. Dominance status structures the relationships among individuals in flocks, a system that reduces strife. Differences in plumage color may serve as badges of dominance status.

Flock formation reduces the risk of predation and improves foraging efficiency. Flock members benefit from one another's vigilance for danger as well as from one another's prospecting for scarce food. Team hunting improves feeding success in some species. Flocks may be loose opportunistic aggregations or highly structured social systems with closed memberships. Mixed-species flocks increase the benefits of mutual protection without the costs of sharing space or food with competing individuals of the same species. The advantages of membership in mixed-species flocks may promote convergence in plumage color patterns and social "mimicry."

Further Readings

Krebs, J.R., and N.B. Davies. 1984. Behavioral Ecology, 2nd ed. Sunderland, Mass.: Sinauer Associates. *The current forefront of a fast-developing field.*

McLean, I.G., and G. Rhodes. 1991. Enemy recognition and response in birds. Current Ornithology 8: 173–211. *Reviews the application of optimality theory and reciprocal altruism theory to the mobbing behavior of birds.*

Morse, D.H. 1980. Behavioral Mechanisms in Ecology. Cambridge, Mass.: Harvard University Press. *A scholarly review of the literature.*

Wilson, E.O. 1975. Sociobiology. Cambridge, Mass.: Belknap Press. *A classic book that helped to establish a new field of study.*

Wittenberger, J.F., and G.L. Hunt, Jr. 1985. The adaptive significance of coloniality in birds. Avian Biology 8: 1–78. *A detailed review of the advantages and disadvantages of colonial breeding and roosting.*

PART V

Reproduction and Development

CHAPTER 15

Reproduction

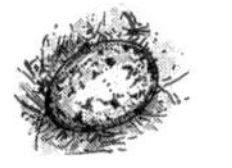

COMPETITION AND TERRITORIALITY reach full expression during the breeding season when birds compete with one another for the chance to breed, for mates, and for the resources they need to raise young. Different balances between competition and cooperation promote the evolution of alternative mating systems—monogamy versus polygyny—and also of nest-spacing systems—competitively dispersed territories versus socially aggregated colonies.

This chapter reviews the fundamentals of sex in birds, including the physiology and anatomy of the gonads, copulation and the fertilization of the ovum, and the production of a fully formed egg in the oviduct. The next chapters examine the nests, which cradle the eggs and chicks of birds; incubation behavior, which ensures the development of the embryo to hatching; the mating systems of birds, which reflect limited reproductive partnerships; the growth and development of young birds, which range from helpless to independent; patterns of parental care; and finally brood parasitism and cooperative breeding, the extremes of parental care systems among birds.

Gonads

The gonads of birds consist of paired testes in the male and usually a single ovary in the female (Figure 15–1A, B; see also Figure 6–17). These sex organs are responsible for both the production of gametes and the secretion of sex hormones. Testes, which are internal bean-shaped organs that are attached to the dorsal body wall at the anterior ends of the kidneys, usually are cream-colored but they may be dark gray or even blackish. Initially only a few millimeters long in small birds, they swell rapidly at the beginning of the breeding season, often reaching 400 to 500 times their inactive mass. The testes of a mature Japanese Quail, for example, increase from 8 to 3000 milligrams in just 3 weeks. Fertility in domestic geese is directly related to the weight of the mature testes, which is an inherited trait (Szumowski and Theret 1965).

The avian ovary resembles a small cluster of grapes. Most birds have only the left ovary with an associated oviduct, but two functional ovaries are typical of many raptors and also of Brown Kiwis (Kinsky 1971); they also occur occasionally in pigeons, gulls, and some passerines. At

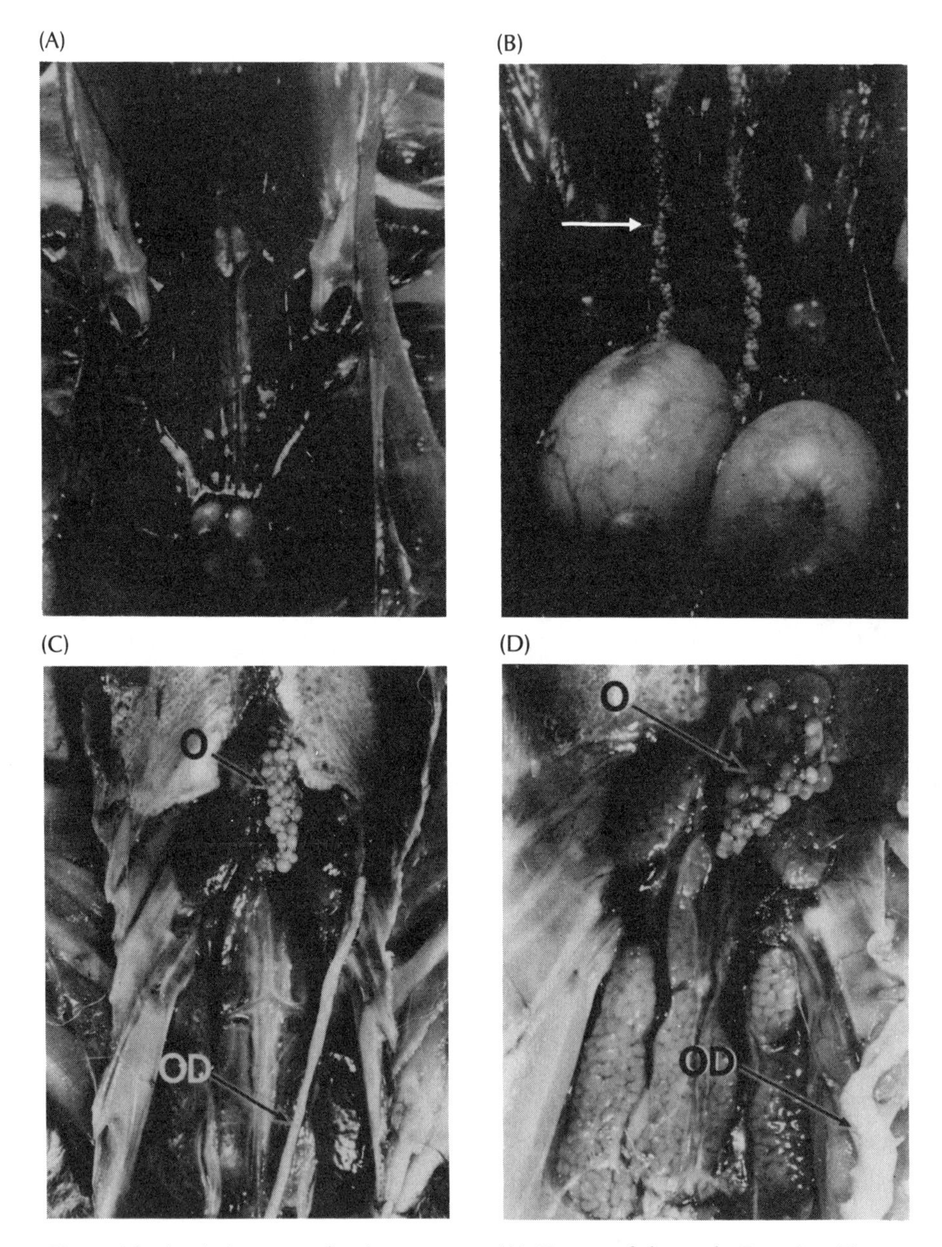

Figure 15–1 Avian reproductive systems. (A) Testes of the male Eurasian Tree Sparrow in winter and (B) at full size during the breeding season. Note also the enlarged vas deferens indicated by the arrow (magnification of A and B, ×5). (C) Ovary (O) and oviduct (OD) of the female Eurasian Tree Sparrow in winter and (D) during the breeding season (magnification of C and D, ×4). (Courtesy of B. Lofts)

maturity, the microscopic ovarian granules of the immature bird have increased in size 10 to 15 times (Figure 15–1C, D). The total number of primary oocytes (see below) in a wild bird is at least 500; more often there are several thousand, certainly many more than are used to produce functional eggs.

During early development of the embryo, primordial germ cells migrate to the site where the gonads will develop. More of these germ cells settle on the left side than on the right, which establishes the asymmetry that persists throughout subsequent gonadal development in both males and females, and leads to the development of an unpaired left ovary in many female birds. The primordial germ cells first generate the medullary tissue, which, although found in both testes and ovaries, is the principal tissue of the testes. A second phase of proliferation creates the cortex, the principal tissue of the ovary. Removal of the left ovary, particularly from young female birds, prompts the right gonadal tissue to develop into an intermediate structure, the ovotestis. Ovarian medullary tissue normally becomes more active with age in females, which in extreme cases causes overt masculinization in older females. With age, somber female Golden Pheasants, for example, acquire the spectacular plumage of males as a result of this phenomenon.

Follicle-stimulating hormone regulates gamete formation in both the testes and the ovary, and luteinizing hormone regulates hormone secretion in the testes and maturation of ova in the ovary. Gonadal secretion of the two principal steroid hormones—testosterone and estrogen—directly activates reproductive behavior and controls the development of secondary sex characteristics. Although testosterone is well known as the male hormone and estrogen as the female hormone, both hormones are present in males and females. The proportion of the two hormones, and the ways that body tissues react to each of them, cause male or female attributes.

Secondary Sex Characteristics

Sexual distinctions in plumage, body size, and voice are influenced by testosterone and estrogen. Acquisition of male breeding plumage in many species results from increasing amounts of testosterone in the blood. Removing the source of this hormone by castration prevents Ruffs, for example, from acquiring their fancy neck feathers. Testosterone causes the bills of European Starlings to turn bright yellow in the breeding season, whereas estrogen causes the red bills of female Red-billed Queleas to turn yellow in the breeding season. The growth of red head ornaments—wattles and combs—on roosters and the development of the bill ornamentations of breeding auklets also depend on testosterone.

Testosterone is responsible for a variety of sex-related peculiarities. Growth of colorful feathers by either sex is triggered by injection of testosterone. Phalaropes, for example, are unusual sandpipers in that the bright-plumaged females defend breeding territories and the less colorful males assume the duties of incubation and parental care. Female phalaropes normally have higher concentrations of testosterone than do males, whose maximum levels of testosterone remain below the threshold required to produce colorful feathers. In a similar case, males of some breeds of chickens have femalelike feathers because the cellular chemistry in the skin actively converts testosterone into estrogen. When castrated, they grow male feathers. Injection of testosterone into these castrated males causes them to revert to the female type of plumage (George et al. 1981).

Testosterone and estrogen are not responsible for all sex differences. Luteinizing hormone controls breeding physiology in both sexes, but its potential influence on female plumage is inhibited by the presence of estrogen. In weavers, the colorful breeding plumages of males result from responses of feather follicles to luteinizing hormone secreted by the pituitary gland.

As in butterflies, but not mammals, females are the heterogametic sex in birds; they have the inactive W sex chromosome and only one Z sex chromosome (ZW), which establishes the patterns of sex-linked inheritance (Buckley 1987; Shields 1987). Male birds, on the other hand, have two Z chromosomes (ZZ). Occasionally, following an aberration in the first mitotic division of the fertilized ovum, half of the embryo becomes ZW (female) and the other half becomes ZZ (male). Such individuals are bilateral gynandromorphs—or half male and half female (Figure 15–2). Internally they have a testis on one side and an ovary on the other; externally they have male and female plumages on the corresponding right and left sides of the body, with a sharp division down the center. In species that exhibit sexual size dimorphism, the size of the left and right sides of the skeleton may also be affected (Lowther 1977). Bilateral gynandromorphs have been reported occasionally among a wide variety of bird species, including an Orchard Oriole, a Black-throated Blue Warbler, Evening Grosbeaks, American Kestrels, and House Sparrows (Kumerloeve 1987; Patten 1993). Nothing is known about the breeding activities of such birds.

Sperm Production

The thick, outer fibrous sheath of the testis encases a dense mass of convoluted tubules—called seminiferous tubules—which are lined by active germinal epithelia that produce sperm and secrete sex hormones. Secretion of steroid sex hormones, notably testosterone, occurs both in the Sertoli cells lining the tubules and, particularly, in the Leydig cells, which are packed between the tubules. These cells undergo well-defined seasonal cycles in the accumulation of lipid and cholesterol used in spermatogenesis. Local transformation of the cells of the germinal epithelia into mature sperm proceeds in synchronous waves down the tubule. The entire length of a seminiferous tubule produces sperm at the same time in

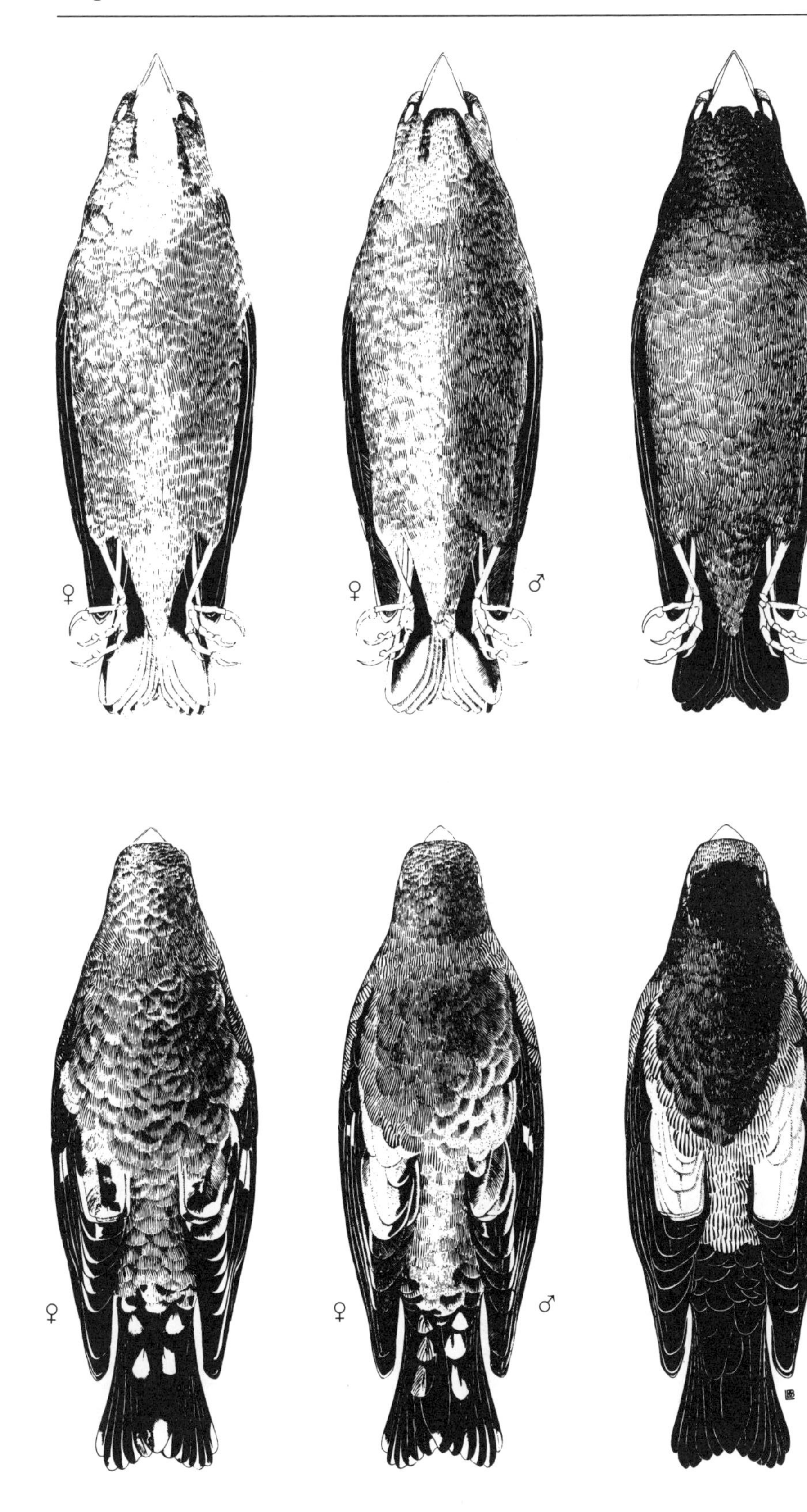

Figure 15–2 Rare individual Evening Grosbeaks are male on one side and female on the other as a result of an aberration in the first cell division. These individuals are called bilateral gynandromorphs (center bird, both top and bottom). Such gynandromorphism is not seen in mammals, in which hormones override the genetic differences between right and left sides. (From Laybourne 1967)

strongly seasonal breeders, such as Arctic shorebirds. Mature sperm quickly leave the testis through a series of thin tubules—rete tubules, vasa efferentia, epididymis, and vas deferens (Figure 15–3).

The testes of most mammals and reptiles are found in cooler external scrota, whereas the testes of birds are housed at body temperature inside the abdominal cavity. To compensate for the extra body heat, spermatogenesis occurs primarily at night when body temperature is slightly lower. New sperm are then stored in swollen seminal vesicles, which are responsible for the conspicuous cloacal protuberance of breeding males.

The seminal vesicles, which are the expanded bases of the two vasa deferentia, swell with accumulated semen awaiting discharge. In mammals, the seminal vesicles and the two other accessory glands add nutritious ingredients to the semen. In birds, the seminal vesicles do not supply many of these nutrients, and the other glands are absent. Both fructose and citrate are absent from bird semen, and chloride concentrations are low (Sturkie 1976). Temperatures in the seminal vesicles, a functional analogue of the mammalian scrotum, are 4°C cooler than internal body temperatures (Wolfson 1954).

Typical vertebrate spermatozoa, including those of birds, consist of three sections (Figure 15-4). The head (acrosome and nucleus) contains the male genetic materials. The midpiece provides metabolic power. The tail (axial filament and tail membrane) propels the sperm forward. Distinctive sperm structures characterize various kinds of birds, such as the oscine passerine birds (McFarlane 1963; Henley et al. 1978). Unlike nonpasserine sperm, which are generally long and straight like those of mammals, passerine sperm have a spiral head and a long, helical tail membrane (Figure 15–4F). Instead of swimming by beating the flagellalike tail, they spin.

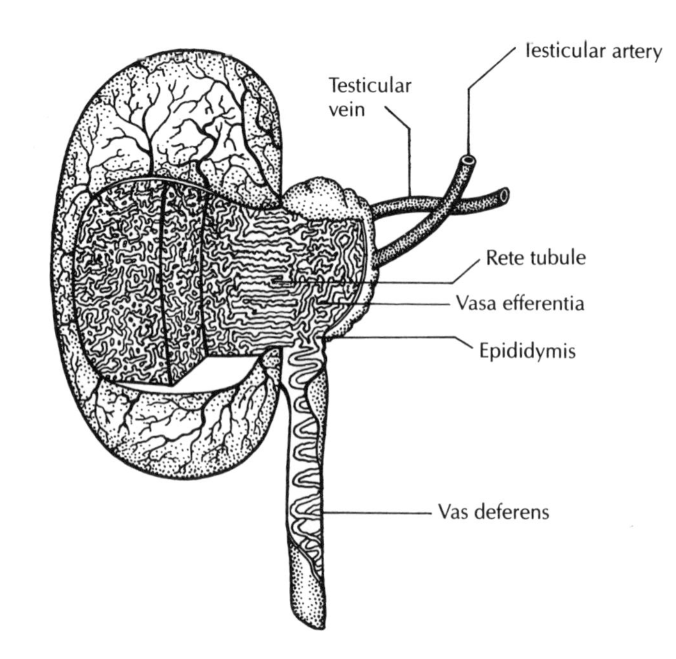

Figure 15–3 Internal anatomy of the avian testis. (From Marshall 1961)

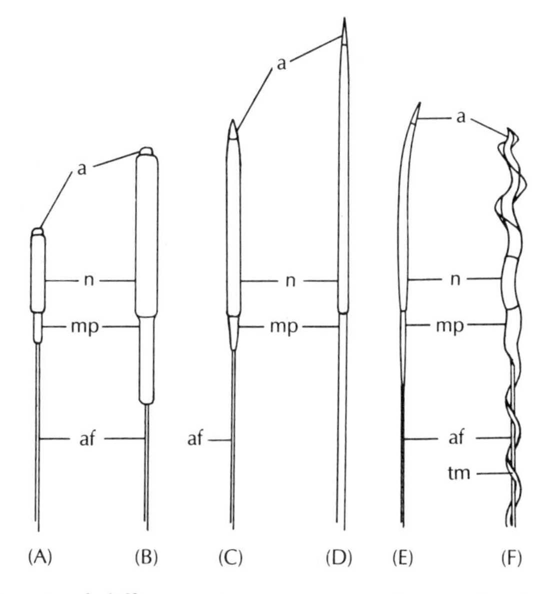

Figure 15–4 Structural differences in spermatozoa characterize the orders of birds. (A) Collared Trogon (Trogoniformes); (B) Great Black-backed Gull (Charadriiformes); (C) Common Eider (Anseriformes); (D) Blue Ground-Dove (Columbiformes); (E) Domestic Chicken (Galliformes); (F) Yellow-rumped Warbler (Passeriformes). a, acrosome; af, axial filament; mp, midpiece; n, nucleus; tm, tail membrane. (From McFarlane 1963)

Maturation of the Ovum

A bird's egg is one of the most complex and highly differentiated reproductive cells achieved in the evolution of animal sexuality. The freshly laid egg consists of (1) the ovum, if unfertilized, or an embryo, if fertilized; (2) a full supply of food to nourish the embryo; and (3) protective layers to ensure the security of the internal environment. Initially only microscopic in size, an ovum swells over 1000 times in volume by the time it is laid. The infusion of yolk, the deposition of egg white—or albumen—and the shell layers all contribute to the enlargement. The yolk is added to the ovum prior to ovulation, and the rest of the components of the egg are added during the egg's passage through the oviduct.

Development of a mature ovum includes two different yet interdependent processes: the formation and deposition of yolk layers, and the differentiation, growth, and maturation of the germ cell itself. Primary oocytes—the cells that give rise to ova—are already present in a hatchling bird, but distinct ova do not appear until the bird is older, when the first small amounts of true yolk are added to the oocyte. Most of the yolk is added to an ovum much later, during the week prior to ovulation, when the ovum swells to its functional size. Each ovum is contained within a follicle, which consists of layers of vascularized cells that are responsible for yolk formation as the ovum matures.

The period of yolk formation—or follicular maturation of the ovum—lasts 4 to 5 days in passerine birds such as the Great Tit, White-crowned Sparrow, and Eurasian Jackdaw, 6 to 8 days in larger birds such as ducks and pigeons, and up to 16 days in some penguins (King 1973; Grau 1982). A study of yolk formation periods among shorebirds and seabirds, using yolk ring patterns, revealed that small shorebirds take 4 to 7 days, small gulls 5 to 8 days, small auks 8 to 10 days, and large gulls 10 to 13 days to form the yolk (Roudybush et al. 1979). Most of the nutrients that supply the energy present in the completed egg are added during this formation period.

The yolk is not homogeneous (Figure 15–5). Rather it comprises alternating layers of yellow yolk in large globules (0.025 to 0.15 millimeter in diameter) and white yolk in smaller globules (0.004 to 0.075 millimeter in diameter). The layers reflect daytime (yellow) versus nighttime (white) yolk deposition. These layers can be counted like a tree's growth

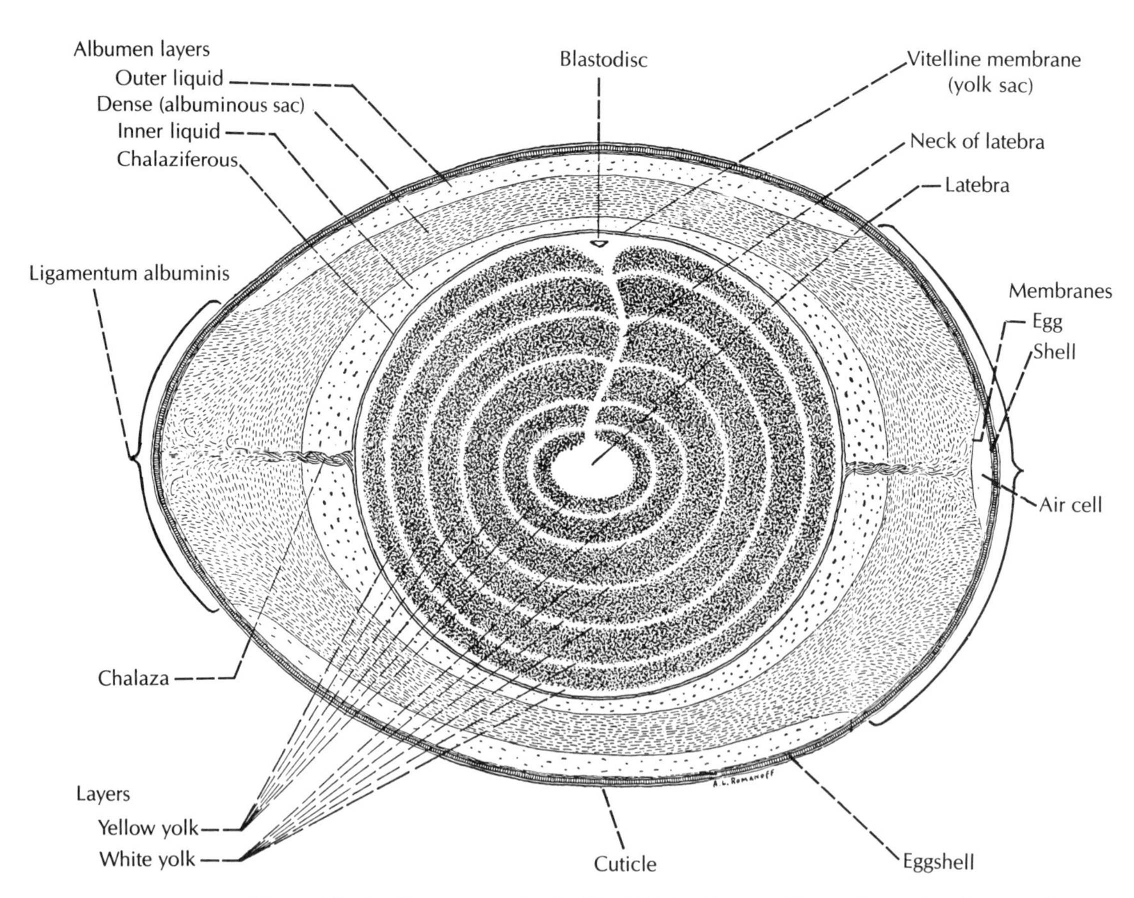

Figure 15–5 Structure of a freshly laid hen's egg. Note alternating layers of white yolk and yellow yolk. The components of egg structure are discussed throughout this chapter. (Adapted from Romanoff and Romanoff 1949)

Figure 15–6 The ovary of a sexually mature chicken showing mature follicles (MF) with basal stalk, or pedicel (P), resorbed or atretic follicles (AF), and a recently discharged follicle (DF). S indicates a stigma, the scarlike area where the follicle will rupture during ovulation. (Courtesy of B. Lofts and R.K. Murton)

rings to determine the time required for yolk formation (Grau 1976; Roudybush et al. 1979). The center of the yolk—or central latebra—is composed of a fluid, white substance called vitellin, which extends to the periphery through a distinct, narrow passage. A thin vitelline membrane encases the yolk, separating it from the albumen that will be added later.

When the full-sized ovum is swollen with yolk, it is ready for its passage down the oviduct. Only a few ova actually make it to this stage. Many follicles—called atretic follicles—stop developing during the early stages of maturation and are resorbed (Figure 15–6). During ovulation, when the egg is released from the ovary, the follicle enclosing the mature ovum ruptures at the stigma—a layer of smooth muscle fibers—and the enlarged ovum pops out and falls into the ovarian pocket—an irregular cavity formed around the ovary by the surrounding organs. Entry into the oviduct is not simply a matter of chance. The open upper end of the oviduct—called the infundibulum—pulses back and forth toward the new ovum, partially engulfing it and then releasing it for up to half an hour, before finally taking it in (Romanoff and Romanoff 1949). When it is finally inside the infundibulum, the ovum is ready for fertilization.

Copulation and Fertilization

Most birds lack external genitalia, so mating normally involves only brief cloacal contact, usually described as a "cloacal kiss." Standing or treading precariously on the female's back, the male twists his tail under hers, and she in turn twists into a receptive position (Figure 15–7). The male may slip off while trying to maintain contact for the few seconds required. Swifts copulate in midair. The prolonged copulation behavior of the Aquatic Warbler is extraordinary (Birkhead 1993). Rather than the normal 1 to 2 seconds, copulation in this species lasts 25 minutes, while the male and female lie together on the ground, male atop the female holding onto her head feathers with his bill. The male inseminates the female repeatedly every few minutes just prior to and after egg laying, a behavior apparently ensuring that his sperm will be positioned to fertilize the next ovum released from the ovary.

Sperm transfer occurs when each partner's cloaca everts and tiny papillae protruding from the posterior walls of the male's sperm sacs into his cloaca are brought into contact with the opening of the female's oviduct, followed by ejaculation. In chickens, average concentrations of spermatozoa are 3.5 million per cubic millimeter of semen. A single ejaculation passes 1.7 to 3.5 billion spermatozoa (with records of 7 to 8.2 billion by Brown Leghorn cocks); however, the concentration of spermatozoa drops rapidly after three or four ejaculations. A minimum of about 100 million spermatozoa is required for the proper fertilization of hens (Sturkie 1976). Little is known about sperm counts in wild birds.

Figure 15–7 Copulation in Common Black-headed Gulls (left) and Little Penguins (right). (Courtesy of E. and D. Hosking and J. Warham)

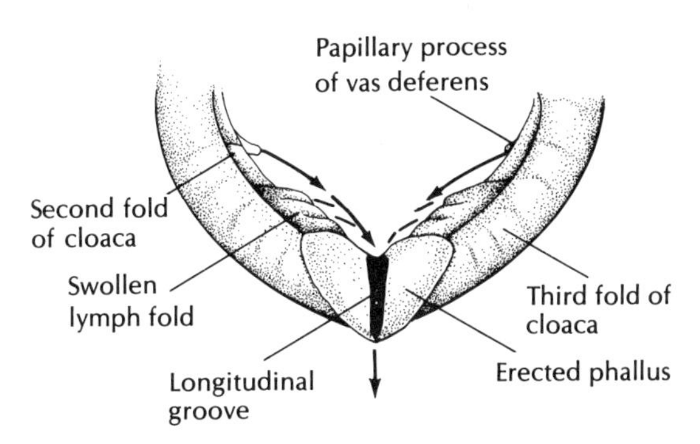

Figure 15–8 Ejaculation of semen by a rooster. Semen, ejected from the papillary process of the vas deferens, combines with transparent fluid from the swollen lymph fold and passes externally via the longitudinal groove of the erect phallus. (After Nishiyama 1955; Sturkie 1976)

A few birds have an erectile, penislike intromittent organ, which is a special modification of the ventral wall of the cloaca (King 1981). The list of species so endowed includes tinamous, most waterfowl, curassows, and Ostriches. The fully extended penis of an Ostrich may be 20 centimeters long and is bright red. Chickens and turkeys have a small penis, which enlarges with lymph fluid that is added to semen in the vas deferens; ejaculation of this fluid occurs through a longitudinal phallic groove (Figure 15–8). Only one passerine bird, the White-billed Buffalo-Weaver, has a phallus, which measures 15 millimeters long by 6 millimeters wide (Bentz 1983). Why such different birds have evolved intromittent organs is not clear, although the nonmonogamous habits of some of them may favor an apparatus that increases the probability of insemination during the brief copulation period. The waterfowl penis probably facilitates sperm transfer underwater.

Some species, including ducks and grebes, have elaborate postcopulatory displays. Immediately after copulation, a male Mallard, for example,

---

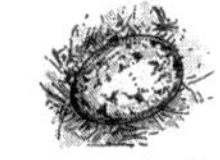

Box 15–1
The Smith's Longspur copulates frequently

A sparrowlike bird of the subarctic tundra, the Smith's Longspur, has the highest copulation rate known for any bird (Briskie 1992, 1993). During the peak week in June, a female longspur copulates frequently with one to three different males. Females solicit copulations an average of seven times per hour and are mounted by their males three times per hour. Each clutch of eggs laid is preceded by an average of 365 copulations. The males themselves have greatly enlarged testes. Such extraordinary sexual effort probably evolved as an adaptation to "sperm competition" in which frequent copulations dilute or displace ejaculates of rival males.

Figure 15–9 Bridling, a postcopulatory display of the male Mallard. (Courtesy of F. McKinney)

suddenly flings his head upward and backward—called the bridling display—and gives a whistled call (Figure 15–9). Then he swims around the female, holding his head low to the water—called nod-swimming. These displays apparently announce successful intromission (Hailman 1977).

Avian sperm swim directly to the upper end of the oviduct, where they may encounter the ripe ovum. They can reach the infundibulum in less than 30 minutes. Normally, eggs are fertilized within a few days of copulation, but some sperm remain viable for weeks. Domestic chickens and turkeys, in particular, can produce fertile eggs 30 to 72 days after copulation. In most birds, the probability of laying fertile eggs decreases rapidly 1 to 2 weeks after copulation. Some female birds have special sperm storage glands, which hold the sperm for up to 10 weeks and then release sperm for transport to the infundibulum for purposes of fertilization (Shugart 1988; Birkhead 1988; Briskie 1992).

As a rule, when female birds mate with several males, mating order and interval between copulations determine paternity of the resulting offspring (Birkhead and Møller 1992). If sequential copulations are separated by more than 4 hours, the sperm that win are likely to be those of the last male to mate. The reasons for this outcome remain uncertain. One possibility is that the last sperm in will form the top layer of sperm stored by the female in her sperm storage tubules and hence will be used first to fertilize the next egg. The evidence for stratification of stored sperm contributions, however, is weak.

An alternative hypothesis is that the sperm of previous males is ejected, removed, or displaced during subsequent copulations. Just before they copulate, for example, the male Dunnock, or Hedge Accentor, pecks at the exposed cloaca of the receptive female, and the female ejects a small droplet of sperm from a previous copulation (Davies 1983). Although lower than those of Smith's Longspurs, the copulation rates of female Hedge Accentors are still high, and they receive sperm one to two times an hour during their 10-day mating period.

Some unfertilized eggs develop normally in domestic turkeys; 32 to 49 percent of infertile eggs may initiate so-called parthenogenetic development, but the embryos usually die at an early stage. Surviving parthenogenetic turkey chicks are always males (because they have the duplicated ZZ sex chromosome combination) and have a full diploid set of chromosomes; they may even be sexually competent (Olsen 1960; Sturkie 1976).

No bird retains and nurtures the fertilized egg inside its body and bears live young—a trait called viviparity—which is a widespread practice among vertebrate animals. All birds quickly form and lay a shelled egg for external incubation—a trait called oviparity. The generally large size of avian eggs would preclude retention of more than a single egg; larger clutches would increase the energetic cost of flight and the vulnerability of gravid females to predators. The benefits of viviparity, it seems, would be offset by reduced numbers of offspring and probably by increased female mortality (Blackburn and Evans 1986). Furthermore, the high body temperatures of birds might dictate against retention of even one egg inside the body cavity (Anderson et al. 1987). The intolerance of embryos for temperatures above 40°C appears to be the reason. Incubation temperatures above 40°C have been found to cause high mortality and malformation of most vertebrate embryos. Thus, the body temperatures of 40° to 42°C that are typical of birds mandate rapid expulsion of the fertilized egg to cooler temperatures outside the body, followed by external incubation in nests.

Egg Production

Fertilization converts an ovum into an embryo, which then begins its passage through the oviduct to complete the process of egg formation. The oviduct is a long, convoluted tube with elastic walls able to accommodate the egg as it enlarges (Figure 15–10). Peristaltic contractions of smooth muscle layers propel the egg from the infundibulum to the vagina through distinct sections in which a glandular epithelial lining adds the albumen, shell membranes, and pigmentation in succession. The passage of the egg through the oviduct usually takes about 24 hours but may require a week. After only a brief stay in the infundibulum (20 minutes), the egg of a chicken enters the main length of the oviduct for 3 to 4 hours, progressing at a rate of 2.3 millimeters per minute. The albumen is added during this period. The membranes of egg and shell are added next, in a 1-hour passage through the isthmus section of the oviduct at a rate of about 1.4 millimeters per minute. Shell formation in the uterus then takes 19 to 20 hours.

Albumen is secreted in the anterior section of the oviduct—called the magnum—where four layers of egg white are added. These layers differ from one another in viscosity and material composition. The innermost liquid layer is literally squeezed from the other albumen by the tightening of a meshwork of microscopic fibers. The yolk rotates gently in response to the slight spiral arrangement of the cellular ridges that line the oviduct's interior. The twisted strands of albumen—called chalazae—that form as the yolk rotates act as small built-in springs that help stabilize the yolk position, thereby keeping the dividing cells that form the embryo oriented upward in the finished egg.

Now covered with albumen, the egg enters the isthmus, where an inner membrane surrounds the albumen and an outer shell membrane is added. This outer membrane is usually firmly attached to the shell itself. It is pliable and tough, and is generally made of felted protein fibers strengthened by albuminous cement, which is riddled with tiny pores that allow the passage of gases and liquids by osmosis and diffusion. The thin inner egg membrane is formed when the cells lining the isthmus apply sticky keratin fibers to the egg surface. Small amounts of pigments, which are also added to the shell membrane, may impart a pinkish hue. Shell colors are added first as pigments deposited during shell formation—the

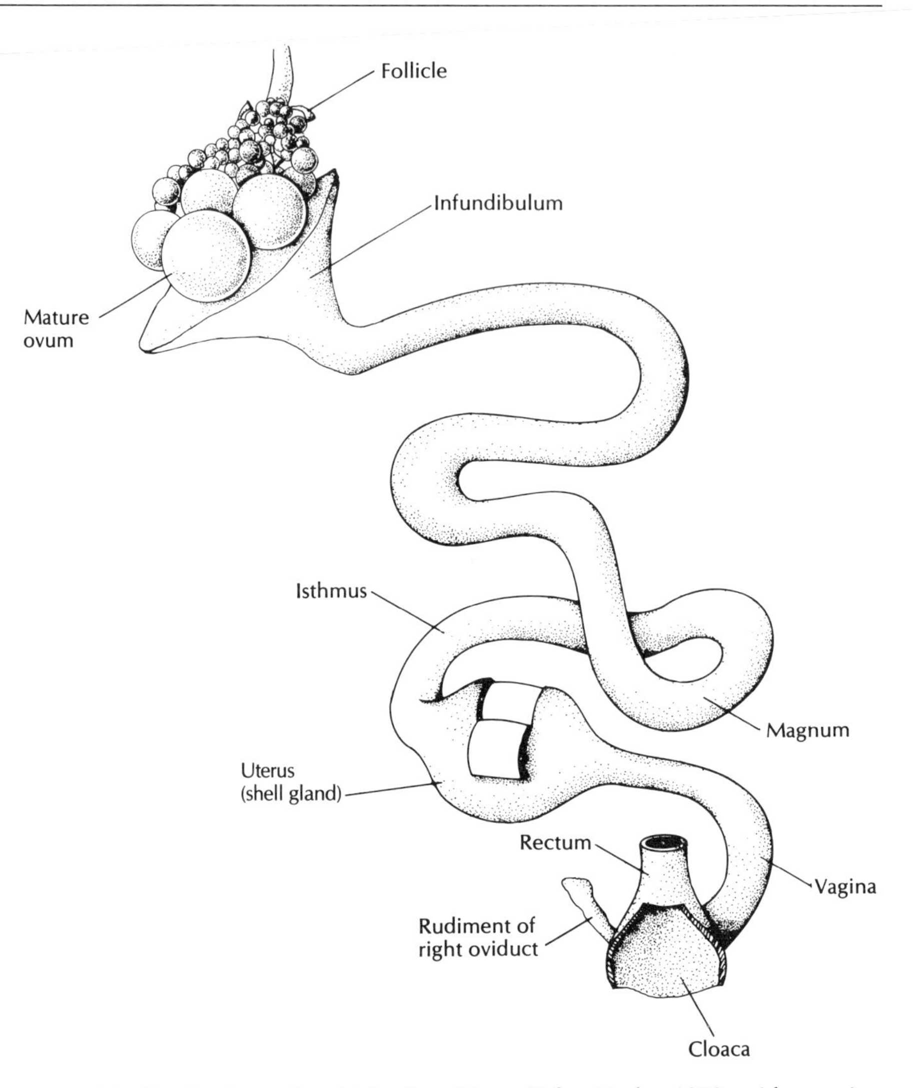

Figure 15–10 Sections of a chicken's oviduct. (After Taylor 1970, with permission of Scientific American)

ground color—and then later as superficial markings in the cuticle, the thin transparent coating of protein molecules that covers the entire shell. The shell pigments are porphyrins (Chapter 4), which derive from the hematin of old blood cells that have been broken down in the liver and transformed into bile pigments.

The final stage of egg production—the addition of a hard shell consisting mostly of calcium carbonate ($CaCO_3$) in the form of calcite crystals—occurs in the region of the oviduct called the uterus. Magnesium and phosphate are minor components of the shell structure, but even slight variations in their concentrations affect the strength and hardness of the shell in dramatic ways. Slight excesses of phosphate may hinder calcite formation by preventing $CaCO_3$ precipitation; slight excesses of

BOX 15-2
Pesticides thin eggshells

Pesticides, such as DDT and DDE, affect normal eggshell formation by increasing magnesium and phosphate levels—with fatal consequences. The normal level of magnesium in Common Tern eggshells is 1.54 percent; the normal phosphate level is 0.25 percent. Exposure to DDT and DDE increases these concentrations to 2.1 percent and over 0.6 percent, respectively, causing denting and developmental failure (Fox 1976). An even higher phosphate level (0.86 percent) has been associated with dead embryos. These pesticides were responsible for widespread eggshell thinning and reproductive failure in the 1960s of Brown Pelicans, several raptors, and penguins. Many eggs were so thin that the weight of the incubating parent crushed them.

magnesium hinder calcite crystal growth (Cooke 1975). Thinning and increased fragility of the eggshell result, and the delicate balance of gas and water required by the embryo may be altered. Magnesium is usually concentrated in a very thin layer of the inner shell, where it plays a role in the reclamation of eggshell salts by the embryo. Fowl (Galliformes) are distinguished from other orders of birds in that they have two layers of magnesium deposition (Board and Love 1980).

When all layers, membranes, and the shell have been added, the egg is ready to be laid. In a few cases, an egg may first rotate 180 degrees so that the blunt end exits first (Figure 15–11). Birds eject an egg voluntarily with their powerful vaginal musculature. Most birds lay their eggs early in the morning, probably to avoid the risks that daytime activity could pose to a bird carrying a heavy, fragile egg in its oviduct.

As a rule, the interval between eggs is 1 to 2 days. Most passerines, ducks and some geese, hens, woodpeckers, rollers, small shorebirds, and small grebes can lay an egg a day. At the other extreme, moundbuilders (Megapodiidae) require 4 to 8 days to produce one of their huge eggs. Ratites, penguins, and large raptors take 3 to 5 days; boobies and hornbills take up to 7 days.

Bird eggs vary in size from the tiny (0.2 gram) eggs of hummingbirds to the enormous (9 kilograms), half-gallon eggs of extinct elephant birds (Aepyornithidae) (Heinroth 1922; Schönwetter 1960–1980). Although egg size increases with body mass, small birds lay much larger eggs relative to their body mass than do large birds. Most birds lay eggs ranging from 11 percent to only 2 percent of body mass. However, there are some dramatic exceptions. Kiwis, for example, lay unusually large eggs. The Brown Kiwi lays two, sometimes three, 500-gram eggs, each of which is 25 percent of the female's own mass; it lays these enormous eggs at 4-week intervals (see Figure 17–1). Mousebirds, swifts, and parasitic cuckoos lay small eggs relative to their body masses. Occasionally, birds lay dwarf—or runt—eggs that are less than half the size of their normal eggs (Ricklefs 1975); most of these lack a yolk and result from aberrant stimulation of the oviduct by an object such as a blood clot.

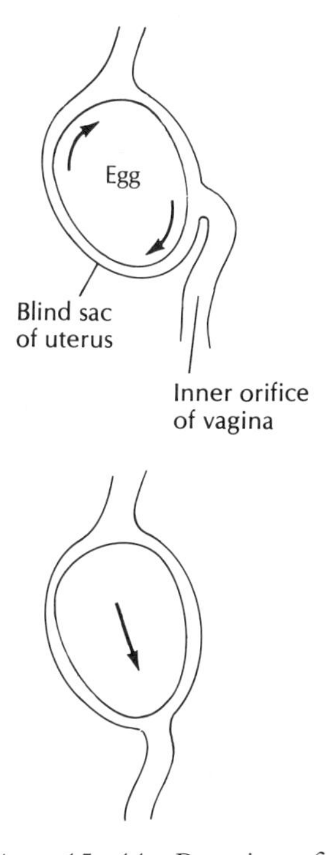

Figure 15–11 Rotation of the egg in the uterus before laying. Some eggs turn 180 degrees and are laid blunt end first (top); most, however, proceed directly through the uterus (bottom). (After Romanoff and Romanoff 1949)

Clutch Size

The number of eggs a bird lays in one set—called the clutch—is subject to short-term constraints, such as energy available for egg formation, and to long-term considerations of lifetime reproductive success. Tube-nosed seabirds lay only a single egg each breeding season, and Arctic sandpipers lay four eggs; but not all birds have such fixed clutch sizes. Average clutch sizes range from 3 to 12 among species of waterfowl and from 2 to 23 among species of gallinaceous birds (Lack 1968).

Clutch sizes can also vary within a single species, for example, from 4 to 14 in the Northern Flicker and from 8 to 19 in the Blue Tit. Some of these variations reflect genetic differences between individuals, but age, food availability, and season also affect how many eggs a female lays (Boag and van Noordwijk 1987). The inheritance of egg-laying ability is well known to poultry farmers, who increase egg production by artificial selection (King and Henderson 1954; Kinney 1969). Great Tits inherit a tendency to lay larger than average clutches, average clutches, or smaller than average clutches, but not a particular number of eggs (Perrins and Jones 1974).

The evolution of such variations in clutch size is an exceedingly complex topic (see Chapter 20). Here we shall consider only the physiological constraints on egg production. The high costs of egg production can strain a female's daily energy balance and slow egg formation (Ricklefs 1974; Walsberg 1983). The daily energy requirements for egg production average 40 to 50 percent of the basal metabolic rate for land birds with helpless—altricial—hatchlings and 125 to 180 percent for birds producing precocial hatchlings—well-developed hatchlings like those of ducks and quails. Expressed simply in terms of the extra daily cost of egg production, land birds with altricial young increase their daily energy budget by 13 to 16 percent, galliformes by 12 to 30 percent, and waterfowl by 51 to 70 percent (King 1973). Also, the lack of protein and minerals, such as potassium and calcium, may limit egg production, especially in birds that eat fruit and seeds, foods poorly supplied with these elements.

The greatest costs of egg formation are incurred during the period of yolk production. The peak daily energy expenditure for total egg production depends on the amount of overlap in growth cycles of separate ova and the number of follicles growing simultaneously (King 1973). In the Fiordland Penguin, for example, the peak occurs on day 20 as the albumen is added to the first egg at the same time as the last of the yolk is added to the second egg (Grau 1982). Generally speaking, the amount of energy transferred to the egg varies from 4.2 kilojoules per gram in passerine birds such as the Eurasian Tree Sparrow to as much as 8.4 kilojoules per gram in the fat-rich eggs of waterfowl. The efficiency of energy transfer is only about 20 percent; a laying female passerine bird, for example, must eat 5 kilojoules of food for every kilojoule that is transferred to her eggs.

The resources required for egg production come from increased daily intake, which is sometimes coupled with reduced activity and with use of a female's stored energy reserves (Blem 1990). Increased foraging by laying female waterfowl and shorebirds is particularly conspicuous. Wood Ducks lay large clutches of about 12 richly provisioned (and, therefore, energy-expensive) eggs at a total metabolic cost of 6000 kilojoules (Drobney 1980). A hen's fat reserves provide most of the energetic requirements of egg production (88 percent). The protein content of these eggs comes from invertebrates that the hen eats during the laying period. The cost of feeding on invertebrates, which is not profitable in terms of energy, is also supported by the hen's fat reserves. Reduced activity in female Willow Flycatchers and Black-billed Magpies enables them to shunt the energy conserved into egg production (Ettinger and King 1980; Mugaas and King 1981).

Use of reserves for egg production by passerine birds is not yet well studied except in the Brown-headed Cowbird of North America and the Red-billed Quelea in Africa. Brown-headed Cowbirds do not use stored reserves despite their great egg production—approximately an egg a day for over a month. Instead they obtain the nutrients for egg production directly from their diet (Ankney and Scott 1980). In contrast, Red-billed Queleas draw heavily on breast muscle for some of the protein they need for their eggs (Jones and Ward 1976). They also use their fat reserves as fuel for the activity required to find additional protein and for the energy required to produce their eggs.

Food shortages can reduce or stall egg production and thus affect clutch size (King 1973; Ricklefs 1974). Year-to-year variations in average clutch size in the Great Tit relate directly to food abundance (Perrins 1965). Both clutch size and egg size of California Gulls reflect variations in food supplies (Tasker and Mills 1981; Winkler and Walters 1983).

Female Snow Geese, among many other birds, depend on their reserves for egg production and also for incubation (see Chapter 16). The number of very large follicles in a female's ovary, and hence her projected clutch size, are directly related to the quantity of protein and fat reserves on her arrival on the Arctic breeding grounds (Ankney and MacInnes 1978). After laying their clutches of various sizes, all females have about the same reserves, a finding that indicates the limits set on clutch size by the amount of reserves (Figure 15–12). Females that arrive with low reserves fail to lay at all.

Historically, a distinction has been made between determinate layers—species that lay a fixed number of eggs—and indeterminate layers—species that lay extra eggs if some are removed from the nest early in incubation. The classic example of an indeterminate layer is a prodigious female Northern Flicker that laid a total of 71 eggs in 73 days to replace those removed as soon as they were laid (Bent 1939). Domestic hens and Japanese Quails can produce an egg a day all year long. In contrast are the determinate layers, which do not replace eggs removed from their nests. Shorebirds and gulls are usually classified as determinate layers, but they

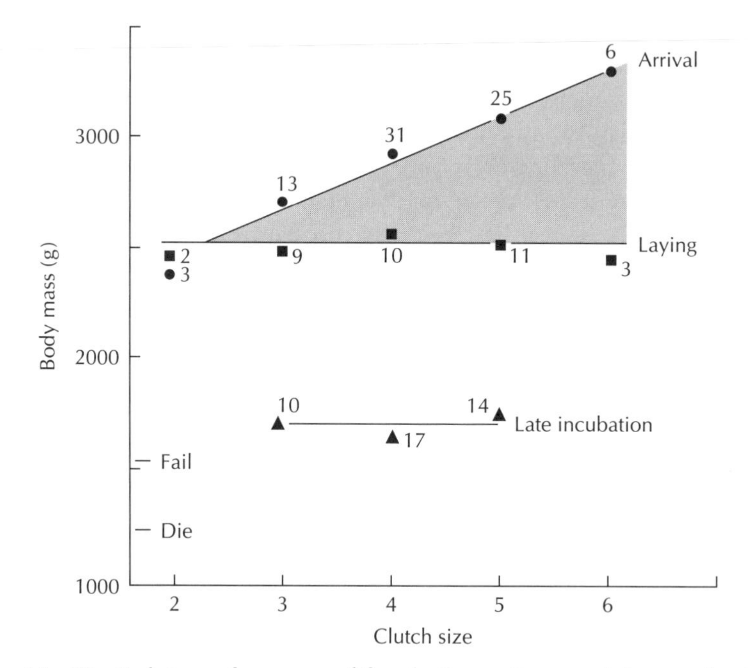

Figure 15–12 Relation of reserves of female Snow Geese arriving on Arctic breeding grounds to their projected clutch size. Females use some reserves (measured by loss in body mass) to produce eggs and then use more reserves during incubation. The number of eggs a female lays is directly related to its reserves. Most females finish laying and start incubating with approximately the same body mass and hence similar reserves. Females that start incubation with inadequate reserves may abandon their eggs to avoid starving, but sometimes they do not do so in time. Dots, squares, and triangles indicate mean values of mass of females weighed upon arrival, while laying, and during late incubation, respectively. Next to each symbol is the number of females included in each sample. (After Drent and Daan 1980; from data in Ankney and MacInnes 1978)

do replace eggs that are removed as soon as they are laid. Pheasants and ducks lay a full complement of replacement eggs if all but one egg of the original clutch is removed as soon as laying is complete. The importance of genetic, nutritional, and psychological factors controlling replacement egg production needs study, and the basic concept of determinate versus indeterminate laying ability needs to be reevaluated (Winkler and Walters 1983). No clear classification of species can be made until all the species being compared are subjected to the same experimental regimen.

The Avian Egg

I think, that, if required, on pain of death, to name instantly the most perfect thing in the universe, I should risk my fate on a bird's egg. (T.W. Higginson 1863, p. 297)

Bird eggs fascinated the earliest ornithologists. Their many sizes, shapes, tints, and textures inspired naturalists to collect them, and, like seashells, rare and beautiful eggs command great prices. In fact, interest in the avian egg influenced the development of ornithology as a comparative science. Nineteenth-century ornithologists published enormous monographs illustrating the eggs of British and African birds (for example, Seebohm 1885); and serious students of oology—the study of eggs—undertook detailed studies of the microscopic structure of eggshells and embryos (Romanoff and Romanoff 1949) (Figure 15–13).

Figure 15–13 Eggshell patterns: 1, Patagonian Tinamou; 2, Bronze-winged Courser; 3, Chilean Tinamou; 4, Lesser Bird-of-Paradise; 5, Cape Crow; 6, Gray-necked Rockfowl; 7, Three-banded Plover; 8, African Jacana; 9, Stripe-backed Bittern; 10, White-throated Laughingthrush; 11, 12, Brown Babbler; 13, 14, 15, 16, Tawny-flanked Prinia; 17, Grayish Saltator; 18, 19, 20, Winding Cisticola; 21, Yellow-green Vireo. (From Winterbottom 1971; painting by A. Hughes)

We have reviewed the physiology of sex and of egg production and now turn to the egg itself as a self-contained chamber that nourishes and protects the growth and development of the embryo. The avian egg is closed—or cleidoic—the type of egg that freed the reptiles from the aquatic mode of life. It contains all the nutrients—especially water—that the embryo requires for its early development. Cleidoic eggs evolved from the naked, amniotic eggs of ancestral reptiles, presumably in response to predation by soil invertebrates and microbes (Packard and Packard 1980). The increased calcification of the avian eggshell provided better protection for eggs laid in the soil. What was sacrificed was the ability of the egg to absorb the water needed by an embryo. The flexible shell membranes of primitive reptilian eggs were water permeable, but the harder, calcified, avian eggshells are less so. To compensate, water was added to the egg contents in the form of albumen (the egg white). Among reptiles, rigid-shelled eggs supplied with albumen and functionally cleidoic are typical of some turtles and crocodilians, but the evolution of the cleidoic egg proceeded even further in birds.

Egg Contents

The embryo inside the avian egg is not isolated from the external environment. Its survival requires an active exchange of oxygen, carbon dioxide, and water vapor through the shell membranes. Its growth and well-being depend on the egg's provisions and also on its temperature. Its chances of hatching depend on the ability of the parents to regulate the egg's immediate environment within narrow limits.

Three extraembryonic membranes support the life and growth of the avian embryo (Figure 15–14). The amnion surrounds only the embryo, which floats in a contained environment of water and salts. The chorion is a protective membrane that surrounds all the embryonic structures. The allantoic sac functions in both respiration and excretion and increases in size as development proceeds, while a growing network of fine capillaries keeps it well supplied with blood.

Pressed tightly against the chorion and the shell membranes, the resulting "chorioallantois" is the site of export of carbon dioxide produced by the embryo and import of oxygen from the outside world. The allantois also acts as a sewer for storage of poisonous nitrogenous wastes.

Although the avian egg provides a secure, self-contained environment for embryonic development, it also imposes restrictions on the kind of nitrogenous waste the embryo can produce. Ammonia is not a suitable waste product, because the embryo, confined in its shell, cannot excrete it, and unexcreted ammonia would rapidly reach toxic concentrations. Nor is the water-soluble compound urea acceptable, because the egg does not have the space required to store large volumes of this dilute waste. Birds (embryonic and adult) have found an excellent solution to

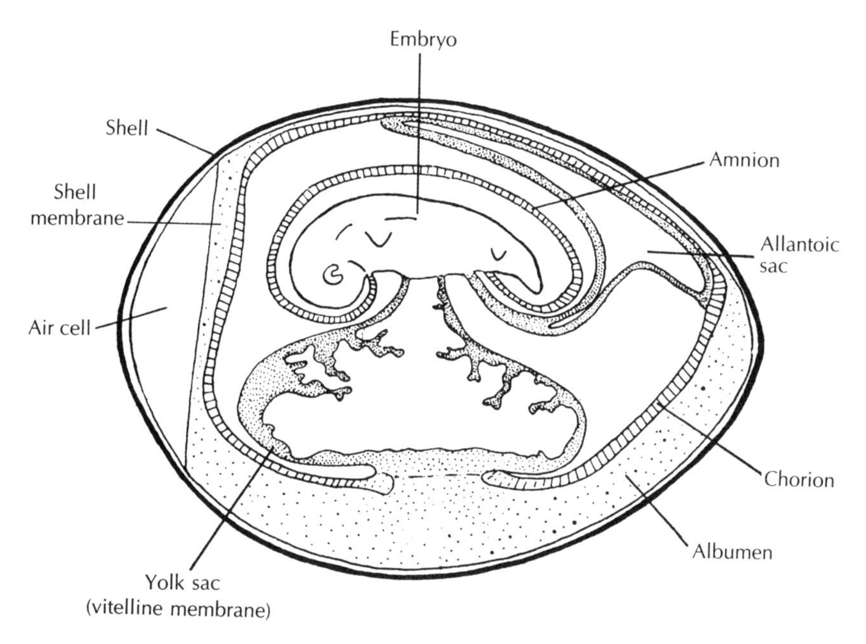

Figure 15–14 The developing embryo and the extraembryonic membranes. (Adapted from Bellairs 1960)

their waste-disposal problem: uric acid. This nonsoluble form of nitrogenous waste can be deposited safely as tiny crystals inside the allantois; it is not toxic and does not require large volumes of water to flush it from the system.

In addition to the blastodisc—the center of growth of the embryo (see Figure 15–5)—and its associated extraembryonic membranes, the freshly laid avian egg consists of three principal components: yolk, albumen, and shell. The yolk is an energy-rich food supply for the embryo. Lipids represent 21 to 36 percent of the yolk materials, and proteins make up another 16 to 22 percent. The rest is primarily water. The yolk sac—or vitelline membrane—which functions as the early analogue of a stomach and intestines, contains the yolk and is ultimately absorbed into the body cavity. In addition to providing nutrition, the yolk initially cradles the tiny embryo in a small pocket. Yolks vary in color from pale yellow or light cream to dark orange-red or even brilliant orange. Within a species, such variations partially reflect diet. Hens that eat red peppers rich in carotenoid pigments, for example, lay eggs with red yolks instead of the normal yellow yolks (Fox 1976).

The albumen—or egg white—consists primarily of water (90 percent) and protein (10 percent). Besides being the embryo's water supply, the albumen is an elastic, shock-absorbing cushion that protects the embryo when the egg is moved or jolted. It also insulates and buffers the embryo from sudden changes in air temperature and slows the cooling rate when the parent is not incubating. Albumen constitutes 50 to 71 percent of the total egg weight.

The external layers of the egg shield the embryo, conserve food and water, and facilitate the respiratory exchange of gases. Above all, the hard shell provides structural support and protection of the egg contents from soil invertebrates and microbes. Eggshells vary in thickness from paper thin in small land birds to as much as 2.7 millimeters thick in Ostriches. They are strong enough to withstand the weight of an incubating adult yet delicate enough to allow the chick to break out. The shell usually constitutes 11 to 15 percent of an egg's total weight—up to 28 percent in extreme cases.

Egg Shapes

The term *egg-shaped* brings to mind a rounded structure, longer than it is wide, and slightly more pointed at one end than at the other, as is the familiar hen's egg. Physiological factors influence the egg shapes of domestic hens, but males do not come from pointed eggs nor do females from more rounded ones, as Aristotle once suggested. Egg shapes vary from the nearly spherical eggs of petrels, turacos, owls, and kingfishers to the pointed (pyriform—literally "pear-shaped") eggs of plovers and murres. Between these shapes are the ellipsoidal—or biconical—eggs of grebes, pelicans, and bitterns.

Egg shapes are a compromise between structural advantages, clutch volume, and egg content (Andersson 1978). Spherical eggs maximize shell strength, conservation of heat, and conservation of shell materials by maximizing volume relative to shell surface. Pointed eggs, in the clutches of three to four large eggs typically laid by shorebirds, further enhance the volume or content of large eggs within the limits set by the area an incubating parent can cover with its body. The pointed eggs of murres and other cliff-nesting birds offer an additional advantage; they roll only in a tight arc, which lessens their chance of falling from their precarious positions on nest ledges (Drent 1975) (Figure 15–15).

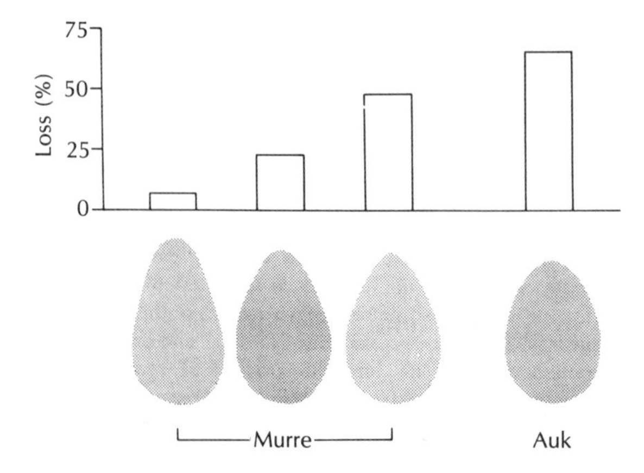

Figure 15–15 Pointed eggs such as those of the Common Murre are less likely to roll off a cliff ledge than are the more rounded eggs of Razorbills. Data presented here are from 400 trial experiments in which eggs of each type were pushed gently on a nesting ledge. (Adapted from Drent 1975; Tschanz et al. 1969)

Eggshell Color and Texture

Many bird eggs, especially those of hole or burrow nesting species, are dull white. The need for camouflage is minimal in such nest sites, and enhanced visibility of the eggs in the dark interior of the nest cavity may reduce accidental breakage by the parents. Most eggs laid in open nest sites, however, are exquisitely colored and patterned. Shaded ground colors, superficial blotches, and fine specklings or scrawls help blend the smooth contours of an egg into its background. In exceptional cases in which the eggs of some ground nesting species, such as nightjars, are conspicuously white, the well-camouflaged incubating parent shields them from the eyes of potential predators. The whitish eggs of grebes are camouflaged by brownish stains from mud and rotting nest vegetation.

A variety of birds, such as American Robins, lay bright blue eggs. The brightest blue eggs of all are those of the Great Tinamou of Central and South America. The function of blue coloration is still not known, but there is a general correlation with nest site. British thrushes that nest in open-forked branches above the ground lay blue eggs, which might look like holes in the vegetation when viewed from above by a predator, whereas related species of thrushes that nest on the ground or in dense thickets lay well-camouflaged, brown-speckled eggs (Lack 1958).

Different shell textures characterize the various families of birds. Accentuating the bright blues, greens, and violets of tinamou eggs is their polished, enamellike texture. The eggs of ibises and megapodes, in contrast, have dull, chalky textures, whereas the eggs of ducks are oily and waterproofed, the eggs of cassowaries are heavily pitted, and the eggs of jaçanas appear lacquered. The functions of such shell textures are not well understood.

Eggshell Structure

Eggshells are made of inorganic calcium and magnesium salts (carbonates and phosphates) embedded in a network of delicate, collagenlike fibers (Board and Scott 1980; Carey 1983). There are two distinct layers of shell microstructure: an inner cone layer with basal protuberances that adhere to the shell membrane, and a palisade layer that makes up most of the shell material (Figure 15–16). Crystalline calcite is the principal construction material. This inorganic salt is gradually mobilized from the shell and used as calcium for bone growth by the embryo. Covering the outer surface of the eggshell is the cuticle, a thin, proteinaceous froth of air bubbles that blocks invasion by microorganisms. The chemical elements that make up eggshells are extremely stable; with proper calibration for past temperatures, Ostrich eggshells can be used to estimate the ages of archeological sites up to 1 million years old (Brooks et al. 1990).

Eggshell textures reflect a porous microstructure that regulates the passage of water vapor, respiratory gases, and microorganisms between the inside of the egg and the external world. The eggshell is permeated by thousands of microscopic pores (Becking 1975; Board and Scott 1980;

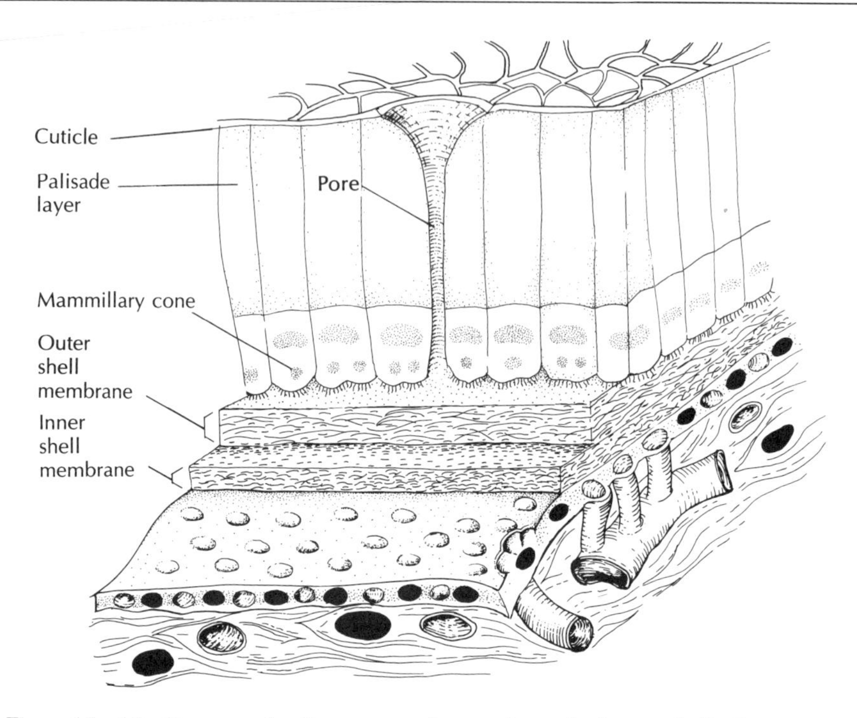

Figure 15–16 Pore canals allow gas exchange through the eggshell. Oxygen enters the eggs via pores that traverse the cuticle and passes through columns of crystals to the permeable shell membranes. Carbon dioxide and water vapor escape to the outside environment through these same pores. Blood vessels in the capillary bed of the chorioallantois link the developing embryo to the gas exchange pathway. (After Wangensteen et al. 1970; Drent 1975; Rahn et al. 1979, with permission of Scientific American)

Tullett and Board 1977). The ordinary hen's egg has over 7500 pores, most of which are at the blunt end of the egg. The shells of most avian eggs have simple, straight canals that widen slightly toward the openings on the exterior surface. The eggshell pores of swans and ratites, however, branch from their origins near the shell membrane into a more complex network (Tyler and Simkiss 1959). Covering the exterior openings of the pore canals of all avian eggshells except those of pigeons and doves are tiny plugs or caps, which may act as pressure-sensitive valves.

Diffusion through the shell membranes allows the exchange of water vapor and gases, which are vital to embryonic life (Ar and Rahn 1980; Carey 1983, 1991). Eggs breathe passively. No active, regulated exchange is known, nor is it required to account for the known rates of gas and water vapor exchange. The density of pores is an exquisite compromise between the optimal high densities that would facilitate rapid gas exchange and the low densities that would minimize water loss. The dynamics of gas exchange vary as incubation proceeds. Removal of calcium from the shell itself for incorporation into the embryonic skeleton promotes progressive thinning of the shell, which increases the rate of gas exchange at a time of increasing respiration by the growing embryo. The permeability of the shell membranes to oxygen increases as they dry out, and the rate of inward movement of oxygen increases as the growing embryo draws increasing amounts of oxygen from the chorioallantois.

The structure of eggshells could limit the altitudinal or geographical distributions of birds (Carey 1991). The high rates of potential water loss in xeric—dry—habitats, where relative humidity is low, or at high altitudes, where barometric pressures are low, might limit the hatchability of eggs without some adjustment in pore density or length. Domestic Chickens change their eggshell microstructure with altitude (Rahn et al. 1982). Some compensation for altitude also occurs in the eggs of swallows (Carey 1980), but how much eggshell microstructures vary as a form of environmental adaptation is not yet known.

In addition to caring for the egg's physiological requirements, parents must also protect the egg from predators. Birds respond to these challenges by incubating their eggs in nests, the topic of the next chapter.

Summary

Birds reproduce sexually. The gonads consist of paired testes in males and (usually) a single ovary in females. Avian testes are located internally, attached to the dorsal body wall at the anterior ends of the kidneys. The avian ovary, which resembles a cluster of grapes, comprises hundreds, sometimes thousands, of oocytes. The gonads are controlled by two hormones secreted by the anterior pituitary: follicle stimulating hormone, which regulates gamete formation, and luteinizing hormone, which regulates hormone secretion by the testes and maturation of follicles in the ovary. Estrogen and testosterone control sexual distinctions in plumage, body size, voice, and behavior.

Because most birds lack external genitalia, copulation normally involves only brief cloacal contact. Sperm swim directly to the upper end of the oviduct, where fertilization occurs, usually within a few days of copulation. After fertilization, the egg with its tiny embryo passes through different regions of the oviduct, a process that generally takes about 24 hours. Albumen, egg and shell membranes, and a hard shell made of calcium carbonate are added during the journey down the oviduct. Egg formation takes from 1 to 7 days, depending on the species. Some birds have fixed clutch sizes, but others do not. Energy requirements, food supplies, egg size, and parental care requirements, as well as genetics, all influence clutch size.

The avian egg is one of the most complex and highly differentiated reproductive cells achieved in the evolution of animal sexuality. It provides not only nourishment for the developing embryo but ventilation, insulation, resistance to rapid heating or cooling, and protection as well. Pores that permit gas exchange and water loss permeate the microstructure of the avian eggshell, which evolved to protect the embryo from soil invertebrates and microbes.

FURTHER READINGS

Birkhead, T.R., and A.P. Møller. 1992. Sperm Competition in Birds. New York: Academic Press. *A forward-looking synthesis of the evolutionary foundations of avian reproductive behavior.*

Carey, C. 1983. Structure and function of avian eggs. Current Ornithology 1: 69–103. *An excellent summary of the physiology of eggs.*

Gilbert, A.B. 1979. Female genital organs. *In* Form and Function in Birds, Vol. 1, A.S. King and J. McLelland, Eds., pp. 237–360. New York: Academic Press. *A detailed review of the anatomy and physiology of the female system.*

King, A.S. 1981. Phallus. *In* Form and Function in Birds, Vol. 2, A.S. King and J. McLelland, Eds., pp. 107–147. New York: Academic Press. *A detailed review of the anatomy and physiology of the male system.*

Lofts, B., and R.K. Murton. 1973. Reproduction in birds. Avian Biology 3: 1–108. *A detailed review of the anatomy and physiology of reproduction.*

Ricklefs, R.E. 1974. Energetics of reproduction in birds. *In* Avian Energetics, Publications of the Nuttall Ornithological Society No. 15, pp. 152–292. *A detailed review of the subject.*

Romanoff, A.L., and A.J. Romanoff. 1949. The Avian Egg., New York: John Wiley & Sons. *A classic reference.*

Shields, G.F. 1987. Chromosomal variation. *In* Avian Genetics, F. Cooke and P.A. Buckley, Eds., pp. 79–104. New York: Academic Press. *A comprehensive review of the chromosome genetics of birds.*

Sturkie, P.D. 1986. Avian Physiology, 4th ed. New York: Springer-Verlag. *A general review with an emphasis on poultry.*

C H A P T E R 16

Nests and Incubation

NO BIRD GIVES BIRTH to live young. Instead, birds prepare a nest to cradle their eggs and young. Caring first for the eggs and then for the young requires a major commitment of time and energy, often by both sexes. The associated risks are great, because eggs and nestlings, as well as the attending, vulnerable parents themselves, tempt a host of predators. These pressures have fostered the evolution of diverse nest architectures that reduce the risks of reproduction. Different nest sites and social arrangements also help to reduce vulnerability to predation.

The incubation behaviors of parents correspond to the survival and growth requirements of their offspring. Successful reproduction requires accommodations to the narrow thermal tolerances of the embryos and later to those of nestlings. Embryos inside the egg require heat from their parent's body to grow to a hatchling; they also must be protected from excessive heat or lethal cold.

This chapter first scans the adaptive architectures of bird nests, which in swallows, for example, reflect the evolutionary history of a taxon. Then follows a more detailed discussion of the ways that nests achieve safety from predators, including the social option of nesting in colonies. The topic of nest structures concludes with reviews of nest construction behavior and the roles of nests in temperature regulation. Finally, incubation of eggs, which requires dedicated parental effort, is discussed.

Nest Architecture

The nesting behaviors of birds mirror their solutions to the local challenges of reproduction. Most birds build isolated, hidden nests, many of which remain unknown to science. At the other extreme are conspicuous, open breeding colonies, some with millions of pairs. On the Peruvian coast, black-and-white Guanay Cormorants pack together at densities of up to 12,000 nests per acre and have attained total colony sizes of 4 to 5 million birds. In Africa, 2 to 3 million pairs of the sparrowlike Red-billed Quelea nest in less than 100 hectares of thornbush savanna. Also providing safety from predators are the burrows of nocturnal auklets and petrels, which riddle the hillsides of oceanic islets, and the woven nests of caciques and weavers, which dangle from crowded tall trees, often over water.

Birds build nests to protect themselves, their eggs, and their young from predators and from adverse weather. Structure and function are in-

(A)

(B)

(C)

(D) (E)

Figure 16–1 The nests of birds vary from simple to elaborate, large to small. (A) Floating platform nest of Western Grebe; (B) sandy scrape nest of Wilson's Plover; (C, D) down-lined, camouflaged nest of Cinnamon Teal; (E) mud nest of Rufous Hornero; (F) mud nests of Cliff Swallows; (G) hole nest (in cactus) of Gila Woodpecker; (H) straw nest of Cactus Wren; (I) stick nests of Great Blue Herons; (J) stick nest of Rufous-fronted Thornbird; (K) cup nest of Broad-tailed Hummingbird; (L) suspended cup nest of Warbling Vireo; (M) suspended nests of Crested Oropendola; (N) intricately woven nest of Black-throated Malimbe. (Courtesy of A. Cruickshank/VIREO, A–D, F–H, K, L; O. Pettingill/VIREO, E, M; T. Fitzharris/VIREO, I; P. Alden/VIREO, J; E. and N. Collias, N)

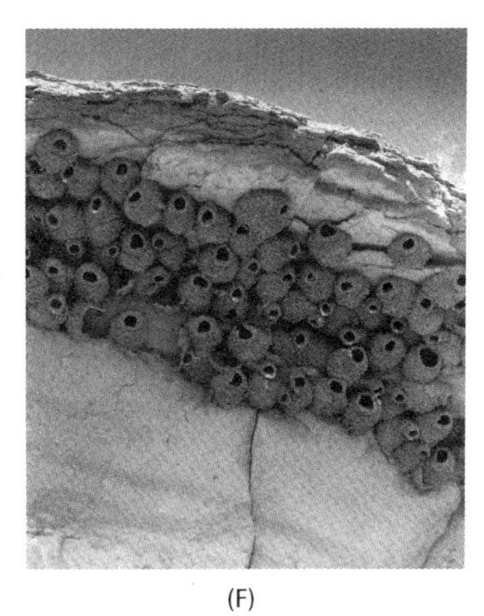
(F)

(G)

(H)

(I)

(J)

(K)

(L)

(M)

(N)

separable features of nest architecture. The more conspicuous features provide protection from predators and from adverse climates, whereas the more subtle features aid in the regulation of temperature and humidity. Other animals also build nests, but birds do so in a greater variety of forms, from a greater variety of materials, and on a greater variety of sites. Bird nests range from precarious sites on bare branches to enormous communal apartments and from simple scrapes on the ground to elaborate stick castles (Figure 16–1). Nests may be casually constructed from ready-for-use pebbles and sticks or laboriously woven from natural fibers. Bird nests range in size from the few sticks assembled by some doves to the gargantuan aeries of eagles. One Bald Eagle aerie weighed over 2 tons when it finally fell in a storm after 30 years of annual use, repairs, and additions (Herrick 1932).

Passerine birds construct the most elaborate nests of all. Baglike—or pensile—nests hang by silky cobwebs or by wiry, black fungal fibers. Some are suspended far below a main branch. Others, such as those of the Northern Oriole, are hung from the thin, outermost branches of large trees. The integrity of pensile nests derives from their tightly woven construction, tough knots, and strong binding materials. The intricately woven, meter-long nest of the Black-throated Malimbe, a West African weaver, may well be the pinnacle of avian nest construction (Figure 16–1N) (Collias and Collias 1984).

Many birds nest in colonies, but only a few actually build compound, communal nests divided into individual compartments. Instead of nesting in excavated cavities or burrows like most parrots, up to 15 pairs of the Monk Parakeet of Argentina occupy huge, communal, stick nests, which also attract nesting pairs of Speckled Teal and Spot-winged Falconets (Martella and Bucher 1984). These nests are now a common sight in Florida, where introduced Monk Parakeets are increasing in numbers. The nests of the Sociable Weaver of southwestern Africa are perhaps the largest and most spectacular of all communal avian nests; they each resemble a large haystack in a thorny tree. The pairs that will occupy the structure share in building the common roof that covers 100 or more separate nest chambers, which are cool by day and warm by night. The geographical distribution of this species is limited to the extremely arid sections of southwestern Africa, probably because rain would saturate the nest and create an insupportable weight.

Nest Materials

Nests made of plant matter may contain twigs, grass, lichens, and leaves. Some plants that birds add to their nests help combat disease and ectoparasite infestations. Green vegetation seems to be particularly useful in this regard. In general, hole nesters incorporate fresh, green vegetation more regularly into their nests than do open nesters (Clark and Mason 1985). European Starlings, in particular, select by odor certain plants, such as

Box 16–1
The nests of swallows reflect their evolutionary history

Nest construction is more diverse among species of swallows, the Family Hirundinidae, than in any other family of songbirds. Some species burrow into hillsides, others adopt tree cavities, and still others build mud nests on cliffs. The use of pure mud to construct hanging nests is unique among all birds. David Winkler and Fred Sheldon (1993) reconstructed a phylogenetic history of 17 swallow species, using DNA hybridization methods (Box 3–1), and then asked whether related species constructed the same kinds of nests (see figure). The answer—nest construction habits reflect the inferred evolutionary history of species—was startling, because modern ornithologists had assumed that nest construction behavior changed easily as an adaptation to local circumstances.

Among swallows, burrowing into the soil appears to be the primitive mode of nesting. Burrowing ancestors gave rise first to a swallow that adopted natural cavities and then to a radiation of descendant, related species, such as the Tree Swallow and Purple Martin. The obligate cavity adoption behavior of such species appears to be tied to their evolution in the rich forest habitats of the New World.

A burrowing ancestor also gave rise to swallows in another major clade, all members of which now use mud construction techniques. Barn Swallows and Cliff Swallows belong to this latter clade. Construction of mud nests originated only once in the evolutionary history of swallows. The mud nesters diversified principally in Africa, where a dry climatic history has favored their mode of nesting. Winkler and Sheldon suggest that mud nests increased in complexity from simple mud cups to fully enclosed, jug-shaped nests, which, in turn, favored the evolution of high-density, colony nesting behavior of species such as the Cliff Swallow.

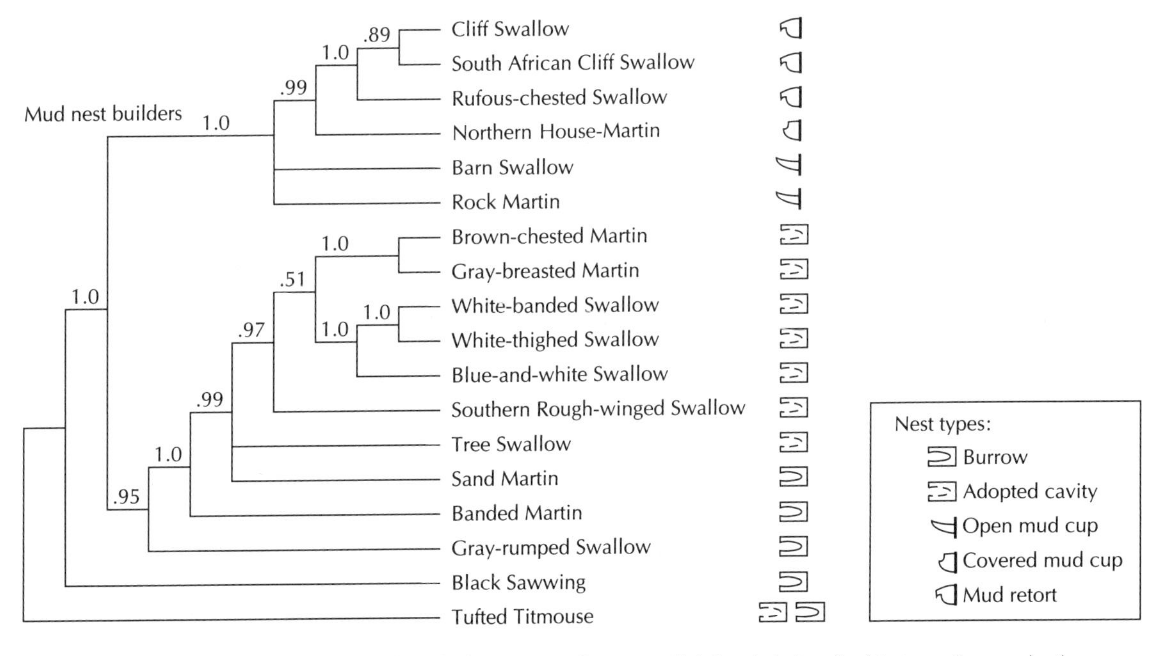

A phylogeny of 17 swallow species with their typical nests. The phylogeny was constructed using DNA hybridization techniques and the cavity nesting Tufted Titmouse as an external reference—or outgroup. The mud nesting species form a well-defined clade of related species; so do the burrow nesting and cavity nesting species, in which the latter apparently evolved from the former.

Red-dead Nettle and Yarrow, which contain volatile chemical compounds that inhibit the growth of bacteria and the hatching of the eggs of arthropod nest parasites. Experimental removal of these green plant materials leads to a dramatic increase in the populations of bloodsucking mites, tiny parasites that each day can drain 20 percent or more of the blood volume of a starling chick (Clark 1991).

Both inorganic materials, including mud pellets, rocks, tinfoil and ribbons, and animal products are used in nest construction. Some birds use spiderwebs for mooring or adhesion, feathers and hairs for the final lining, and snake skins for external embellishment. In one remarkable example from South Africa, Black-eared Sparrow-Larks adorn the edges of their nests with the scarlet-colored lids that cover the burrows of a particular species of trapdoor spider (P. Ryan and W.R.J. Dean, personal communication). The geographical distribution of this sparrow-lark coincides closely with that of the spider. The nests of Long-tailed Tits and Common Goldcrests in Europe may contain 2000 or more feathers. Waterfowl pluck down from their own breasts, and the Superb Lyrebird plucks down from its flanks to line the nest. Birds go to extremes to get prime materials, which may be in short supply. Thievery is common, especially in large seabird, heron, and penguin colonies; it is often much easier to steal than to collect fresh materials. Many birds pluck hair, a prized nest-lining material, from livestock. Galápagos Mockingbirds occasionally snatch hair from the heads of surprised tourists.

Nest Safety

Predation on nests and their contents, including the incubating parents, severely reduces breeding success—more chicks may leave the nest via the stomach of a predator than on their own volition. Predation accounted for 88 percent of nest losses in deciduous scrub vegetation in Indiana (Nolan 1963), 75 percent of nest losses in Britain (Lack 1954), and 86 percent of nest losses in the tropical White-bearded Manakin (Snow 1962). Nest predation is a powerful force that influences nest architecture and placement (Skutch 1985), the evolution of life history traits such as clutch size (Martin 1988a), and possibly also local patterns of coexistence of different species, which must each find safe nest sites (Martin 1988b; Ricklefs 1989b).

Invisibility, inaccessibility, and impregnability all contribute to nest safety (Skutch 1976). The camouflage of incubating nightjars and of shorebird eggs renders them nearly invisible, as do the lichen decorations on the sides of a hummingbird's nest. Cryptic sites in dense clumps of grass, vine tangles, or hidden crevices also minimize the chance of discovery. Seabirds that nest on sheer cliffs and swifts that nest in deep caves or behind waterfalls achieve safety through inaccessibility. Horned Coots pile up stones in the middle of high Andean lakes to build their own

Box 16–2
Kittiwakes adapt to cliff nesting

Most gulls nest on the ground, where they are vulnerable to predation by other gulls, crows, and mammals. Black-legged Kittiwakes, however, nest on narrow, predator-free ledges on windswept, seaside cliffs. The selection of these relatively safe nest sites broadly molds the behavior and morphology of the kittiwake. They cling to their safe nest sites, using strong claws and toe muscles (see figure). Reduced nest predation has also fostered the loss of antipredator behaviors evident in other gulls, including alarm calls, predator mobbing, and removal of eggshells from the nest site. Young kittiwakes, which are a conspicuous silvery white, stand still and hide their beaks when frightened, rather than running and hiding as the cryptically colored chicks of ground nesting gulls do.

The physical restrictions of a kittiwake's narrow nesting ledge have also favored aggressive and courtship displays that differ from those of other gulls. Instead of using the long call (see Chapter 9), for instance, kittiwakes announce territorial ownership with a modest choking display. Aggression between males is expressed by bill jabs from a fixed position; it does not extend to the flamboyant charges of other species. Females commonly hide their beaks to minimize the likelihood of attack and physical displacement from the narrow ledge. Courting males do not regurgitate food onto the ground in front of the female (there is no place to do so) but instead give it to her directly.

Unlike most other gulls, Black-legged Kittiwakes nest on cliffs. (Courtesy of E. Hosking)

nesting islands out of reach of terrestrial predators (Figure 16–2), and many grebes build nests of floating vegetation.

Nests on the ground are more vulnerable to mammalian predators than nests in trees or bushes. In Oklahoma, 71 percent of 130 Mourning Dove nests placed on the ground were destroyed, whereas only 51 percent of 167 dove nests in trees in the same area were destroyed (Downing 1959). Ospreys and American Robins usually nest in trees, but on Gardiners Island, New York, where there are no predators, these species nest on the ground. Conversely, Tooth-billed Pigeons once nested on the ground on Samoa, but they shifted to tree nesting after cats were introduced to this South Pacific island by whalers (Austin and Singer 1985).

Many birds build enclosed or pensile nests to discourage predators. Tropical passerines build globular or enclosed nests, often with an entrance tube on the sides (Figure 16–1H, M). Eggs in a covered nest are

(A)

(B)

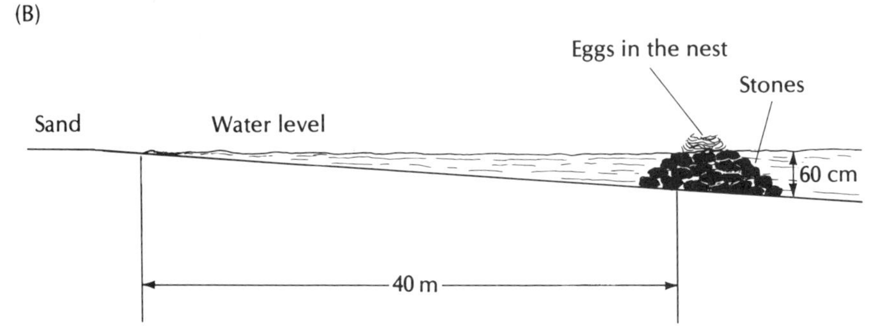

Figure 16–2 Horned Coots build their nests on stones, which they assemble in piles in high-altitude Andean lakes. (A) Horned Coot on nest; (B) diagram of nest structure. (A courtesy of P. Canevari/VIREO; B after Ripley 1957)

less visible to potential predators when the parent is absent than are eggs in an open nest. Snakes cannot easily reach pensile nests or easily crawl inside protruding entrance tubes.

Cavity nesting is safer than open nesting. Half of the avian orders, among them all parrots, trogons, and kingfishers and their relatives, nest in cavities or holes. Owls, parrots, and Australian frogmouths nest in natural cavities; trogons, titmice, and piculets excavate cavities in the soft or rotten wood of old trees. Kingfishers, small petrels, and some penguins dig burrows into the ground. The most highly developed cavity nesting is found among woodpeckers, which chisel cavities into living hardwoods. Other birds unable to make their own holes compete intensely for abandoned woodpecker holes. Indeed, the availability of nest holes limits the population sizes of some species (see Chapter 21).

Some birds nest in an area protected by large animals or stinging insects. European Starlings and House Sparrows nest on the fringes of Imperial Eagle aeries. The Eurasian Thick-knee, a ploverlike bird of Africa, nests

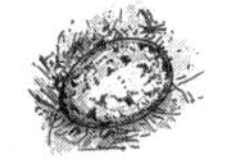

Box 16–3
Forest fragmentation increases nest predation

David Wilcove (1985) studied the patterns of nest predation by placing artificial nests with fresh quail eggs in forests of different sizes in Maryland and Tennessee in the eastern United States. Nests in small woodlots suffered more predation than those placed in large forest tracts. Predation was much more intense in woodlots near suburban neighborhoods than in isolated, rural woodlots. Open-cup nests were more vulnerable to predators when placed on the ground than when placed 1 to 2 meters above the ground; and open-cup nests were generally more vulnerable than experimental cavity nests. These experiments suggest that migratory songbirds, many of which construct open-cup nests, sometimes on the ground, will suffer higher predation rates in the small forest tracts on which they must rely as a result of forest fragmentation throughout their breeding ranges.

on sandy shores beside nesting crocodiles. Bananaquits of the West Indies reduce rates of nest predation by nesting in association with wasps (Wunderle and Pollock 1985). Descriptions of such associations are widespread. In one experimental study of Rufous-naped Wrens in Costa Rica, Frank Joyce (1993) demonstrated that the presence of wasp nests significantly reduced nest losses to monkeys, the primary predator. He did this by relocating wasp nests to new associations with previously unprotected nests. Tropical birds often associate with termites. At least 49 species, including 25 percent of all kingfishers, excavate nest cavities inside the mounds—called termitaries—of social termites (Hindwood 1959). The Orange-fronted Parakeet of Central America nests exclusively in the termitary of one species, *Eutermis nigriceps*. The geographical distribution of this parakeet is restricted to that of its termite host.

Birds rely on cryptic coloration in combination with distracting and surreptitious behavior to thwart predators. Many incubating parents avoid discovery by not flushing until the last possible moment, relying on their own camouflage and that of the nest for protection. Female Common Eiders, for example, are camouflaged from Arctic Foxes and weasels by their finely patterned plumage. If discovered, however, they defecate a noxious repellent fluid over their eggs as they flush (Swennen 1968). Many birds directly attack trespassers. Eastern Kingbirds chase anything that violates nearby airspace. Northern Mockingbirds, Blue Jays, and Arctic Terns can draw blood and bits of fur from cats that come too close to their nests or young. They may attack people as well. Large owls and eagles with powerful feet and sharp talons can seriously wound approaching climbers.

A parent flushed from the nest may distract a predator's attention away from the nest site with distraction displays. The two most common displays are the injury flight and the rodent run (Simmons 1952, 1955; Brown 1962) (Figure 16–3). Using the injury-flight display—feigning a broken wing and calling in great alarm—an adult sandpiper can easily

(A)

(B)

Figure 16–3 Distraction displays. (A) By feigning injury, a Common Nighthawk distracts predators and thus protects its young. (B) The rodent-run display (bottom) in the Tasmanian Native-Hen contrasts with its normal upright walking posture (top drawing). (A courtesy of S.A. Grines; B from Ridpath 1972)

draw a fox away from its nest. To keep the fox's attention, the sandpiper may then switch to the rodent-run display—running in a low crouch—an action that appeals to the mouse-catching instincts of the fox. Distraction displays are risky, but more often than not the parent escapes and the predator loses track of the original nest location.

To avoid attracting attention when returning to the nest after a recess, parents adopt surreptitious behavior. Meadowlarks land some distance from the nest and sneak back to it through the grass, using one of several indirect routes. Bearded Parrotbills pretend to look for food as they get near their nests and then enter rapidly if they perceive that the coast is clear. The female Long-tailed Hermit, a tropical hummingbird, behaves similarly. Upon returning from foraging, she searches intensively for spiders on the buttresses of large trees before quickly slipping onto her nest and sitting very still.

Colonial Nesting

About 13 percent of bird species, including most seabirds, nest in colonies. Colonial nesting evolves in response to a combination of two environmental conditions: (1) a shortage of nesting sites that are safe from predators and (2) abundant or unpredictable food that is distant from safe nest sites (Siegel-Causey and Kharitonov 1990). Colonial nesting has both advantages and disadvantages. First and foremost, individuals are safer in colonies that are inaccessible to predators. Also, colonial birds detect predators more quickly than small groups or pairs and can drive them from the vicinity of the nesting area; in one classic example, the effectiveness with which Common Black-headed Gulls mobbed predators increased with the number of participants (Kruuk 1964). Because nests at the edges of breeding colonies are more vulnerable to predators than those in the centers, the preference for advantageous central sites promotes dense centralized packing of nests, even in ample areas.

Coordinated social interactions tend to be weak in the initial evolutionary stages of colony formation, but true colonies provide extra benefits. Synchronized nesting, for example, produces a sudden abundance of eggs and chicks that exceeds the daily needs of local predators. Also, colonial neighbors can improve their foraging by watching others. This behavior is especially valuable when the off-site food supplies are restricted or variable in location, as are swarms of aerial insects harvested by swallows or schools of small fish harvested by seabirds. The colonies of Cliff Swallows, for example, serve as information centers in which unsuccessful individuals locate and then follow successful neighbors to good feeding sites (Brown 1988). As a result of their enhanced foraging efficiency, parent Cliff Swallows in large colonies returned with food for their nestlings more often and brought more food each trip than did parents in small colonies.

To support large congregations of birds, suitable colony sites must be near rich, clumped food supplies. Colonies of Pinyon Jays and Red Crossbills settle near seed-rich conifer forests, and Wattled Starlings nest in large colonies near locust outbreaks. The huge colonies of Guanay Cormorants and other seabirds that nest on the coast of Peru depend on the productive cold waters of the Humboldt Current. The combination of abundant food in the Humboldt Current and the vastness of oceanic habitat can support enormous populations of seabirds, which concentrate at the few available nesting locations. The populations crash when their food supplies decline during El Niño years.

On the negative side of the balance sheet, colonial nesting leads to increased competition for nest sites, stealing of nest materials, increased physical interference, and increased competition for mates. In spite of food abundance, large colonies sometimes exhaust their local food supplies and abandon their nests (Payne 1969b; Brown and Urban 1969; Johnson and Sloan 1978; Jones and Ward 1979). Large groups may actually attract predators, especially aerial, avian predators, or may facilitate the spread of parasites and diseases. High densities of ticks cause cormorants, boobies, and pelicans to desert their nests in the huge seabird colonies on the coast of Peru (Duffy 1983). The globular mud nests in large colonies of the Cliff Swallow are more likely to be infested by fleas or other bloodsucking parasites than are nests in small colonies (Brown and Brown 1986). Experiments in which some burrows were fumigated showed that these parasites lowered survivorship by up to 50 percent in large colonies, but not significantly in small ones. The Cliff Swallows inspect and then select parasite-free nests, or, in large colonies, they tend to build new nests rather than use old, infested ones. On balance, however, the advantages of colonial nesting outweigh such disadvantages.

Box 16-4
Coloniality in Yellow-rumped Caciques reduces predation

The Yellow-rumped Cacique nests in colonies in Amazonian Peru. These tropical blackbirds defend their closed, pouchlike nests against predators in three ways (Robinson 1985). First, by nesting on islands and near wasp nests, caciques are safe from arboreal mammals such as primates, which destroy more accessible colonies of other birds. Caimans and otters also protect the island colonies by eating snakes that try to cross the open water surrounding the colony. Second, caciques mob predators as a group. The effectiveness of mobbing increases with group size, which increases with colony size. Third, caciques hide their nests from predators by mixing active nests with abandoned nests. Overall, nests in clusters on islands and near wasp nests suffer the least predation. Females switch colonies after losing nests to a predator, usually to sites that offer better protection against that predator. By such mechanisms, the best colony sites accumulate the largest numbers of nests.

Nest Building

A nest may be built by either member of a pair of birds, or it may be built jointly during courtship and pair formation. Nest site selection, accompanied by displays, is an integral component of pair formation. Wrens and weavers construct nests for evaluation by prospective mates. If a prospective mate rejects the nest, Village Weavers tear it down and build a new one. Male Marsh Wrens may build more than 20 nests for comparison by prospective mates; bigamous males build an average of 24.9 nests, monogamous males 22.1, and unmated males 17.4 (Verner and Engelson 1970). As in the tropical caciques (Box 16–4), unused nests serve as dummy nests that help to confuse nest predators.

Nest building itself can be strenuous. Female Great Tits and Blue Tits devote 1 to 3 hours a day to building their nests, time that could be devoted to feeding for self-maintenance and egg production. Northern House-Martins that build new nests lay one less egg than neighbors that recondition old nests, apparently because they devote metabolic reserves to nest building instead of to egg production (Lind 1964). Most monogamous male North American passerines contribute to the nest-building effort (Verner and Willson 1969). A male's presence at the nest site in the earliest stages of nesting, however, may be primarily to protect his female from insemination by other males (to guard his paternity).

(A)

(B)

Figure 16–4 Lovebirds carrying nest strips. (A) The Rosy-faced Lovebird tucks them into its rump feathers, whereas the Yellow-collared Lovebird (not shown) carries them in its bill. (B) Hybrids of these two species try to tuck strips but usually fail. (From Dilger 1962, with permission of Scientific American)

Nest-Building Behavior

The behavior associated with nest building varies from the simple accumulation of materials to elaborate construction. The nonincubating parent may simply toss materials in the direction of the nest site, thereby creating a mound of debris or a conspicuous rim near the eggs and leaving the incubating parent to delineate the nest site by drawing the materials toward itself. Deliberate transport of suitable materials to the nest site was a major step in the evolution of nest-building behavior among birds (Collias and Collias 1984); it led to the modification and design of the nest site and to more complex nest architecture.

Birds usually carry nest materials in their bills or feet. Some lovebirds, which are small African parrots, transport their nest materials in an unusual way that apparently is genetically determined (Dilger 1962). The Yellow-collared Lovebird carries one strip of nesting material at a time in its bill, but the related Rosy-faced Lovebird tucks the ends of several strips beneath its rump feathers and flies to the nest with the strips in tow (Figure 16–4). Hybrids between these two species try to tuck strips into their rump feathers but cannot do so correctly. Sometimes the hybrids fail to complete the tuck; more often they hold the strip by the middle instead of the end, fail to let go of the strip after tucking it, or tuck it into the wrong place; many strips do not reach the nest box. The hybrid's genetic program for carrying nesting material apparently contains conflicting instructions.

Bills and feet are the essential nest-building tools. Bills serve as wood chisels and drills, as picks for digging into the ground, as shuttles for weaving, as needles for sewing, as trowels for plastering, and as forceps for plucking and inserting (Skutch 1976). In general, bill structures have evolved toward efficiency in food gathering (Chapter 7) rather than nest construction, although the Finch-billed Myna may be an exception; its heavy bill seems unnecessary for consuming the soft fruit it eats but is essential for digging its nest into the trunks of old trees (Gilliard 1958). Nests are also built by stamping, scraping, kneading, and scratching as reptilian ancestors did. Burrow nesters dig by kicking loose soil backward; they then mold the internal nest dimensions by using their bills, breasts, and feet.

The cup nests of small arboreal land birds are usually built from the bottom up. Others, such as the open-cup nests (suspended by the rim) of vireos, are built by wrapping nest materials around the supporting twigs first and then by looping strands of material from side to side to form the framework of the cup. The long, hanging nests of tropical flycatchers, such as the Sulphur-rumped Flycatcher, begin as an accumulation of materials stuffed into a tangled mass (Figure 16–5). The flycatcher forces its

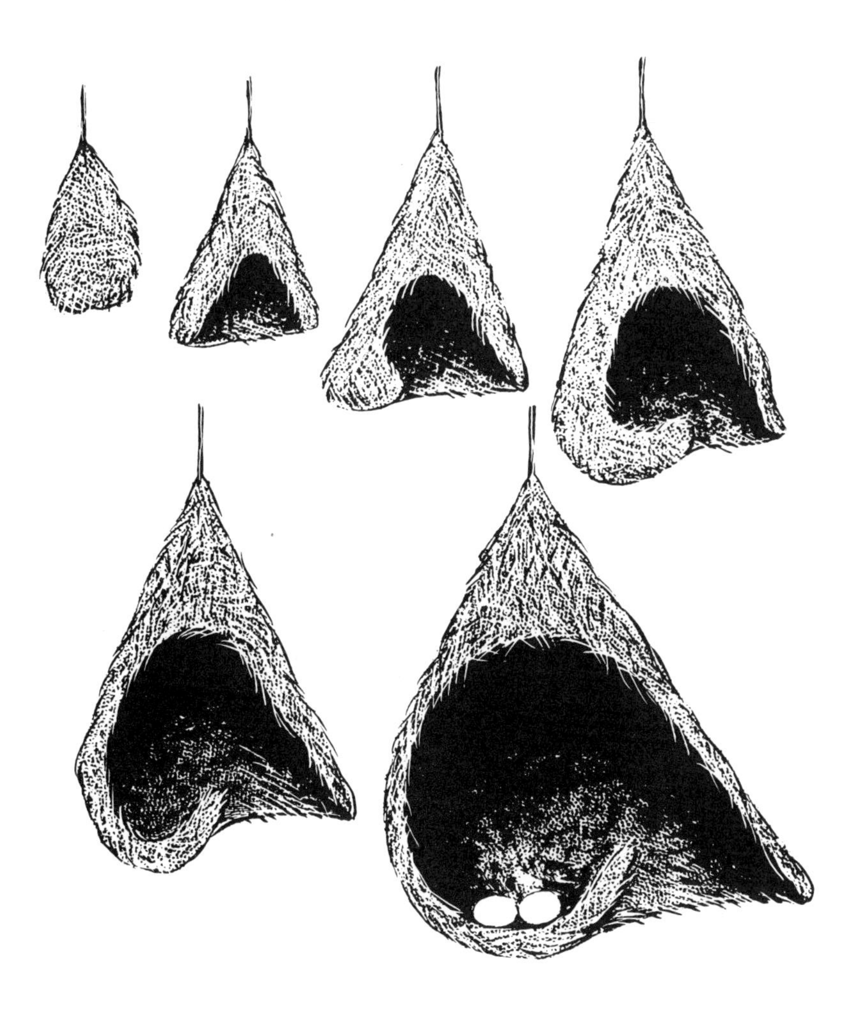

Figure 16–5 Stages of nest building by Sulphur-rumped Flycatchers. (From Skutch 1960, 1976)

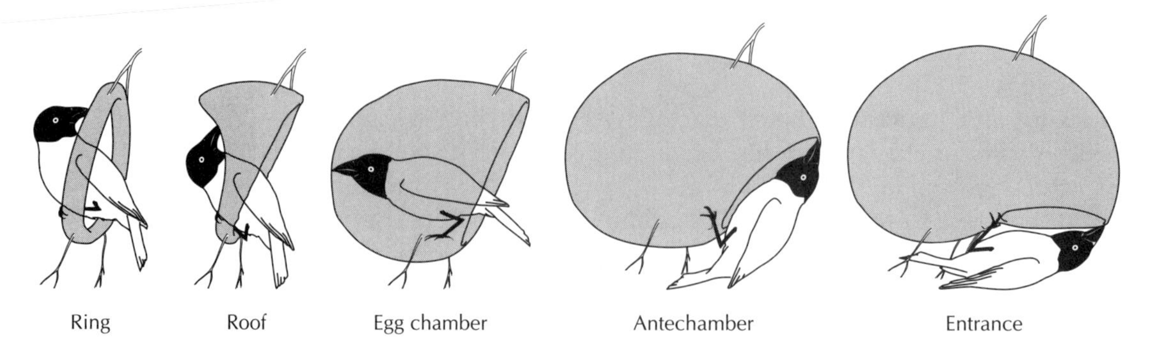

Figure 16–6 Stages of nest construction by the male Village Weaver. (Adapted from Collias and Collias 1964)

way into the center and gradually expands the nest cavity from the inside out; then it reinforces and lines the hollowed-out cavity (Skutch 1976).

Weavers and New World orioles weave elaborate hanging nests. The male Village Weaver, for example, begins with a vertical ring, to which it adds in succession a roof, the walls of the main nest chamber, an antechamber, and finally the finished entrance (Figure 16–6). The structural features of these nests are woven into their final positions using special knots. The types of knots used are species specific. Some weavers tie simple knots; others tie half hitches and slipknots (Figure 16–7).

Role of Experience

Most passerine birds build nests with architectural features so typical that we can identify the builder to genus or species. How does a young bird know how to build a complex nest similar to that built by its parents? A male Village Weaver, hand-raised in isolation without ever seeing a nest, can build a nest that is typical of its species, an ability suggesting a strong genetic control of this behavior. Early experiences also play a role. Improvement in nest construction is particularly evident in species that build elaborate nests, such as the Village Weaver (Collias and Collias 1964). Although immature males build crude structures at first, they become more skilled in the arts of knot tying and weaving. Older males can build refined products. When nesting for the first time, the Eurasian Jackdaw, a small European crow, rapidly improves its skills from clumsy movements with inappropriate nest materials to efficient construction with a range of suitable nest materials. At first, the inexperienced young jackdaw tries to shove almost anything into the nest platform. Sticks of the right size and texture insert easily and firmly into the matrix, but objects such as light bulbs do not. By the time the nest is complete, the range of materials gathered has narrowed to the types of twigs that are most suitable for nest construction (Lorenz 1969).

Raptors imprint on their natal nest sites; consequently, they choose a similar situation several years later when they reach maturity (Temple

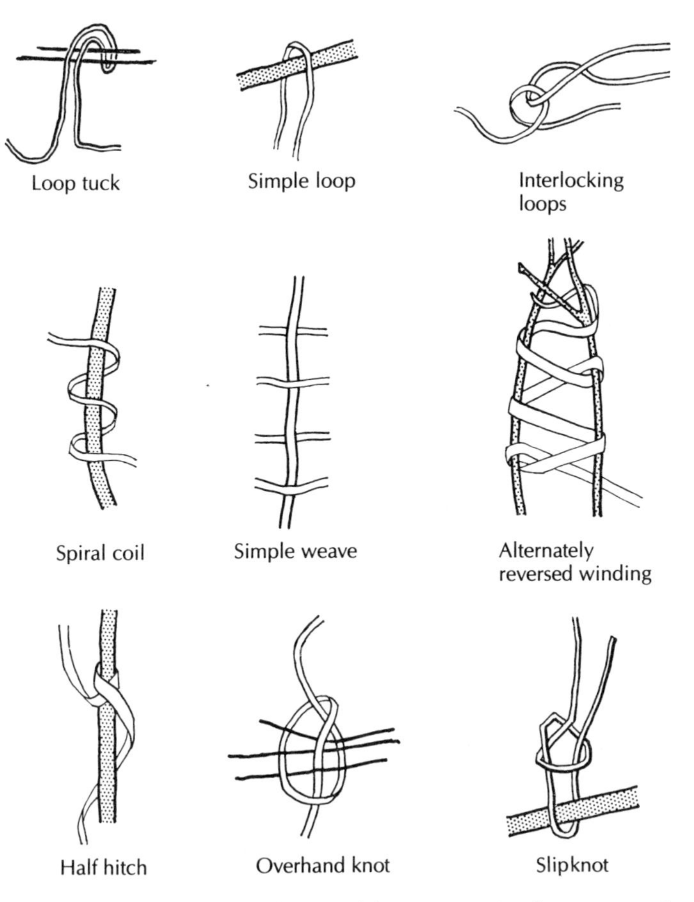

Figure 16–7 Some knots and stitches used by weavers in the construction of their nests. (From Collias and Collias 1964)

1977). This ability has proved important to the conservation of endangered species. The Mauritius Kestrel, for example, nested in tree cavities that became vulnerable to predation by introduced monkeys, causing the kestrel population to decline to only a few pairs in the 1960s. One pair of kestrels then nested on a cliff ledge, out of reach of the monkeys, and started a new tradition that provides hope for the survival of the species. Similarly, the Peregrine Falcon reestablishment program relies on the tendency of young Peregrines to imprint on special nest boxes for future breeding (Chapters 18 and 24).

Nest Microclimates

The microclimate of a nest is crucial to the successful incubation of the eggs and to the health of baby birds. Nest microclimate also influences the daily energy requirements of the adult and the amount of time they must spend on the nest incubating eggs and brooding young. Nest warmth, like that of a house, is determined by the thickness of insulation (Skowron and Kern 1980). In addition to keeping the eggs warm, nest insulation reduces the energy requirements of the incubating parent (Walsberg and King 1978). The outstanding insulating properties of breast down used by eiders and other waterfowl greatly reduce the cooling rate of eggs not covered by the parent. The eggs of Mallard ducks cool twice as fast uncovered as when they are buried in down. Experimental reduction of the lining thickness of a Village Weaver nest results in a corresponding increase in the amount of time the female stays on the nest to keep the eggs warm and, thereby, a reduction in the time she can feed (White and Kinney 1974). The weavers adjust the amount of nest insulation in relation to need, increasing it with the altitude of their nests (Collias and Collias 1971). Seasonal differences in the thermal conductance of songbird nests result from variations in the tightness of the weave (Skowron and Kern 1980).

Nest placement in or out of the sun, shade, or wind has a major effect on the nest microclimate and, therefore, on a pair's breeding success. Early in the season in Arizona, Cactus Wrens build nests where they are protected from cool winds and are bathed in the warm morning sun. Later in the season, when it is hot, they build well-shaded nests that are exposed to cooling breezes (Ricklefs and Hainsworth 1969; Austin 1976). The enormous communal nests of the Sociable Weaver have great thermal inertia, which keeps them cool in the daytime and warm at night. Temperatures inside the nest at night remain 18° to 23°C above external temperatures, owing in part to heat absorbed during the day and in part to heat generated from the bodies of large numbers of roosting birds. F.N. White and his colleagues (1975) estimated that communal roosting by Sociable Weavers reduces their metabolic costs by 43 percent.

Thick nest insulation and the careful selection of sites reduce heat loss by incubating hummingbirds. Their energy requirements can drop 13 percent as a result of a minor 0.05-centimeter increase in nest thickness (Smith et al. 1974). Heat loss from the exposed upper surface of the hummingbird's body can be substantial, especially at night. Consequently, hummingbirds choose nest sites beneath branches, a location that reduces heat loss to the air by half (Calder 1974; Southwick and Gates 1975). Careful selection of nest sites to reduce heat loss is especially important in species such as the tiny Calliope Hummingbird, which nests at high elevations in the Rocky Mountains, where night temperatures are cool (Calder 1971).

The placement of nests in cavities and burrows also conserves energy. Like the haystack nests of Sociable Weavers, cavity nests and burrow nests buffer eggs, parents, and young against fluctuations in external temperatures. The temperatures inside the burrows of Jackass Penguins, for example, stayed between 17° and 20°C despite an outside temperature range of 12° to 36°C; the burrow nests of European Bee-eaters remained close to 25°C despite an outside range of 13° to 51°C (White et al. 1978).

Deep, cool burrow nests have their drawbacks, however. Poor ventilation limits the amount of time parents can spend inside with growing young. Diffusion of gases through the soil and the nest tunnel usually keeps the air in the nest chamber adequately ventilated. The movements of adults in and out of the nest pump air much as a moving piston would. On windless days, however, ammonia and carbon dioxide tend to build up as a result of decay of excreta amid unsanitary nest conditions, and oxygen levels occasionally decline until the occupants have difficulty breathing.

Incubation

No bird incubates its eggs internally as do animals that bear live young. Consequently, the narrow temperature tolerances of embryos inside the egg commit parents to rigorous external incubation patterns.

The incubation behavior of birds is mediated by the hormone prolactin (Goldsmith 1991). The levels of this hormone circulating in the blood are high during the incubation period (and also brooding of chicks) in both free-living and captive individuals of many bird species. The incubation roles of males and females are mirrored by the circulating levels of prolactin—that is, where one sex contributes most of the parental care, it has relatively high levels of prolactin. Conversely, the sex hormone testosterone, which mediates aggressive and sexual behavior, inhibits the expression of parental behavior in birds (Wingfield 1991). Blood levels of testosterone in male birds that incubate drop sharply once egg laying begins (see Figure 17–7).

Most birds delay the onset of incubation until the clutch is complete. This behavior ensures that the embryos begin to develop and later hatch at the same time even though some eggs are laid earlier than others. Pigeons and doves, for example, sit on the first egg before the second is laid but do not bring it up to the temperatures required for incubation. Owls and raptors, on the other hand, begin incubation before the clutch is complete, with the result that young hatch asynchronously at intervals.

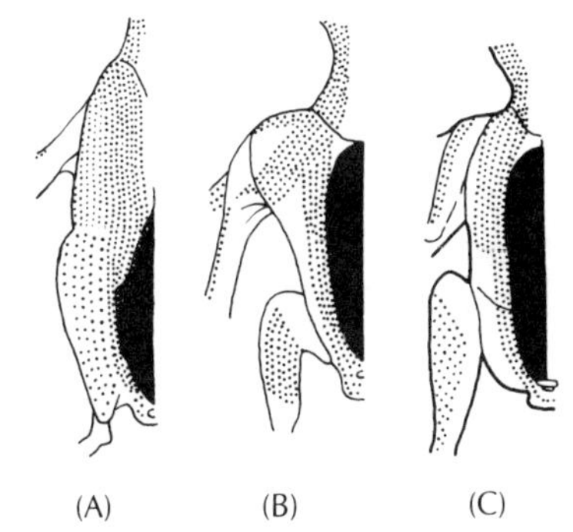

Figure 16–8 Incubation patches (in black) of (A) a grebe, (B) a hawk, and (C) a passerine bird. Stippling indicates feather tracts. Clear areas indicate areas without feathers—called apteria. (Adapted from Drent 1975)

Brood Patches

Birds transfer body heat to their eggs through brood patches—or incubation patches—which are bare, flaccid sections of skin on the abdomen or breast. This area may be a single median patch, as in most passerines and pigeons, or two lateral patches, as in most shorebirds, gulls, and quails (Figure 16–8). Most birds lose the feathers to form an incubation patch for the purpose of brooding; pigeons and doves use a normally bare apterium—or featherless region (Chapter 4). The accumulation of fluids—edema—and the infiltration of white blood cells swell and soften the skin, a condition that allows better contact between the surfaces of the incubation patch and the egg. The epidermis itself thickens into a callused surface that is not damaged by sustained contact or friction with the eggs. Finally, blood vessels, which deliver body heat to the eggs, proliferate throughout the patch. The arterioles in the network of blood vessels have well-developed musculature that directs the flow of warm blood to the skin surface during incubation and stops it when the parent is not actively incubating.

Incubation patches develop just before the incubation period and regress after hatching. If both parents incubate, then the patches occur in both sexes; otherwise, individuals of the nonincubating sex usually have the potential for brood patch development in the event that they should have to incubate for some unusual reason, such as a mate's death.

Seasonal development of incubation patches is under direct hormonal control (Bailey 1952). Prolactin or estrogen, or both, depending on the species, stimulate defeathering and vascularization of the incubation patch. Progesterone stimulates thickening and increased sensitivity of the epidermis. Most birds develop brood patches in response to experimental hormone treatment, except brood parasites, such as Brown-headed Cowbirds, which never incubate (Selander and Kuich 1963).

Although incubation patches are the typical mode of heat transfer between parent and eggs, some birds (waterfowl, penguins, and pelicans) lack them. By plucking down from the breast, waterfowl at once produce an insulated nest lining and barer skin regions. Gannets and boobies lack a brood patch and, instead, incubate with their feet. They may grasp a single egg in their well-vascularized, webbed feet or even hold two eggs, one in each foot. Murres and penguins incubate their eggs on the top surfaces of their feet. Some penguins have a muscular pouch of belly skin that holds the single egg in this position.

Keeping Eggs Warm

The first priority of incubation is to keep the eggs close to the optimum temperature for development, that is, 37° to 38°C. Serious problems result if the embryo is exposed to temperatures outside the range of 35° to 40.5°C (Dawson 1984). Exposure to higher temperatures is lethal, and even a short exposure to cool temperatures between 26° and 35°C can disrupt normal development. Below 26°C, the development of young embryos simply stops. Thus, frequent or continuous warming is necessary unless ambient air temperatures are hot.

Internal egg temperatures are low at first, but they increase steadily during incubation as a result of parental incubation, further incubation patch development, and heat generated internally by the metabolism of the growing embryo. The temperature increase has been demonstrated by regularly measuring the temperature of naturally incubated eggs of the Herring Gull (Figure 16–9).

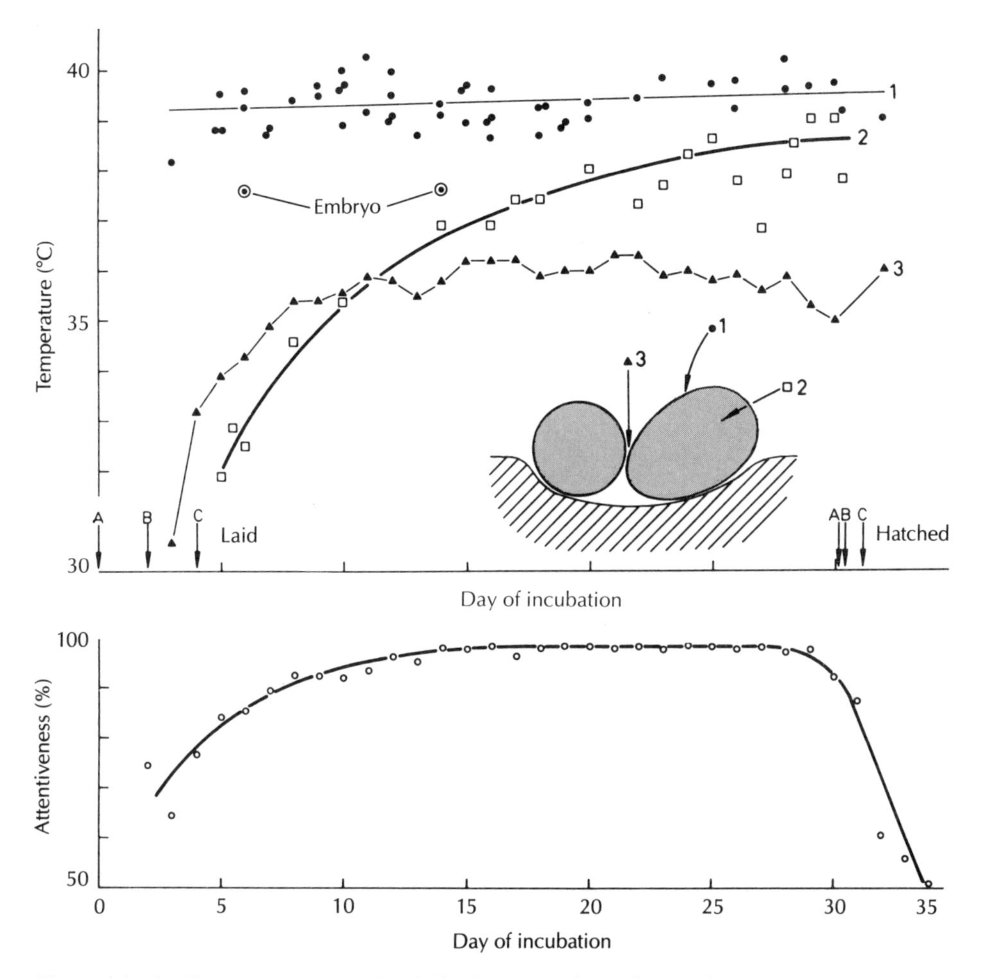

Figure 16–9 Egg temperatures (top) during natural incubation by a Herring Gull. Sites of measurement included (1) the egg surface, (2) inside the egg, and (3) air temperature between eggs A, B, and C. Points labeled "Embryo" indicate measurements taken beside the embryo on days 6 and 14. The constancy of incubation (attentiveness) of adults increased steadily during the first 2 weeks of incubation (bottom). (Adapted from Drent 1975)

Incubating parents are able to keep the internal temperatures of their eggs remarkably stable, despite the fact that incubation behavior itself comprises many conflicting options. Central to the question of egg temperature regulation is the pattern of attentiveness—or incubation sessions versus incubation recesses (Kendeigh 1952; Haartman 1956; Drent 1975; Haftorn 1978a, b). The natural incubation rhythm of a species (Figure 16–10) is geared directly to the maintenance of critical egg temperatures. At cooler air temperatures, sessions on the eggs are longer and recesses for food and drink are shorter. As was shown in experiments, increases in the air temperature inside the nest boxes of European Pied Flycatchers cause the adults to shorten their sessions on the eggs (Haartman 1956).

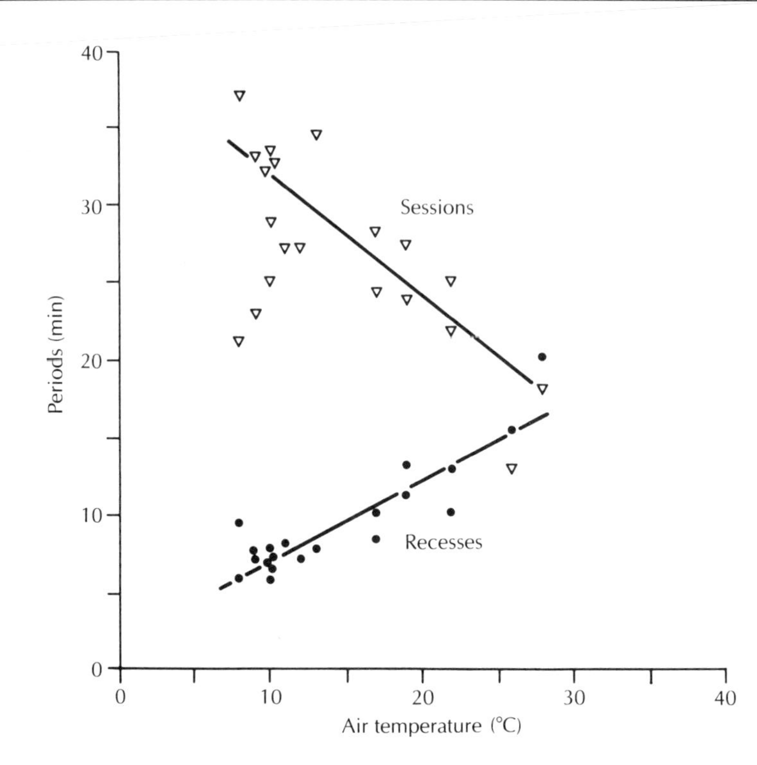

Figure 16–10 Incubation rhythms of the Great Tit are directly related to the air temperature in the nest box. Time on the eggs (sessions) increases and time off the eggs (recesses) decreases when the air is cooler. (Adapted from Drent 1972; Kluijver 1950)

Similarly, raising or lowering the temperature of artificial eggs placed in the nests of Ringed Turtle-Doves and Savannah Sparrows decreases or increases the length of the incubation sessions, respectively (Franks 1967; Davis et al. 1984).

Experiments with Crested Mynas and European Starlings on Vancouver Island, British Columbia, demonstrate the effects of inadequate incubation behavior (Figure 16–11). Crested Mynas, a type of starling

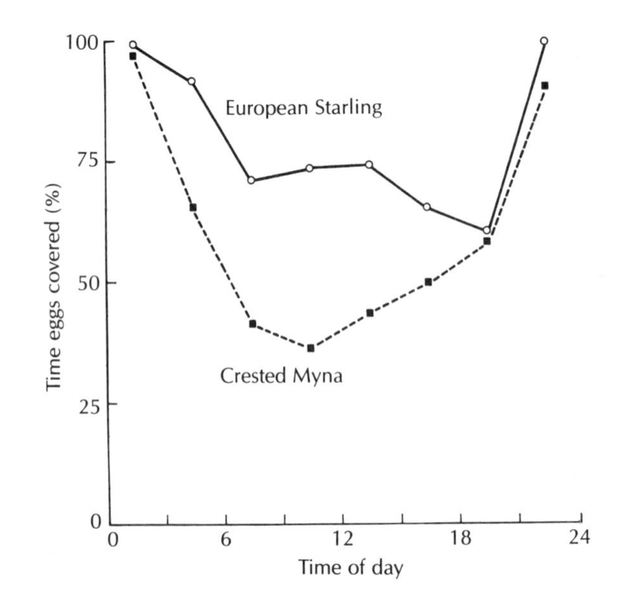

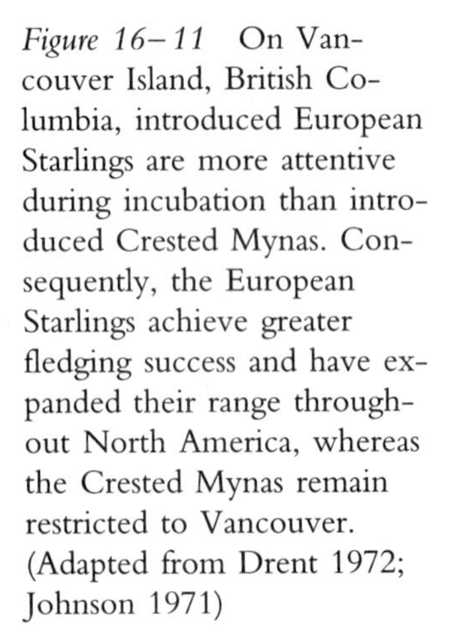

Figure 16–11 On Vancouver Island, British Columbia, introduced European Starlings are more attentive during incubation than introduced Crested Mynas. Consequently, the European Starlings achieve greater fledging success and have expanded their range throughout North America, whereas the Crested Mynas remain restricted to Vancouver. (Adapted from Drent 1972; Johnson 1971)

Box 16–5
Moundbuilders use natural heat

The megapodes—or moundbuilders—of Australasia are fowllike birds that use heat from the sun, volcanic steam, or decomposing vegetation to incubate their eggs. Reptilian as it may seem, this behavior appears to have evolved secondarily from normal avian incubation behavior. Some species, most notably the Malleefowl, regulate the temperatures of the nest heap with great sensitivity (Frith 1962). The Malleefowl hen lays her eggs in a large mound, up to 11 meters in diameter and 5 meters high, made of decaying vegetation and sand, not unlike a compost heap (see figure). The male, which tends the mound alone for 10 to 11 months a year, spends an average of 5 hours a day digging to manipulate the amount of material covering the eggs.

Incubation temperatures inside the mound remain at 32° to 35°C as external air temperatures range from 0° to 38°C. The male regularly checks the temperature inside the mound by testing the soil in his mouth. In the spring and summer, an incubating male must cool the mound, which he does by opening it to release accumulated heat and by replacing hot sand with cooler sand. Summer nests are placed in deep pits, which protect the eggs from the hot sun. Heating the eggs in the fall becomes more difficult because there is less sun and less decay. The pit for the eggs, therefore, is a shallow one that takes advantage of daytime solar heating. At night, the male adds extra insulation to seal in the heat.

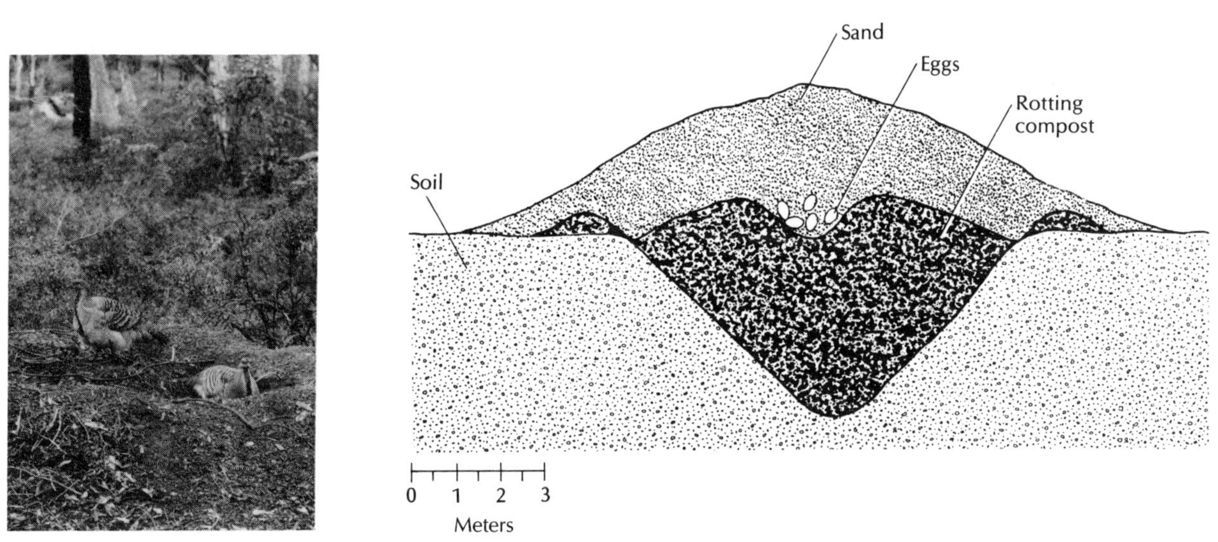

(Left) Moundbuilders do not incubate their eggs directly but regulate the temperatures inside their nests by varying the rate of natural heat loss and gain. (Right) Cross section of incubation mound with eggs. Underneath the egg chamber is a pit full of decaying vegetation. Sandy soil covers the eggs. (Photograph courtesy of J. Warham; drawing after Frith 1962, with permission of Scientific American)

introduced to Vancouver from Hong Kong, hatch and fledge young from only 38 percent of their eggs because they persist with an incubation rhythm that is suitable for the tropical climates of their native Hong Kong but unsuitable for the cool Vancouver climate (Johnson 1971). Unlike

Great Tits, they do not regulate incubation time by air temperature. However, they hatch and fledge more young when their nest boxes are heated artificially. A low rate of reproduction is part of the reason Crested Mynas remain restricted to the vicinity of Vancouver and are declining in number. The related European Starling, in contrast, expanded its range rapidly after its introduction to North America. This starling is more attentive during incubation and therefore fledges more young (68 percent). Crested Myna eggs that are incubated by the European Starlings usually hatch, thus showing that the normal incubation pattern of Crested Mynas, not the quality of their eggs, is at fault.

Keeping Eggs Cool

Birds that nest in hot places face demanding incubation challenges of a different nature. The temperatures of unprotected eggs quickly rise to lethal levels. Just leaving the nest to chase predatory gulls caused a Forster's Tern's egg temperature to rise to 46°C in 10 minutes, and in a separate response to a shorebird's alarm call to 50°C during a 25-minute absence (Grant 1982). Shading of eggs, therefore, is a critical part of incubation behavior. Gray Gulls that nest in the deserts of northern Chile incubate their eggs at night, when it is quite cold, but shade them during the day, when air temperatures are 38° to 39°C (Howell et al. 1974).

Wetting the nest or eggs, which counteracts extreme heat with evaporative cooling, is a common practice among shorebirds, gulls, and terns (Grant 1982). Killdeer, for example, cool their eggs by transferring water from wet belly feathers (Schardien and Jackson 1979). The Crocodile-bird, an Egyptian plover that nests on hot sandbars of the Nile River, cools its eggs by covering them with a thin layer of sand and then sprinkling water on top. The nest temperature stays near 37.5°C as a result (Howell 1979).

Heat and water problems impose stress on the parent trying to cool its eggs directly in a hot environment. To protect eggs from the hot sun, the incubating parent must absorb and dissipate enormous amounts of radiant energy without overheating itself. Sooty Terns dissipate heat by extending their legs fully, erecting their feathers, and panting (see Figure 6–13). The heat that their black backs absorb is removed by the breeze. The more sunlight that incubating Herring Gulls absorb, the more they must pant (Drent 1970). The stress on an individual's water balance is so great and the consequences of even temporary absences so severe that mates must take turns to provide continuous egg coverage.

Conservationists, sightseers, and research scientists should be aware of the dangers of egg exposure; the unwitting disturbance of seabird nesting colonies destroys embryos because disturbed parents leave their nests and expose their eggs to the sun. Disturbance by people also increases the risk of predation or desertion (Götmark 1992).

Turning Eggs

Incubating birds rise periodically to peer sharply down at their eggs and then draw each egg backward with a sweeping motion of the bill, rearranging the clutch and turning the eggs. Parents rearrange the eggs so that those that have been on the outside of the clutch become more centrally situated, where the temperature is several degrees higher (a fact that is incorporated into the design of artificial incubators). Regular turning of eggs during early incubation prevents premature adhesion of the chorioallantois (Chapter 15) to the inner shell membranes. Premature adhesion interferes with albumen uptake by the embryo and obstructs its ability to attain the tucking position essential for hatching. Hence, the turning of eggs seems crucial for normal embryonic development, but not in all species. Fork-tailed Palm-Swifts glue their eggs to palm fronds, where they remain fixed for the full period of incubation. Perhaps the movement of the palm frond prevents adhesion of the membranes to the eggshell.

Costs of Incubation

Incubation at normal temperatures consumes 16 to 25 percent of a bird's daily productive energy. Yet, the sedentary incubating adult actually saves energy compared with the alternatives off the nest. Red-winged Blackbirds benefit from the nest insulation and the favorable microclimate of the nest site (Walsberg and King 1978) (Figure 16–12). Because their foraging time is limited, incubating birds sometimes fast or depend on

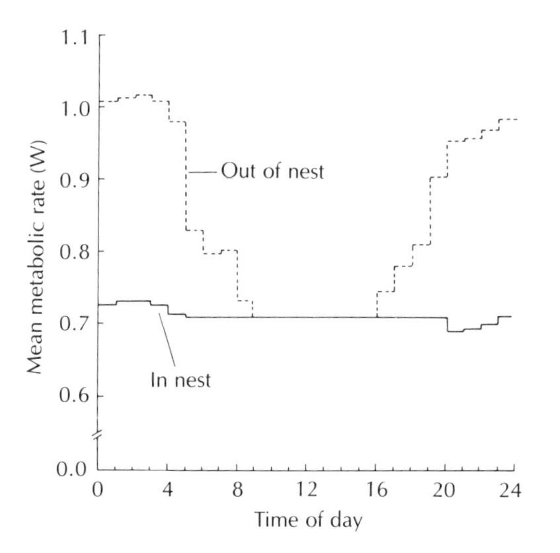

Figure 16–12 The energy expenditures of a Red-winged Blackbird are higher while perching near the nest than while in the nest incubating eggs. (From Walsberg and King 1978)

their mates for supplementary food. A female Snow Goose, for example, subsists on the reserves remaining after egg production. Inadequate reserves at this stage in the nesting cycle cause some females to desert their eggs during incubation; others actually die of starvation (Ankney and MacInnes 1978) (see Figure 15–12). A male Emperor Penguin starts incubation with substantial reserves that allow it to fast for up to 4 months in the Antarctic winter—and loses substantial weight in the process. At the other extreme, a male hornbill provides all the food for its mate, which is imprisoned in the nest cavity. A female Red Crossbill also receives all its food from its mate, an arrangement that enables continuous incubation in the middle of winter. A female Yellow-billed Magpie, which forages 40 percent of the day while laying, forages only 1.7 percent of the day while incubating. To compensate, the foraging time of the male doubles to 78 percent of the day (Verbeek 1972).

Males feed their mates during the final days of incubation. Supplementary feeding at this time could be essential to mates that have exhausted their reserves during incubation or that need to build their own reserves for the demanding efforts that follow hatching. Males may call their mates off the nest to feed them or to escort them during a foraging recess. This habit is common among tyrant-flycatchers, tanagers, vireos, wood warblers, and finches.

Birds abandon their eggs during incubation when foraging becomes difficult or when their reserves dwindle to critical levels. When forced to choose between self-maintenance and caring for progeny, birds usually take care of themselves first. Desertion of eggs is fairly common, but in only a few cases have ornithologists correlated it with food availability. Common Kestrels and Tawny Owls readily desert their eggs when hunting becomes difficult (Cavé 1968; Southern 1970). Red-footed Boobies, Laysan Albatrosses, and Common Wood-Pigeons desert their eggs if their mate fails to return on schedule to relieve them (Nelson 1969; Fisher 1971; Murton 1965).

Both sexes incubate in a majority of avian families. The female incubates alone in about 25 percent of the families, the male incubates alone in only 6 percent. Regular alternation of incubation shifts accomplishes nearly continuous coverage of the eggs in many groups, including some penguins, woodpeckers, doves, trogons, hornbills, hoopoes, and antbirds. Incubation shifts by alternating mates may last for 1 or 2 hours; for 12 hours when one sex incubates by day and the other by night; for 24 hours when each sex takes a day at a time; or several days, as in many pelagic seabirds.

Most small land birds lack conspicuous relief ceremonies; they slip on and off the nest surreptitiously to avoid detection by predators. Other birds have ritualized relief ceremonies. When changing the guard, Pied-billed Grebes touch bill tips lightly. Least Bitterns erect their crown feathers and rattle their bills. Some herons present a stick for the nest to their mates, and terns may offer a freshly caught fish, as the males do during courtship.

Penguins have elaborate changeover rituals that facilitate individual recognition and reinforce the pair bond:

> As a Yellow-eyed Penguin approached his incubating partner, she broke into an "open yell." He ran up with arched back and beak to the ground. Then both put their heads together to perform a hearty welcome ceremony, in which a great volume of sound issued from their widely opened mouths as they faced each other, standing erect close together. After several less-intense displays of mutual affection and three repetitions of "welcome," the female resumed her position on the eggs, then rose to relinquish them to her mate. (Skutch 1976, p. 171, from Richdale 1951, pp. 227–228)

Incubation Periods

The incubation period is the time required by embryos for development in a freshly laid egg that is given regular, normal attention by incubating parents. It is defined as the interval between the laying of the last egg of a clutch and the hatching of that egg (Drent 1975). For most birds, the incubation period is fairly fixed, but in some birds, such as swifts, the incubation period varies in length because of weather-induced interruptions. Incubation periods, which vary from as short as 10 days in some woodpeckers, cuckoos, and small songbirds to as long as 80 to 90 days in albatrosses and kiwis, often are characteristic of a particular taxonomic family or order of birds (Rahn and Ar 1974). The short incubation periods of woodpeckers produce underdeveloped young at hatching, which then require more time to complete their development in the nest. Long incubation periods produce active, precocial chicks with advanced muscular and sensory development.

Reflecting adaptive, genetically controlled differences in the development programs of their embryos, similar-sized eggs of different taxa differ greatly in the amount of time they take to hatch and in the chick's state of development upon hatching. A broad survey of 47 families or subfamilies of birds revealed that incubation periods relate directly to how long adult birds live (Ricklefs 1993). Incubation periods also reflect the probability of predation. Species that nest in holes tend to have longer incubation periods than do species that nest in less safe, open sites (Skutch 1976).

Hatchability of eggs depends on their rate of water loss, which is inevitable because of differences between the water-saturated interior atmosphere of the egg and its unsaturated external environment. During incubation, eggs lose from 10 to 23 percent of their weight, primarily as a result of loss of water vapor. It is greatest in deserts and at high altitudes. The net loss of water from the egg has both positive and negative effects. On the positive side, the space vacated inside the egg becomes the air cell at the blunt end of the egg. This is the source of air for a chick's first breath as it starts to break out of the egg. An adequate volume of air must be available for that critical inhalation. On the negative side, excessive water loss may fatally dehydrate the embryo.

Incubation ends when the young hatch, but the parents then enter a new period of strenuous effort to feed their young as well as themselves. The conflicts between maximizing mating opportunities and ensuring the survival of personal offspring govern the social relationships of birds during the breeding season. The next chapters review first the evolution of avian mating systems, then the growth and development of chicks, patterns of parental care, and finally two elaborations of parental care systems—brood parasitism and cooperative breeding.

Summary

Reproduction in birds requires the nurturing of eggs and young outside the body. Nests, which provide a receptacle for eggs during incubation and for baby birds until they fledge, vary in construction from simple accumulations of sticks or scrapes in the earth to major architectural achievements. Woven, pensile nests and the huge, apartmentlike, compound nests of certain weavers represent the pinnacles of nest construction. Nest materials may include specific plants with pharmacological properties, or feathers, hair, and spiderwebs. Particular methods of gathering nest materials and of constructing nests, which characterize each species, may have a genetic basis.

Nests have four primary purposes: protection from predators, provision of a microclimate suitable for egg incubation, cradles for dependent young, and roosting chambers for adults attending their eggs and young. Various forms of camouflage, inaccessible locations, and fortresslike structures reduce the vulnerability of nests to predators. Roofed nests and cavity nests have many protective advantages. Birds also nest near stinging insects for protection and are able to defend their nests or lure would-be predators away with great skill. The choice of safe nest sites may influence the evolution of other aspects of the morphology and behavior of a species.

Colonial nesting reduces predation and provides access to nearby abundant or spatially unpredictable food supplies. About 13 percent of all bird species nest in colonies, most of them marine seabirds. Slightly offsetting the advantages of colonial nesting are the potentials for social confusion and for dissemination of diseases and parasites, potentials that are directly correlated with colony size.

The nest microclimate is governed in part by the thickness of insulation, which reduces the rate of egg cooling when the adult leaves the nest to feed. Nest location with respect to sun, shade, prevailing breezes, or sheltering objects has a major effect on incubation behavior. Burrows and cavity nests tend to buffer birds and their eggs from daily temperature cycles. Precise incubation schedules keep egg contents within the narrow limits of embryo temperature tolerances. Most birds transfer body heat directly to their eggs through the bare, flaccid, and highly vascularized brood or incubation patch. Moundbuilders, however, incubate their eggs by using the heat generated in heaps of decaying vegetation. In hot desert environments, cooling of eggs by means of midday shading or wetting may be necessary. Incubation not only exposes parents to temperature stresses and predators but also reduces the time they have for feeding themselves. Male birds may relieve their mates by sharing incubation or may help provision them with food.

Further Readings

Carey, C. 1983. Structure and function of avian eggs. Current Ornithology 1: 69–103. *A modern summary of the physiology of eggs.*

Collias, N.E., and E.C. Collias. 1984. Nest Building and Bird Behavior. Princeton, N.J.: Princeton University Press. *A comprehensive review of the natural history of avian nests.*

Drent, R.H. 1975. Incubation. Avian Biology 5: 333–419. *A summary of bird nests and incubation.*

Drent, R.H., and S. Daan. 1980. The prudent parent: Energetic adjustments in avian breeding. Ardea 68: 225–252. *A well-illustrated summary of energetic aspects of parental care.*

Siegel-Causey, D., and S.P. Kharitonov. 1990. The evolution of coloniality. Current Ornithology 7: 285–330. *An excellent synthesis of the growing literature, including key Russian contributions.*

Skutch, A.F. 1976. Parent Birds and Their Young. Austin: University of Texas Press. *A readable account of the natural history of avian reproduction, with chapters on nests and incubation behavior.*

Webb, D.R. 1987. Thermal tolerance of avian embryos: A review. Condor 89: 874–898. *A thorough review of the temperature considerations of incubation.*

Wittenberger, J.F., and G.L. Hunt, Jr. 1985. The adaptive significance of coloniality in birds. Avian Biology 8: 1–78. *A thorough review of the advantages and disadvantages of colonial breeding.*

C H A P T E R 17

Mates

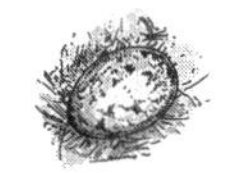

MALE AND FEMALE BIRDS differ in their reproductive options and in their potential reproductive success. The reproductive success of males—defined as the number of offspring they sire—tends to be limited by the availability of mates. Male reproductive output is potentially greater than that of females, which tends to be constrained by the production and care of limited numbers of large eggs. The Brown Kiwi is an extreme example of the demands of reproduction in female birds. Its eggs are huge, each being 25 percent of body size (Figure 17–1). In contrast, males produce vast numbers of tiny sperm, capable of fertilizing many such eggs. The sexual differences in gamete production control the ways males and females compete for mates and how much time and energy they devote to parental care. Inseminated females usually concentrate on the care of their limited zygotes to maximize their reproductive contribution to the next generation, but

Figure 17–1 The Brown Kiwi, which produces an enormous egg relative to its body size, provides an extreme example of the great investment of reproductive energy that female birds put into egg production. This X-ray reveals the egg in the oviduct just prior to laying. (Copyright 1978 by the Otorohanga Zoological Society, used with permission)

males may benefit more from additional matings than from caring for offspring.

Until recently, ornithologists have focused on the different gametic investments of males and females and the correlations between mating systems and ecology, mediated by parental care requirements. Current studies of avian mating systems focus on the antagonisms between male and female options for maximizing mating opportunities and on the nature of monogamy itself. Instead of a spirited, cooperative partnership, monogamy may be a temporary truce between selfish, competitive individuals. Males must balance the alternatives of competing with other males to mate with extra females versus caring for their own young. As a result, extra-pair copulations are most common in the early stages of the nesting cycle before the males start helping to raise their nestlings. Females are tempted also to improve the quality of their offspring through extra-pair copulations with high-quality males, but they risk losing the parental assistance of their mate by diluting the paternity of the brood.

This chapter considers the varied mating systems of birds. Most birds pair with a single mate and raise the offspring together, both parents being needed to provide adequate care for their helpless nestlings. A minority of bird species adopt other mating systems that involve multiple mates and uniparental care.

Kinds of Mating Systems

Pair bonds vary from brief sexual unions to sustained mutual efforts. Both the duration of the association and the number of sexual partners help to define various mating systems (Oring 1982). The reproductive success of males relative to that of females also varies from system to system. The principal kinds of mating systems, as defined in Box 17–1, are

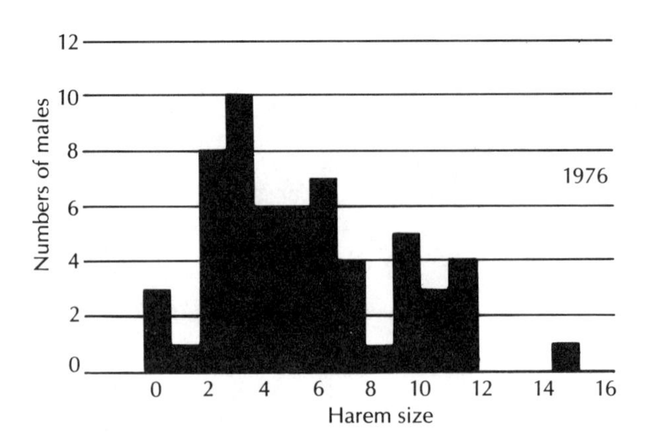

Figure 17–2 Variation in the sizes of harems of male Red-winged Blackbirds in Washington State. (Adapted from Searcy and Yasukawa 1983)

Box 17–1
The type of pair bond defines the mating system

Monogamy (Greek: *mono,* single; *gamos,* marriage) The predominant avian mating system in which there is a prolonged and essentially exclusive pair bond with a single member of the opposite sex for purposes of raising young.

Polygamy (Greek: *poly,* many; *gamos,* marriage) Any mating system involving pair bonds with multiple mates of the opposite sex. Only 3 percent of all birds practice polygamy.

Polygyny (Greek: *poly,* many; *gyna,* woman) That kind of polygamy in which a male pairs with two or more females. It is called bigamy if the male pairs with only two females. For polygynous birds, breeding success of males is more variable than that of females. Sustained associations distinguish this mating system from promiscuous behavior. About 2 percent of all birds are polygynous.

Polyandry (Greek: *poly,* many; *andros,* man) That kind of polygamy in which a female pairs with several males (that is, the opposite of polygyny). Each male may tend a clutch of eggs. For polyandrous birds, female breeding success is more variable than that of males. If a female lays full clutches of eggs for successive mates, the system is called sequential (or serial) polyandry; if she has two or more mates at once, the system is called simultaneous polyandry. Fewer than 1 percent of all birds are polyandrous.

Polygynandry (Greek: *poly,* many; *gyna,* woman; *andros,* man) That kind of polygamy in which a female pairs with several males, each of which also pairs with several different females. This mixed mating system, which is common among fishes, is characteristic of tinamous, the flightless ratites (Ostrich, rheas, Emu), and also some unusual songbirds, such as Smith's Longspurs and Hedge Accentors. Male ratites and tinamous incubate mixed clutches of eggs from several females, which deposit eggs successively with different males.

Promiscuity (Latin: *pro,* forth; *miscere,* to mix) Indiscriminant sexual relationships, usually of brief duration. Male hummingbirds, manakins, and grouse, which mate with any receptive visiting female, are technically promiscuous. Variance in male reproductive success reaches its maximum value in this system. About 6 percent of all birds are promiscuous.

---

monogamy, polygamy, polygyny, polyandry, polygynandry, and promiscuity. One notable group, the sandpipers and their relatives in the Family Scolopacidae, have evolved diverse mating systems (Pitelka et al. 1974; Myers 1981a, b) (see Box 17–4). Arctic breeding sandpipers are monogamous, polygynous, polyandrous, promiscuous, and mixtures of these alternatives. The three phalaropes also are polyandrous, whereas woodcocks and some snipes are promiscuous.

Tied to alternative mating systems is the process of sexual selection. Males rather than females tend to compete for mates or are subject to selection by females. The reproductive success of males varies more than that of females (Payne 1979). The reproductive success of monogamous males varies from zero to the few offspring produced by one female. At the other extreme are promiscuous males and masters of harems with a dozen or more females (Figure 17–2). As potential male reproductive success increases, so does the value of the characteristics—such as large size, fancy plumage, intricate songs, and striking displays—that are responsible for the success. The resulting evolutionary process of sexual selection leads to differences between the sexes in size and ornamentation—

sexual dimorphism (Chapter 9). Polygamous mating systems and, particularly, behavior on communal display grounds—or leks—are the most elaborate results of this process.

The large sizes and conspicuous plumages favored in reproductive displays may be liabilities in other regards (Figure 17–3). The same bright colors that announce a male's presence to potential rivals or mates may attract predators. Fancy display plumage, such as that of a male bird-of-paradise, may also hinder escape; and large size carries the liability of greater energy expenditure. There is some evidence in Red-winged Blackbirds that large males are at a disadvantage because they must sacrifice display time for feeding. Among species of North American blackbirds, males that are much larger than females tend to suffer greater mortality (Searcy and Yasukawa 1983). Male Western Capercaillies, huge European grouse, grow twice as fast as females to reach full adult size by the end of their first summer. As a consequence, their higher energy requirements render them more vulnerable than the females to starvation when food is scarce (Wegge 1980).

Figure 17–3 Strutting male Wild Turkeys compete for mates. (Courtesy of A. Cruickshank/VIREO)

Some male birds are smaller than the females of their species. The combination of small males and large females—called reverse sexual size dimorphism—is typical of shorebirds, raptors, and hummingbirds. Historical explanations have centered on reduced sexual competition for food and metabolic advantages of large size in incubating females (see Jönsson and Alerstam 1990), but Joseph Jehl and Bertram Murray (1986) suggest instead that acrobatic aerial displays favor small males in these groups.

Monogamy

Monogamy refers to a prolonged and essentially exclusive pair bond with a single member of the opposite sex for purposes of raising young. The practices of biparental care also serve to reduce energetic strain on adults. Pair bonds may last for a breeding season or for life. Pairs of parrots, albatrosses, eagles, geese, and pigeons all sustain lifelong associations, although separations do occur. Among Mute Swans, for example, roughly 5 percent of breeding pairs and 10 percent of nonbreeding pairs separate each year (Minton 1968). A 16 percent separation rate is typical of old pairs of Adelie Penguins, and nearly half (44 percent) of young pairs do not stay together more than one breeding season (Ainley et al. 1983). In general, the probability of divorce increases if a breeding attempt fails.

Still, birds are among the most monogamous of organisms. Less than 10 percent of all birds engage in other kinds of mating relationships. Traditionally, ornithologists have viewed monogamy as the mating system of choice; a pair can raise more young than can a female without a mate (Lack 1968). Most birds spend weeks or months tending their eggs and young; in contrast, most reptiles simply lay their eggs and leave them. Not only do avian eggs and chicks require more parental care than do the offspring of most vertebrates, but the participation of both sexes frequently appears to be essential (Chapter 18). Defense of territorial space, which often provides adequate food supplies for the female and young, generally falls to the male. Most monogamous males also help their mates build nests and feed young; some share incubation (Verner and Willson 1969).

Pair Formation and Assessment

The parental roles of monogamous males often are substantial. Hence, mutual assessment of prospective partners is a subtle, but vital, aspect of the early stages of courtship and pair formation. Female pigeons, for example, prefer experienced but not too elderly males (Burley and Moran 1979). Recall previous discussions of how females might judge health and parasite loads of a male by the quality of its plumage ornamentation (Chapter 9). A female must also assess her prospective mate's commitment and ability to sustain efforts in raising young. Ornaments that reliably reflect the superior condition of certain males and that enable females to select the best possible mates will be favored and maintained by sexual selection (Kodric-Brown and Brown 1984). The quality of combs of Red Junglefowl and the length of the tail streamers of Barn Swallows are two examples of such reliable cues (Box 9–2). Display effort also demonstrates a male's physical condition. The familiar flight displays of male Bobolinks over lush fields, for example, serve to advertise their condition. Monica Mather and Raleigh Robertson (1991) demonstrated that females favored the males that displayed longer; they had larger fat reserves and consequently fledged more young than their neighbors did. Experiments confirmed the observed correlations; males with clipped wings had shorter flight displays and acquired fewer mates than did control males.

The ritualized display of food or nest materials during courtship may reveal an individual's skill in gathering these essential items. Courtship feeding by the male not only helps a female build the nutritional reserves needed for egg production but also may serve to demonstrate a male's food-gathering abilities. In Common Terns, a male's ability to feed young correlates well with the intensity of his courtship feeding efforts (Nisbet 1973) (Figure 17–4).

Courtship feeding is not only a ritual of pair formation; it is of considerable energetic benefit to the female. Courtship feeding of their mates by male Common Terns during egg laying directly affects the timing of laying and the size of the eggs laid (Nisbet 1973). Male Dot-winged Antwrens provide 40 percent of their mate's daily intake during egg laying and incubation (Greenberg and Gradwohl 1983). Male European Pied Flycatchers contribute nearly half of the incubating female's food (Curio 1959; Haartman 1958). When the male is experimentally

Figure 17–4 A male Common Tern feeds its newly hatched young while the female continues to incubate the second egg. (Courtesy of E. and D. Hosking)

removed, the female spends less time incubating (58 percent versus 79 percent with a male present) and more time foraging, but she still loses weight. Male Great Tits and Blue Tits feed their mates more frequently during egg laying and incubation than during courtship, up to 160 feedings per day for the Great Tit and over 1000 for the Blue Tit, which eats smaller prey (Royama 1966a). Such supplementary feeding more than doubles a female's rate of energy intake and helps her maintain body weight while producing eggs in a timely fashion (Krebs 1970). Male European Robins feed their mates nearly every 5 minutes during incubation (East 1981).

Conversely, a male must judge during courtship how receptive the female is, and in particular, he must be certain that he alone is the father of the chicks he will be caring for. Unreceptive females do not tolerate the male's initial aggression or at least do not return his attentions readily. In some birds, courtship stimulates the last phases of gonadal activity and helps pairs to synchronize their readiness to mate. Prolonged courtship, even of receptive females, may also help to ensure paternity. Experiments with Ringed Turtle-Doves show that aggression by the male delays ovulation in its prospective mate. Because most sperm die within 6 days of their release, delayed ovulation helps ensure the demise of any sperm remaining from earlier inseminations and increases the likelihood that the current suitor will be the father (Erickson and Zenone 1978; Zenone et al. 1979). High copulation rates, as in Smith's Longspurs (Box 15–1), also help to displace the sperm of other males.

Extra-Pair Copulations

Monogamy, as we usually see it, is a social relationship between members of the opposite sex that is built on the assumption that the offspring are truly their genetic offspring. Mixed genetic paternity of broods due to forced copulations or to copulations with additional males—both called extra-pair copulations—is proving to be common (reviewed by Westneat et al. 1990) and is threatening the working value of the traditional classification of mating systems. Forced copulations, the avian analogues of rape, are a way of life among waterfowl (McKinney et al. 1983; Evarts and Williams 1987). The proportion of chicks sired through extra-pair copulations may be 30 to 40 percent in broods of Indigo Buntings and White-crowned Sparrows (Westneat et al. 1987; Sherman and Morton 1988) and 50 percent or more in some Tree Swallow populations (Lifjeld et al. 1993). Territorial male Red-winged Blackbirds realize more than 20 percent of their reproductive success through extra-pair copulations (Gibbs et al. 1990), and female Red-winged Blackbirds whose mates are vasectomized still manage to achieve a high rate of fertility in their clutches through liaisons with other males (Bray et al. 1975). DNA fingerprinting analyses have revealed that most blackbird broods contain chicks fathered by neighbors.

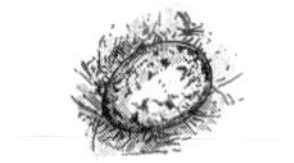

BOX 17–2
Purple Martins practice cuckoldry in the condo

Native Americans used to attract Purple Martins by hanging hollow nesting gourds. Elaborate, multistory, white condominiums—sometimes with hundreds of nest chambers—now attract colonies of martins to the backyards of lucky homeowners. Appreciation of his backyard martins piqued the curiosity of Smithsonian ornithologist Eugene Morton to learn more about social life inside his 24-room martin mansion (Morton et al. 1990; Morton 1991). Among his findings—rampant cuckoldry. Older experienced males arrived first and took charge of the top floors, where the nests are safest from predators. After establishing themselves with their mates in the best available condos, the experienced male martins sang a special song high in the dark predawn sky to attract late-arriving yearling males to the colony. The older males then concentrated on copulating with the mates of their naive, young neighbors—with much success.

Adult males added, through extra-pair copulations, an extra 3.6 fertilized eggs to the 4.5 eggs produced by their own mates. In contrast, yearling males fathered only 29 percent of their young; most of the nestlings they fed were sired by another male. Some offspring, however, are better than none at all, and with time yearling males inherited the prime nest chambers and also sang their predawn songs to attract junior neighbors to the suites below.

Unpaired birds mate on migration. In one set of studies, the presence of viable cloacal sperm revealed that at least 25 percent of the females of North American migrant passerines had copulated before they reached the breeding grounds and paired with a territorial male (Quay 1985, 1989). Adding uncertainty to genetic maternity as well as paternity of chicks in a single nest is the possibility of intraspecific brood parasitism (see Chapter 19).

Close association with mates to ensure paternity—mate guarding—is a major feature of early stages of nesting and repeat nesting—called renesting. Competition for and control of receptive females are especially keen in the colonial Yellow-rumped Cacique of South America (Robinson 1986). Male caciques gather at the colonies to establish their status in the dominance hierarchy through mutual displays. High-ranking males closely guard and mate with females that have chosen the safest nest sites with the least risk of predation. But the costs of rank and consortship—or close association with a mate—are high. Dominant males lose weight with age and soon are displaced to peripheral, secondary status in the colony by more vigorous, younger males.

Mating Systems and Ecology

Like variations in social behavior, pair bonds and avian mating systems reflect the availability of key ecological resources—space, food, and protection—as well as the availability of mates and the feasibility of uniparental care. The relations of mating systems to ecology are conspicuous in the weavers of Africa and the birds-of-paradise of New Guinea.

African weavers include both monogamous and polygynous species (Crook 1964) (Figure 17–5). Those, such as the Forest Weaver, that are adapted to stable forest environments with uniform food distributions tend to be territorial—solitary—and monogamous. Savannah-living weavers, such as the Golden-backed Weaver and the Red Bishop, which exploit ephemeral and unpredictable habitats that occasionally have superabundant foods, tend to be polygynous species in which males control

Figure 17–5 African weavers have different mating systems. (A) Forest Weaver is a territorial, monogamous species; (B) Red Bishop, a territorial, polygynous species; (C) Golden-backed Weaver, a colonial, polygynous species; (D) Red-billed Quelea, a colonial, monogamous species.

limited safe nest sites near good food supplies. The most abundant of the colonial savannah weavers, the Red-billed Quelea, is an instructive exception. It is monogamous even though it nests in huge colonies near abundant food. Quelea colonies are so large that their members deplete nearby food stores during nesting and must commute farther and farther to gather food for their young. Male assistance becomes essential to ensure that older nestlings are fed (Ward 1965).

Fruit diets favor the evolution of polygynous mating systems in birds. Details of fruit dispersion patterns, abundances, and nutritional value are basic considerations (Ricklefs 1980a). Males of 34 of the 43 birds-of-paradise are known or presumed to be promiscuous (Beehler and Pruett-Jones 1983; Beehler 1989). The two best-studied species—Magnificent Bird-of-Paradise and Count Raggi's Bird-of-Paradise—feed on assorted predictable, nutritious fruits. Females are able to raise their young alone. Monogamous species such as the Trumpet Manucode, however, specialize on locally abundant figs that have little nutrition. Females need the assistance of a mate to help deliver the large quantities of such low-quality food that are required to sustain the nestlings (Beehler 1985). In birds-of-paradise, therefore, food quality overrides food distribution in determining mating systems.

Generally, when male participation is essential for raising young or when males cannot commandeer the resources necessary for supporting extra mates, monogamy may be the only option. But as a female's ability to take care of young by herself increases, polygyny becomes a more viable option for those males that are able to control the best territories or to attract the most females. Also, if males become superfluous or even liabilities during parental care, they no longer are limited by their pair bonds and can compete with other males for extra matings.

What conditions release a parent from parental care responsibilities? One of the most obvious is mode of development of the young. Young of species, such as the sandpipers, that leave the nest soon after hatching and that can feed themselves require less care than those that remain dependent in the nest. Seventeen percent of birds with young that leave the nest soon after hatching—nidifugous young—are polygamous or promiscuous, whereas only 7 percent of birds with young that remain in the nest—nidicolous young—are nonmonogamous.

Abundant or easy-to-find food may also relieve parental pressures. Savannah Sparrows are sometimes bigamous where food is abundant and are monogamous elsewhere (Welsh 1975). Female blackbirds, wrens, warblers, and sparrows that nest in marshes care for young alone by exploiting emerging aquatic insects on prime territories. Temporary food abundance on the Arctic tundra and on the African savannah also eases the need for dual parental effort.

Like male birds-of-paradise, males of many tropical, fruit-eating birds do not help care for their young. Fruit and floral nectar are conspicuous food sources that require little searching; once a bird locates them, regular

Box 17–3
Waterfowl have unusual breeding systems

Three features distinguish the unusual breeding systems of most ducks and other waterfowl—winter pairing, monogamy, and female fidelity to the natal locations where they were hatched and raised. "Emancipated" drakes, which do not help incubate or raise their precocial young, are logical candidates for the polygynous mating systems that characterize many marsh nesting species. Frank Rohwer and Michael Anderson (1988) suggest that waterfowl do not conform to the system because the drakes cannot attract harems on the breeding grounds through control of high-quality marsh territories. Instead, the females—or hens—initiate a monogamous pairing as early as possible on the wintering grounds.

Drakes typically outnumber females, which suffer high mortality while they alone care for the ducklings. A drake has little choice but to pair monogamously, if chosen by a female on the wintering grounds, and then to follow her back to the place where she herself was raised and will nest. A drake cannot follow two females back to their respective nesting grounds! Protection by her drake on the wintering grounds increases a hen's ability, through increased foraging efficiency and energy conservation, to build up the substantial nutrient reserves she requires for migrating and nesting. Once back on the breeding grounds, a drake enhances its reproductive efforts outside the primary pair bond through forced extra-pair copulations and through fertilizations of renesting females.

revisitation minimizes foraging effort. Incubating females can easily slip off the nest to feed quickly. As long as the energetic requirements of nestlings can be partially satisfied with fruit or nectar, one parent can raise them successfully. Thus, males of these species devote themselves to display to attract additional mates.

Polygyny

Careful study of color-marked individuals often reveals a few bigamous males in an otherwise monogamous species, but only 2 percent of all birds are regularly polygynous. In North America, these include 14 of the 278 breeding songbird species, 11 of which nest in marshes or grasslands (Verner and Willson 1969; Ford 1983). Throughout the Tropics, birds that nest colonially in "safe" trees or in marshes tend to be polygynous.

Variation in territory quality leads to polygyny. Polygynous male Marsh Wrens, Red-winged Blackbirds, and Indigo Buntings, among others, all control better quality territories than do unmated or monogamous males in the same area (Verner 1964; Searcy and Yasukawa 1983; Carey and Nolan 1975). The mating success—number of mates—of Marsh Wrens, for example, reflects the proportion of the territorial area with emergent marsh vegetation that provides good nest sites: 54 percent of the area on territories of unmated males, 80 percent on territories of monogamous males, and 95 percent on territories of polygynous males.

Assuming no shortage of potential males, why should two or more females share a male? Females that join a harem presumably do so because they can do as well as or better than when paired alone with a male on a territory of poorer quality. In polygynous Lark Buntings, a grassland bird of the western United States, the primary female gets all or most of the male's help in raising the young, whereas the secondary female gets some protection and use of the territory but no direct assistance (Pleszczynska and Hansell 1980). The secondary female's main advantage derives from the availability of shaded nest cover on better quality territories. Nestling Lark Buntings easily overheat and die in poorly shaded nests on poor territories. Therefore, some females choose to raise young by themselves in shaded nests on good territories.

In other cases, primary and secondary females share male help—at a cost. In Great Reed-Warblers, nestlings of polygynous males die more often from starvation than do those of monogamous males (Dyrcz 1977). Starvation is most frequent during cold, wet spells when food is scarce and the young depend on food delivered by the male as well as by the female parent. Offsetting losses to starvation on polygynous territories, which have better nesting sites, are reduced losses to predators.

Control or monopolization of quality resources leads to the evolution of what is called resource defense polygyny (Emlen and Oring 1977). Clumped resources are easier to monopolize than are uniformly distributed resources. Extending Brown's concept of economic defensibility (Chapter 14), the environmental potential for polygyny increases with clumped resource distributions. Male Red-winged Blackbirds compete with one another for defensible, high-quality territories that attract females. Defense of specific feeding sites by males takes resource defense polygyny one step further. Male Yellow-rumped Honeyguides, for example, commandeer the best bee nests in an area and await females that come to feed on the wax (Cronin and Sherman 1976). Dominant males that control the largest bee nests in an area attract the most mates. Honeyguides are specialized brood parasites (Chapter 19); therefore, the male honeyguides have no parental duties and occupy themselves fully with control of prime sites and promiscuous mating behavior.

To acquire extra females, males of the well-studied, hole nesting European Pied Flycatcher of northern Europe set up side territories where they advertise for an additional late-arriving female (Alatalo and Lundberg 1984). The male flycatchers sing on their secondary territories after their first mate starts laying and incubating the eggs. Their extra territories are typically far enough away from their primary territory to allow them to project themselves as unmated males, thereby tricking secondary females into an unfavorable situation. Secondary females get less male assistance than the primary females and fledge fewer young as a result. Conversely, females widowed during incubation solicit copulations from neighboring males and thereby entrap them into helping raise offspring that could be their own, but actually were fathered by a deceased mate (Gjershaug et al. 1989).

BOX 17–4

Varied mating systems of shorebirds: Are they escapes from the constraints of a fixed four-egg clutch?

The diversity of mating systems exhibited by sandpipers of the shorebird family Scolopacidae continues to puzzle ornithologists (Ligon 1993). Fifteen species are monogamous, with shared incubation at a single nest; two or three species are socially polygamous or polyandrous, with incubation of successive clutches by different individuals; two or three other species have polygynous males that provide no parental care; and three species are promiscuous, lek species. There is neither rhyme nor reason for using one mating system rather than another. Early attempts to relate these varied mating systems to factors such as differences in ecology, length of breeding season, or predictability of food resources were not convincing.

Despite their diversity of body sizes and feeding habits as well as mating systems, shorebirds of the Order Charadriiformes typically lay four eggs in a clutch. Why they do not lay more—and rarely fewer—eggs itself is a puzzle, but the answer seems to lie deep in the evolutionary history of these birds, perhaps tied to the physiology of incubation. Given their fixed clutch size, however, David Ligon (1993) suggests that there are advantages to increasing reproductive success in other ways. Increasing the number of clutches produced and tended through nonmonogamous mating systems is the result. The details of the solutions may be as much due to evolutionary accidents as to well-defined ecological controls.

Polyandry

In only a few birds do females pair with several males, which then incubate the eggs and take care of the young. Such females defend territories, compete for males, and take the lead in courtship. Classic polyandry, in which females divide their attention among two or more mates, is distinguished from cooperative polyandry (Oring 1986; Ligon 1993), in which several males cooperate to assist a female. The latter system is restricted to two hawks (Galápagos Hawk, Harris's Hawk), two gallinules (Dusky Moorhen, Tasmanian Native-Hen), a woodpecker (Acorn Woodpecker), and a songbird (the Hedge Accentor). Classic polyandry has evolved primarily in two orders of birds. In the Order Gruiformes, the buttonquails, roatelos, and some rails are polyandrous; in the Order Charadriiformes, the jacanas, painted-snipes, the ploverlike Eurasian Dotterel, the buttonquail-like Plains-Wanderer of Australia, a plover, and a few sandpipers are polyandrous. Sex-role reversal has led to the evolution of large and brightly colored females. Polyandrous female phalaropes, for example, compete for males in congregations at productive feeding sites and initiate courtship with males. Males incubate the resulting clutch of eggs by themselves and do not tolerate the female near the nest after the clutch is complete. Females then lay additional clutches for other males.

The Spotted Sandpiper of North America provides one of the best-studied cases of classic avian polyandry (Oring and Knudsen 1972; Lank et al. 1985; Oring 1986). Females, which are 25 percent larger than males, defend large nesting territories and fight each other for the available males (Figure 17–6). During each breeding season, they lay a clutch of four eggs for their primary mate and for one to three secondary males. A female's reproductive success directly reflects her ability to obtain extra mates. Each male incubates its clutch of eggs and defends a surrounding territory against other males. When a male loses its clutch of eggs to a predator, the female quickly replaces the clutch with a new set of eggs.

A closely related species, the Common Sandpiper of Eurasia, is strictly monogamous with biparental care, as is the case of other related sandpipers in the genus *Tringa.* Also, in low-density populations of the Spotted Sandpiper, both monogamous and polyandrous females share parental care, including incubation. In dense populations, where competition with other females is more intense, polyandrous females only help their secondary males. Polyandry in the Spotted Sandpiper, therefore, represents the emancipation of females from parental care, rather than a novel assumption of incubation responsibilities by the male.

Changes in the hormones that mediate aggression and parental behavior correspond to the reversal of sex roles in this sandpiper (Oring and Fivizzani 1991) (Figure 17–7). Levels of the sex hormone testosterone, which inhibits incubation, drop more sharply in male Spotted Sandpipers than in other sandpipers that share incubation behavior, and drop lower than the testosterone levels found in their aggressive females. Conversely, levels of the hormone prolactin, which mediates incubation and other parental behavior, rise higher in male Spotted Sandpipers than in females.

Why should such systems ever evolve? And why not more often? The development of a theory of the evolution of polyandry has lagged far behind that for the evolution of polygyny. Ornithologists are still groping for explanations. "No single explanation of avian polyandry is satisfactory,

Figure 17–6 Female Spotted Sandpipers, which are polyandrous, defend territories and lay clutches for each of several males, which tend them. (Photography by S.J. Maxson, courtesy of L. Oring)

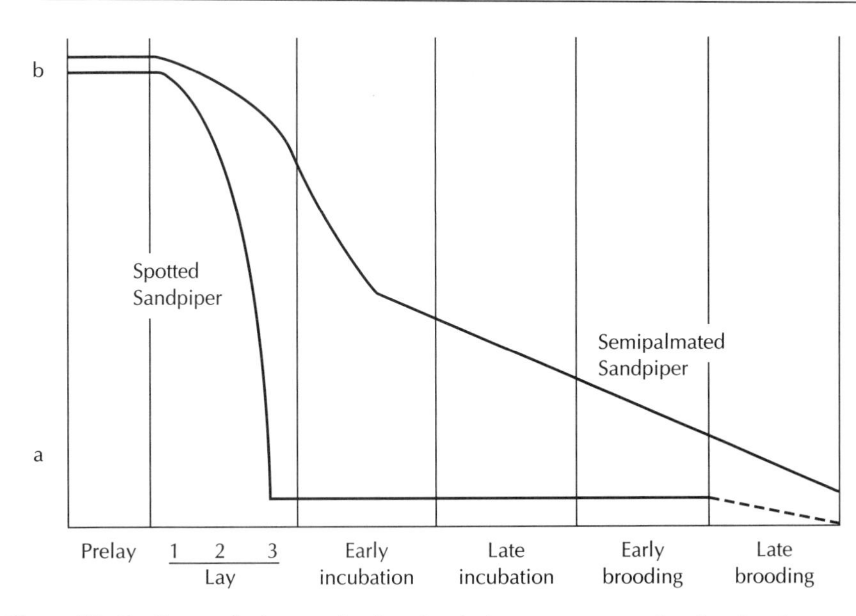

Figure 17–7 Seasonal changes in the circulating testosterone levels of male Spotted Sandpipers and Semipalmated Sandpipers. In male Spotted Sandpipers, which perform all or most of the parental care, testosterone levels drop sharply from (b) the elevated levels that support sexual activity before eggs are laid to (a) basal levels at the onset of incubation. Numbers over the word "Lay" indicate first, second, and third eggs of the four-egg clutch. In male Semipalmated Sandpipers, which share incubation with their mates, testosterone levels decline gradually throughout the parental care period. (After Oring and Fivizzani 1991)

and its evolution continues to puzzle behavioral ecologists" (Clutton-Brock 1991, p. 259). The diversity of systems called polyandry requires different explanations of classic and cooperative polyandry. Past thinking stressed responsibilities for incubation and risks of predation (Jenni 1974; Oring 1982). Cases of double clutching—incubation of separate clutches by each member of a monogamous pair—were once thought to represent the first step toward uniparental care by the male. But none of the double-clutching species was closely related to a classic polyandrous species; and, if Spotted Sandpipers point the way, then the key issue is female desertion of parental care to increase mating opportunities. It could even be in the best interest of both sexes for the male to assume responsibility for incubating the early clutches, thereby freeing the female to lay additional clutches, especially if the male gets to fertilize some of the eggs in successive clutches (Ligon 1993). Perhaps past preoccupation with different gametic investments (eggs versus sperm production) misled thinking about the evolution of these mating systems. Instead, attention might be directed more productively toward the ways birds increase mating opportunities, especially when constrained by physiological or historical limitations, such as the fixed, four-egg clutch of shorebirds (Ligon 1993).

J. Maynard Smith (1977) suggested that the options for parental care could be likened to an evolutionary game in which each parent has to

Box 17–5
Hedge Accentors do it all

Careful scrutiny of a species' mating system, and the discovery of frequent extra-pair copulations or occasional bigamy, thwart the simple labeling of mating systems. The labeling challenge reaches an extreme in the complex mating system of the Hedge Accentor, a drab, sparrowlike songbird that lives in the dense hedgerows of England. Hedge Accentors show a range of relationships including monogamy, polygyny, polyandry, and polygynandry (Box 17–1). Only the female incubates. Detailed studies of the social behavior of Hedge Accentors near Cambridge by Nicholas Davies (1992; see also Oring 1986; Gibson 1993) suggest that there really is an underlying structure to the Hedge Accentor's sexual complexities.

Hedge Accentors eat the tiniest soil arthropods in dense cover, where exclusive nesting territories are difficult to maintain. The sexes establish independent but overlapping territories in relation to food density. The patterns of overlap prescribe the varied pair bonds. Where food is dense, one male may overlap the small territories of one female (monogamy) or two (bigamy). At lower food densities, however, males cannot monopolize the food resources required by females and their chicks. Females then have large territories that overlap with two (unrelated) males. Because they fledge the most chicks when they have two males helping them, such females solicit copulations from both males, which increases the commitment of each male to the parental care of her chicks. Experimental addition of food causes reductions of territory size and can promote polygyny rather than polyandry. More complex groups of two or more females may share two or more males, depending on the arrangements of their feeding territories.

Contrary to the female's best course, males fare best as bigamists. The resulting conflicts lead to intense copulation rates (as in Smith's Longspur; see Box 15–1) and sperm competition. Preceding copulation is an elaborate display in which the male pecks the female's cloaca, thereby stimulating her to eject sperm from previous matings and simultaneously increasing the probability of egg fertilization by the latest—namely, his—ejaculate. Underneath the complexity, each sex adjusts its personal and parental relationships in ways that optimize reproductive success. To tell whether they are the fathers of certain young, and therefore, how much they should feed them, the males monitor the appearance of eggs in nests of the females with which they have mated!

---

choose between taking care of the young or abandoning them to seek additional mates. The relative payoffs determine each parent's decision. This game-theory approach successfully predicted the relative parenting efforts of male and female Hedge Accentors that breed either in pairs or in trios of two males with one female (Box 17–5).

Lek Displays of Promiscuous Birds

Promiscuous males of some species display in courtship arenas that contain no resources. Females visit for one purpose only—fertilization—and then build their nests and raise their young elsewhere by themselves. The display grounds of promiscuous birds vary from solitary courts to large, communal display grounds—or leks (Figure 17–8). At the one extreme, Great Argus males in Malaysia and Superb Lyrebird males in Australia hold forth on isolated deep-forest courts. At the other extreme, Cock-of-the-Rock males gather like glowing orange ornaments on leks in the

Figure 17–8 The mating grounds of promiscuous birds include (A) communal leks of Black Grouse and (B) isolated display courts of the Great Argus. (Adapted from Lack 1968)

understory of South American rain forests, golden-plumed Lesser Birds-of-Paradise dance in the open-forest canopy of New Guinea, puffed-up Sage Grouse strut on the open plains of the western United States, and Black Grouse face off on the moors of northern Eurasia.

Lek displays have been incorporated into the ceremonial dances of human societies (Armstrong 1942). The Jivaro Indians of South America copy the Andean Cock-of-the-Rock in a sensual dance ceremony. Siberian Chukchees mimic the notes of the Ruff. Blackfoot Indians of the western United States mimic the foot stomping, bowing, and strutting of the Sage Grouse while wearing costumes that mimic the grouse's spread tail.

Promiscuous males, and especially those on large leks, vary greatly in their mating success each season. Intense competition ranks males in a dominance hierarchy that determines who sires most of the next generation. In the well-studied Black Grouse of Europe and the Sage Grouse of the western United States, less than 10 percent of the males on large leks achieve 70 to 80 percent of all matings (Kruijt et al. 1972; Wiley 1974). This is also the case for the White-bearded Manakin, Red-capped Manakin, and Lesser Bird-of-Paradise (Lill 1974a, b; Beehler 1983). One male Lesser Bird-of-Paradise displaying on a lek of six other males made 24 of the 25 observed copulations. Dominance is a matter of age, experience, and ability. By mating with a dominant male, females obtain for their offspring the genes responsible for the male's superior traits. The dominance hierarchy, in effect, selects among males and thus simplifies the selection of a good male. Recall that females of some species may judge male health from the quality of their plumage or displays (Chapter 9). Female Sage Grouse preferred males that were free of lice and malaria (Johnson and Boyce 1991; Spurrier et al. 1991). Females could identify males with louse infestations by the red blood spots on their air sacs, which the males inflated while strutting.

Evolution of Leks

Why should promiscuous males gather in leks, in which a few dominant individuals mate most frequently? What are the conditions that favor clustered males in leks over dispersed nonmonogamous males trying to attract the same females? No one knows for sure, but several possibilities stand out. Reduced risk of predation is one possibility. As in the case of flocking (see Chapter 14), predators may have a smaller chance of surprising a group of males that display together than of grabbing an isolated male preoccupied with sexual display. R. Haven Wiley (1974) suggests that this is the primary reason that open-country grouse display in leks whereas forest grouse tend to display solitarily. Birds-of-paradise that display in leks are species that inhabit forest borders and second-growth forest, where predation risk tends to be higher than in primary forest (Beehler and Pruett-Jones 1983). Conspicuous display on traditional sites, however, may attract predators and thereby counter any possible advantages. The Tiny Hawk of Central America, for example, seems to specialize on lekking hummingbirds as prey (F.G. Stiles 1978).

The two favored hypotheses for the evolution of leks are that males gather at sites—"hot spots"—where they are most likely to encounter roaming females, or that females prefer to choose mates—"hot shots"—from large aggregations of males because group displays facilitate comparisons (Vehrencamp and Bradbury 1984; Beehler and Foster 1988). In the male-initiated model, good regional positioning more than offsets competition within the lek, especially if a male is dominant or has a chance of attaining dominant status. In the female-initiated model, grouping of males allows more rapid, efficient comparison than that possible with scattered individuals. Young females also might avoid naive mistakes by learning from the choices of older females.

In some of the best field studies to date, David Lank and Constance Smith (1987, 1992) demonstrated that mating behavior of the Ruff reflects both models. Males directly pursue females (Followers), wait for them at resource-rich feeding grounds (Interceptors), and also wait in groups for females to visit classic leks away from feeding grounds (Lekkers). Males switched their tactics as they tracked the behavior of the females, but committed Lekkers proved to be most successful in the mating game. Lank and Smith suggest that white-ruffed satellite males evolved as specialized Followers adept at tracking the movements of females among neighboring leks. In related, controlled experiments, Lank and Smith (1992) learned that the preference by female Ruffs—called Reeves—for large-sized leks significantly increased the average per capita visitation rates of males on those leks, and thereby favored aggregation rather than separation of displaying males.

Equally demanding of an explanation are the intermediate dispersion patterns of promiscuous male display courts—or exploded leks—in which small numbers of males display 50 to 150 meters apart out of sight of one another but usually within earshot. More numerous than classic lek aggregations are the exploded leks of some promiscuous sandpipers, manakins, flycatchers, parasitic finches, bowerbirds, birds-of-paradise, and hummingbirds. The dispersion of the display courts of birds-of-paradise, for example, is clearly correlated with diet. Fruit-eating species aggregate, insect-eating species disperse, and species with mixed diets are intermediate (Beehler and Pruett-Jones 1983). The reasons for such dispersion patterns remain obscure but probably have an ecological basis.

Failure to consummate copulation because of disruption is one of the major liabilities of joining an aggregation of eager males (Foster 1983; Trail 1985). Destruction of neighboring, rival bowers clearly discourages a clustering of the display courts of the Satin Bowerbird (Chapter 9). Thus, the advantages of dispersed display with little disruption counter the potential advantages of displaying together. The final dispersion of male display grounds probably represents an equilibrium between such competing alternatives. The intermediate dispersion patterns of exploded leks are to be expected when either female preferences for aggregations or the advantages for males are not too strong.

Cooperative Lek Displays

Clustering of males on leks may be partly a result of a tendency of young, inexperienced males to gather near older or successful males. In this way, the young males may get occasional matings and gradually achieve a controlling position in the system. Extreme cases of such associations are seen in *Chiroxiphia* manakins. Several species of these blue-backed, red-capped manakins engage in circuslike, cooperative routines (Foster 1977, 1981). When an interested female visits the lek, two or three males line up on a single branch and perform the cartwheel dance (Figure 17–9), described for the Swallow-tailed Manakin by Helmut Sick:

The males (blue with red crown, black wings and tail) perch closely side-by-side, in a row, on a slightly sloping (or horizontal) twig, face the same direction, all crouched, tripping [moving back and forth with tiny steps], forming a vibrating mass. They call in the recurrent rhythm of a perfectly synchronized "frog chorus." Suddenly the lowest male on the twig rises straight into the air one to two feet and hangs momentarily suspended facing the female. He delivers a sharp *dik dik dik,* then lands at the upper end of the row of males at the side of the motionless female. He pivots immediately in the direction of the other males and joins the other males in tripping. Now the lowest bird performs in a similar manner and so on. The entire performance occurs rapidly, giving the impression of a turning wheel; the speed varies. (1967, p. 17)

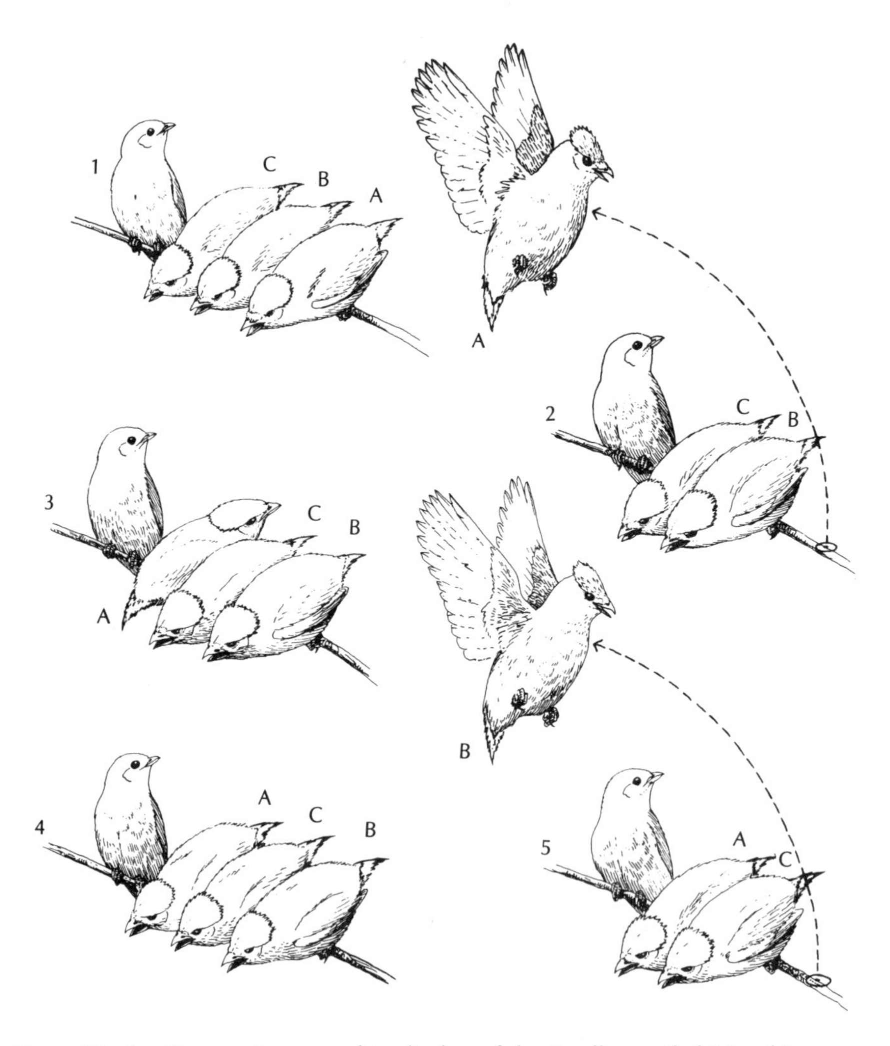

Figure 17–9 Cooperative courtship display of the Swallow-tailed Manakin. Males leap over one another in rapid succession before a waiting female, which may then copulate with the oldest, dominant male. (Adapted from Sick 1967)

This team performance becomes more and more frenzied, then suddenly stops. The oldest, dominant male does a brief, solo precopulatory display and then mounts the female (Foster 1981). Only cooperative group displays attract and excite females, but subordinate males are not being altruistic. They occasionally copulate when the dominant male is absent. They also develop their expertise, and some of them eventually achieve a dominant status.

Male Long-tailed Manakins, a long-lived, cooperative lek display species that is closely related to the Swallow-tailed Manakin, work their way slowly up the social ladder. Young males take at least 8 years to attain the top dominance status that is required to win access to females (McDonald 1989, 1993a). Young males take 4 years to acquire their definitive adult plumage; distinct subadult plumages indicate their first years. The young males queue for dominance rank on the basis of their age and participate in cooperative displays with the top-ranked—alpha—male. They do not actually mate with visiting females, however, until they rise to the second highest position—beta status—in the hierarchy. Mating opportunities remain rare until the manakin achieves alpha status. In a 4-year study, only 8 of 85 marked males achieved any of 117 observed copulations, 90 percent of which were credited to just four males and 67 percent to one alpha male. The differences among male mating success, biased so strongly in favor of alpha-ranked individuals, favor long-term behavioral strategies such as cooperation with older males and delayed plumage maturation, which increase prospects for future success.

Summary

Over 95 percent of birds are socially monogamous. However, the potentially high incidence of chicks fertilized through extra-pair copulations adds uncertainty to the paternity of a brood. Biparental care of young provides the main advantage of monogamy. Monogamy serves the needs of chicks and helps parents (especially females) to reduce their own energy costs. Different reproductive courses are available to males and females, and uncertainties about genetic parentage put stress on the pair bonds and each partner's commitment to care of the young.

Polygyny becomes a viable system when females can take care of young without the assistance of males. Species with precocial young and those that feed on easily accessible resources tend to be polygynous. Polyandry, found primarily in the Orders Gruiformes and Charadriiformes, is a system with competitive, territorial females, which are generally larger than their male counterparts. Promiscuity, characterized by brief matings and immediate separation, often leads to spectacular display behavior, especially among males that establish themselves on leks. We are not entirely sure why leks evolved. The advantage from the male viewpoint is that leks may increase their chances of encountering females; from the female viewpoint, leks facilitate comparisons of prospective mates.

Further Readings

Birkhead, T.R., and A.P. Møller. 1992. Sperm Competition in Birds. New York: Academic Press. *The new view, with many current research examples of the sexual undercurrents of avian mating systems.*

Davies, N.B. 1992. Dunnock Behavior and Social Evolution. New York: Oxford University Press. *A fascinating study of one of the most complex avian mating systems.*

Gowaty, P.A., and D.W. Mock. 1985. Avian Monogamy. Ornithological Monographs No. 37. *A collection of important papers on the advantages of monogamy in birds.*

Lack, D. 1968. Ecological Adaptations for Breeding in Birds. London: Methuen. *A classic review of the diversity of avian mating systems.*

Ligon, J.D. 1993. The role of phylogenetic history in the evolution of contemporary avian mating and parental care systems. Current Ornithology 10: 1–46. *A provocative review of unsolved puzzles.*

Oring, L.W. 1982. Avian mating systems. Avian Biology 6: 1–91. *A comprehensive review of the literature.*

Oring, L.W. 1986. Avian polyandry. Current Ornithology 3: 309–351. *A critical analysis of current ideas about this perplexing mating system.*

Searcy, W.A., and K. Yasukawa. 1989. Alternative models of territorial polygyny in birds. Am. Nat. 134: 323–343. *A timely review of changing theories.*

Snow, D.W. 1976. The Web of Adaptation: Bird Studies in the American Tropics. New York: Quadrangle/New York Times Co. *A lively introduction to the mating systems of cotingas and manakins.*

C H A P T E R 1 8

Growth and Development

THE DEVELOPMENT OF AN INDIVIDUAL begins with embryonic cell divisions and ends with the learning of the complex behavioral skills of a capable young adult. Baby birds undergo part of their development inside the egg, then hatch from the egg, leave the nest, join flocks, and sometimes migrate to distant places. They learn to fly, feed, and sing. They soon distinguish predators from prey and potential mates from potential rivals. This chapter follows the life of a baby bird from hatchling to fledgling. A central theme is the contrast between the altricial mode of development, represented by songbird chicks that are blind and immobile when they hatch, and the precocial mode of development, represented by mobile ducklings or quail chicks that hatch in a more advanced physical state. Such different modes of development affect not only the way that fledglings leave the nest but also the patterns of parental care that are permissible.

The selfish interests of chicks inevitably conflict with those of their parents (Trivers 1974). Chicks vie with one another for parental attention, protection, and extra portions of food. But parents should raise as many, equally vigorous young as possible. Chicks also compete directly with their parents' own requirements for self-maintenance—time, energy, and limited food supplies, for example. Sibling competition leads routinely to siblicide, especially when staggered hatching of the eggs produces nestmates of different ages and abilities. Chicks benefit from continuous, demanding care that exposes their parents to increased predation risk and physiological stress and that limits their parents' abilities to raise additional offspring. Whereas this chapter concerns the chick, the next chapter switches to the viewpoint of the parent.

Embryonic Development

From fertilization to hatching from the egg, the avian embryo undergoes a standard sequence of 42 stages of development, regardless of the length of the incubation period (Starck 1993) (Figure 18–1). The first 33 stages, during which the body plan develops and the initial phases of organogenesis and tissue differentiation take place, vary little among different species. The basic systems of life are established, including a feathered integument, a skeleton made first of cartilage and then gradually calcified, a brain that may continue to enlarge and build internal neural networks, and a digestive system that will set limits to energy intake. Species-specific features then introduce taxonomic variations in the durations of the final stages of functional development. Stage 39, for example, is prolonged in species, such as moundbuilders, that hatch in advanced physical condition but is abbreviated in songbirds and other species that hatch in helpless physical condition.

Hatching

Hatching—breaking the eggshell and emerging from it—is a physical challenge. In its final stages of development, the folded and compact chick fills the limited space inside the egg that was once occupied by yolk and albumen; the chick barely seems to fit inside the tight confines of the shell. To break out of the egg, the hatchling-to-be withdraws its head so

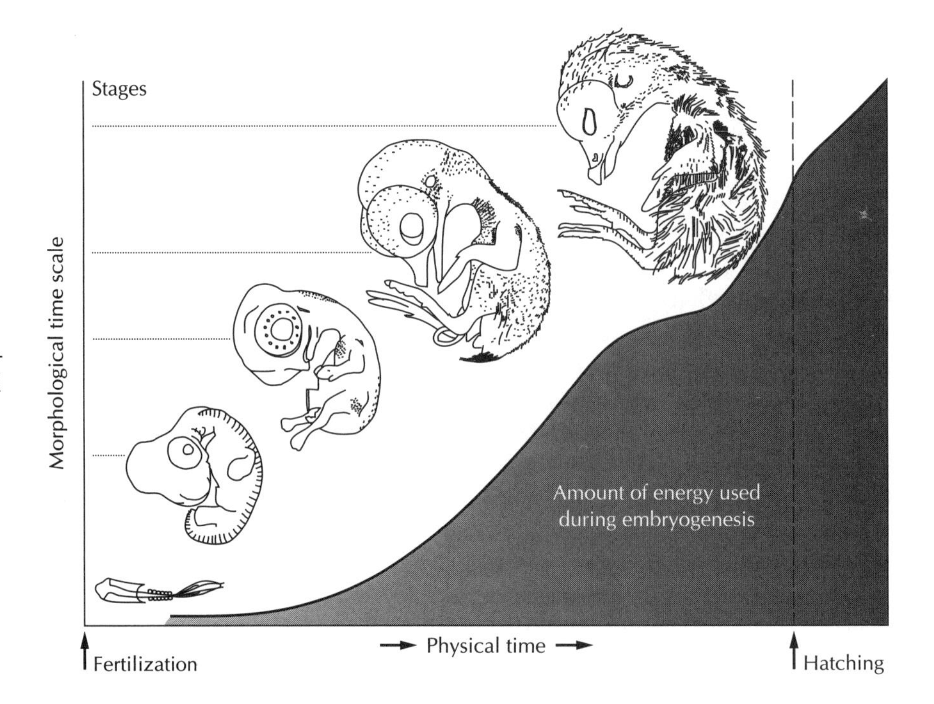

Figure 18–1 The development of the avian chick proceeds through a well-defined sequence of morphological stages from fertilization to hatching. Although the sequence of stages is the same among species, the rate of morphological change and the amount of energy used in each stage vary among species. (From Starck 1993)

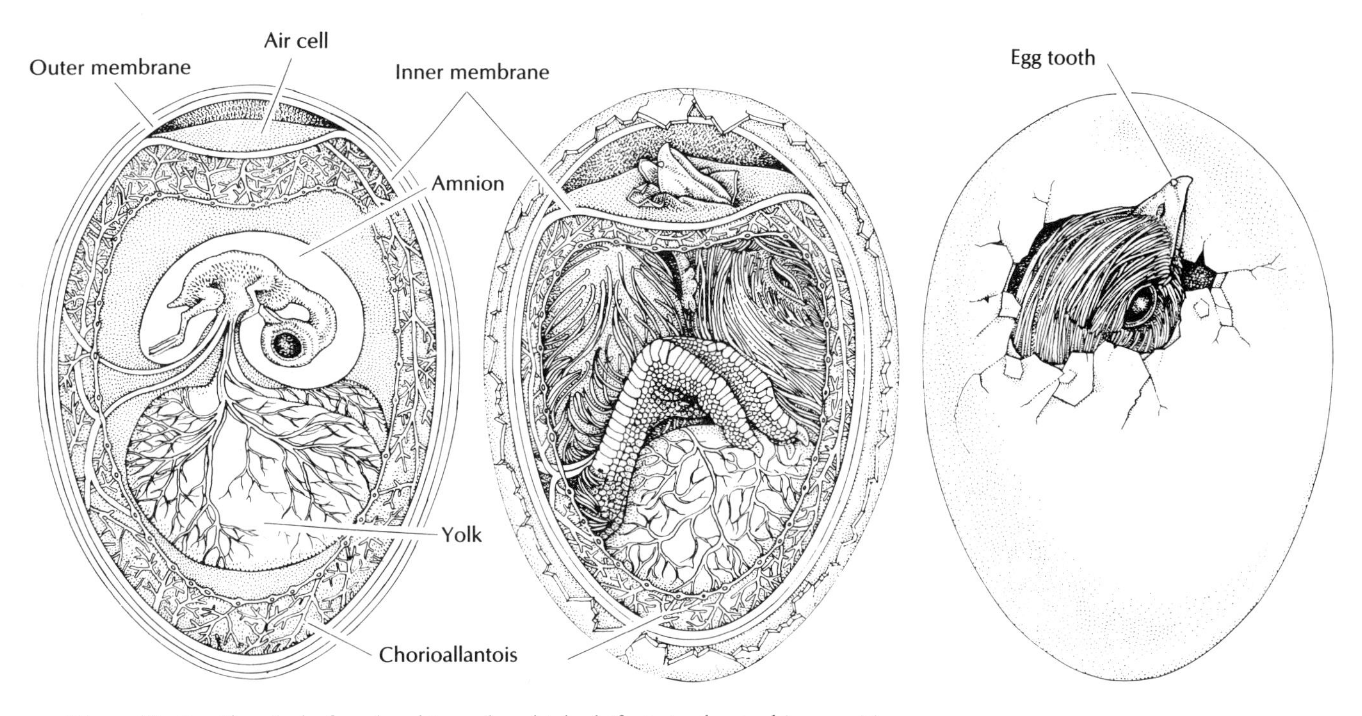

Figure 18–2 Shortly before hatching, the chick shifts into the tucking position, breaks into the air chamber with its beak (middle), and inflates its lungs for the first time. Prior to this event, the developing chick depends on oxygen exchanged through the capillary network of the chorioallantois. The chick chips its way through the eggshell with the aid of an egg tooth. (After Rahn et al. 1979, with permission of Scientific American)

that its bill passes between the body and its right wing. The so-called tucking position increases the efficiency of pipping—or breaking the eggshell—and, thereby, the chances of hatching successfully (Brooks 1978). To hatch, the chick first punctures the membrane that encloses the air chamber at the large blunt end of the egg. Then the chick pecks feebly but regularly at the shell while slowly rotating in a counterclockwise direction by pivoting its legs. After 1 to 2 days of "bumping," the chick leaves a circular series of fractures on the eggshell, and finally penetrates through the eggshell to the world outside (Figure 18–2). The power for the first pecks comes from the hatching muscle on the back of the neck (Figure 18–3). The hatching muscle withers once its task is done.

A special, calcified egg tooth on the tip of the bill helps the chick to break the shell. The hard, sharp-edged egg tooth is generally located just before the bill tip where the tip curves downward. The sheath of the egg tooth includes the lower mandible in loons, rails, bustards, pigeons, shorebirds, auks, hornbills, and woodpeckers. Moundbuilders have an egg tooth early in their development but lose it by hatching time; they kick rather than peck their way out of the egg. Egg teeth drop off the bills of

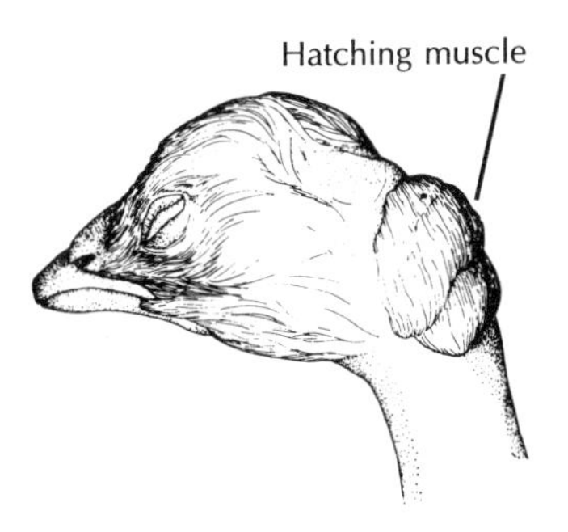

Figure 18–3 The hatching muscle is a transient feature of chick anatomy that helps the chick break out of the egg. (From Bock and Hikida 1968)

most baby birds soon after hatching: 1 to 3 days in shorebirds and fowl, up to 3 weeks in petrels (Clark 1961). Songbirds gradually absorb the egg tooth.

Most birds chip a big hole out of the eggshell and finally shatter it with their body movements. Emerging woodcocks and Willets, however, split the eggshell longitudinally, ripping open a seam rather than breaking the eggshell into pieces (Wetherbee and Bartlett 1962). After hours or even days of struggling, Ostrich chicks virtually explode from their thick-shelled eggs, shattering the shell into many pieces (Sauer and Sauer 1966). Sometimes a parent Ostrich will help crack the shell with its breastbone and pull the chick out by the head. Parents also may help their chicks to hatch by enlarging the initial hole.

Prompt removal of eggshells after hatching protects the camouflage of a nest site. Parents may eat the shell, feed it to their chicks, or take it away from the nest for disposal. Niko Tinbergen (1963) demonstrated that removal of eggshells from nests of Herring Gulls reduced predation by crows from 65 to only 22 percent.

Eggs in a clutch may hatch almost synchronously or asynchronously at intervals that range from a few hours to more than a week (Clark and Wilson 1981). Staggered, asynchronous hatching reflects the onset of incubation before the clutch of eggs is complete. As a rule, the first hatched young have an advantage over their younger siblings, which succumb first to shortages of food and sometimes to physical abuse. Highly synchronized hatching is a feature of species, including many waterfowl and quails, which move their large broods from the nest to safer sites soon after hatching (Johnson 1969).

The 11 to 13 eggs that constitute the clutch of a Mallard duck all hatch within 2 hours despite having been fertilized and laid over a 2-week period. The stages of development of the eggs are brought closer together by differences in their rates of development, but coordinated adjustments during actual emergence are responsible for the final synchrony of hatching. Vocal communication among chicks within eggs enables them to synchronize the time of hatching. Older chicks about to hatch "click" slowly (1.5 to 60 times per second), causing younger siblings to accelerate their hatching effort. Conversely, younger chicks click rapidly (over 100 times per second), causing their older siblings to delay emergence as long as 33 hours (Vince 1969; Driver 1967; Forsythe 1971). Jarring of adjacent eggs by the first hatchling is the final signal; it stimulates nestmates to make their final hatching moves and to escape synchronously 20 to 30 minutes later.

After chicks hatch, they impose two pressing demands on their parents: brooding and feeding. In their first week of life, baby birds cannot regulate their body temperature or generate adequate heat. They need protection from cool air temperatures and from the hot sun. Consequently, parents routinely brood their young, by sitting on them, usually in the nest. Brooding parents protect their young from the rain and predators as well as keeping them warm. Regardless of climate, single parents face potential conflicts between time for brooding on the nest and time for gathering food away from the nest. Within a week or so, however, the young develop thermal independence and greater tolerance for exposure, developments that allow single parents to gather the required food. In cool climates, large broods need less parental care than small broods do (Clark 1983). Decreased individual exposure, pooled metabolic heat, and the greater thermal inertia of their combined mass enable the chicks in large broods to keep one another warm.

Early Experience

Heritage and experience both affect the behavior of birds. Brief imprinting exposures and prolonged learning link the individual's genetic heritage of nerves, hormones, muscles, and bones to social and ambient environments. Ended now are the intense debates of past decades as to whether a particular behavior is innate or learned. The dichotomy was a false one. Instead, behavioral patterns of birds range continuously from those modified slightly by experience to those derived entirely from experience. The innate capacity to learn is evident in a bird's recognition of its own species, its mating and nesting, its choice of habitats and search for food, its social relationships, its escape from predators, and, perhaps most striking, its development of song.

The natural responses of young birds include positive responses of clear adaptive value. Hand-raised Great Kiskadees and Turquoise-browed Motmots, for example, are frightened by sticks painted with black, red, and yellow bands to look like coral snakes (Smith 1975, 1977). Such a reaction is clearly adaptive—coral snakes are dangerous. Rather than having to learn to associate this color pattern with danger by direct experience, birds are programmed from the outset to avoid the risk.

Young birds also learn to recognize predators by observing the mobbing behavior of other birds (Curio et al. 1978). Owls and snakes are scolded vocally and attacked when discovered. Inexperienced birds quickly associate potential danger with this commotion. Eurasian Blackbirds that watch others mob an owl in an aviary will later copy the behavior. Blackbirds will mob a harmless stuffed songbird or even a Clorox bottle if, in experiments, they have seen other birds appear to mob them.

As soon as they are physically able, hatchling Herring Gulls peck at the red spot on the bill tip of a parent to receive food (Tinbergen and Perdeck 1950). The apparently simple stimulus of red near the end of the bill is in reality quite complicated, involving several ingredients such as shape and color contrast. Experiments with color-patterned bill-like sticks in this species and in the Laughing Gull revealed that the most effective stimulus for eliciting pecking was a red or blue, 9-millimeter-wide, oblong rod, held vertically at the chick's eye level and moved horizontally 80 times a minute (Hailman 1967). Hatchling Herring Gulls react faster

(A)

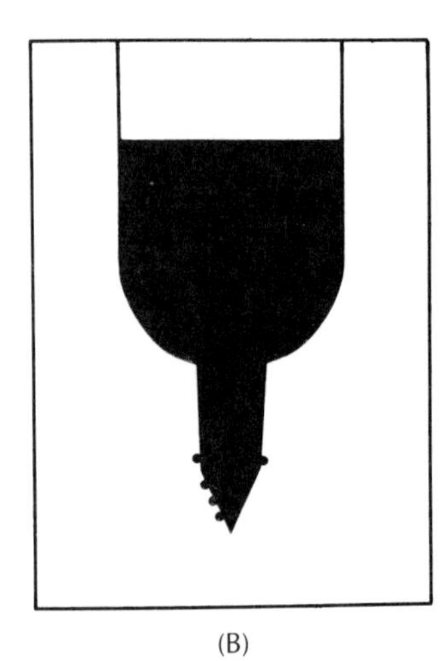

(B)

Figure 18–4 The accuracy of a Laughing Gull chick's pecking at an adult's bill improves with age, as shown in this experiment using a painted card. Dots indicate pecks. (A) The record of a newly hatched chick and (B) 2 days later. (From Hailman 1969, with permission of Scientific American)

to a red knitting needle with three white bands near the tip than they do to a parent's bill. The contrasting red and white borders of this stimulus enhance the most important stimulus features of a real bill.

All behavior shows some refinement with age. Accuracy in pecking increases with age as a Laughing Gull chick's depth perception, motor coordination, and ability to anticipate the parent's position improve (Hailman 1967). Older, more experienced chicks restrict their pecking to stimuli most similar to the head and bill of a real adult (Figure 18–4).

The process of imprinting is fundamental to the development of behavior in many birds (Gottlieb 1968, 1971; Bateson 1976; Hess 1973; Immelmann 1975; Smith 1983). Imprinting is a special kind of learning that occurs only during a restricted time period called the critical learning period; it is irreversible, that is, something once learned during this period persists and cannot be forgotten (Smith 1983). The moving objects that ducklings follow in the first 24 hours after hatching define their future acceptance of comrades and mates. Imprinting determines adult habitat preferences (Klopfer 1963), the prey-impaling behavior of the Loggerhead Shrikes (see Figure 7–20) (Smith 1972), the selection of nest materials and sites by adult Zebra Finches (Sargent 1965), and mate preferences (Chapter 17).

A well-defined period of critical learning is a distinguishing aspect of imprinting (Bateson 1976). An early sensitive period enables young precocial birds to establish the critical concept of "parent," on which their survival depends. Ducklings, for example, imprint most strongly on a moving and calling object when they are 13 to 16 hours old (Hess 1959a, b). Young of species that leave the nest shortly after hatching, such as ducklings, must learn to distinguish their parents from inanimate or inappropriate objects. Two particular stimuli help define a parent to ducklings: movement and short, repetitious call notes. Imprinting is enhanced when both stimuli are present, but movement alone is sufficient. Chicks, ducklings, and goslings will follow and imprint on a human, a moving box containing a ticking alarm clock, or even a moving shadow on a wall. The strength of imprinting increases with the conspicuousness and variety of stimuli presented by the parent (Smith 1983).

The next step in the behavioral development of a chick is to learn to distinguish its parents from other adults. The parents' visual appearance alone may be an important distinguishing factor (Collias 1952). In experiments with various breeds of hens, baby chicks followed the hen that looked most like their mother, on whom they had imprinted initially. Aggressive rebuffs by other adults may reinforce this process; Common Coots learn to avoid menacing adults when 8 to 11 days old.

Baby birds may also imprint quickly on a parent's voice, which is one of the first sounds they hear, perhaps even while they are in the egg. Common Murre chicks recognize their parents' voices upon hatching (Tschanz 1968). Accurate parent-chick recognition is most important in birds that gather in large colonies and yet whose chicks require parental attention (Chapter 17). Young Bank Swallows in large colonies are apt to wander into the wrong burrow and perish because they are not fed. Adults learn to recognize their own young by means of individually distinctive calls and do not accept strange young (Beecher 1982) (Figure 18–5). In contrast, Northern Rough-winged Swallows, a related but solitary nesting species, do not discriminate between their own offspring and those of others placed in their nests. They feed whatever young occupy their nest.

(A)

(B)

Figure 18–5 (A) Colonies of Bank Swallows riddle dirt embankments with their nesting tunnels. (B) A brood of three young swallows, almost ready to fledge, waits for food at an entrance. (Courtesy of A. Cruickshank/VIREO)

Altricial versus Precocial Modes of Development

> Perhaps the single most striking feature of postnatal growth in birds is the dichotomy between precocial and altricial development. (Ricklefs 1983a, p. 3)

Some hatchlings are helpless and depend entirely on their parents. Others are mobile and able to find their own food. A 3-day-old Lesser Scaup duckling, for example, can dive, catch a minnow, and return to the surface. Even more precocious are the moundbuilders, whose chicks are independent at hatching. The terms *altricial* and *precocial* refer to the extremes of this spectrum of increasing maturity at hatching and decreasing dependence on parental care (Nice 1962; Starck 1993) (Table 18–1).

Although most birds are altricial or precocial, the variety of conditions among hatchlings requires an artificial classification with at least six major categories of hatchlings based on primary criteria of mobility, open or closed eyes, presence or absence of down, and the extent of parental care (Box 18–1; Figure 18–6)

Altricial birds are naked, blind, and virtually immobile when they hatch and thus are completely dependent on their parents (Figure 18–7). The helpless, grublike nestlings of altricial birds appear to have hatched prematurely. Altricial hatchlings have huge bellies and large viscera, which reflect their need for food and fast growth. In contrast, precocial chicks are well-developed little birds, usually covered with fuzzy down. They

TABLE 18–1

Comparison of altricial and precocial modes of development

Character	Altricial	Precocial
Eyes at hatching	Closed	Open
Down	Absent or sparse	Present
Mobility	Immobile	Mobile
Parental care	Essential	Minimal
Nourishment	Parents	Self-feeding
Egg size	Small (4–10%)[a]	Large (9–21%)[a]
Egg yolks	Small	Large
Brain size	Small (3%)[a]	Large (4–7%)[a]
Small intestine	Large (10.3–14.5%)[a]	Small (6.5–10.5%)[a]
Growth rate	Fast (3–4 times precocial rate)	Slow

a. Percentage of adult weight.

Condition		Down	Sight	Mobility	Parental nourishment	Parental attendance	Examples
Superprecocial		○	○	○	○	○	Moundbuilders
Precocial		○	○	○	○	●	Duck, shorebirds, quails, grouse, murrelets
Subprecocial		○	○	○	◒	●	Grebes, rails, cranes, loons
Semiprecocial		○	○	◒	●	●	Gull, terns, auks, petrels, penguins
Semialtricial	1	○	○	●	●	●	Herons, hawks
	2	○	●	●	●	●	Owls
Altricial		●	●	●	●	●	Songbirds, woodpeckers, parrots

○ **Precocial characters** ● **Altricial characters**

Figure 18–6 Development characteristics of baby birds at hatching, according to Margaret Nice's (1962) classification. Down: present or absent; Sight: open or closed eyes; Mobility: ambulatory or nestbound; Parental nourishment: no or yes; Parental attendance (for brooding or defense): absent or present. (After Ricklefs 1983a)

can feed themselves, run about, and regulate their body temperature soon after they hatch. Their brains are quite large compared with those of altricial chicks.

Precocial birds, such as quails and plovers, lay larger eggs than do altricial birds of the same size. Precocial birds lay eggs that are composed of 30 to 40 percent yolk; altricial birds' eggs are 15 to 27 percent yolk. The eggs of moundbuilders have extremely large yolks (62 percent of the egg mass). The larger eggs of precocial birds yield larger chicks that are well advanced in their development at hatching or that have large food stores to increase their initial chances of survival out of the egg. They absorb their substantial yolk reserves as a supplement to their feeding for several days after hatching (Ricklefs 1974). The total energy requirements of the embryos of precocial birds are greater because their incubation periods are longer and their rates of metabolism are higher.

The precocial chicks of quails and waterfowl leave the nest soon after hatching. Although they do not depend on their parents for delivery of food, they rely on them for food location and for protection. Quail chicks quickly learn what is and is not edible by pecking at objects shown to them by their parents. Parent ducks and geese also guide their chicks to food, with one exception: The Pied Goose of Australia transfers food from its bill tip to the chicks' bill tip (Kear 1963).

Precocial chicks of quails and shorebirds require more parental care than is usually recognized (Safriel 1975; Walters 1984). In addition to brooding, protection from predators is a demanding effort that requires constant parental vigilance. Males and females of a large species of plovers called lapwings, for example, face major time constraints while taking care of their mobile young (Walters 1982). They alternate "tending" behavior in order to feed.

Each mode of development has distinct advantages, some of which reflect food habits (Nice 1962; Ricklefs 1983a). The food of most precocial

BOX 18–1

Ornithologists recognize six development categories of hatchlings

Superprecocial Wholly independent. Examples: moundbuilders and Black-headed Ducks

Precocial Leave the nest immediately (nidifugous) and follow their parents; pick up their own food soon after hatching, although parents help to locate food. Examples: ducks and shorebirds; quails, grouse, and murrelets; also Ostrich and kiwis

Subprecocial Leave the nest immediately and follow their parents; are fed directly by their parents. Examples: rails, grebes, cranes, loons; also guans and some pheasants

Semiprecocial Capable of body temperature regulation; mobile but stay in the nest; fed by their parents. Examples: gulls, terns, auks, petrels, and penguins

Semialtricial Stay in nest (nidicolous), although physically able to leave the nest within a few hours or the first day; fed and brooded by parents. Examples: herons, hawks; also nightjars, albatrosses, and seriemas

Altricial Naked, blind, and helpless at hatching. Examples: songbirds, woodpeckers, hummingbirds, swifts, trogons, kingfishers, pigeons, parrots

birds—small invertebrates or, occasionally, seeds—can be procured by young chicks. Semiprecocial species and many altricial species live on food that must be located and captured with adult strength and skill. For these chicks, dependence on parents during a period of growth, maturation, and learning is essential before they can feed themselves. Semiprecocial chicks of gulls, terns, auks, and petrels are fed at the nest, but they can regulate their body temperature, thus freeing their parents to commute to distant feeding grounds (Ricklefs 1979a; Ricklefs et al. 1980). The subprecocial chicks of grebes and loons, which cannot dive and chase prey skillfully, are also fed by parents, but the young of these species must leave the nest to avoid predation and to free parents from costly flights between feeding areas and nests. Consequently, rails and coots brood their young away from the nest on special platforms that they build above the water. Rails regularly pick up chicks with their bills and move them to safety (Turner 1924; Bent 1926). Grebes carry their young on their backs, often under their wings, diving and feeding relatively undisturbed. Sungrebes have special pouches under each wing for carrying new hatchlings (Alvarez del Toro 1971).

There is no simple evolutionary sequence from the precocial to the altricial condition. Precocial development was the original mode among birds, being typical of many primitive groups of birds, including the ratites, waterfowl, and chickenlike birds. The altricial condition has evolved independently in many groups of birds; semialtricial or semiprecocial modes of development evolved secondarily from altricial modes of development (Ricklefs 1983a).

The primary advantage of altriciality relates to specialized nutritional adaptations that require food delivery by parents and learning of feeding skills. Altricial development appears to be a necessary prerequisite for the evolution of a large adult brain volume. The altricial mode of development

(A)

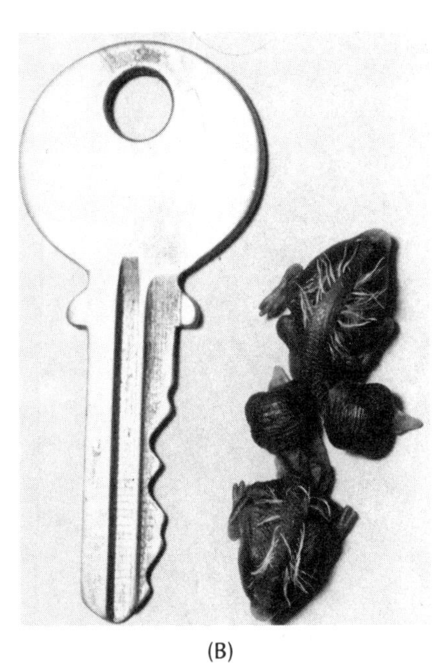

(B)

(C)

(D)

(E)

(F)

Figure 18–7 Baby birds and their states of development at hatching. (A) Cedar Waxwing, altricial; (B) Ruby-throated Hummingbird, altricial; (C, D) Least Bittern, semialtricial; (E) Leach's Storm-Petrel, semiprecocial; (F) Whimbrel, precocial; note the egg tooth, the white structure at the tip of the bill, which the chick uses to break the eggshell. (Courtesy of O. Pettingill, Jr./VIREO; A. Cruickshank/VIREO; W. Conway; and D. Hosking)

substitutes parental care for functional differentiation of the brain, thereby allowing the undifferentiated brain to continue to grow after hatching (Starck 1993). The enlarged forebrain of altricial birds differentiates its control functions at a later stage than does the brain of precocial birds. At the other extreme are the superprecocial moundbuilders and Black-headed Ducks. Their reptilelike development patterns are breeding specializations derived from precocial ancestors rather than primitive states.

Temperature Regulation

Homeothermy—the ability to generate metabolic heat (endothermy) and to maintain a high, constant body temperature—is a major step of early development. Homeothermy releases a chick from its absolute depen-

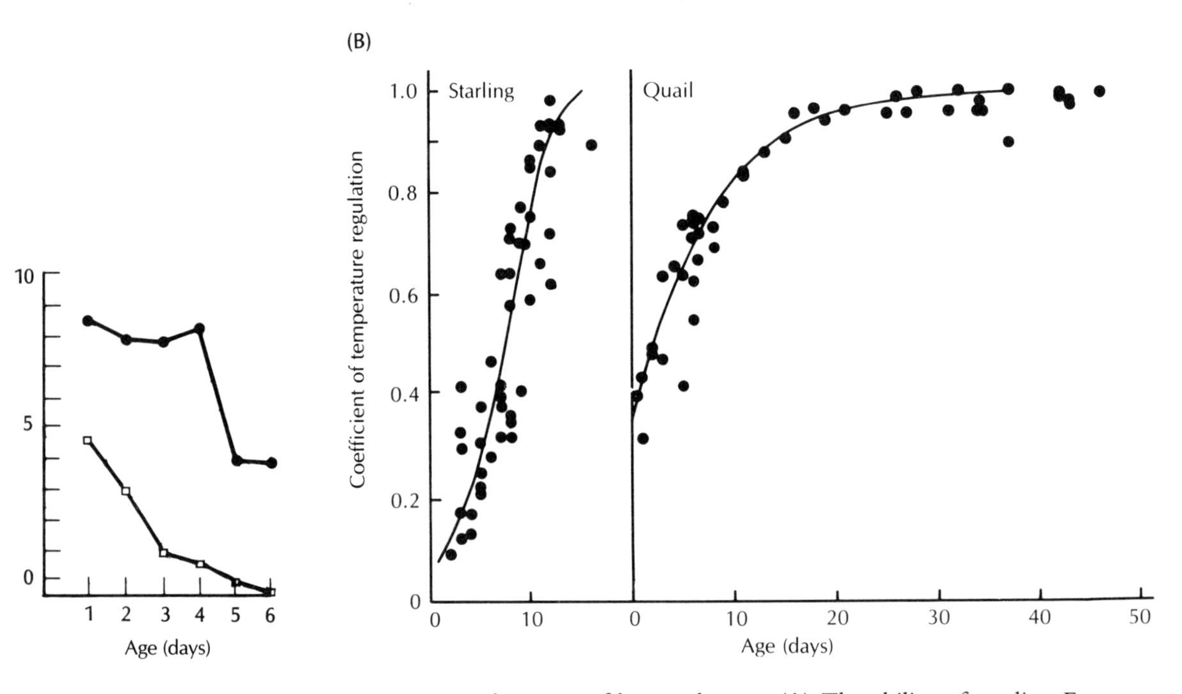

Figure 18–8 Development of homeothermy. (A) The ability of nestling European Starlings to maintain a body temperature of 39°C increases with age and number of broodmates. The body temperature of a single nestling (solid circles) 1 to 4 days old drops 8° to 9°C after 1 hour of exposure to an ambient temperature of 20°C, whereas the body temperature of a 5-day-old nestling drops only 4°C under the same conditions. The presence of broodmates together (seven in this experiment; open squares) greatly reduces individual loss of body temperature. (B) Precocial chicks such as those of the Japanese Quail can maintain a high body temperature better on hatching than can European Starlings; this ability improves with age in both species. The coefficient of temperature regulation is the percentage of the difference between adult body temperature and air temperature at 20°C that a chick maintains after 30 minutes. (A after Clark 1983; B from Ricklefs 1979b)

dence on parental brooding and enables it to tolerate exposure. Precocial and semiprecocial chicks such as those of quails, gulls, and terns have partially developed capabilities for homeothermy when they hatch and then achieve 90 percent of their adult thermoregulatory capability within 1 week. Most small, altricial songbirds take a week to develop the initial stages of homeothermy. Rapid changes in the metabolism and temperature responses of hatchlings relative to those of advanced embryos mark the process of hatching as a major gateway for the development of homeothermy in the Brown Noddy, a large, tropical tern (Mathiu et al. 1991). Shutdown of the chorioallantoic circulation in the days preceding hatching reduces heat loss by the unhatched chick. Pipping through the shell membranes improves the chicks access to oxygen, which supports increased metabolism. Emergence from the shell itself allows increased movement, and ventilation and shivering. Once out, the chick's down dries to form functional insulation. The stage is then set for the rapid development of endothermy and refined thermoregulation through maturation of muscular tissue and endocrine control systems.

Figure 18–9 Gray Gull shading its young from the intense desert sun. (Courtesy of T.R. Howell)

Regulation of temperature by both precocial and altricial chicks improves during development as a result of increased mass relative to surface area, improved insulation (due to feather growth and fat deposition), increased metabolic heat production, and the development of nervous and endocrine system control. The functional mass and thermal inertia of each chick in a brood increase with brood size. As a result, altricial chicks in large broods achieve homeothermy earlier than those in small broods (Figure 18–8).

Skeletal muscle is the main source of heat production. The large leg muscles of a young chick are of primary importance in early thermogenesis, followed by the pectoral muscles. Early development of large pectoral muscles in the Willow Ptarmigan (Aulie 1976) and Leach's Storm-Petrel facilitates heat production (Ricklefs et al. 1980). The pectoral muscles of nestling Leach's Storm-Petrels mature by 2 weeks of age, even though the chicks do not fly for 9 to 10 weeks.

Figure 18–10 A young Laysan Albatross loses heat by leaning back and exposing its feet to the breeze. (Courtesy of T.R. Howell and G.A. Bartholomew)

A chick's ability to retain metabolic heat also improves as its feather coat thickens. Oxygen consumption of nestling Great Tits and European Pied Flycatchers, for example, increases by 25 and 15 percent, respectively, if they are shaved (Shilov 1973). Down insulation enhances the thermoregulation abilities of hatchling precocial birds.

The chicks of seabirds that nest in the hot sun are vulnerable to overheating; shading by their parents may be essential (Figure 18–9). However, young Least Terns and Sooty Terns can prevent overheating better than they can prevent chilling (Howell 1959; Howell and Bartholomew 1962). Laysan Albatross chicks can also thermoregulate at an early age by dissipating excess heat from their large feet, which they expose to the breeze by leaning back on their ankles (Figure 18–10). Dehydration and poor thermoregulation are the primary causes of death among Laysan Albatross chicks on Midway Island in the Pacific Ocean (Sileo et al. 1990).

Energy and Nutrition

Chicks require energy for maintenance, temperature regulation, activity, excretion, and growth. Growth accounts for a major fraction of total energy expenditures early in development—31 percent in the Common Tern, 46 percent in the Sooty Tern, and 56 percent in Leach's Storm-Petrel—but a small fraction in the later stages (see Ricklefs and White 1981; Ricklefs et al. 1980). Peak needs of total energy expenditures occur late in development. The energy channeled into growth constitutes roughly 21 to 40 percent of a chick's energy budget for the entire developmental period. Important as energy is, it may be less important in determining rates and patterns of development than is nutrition.

Production of new tissues during growth requires nutrients such as the amino acids that the body cannot manufacture and the sulfur-containing amino acids cysteine and methionine, which are used in feather production. To get the calcium they need, some chicks are fed fragments of teeth, bone, snail shells, and eggshells as dietary supplements (Löhrl 1978; Walkinshaw 1963). The bone growth of Lapland Longspurs cannot depend on the meager amount of calcium (0.1 percent by dry weight) in the craneflies and sawflies they eat; their parents must also feed them lemming bones and teeth (Seastedt and Maclean 1977).

Chicks require protein, especially in the early stages of their development. Parents of many species of songbirds supply mostly small, soft-bodied insects at first, gradually increasing the proportion of fruits and seeds. Protein-rich aquatic insects constitute 90 percent of the diet of baby American Black Ducks during the first 5 days after hatching but later drop to 43 percent of the diet (Reinecke 1979). Fruits do not usually provide an adequate diet for nestling growth (Morton 1973; White 1974; Foster 1978; Ricklefs 1976b). The chicks of Bearded Bellbirds and Oilbirds, which eat only fruits, grow half as fast as those of other tropical birds. The Resplendent Quetzal, primarily a fruit eater, feeds its young only insects for the first 10 days and so attains a more normal growth rate.

Pigeons, flamingos, and Emperor Penguins feed nutritious esophageal fluids to their young (Table 18–2). Pigeon milk, the best known of these, is full of fat-laden cells sloughed off the epithelial lining of the

TABLE 18–2
Composition of avian esophageal fluids

Bird	Protein (%)	Lipid (%)	Carbohydrates (%)
Pigeon	23	10	0.0
Flamingo	8	18	0.2
Penguin	59	29	5.5

From Fisher 1972.

crop. Pigeon crop milk often contains some small food fragments as well (Ziswiler and Farner 1972). Like the milk of marine mammals, this fluid is rich in protein (23 percent) and fat (10 percent). It also includes essential amino acids (Fisher 1972). Esophageal fluid is initially the sole source of nutrition for the chicks of Greater Flamingos. Flamingo milk has more fat and less protein than pigeon milk. The rich esophageal fluid on which Emperor Penguin chicks feed is rich in both fat and protein and contributes to a doubling of their body weight during the first week of life.

Growth Rates

The growth of body mass of a baby bird during development follows an S-shaped—or sigmoid—curve (Figure 18–11). At first the chick grows slowly, then the growth rate accelerates and mass increases rapidly, and finally, growth decelerates as the chick's weight approaches that of the adult. The sigmoid curve allows comparison of species that differ in size and growth strategies because it is defined mathematically by only a few variables: initial size, growth rate, and final (asymptotic) value.

There are two major exceptions to the typical shape of the growth curve. The mass of young ground feeding birds, such as doves and Curve-billed Thrashers, levels off below adult mass. Chicks gradually

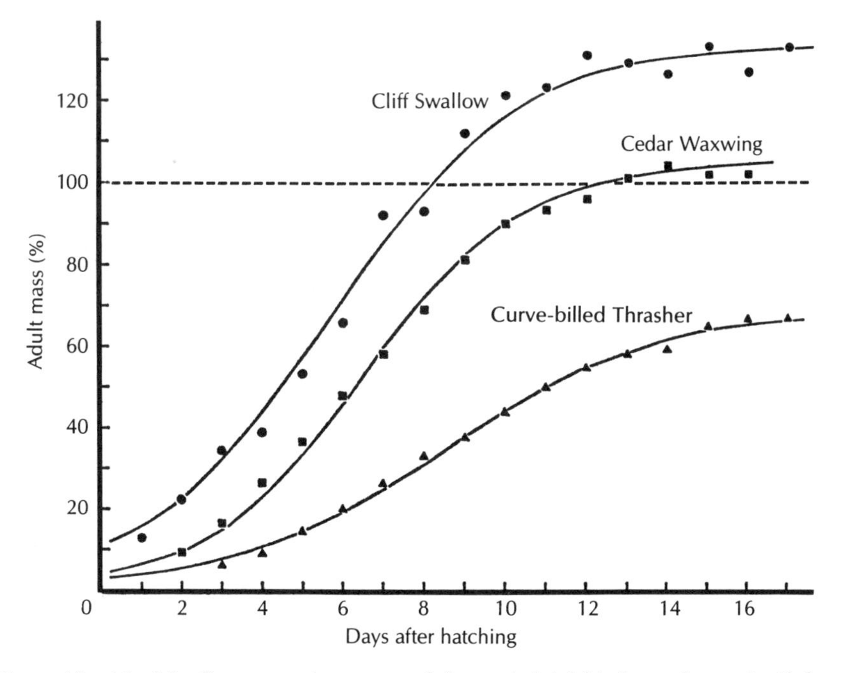

Figure 18–11 Nestling growth curves of three altricial birds, as determined from initial size, growth rate, and final maximum value. Data are standardized to the maximum values of the growth curve in order to directly compare birds of different species size. (Adapted from Ricklefs 1968)

achieve full size after fledging, when their muscle and plumage development catches up with their skeletal development. In aerial species, such as swallows, swifts, and oceanic seabirds, chick mass overshoots that of adults in the final stages of development and then declines as a result of metabolism of fat deposits and loss of water from maturing tissues, especially in the skin and at the bases of the feathers. In some of the latter species, then, chicks near fledging can temporarily outweigh adult birds.

Some chicks store excess energy as fat to insure themselves against poor food delivery by parents or as reserves for the days just after fledging when the chick learns to feed itself. Aerial passerines such as swallows deposit more fat than other species, apparently as an adaptation to the irregularity of their food supply (O'Connor 1977). The accumulation of fat is most striking in petrels; their obese chicks reach masses twice those of the adults. Young Oilbirds, which are raised on the oily lipid-rich fruits of palms and other tropical trees, also accumulate large lipid stores. These stores are in large part excess energy that must be stored on the side in order for the chicks to extract adequate amounts of protein from their specialized, protein-poor and lipid-rich foods. Such excess lipid supplies also act as reserves for bad times rather than as energy stores needed to complete development (Ricklefs et al. 1980).

The growth rates of chicks of different bird species vary 30-fold. Over half of the variation in growth rate relates directly to adult body weight: big birds grow more slowly than little birds, the rate decreasing roughly as the cube root of adult body weight increases. Contrasting with the fast growth of small songbirds, which have short nestling periods of 10 to 12 days, are the slow growth rates of the Wandering Albatross, one of the largest seabirds. It has the longest known nestling period of any bird—up to 303 days. Hole nesters also grow more slowly than open nesters, and tropical region land birds grow more slowly than temperate region land birds. Rapid growth is the advantage of altricial development: Altricial nestlings grow three to four times faster than precocial chicks. The brevity of the altricial development period more than compensates for the chicks' vulnerability to predators and bad weather.

The evolution of different growth rates is a major research topic with several competing hypotheses. David Lack (1968) regarded growth rate as a balance between selection for rapid growth to escape predation and selection for slow growth to reduce food requirements; parents might be able to rear more slow-growing offspring when food is limited. Ricklefs' (1983a) models of the energetics of growth suggest that these considerations are important to altricial birds but cannot explain the difference in growth rates between altricial and precocial birds.

The tissue allocation hypothesis is a unifying and simplifying explanation of the differences in growth rates between altricial and precocial species (Ricklefs 1979a, b). According to this hypothesis, growth rates reflect the channeling of limited resources either into increased tissue mass or into maturation of tissue functions required for survival. In other words, growth of tissue mass and maturation of tissue functions (such as muscle contraction) are mutually exclusive, so growth should slow as the individual matures.

Comparison of the altricial European Starling, the semiprecocial Common Tern, and the precocial Japanese Quail illustrates the interaction between precocity of tissue maturation and overall growth rate (Figure 18–12). Of these three species, the starling grows fastest, the tern grows nearly as fast as the starling, and the quail grows relatively slowly. The rapid maturation of the quail's large leg muscles, essential for precocial locomotion, detracts from the chicks potential growth rate. The tern also exhibits rapid leg development, but the material and energy needed for growth of its tiny legs are only minor investments relative to its overall growth. The starling puts energy into growth before tissue maturation.

Individuals of a species can also exhibit markedly different growth rates. Growth rates of individuals are affected by variations in quality and quantity of food, temporal pattern of feeding, and temperature, all of which vary according to locality, season, habitat, and weather. For example, the average fledging weights of Rhinoceros Auklets in Puget Sound vary from 339 grams in bad seasons to 521 grams in good seasons (Summers and Drent 1979). The effects of food supply on growth rate are perhaps best known in swifts and martins (Lack 1956; Bryant 1975,

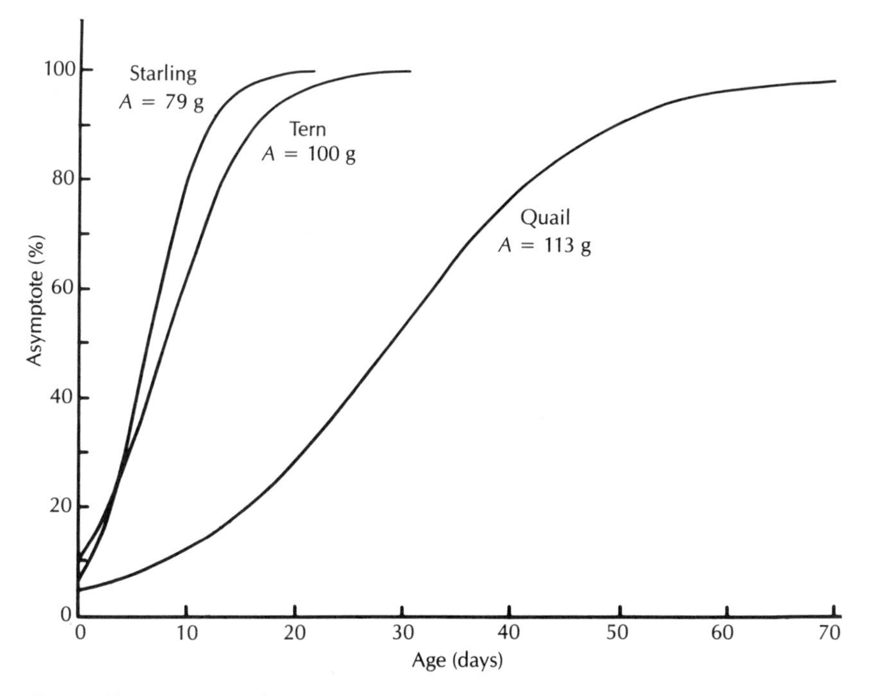

Figure 18–12 Growth curves for an altricial bird (European Starling), a semiprecocial bird (Common Tern), and a precocial bird (Japanese Quail). A is the bird's mass at the maximum value of the growth curve. (From Ricklefs 1979b)

1978a, b). The maturation of Common Swifts, for example, varies from 37 to 56 days, reflecting feeding conditions. Chicks of these swifts can survive up to 21 days of starvation by ceasing to grow and becoming hypothermic (Koskimies 1948). Domestic poultry chicks can remain at a physiological age of 10 days for months if their diets are deficient (McCance 1960; Dickerson and McCance 1960); with access to an adequate diet, normal growth resumes.

Brood size mediates the effects of food limitation on nestling growth rate (Royama 1966b; O'Connor 1975; Mertens 1969). Nestling growth rates often decrease as brood size increases, a relation that suggests that parents cannot deliver enough food to all nestlings to ensure maximum growth (Klomp 1970). Large broods enhance the differences in growth rate between older and younger siblings of nestling Northern House-Martins (Bryant 1978a, b). The predictable starvation of small chicks when food is insufficient tends to adjust brood size to food availability (Lack 1954). Competition among nestmates for food may also promote siblicide.

Sibling Rivalry

Vicious rivalry seems to be normal among the chicks of some birds (O'Connor 1978; Stinson 1979; Mock 1984a, b), especially if the eggs hatch at staggered intervals, thereby producing different aged young that must coexist in the nest and compete for parental care. Larger siblings bully their nestmates and thereby get the first choice of food delivered by their parents. As a rule, parents react passively to the deeply rooted, destructive behavior of their offspring. Siblicide—or cainism—is a standard method of brood reduction in the nests of some eagles, skuas, herons, and boobies. In the well-studied Verreaux's Eagle, for example, only once in 200 records did both siblings survive to the fledging stage. In most cases, the older sib deliberately killed the younger eaglet (Brown et al. 1977; Gargett 1978). Experimental size matching of brood mates in the Lesser Spotted Eagle did not prevent the inevitable dominance and eventual killing of one by the other (Meyburg 1974).

In the South Polar Skua, however, the younger of two siblings has a good chance of surviving if it is nearly the same size as the older chick but a poor chance of survival if it is as little as 8 grams (10 percent) lighter than its older nestmate (Spellerberg 1971; Procter 1975). Thus, small delays in hatching time place a younger chick at a competitive disadvantage with respect to its nestmates, particularly when food is in short supply. In general, sibling competition within broods may be a powerful selective force that affects the length of the incubation period in birds (Ricklefs 1993).

Sibling rivalry is a way of life in some colonial herons, such as the Great Egret, but not in others. In Texas breeding colonies, Great Egret nestlings

TABLE 18–3
Fate of the youngest chick in natural and experimental foster broods

Brood[a]	Number alive by day 25	Number of siblicidal deaths	Number of other deaths	Number of broods studied[b]
Great Egret chicks				
Natural	5	8	4	17
Foster	4	6	0	10
Great Blue Heron chicks				
Natural	8	1	10	19
Foster	1	6	2	9

a. Natural broods were raised by parents of the same species. Foster broods were experimentally switched so that they were raised by parents of the other species; for example, foster Great Egret chicks were raised by Great Blue Heron parents.
b. Typical brood size in all cases was three or four chicks.
From Mock 1984a.

were often killed by an elder sibling, but Douglas Mock (1984a) rarely saw siblicide in the Great Blue Heron. Why should two such similar species differ in this way? The type of food the parents bring their nestlings is part of the answer. Great Egrets bring small fish, which are easily monopolized by the aggressive older sibling, whereas Great Blue Herons bring larger fish, which cannot be easily monopolized. When experimentally cross-fostered in Great Egret nests, young Great Blue Herons adopted the siblicidal tactics typical of the egret, in response, it seems, to the opportunities presented by the smaller food. Surprisingly, the converse result did not take place, for Great Egrets cross-fostered in Great Blue Heron nests did not become more tolerant of their nestmates. Sibling aggression in the Great Egret is a deep-seated, obligatory behavior similar to that in raptors; it is optional in the Great Blue Heron (Table 18–3).

Feeding the Nestlings

Normal feeding behavior requires the interaction of chick and parent, using a range of auditory, visual, and tactile cues. For example, the begging cries of young stimulate parents to deliver food to the nest. Experimental increases in the volume and continuity of begging cries at the nest have prompted greater activity. In a classic experiment, Lars von Haartman (1953) hid extra young European Pied Flycatchers behind the wall of a nest box. In response to their cries, the parents brought more food to the nest than was required for their nestlings. Usually it seems as if the largest young and biggest mouth gets most of the food, but females of at least two species—European Pied Flycatchers and Budgerigars—preferentially feed runt nestlings (Stamps et al. 1985; Gottlander 1987).

The visual stimuli of the gaping mouths of nestlings facilitate or in some instances allow food delivery by the parents. The chicks of some cavity nesting species have brightly colored mouth markings to attract parental attention and serve as targets (and as stimulants) for food delivery. Gouldian Finches of Australia have black spots inside their mouths and three opalescent green and blue spots in the corners. The arrangement of spots on the insides of the mouths of some African finches also is unique to the species and apparently protects the young against competition from brood parasites (see Chapter 19).

Nestlings receive food by direct insertion, sometimes deeply into the digestive tract. Young hummingbirds receive an injection of nectar and insects through their mother's long, hypodermiclike bill (Figure 18–13). Regurgitation of a meal either directly into a nestling's mouth or onto the ground for the nestling to pick up is common among seabirds. Young penguins and pelicans plunge their heads deeply into their parents' gullets. Spoonbills and albatrosses cross their large bills with those of their young, like two pairs of open scissors, so that the chicks' mouths are in position for food transfer.

Food delivery rates to nestlings range from every second or third day in albatrosses to once or twice daily for seabirds, swifts, and large raptors to once per minute for some small land birds with large broods. Normal rates of food delivery by small and medium-sized land birds average 4 to 12 times per hour. Trogons bring food to the nest once per hour, Bald Eagles 4 to 5 times per day, and Barn Owls 10 times a night. Recorded extremes of rapid food delivery to large broods include 990 trips per day by the Great Tit and 491 trips per day by the House Wren.

Food delivery rates vary according to the age of the young. Hatchlings require only small amounts of food, but as they develop, their appetites grow. The European Pied Flycatcher brings food to the nest every 2 minutes, making about 6200 feeding trips to nourish its young from hatching to fledging. In general, parents must gather two to three times as much food as they need for themselves to cover the energy needs of their nestlings (Walsberg 1983). To meet such demands, the Common Swift of Europe flies 1000 kilometers a day, scooping insects from the sky.

Nest Sanitation

Some young birds instinctively eject liquid feces away from the nest; others eliminate feces accurately through nest hole openings. For example, female hornbills, sealed in their nest holes, defecate through the narrow slit remaining in the mud-sealed opening. But fouling of the nest is common. Nests of many pigeons, raptors, and carduelline finches, such as the House Finch, are well known for their filthy conditions. Many birds, however, are fastidious, regularly removing feces and other debris

Figure 18–13 Parent birds feeding young. (A) Anhinga young begging for food; (B) parent Anhinga feeding one of the young; (C) Ruby-throated Hummingbird nestlings begging for food; (D) parent hummingbird feeding one of the nestlings. (Courtesy of A. Cruickshank/VIREO)

to prevent the nest from becoming a breeding ground for disease and insects and other parasites. The larvae of a particular moth species help to clean the nests of the Golden-shouldered Parrot of Australia, and a beetle provides nest sanitation services for Australian finches.

Nest sanitation is made easier for most passerine birds and woodpeckers because their young excrete fecal sacs, which are packages of excrement surrounded by a gelatinous membrane. The parent can easily pick up the sac and drop it away from the nest (Figure 18–14). Incomplete digestion by nestlings leaves some residual food in their fecal sac, which is often eaten by parents for nutrition as well as sanitation purposes. In one study, fecal sacs provided 10 percent of the daily energy requirements of adult White-crowned Sparrows (Morton 1979).

Figure 18–14 Bearded Parrotbill removing fecal sac from nest. (Courtesy of E. Hosking)

Fledging from the Nest

As a naked, blind hatchling transforms into a feathered juvenile, the young bird approaches a pivotal event in its life—leaving the nest. Technically speaking, the nestling period is the interval between hatching and departure from the nest; the fledging period is the interval between hatching and flight (Skutch 1976). The nestling and fledging periods may be the same for altricial birds such as hummingbirds but different for sub-precocial and precocial birds, which have short nestling and long fledging periods. The moment of departure from the nest by altricial birds is commonly termed fledging even though the young birds may only flutter and scramble about for a few days before their first flight. Departure from the nest increases vulnerability to predators and the weather. Unable to fly well, the baby bird cannot easily escape predators, and the mortality rate during this period is high. Once past the first dangerous days, however, the fledged chick is safer than it would have been back in a vulnerable nest. Fledglings respond to the warning calls of their parents by hiding or by staying still. Immobility combined with camouflaging plumage can render chicks extremely difficult to find.

Mobile young birds move with their parents closer to good feeding grounds, a tactic that reduces the strain on the parents. The initial journey away from the nest is often a heroic one. Wood Ducks and other tree nesting waterfowl have been known to carry their young to and from the ground in their bills or on their backs (Johnsgard and Kear 1968), but usually they just jump. One brood of Wood Ducks jumped 2 meters to the ground from their nest in a tree cavity and then followed their mother down a bluff and across a railroad track before swimming three-quarters of a mile across the Mississippi River to feeding grounds in good bottomland (Leopold 1951). More amazing still is the pair of Egyptian Geese that bred for several years on the roof of a three-story building in Johannesburg, South Africa. Once the chicks hatched, the female herded them toward the drain inlet on the roof, and after a little pushing and shoving, they fell three stories down the drain pipe to be shot out parallel to the ground by the curved end of the drain pipe (P. Ryan, personal communication).

Under more natural circumstances, precocial chicks that leave nests in tall trees or high cliffs must leap to the ground below, bouncing off soft earth if they are lucky or off jagged rocks if they are not. Torrent Ducks, for example, live in the dangerous waters of fast-flowing streams high in the South American Andes. To leave their nests in cliff crevices or holes above the streams, ducklings plunge as much as 20 meters into the turbulent water of the rocky stream below. Only rarely do they hurt themselves. Their light weight, buoyancy, and downy cushioning protect them from severe impact. Young seabirds that grow up on tiny cliff ledges overlooking the sea must also leap into space, fall 150 meters, sometimes bouncing off jagged rocks, and then flutter to the water below.

In general, chicks of auks and their relatives in the Family Alcidae face a trade-off between staying in their inaccessible, safe nests, which often are on cliff ledges, and fledging to the ocean below, where growth rates are faster (Ydenberg 1989). At sea, parents do not have to commute long distances between feeding grounds and their hungry chicks. Better feeding rates at sea combined with reduced parental risk of predation favor early fledging. Young murrelets, for example, leave the nest shortly after hatching and swim rapidly out to sea; chicks only 2 days old have been found 15 miles from land.

Most dramatic is the emergence of a baby Malleefowl from the compost heap where it was deposited as an egg (see Chapter 17). The newly emerged, exhausted, and weak hatchling must work its way up through several feet of sand and debris to the surface of the mound. This task takes 2 to 15 hours:

> Suddenly the back of its neck appears at the mound's surface. After the neck is free, the head quickly follows. The chick opens its eyes for the first time and rests briefly. Then it resumes its struggles, freeing one wing and then the other. Soon the whole body follows. Temporarily exhausted, the young Mallee-fowl may lie exposed on the surface for some time, an easy prey to predators; but more often it tumbles down the side of the mound and staggers to the nearest bush to collapse in the shade, where it recuperates its strength after such prolonged exertion. Its recovery is swift: within an hour it can run firmly; after two hours it runs very swiftly and can flutter above the ground for thirty to forty feet. Twenty-four hours after its escape from the mound, it flies strongly. (Skutch 1976, p. 234)

Long before they are ready to leave the nest or to fly, young birds develop essential strengths through exercise. Young pelicans jump up and down and flap their growing wings with increasingly effective strokes. Young hummingbirds grip nest fabric with their feet as they practice beating their new wings, anchoring themselves so as not to take off. When first airborne, the young bird responds to the new experience with astounding ability and control. When a young Osprey launches itself on its first flight over a northern lake, it wobbles and flaps uncertainly, loses altitude, and seems certain to splash into the lake. In the last possible moments, it flaps more effectively and gains altitude, climbing steadily until it is high above the lake. It then glides in circles and practices steering and control. Successfully launched from the nest, the young Osprey enters the local population of its species and may someday become a breeding adult.

Learning Essential Skills

After the chicks leave the nest, they enter a period of intense learning and practicing essential skills, including foraging, predator avoidance, and dominance behavior. Their next challenge is to find a mate and attain breeding status. Fledglings of most small passerines stay with their parents for 2 to 3 weeks after they have left the nest (Nice 1943). Young Bewick's Swans, for example, stay with their parents for 1 to 2 years, through several long-distance migrations. Young boobies and terns depend on their parents for up to 6 months after they have fledged, that is, until they have mastered the art of plunging after fish. In the Tropics, where long apprenticeships also seem necessary to develop feeding skills, some young passerines stay with their parents for 10 to 23 weeks.

Prolonged learning as well as early imprinting exposures link an individual's genetic heritage to its social and ambient environments. Young birds must first develop their skills at food finding and capturing prey. Seemingly inept juvenile Royal Terns, for example, drop caught fish almost 14 times more often than adults do (Buckley and Buckley 1974). Among fledgling Yellow-eyed Juncos, rates of pecking and feeding as well as scanning for predators increase dramatically with age, reaching adult levels of performance as the juveniles reach independence (Sullivan 1988). Reflecting their inexperience, independent juveniles spend more time exposed to predators while foraging than adults do. Combined with greater risk of starvation when severe weather limits foraging time, inefficient foraging leads to increased mortality. By feeding together in flocks, newly independent juveniles increase their foraging time by sharing vigilance tasks.

Other essential skills also develop with age and social experience. Orientation and navigation skills require calibration of compasses and definition of goals. Young songbirds acquire their vocal repertoire and learn to communicate through social interactions. The extraordinary Gray Parrot named Alex required social exchange with his tutor to learn words and concepts; he could not learn from a television video program (Pepperberg 1991). Social skills and dominance also improve with age. Mating status itself may require years of apprenticeship as does successful parenting.

Play can be a form of practice for development of essential locomotory and social skills (Ficken 1977; Fagen 1981; Smith 1983). When young Garden Warblers discovered that dropping pebbles into a glass made a ringing sound, it became a group activity. Young crows, ravens, jackdaws, and their relatives frequently play and even create elaborate social games similar to "king of the mountain" or "follow the leader." Stick balancing and manipulation or exchange of sticks, sometimes while upside down, and taking turns sliding down a smooth piece of wood in a cage are among the many games that these intelligent birds play (Gwinner 1966). This seemingly frivolous play activity can have adaptive (and survival) value for young birds.

Peregrine Falcons develop their professional skills through practice and social interactions. After they fledge, young Peregrines depend on their parents for food for 1 to 2 months. They develop their flying and hunting skills through aerial interactions that occur while playing with their siblings. In aerial dogfights, they chase and dive at each other—called stooping—and roll over to grapple each other's talons. As such acrobatic skills improve, they take food directly from their airborne parent by rolling over and snatching food directly from the parent's talons. When hungry, young Peregrines also try to chase their parents, which do their best to avoid the hassle by perching inconspicuously on tall city buildings (H.B. Tordoff, personal communication).

Programmed to chase, juvenile peregrines can develop good hunting skills without much help from their parents. Initially they chase anything large that flies nearby, including herons and vultures, as well as each other. After 2 or 3 weeks of this, they start to focus on smaller, potential prey in the right size range. Their first captures seem almost to be accidental, surprising contacts. Kills soon become more deliberate, usually directed at first at large easy-to-catch insects, such as butterflies and flying beetles, which they may eat on the wing. The adolescent Peregrines then graduate to bird prey, which they kill with increasing efficiency.

Because yearling Peregrine Falcons usually are not good parents, they typically wait 2 to 3 years to breed. First year pairs that try to breed usually fail, but 1-year-old females nest successfully more often than 1-year-old males; their success is enhanced by having an older mate. If they chance to breed, young males may kill their own young. A 1-year-old male in Milwaukee fertilized his mate's eggs and helped to fledge the young, but then he killed them by aggressively diving at them and accidentally breaking their wings. This male was not as aggressive the next year toward his offspring, which survived. In another case, a young female joined an experienced 4-year-old male in Minnesota. She dropped her first egg into the nest box by accident while sitting on the front ledge and didn't know what do with it. She was unable to roll it from the edge of the box where it was lodged to the central nest scrape for incubation. The older male came to the rescue and promptly rolled it expertly from the edge of the box to the middle of the scrape. Once the egg was in its proper place, thanks to the male, the female settled on it right away and laid the rest of the clutch in the scrape. In this case, incubation was successful, but often young females are inattentive and haphazard in their incubation behavior, thus leading to nest failure.

Summary

A chick's first challenge is to break out of its shell. The egg tooth, a sharp-edged structure on the top of the bill, is a special feature for breaking the eggshell. Synchronized hatching of precocial young enables them to leave the nest together. Most nestlings fall into one of two categories, altricial or precocial. The former is a state of almost complete dependence on parents; the latter, a state of greater independence. Ornithologists recognize the intermediate categories of semialtricial, semiprecocial, and subprecocial, plus an extreme category of superprecocial for young that are wholly independent when they hatch.

The growth in mass of chicks follows a sigmoid curve. There is a 30-fold variation in the growth rate of chicks of the various species. One school of thought regards growth rate as moderated by selection for slow growth to reduce food requirements. Another theory suggests that growth rates relate directly to precocity of development and adult body proportions. If growth of tissue mass and maturity of function are mutually exclusive, growth should slow as the individual matures. Growth rates of individuals of the same species may also vary considerably, owing to diet quality, food availability and reliability, and temperature. Brood size is also a factor here. Older, larger chicks have a greater chance of survival than smaller, younger siblings. Asynchronous hatching contributes to siblicide because it results in chicks of unequal size and strength. Food type is also a factor; food that can be monopolized, such as small fish, promotes physical and sometimes mortal competition between siblings.

Young birds show extraordinary skill and daring when they leave the nest. For example, the flightless young of seabirds that nest on cliff ledges sometimes must leap great distances into turbulent waters below. How long young stay with their parents depends on the difficulty of skills that must be acquired. Both heritage and experience affect the behavior of birds. The nature of formative experiences varies from brief imprinting exposures during critical sensitive periods early in life to prolonged learning and cultural exchanges of information. Imprinting affects many aspects of avian behavior, from recognition of species to choice of nest sites and habitats. However, young birds must use experience to learn to find appropriate food efficiently and to avoid danger. Avian play is an important way for young birds to practice the essential locomotory and social skills they need to survive on their own.

Further Readings

Kroodsma, D.E., and E.H. Miller. 1982. Acoustic Communication in Birds. New York: Academic Press. *Includes excellent review papers on song learning.*

O'Connor, R. 1984. The Growth and Development of Birds. Chichester: John Wiley & Sons. *A comprehensive review of the evolutionary ecology of avian development.*

Ricklefs, R.E. 1983. Avian postnatal development. Avian Biology 7: 1–83. *A powerful review of an emerging research topic.*

Skutch, A.F. 1976. Parent Birds and Their Young. Austin: University of Texas Press. *A readable review of the natural history of avian reproduction, stressing the topics covered in this chapter.*

Smith, S.M. 1983. The ontogeny of avian behavior. Avian Biology 7: 85–159. *A fine review of the subject.*

Starck, J.M. 1993. Evolution of avian ontogenies. Current Ornithology 10: 275–366. *A powerful new integration of embryology and chick development, with excellent illustrations.*

C H A P T E R 1 9

Parental Care

THIS CHAPTER SWITCHES from the focus of the previous chapter, the growth and development of baby birds, to the efforts of their parents. The raising of young requires a conspicuous and consuming effort by adult birds and mammals. Males and females may share the effort equally or not at all; their individual selfish best interests may conflict with each other and, at times, with the selfish interests of their offspring. Some individuals cheat, but others cooperate.

The diversity of parental care systems is rooted in the evolutionary history of major taxa and is refined by local ecological opportunities (Silver et al. 1985; Ligon 1993). Guiding the evolution of alternative breeding behaviors are trade-offs between current and future efforts, conflicts between parents and their offspring, and uncertainties about parentage. Cuckoldry and brood parasitism are normal threads of the complex fabric of reproductive effort. In a different direction, the challenges of adequate parental care foster cooperative societal interactions that appear to be altruistic but that are actually selfish.

Parental Care and Monogamy

Most birds raise young that require extended and substantial amounts of parental care in terms of both time and energy. Consider the energy costs of reproduction in Phainopeplas (Walsberg 1978) (Figure 19–1). This unusual songbird breeds twice each year, once in the summer in the California coastal woodlands and then again in the winter in the Arizona deserts. Average daily energy expenditures are higher in the woodlands than in the desert. Breeding adults feeding two woodland nestlings expend 12.5 percent more daily energy than do nonbreeding birds. Breeding adults spend 18.6 percent more energy than do nonbreeding birds in the desert. Woodland broods of three instead of the usual two increase adult daily energy expenditures by an additional 7 percent. A breeding Phainopepla must harvest 52 percent more food in the woodlands and 72

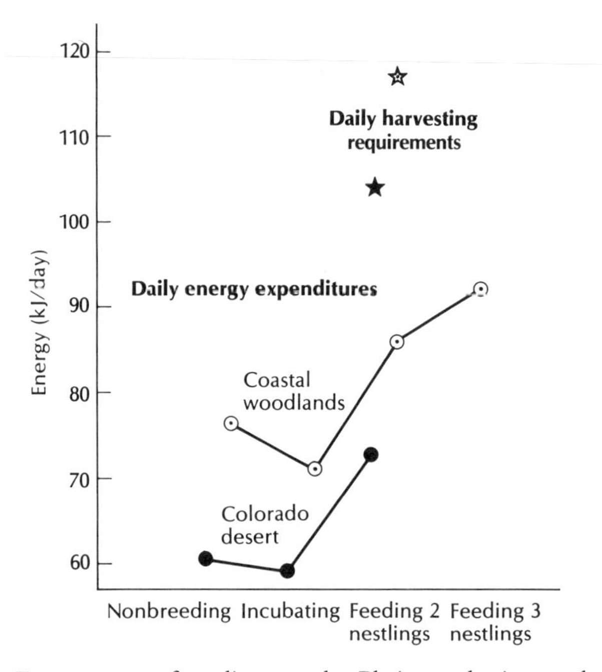

Figure 19–1 Energy costs of nestling care by Phainopeplas in two habitats, compared with the energy costs of its incubation and nonbreeding activity. Also shown (stars) is the amount of food energy an individual must harvest to feed both itself and two young. The black star indicates food-harvesting requirements in the desert; the white star indicates harvesting requirements in the woodlands. (After Walsberg 1978)

percent more food in the desert to meet its own requirements and those of its nestlings.

Peak breeding activity increases total daily energy expenditures by up to 50 percent, the fueling of which requires some combination of increased foraging time or food supplies, use of accumulated reserves, or help by mates or fully grown offspring (Ricklefs 1974; Walsberg 1983). In Little Penguins, the seasonal rearing of chicks consumes 31 percent of an adult's total annual expenditures of energy (Gales and Green 1990). These costs become extreme as the chicks reach full size, when the daily food consumption by hardworking parents exceeds 60 percent of their body mass.

The substantial costs of parental care favor participation by both parents and sometimes even the participation of other individuals, called helpers. Biparental care by monogamous mates enhances fledging success of larger broods. This effect is particularly apparent in Temperate Zone species, in which large clutch sizes seem to be tied to effort by both parents. Deprived of their mates, which share equally in the feeding of the nestlings, female Dark-eyed Juncos doubled their feeding rates, but at the cost of reduced brooding time and nest guarding (Wolf et al. 1990). Unaided females lost more chicks to exposure, starvation, and predators than did controls with helpful mates. Males not only help to feed young but also bring food to the female, food that is essential during egg formation, incubation, or early brooding. The conflict between the time required for parental care and the time required for self-maintenance emerges as one of the key constraints on solo parenthood in birds (Walters 1984).

Cold temperatures and risk of predation may discourage and penalize even brief exposure of the eggs or young and mandate a shared vigil. Mandatory continuous incubation and prolonged close care of their young by penguins in Antarctica practically preclude solo parenthood; South Polar Skuas quickly consume unguarded eggs or chicks. For seabirds in general, both parents are needed to avoid the triple dangers of extended feeding absences, harsh climates, and predators.

Incubation may be a difficult period because of the constraints on foraging time. An individual must find adequate food during brief absences from the nest. Eggs cannot be allowed to cool, nor can they be exposed to predators. Raptors cannot be certain of feeding quickly, because they cannot be sure of finding and capturing suitable prey; thus male raptors, such as the Northern Goshawk, regularly feed their incubating mate and may be responsible for almost all of her food intake. Males of most North American passerine birds (211 of 250 species for which information is available) help incubate or feed their mates during incubation (Verner and Willson 1969). Male Snow Buntings frequently feed their incubating females in the high Arctic, whereas nearby male Lapland Longspurs do not (Lyon and Montgomerie 1987). The colder microclimates of the buntings' cavity nests prevented long feeding absences by the females, whereas direct sun exposure warmed the eggs in the longspurs' nests while the female left to feed.

Biparental care may be more an assist than a requirement in the case of the usually monogamous Willow Ptarmigan (Martin and Cooke 1987). Lone females, both naturally and experimentally widowed, raised equivalent numbers of chicks to fledging and survived equally well during the breeding season and in subsequent years. The help normally provided by monogamous males appears to be more than is required to raise the young successfully and to maintain their mates' good health. Kathy Martin and Fred Cooke suggest that males stay with their mates primarily to capitalize on renesting attempts.

Trade-offs and Conflicts in Parental Care

How much effort a bird theoretically should invest into the protecting and nurturing of its chicks reflects a basic trade-off between increased survival rate of chicks at the cost of a parent's survival or additional mating opportunities (Figure 19–2).

David Winkler (1991) examined the relationship between past investments and expected benefits in parental "decision making" by Tree Swallows. In a series of carefully planned experiments that simulated losses to predators, Winkler reduced clutch size in some nests from five

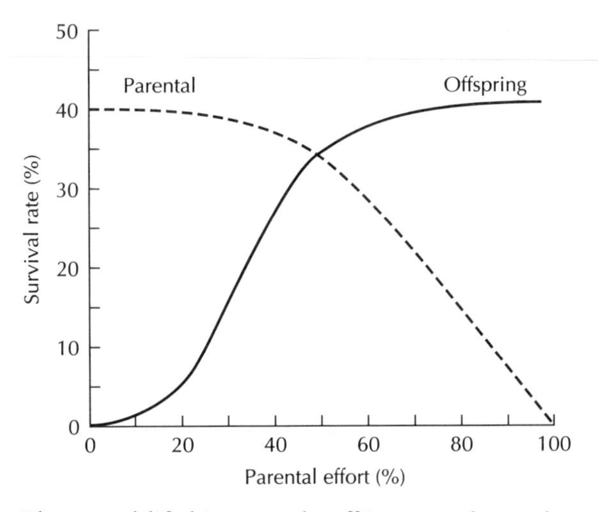

Figure 19–2 The central life history trade-off in parental care theory. Parental survival rate decreases with increasing parental effort required to increase the survival of offspring. The shapes and values of the curves in this figure are hypothetical because they have not been measured for any bird. (After Winkler and Wilkinson 1988)

(sometimes six) eggs to three eggs or one egg early in incubation. Reduction of clutches to three eggs caused only 3 of 14 pairs to renest. All 19 pairs abandoned clutches that were reduced to one egg; 14 of these pairs renested in the same or nearby boxes. Winkler also reduced brood sizes to three chicks or to one chick in other sets of nests late in the hatchling period and looked for changes in parental defense behavior. Because these swallows could not renest, Winkler hypothesized that they should be more inclined to defend what remained of their investment. There were, however, no detectable effects on parental defense behavior. These results suggest that immediate future prospects for producing additional young by renesting are the most natural currency for comparing the costs and benefits of continued parental care.

The trade-off between the selfish interests of the parent and those of the chicks sets the stage for a natural conflict between the parents, which should maximize their own lifetime reproductive success, and the chicks, which should capture as much personal attention from their parents as possible (Trivers 1974). Also tempering investments into parental care are the uncertainties of parentage. Both cuckoldry due to extra-pair copulations and eggs added to the nest by other females—intraspecific brood parasitism—dilute the potential genetic relationship between a parent and the chicks it nurtures. Decreasing confidence of genetic relationship reduces the allowable sacrifices of parental care and promotes desertion or infanticide.

A parent that has replaced a lost mate faces a pair of challenging alternatives. Should it adopt and care for its predecessor's offspring, or should it kill and replace unrelated chicks with its own? Infanticide, which is quite normal under these circumstances among mammals, is rarely documented in birds (Rohwer 1986). One exception concerns the long-toed, tropical shorebirds called lily-trotters or jacanas, which are polyandrous (Chapter 17). In Central America, female Wattled Jacanas defend territories and fight with one another for males, which incubate their mate's eggs and care for her young. When Stephen Emlen and his colleagues (1989) removed resident females—thereby creating opportunities for new females to take over the undefended territories and associated males—the new, replacement females killed or evicted three of four existing broods of chicks and solicited copulations from four of the five "widowed" males.

Infanticide also occurs in species that pair monogamously (Rohwer 1986). Tree Swallows, which are monogamous and compete for limited nest sites, practice deliberate infanticide. Intense competition among males for nest sites results in a floating population of unmated males. In one study, five of seven such males that replaced experimentally removed predecessors killed the unrelated nestlings (Robertson and Stutchbury 1988). One of the killers mated with the widowed female, but two others brought in new mates. Raleigh Robertson and Bridget Stutchbury suggest that infanticide by swallows is favored by sexual selection because it enables unmated males to obtain a nest site and breed.

Cheating and Cooperation

The demands and conflicts of parental care invite both cheating and cooperation. Brood parasitism—the surreptitious addition of eggs to another female's nest—is a common form of cheating. It allows females to increase the number of eggs they lay without increasing the costs of parental care. It also dilutes the genetic relationship between parents and their dependent offspring. Brood parasitism among birds is manifest in two major forms. Many species practice intraspecific brood parasitism, that is, they lay extra eggs in the nests of other females of the same species as a supplement to those tended in their own nest. Some species—called obligate brood parasites—never build their own nests or raise their own young. Instead they depend on other species for the services of parental care.

The next sections of this chapter introduce, first, avian brood parasites—their natural history, mimicry of hosts, and effects on their hosts—and, second, cooperatively breeding birds, in which helpers increase their own prospects for reproduction by aiding their parents and stepparents. Brood parasitism and cooperative breeding lie at opposite ends of the spectrum of parental care practices among birds. Obligatory brood parasites are selfish cheaters whose evolution is consistent with Darwin's theory of natural selection. In contrast, the apparent altruism of cooperative breeders seems to challenge the basic tenets of evolutionary theory. Underneath the apparent cooperation, however, is a strong undercurrent of conflict and self-interest.

Intraspecific Brood Parasitism

Intraspecific brood parasitism is widespread among birds. The habit is most prevalent among waterfowl, but the long list of practitioners includes grebes, fowl, gulls, the Ostrich, pigeons and doves, and songbirds. Most nests in some duck populations contain parasitic eggs (McCamant and Bolen 1979; Andersson and Ericksson 1982). Ten percent of House Sparrow clutches in Australia contain eggs of other females (Manwell and Baker 1975). Intraspecific nest parasitism increases when there is a shortage of nest sites and when population density is high. For example, Cliff Swallows nesting in large, dense colonies in southwestern Nebraska regularly lay their eggs in each other's nests (Brown 1984). Careful daily monitoring of the number of eggs in nests revealed that at least 24 percent of the nests in colonies of over 10 pairs of swallows received eggs from neighbors. Parasitic females quickly deposited an egg in a host nest when the host was away—in one instance it took only 15 seconds to do so. Such parasitism reduced the reproductive success of host females, which acted as though the parasitic eggs were their own and laid fewer eggs.

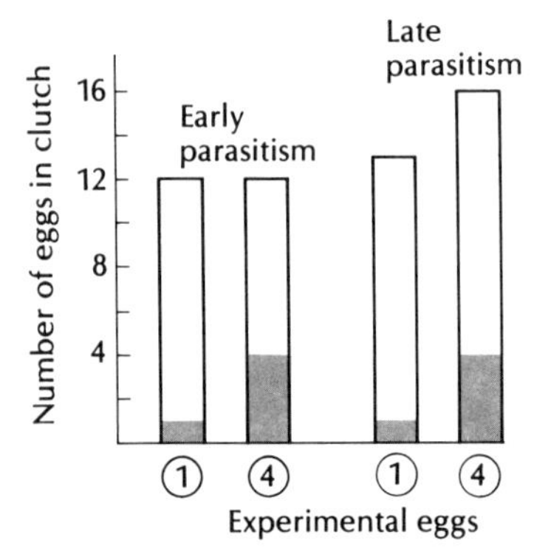

Figure 19–3 Female Common Goldeneyes respond to the experimental addition of eggs to their nests early in the laying cycle by laying fewer eggs. In this experiment, females laid significantly fewer eggs when four eggs were added than when only one egg was added. Shaded portion of bar denotes number of eggs added, and white portion denotes eggs laid by incubating female. The final result in both cases was a clutch of 12 eggs for incubation. Female Common Goldeneyes laid their usual number of eggs when other eggs were added late in the laying cycle, a response resulting in larger clutch sizes. (After Andersson and Ericksson 1982)

A similar result was obtained by Malte Andersson and Mats Eriksson (1982). They showed that female Common Goldeneyes, a sea duck that nests in tree holes, produced fewer eggs when parasitic eggs were added early in the laying period and contributed to the normal final clutch size. However, the goldeneyes did not compensate for eggs added late in the laying period and consequently incubated unusually large clutches (Figure 19–3).

European Starlings commonly lay eggs in nests other than their own. One of every four early nests in both New Jersey and Great Britain acquires foreign eggs (Evans 1988; Lombardo et al. 1989). Breeding females guard against such brood parasitism by removing foreign eggs that are deposited before they themselves start to lay. Once a female starts its own clutch, however, it cannot distinguish the parasitic eggs, and like the goldeneye ducks, does not compensate for additional parasitic eggs. Roaming, parasitic females often remove one of the host's eggs and replace it with their own. In addition to making detection more difficult, egg removal keeps the clutch size closer to the optimal number for nest success. Clutches with six eggs produce the most fledglings; seven-egg clutches suffer losses as a result of overcrowding (Power et al. 1989). Noting that host starlings usually lay five, not six, eggs, Harry Power and his students suggest provocatively that this is the best course in situations with a high probability of parasitism; one parasitic egg will bring the clutch size up to the optimum, instead of having a negative effect on the whole brood. That same female, having laid five eggs in her nest, may lay an egg in another nest, thereby guarding against complete reproductive failure in the event her nest is destroyed.

Intraspecific parasitism could be the first step in the evolution of obligatory brood parasitism. Occasional parasitism—called facultative parasitism—of the nests of related species probably is the next step. The Black-billed Cuckoos and Yellow-billed Cuckoos of North America, for example, occasionally parasitize each other, particularly when abundant food encourages the production of extra eggs (Nolan and Thompson 1975). The acceptance of a parasite's eggs and successful raising of its young then lead logically to increasing frequencies of parasitism and eventually to obligatory interspecific brood parasitism.

Obligate Brood Parasites

Cowbirds and cuckoos are the most familiar birds that relinquish care of all their young to foster parents of other species. Such obligate brood parasites always lay their eggs in the nests of other birds (Figure 19–4). This is a most unusual breeding strategy; a few social insects but no mammals are brood parasites. Among birds, the practice has evolved independently at least seven times: in cowbirds (Icteridae), honeyguides (Indicatoridae), Old World cuckoos (Cuculidae, Subfamily Cuculinae), New World cuckoos (Estrildidae, Subfamily Neomorphinae), whydahs and indigobirds (Ploceidae, Subfamily Viduinae), the Parasitic Weaver (Ploceidae), and the Black-headed Duck (Anatidae).

These obligatory parasitic birds take advantage of the parental care of other birds. By reducing their costs and risks, parasitic birds are able to lay more eggs each season. Among cuckoos, some species are obligate parasites, other nonparasitic species build nests and care for their own young. Parasitic species of cuckoos lay more eggs than do the nonparasitic cuckoo species. By not putting all their eggs into one nest, brood parasites improve the chances that some of their offspring will escape predation (Payne 1977a).

Figure 19–4 Foster parent wagtail feeding a parasitic young Common Cuckoo. (Courtesy of E. and D. Hosking)

Brood parasites lay eggs in clutches, like most birds (Payne 1977b). Female Brown-headed Cowbirds lay 30 to 40 eggs per season in weekly sets of 2 to 5 eggs (Scott and Ankney 1983). These cowbirds deposit their eggs at random; most host nests have only 1 cowbird egg, but some may have up to 12 (Lowther 1993). African cuckoos of several species lay 16 to 25 eggs per season in clutches of 3 to 6 eggs, but they lay only 1 egg per nest.

As a rule, the eggs of brood parasites require 2 to 4 days' less incubation time than do those of the host. This property ensures earlier hatching and dominance by the young parasite. The Pied Cuckoo and Common Cuckoo incubate eggs in their oviduct for up to 18 hours prior to laying (Payne 1973a). Hatchling parasites also grow faster than nonparasites. Such advantages enable the young parasite to garner most of the parental attention.

Mimicry and Other Adaptations of Parasites

Obligatory brood parasites are highly specialized birds that use specific hosts (Table 19–1). Hard-shelled eggs and deliberate destruction of host eggs or host young also enhance a brood parasite's self-serving trade.

The eggs of parasitic cuckoos are thick-shelled and resistant to cracking; females drop their eggs into deep nests, sometimes damaging the hosts'

TABLE 19–1
Host specializations of African honeyguides and Japanese cuckoos

Brood parasite	Primary host(s)
African honeyguides	
Greater Honeyguide	Rollers, starlings, bee-eaters
Lesser Honeyguide	Large barbets, woodpeckers
Scaly-throated Honeyguide	Woodpeckers
Least Honeyguide	Tinkerbirds, small barbets
Cassin's Honeyguide	Rock-sparrows
Wahlberg's Honeyguide	White-eyes, small warblers, flycatchers
Japanese cuckoos	
Common Cuckoo	Great Reed-Warbler, Bull-headed Shrike, Meadow Bunting
Oriental Cuckoo	Eastern Crowned-Warbler
Hodgson's Hawk-Cuckoo	Chats
Little Cuckoo	Wren, Japanese Bush-Warbler

From Lack 1963, 1968.

Figure 19–5 Baby brood parasites dispose of their competitors: Hatchling Common Cuckoo (left) pushes the eggs of the host from the nest; a hatchling Greater Honeyguide (right) kills host nestlings with the hooklike tip of its bill. (Adapted from Lack 1968)

eggs rather than their own. The eggs and subsequent chicks of brood parasites are normally the same size or larger than those of their hosts, thereby increasing their dominance over the host's chicks. Some baby brood parasites are aggressive; it is not uncommon for a hatchling cuckoo to shove the unhatched eggs of its host out of the nest. Baby honeyguides have special fanglike hooks at the ends of their bills for killing their foster nestmates (Figure 19–5).

The adaptations for brood parasitism include egg mimicry. To minimize detection and destruction by the host, cuckoo eggs have come to resemble or mimic those of their primary hosts. In Africa, the eggs of the Dideric Cuckoo are so similar to those of its host, the Vitelline Masked-Weaver, that one ornithologist resorted to chromosome analysis to distinguish them (Jensen 1980). In Europe, Common Cuckoos parasitize and mimic the eggs of a variety of host species that differ in egg color patterns (Figure 19–6). The blue eggs of the Common Cuckoo in Finland match those of its primary hosts, the Common Redstart and the Whinchat, whereas in Hungary, Common Cuckoos lay greenish eggs with dark markings, similar to those of the Great Reed-Warbler. Such egg "races" of the Common Cuckoo correspond to major habitats in Scandinavia and central Europe, a correspondence suggesting that they evolved in isolation but now coexist in some parts of central Europe.

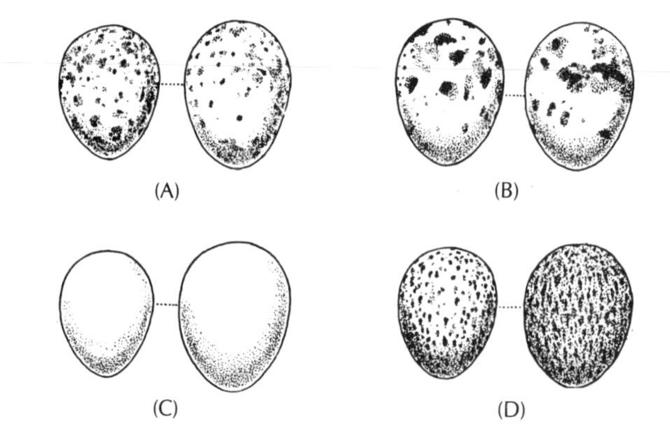

Figure 19–6 Common Cuckoo eggs (right) of particular types—egg "races"—closely match the color pattern of the eggs of their hosts (left). Identity of hosts: (A) Garden Warbler; (B) Great Reed-Warbler; (C) Common Redstart; (D) White Wagtail. (From Rensch 1947)

Plumage, mouth, and song mimicries—all forms of host mimicry—are other traits that accompany host specializations of brood parasites. Feathered young Common Koels, a kind of cuckoo, for example, look like nestling crows. Nestling whydahs have evolved mouth markings that mimic those of the nestlings of the estrildine finch hosts (Nicolai 1974; Payne 1982a) (Figure 19–7). The evolution of host-specific whydahs in Africa has produced species adapted precisely to particular host finches, which have distinctive mouth color markings and do not feed nestlings with the wrong markings. Nestlings of the Northern Paradise-Whydah, for example, have the single spot resembling the nestlings of their host, the Green-winged Pytilia. Straw-tailed Whydah nestlings have three spots like those of the nestlings of their host, the Purple Grenadier. Young whydahs also imitate their nestmates by begging with their heads upside down (Nicolai 1964). Nestlings of the Village Indigobird have a ring of five spots like those of the nestlings of their host, the Jameson's Firefinch. Both host and parasite of this latter pair of species also have the same mouth color pattern: white spots, blue corners, and yellow roof. Color pattern may be a more important parental cue than the arrangement of the spots in some pairs of species.

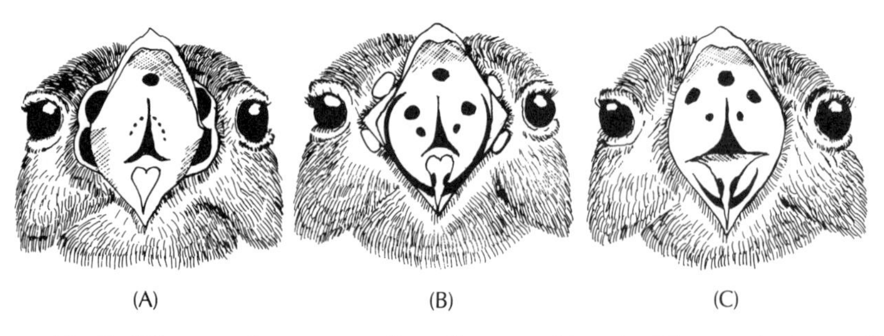

Figure 19–7 The mouth patterns of nestling parasitic whydahs match those of their hosts. (A) Northern Paradise-Whydah and Green-winged Pytilia; (B) Straw-tailed Whydah and Purple Grenadier; (C) Village Indigobird and Jameson's Firefinch. (After Nicolai 1974; Lack 1968)

How do brood parasites develop a sense of identity and association with other members of their own species? What little is known suggests that female Brown-headed Cowbirds rely largely on their genetic heritage. Females raised in isolation are fully responsive to songs of males of their own species (King and West 1977); when they come into breeding condition and hear male songs, they solicit copulation. They do not respond to the songs of other species, even those of their foster parents. The calls of the Common Cuckoo are also inherited. Other parasites, however, imprint on the vocalizations of their foster parents. Young Village Indigobirds imitate the songs of their foster fathers, including dialect variations (Payne 1982a). Thus, geographical patterns of song dialects correspond in host and parasite. The odd male Village Indigobird raised in the nest of a different host species acquires different songs, and the female Village Indigobird raised in such a nest becomes responsive to these different songs. Host vocalizations enable female indigobirds to recognize potential mates with the same host heritage. In the short run, this response ensures that young parasites will mimic the host's mouth markings, which reduces the chances of rejection. Host-specific lineages may then perpetuate themselves and evolve into separate species (see Box 22–5).

Effects of Brood Parasites on Their Hosts

Parasitized nests rarely fledge young of the host, so severely do brood parasites limit their host's breeding success. Parasitic adults or young may toss out or eat the host's eggs. The young may also kill nestlings or cause the host to desert its nest. Feeding a large, insatiable parasite undoubtedly exhausts host parents and reduces their survivorship and ability to renest. Thus, the impact of brood parasitism can be substantial, especially in species with small populations. The Yellow-shouldered Blackbird, a species found only on Puerto Rico, is a case in point (Post and Wiley 1977). This island blackbird recently became a host for the parasitic Shiny Cowbird, which expanded throughout the West Indies from 1950 to 1970. Now endangered, the Yellow-shouldered Blackbird breeds successfully only on tiny nearby islets where there are no Shiny Cowbirds.

Parasitism by Brown-headed Cowbirds can overwhelm local populations of Neotropical migrants (Robinson 1992). Abundant cowbirds parasitize three-fourths of the nests of Neotropical migrants in small forest fragments in Illinois. Combined with high rates of nest predation, few nests succeed. Cowbirds were responsible for the precipitous decline of the endangered Kirtland's Warbler in Michigan (Mayfield 1960). In 1957, parasitism was high (about 55 percent); 75 percent of the nests examined between 1957 and 1971 were parasitized (Walkinshaw 1972). In just one decade, the number of singing male Kirtland's Warblers dropped from 502 (1961) to 201 (1971); parasitized nests produced nearly 40 percent fewer young than unparasitized nests. Control of cowbirds is now central to the management of Kirtland's Warbler populations.

Host Responses to Brood Parasites

Some host birds accept the eggs of a brood parasite, but others do not. Stephen Rothstein (1975) placed artificial cowbird eggs in 640 nests of 30 species of North American birds. Twenty-three of these species usually accepted the eggs (meaning that they threw them out less than 30 to 40 percent of the time), whereas seven species usually rejected the different eggs. "Rejectors" typically threw out the parasite eggs as a natural extension of nest sanitation behavior. It appears that rejectors acquire true recognition of their own eggs through some form of imprinting that is less developed in "acceptor" species (Rothstein 1982). The rejection defenses of American Robins against cowbird parasitism vary with location (Briskie et al. 1992). In Churchill, Manitoba, north of the range of cowbirds, American Robins are more likely to accept a parasitic egg than in southern Manitoba, where cowbirds have parasitized local birds for centuries. One-third of the southern robins rejected an experimental egg placed in their nests by "parasitic" ornithologists, whereas all Churchill robins accepted them.

Some birds, such as the Yellow Warbler, respond to the discovery of a cowbird egg by deserting the nest or by burying the entire clutch in additional nest materials and laying a fresh clutch of eggs on top. A Yellow Warbler's method of rejection depends on the point in its laying cycle at which the cowbird adds the egg (Clark and Robertson 1981). Egg burial occurs most often when the warbler has just started its own clutch. This behavior enables the warblers to renest without rebuilding the entire nest. Nests in which eggs were buried produce more young of the hosts than nests in which cowbird eggs are accepted (Table 19–2).

One unusual host, the Chestnut-headed Oropendola, tolerates young brood parasites because they may help their nestmates survive. In Panama, nestling Giant Cowbirds in Chestnut-headed Oropendola nests pluck fly maggots from nestmates (Smith 1968; Ricklefs 1979c). Flies cause a high level of mortality in oropendola nestlings; hence, adult

TABLE 19–2
Nesting success of parasitized Yellow Warblers

Nest status	Number of nests	Nest success[a]
Parasitized		
Buried	13	0.78
Deserted	10	0.00
Accepted	12	0.53
Not parasitized	64	0.80

a. Average number of fledged young per egg laid, including buried eggs.

After Clark and Robertson 1981.

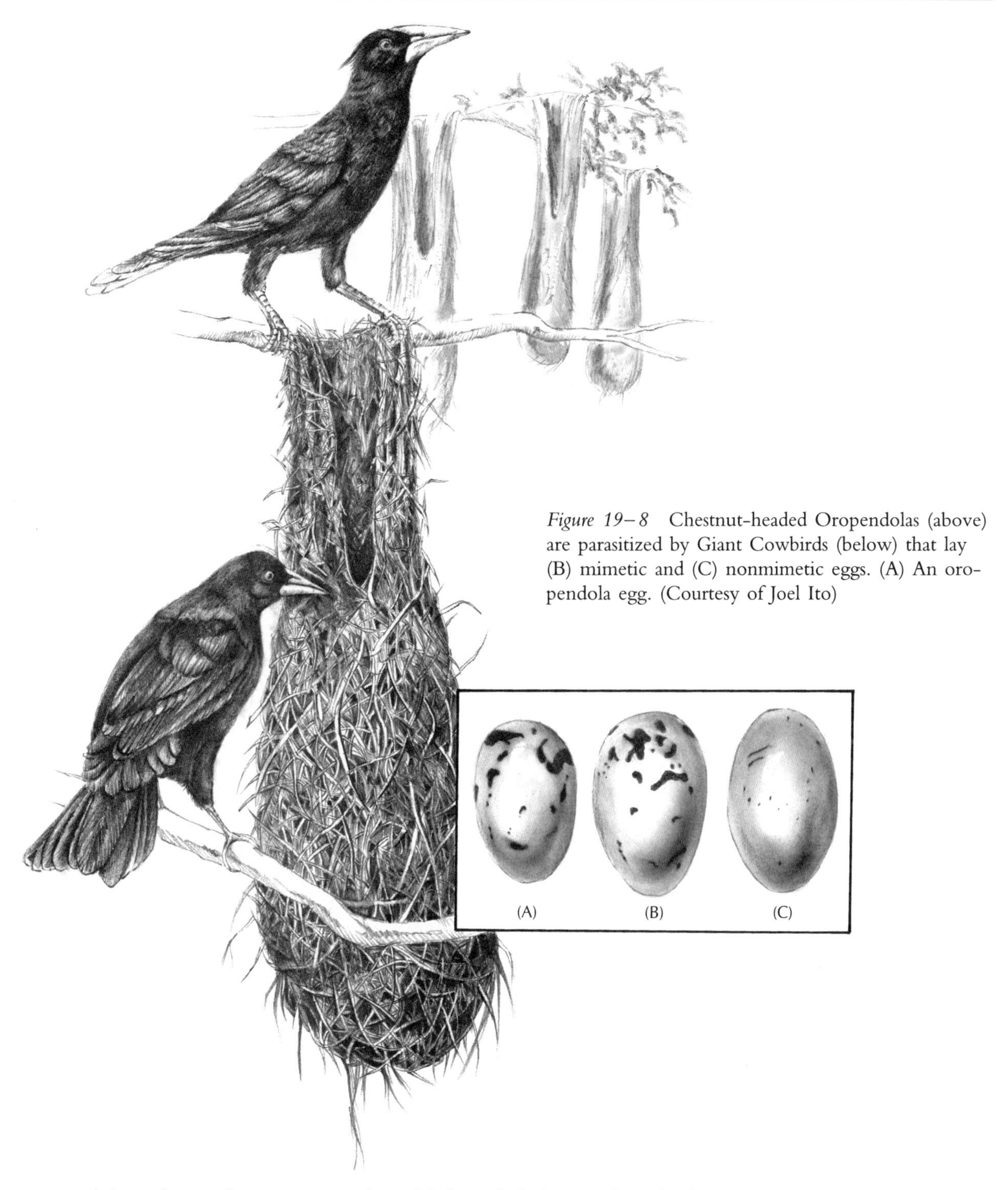

Figure 19–8 Chestnut-headed Oropendolas (above) are parasitized by Giant Cowbirds (below) that lay (B) mimetic and (C) nonmimetic eggs. (A) An oropendola egg. (Courtesy of Joel Ito)

oropendolas tolerate the presence of cowbirds and their eggs in colonies with serious infestations of flies but not in colonies with few flies. Cowbirds facing rejection have deceptive, mimetic eggs, whereas beneficial cowbirds in other colonies have plain, easily detected eggs (Figure 19–8).

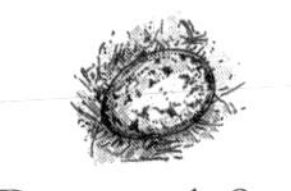

Box 19–1
Obligate brood parasites increase the variety of hosts used

The seven known species of cowbirds (*Molothrus*) differ in the number of species of hosts they parasitize. At one extreme, the Screaming Cowbird of Argentina parasitizes only the Bay-winged Cowbird, a nonparasitic species that raises its own young in the old nests of other birds (Hudson 1920). At the other end of the spectrum is the Brown-headed Cowbird of North America, which parasitizes more than 200 species of small land birds. DNA sequencing studies of the relationships among the cowbirds indicate a progressive increase in the number of hosts parasitized by more recently evolved species (see figure). In this phylogenetic analysis, the specialized Screaming Cowbird assumed the oldest, primitive position in the evolutionary tree, followed by the Giant Cowbird, which parasitizes seven species of tropical caciques and oropendolas, and then by the more generalized species. The Bay-winged Cowbird is more closely related to nonparasitic blackbirds than to the parasitic cowbirds.

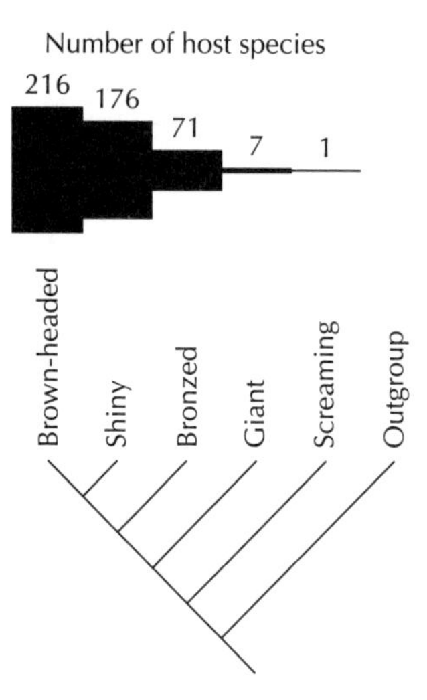

Phylogeny of the brood parasitic cowbirds. A single most parsimonious tree based on the analysis of DNA base pair sequence characters indicates a trend from host specialist to host generalist. Numbers of host species known for each cowbird species are presented above the phylogeny.

Cooperative Breeding

Brood parasitism evolves because it clearly favors the reproductive success of some individuals that exploit others. Individual advantages are not as obvious in cases of cooperative breeding, which occurs when "helpers" provide parental care for young that are not their own. In Florida Jays, one of the best-known cases, helpers are an integral part of the social system (Figure 19–9). The basic social unit comprises a breeding pair with up to six helpers. They defend a territory throughout the year. The helpers, which often are young, nonbreeding individuals, also protect and feed the nestlings (Woolfenden and Fitzpatrick 1984). More generally, helpers contribute to parental care in a diverse array of cooperative breeding systems that range from those of Hedge Accentors (Box 17–5), through the social units of Mexican Jays and Groove-billed Anis, which have multiple breeding females, to the complex societies of White-fronted Bee-eaters and Galápagos Mockingbirds, in which helping behavior changes from one season to the next.

Figure 19–9 The Florida Jay is one of the most thoroughly studied species of cooperative-breeding birds. (Courtesy of A. Cruickshank/VIREO)

The phenomenon of "helpers at the nest" was first reviewed by Alexander Skutch (1961). We now know of cooperative breeding in over 220 avian species, and the list is growing (Stacey and Koenig 1990). Cooperation is a major aspect of the breeding biology of birds, and one that has received intensive field study in the past decade. The apparent altruism of cooperative breeding challenges the basic tenets of evolution by natural selection: Charles Darwin himself offered the discovery of altruistic behavior as a way to disprove his theory. A century later, V.C. Wynne-Edwards (1962) shocked the establishment of evolutionary biology when he concluded that individuals place the good of their populations or species above their individual well-being. In particular, helpers at the nest seemed to offer the most compelling cases of altruism.

Cooperation or Opportunism?

Do helpers make positive contributions, or do they interfere? Do helpers serve their own interests in some way, or do they sacrifice their reproductive potential to help others? Answers to these questions could help to reconcile cooperative breeding with evolutionary theory.

The simplest avenue of reconciliation is based on discovering how helpers directly enhance their reproduction by delaying their own dispersal to breed independently. Helpers may also obtain indirect benefits, either by enhancing their own lifetime reproductive success through production of genetic relatives—called kin selection—or by obtaining help in return—called reciprocal altruism. Kin selection is one route to understanding complex social behavior in ants, bees, and wasps, in which sterile castes help their mother produce sisters (Hamilton 1964; West-Eberhard 1975). Alternatively, help might be based on a principle of compensation. Reciprocal altruism—or mutualism—could be in an individual's best interest as long as cheating does not occur (Trivers 1971; West-Eberhard 1975; Axelrod and Hamilton 1981). Field studies of cooperative breeders show increasingly that helpers achieve direct benefits, supplemented in some cases by such indirect benefits.

Ian Rowley (1965) pioneered the study of the details of cooperative nesting in Australian birds, specifically in the Superb Fairywren. His discovery that helpers are young from previous broods has been confirmed in other species: Most helpers help either genetic parents or stepparents to raise siblings or half-siblings. Of the 199 Florida Jays that Glen Woolfenden recorded as helping from 1969 to 1977, 118 (59 percent) helped both their mothers and their fathers, 49 (25 percent) helped one parent and one stepparent, and 32 (16 percent) helped distant kin as well as nonkin (Emlen 1978). The functional social units are essentially extended families, and the production of genetically related siblings seems consistent with evolutionary theory. The apparent importance of kin selection, however, is weakened by observations that helpers continue to help unrelated stepparents after their own parents die. In the Mexican Jay, helpers may be only distantly related to the nestlings they help feed because these jays regularly immigrate from other family units (Brown and Brown 1981a).

Parental tolerance of grown offspring on their natal territories, despite the fact that they may compete for resources, is a key step in the evolution of cooperative breeding systems (Brown 1974; Gaston 1978). Reasons to tolerate the continued presence of young from previous broods center on their helpful contributions to reproduction as well as survival of the breeding pair itself.

Most studies show that helpers truly help, rather than hinder, the parents in their social unit. In some cases, the number of young fledged increases with the number of helpers. Breeding pairs of Florida Jays with helpers fledge more young per season than do groups without helpers, principally as a result of better group defense against snakes, the primary predator on young jays (Woolfenden and Fitzpatrick 1984). Ronald Mumme (1992) demonstrated the effects of helper jays by temporarily removing them from some territories at the beginning of the breeding season. Breeding pairs with helpers produced more young that fledged from the nest and that lived longer after they left the nest than did breeding pairs without helpers. In addition to increasing the production of surviving fledglings, helpers improved the survival of the breeding parents. Breeding Florida Jays with helpers survived more often than did those without helpers (Woolfenden and Fitzpatrick 1984). Hence, helpers also increased lifetime reproductive success of the breeders in this system.

Jerram Brown and his colleagues (1982) removed helpers from family units of the Gray-crowned Babbler, an Australian bird, and observed results similar to those for jays. The average number of fledglings declined from 2.4 in the first broods (with helpers) to 0.8 in the second and third broods (without helpers) (Brown and Brown 1981b) (Figure 19–10). Helpers also increased the ability of the parents to start a second or third brood. Larger breeding groups of Gray-crowned Babblers renested sooner and started more clutches than did smaller breeding groups with few or no helpers (Brown and Brown 1981b).

Uli Reyer's (1980) studies of Pied Kingfishers on Lake Naivasha and Lake Victoria in Kenya shed more light on the importance of local ecology in the acceptance of helpers. The acceptance by breeding male kingfishers of unrelated male helpers is directly related to their need for help in delivering fish to their young. On Lake Victoria, where fishing is difficult, a single helper doubles the average fledging survival rate from 1.8 to 3.6 young per nest. On Lake Naivasha, where fishing is easier, helpers have less effect; the rate increases from 3.7 to 4.3 young per nest. Most breeding pairs on Lake Victoria have helpers, at least one of which is not their own progeny. On Lake Naivasha, however, few pairs have helpers and of those that do, almost all are their own young. These auxiliary male helpers increase their own chances of eventually pairing with a female by helping her raise the brood, and they also may recruit the young they helped to raise as their own assistants in the future.

Ecological Constraints and Delayed Dispersal

It may be in the parents' best interests to have helpers, but why do the helpers themselves stay at home, and help, rather than disperse from their natal territory and breed elsewhere on their own? Even though they may

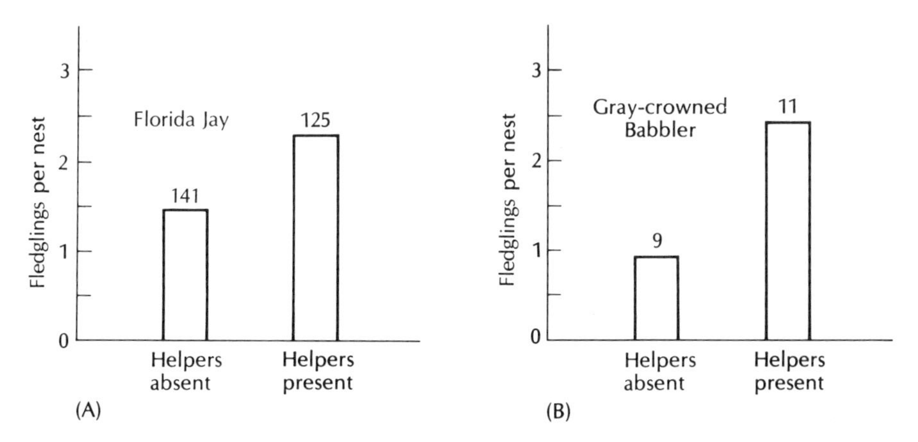

Figure 19–10 Groups with helpers fledge more young. (A) Groups of Florida Jays with helpers produce more fledglings per nest than do pairs without helpers. (B) Experimental removal of helpers from breeding groups of Gray-crowned Babblers reduces the average number of young fledged per nest. Numbers of unassisted pairs and pairs with helpers studied are indicated over the bars. (A adapted from Woolfenden 1981; B adapted from Brown et al. 1982)

be physiologically capable of breeding, these young birds delay dispersal and reproduction on their own for several years, apparently sacrificing their potential reproductive output.

Dispersal is risky because young birds must explore unfamiliar habitats with uncertain food supplies and new predation possibilities. Mortality increases during dispersal. Offsetting the higher risk of mortality, however, is the potential for personal reproductive success. In the role of a helper, a Gray-crowned Babbler rears an average of 0.46 fledgling rather than the average of 3.6 fledglings it could produce as a breeder, with help of its own (Brown and Brown 1981b). The per capita reproduction of extended family units usually decreases with group size and is less than that of pairs without helpers. But such figures assume that helpers could breed successfully on their own. This is not always the case; there is no guarantee that a dispersing young bird will find a place as a breeder on a territory of high quality.

Breeding status on an exclusive territory may be difficult to obtain when most of the available habitat is controlled by established pairs. This situation is most common in species that reside in stable environments and that have specialized habitat requirements and high adult survivorship. Florida Jays, for example, are restricted to undisturbed oak-palmetto scrub habitat in central Florida, which exists only as small islands of scrub surrounded by other vegetation. The available habitat seems saturated by occupied territories. Female helpers wait for openings. They monitor nearby groups and move quickly to replace breeding females that disappear. Males, however, wait to inherit breeding positions on their natal territories in relation to their age and status. The dominant (usually oldest) son replaces its (deceased) father, stepfather, or brother. Alternatively, helper males may take over a separate portion of the family territory for their own breeding purposes. The territories tend to expand with group size. When the expanded territory is large enough, one part is ceded to the oldest male, which then recruits a female from outside the family unit, apparently to lessen inbreeding.

Stephen Emlen (1982b, 1984) proposed a general hypothesis that ecological constraints limit successful dispersal and reproduction of young birds entering the breeding population (Figure 19–11). The constraints may be a shortage of quality territories, which discourages dispersal, or food stress during dry years, which favors recruitment of assistance for feeding the young. More recent development of this theory (Stacey and Ligon 1991; Koenig et al. 1992) suggests that delayed dispersal and group living reflect a mixture of extrinsic ecological constraints, such as habitat saturation, and intrinsic social benefits, such as improved survivorship and the acquisition of essential skills.

There are many potential intrinsic benefits to delayed dispersal, to participating in a large cooperative social unit for several years, and to assisting parents in the raising of siblings or stepsiblings (Stacey and Ligon 1991; Koenig et al. 1992). In addition to waiting for a territorial opening in saturated habitats, a young bird might stay in its home territory that

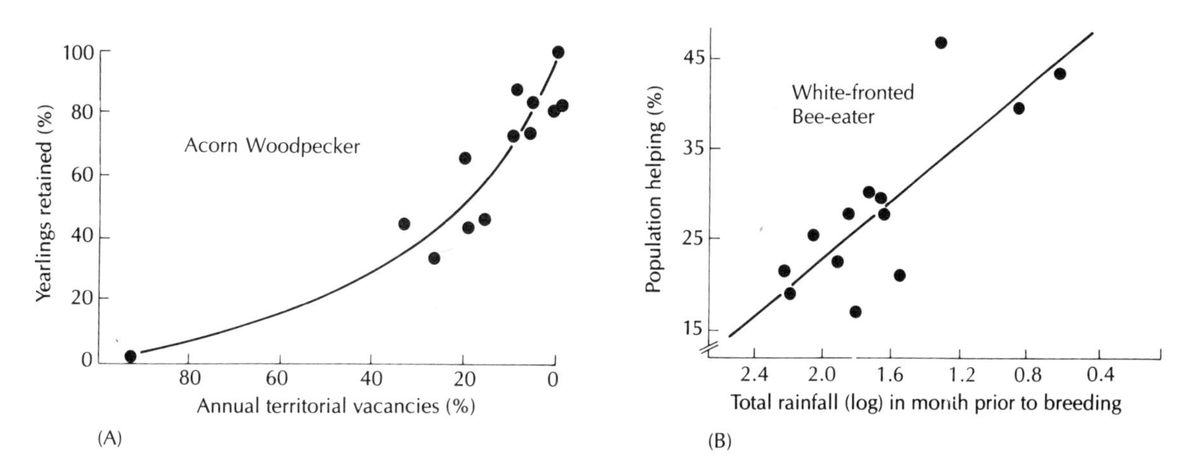

Figure 19–11 The retention of young as helpers may result from ecological constraints: (A) territory shortage, Acorn Woodpecker; (B) environmental harshness (lack of rain), White-fronted Bee-eater. (Adapted from Emlen 1984)

contains key resources not readily available elsewhere. Acorn storage granaries are the key resource for Acorn Woodpeckers (see Figure 14–5), and tree holes for nesting and roosting are a key resource for the endangered Red-cockaded Woodpecker (Box 21–2). Both of these species of woodpeckers are cooperative breeders with young helpers from previous broods.

Young birds might not be able to breed on their own until they achieve the behavioral skills and social status required to control territorial space, acquire mates, and feed young. Acquisition of such critical skills during a period of apprenticeship may favor delayed dispersal and prolonged membership in an extended family unit (Brown 1987). In

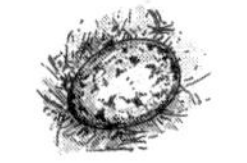

Box 19–2

An endangered island species uses all available habitat

The Seychelles Brush-Warbler is an endangered species, confined until recently to the tiny island of Cousin in the Seychelles islands of the western Indian Ocean. The territories of this drab island species consumed all the available habitat; young warblers had no choice but to wait as helpers until a breeding adult died.

As a part of the program to prevent the extinction of the Seychelles Brush-Warbler, Komdeur (1991) transplanted breeding adults from 16 territories to the nearby and much larger island of Aride, in hopes of establishing a new population. All the vacancies created on Cousin were quickly filled by individuals that had been helpers, sometimes within hours. The transplanted pioneers not only established themselves successfully on Aride and started to breed without help, but their 61 young dispersed and bred independently on territories of their own the next year. They did not serve as helpers. In addition to being a conservation success story, this experiment demonstrated that habitat saturation can prevent dispersal behavior and lead to cooperative group living.

White-winged Choughs, an Australian crowlike bird, the acquisition of the foraging skills required for independence takes almost 7 months (Heinsohn 1991). Young choughs depend on food provided by helpers while acquiring essential skills, and conversely helpers may develop their own breeding skills by observing and practicing on their siblings.

The recruitment of help from younger siblings is sometimes a side benefit of cooperative breeding systems. Green Woodhoopoes, medium-sized, hole nesting birds of the African savannas, typically live in extended family groups of helpers (Ligon and Ligon 1978; Ligon 1981) (Figure 19–12). Large roost holes in dead trees are a key resource for these cold-sensitive birds, which stay warm at night by sleeping together inside a deep hole (Ligon et al. 1988; Williams et al. 1991). Morne DuPlessis (1990) discovered that, where suitable roost holes abound in some habitats in South Africa, young woodhoopoes disperse readily to new territories, leaving pairs of adults to breed on their own. In the lakeside forests of the Rift valley of East Africa, however, roost holes are scarce and competition for territories containing them is keen. There, pairs of young woodhoopoes, usually an older and younger sibling or half-sibling, cooperate to secure new breeding space. Young male woodhoopoes recruit the help of their former charges to take control of a quality territory. Thus, the initial cooperative breeding effort leads to long-term working partnerships between siblings. The alliance is in the younger woodhoopoe's interest because it will eventually replace its partner as the breeding male of the new unit.

Unpredictable or difficult breeding conditions are another ecological feature of cooperative breeding in some birds (Emlen 1982b). Many species that live in the dry forests of Africa and Australia breed cooperatively. Some are nomadic. Others, such as the White-fronted Bee-eater of East Africa, are resident and colonial. Nestling bee-eaters often starve when adequate rains and good supplies of insects fail to materialize. Helpers increase the rate of food delivery. Helpers could start their own nests but can only raise young successfully by themselves in good years. Consequently, cooperative group size increases with environmental harshness as measured by low rainfall and poor food availability in the month preceding the onset of breeding (see Figure 19–11).

Conflict and Competition

Behind the facade of a cooperative social order is a swirl of competition, strife, and harassment. Helpers may deliberately interfere with parental reproduction to increase turnover and thereby increase their own chances of breeding (Zahavi 1974). Conversely, adults may sabotage the initial breeding efforts of young to increase the incentives for the young to stay at the nest as helpers. Young helper males sometimes mate with their stepmothers (Emlen 1982a), and helper females sometimes slip an egg of their own into the parental clutch. In cases of communal nesting, females may destroy one another's eggs.

Individual selfishness prevails beneath the surface of communal breeding by Groove-billed Anis, large black cuckoos of the New World Tropics that form social units of one to four monogamous pairs. All members of the unit lay their eggs in a single nest, and all the individuals in the unit help incubate and feed the communal brood. The main advantage of communal nesting in this species is in sharing the high nocturnal predation risks during incubation and brooding, thereby improving individual survivorship (Vehrencamp 1978; Vehrencamp et al. 1986). Female anis,

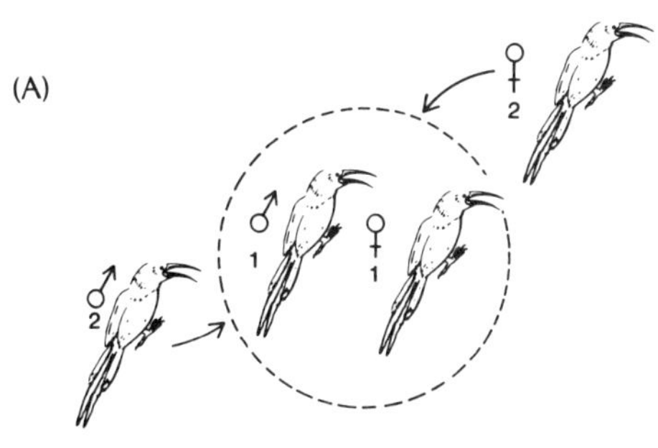

♂$_1$ and ♀$_1$ live at "seams" of other territories, unable to defend space from neighboring groups. An unrelated, younger, ♂$_2$ and ♀$_2$ are allowed to join the original pair.

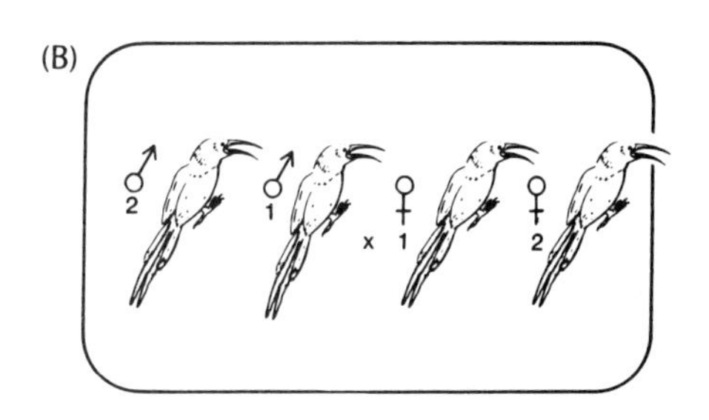

This foursome is now able to defend a territory from neighboring flocks. ♂$_1$ and ♀$_1$ breed when environmental factors permit.

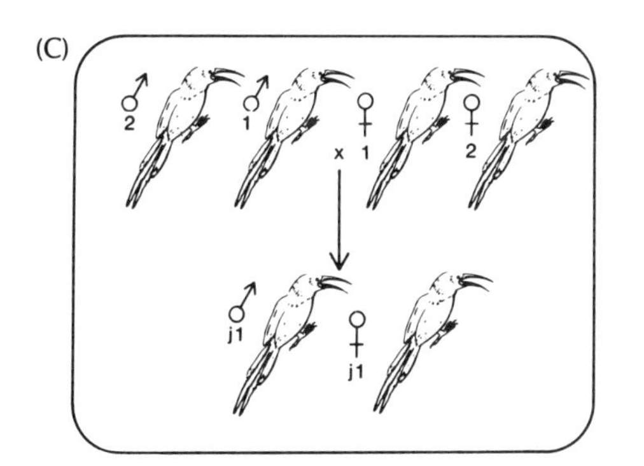

Two juveniles produced. Cared for by all four adults. Territory boundaries expand with increase in group size.

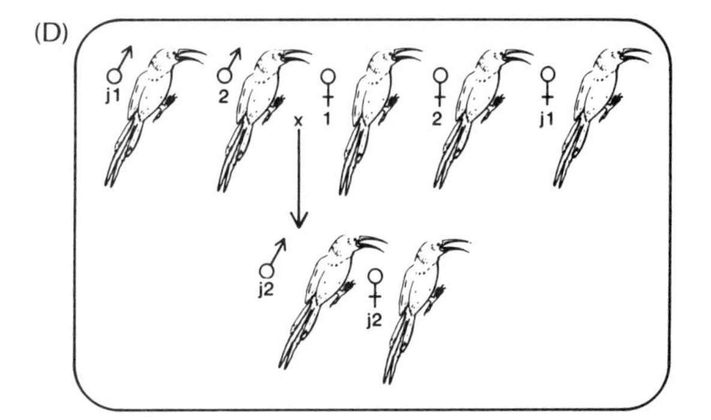

Original ♂ (1) dies; ♂$_2$ and ♀$_1$ breed, producing juveniles ♂$_{j2}$ and ♀$_{j2}$.

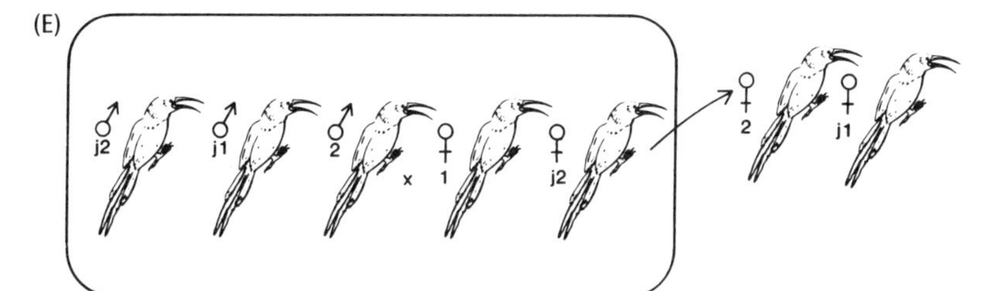

♀$_2$ and ♀$_{j1}$ move together to a nearby territory where prior ♀ occupants have died. ♀$_2$ now becomes the breeder in this new territory, with ♀$_{j1}$ as an ally in territorial defense against other ♀♀, and as a helper for ♀$_2$'s offspring.

Figure 19–12 Scenario of cooperative partnerships and reciprocity in the Green Woodhoopoe, which suffers high and unpredictable mortality. (A–E) Offspring and their older helpers ultimately move together to take over a nearby breeding territory. (From Ligon and Ligon 1983)

however, compete among themselves to ensure the success of their respective contributions to the clutch. Because one nest cannot hold all the eggs, the females throw one another's eggs out to make room for their own. Young subordinate females start laying first. The older females toss out some of these eggs to make room for their own eggs, which make up most of the clutch. Subordinate females counter these actions by increasing the total number of eggs added to large clutches, by prolonging the interval between eggs laid, and by producing a "late egg" as the clutch size nears completion. There are natural limits to the subordinate female's attempts, because the last-born nestling is the smallest and most vulnerable member of the brood. The genetic relationships among members of the ani's social units remain unknown, except that, as in Florida Jays and other cooperative breeders, young males tend to stay home and cooperate with their parents (Bowen et al. 1989).

The Complex Society of a Bee-eater

The potential for complex social relationships may be greatest where contacts with large numbers of individuals are frequent and predictable, as in colonial breeding birds. The White-fronted Bee-eater is a case in point. These bee-eaters breed in large colonies in East Africa but function on a daily basis in clans of two to seven individuals that feed together and defend a group territory within 20 miles of the colony (Hegner et al. 1982). Members of each clan feed, roost, and breed cooperatively.

Steve Emlen's studies of color-marked White-fronted Bee-eaters in Kenya's Rift valley reveal that the fabric of the complex bee-eater society is a "mixture of openness and fluidity of group memberships on the one hand, with stability and fidelity of certain social bonds on the other" (Emlen 1981, p. 224). Individuals appear to remember past associations. They leave groups to join other groups but return months or years later to roost or nest with old associates. They preferentially help close relatives (Emlen and Wrege 1988; Emlen 1990). Despite their flexibility and fluidity, personal relationships based on individual recognition and long-term memory are the social foundations of subtle forms of reciprocal altruism, social manipulation, and kinship responses, all extraordinary levels of social complexity (Figure 19–13).

The open cooperative breeding system of the White-fronted Bee-eater is adapted to the unpredictable environment of the Rift valley. In some years pairs can breed successfully by themselves; in other years they cannot. Unlike closed cooperative breeding systems in saturated stable environments, where young cannot disperse and must compete with established individuals for breeding status, adult bee-eaters have less control over the breeding options of potential helpers. Emlen suggests that, as a result, potential breeders must sometimes allow helpers to share paternity or maternity of group clutches to attract their assistance. Complex social bonds result.

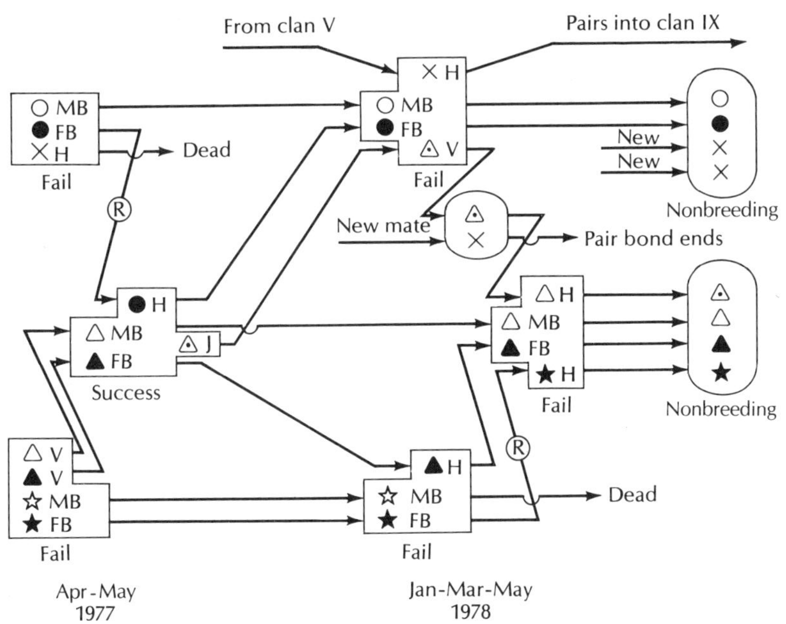

Figure 19–13 Clan relations in White-fronted Bee-eaters. Core members of the clan are identified individually by symbols (circles, triangles, stars), and connecting lines trace their social movements over time. Temporary associates are indicated by ×. In 1977 the clan consisted of three monogamous pairs and their associates. Two of the pairs failed in their breeding attempts and one succeeded. Each box represents a breeding or roosting chamber in the colony. MB, male breeder; FB, female breeder; H, helper; V, visitor, that is, a bee-eater that roosted in the chamber but did not help in the nesting effort; J, juvenile. Note the rearrangements of the associations within the clan: redirected helping (R) by breeders whose own efforts failed; and the reciprocal helping between females represented by the solid star and solid triangle. (After Hegner et al. 1982; photograph courtesy of S. Emlen and N. Demong)

Complex social bonds biased by kinship may prove to be more widespread among birds than we realize. Other studies of cooperative birds suggest that flexible helping and complex social relationships are not restricted to colonial bee-eaters. Like Emlen's bee-eaters, species such as Galápagos Mockingbirds and Splendid Fairywrens also are subject to unpredictable climatic variations from year to year (Curry and Grant 1990; Rowley and Russell 1990). Unable to relocate to optimal conditions, such birds, which reside on and defend permanent, fixed territories, must respond flexibly to different circumstances each year. Social flexibility with continuity enables them to keep their options open and cooperate with kin as needed.

Summary

Most birds raise young that require extended and substantial amounts of parental care in terms of both time and energy. The substantial costs of parental care favor participation by both males and females, sometimes aided by helpers. Biparental care by monogamous mates enhances fledging success of larger broods. This outcome is particularly apparent in Temperate Zone species, in which large clutch sizes seem to be tied to effort by both parents. Guiding the evolution of alternative breeding behaviors are trade-offs between current and future efforts, conflicts between parents and their offspring, uncertainties about parentage, plus opportunities to cheat or to cooperate. How much effort a bird theoretically should invest into the protecting and nurturing of its chicks reflects a basic trade-off between increased survival rate of chicks at the cost of a parent's survival or additional mating opportunities. Deliberate infanticide is known in some bird species, but others adopt foster chicks.

Brood parasitism and cooperative breeding lie at opposite ends of the spectrum of parental care practices among birds. Brood parasites lay their eggs in the nests of other birds. By allowing foster parents to raise their young, brood parasites reduce their own costs and avoid the risk of losing all the young in a single nest. Cooperative breeders help other breeders (usually parents or stepparents) to raise young while waiting for an opportunity to breed themselves.

Cowbirds and cuckoos are the most familiar brood parasites, but this practice has evolved along at least seven separate lines among birds. Adaptations for brood parasitism include egg mimicry, nestling mimicry, host mimicry, egg size and hardness, and destruction of host eggs and young. A high incidence of brood parasitism is responsible for the decline of some host populations. Countermeasures evolved by hosts include egg recognition and nest abandonment.

Intraspecific brood parasitism—the leaving of eggs in nests of other females of the same species—is proving to be quite common among birds. Facultative parasitism of other species is an intermediate step in the evolution of obligatory brood parasitism.

Although cooperative breeders may appear to act altruistically, they actually act in their own best interest. Cooperative breeding evolves under conditions of ecological constraint, such as lack of breeding territories that prevent birds from breeding on their own. Intrinsic benefits, such as learning behavioral skills may also favor delayed dispersal. By helping to raise other broods, these birds enhance their own chances for breeding through inheritance of a territory or through other forms of territory acquisition. Breeding pairs with helpers fledge more young than those without helpers, primarily because they suffer less stress and hence survive longer and are more likely to renest. Strife abounds in the social relations of cooperative breeders such as Groove-billed Anis, which throw one another's eggs out of their communal nests. White-fronted Bee-eaters are remarkable for the complex, cooperative social systems that help them adapt to the unpredictable environments of East Africa.

Further Readings

Brown, J.L. 1987. Helping and Communal Breeding in Birds: Ecology and Evolution. Princeton, N.J.: Princeton University Press. *A broad review of evolving hypotheses in a fast-moving field of research.*

Emlen, S.T. 1991. Cooperative breeding in birds and mammals. *In* Behavioral Ecology: An Evolutionary Approach, 3rd ed., J.R. Krebs and N.B. Davies, Eds., pp. 301–337. Sunderland, Mass.: Sinauer Associates. *A major review of conceptual issues.*

Koenig, W.D., and R.L. Mumme. 1987. Population Ecology of Cooperatively Breeding Acorn Woodpeckers. Princeton, N.J.: Princeton University Press. *A synthesis of long-term studies of an unusual breeding system.*

Koenig, W.D., F.A. Pitelka, W.J. Carmen, R.L. Mumme, and M.T. Stanback. 1992. The evolution of delayed dispersal in cooperative breeders. Q. Rev. Biol. 67: 111–150. *A major review of conceptual issues and challenges for future research.*

Lack, D. 1968. Ecological Adaptations for Breeding in Birds. London: Methuen. *A classic.*

May, R.M., and S.K. Robinson. 1985. Population dynamics of avian brood parasitism. Am. Nat. 126: 475–494. *Theoretical models of the consequences of brood parasitism in birds.*

Payne, R.B. 1977. The ecology of brood parasitism in birds. Ann. Rev. Ecol. Syst. 8: 1–28. *A critical review of the literature.*

Stacey, P.B., and W.D. Koenig, Eds. 1990. Cooperative Breeding in Birds. Cambridge: Cambridge University Press. *A collection of important reviews that update major studies.*

Woolfenden, G.E., and J.W. Fitzpatrick. 1984. The Florida Scrub Jay: Demography of a Cooperative-Breeding Bird. Princeton, N.J.: Princeton University Press. *Details of an extraordinary long-term study of one species.*

PART VI

Population Dynamics and Conservation

CHAPTER 20

Demography

DEMOGRAPHERS STUDY THE characteristics of populations, that is, age composition, reproductive output, and mortality. Comparative avian demography—the study of life history patterns—is a field of study that explores the links between parental efforts and life span. Some birds lay many eggs; others lay just one. Some birds typically live for decades; others for just 2 or 3 years. To maximize their contribution to the next generation, individuals must achieve optimal combinations of four variables that affect lifetime reproductive output, namely, the age at which they first reproduce, the number of young they fledge each year, the survival of those young, and their own longevity as adults. An even more specific topic, the evolution of optimum avian clutch sizes—how many chicks should a bird attempt to raise at one time—is one of the vital intellectual issues of avian biology. Under investigation are the ways that clutch sizes maximize lifetime reproductive success, plus the reasons why clutch sizes vary among populations of the same species.

This chapter starts with an introduction to life history patterns and life table analysis and then proceeds to their two main components, survival and fecundity. The term *survival* refers to the probability of living to a particular age. The term *fecundity* refers to the number of young successfully raised or fledged per year. The costs and often poor results of attempts to breed by yearling or inexperienced birds favor delayed maturity in some species. After this introductory foundation, the chapter proceeds to review competing hypotheses for the evolution of avian clutch sizes. Concluding the discussion of demography is the topic of brood reduction—or retrospective adjustment of brood sizes and the strategic apportioning of parental care to the chicks.

Life History Patterns

Several patterns characterize avian life histories (Table 20–1). First, mortality rates are initially high among young birds less than 1 year old and then decline to nearly constant lower levels among adults. Second,

TABLE 20–1
Extremes of avian life history patterns

Species	Survival before breeding	Age of first reproduction	Fecundity	Adult mortality rate
Albatrosses and eagles	Moderate (30%/year)	Late (8–10 years)	Low (0.2 young/year)	Low (5%/year)
Ducks and small passerines	Low (15%/year)	Early (1 year)	Moderate (3 young/year)	High (50%/year)

reproductive success and effort improve with age and experience. Third, long-lived species, such as albatrosses, penguins, and eagles, tend to have low fecundity, whereas short-lived species, such as songbirds and ducks, tend to have high fecundity. Finally, adaptation to local environments molds life history patterns in predictable ways. In the north and in arid environments, clutches are generally larger than those in the south and in wet environments. Direct relationships between early development patterns, particularly long incubation periods and sibling competition, and long life spans in birds such as albatrosses (Figure 20–1) also suggest a high level of evolutionary integration of life history traits in birds, from fledging to old age (Ricklefs 1993).

Illustrating the extremes of avian life history patterns are those of a small, short-lived bird, such as a sparrow, compared with a large, long-lived bird, such as a penguin or an albatross (Figure 20–2). A sparrow starts breeding after 1 year and concentrates high reproductive achievements into a few consecutive years, whereas a penguin starts breeding after 4 years and spreads lower annual fecundity over a long span of years. The Wandering Albatross starts breeding at 8 to 11 years, and produces one chick every 2 years for up to 50 years.

Life Tables

Survival and fecundity change with age. Life tables summarize the vital population statistics of age-specific survivorship and age-specific fecundity under a given set of conditions (Table 20–2). Life tables usually are based on the statistics of females because these are more reliably measured than are those of males. Although subject to error, it is more pragmatic to associate eggs in a nest with the female(s) that laid them than with the male or males that might have fertilized them. From these statistics, one can project life expectancies and family sizes of other individuals facing the same conditions. From life table data, one can also project rates of population growth, which are the net result of survival and fecundity. Life

Figure 20–1 Albatrosses are long-lived birds that raise only one offspring at a time. (Courtesy of J. Warham)

table data thus project future population trends. Life table analyses also reveal which stages of the life cycle are most sensitive to change and hence should be targeted by conservationists to maintain healthy, stable populations.

To study the life history patterns of a particular kind of bird, the ornithologist follows the annual progress of a cohort of eggs, nestlings, or fledglings until the last one dies. The proportion of the cohort that survives each year defines the annual survivorship (S_x). The probability of survival to a particular age (L_x), therefore, is the product of the preceding annual survival rates. The number of young produced each year by adults in the cohort defines age-specific fecundity (B_x). The product L_xB_x specifies an individual's expected annual fecundity, which is to say fecundity at a certain age discounted by the chance of dying before reaching that age (Figure 20–2).

TABLE 20–2
Time-specific life tables for female Eastern Screech-Owls in either a suburban or a rural study area, 1976–1991

Age classes[a]	S_x[b]	L_x[c]	Average number of fledglings per individual	B_x[d]	L_xB_x[e]
			Suburban		
Fledglings	0.36	1.00	0.0	0.0	0.00
Adults					
1	0.49	0.49	1.6	0.8	0.39
2	0.58	0.18	2.6	1.3	0.23
3	0.61	0.10	3.1	1.5	0.15
4	0.67	0.06	3.2	1.6	0.10
5	0.75	0.04	2.7	1.3	0.05
6	0.75	0.03	2.7	1.3	0.04
7	0.75	0.02	2.7	1.3	0.03
8	0.75	0.02	2.7	1.3	0.03
9	0.75	0.01	2.7	1.3	0.00
10	0.75	0.01	2.7	1.3	0.00
			Rural		
Fledglings	0.30	1.00	0.0	0.0	0.00
Adults					
1	0.36	0.30	1.6	0.8	0.24
2	0.60	0.11	2.3	1.1	0.12
3	0.67	0.06	3.2	1.6	0.10
4	0.53	0.04	2.0	1.0	0.04
5	0.50	0.02	2.0	1.0	0.02

a. All age classes present in each study area are included. Numbers in column represent age (in years) of adult.

b. S_x, survivorship.

c. L_x, probability of survival.

d. B_x, number of female offspring per female subject, based on known 1 : 1 sex ratio.

e. Summation of values in this column yields an R_0 (population replacement rate) = 1.01 for the suburban population and R_0 = 0.52 for the rural population.

From F.R. Gehlbach, unpublished data.

The values of L_xB_x for all age categories add up to R_0, which is the net reproductive rate—or the expected rate of recruitment of new individuals into the population. If one female is replaced by one other during her lifetime, R_0 equals 1. A population composed of many such individuals is stable in size. Larger values of R_0 are expected in growing populations and smaller values in declining populations. Thus, if $R_0 = 1.5$, the population will increase 50 percent in one generation. Conversely, a value of 0.8 indicates a declining population.

Consider a life table example for Eastern Screech-Owls, which nest commonly in cavities in wooded habitats. These small owls are monogamous, permanent residents that live between 7 and 13 years, and produce one brood of two to three young each year. Fred Gehlbach (1989; unpublished data) compiled life table statistics for two study populations in Texas, one in the suburbs, the other in rural woodlands (Table 20–2).

In the suburbs, annual survivorship (S_x) increased with age to a maximum of 75 percent per year. Eastern Screech-Owls achieve full breeding potential by the age of 2 years, by which time each female produces an average of 1.3 female offspring each year (B_x). Actually, reproductive output varied greatly among females; a minority (less than 25 percent) of long-lived females produced the majority of fledglings (Gehlbach 1989). The net reproductive rate R_0, obtained by adding the annual products of survivorship and fecundity (L_xB_x), was 1.01, a value indicating simple lifetime replacement of a female by one daughter and thus a stable population size. Screech-owls do not fare as well in the rural woodlands of central Texas. Both annual survival and fecundity are lower. The net reproductive rate is 0.5, and thus the population is declining.

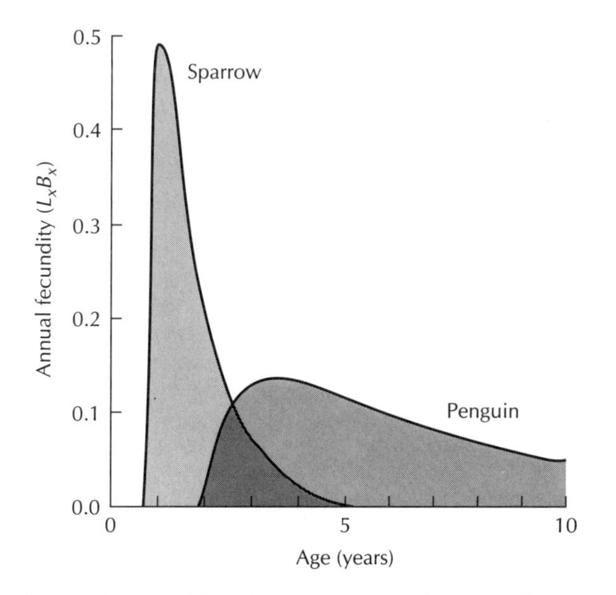

Figure 20–2 Reproductive efforts by a sparrow and a penguin, expressed in terms of expected annual fecundity (L_xB_x), where L_x is the probability of survival to a particular age, and B_x is age-specific fecundity. The short-lived sparrow produces more young every year than does the long-lived penguin, but their total lifetime fecundities (areas under the curves) are roughly the same. (Adapted from Ricklefs 1973b)

Annual Survival and Mortality

Rates of annual survival, a main ingredient of demographic analyses, change conspicuously with age in the first years of life. Survival rates may also differ between the sexes, among species, and in various geographical regions.

Effects of Species, Age, and Sex on Survivorship

In general, large species survive better than small species, seabirds survive better than land birds, and tropical species may survive better than Temperate Zone species (Table 20–3). The survival rates of adult birds range from as little as 30 percent per year in Blue Tits and Song Sparrows to over 95 percent in Royal Albatrosses, Bald Eagles, and Atlantic Puffins. Most birds die inconspicuously of natural causes—starvation, disease, predation, and disasters related to the weather.

A young bird's annual chance of survival from fledging to breeding age typically is about half that of an adult. Small land birds are especially vulnerable in their first year. Mortality during the first few weeks out of the nest is generally high. In the case of Yellow-eyed Juncos of the Chiricahua mountains of southeastern Arizona, nestlings and fledglings experience two early episodes of high mortality; only 11 percent of banded nestlings reappear the following spring (Sullivan 1989). Grown nestlings and fledglings incapable of extended flight easily succumb to predators, which take about 50 percent of the available young juncos in a 9-day risk period. Survivorship then improves for 3 weeks while parent juncos care

TABLE 20–3
Annual survival of adults

Group	Survival (%/year)
Fowl	20–50
Small land birds	30–65
Ducks	40–60
Raptors	50–96
Herons, gulls, waders	60–80
Seabirds	80–95
Tropical land birds	50–90
Albatrosses	95

From Ricklefs 1973b; Welty 1982; Perrins and Birkhead 1983.

for their mobile fledglings. With independence comes a second episode of high mortality due to starvation. Newly independent young find insects slowly and inefficiently and spend almost all day feeding. Approximately 42 percent of them die, most by starvation. It takes about 2 weeks for the juveniles to develop adequate foraging skills.

In general, the more physically developed the young are when they leave the nest, the greater are their chances of survival. This is one of the advantages of longer nestling periods and of fast growth in altricial nestling (Box 18–1). A fledgling's chance of survival (measured by ornithologists in terms of future recaptures) increases in proportion to its mass at fledging (Figure 20–3). Food availability, the quality of parental care, the number of siblings competing for that care, and the timing of fledging are also important factors.

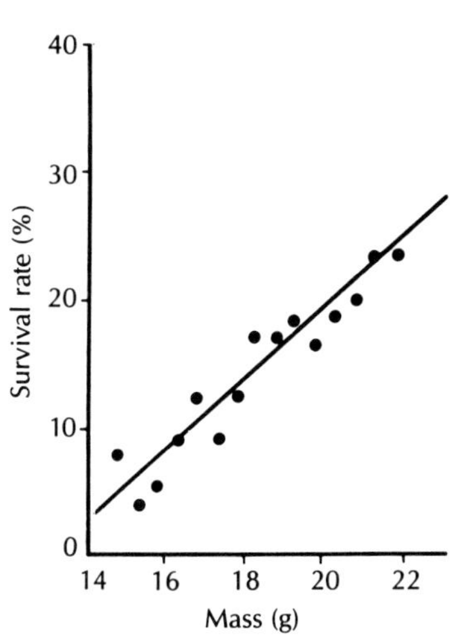

Figure 20–3 The probability of survival (and hence recapture by ornithologists) increases directly in relation to the mass a young Great Tit attains prior to leaving the nest. (Adapted from Perrins 1980)

Some of the best field data of predation on young birds were obtained by the systematic collection of the metal bands ("rings") on songbird carcasses that accumulated at the nests of Eurasian Sparrowhawks that fed on the research populations of Great Tits and Blue Tits in Wytham Wood at Oxford (Perrins and Geer 1980). These predators took 922 ringed tits in 1976, 759 in 1977, and 1220 in 1978. The largest losses were in juveniles; 18 to 34 percent of the Great Tit juveniles and 18 to 27 percent of the Blue Tit juveniles were lost to these hawks each year.

Once birds reach adulthood, their chances of survival increase and stay essentially constant. Survivorship in juvenile Florida Jays, for example, is extremely low during the first few months after they leave the nest (Woolfenden and Fitzpatrick 1984) (Figure 20–4). At ages of 2 to 3 months, their mortality rate is still four times that of breeding adults. Only 40 percent of Florida Jays survive their first year, after which they "graduate" to the high adult survival rates.

Survival of experienced breeders has the greatest potential impact on population growth rate (McDonald and Caswell 1993). Protection of nests per se has relatively little impact because adult jays will breed again, many times if they have access to quality habitat that ensures their long-term survival. Carefully controlled analysis of mortality among high-quality, early breeding Florida Jays also revealed increases in mortality at an average rate of 2 percent per year as a result of the degenerative effects of old age—often called senescence (McDonald et al. 1994). This finding challenges the traditional view that mortality in adult birds is independent of age.

Males and females often differ in their survival rates. Only 20 percent of fledgling Great Tits survive their first year, but 48 percent of breeding females and 56 percent of males survive each 12-month interval thereafter (Bulmer and Perrins 1973). Male birds generally survive better than do females, a situation that leads to a male-biased sex ratio in many populations. The causes and timing of greater female mortality and their relations to parental investment remain uncertain (Breitwisch 1989), but these results support the long-held viewpoint that females have higher costs of reproduction than males do.

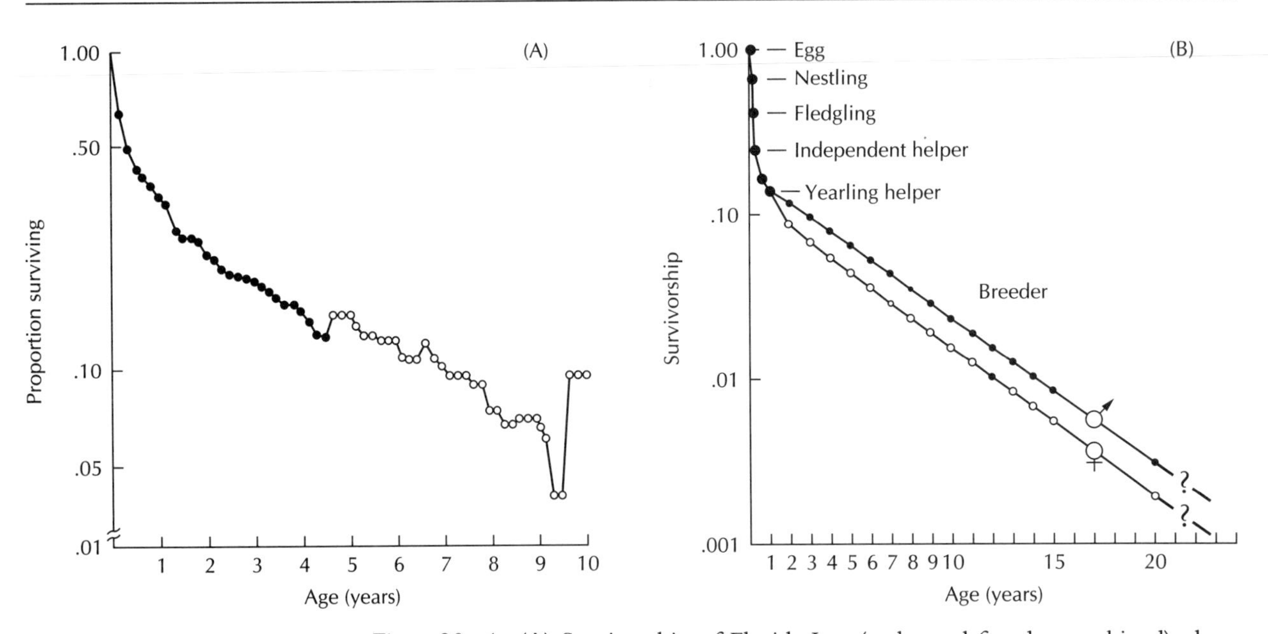

Figure 20–4 (A) Survivorship of Florida Jays (males and females combined) observed in a population in central Florida, from fledging through age 10 years. Black symbols indicate sample sizes of cohorts containing more than 100 potential surviving individuals. Data become more irregular and unreliable at the end because of small sample sizes. Note that the proportion surviving drops sharply in the first year of life to about 40 percent of the initial cohort. (B) A complete survivorship curve from beginning of incubation through possible age of senescence (question marks). Males and females diverge slightly after age 1 year because of the greater mortality of females, which disperse from their natal territory. Survivorship of breeders is identical between sexes. (From Woolfenden and Fitzpatrick 1984)

Tropics versus Temperate Zone

Also debated is the issue of whether survival rates of tropical forest birds are higher than those of temperate forest birds. Much life history theory has been based on this assumption, but few data have been available to test it. Accurate estimates of survival rates require careful statistical analysis of large sample sizes (Dobson 1990; Clobert and Lebreton 1991; Lebreton et al. 1992). Hence, many early estimates of bird survival rates based on scattered, poorly analyzed data were inaccurate. Comparisons of large samples of certain tropical African birds and temperate region European birds suggest that at least some African species attain older ages than their European counterparts do (Oatley and Underhill 1993).

The first detailed comparison of survival rates of comparable sets of birds, however, did not support the conventional wisdom (Karr et al.

1990). Instead, the study revealed that the mean annual survival rate of adult birds is about 55 percent in both Maryland and Panama (Figure 20–5). Both sets of estimates range from 30 to 70 percent. If confirmed, this finding will force a reevaluation of current models of the evolution of life history traits in tropical birds. A prevailing hypothesis for the prevalence of small clutch sizes in the Tropics, for example, is based on the assumption that tropical birds live longer and have more breeding opportunities than do temperate region birds.

Longevity

Reflecting their high annual mortality rates, the life expectancy of most small birds is only 2 to 5 years. In contrast, large birds such as the Adelie Penguin live an average of 20 years. Many individuals die young, whereas others live longer. The maximum ages recorded in wild birds average 10 to 20 years for various species of passerines and 20 to 30 years for water birds and raptors. Among the records are a 53-year-old Laysan Albatross, a 36-year-old Eurasian Oystercatcher, and a 34-year-old Great Frigatebird. One record holder is a female Royal Albatross from New Zealand named "Blue-White"; she was banded as a breeding adult in the summer of 1937 and still laid eggs there 48 years later in 1985. If she was 10 years old in 1937 as estimated, this female lived to an age of at least 58 years old. Captive birds tend to live even longer than their wild relatives; some captive parrots have lived to be 80 years old (Flower 1938).

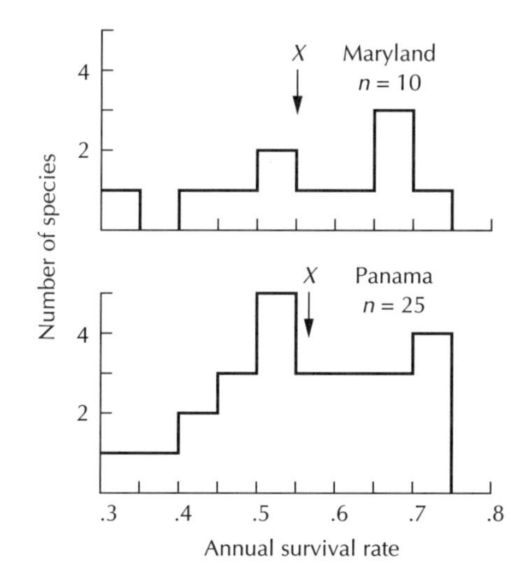

Figure 20–5 Temperate and tropical forest birds have virtually the same mean annual survival rates (X), namely, 0.54 ± 0.03 ($n = 10$) and 0.56 ± 0.02 ($n = 25$), respectively. (After Karr et al. 1990)

Fecundity

Fecundity—the number of young successfully raised—is a measure of an individual's reproductive success. Total lifetime reproductive success depends on the age at which a bird starts to breed, how long it lives, and on the cumulative result of the bird's annual reproductive successes and failures (Newton 1989). Annual fecundity, in turn, reflects the number of nesting attempts and the success of each attempt, the number of eggs laid each time (clutch size), and the age and experience of a breeding individual.

Multiple Broods

The number of broods a pair can raise depends on the length of the breeding season. Long nesting cycles or restricted breeding seasons such as those in temperate and arctic latitudes tend to preclude extra broods. Hence, many temperate and arctic birds attempt only one brood, unless they are trying to replace one that was lost early. Single broods also characterize many tropical and island species that practice prolonged parental care. Within species there may be variation in numbers of broods attempted because of the local geography. The Song Thrush, for example, attempts only one brood in northern Europe (65 to 70 degrees north latitude) but attempts two to four broods in southern Europe (45 to 50 degrees north latitude) (Haartman 1971). Tropical birds generally attempt more broods than temperate birds, owing, in part, to prolonged breeding seasons. In addition, because losses of nests and young to predators are much higher in the Tropics than in the Temperate Zone, renesting is often necessary to replace lost clutches. Two to four clutches are not unusual in the Tropics, and even six are not rare. The White-bearded Manakin, for example, typically lays three to five clutches per season in Trinidad (Snow 1962).

Tropical birds require slightly longer than temperate species to renest, whether they fledge their first brood or whether they replace a lost clutch or brood (Table 20–4). It is possible that tropical birds mobilize reserves more slowly and take longer to accumulate the additional energy needed for egg formation. The fact that tropical species lay on alternate days suggests that food resources in the Tropics may be more limited than in the Temperate Zone. Also, tropical species, unlike temperate species, cannot renest soon after their young have left the nest because parents take care of their fledged young for longer periods of time.

Some birds increase their seasonal productivity by overlapping successive clutches (Burley 1980). Overlapping small clutches can be a better way of increasing fecundity than enlarging a single clutch, because it reduces stress by separating periods of peak parental care into several smaller peaks. Clutch overlap is practiced by Rock Doves, which lay only two

TABLE 20–4
Intervals between broods of temperate and tropical passerine birds

Bird habitat	Interval (days) between broods after nest failure		Interval (days) between broods after successful fledging	
	Mean	Range	Mean	Range
Temperate Zone	7.8	3–11	8.2	1–20
Tropics	13.3	6–24	25.8	8–59

From Ricklefs 1969a.

eggs per clutch. Proficiency in parental care by pigeons is a prerequisite to their management of overlapping clutches; the extent of overlap of clutches increases with a mated pair's combined experience as parents. In Common Goldcrests, which sometimes have two overlapping broods, the male builds the second nest alone (Haftorn 1978c) and assumes primary responsibility for the young in the first nest when the female begins to lay and incubate the second clutch. Faced with two broods of young of differing ages, the male puts his initial effort into the older brood, for which the need is greatest, and then shifts his attention to the second brood after the first has achieved independence.

Success in Nesting

In general, nesting success increases in northern latitudes, in hole nesting species, and in large species with hardy young. The principal causes of failure in nesting are, in descending order, predation (see Chapter 16), starvation, desertion, hatching failure, and adverse weather (Ricklefs 1969b). Starvation of young is most common in marsh and field nesting birds and in raptors. Desertion by parents occurs most often in hole nesting birds. In colonial seabirds, separation from parents often results in the death of chicks, particularly in semiprecocial gulls and terns. Brood parasitism, nest-site competition, ectoparasites, and disease also contribute to nesting failure.

Another cause of reproductive failure—inbreeding (mating of close relatives)—is well documented in domestic animals and plants but has rarely been measured in wild birds (Greenwood 1987). The Great Tit has an average nestling mortality of 28 percent among inbreeders and 16 percent among outbreeders. Fertility of offspring, especially of males, also declines with increasing genetic similarity of the parents. Inbreeding levels tend to be very low in wild bird populations because of dispersal, but it may increase substantially in small, island or endangered populations (van Noordwijk and Scharloo 1981).

Clutch Size

The number of eggs a female bird lays is an essential and heritable component of fecundity (Boag and van Noordwijk 1987; Cooke et al. 1990). Clutch size is an adaptation molded by selection over evolutionary time, but it is also sensitive to immediate environmental conditions (see Chapter 15). Passerines and other small land birds that feed their young lay clutches of 2 to more than 12 eggs, the exact number of which varies among species and, within a species, with latitude, climate, age, and quality of territory. Waterfowl, pheasants, rails, and many other precocial birds have clutches of up to 20 eggs. Other birds have less variable clutch sizes; precocial shorebirds typically lay four eggs, and oceanic birds in Orders Pelecaniformes and Procellariiformes lay only one egg. Hummingbirds and doves always lay two eggs.

Variations in clutch size pose intriguing questions. Why do some birds lay 1 egg and other birds 20? Which particular number of eggs maximizes

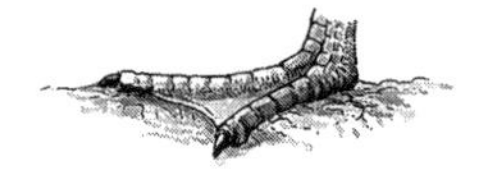

Box 20–1
Penguins and boobies lay an extra egg for insurance

To offset chronically poor hatching success due to infertility, early embryonic failure, or predictable loss to predation, certain seabirds lay an extra "insurance" egg to back up the first egg. Penguins in the genus *Eudyptes*, which includes the Rockhopper Penguin and the Macaroni Penguin, typically lay two eggs of different sizes (Williams 1980; del Hoyo et al. 1992), with features that are unusual among birds. The second egg is 20 to 70 percent larger than the first and tends to hatch first despite being laid up to a week later. The penguins do not start incubation until both eggs are laid.

The smaller, later hatching chick from the first, smaller egg usually starves to death within a few days after hatching. There are two intriguing hypotheses for this situation. The first is that the first eggs laid by these penguins often are lost to predators or pushed out of the nest during fights with neighbors before the parents settle down to incubation; 54 percent of the pairs of Macaroni Penguins lose their first egg this way before they lay their second egg. Such losses combined with high mortality during incubation favor a two-egg clutch containing one insurance egg. Alternatively, *Eudyptes* penguins could be evolving from a two-egg clutch to a one-egg clutch, and what we see is a snapshot of evolution in action.

Another seabird, the Masked Booby, a large goose-sized, tropical seabird, lays one extra egg as insurance against the hatching failure of their first egg (Anderson 1989, 1990). Most gannets and boobies lay only a single egg and seem to be able to raise only one young. Asynchronous hatching of the two eggs ensures predictable siblicide in the Masked Booby: The first chick pushes the smaller, second chick out of the nest shortly after it hatches. Hatching asynchrony followed by siblicide ensures that the peak food demands of two growing young never compromise the parents' ability to secure the survival of at least one chick.

TABLE 20–5
Condition correlated with variations in average clutch sizes

Variable	Conditions correlated with small clutches (2–3 eggs)	Conditions correlated with large clutches (4–6 eggs)
Latitude	Tropics	Temperate/Arctic
Longitude	Eastern Europe	Western Europe
Altitude (temperate)	Lowlands	Highlands
Nest type	Vulnerable	Secure (cavity)
Body size	Large species	Small species
Habitat	Maritime, island, and wet Tropics	Continental, mainland, and arid Tropics
Feeding place	Pelagic seabirds	Inshore seabirds
Development mode	Altricial	Precocial

reproductive success for a particular species? A general answer is that nutritional requirements for egg formation seem to limit clutch sizes of precocial birds, whereas the feeding abilities of parents limit the clutch sizes of altricial birds. This summary opens, rather than closes, further inquiry. No single topic has so occupied the attention of students of avian life history patterns as has the evolution of clutch size. The literature summarizing conspicuous variations in clutch size (Table 20–5) is formidable, and the interpretations of the data are controversial.

Age and Experience

Birds that breed for the first time typically produce fewer eggs and offspring than older, more experienced birds (Saither 1990). European Starlings lay an average of 4.9 eggs per clutch in their first year and 5.9 eggs per clutch thereafter (Kluijver 1935). Older California Gulls produce more young than younger gulls (Pugesek 1981). The oldest gulls (12 to 18 years old) produce 1.5 young per year, whereas middle-aged gulls and the youngest gulls (3 to 5 years old) produce 0.8 young per year. Older gulls feed their young more frequently, spend more time looking for food, and leave the nest unattended less often than do the younger members of the colony. It seems that older gulls invest heavily in reproduction during each season, even though doing so may decrease their chances for future reproduction. Young gulls invest less in their initial efforts (Figure 20–6).

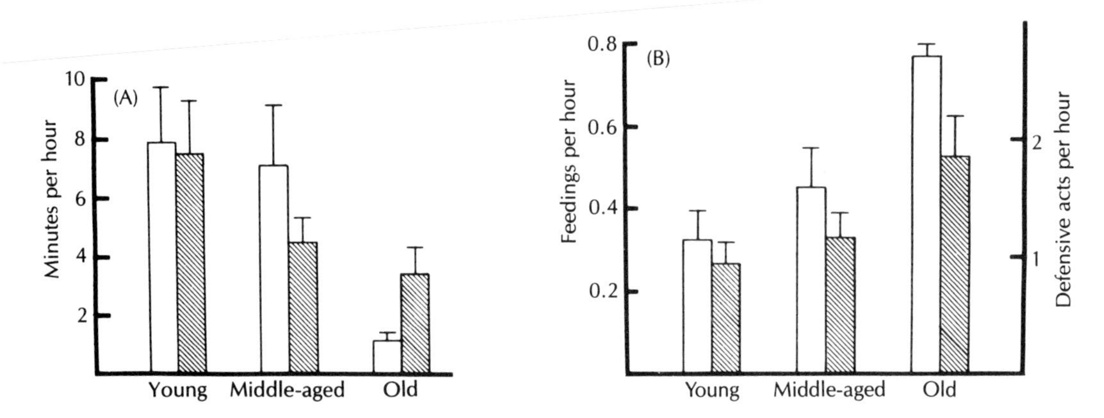

Figure 20–6 Older California Gulls put more effort into nesting than do young and middle-aged gulls. (A) During each hour, older gulls spend less time away from the nest and more time foraging. Open bars denote time nest was unattended; shaded bars denote time that neither parent foraged. (B) Older gulls feed their young more often and defend their nest territories more often. Open bars denote frequency of feeding; shaded bars denote frequency of defense. (From Pugesek 1981; copyright 1981 by the AAAS)

Delayed Breeding

Many birds, although by no means all, breed when they are 1 year old. Swifts breed at 2 years, parrots at 2 or 3 years, and raptors at 3 or more years. Water birds and seabirds generally take 4 or more years to breed for the first time and large albatrosses and condors take 8 to 12 years. Among the species with late maturity, there is a distinct correlation between longevity and age at first breeding.

Why should a bird delay breeding? Every extra year of nesting would seem to increase the chances of leaving some offspring, and theoretically, an early starting age has the greatest impact of all the variables on a bird's potential reproductive contribution to succeeding generations. The age at first breeding controls the interval between generations—or mean generation time—which, in turn, drives the potential growth rate of a population. Individuals that can breed in their first year should soon replace others that delay breeding for several years unless the costs of early reproduction are too severe.

Delayed maturity may contribute to maximizing lifetime reproductive success in long-lived birds when reproduction enhances the risk of death. The factors favoring delayed maturity are well documented in Adelie Penguins (Ainley and DeMaster 1980; Ainley et al. 1983). First, breeding entails greater risk than not breeding; mortality of breeders (39 percent) is greater than mortality of nonbreeders (22 percent). The greatest mortality occurs the first time young Adelie Penguins try to breed. An amazing 75 percent of 3-year-old females die during their first attempt to breed, because they are less efficient at obtaining the food necessary to sustain the costly breeding effort and also may be less adept at escaping predation by leopard seals at the edge of the pack ice. Mortality then declines with age to 10 percent in 11-year-old breeding females.

Offsetting the risks of initial reproduction in Adelie Penguins are improved prospects for raising young in subsequent attempts. Adelie Penguins that breed for the first time when they are 3 to 4 years old (and survive that effort) are less likely to lose their eggs or young in subsequent nesting seasons than those penguins that breed first at a later age. Whether these early starters are inherently better breeders or whether the early start somehow enhances subsequent breeding success is not known.

Three to four years seems to be the minimum possible age for reproduction in small penguins such as the Adelie. Three main factors are responsible for this age requirement. First, studies of another species, the Yellow-eyed Penguin, suggest that 2-year-old penguins usually are not reproductively mature; 65 percent of their eggs are infertile (Richdale 1957). Second, 2 to 3 years of experience seem to be essential for young penguins to develop the foraging efficiency that enables them to accumulate the large energy reserves required for egg production and the long fasts while breeding. Third, at least 1 year of social experience is necessary to develop the behavioral skills required for successful pairing and defense of nest, eggs, and young. Given the increased risks of mortality associated with breeding even when well prepared, there is a clear advantage to the 3- to 4-year delay typical of this penguin.

Males of some songbirds do not acquire full adult breeding plumage in their first breeding season, even though they are capable of breeding. Delayed plumage maturation reaches extremes in cases such as the lek-displaying Long-tailed Manakin, in which young males wait 8 years before reaching breeding status (see Chapter 17). Indeed, delayed maturation or acquisition of adult features is widespread among birds (Lawton and Lawton 1986; Studd and Robertson 1985). Thirty-one of the one hundred and five sexually dimorphic passerines of North America, including Red-winged Blackbirds, Northern Orioles, Scarlet Tanagers, and American Redstarts, do not attain adult male plumage for a year.

Among the unresolved hypotheses to explain this delay in adult plumage are some that propose that young males in femalelike plumage are less conspicuous (and less vocal) and thus less vulnerable to predation (see Rohwer and Butcher 1988; Thompson 1991). Also, because these young males resemble females more than they do rival males, they may avoid attack and eviction by older males and thereby occupy parts of established territories surreptitiously, establish some control in and access to that space, and ultimately gain priority in the use of the territory for breeding. In this way, perhaps, evolution has favored first-year males that postpone the acquisition of full breeding coloration.

In one study, Elizabeth Proctor-Gray and Richard Holmes (1981) found that first-year male American Redstarts with female plumage are sexually mature, establish territories, pair successfully, and fledge as many young as do fully mature males with black-and-orange plumage. By

Box 20–2
Young female Tree Swallows delay breeding

Floating populations of nonbreeding males, but not of females, are typical of birds (Chapter 17). In an unusual case, many female Tree Swallows, especially yearlings, are unable to breed because of intense competition for nest holes. Early exploring of nest holes for vacancies subjects yearling females to attacks by older residents, both males and females. Also counter to the general trend among birds, female, not male, Tree Swallows delay acquiring full adult breeding plumage for one year.

Bridget Stutchbury and Raleigh Robertson (1987) conducted experiments with models of yearlings and adults to test alternative hypotheses about the adaptive significance of subadult plumage in these swallows. The yearling females did not gain any reprieve from adult females, which attacked models with full breeding or subadult plumages with equal intensity. Adult males, however, were always more aggressive toward the adult model. These ornithologists concluded that the subadult plumage of yearling females was advantageous because it signaled their inactive sexual status to resident males and not because it signaled their subordinate status to resident females.

breeding in nonpreferred habitats away from older males, however, these young males avoid investment in bright territorial signaling plumage and use the more cryptic female plumage that helps to reduce the risk of predation.

Evolution of Clutch Size

Theoretically, there is a clutch size for each bird in an average year that produces the maximum number of its young capable of surviving to sexual maturity, and theoretically, an average optimal clutch size should prevail in local populations. However, understanding the evolutionary forces responsible for the evolution of a particular clutch size remains one of the most controversial and unresolved challenges for ornithologists, despite nearly a half-century of intense research. The debate about the evolution of clutch sizes among birds centers on applications and extensions of Lack's food limitation hypothesis (Winkler and Walters 1983; Winkler 1985; Murphy and Haukioja 1986). Bertram Murray, Jr. (1992a, b), has challenged these hypotheses with an alternative theory and a set of predictive equations (Murray and Nolan 1989).

Lack's Food Limitation Hypothesis

The clutch size in birds is adjusted by natural selection to the maximum number of nestlings the parents can feed and nourish. This fundamental postulate by David Lack (1947b, 1948) has guided research for nearly half a century.

The hypothesis assumes that individuals will be disadvantaged if they lay fewer eggs each year than they can raise. Observations, experiments, and theory (Royama 1969; Hussell 1972; Ricklefs 1977) all support this general concept, with some caveats and modifications.

The strongest support for Lack's hypothesis comes from observations of the relative success of various sizes of clutches and from experiments designed to test the ability of parents to feed extra young. For example, Christopher Perrins and Dorian Moss (1975) experimentally increased and decreased clutch sizes and found that clutches of 10 to 12 eggs produced the most surviving young Great Tits in Wytham Wood, near Oxford, England (Figure 20–7). The probability of a chick's survival in a small brood is greater than in a large brood because the nestlings in a small brood are better fed and heavier when they fledge; but the number of potential fledglings from small broods is necessarily low. Above a brood size of 12, chicks tend to be underfed and to die, especially in "bad" years of poor food availability. In 6 of 13 consecutive years, the average natural brood size in the population was 10—that is, close to the most productive number—but the average brood size was slightly lower than predicted in other years, an outcome resulting in an overall average across years of 8.5. Thus, birds seem to err on the side of caution. The vulnerability of large clutches in bad years will favor moderate clutch sizes in the long run (Boyce and Perrins 1987).

Lack's hypothesis seems to explain individual variations as well as, roughly, the average clutch size in a population. In southern Sweden, for

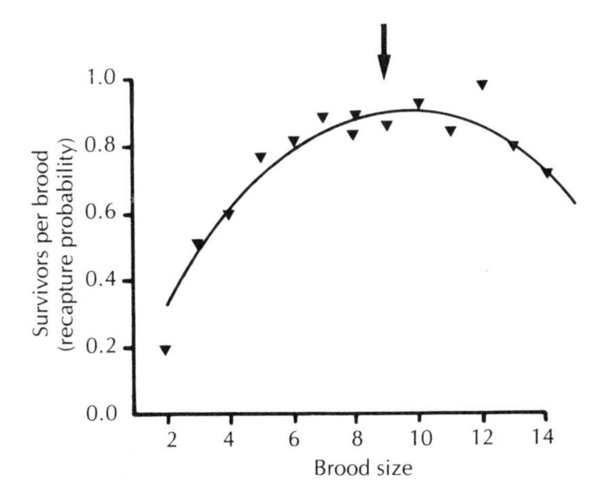

Figure 20–7 Lack's hypothesis of optimal clutch size projects a maximum number of surviving young as a result of the balance between the number of young hatched and the probability of survival. In the population of Great Tits in Wytham Wood, broods of 10 to 12 chicks are the most productive. The average clutch size in this species is 8.5 (arrow). (From Perrins and Moss 1975)

example, female Black-billed Magpies lay clutches of optimal size (Högstedt 1980). Experimental additions of eggs increase losses to predation and starvation and thereby reduce a pair's reproductive output. Experimental removal of eggs simply cuts breeding potential (Table 20–6). Most of the variations in clutch size among females in any 1 year relate to differences in food availability on their territories. The mean clutch size in the population reflects the average quality of territories. Thus, no single, optimal clutch size exists, and the variations are adaptive.

The strengths and weaknesses of Lack's hypothesis can be seen in its application to the well-documented increases of clutch size with latitude. Lack (1947b) and others after him have tried to explain why the average clutch size of many passerines, owls, hawks, herons, terns, gallinules, some fowl, and some grebes increases with latitude. Lack suggested that longer day length at high latitudes would enable birds to have larger broods. Birds nesting during the temperate summer have more time to find food for their young and themselves. However, the potentially positive effects of increasing day length do not explain why clutch sizes increase with longitude from east to west in Europe, with altitude in the Temperate Zone but not in the Tropics, on the mainland compared with adjacent islands, or with latitude in owls that feed at night and thus have less foraging time, not more (Table 20–5).

One problem with Lack's hypothesis is that the commonest clutch size observed in a population may be smaller than that which appears to be the most productive. This is the case in Great Tits in Wytham Wood (Boyce and Perrins 1987). Another problem is that some species of birds, but not others, can raise additional young, when they are provided experimentally (Hussell 1972; Lessells 1986; VanderWerf 1992). Even some large seabirds, such as gannets that normally lay only a single egg, can raise

TABLE 20–6
Reproductive output by Black-billed Magpies with natural and altered clutch sizes

	Output in experimental clutches of various sizes				
Initial clutch size	4	5	6	7	8
5	0.3	0.7	0.5	0.3	0
6	1.7	1.9	2.8	0.8	1.2
7	3.5	2.3	3.1	3.6	2.4
8	2.5	3.5	3.5	4.3	4.5

After Högstedt 1980.

two young when an extra egg is added to the nest (Nelson 1964). Thus, Lack's hypothesis cannot be fully generalized.

Among the many hypotheses that attempt to respond to the shortcomings of Lack's food limitation hypothesis are three principal ones: (1) the resource allocation or "trade-off" hypothesis; (2) the nest predation hypothesis; and (3) the seasonality hypothesis. The resource allocation hypothesis states that prospects for future survival temper annual clutch size. The nest predation hypothesis states that nest predation selects for smaller clutches because they are less conspicuous and minimize short-term losses. The seasonality hypothesis states that clutch size reflects the seasonal availability of resources relative to population size.

The Trade-off Hypothesis

This hypothesis incorporates the long-term costs of maximizing immediate reproductive effort and the trade-offs inherent in the optimal allocation of energy (Williams 1966; Cody 1966; Boyce and Perrins 1987). The total reproductive effort of a parent (which requires its continued survival) is an important variable. For repetitive breeders, an individual's current reproductive effort should take into account its long-range reproductive interests. We discussed this phenomenon with reference to age at first breeding. To the degree also that adult mortality increases with clutch size, as a result of the stress of caring for more young or of the heightened risk of predation, there may be advantages to reducing clutch size, especially in long-lived species.

Larger broods may be more stressful physiologically to produce than small broods. Individual Snow Buntings, European Pied Flycatchers, and Northern House-Martins that raise large broods sustain greater weight losses during the period of reproductive effort than do those individuals of their respective species that raise small broods (Hussell 1972). Eastern Bluebirds that raise large first broods are less likely to initiate a second brood (Pinkowski 1977). Such observations, however, have not determined the long-term survival of the adults.

Contrary to the trade-off hypothesis, it appears that adult survival may play only a minor role in shaping the life histories of short-lived bird species (Ricklefs 1977). Diane De Steven (1980), for example, was unable to demonstrate an inverse relationship between clutch size and adult survival in experiments that were specifically designed to test this hypothesis. Female Tree Swallows that raised experimentally enlarged broods did not lose significantly more weight or survive less well (from year to year) than females that raised normal-sized broods. Also, inexperienced yearling females were no more susceptible to stress (measured as reductions in weight or survival) during reproduction and brooding than were older females. Thus, negative trade-offs between clutch size and future reproductive effort—demographic optimization—do not help to explain the evolution of swallow clutch sizes.

The Predation Hypothesis

Predation is often postulated to be an important force in the evolution of clutch size (Perrins 1977). Increased predation could select for smaller clutches in at least three ways. First, it takes longer to lay a large clutch than a small one, which means that the eggs and chicks of a large clutch are at risk longer than those of a small clutch at the same site. Second, larger broods of young in a nest are noisier and more conspicuous and therefore more likely to attract predators (Skutch 1949). Third, it may be advantageous to risk fewer eggs at a time in the Tropics, where most nests are lost to predators and where a pair of birds has the opportunity to renest several times. Mercedes Foster (1974) suggests that tropical birds can maximize their reproductive output by laying small clutches and by molting while they are breeding, practices that extend the breeding season and, therefore, the renesting opportunities.

Greater safety from predators is probably the reason that hole nesting birds lay larger clutches than do open nesting birds. Selection for small, inconspicuous nests with few eggs also seems to prevail in predator-rich tropical forests and may even discourage an active role by the male in some species (Lill 1974c; Snow 1978). However, evidence that predation increases consistently as a function of clutch size in a way that would explain trends in clutch sizes remains equivocal (Ricklefs 1977; Slagsvold 1982). Furthermore, predation on nests and adult mortality during the breeding season account for only 10 to 25 percent of the variation in clutch size with latitude and as evolutionary forces seem to be too weak to counter the clear advantage of those larger clutches that escaped predation.

The number of fledged young that parents can attend or guard from predators may limit clutch size in some birds, particularly shorebirds, which lay no more than four eggs. Constraints of egg formation, incubation, and phylogenetic history have been suggested (Safriel 1975; Winkler and Walters 1983; Walters 1984; Ligon 1993). Even though the parents of many shorebirds do not feed their precocial young, they brood and tend them actively and guard them from predators. Physical distance between parents and their mobile young increases with brood size and potentially sets an upper limit on brood size. Active tenders—those species that follow their young closely—have smaller clutches than inactive tenders—those species that monitor their young from a distance.

The Seasonality Hypothesis

This hypothesis stresses the seasonality of resources (Ashmole 1963b; Ricklefs 1980b). Resources that are available during the breeding season depend on local demands by consumers, which in turn depend on population density. The population densities of resident birds are regulated by low resource availability during the nonbreeding season. Seasonal increases above this baseline thus control the resources available for breeding on a per capita basis (Figure 20–8).

Clutch sizes in some birds relate directly to seasonal increases in food production rather than to absolute level of production (Ricklefs 1980b). Birds of arid habitats in both Africa and Ecuador have larger clutches than do those that live in less seasonal, humid habitats at the same latitude (Lack and Moreau 1965; Marchant 1960). These results, as well as theoretical considerations, suggest that variation in the seasonability of resources experienced by a species is the ultimate cause of geographical variations in clutch size. In the broad context of the seasonal availability of resources, the immediate environment, such as territory quality, then molds the local clutch size for each species.

The pattern of variation of clutch size of the Northern Flicker supports the seasonality hypothesis (Koenig 1984). Clutches of this widespread North American woodpecker range from 3 to 12 eggs and increase by an average of 1 egg per 10 degrees of latitude. Variation in clutch size is directly correlated with the availability of resources per breeding woodpecker, which Walter Koenig has estimated as the ratio of local summer productivity (in terms of actual evapotranspiration, an index of plant productivity) to the breeding density of all woodpeckers. Local breeding

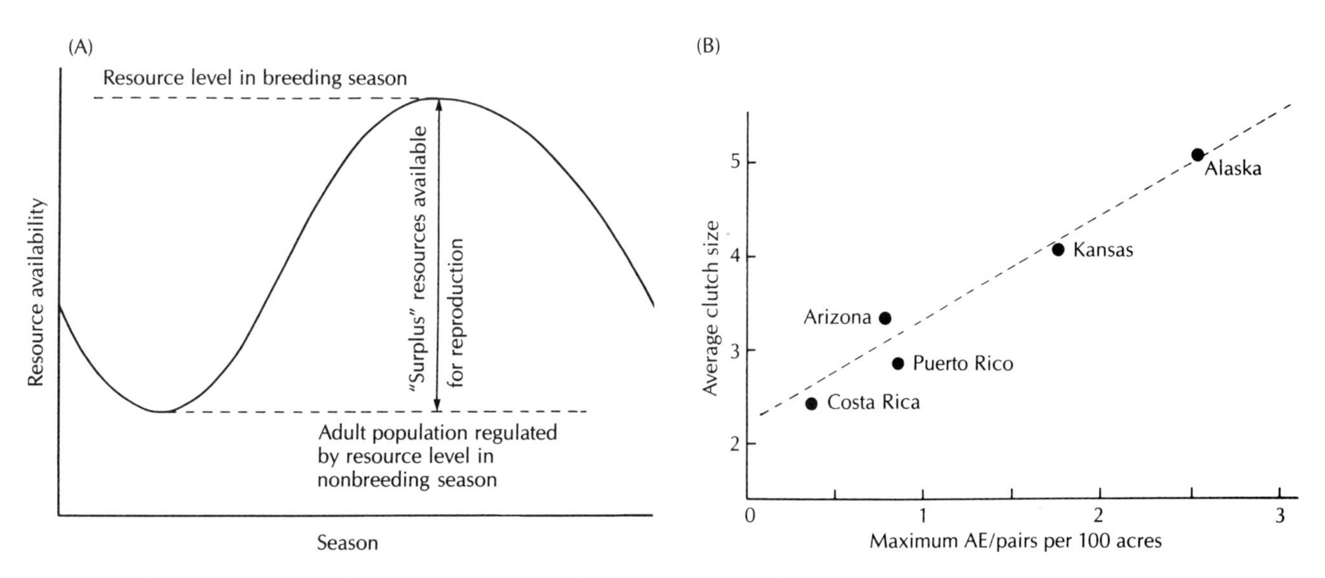

Figure 20–8 The seasonality hypothesis for geographical variation in clutch size. (A) Model of the seasonal increase in resources available for reproduction, measured in some months as the "surplus" above those resources that limit population size in the nonbreeding season; clutch size varies in relation to the ratio of the breeding season surplus to the adult population. (B) Clutch size increases with resource seasonality, measured as the ratio of maximum actual evapotranspiration (AE, an index of plant productivity) to the density of breeding pairs of birds. (From Ricklefs 1980b)

densities of woodpeckers in turn are set by winter productivity, which apparently determines how many woodpeckers survive until the breeding season (Figure 20–9).

Murray's Theory

Bertram Murray, Jr., an iconoclastic ornithologist, proposes that natural selection favors those females that *minimize* egg production consistent with replacing themselves in their lifetime. His controversial theory and predictive equations derive from the application of strict "hypothetico-deductive" logic (Popper 1972) to a small set of explicit assumptions (Wooten et al. 1991; Murray 1992b). First, Murray assumes that females maximize survival probability and lifetime success by tempering and actually minimizing short-term reproductive efforts to the smallest possible level consistent with self-replacement—the trade-off hypothesis. Second, he assumes that young from small broods have the highest probability of surviving to breed—Lack's food limitation hypothesis—and that a population will maintain a steady-state condition, that is, will neither increase nor decrease in the absence of evolutionary change. When life table measurements from natural populations are used, the equations developed from Murray's theory (based on these assumptions) closely predict the clutch sizes of Prairie Warblers (3.5 versus actual 3.9; Murray and Nolan 1989), Florida Jays (3.4 versus actual 3.3; Murray et al. 1989), and House Wrens (6.0 versus 5.8; Kennedy 1991). They also predict the basic trends in clutch size summarized in Table 20–5.

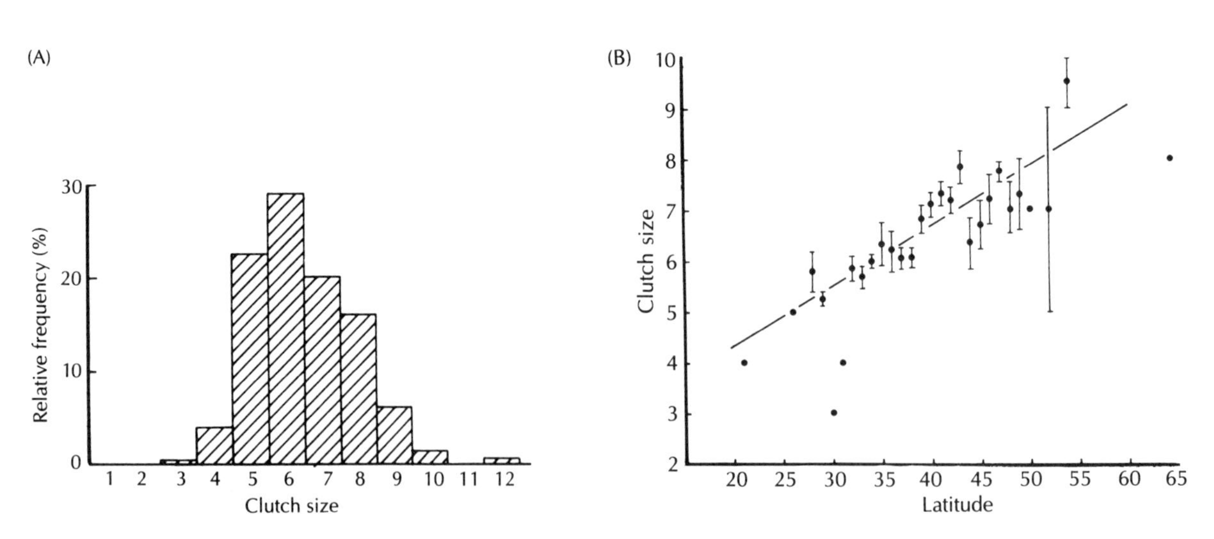

Figure 20–9 (A) The clutch sizes of Northern Flickers vary from 3 to 12 eggs. Relative frequency equals the percentage of total sample that had x number of eggs. (B) The increase in average clutch size with latitude supports the seasonality hypothesis. (From Koenig 1984)

Brood Reduction: A Retrospective Adjustment

One way that birds can cope with uncertainties about the maximum number of young they can raise in any particular year is to lay the number of eggs that should be successful in good years and then to sacrifice some of these eggs if necessary. Brood reduction, as this strategy is called, protects parents against loss of the entire brood should conditions for raising young be poor (O'Connor 1978).

Parents can apportion their reproductive effort selectively among the offspring in a clutch in several ways, including laying eggs of different sizes or nutritional qualities, or selective feeding of certain offspring. Parents may also start incubation before the last egg is laid, a strategy that promotes asynchronous hatching and sacrifices the younger siblings.

In addition to brood reduction, asynchronous hatching provides several additional potential advantages to parents. For example, hatching asynchrony staggers the peak feeding demands of the chicks, which reduces stress on the parents and increases fledging success when food is in short supply. By establishing a feeding rank hierarchy within the brood that is based primarily on age and size, and not on sex, asynchronous hatching also tends to equalize the survival of the sexes, especially in species with pronounced sexual size dimorphism (Slagsvold 1990).

Brood reduction ranges from overt siblicide, as discussed in Chapter 18, to more subtle, selective elimination of some members of a brood. Consider the details of subtle brood reduction by the Common Grackle (Howe 1978). Female grackles that lay large clutches of five or six eggs start incubation before they complete the clutch; but they rarely manage to raise the whole brood. The older siblings are fed first, grow faster, and are more likely to survive than their younger siblings. Females that attempt large clutches put extra provisions of yolk and albumen into the last, larger eggs to partially compensate for their late hatching disadvantages, but these strategies do not prevent starvation of the younger members of the brood when food is scarce. Some female grackles only lay two, three, or four eggs per clutch. They apportion their reproductive investment differently. Those that lay small clutches lay eggs of uniform size and provisions, hatch all the eggs at the same time by waiting until the last egg is laid before they begin incubation, and usually raise the whole brood.

Brood reduction in Red-winged Blackbirds depends on the age of the breeding female and also affects the relative numbers of male and female offspring that are fledged (Table 20–7). Young females fledge more daughters than sons, whereas old females fledge more sons than daughters. Although equal numbers of sons and daughters hatch in the broods of young females, starvation is common and sons starve more often than daughters. Young females lay poorly provisioned final eggs in the clutch, which causes those nestlings to be most vulnerable to starvation. Young females also tend to lack the experience required to feed their nestlings

TABLE 20–7
Surviving offspring of three age classes of female Red-winged Blackbirds

Age class of female parent	Number of surviving offspring		
	Male	Female	Significant difference
Young	28	50	Yes
Middle-aged	53	66	No
Old	54	34	Yes

From Blank and Nolan 1983.

adequately. A sex bias exists in the probability of starvation because male offspring need more food than their sisters do; they grow faster to a larger size and hence are more likely to starve. Older females, however, do not lay inferior final eggs, and they supply their young with better provisions. Hence their large, fast-growing sons are less likely to starve. Ornithologists do not know why older females hatch more sons than daughters, but apparently the female offspring are more likely to die as embryos as a result of unknown causes. Similar sex ratio differences are seen in humans, which give birth to 105 males for every 100 females.

In conclusion, this chapter has reviewed the trade-offs between survival and reproduction by birds. When birds attempt to breed and how they invest and apportion their parental time and energy guides the evolution of different life history patterns among species. Rates of reproduction and survival also combine to define each individual's relative contributions to succeeding generations, the heart of evolution by natural selection. Finally, the rates of reproduction and annual survival of individual birds combine to establish the dynamics of population growth or decline, the topic of the next chapter.

Summary

This introduction to the evolution of life history traits of birds presents the basics of life table analysis of demographic patterns, including the nature of survival and fecundity in birds and the typical patterns of avian life histories. A bird's lifetime reproductive output reflects the age at which it first reproduces, the number of young it fledges each year, the survival of those young, and its own longevity as an adult. Most small birds live 2 to 5 years, whereas large birds may live 20 to 40 years. Although many young birds die during their first year as a result of predation and starvation, the survival rates of adults are much higher and remain the same each following year.

In general, short-lived species usually breed when 1 year old and produce many young each year. Long-lived species tend not to breed until they are several years old and to produce few young each year. Reproductive success and effort usually improve with age and experience. Delayed maturity is a way of maximizing lifetime reproduction. Birds also increase the number of young produced by raising several broods sequentially in a season and in some cases by overlapping successive broods. Because it is difficult to predict the conditions that may favor a particular clutch size, some birds practice retrospective brood reduction, sacrificing surplus, disadvantaged young when necessary.

The evolution of clutch sizes is a major, still unresolved research topic in ornithology. Competing hypotheses extend Lack's food limitation hypothesis, which states that birds raise as many young as they can feed. The trade-off hypothesis states that birds temper short-term reproductive effort to increase long-term survival. The predation hypothesis states that the risk of predation may favor small clutches in the Tropics and in open nests, but not in cavity nests.

The seasonality hypothesis proposes that geographical trends in average clutch size—large clutches in the north and in arid environments, small clutches in the south and in mesic environments—seem best explained by seasonal differences in food availability. Bertram Murray has challenged these hypotheses with an alternative theory and a set of equations that predict not only the general trends in clutch size but also the average clutch sizes of populations of three species of North American songbirds.

Further Readings

Ainley, D.G., R.E. LeRosche, and W.J.L. Sladen. 1983. Breeding Biology of the Adelie Penguin. Berkeley: University of California Press. *A detailed, long-term study of one species.*

Clobert, J., and J.-D. Lebreton. 1991. Estimation of demographic parameters in bird populations. *In* Bird Population Studies, C.M. Perrins, J.-D. Lebreton, and G.J.M. Hirons, Eds., pp. 75–104. New York: Oxford University Press. *A review of methods used to estimate survivorship and fecundity in natural bird populations.*

McDonald, D.B., and H. Caswell. 1993. Matrix methods for avian demography. Current Ornithology 10: 139–185. *An introduction to ornithological applications of an approach to life table analyses that accommodates different social classes such as helpers.*

Murphy, E.C., and E. Haukioja. 1986. Clutch size in nidicolous birds. Current Ornithology 4: 141–180. *A detailed review of one group of birds.*

Murray, B.G., Jr. 1992. Sir Isaac Newton and the evolution of clutch size in birds: A defense of the hypothetico-deductive method in ecology and evolutionary biology. *In* Beyond Belief: Randomness, Prediction, and Explanation in Science, J.L. Casti and A. Karlqvist, Eds., pp. 143–180. Boca Raton, Fla.: CRC Press. *A provocative challenge to prevailing thought about avian clutch sizes and to how ornithologists practice science.*

Newton, I., Ed. 1989. Lifetime Reproduction in Birds. New York: Academic Press. *A collection of important papers based on long-term studies of color-marked birds.*

Ricklefs, R.E. 1983. Comparative avian demography. Current Ornithology 1: 1–32. *A powerful summary of the theoretical foundations of this subject.*

Ricklefs, R.E. 1993. Sibling competition, hatching asynchrony, incubation period, and life span in altricial birds. Current Ornithology 11: 199–276. *A stimulating integration of different components of avian life histories.*

Winkler, D.W., and J.R. Walters. 1983. The determination of clutch size in precocial birds. Current Ornithology 1: 33–68. *A detailed and timely review of perennial issues.*

Woolfenden, G.E., and J.W. Fitzpatrick. 1984. The Florida Scrub Jay: Demography of a Cooperative-Breeding Bird. Princeton, N.J.: Princeton University Press. *A detailed analysis of the demography of a cooperative-breeding bird.*

C H A P T E R 2 1

Populations

THE GROWTH OF POPULATIONS is a geometric function of individual reproductive success because a population potentially can double every few years. But at some point, population growth slows because resource needs begin to exceed availability. Growing populations also become increasingly vulnerable to predation and disease. Determining the ecological, social, and competitive forces that limit the continued growth of bird populations is a central challenge in the field of avian ecology. The conservation management and restoration of endangered bird species, based on sound understanding of their population ecology, are now urgent priorities, as discussed in Chapter 24.

This chapter examines the phenomenon of population size, starting with the principle that populations are dynamic, not static, entities. Then, extending the discussions of life tables and population growth from Chapter 20, the chapter considers the growth potential of populations and the factors that act to limit population growth. The ways in which fecundity and survival depend on local population density are illustrated by research on the Great Tit in England and Holland. Studies of European birds provide much of the best information on long-term population trends and on the dynamics of bird populations and hence provide a majority of examples.

Dynamically Changing Population Sizes

Bird populations are rarely stable and static. Instead, they fluctuate in size from year to year in relation to breeding success and mortality. Populations of European Pied Flycatchers, a hole nesting European bird, fluctuate annually by 50 percent. Densities of the Great Tit population in Marley Wood, near Oxford, England, can vary fourfold among years (Figure 21–1). Great Tit populations throughout Holland and England have grown in average size since 1946 (Kluijver 1951; Lack 1966). The general increase is due to a gradual amelioration of the European climate during

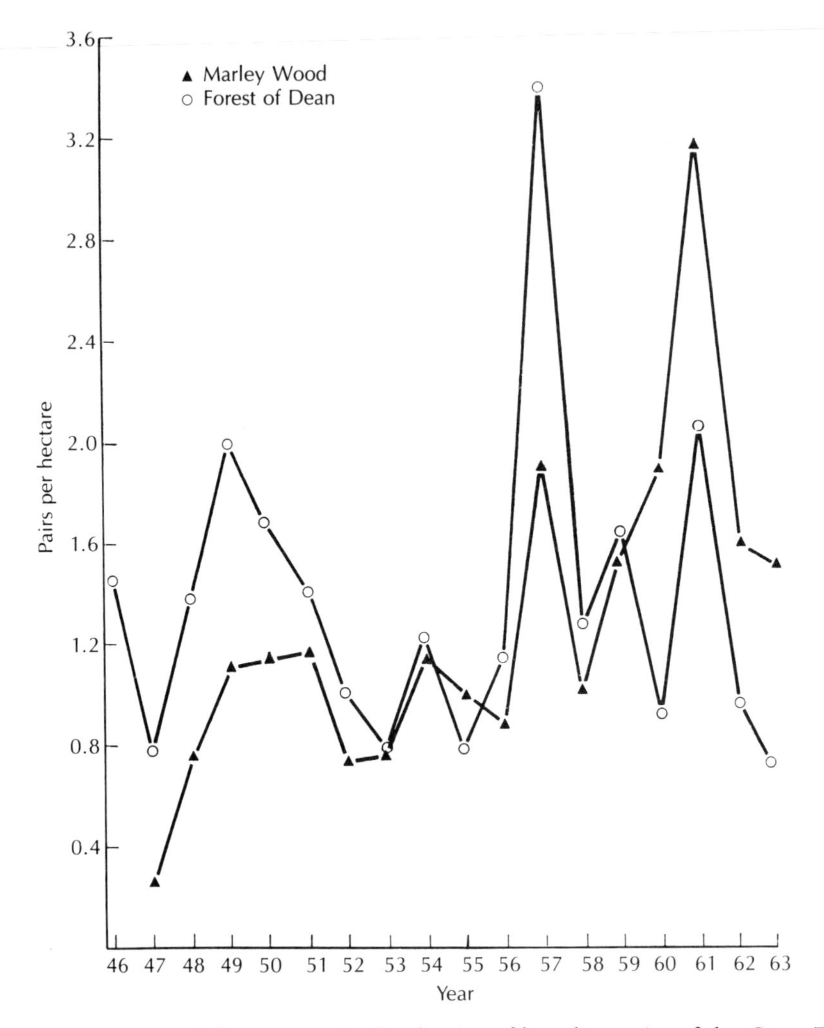

Figure 21–1 Annual fluctuations in the density of breeding pairs of the Great Tit in two British forests, Marley Wood and the Forest of Dean. (From Lack 1966)

this period, whereas the annual fluctuations are caused by winter food limitations.

The general abundances and geographical distributions of certain birds have undergone great changes in recent years. Many species, including the Brown-headed Cowbird and Glossy Ibis, for example, have gone from being rare to being abundant in the northeastern United States in the last 30 years. Others, such as the Red-headed Woodpecker and American Black Duck, have declined during this same period. In Finland, 34 percent of the 233 breeding species have enlarged their geographical distribution or have become more numerous during the past 100 years, 25 percent have declined in numbers, 22 percent have fluctuated, and 20 percent have been fairly stable (Haartman 1973; Järvinen 1980).

Population Growth Patterns

The typical S-shaped (sigmoid) growth pattern of a population in a new environment is similar to that of a baby bird (Figure 18–12). The rate of growth increases slowly at first, accelerates, and then declines because of negative feedback that lowers reproduction and survival. Finally, the growth curve levels off as the population gradually approaches the maximum number that the environment can support.

The two key phases of this S-shaped growth curve are the period of maximum growth and the final period of limited growth. The changing population growth rates derive from the demographic parameters of survival and fecundity reviewed in the previous chapter. Recall that values of per capita replacement, R_0, greater than 1 reflect a growing population and values less than 1 reflect a declining population. The instantaneous growth rate of a population is usually expressed as r, which is the logarithmic form of R_0. During the phase of accelerating growth, the rate of change in the number of individuals with time—dN/dt, from differential calculus—is the product of the instantaneous growth rate r and the population size N at time t. This is expressed mathematically as

$$dN/dt = rN$$

Bird populations have tremendous growth potential in this acceleration phase. For example, the 120 European Starlings that were introduced into the United States in 1890 increased a millionfold in 50 years (Davis 1950). In general, large-bodied species with low reproductive rates have annual growth potentials of 10 to 30 percent, and small-bodied species with large brood sizes and high reproductive potentials have an annual growth potential of 50 to 100 percent in favorable years (Ricklefs 1973b). Strong population growth enables species to recover quickly from short-term setbacks.

Newly established populations of the Cattle Egret and the House Finch grew exponentially (Bock and Lepthien 1976a, b). The Cattle Egret colonized North America in the early 1950s and then spread dramatically, increasing 2000-fold from 1956 to 1971 (Bock and Lepthien 1976a). The calculated value of r was 0.21 for the 16-year period from 1956 to 1971 but was as high as 0.84 during the initial 5 years of this expansion. This high value may be close to the species' maximum potential. More recently, however, as a result of factors explained in the next section, the growth rate of the Cattle Egret population in North America has been dropping (Bock and Lepthien 1976a).

Following the release of caged birds on Long Island in 1940, the eastern population of the House Finch also grew rapidly (Figure 21–2). The average value of r for eastern House Finches from 1962 to 1971, which included some temporary declines, was 0.23, a figure nearly identical to the average value for the egret.

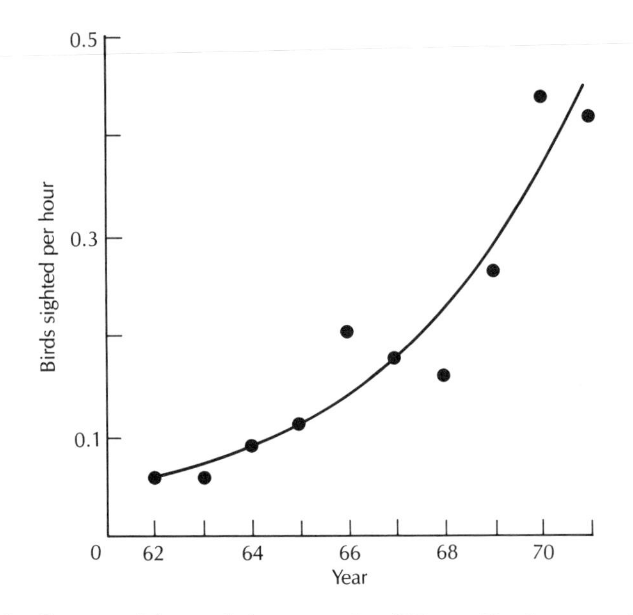

Figure 21–2 Exponential population growth of House Finches east of the Mississippi; curve based on annual Christmas counts. (Adapted from Bock and Lepthien 1976b)

Population Regulation

As the size of a growing population approaches the maximum, that is, the environment's carrying capacity, the population growth rate decelerates. The population then fluctuates in size about an equilibrium value that an environment can support in the typical—or average—year. The carefully monitored growth of a population of Ring-necked Pheasants introduced in 1937 on Protection Island, Washington, illustrates the two phases of population growth (Einarsen 1942, 1945; Ricklefs 1979c). In just 5 years, the population multiplied from the initial 8 pheasants to a total of 1325 birds. The rate of increase then dropped steadily as the population reached the environmental carrying capacity of the island.

Chapter 20, on demography, discussed causes of mortality and variations in individual fecundity. In this chapter the major ecological forces that set limits on the population as a whole are identified after first reviewing the role of population density as a force that helps to maintain a more or less constant population size.

Regulatory Role of Population Density

The growth rate of a population depends directly on the density of individuals whenever mortality and reproductive failure increase with density (Lack 1954, 1966). However, annual variations in mortality and reproductive success can be unrelated to population density (Andrewartha and Birch 1954; Murray 1979). For example, food supply can limit numbers

BOX 21–1

Too many geese spoil the grass

Long-term studies of the demography of the Snow Goose colony at La Pérouse Bay, Manitoba, have documented a steady growth of the population followed by changes that can be attributed to the degradation by the geese of the Arctic salt marshes on which they depend (Cooch and Cooke 1991). This goose population increased steadily in size by approximately 8 percent per year after 1973, because fecundity routinely exceeded mortality. The growing population, however, damaged the quality and availability of nutritious tundra grass that breeding geese and their goslings require. Degradation of food resources caused the average clutch size to decline from 4.7 in 1973 to 4.0 eggs in 1981, gosling weight at fledging and subsequent adult weight to decline 11 percent in the same period, and first-year survival to decline from 50 to 35 percent. These changes in fecundity and survival have caused the population growth rate to slow down, consistent with the operation of density-dependent processes.

in either a density-dependent (Lack) or a density-independent way (Andrewartha and Birch). Although, as noted in earlier chapters, seabirds and some other species frequently fail to raise the youngest and weakest chicks when food supplies are low, the proportion of individuals that starve may be independent of population density if starvation is due to a major ice storm or a sudden drought.

Definitions

A distinction is made between the terms *regulation* and *limitation*, both of which refer to population sizes. Regulation implies that maintenance of the average size of a population depends on population density, which is determined by birth and death rates, whereas limitation refers to any ceiling on population growth. Fecundity or survival may be density-dependent but still not limit population sizes. Density-dependent clutch size or adult survival, for example, might not affect population size if hurricanes kill most of the juveniles each year, thereby causing low recruitment into the breeding population. The following sections consider the ecological forces that limit population sizes and the ways in which population density can influence these forces to create patterns of population regulation.

Natural Factors That Limit Populations

Four principal ecological factors set upper limits to bird populations: habitat, climate, food supply, and disease (including parasites). Social behavior mediates the effects of food and habitat limitation, and perhaps also disease.

Habitat

Habitat availability is the first ecological variable that influences population size. Rail and bittern populations, for example, are declining throughout the United States as wetlands are drained for industrial and suburban development. In recognition of this loss of habitat, the emphasis of many conservation efforts has shifted in recent years to the preservation of critical habitats, and wildlife refuges are being designed to fulfill the requirements of particular species. Endangered Northern Spotted Owls depend on the remnant old-growth forests in the Pacific Northwest, as the entire nation knows all too well (Dawson et al. 1987; Doak 1989) (see Chapter 24). These old-growth forests have been reduced to less than 10 percent of their original extent.

The essential resources provided by a particular habitat range from food to nest sites. For some birds, the limited availability of nest holes limits population size. Woodpeckers can dig their own nest holes, but other birds must either use abandoned woodpecker holes or dig their own in soft dead wood. In the managed forests of Britain and Europe, where dead trees and branches are routinely removed, the shortage of nest sites clearly limits the population densities of species such as the Great Tit and the European Pied Flycatcher and has already caused the extirpation of the White-backed Woodpecker (Haartman 1951; Sternberg 1972). Larger birds such as the Common Kestrel (Cavé 1968) and the Wood Duck (McLaughlin and Grice 1952) also face shortages of nest sites.

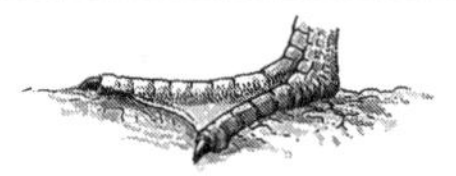

Box 21–2
A southern woodpecker needs old, rotten pine trees

The Red-cockaded Woodpecker is an endangered species that is intimately tied to old-growth southern pine forests (Lennartz and Henry 1985; Costa and Escano 1989; Walters 1991; Jackson 1994). Officially listed as Endangered in 1968, the Red-cockaded Woodpecker is now the controversial focus in the conflict between protection and logging of old growth, long-leaf pine forests in the southeastern United States. Frequent fires started by summer lightning naturally maintain the extensive mature open pine forests. Removal of the forests, including old, rotten trees, by the timber industry, however, has caused severe population declines of this specialized woodpecker.

Red-cockaded Woodpeckers require pine trees that are 80+ to 100 years old and have been infected by the red heart fungus. The fungus rots the old pine tree's heart wood just enough to allow excavation by the woodpeckers. Unlike most woodpeckers, Red-cockaded Woodpeckers excavate cavities in living pine trees. They also excavate tiny holes above and below each cavity, thereby causing an accumulation of sticky resin that protects the woodpeckers and their young from climbing rat snakes. Excavation of the cavities and resin holes requires a major investment of time and energy. No wonder that clans of this cooperative-breeding woodpecker reuse the same cavities for years.

Populations of migrant birds are vulnerable to threats on their wintering grounds. For example, the numbers of the Greater Whitethroat that arrive on its breeding grounds in Britain reflect winter survival in Africa (Winstanley et al. 1974; Batten and Marchant 1977). The breeding population of this species dropped 77 percent one year because of drought south of the Sahara. When wetter winters followed, the breeding population rebounded to its previous level. In this case annual climate probably affected both habitat and the food requirements of the wintering warblers.

The full story of the impact of human activities on the availability of habitats for birds cannot yet be told. Some species benefit from human interference, but many do not. Widespread deforestation has aided species that inhabit open country but has hurt species tied to large timber. Many species of the open grasslands of the midwestern United States have moved into the agricultural fields of the East. Familiar birds of cleared and second-growth habitats, such as the Chestnut-sided Warbler, the American Robin, and the Indigo Bunting, were once scarce. In Finland, Blackcaps and Hedge Accentors have greatly increased in abundance since 1956, when cattle were no longer permitted to graze in the forests. Similar destruction of forest habitat is now occurring in eastern North America as a result of the population explosion of the white-tailed deer. This large, browsing mammal changes the normal vegetation structure of a maturing forest by eating and killing young plants that provide the low cover and shrub layers required by many ground nesting forest birds.

The high densities of migrants on tropical wintering grounds make them especially vulnerable to the destruction of natural habitats (Terborgh 1980). Clearing 1 hectare of forest in Mexico eliminates the same number of warblers as clearing 5 to 8 hectares in the United States. Worse, many migrants congregate in tropical highland areas, which is prime agricultural land. Conservation of these tropical habitats may be essential if the large numbers of migrants returning northward in the spring are to be maintained.

Food and Climate

Food supply, which often depends on climatic conditions, limits population growth and influences population size (Lack 1954, 1966; Newton 1980). A classic example of food limitation is that of the millions of Peruvian seabirds that starve when their main food—the anchovy, a small fish—periodically disappears as a result of changes in surface water temperatures during El Niño (Idyll 1973). The total population of cormorants, pelicans, and other seabirds went from 27 to 6 million birds in 1957 and 1958, increased to 17 million as food supplies returned, and then plummeted again to 4.3 million birds in 1965. In recent years the maximum number of seabirds during good years has been declining as overfishing depletes the anchovy populations. Seabird populations generally seem to be limited by their food supplies (Cairns 1992).

The warming of the European climate, which was responsible for the gradual increase in Great Tit populations, also favored the expansion of southern species into Scandinavia and forced the retreat of some northern species. For example, the Thrush Nightingale occurred as far north as central Sweden in the eighteenth century, but it receded to southern Sweden as the climate turned colder in the nineteenth century. Owing to recent warming, it has expanded again during the twentieth century to its earlier distribution. Similarly, in North America, southern species are expanding northward. The general amelioration of winter climates combined with the growing tending of backyard bird feeders has increased the number of Great Tits that survive the winter in Europe and the number of Northern Cardinals, House Finches, and Evening Grosbeaks that survive the winter in North America.

Most of the evidence of starvation among Temperate Zone birds comes from losses of songbirds, waterfowl, and waders during hard winters. The very cold winter of 1981–1982, for example, was hard on British birds (O'Connor and Cawthorne 1982). During that winter, mortality rates in several species were 2 to 10 times the normal rate. Common Redshanks were unable to feed because their main food—shrimplike amphipods—remained deep in their burrows when intertidal areas froze. White Wagtails, searching for insects along frozen shorelines, could no longer find one every 4 seconds, the average rate needed for their subsistence.

Widespread food shortages can cause irruptions—mass dispersals—of populations, especially of populations of birds in the arctic and subarctic regions. One of the most familiar mass dispersals is the periodic southward invasions by Snowy Owls, presumed (but not demonstrated) to reflect the abundance of lemmings, which are mouselike rodents of the tundra (Parmelee 1992). Over 14,000 Snowy Owls were counted in southeastern Canada and New England during the great Snowy Owl invasion of 1945 to 1946. Because they were away from their usual habitat, many of the owls were killed or died of starvation.

Irruptive invasions of dispersing populations of the seed-eating birds of northern coniferous forests are also dramatic ornithological events in both Europe and North America (Bock and Lepthien 1976c). Invasion years, which are often the same in the New and Old Worlds, correspond to years of poor boreal forest seed production. During invasion years, flocks of northern finches appear along roadsides and at backyard feeders. Eight North American species—the Pine Siskin, the Red-breasted Nuthatch, Red Crossbill, White-winged Crossbill, Purple Finch, Pine Grosbeak, Evening Grosbeak, and Common Redpoll—tend to invade during the same years (Figure 21–3).

Detailed local studies have documented the correlation between food abundance and population size. For example, Daphne Major, one of the small islands in the Galápagos archipelago, suffered severe drought in 1977, a condition that caused a critical shortage of the seeds that sustain

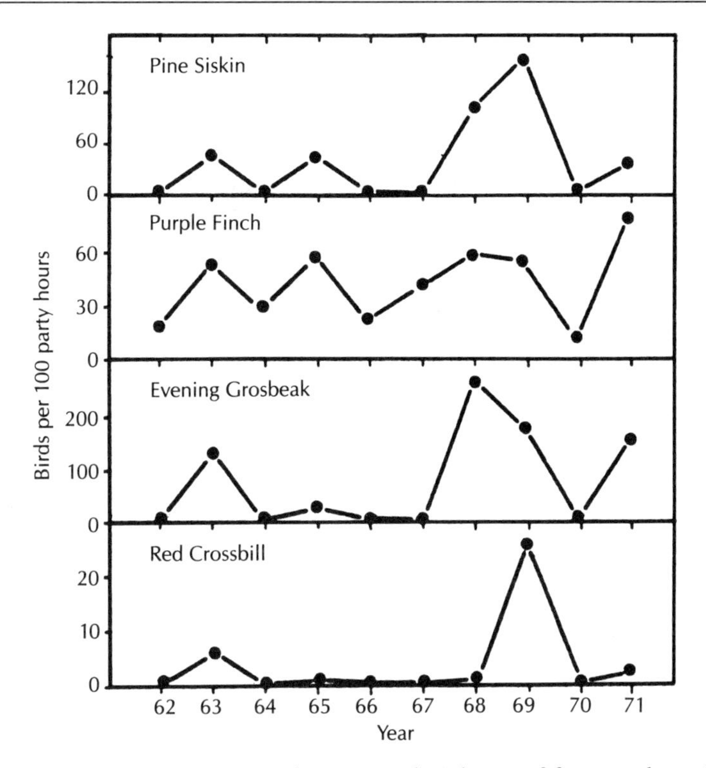

Figure 21–3 Annual variations in the winter abundance of four northern finches in the Chesapeake Bay region; curves based on annual Christmas counts. (Adapted from Bock 1980)

the resident ground-finches. When seed abundance, both in numbers and volume, plunged sharply relative to the number of seeds available in 1973 (a wet year), finch abundance declined by a similar order of magnitude in both numbers and total biomass (Grant 1986) (Table 21–1). The effect of this event on average bill sizes in the population is described in Figure 7–5.

TABLE 21–1
Effects of seed availability on ground-finch abundance on Daphne Major in the Galápagos

	Seeds		Finches	
Year	Total number per m^2	Total volume (cm^3 per m^2)	Total number	Biomass (kg)
1973 (wet)	4821	15	1640	26
1977 (dry)	295	5	300	6

From Grant and Grant 1980.

Populations of sparrows that winter in southeastern Arizona are also limited by the availability of seeds, which the rainfall of the previous summer determines (Pulliam and Parker 1979; Dunning and Brown 1982). Populations of resident sparrows such as Grasshopper Sparrows are limited by food supplies in poor years when most of the seeds are gone by the end of the winter (Figure 21–4). Migrant sparrows such as Chipping Sparrows use the same grasslands as Grasshopper Sparrows when winter food is abundant; and, although many competing sparrows are present, they have little impact on seed abundance. In poor winters, however, virtually all of the Chipping Sparrows move further south to wintering grounds in Mexico, where the resulting high densities of sparrows can exhaust the usually abundant local food supplies.

Disease and Parasites

Diseases and parasites can devastate bird populations. Lowland populations of the Hawaiian honeycreepers, for example, were destroyed by bird pox and malaria when mosquitos that carried these diseases were accidentally introduced by Captain Cook and then by European settlers in the early 1800s (Warner 1968). Island bird populations are particularly vulnerable because of their tendency to lose their resistance to mainland diseases. Until recently, however, the frequency of diseases and their impact on natural bird populations were not well known. Renewed interest

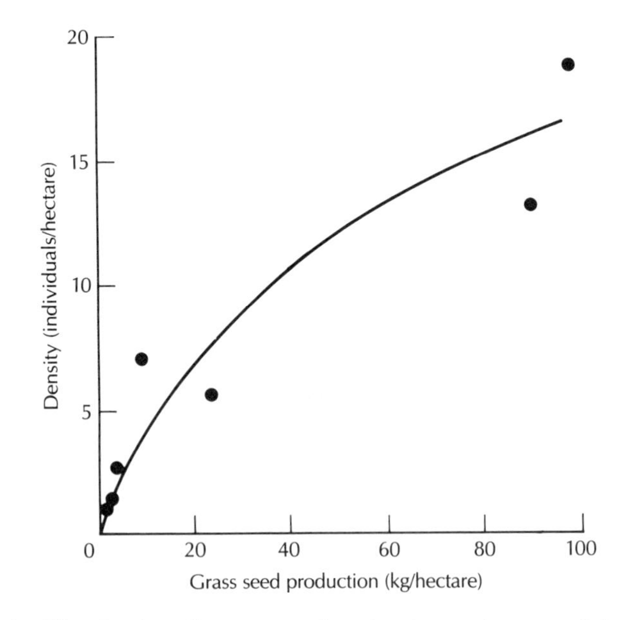

Figure 21–4 The density of sparrows wintering in southeastern Arizona increases with the abundance of the seeds that make up their food supply. In poor years, these sparrow populations may be limited by food supplies; in good years, food supply is not a limiting factor. (Adapted from Pulliam and Parker 1979)

in this topic has led to a surge of discoveries, including reevaluation of previous explanations for changes in population size (Loye and Zuk 1991). The population cycles of the Red Grouse provide one such case.

On the moorlands of Scotland, local populations of the Red Grouse cycle every 6 years from lows of 30 birds to highs of 120 birds per square kilometer. Different explanations for these cycles ranged from food availability, to cycles of genetically controlled social behavior, to intestinal parasites (Hudson and Dobson 1991a, b). The leaves of heathers are the primary food of this grouse. The grouse select nutritious leaves, and in the spring the leaf quality affects maternal nutrition, egg quality, brood size, chick survival, and adult summer survival. Experimental addition of fertilizer to selected moorland plots increases heather growth and the quality of the leaves and, predictably, improves the breeding success and survival of the grouse on those plots. In the 1970s, however, a major decline of the Red Grouse population occurred despite an abundance of nutritious heather leaves.

An intestinal parasite, a nematode worm named *Trichostrongylus tenuis*, burrows into the soft walls of the intestinal ceca, causing local damage, internal bleeding, decreased absorptive function, and mortality in the Red Grouse. Infection levels of individual grouse can be severe, and it is now clear that this worm was responsible for "grouse disease," as evidenced by moribund and emaciated birds, which devastated grouse populations in the nineteenth century. Experiments that purged the parasites from some grouse demonstrated that the nematode reduces rate of weight gain in females prior to incubation, clutch size, hatching success, and chick survival. Adult survival also is affected. Secondarily, the parasites may increase vulnerability to predation by reducing ability of the grouse to control scent emission from the intestinal caeca; hunting dogs and foxes both use these bird odors to locate grouse. Finally, worm infections reduce aggressiveness and the ability of males to compete successfully for territories, behavior that was once thought to be the root cause of the population declines. The density-dependent effects of the parasites on breeding production and survival increase with population size, ultimately causing short-term population declines.

Social Forces

Subtle social forces mediate the availability of habitat and, therefore, local population size. The spacing of territorial individuals in prime habitat may exclude some individuals from the breeding population or force them to occupy secondary habitats where nesting is less successful and the risk of mortality is greater. Dispersal increases with population density. In one well-studied case, young male Great Tits dispersed as little as 354 meters (median) in years of low population density and up to 1017 meters (median) in years of high population density (Greenwood et al. 1979). Young male Great Tits disperse farther in populous years to find unoccupied territories, which are scarce because of the high level of survival of established males. Young that fledge late in the season usually must disperse farther because young fledged earlier in the season occupy the nearest territorial openings.

The occupation of available habitat has three stages (Brown 1969) (Figure 21–5). First, prime habitat is filled. Then, unable to find vacancies in prime habitat, surplus birds move to suboptimal habitat, and wait for vacancies in better habitat. Finally, as suboptimal habitats are filled, remaining birds, often large numbers of them, simply must wait, usually as floaters, for vacancies in either habitat.

Geoffrey Hill (1988) studied the occupation of different quality territories by male Black-headed Grosbeaks, a species that exhibits delayed plumage maturation. Young males wear duller breeding plumages than do old males. Hill found that the breeding territories defended by males varied in vegetation structure and thereby the abundance of Scrub Jays and Steller's Jays, the grosbeaks' principal nest predators. Males that were 3 or more years old defended territories with mixed vegetation and few jays; these males achieved the highest reproductive success. Yearling and 2-year-old males settled on territories with denser vegetation and more jays; these males suffered more nest predation. Among these young males and irrespective of their age, the brightly colored individuals occupied slightly better quality territories than the dull-colored individuals. Each year, the older males shifted to better quality territories.

Floaters either form flocks in areas that are not occupied by territorial breeders (Carrick 1963; Holmes 1970), or they live singly on home ranges that overlap the breeding territories of established pairs. About 50 percent of a population of the Rufous-collared Sparrow was made up of nonterritorial floaters (Smith 1978). This tropical bunting, which is

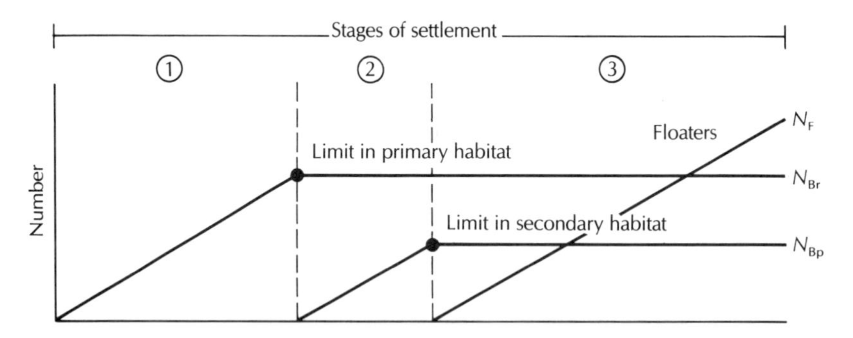

Figure 21–5 Stages of settlement in a local population of breeding birds. The first breeding birds to arrive in an area occupy primary habitat (stage 1). Individuals unable to establish territories in primary habitat settle in secondary—or poorer—habitat (stage 2). Floaters are individuals unable to establish territories because they arrive after all the breeding habitat is filled (stage 3). N_F, number of floaters; N_{Br}, number breeding in primary habitat; N_{Bp}, number breeding in secondary habitat. (Adapted from Brown 1969)

Figure 21–6 Rufous-collared Sparrow, a species with a well-developed "underground" of individuals waiting for a breeding opportunity.

closely related to the White-crowned Sparrow of North America, defends territories and breeds throughout the year (Figure 21–6). Floaters—or members of the "underworld"—live in well-defined, small home ranges. The ranges of young females were restricted to a single territory, whereas the ranges of young males encompassed three to four established territories. Males and females of the underworld whose home ranges overlapped had well-defined, intrasexual dominance hierarchies. The dominant individuals of the appropriate sex filled new vacancies.

Floaters quickly replace established males that disappear or that are experimentally removed (Stewart and Aldrich 1951; Hensley and Cope 1951). The dynamics of control and attempted takeover of limited territorial spaces are illustrated by Susan Smith's (1978) description of what happened when a territorial male Rufous-collared Sparrow (color banded RO) disappeared for 9 days after capture and banding on August 10:

> Less than one hour after his capture, two banded underworld males were courting his mate, GY, but she actively chased both throughout the day. Also, at least four neighbor male owners invaded the territory repeatedly and were driven out by GY. By August 15 one of these, YO, had formed a stable pair with GY, and two other underworld males . . . had established small territories at each end of YO's former territory. Both actively courted YO's former mate, RRO, who, unlike GY, readily associated with both. On August 17 I saw RRO copulating with the one that sang more, RBO, and by August 18 they were established as a pair in her territory. Yet less than 24 hours later RO had returned and regained his territory and mate, and YO had reclaimed most of his old territory with RBO, holding a small corner, forming a trio of one female (RRO) and two males (YO and RBO). Five weeks later YO had regained all his territory, and RBO rejoined the underworld. (p. 577)

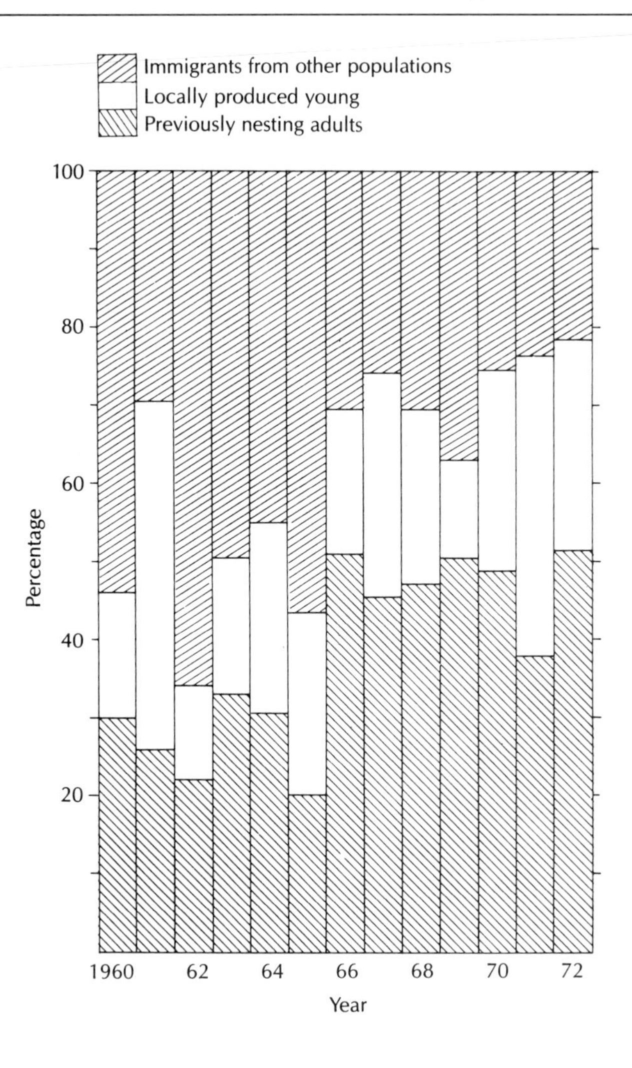

Figure 21–7 Composition of Great Tits breeding on the island of Vlieland off the coast of Holland. Opportunities for immigration depend on the overwinter mortality of the resident adults. (Adapted from Van Balen 1980)

Recruitment

Recruitment of young birds into a local population every year reflects the number of young produced during the breeding season and, particularly, the proportion of those that survive their first 6 months of life. The yearlings recruited into the population include some that are produced locally and some that immigrate from other places. In one study on an island off the coast of Holland, local recruits averaged about 20 percent of the breeding population each year. Recruitment varied in relation to adult mortality over the winter (Van Balen 1980) (Figure 21–7). Recruitment of young birds into the breeding population dropped when winter food was provided at bird feeders, starting in 1966, decreasing adult mortality and, consequently, the number of vacancies available. Such observations demonstrate once again that the density of established adults influences dispersal and recruitment patterns. Some form of population regulation is in effect.

Recruits can be an important fraction of a growing population. The population of Atlantic Puffins on the Isle of May, off eastern Scotland, for example, increased by 19 percent a year from 1973 to 1981 (Harris and Wanless 1991). The growth rate of the population calculated as the per capita replacement, R_0, from its life table parameters was only 16 percent. The difference reflected recruitment of immigrating young birds from another population on the Farne Islands 80 kilometers away. After 1981, both adult and immature survival declined, apparently as a result of widespread declines of winter food in the North Sea. The lower survival rate caused the population to cease growing and stabilize in size.

Regulation of Great Tit Populations

Ornithologists have monitored populations of the Great Tit in Holland since 1912 and in England, especially in Wytham Wood, near Oxford, since 1947. This species is quite sedentary and nests readily in boxes, especially in managed woodlands where natural cavities are scarce. Inspection of the nest boxes yields accurate censuses of breeding pairs, clutch sizes, and young raised. Deciphering the dynamics of population regulation has been a primary goal of this research (Perrins 1979; McCleery and Perrins 1991). The main finding is that population regulation in Great Tits is a density-dependent phenomenon (Figure 21–8), which is evident mainly in the effects of food limitation on juvenile survival during the winter. Habitat limitation, territorial behavior, and dispersal also play mediating roles. Although fecundity and survival in the breeding season

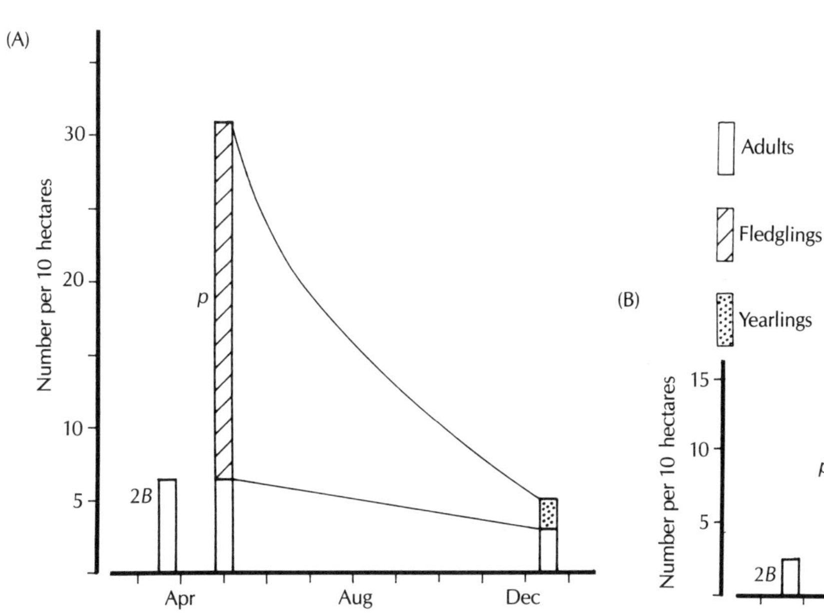

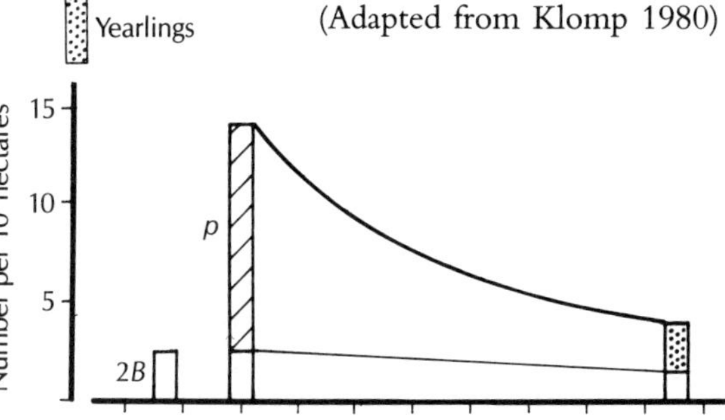

Figure 21–8 Great Tit production (p) and survival to the winter in years with (A) high and (B) low breeding densities in a model population. Nesting in the spring adds many young birds to the population, but most of these die by winter. Averages of 6 and 3.5 fledglings are produced per breeding pair (2*B*) in low- and high-density years, respectively. (Adapted from Klomp 1980)

are density dependent, their impact on overall population size is minor. This section examines these effects in detail.

Seasonal changes in Great Tit populations are influenced by reproduction, mortality, and migration. Each year, the population increases rapidly with the production of fledglings by the breeding adults. Both mean clutch size and number of fledglings (fecundity) depend on local population density. Great Tits lay fewer eggs when population density is high than when it is low. Sixty percent of the variation in annual mean clutch size is directly related to population density. Success in rearing nestlings also decreases as population density increases because of increased predation and because fewer females attempt second broods (Figure 21–9).

The rapid growth of the population following breeding is short lived. Heavy mortality of young birds and the loss of some adults then cause a steady decline. An average of only 22 percent of the juveniles survive their first year. Survival of both juvenile and adult Great Tits is density dependent. Recoveries of banded individuals throughout Britain showed that females are less likely to survive in a high-density population year than in a low-density population year. Most telling were experiments by Kluijver (1966) in a population on an isolated Dutch island in the North Sea. He removed 60 percent of the eggs and nestlings in some years but not in others. Both juvenile and adult survival doubled in the years when he removed eggs and nestlings. Juvenile survival rose from 11 to 20 percent and that of adults rose from 26 to 54 percent. Immigration and emigration did not affect these experimental results because they did not occur on the tiny island. In other experiments, the survival rate of juveniles in autumn and winter was positively correlated with the percentage of breeding birds removed in the summer.

Although annual variations in reproductive success, adult survival, and juvenile survival all potentially influence the density of the population during the breeding season, survival outside the breeding season, particularly of juveniles, actually controls population size the following year (Klomp 1980). Losses to Eurasian Sparrowhawks, although substantial, have little final impact on the population as a whole (McCleery and Perrins 1991). Instead, winter food supplies, especially the seeds of beech trees, control juvenile survival in both Oxford and Holland. Young Great Tits in Holland, for example, depend on beechnuts from November to late February when other food sources are scarce. Beechnut crops vary greatly from year to year, and this essential reserve food supply is easily exhausted in poor crop years.

The local population density of Great Tits also varies with habitat. Deciduous oak forests, for example, support 10 times as many breeding pairs as do pine forests. Mixed oak–pine forests support intermediate population densities. As the composition of trees in a local forest shifts from pine to oak, the density of tits increases (Figure 21–10). There is less food in pine forests than in oak forests, and, therefore, territories are larger in pine forests and more nestlings starve. The amplitude of annual population fluctuations is greater in thinly populated pine habitats than in densely populated deciduous woodland. The pine forest is a secondary or

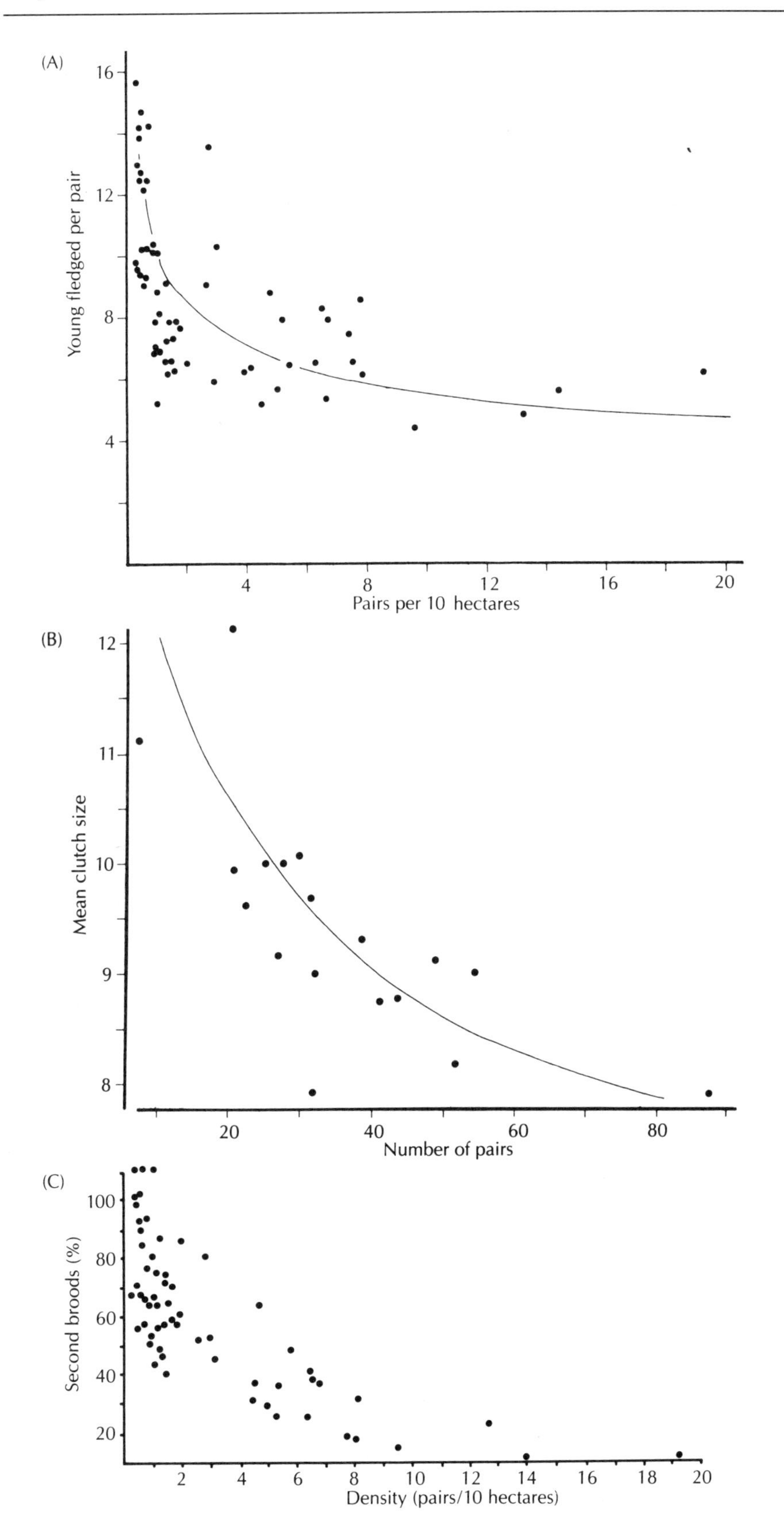

Figure 21–9 (A) Reduced fecundity at higher densities in the Great Tit reflects (B) smaller clutches at high population densities and (C) less frequent attempts to raise second broods. (Adapted from Klomp 1980; Kluijver 1951)

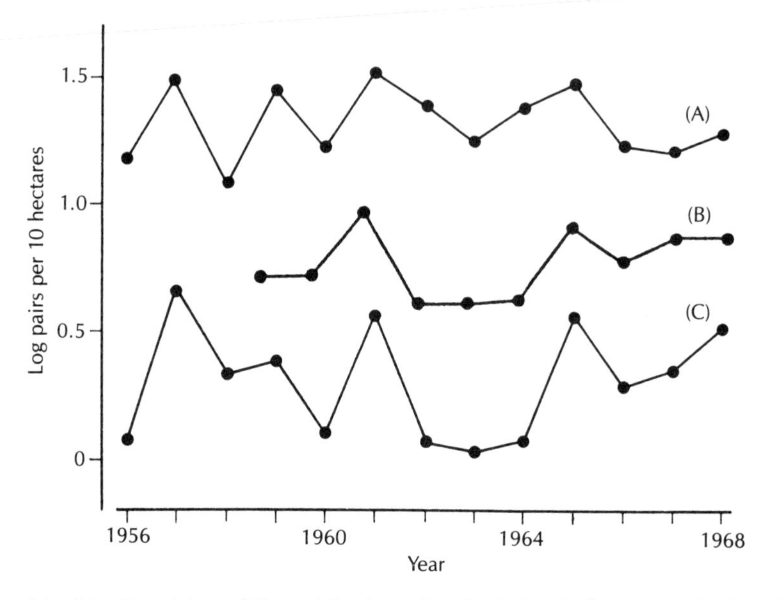

Figure 21–10 Densities of Great Tits breeding in (A) oak forests are higher than in (B) mixed oak–pine forests and much higher than those in (C) pine forests. (Adapted from Klomp 1980)

suboptimal habitat occupied by an overflow of Great Tits that are excluded from deciduous woods, and individuals from pine forest habitats quickly fill vacancies in the deciduous forest. Farmlands and other nonforest habitats act as suboptimal buffers to woodland populations, which are regulated by density-dependent factors.

The patterns of population regulation seen in the Great Tit have broad application to other birds. As a rule, bird populations seem to be limited by food scarcity during the nonbreeding season. Food abundance fluctuates from year to year. Survival of juveniles in a year of scarcity has a major effect on recruitment levels the following year, and hence on population size.

This chapter started with the point that bird populations are dynamic, fluctuating in size from large to small and back to large again. Population sizes may change dramatically in size over periods of several years, reflecting annual variations in survival and fecundity. Recent analyses of genetic variability in abundant and widespread North American bird species, such as Red-winged Blackbirds and Black-capped Chickadees, suggest that they also passed through historical "bottlenecks" of severely reduced size (Avise et al. 1988; Gill et al. 1993a). In general, most populations probably pass through periods of small size. Some do not recover and go extinct. Others rebound. Periodic bottlenecks reduce genetic variability in local populations and, more important, provide the principal theaters for evolutionary change and speciation, the topic of the next chapter.

Summary

Bird populations range in size from hundreds of millions of individuals to just a handful of survivors. In general, population sizes fluctuate dynamically from year to year as a result of changes in breeding success and mortality. Growth of a population in a new environment usually follows a pattern of a slow initial rate of increase, followed by accelerated growth rates, and finally, a decline in growth rates in response to factors that lower reproduction and survival. Established populations tend to stay close to a long-range average size.

The availability of quality habitat and food limits the sizes of bird populations. Diseases and parasites probably play a more substantial role in limiting bird populations than has been appreciated until recently. Habitat loss and pesticide poisoning of the food chain severely affect some populations. Some bird species, especially those that require cleared or shrubby habitats, benefit from human expansion. Social forces, such as territoriality and aggressiveness, also limit populations.

Although populations may grow as a result of recruitment, population sizes depend on the balance between rates of dispersal and recruitment, which, in turn, are often density dependent. Detailed studies of Great Tit populations in Holland and England illustrate the nature of density-dependent regulation of population size.

Dispersal patterns together with changes in size affect the genetic composition and potential for evolutionary change of local populations.

Further Readings

Cairns, D.K. 1992. Population regulation of seabird colonies. Current Ornithology 9: 37–61. *A review of recent literature on random versus density-dependent forces that regulate marine seabird populations.*

Cooke, F., and P.A. Buckley. 1987. Avian Genetics. New York: Academic Press. *A collection of significant papers on the relation of genetics to the population ecology of birds.*

Grant, B.R., and P.R. Grant. 1989. Evolutionary Dynamics of a Natural Population. Princeton, N.J.: Princeton University Press. *The results of a fascinating, long-term study of one species of Galápagos finch.*

Lack, D. 1954. The Natural Regulation of Animal Numbers. Oxford: Oxford University Press. *A classic.*

Loye, J.E., and M. Zuk, Eds. 1991. Bird–Parasite Interactions. Oxford: Oxford University Press. *A landmark collection of papers that establish the foundations for future study of the role of diseases and parasites in bird populations.*

Perrins, C.M., J.-D. Lebreton, and G.J.M. Hirons, Eds. 1991. Bird Population Studies. Oxford: Oxford University Press. *A collection of papers summarizing some of the best population studies of birds, with recommendations for conservation management.*

Terborgh, J. 1989. Where Have All the Birds Gone? Princeton, N.J.: Princeton University Press. *A well-written, personalized view of the problems facing migrant birds in the Western Hemisphere.*

C H A P T E R 2 2

Species

THE FIRST BIRD SPECIES of the Mesozoic era 150 million years ago multiplied themselves manyfold in the succeeding epochs of Earth history. That multiplication and the associated enrichment of avian diversity on Earth were due to repeated speciation—the separation of one species into two or more derived species.

This chapter concerns the process of divergence and diversification of species of birds. It addresses the questions, What are species, and how do they arise? The chapter begins with an introduction to species concepts and a review of the patterns of geographical and genetic variation in birds, including population structure, gene flow, and clines. Then follow discussions of the evolutionary divergence of geographically isolated populations, assortative mating, secondary contact and hybridization, plus ecological and behavioral aspects of speciation.

What Is a Species?

Species are the primary units of systematic biology; they serve as the basis for describing and analyzing biological diversity. Although they are seemingly fundamental units of biology, species are not fixed or easily defined entities. Competing species concepts and definitions range from the practical to the philosophical. The species concept that prevails in ornithology and in this textbook is called the Biological Species Concept. The reproductive compatibility of individuals serves as the ultimate criterion for inclusion in a unit called species. According to the Biological Species Concept, "Species are groups of interbreeding natural populations that are reproductively isolated from other such groups" (Mayr 1970, p. 28).

Each species has a characteristic size, shape, color, behavior, ecological niche, and geographical range. In each species, sexual reproduction links males, females, and their offspring into cohesive populations. Mating of like individuals with each other—called assortative mating—unites some sets of cohesive populations and reproductively isolates these from other sets of similarly cohesive populations. White-crowned Sparrows mate with each other, but they do not interbreed with coexisting Song Sparrows. White Ibises mate with each other, but they do not interbreed with the Glossy Ibises that nest in the same colonies.

Such species evolve as a result of the divergence of isolated populations (Figure 22–1). After populations separate in geographical space, the sister populations diverge gradually, or sometimes rapidly, thus enabling separate taxonomic diagnoses and ultimately leading to reproductive isolation. Sister species may later come back into contact if geographical barriers disappear. When this renewed contact occurs, sister species either coexist or hybridize with each other, depending on the extent of their divergence. The interactions of divergent sister taxa when they again occupy the same area—or sympatry—test their reproductive, ecological, and behavioral compatibility. Although the general patterns of geographical speciation in birds are well known, the details of the process itself are not. Still to be resolved are the roles of ecological and social adaptations, as well as the timing and nature of the genetic changes related to them.

How best to define species also continues to be one of the perennial debates in biology. The debate has intensified on both methodological and philosophical grounds in recent years, producing as many alternative definitions as there are viewpoints (O'Hara 1993). Studies of hybridizing populations permit direct application of the Biological Species Concept, discussed in later sections of this chapter, but the vast majority of recently isolated and divergent populations do not come into contact. They remain geographically separated, thus forcing ornithologists to guess what might happen if contact should be established in the future. This limitation and other concerns about practical application prompt some ornithologists to question the modern utility of the Biological Species Concept and to adopt instead the Phylogenetic Species Concept.

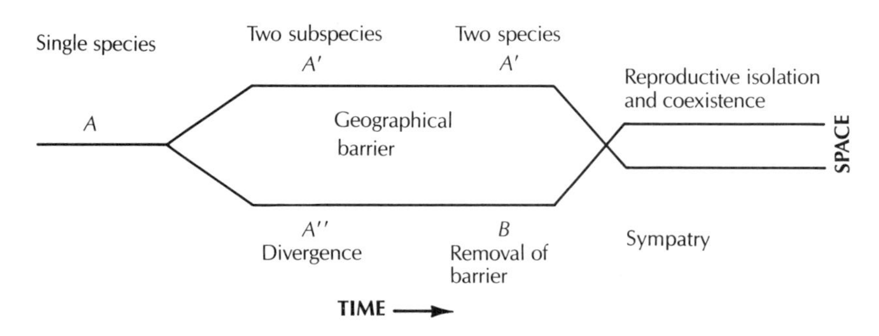

Figure 22–1 Geographical speciation proceeds via the divergence of populations in space and time. Letters designate genetically discrete populations. The separation of a population facilitates genetic divergence (A yields A′ and A″). Reproductive incompatibility of populations (A′ and B) can result from sustained isolation. Reversal of geographical isolation and range expansions can lead to coexistence as separate species.

Each historically fragmented population with distinct characteristics serves as the working unit of the Phylogenetic Species Concept (McKitrick and Zink 1988; Cracraft 1989). According to the Phylogenetic Species Concept, a species is the smallest aggregation of populations diagnosable by a unique combination of character states in comparable individuals (Nixon and Wheeler 1990).

The goal of the Phylogenetic Species Concept is to define indivisible taxa that can be used in cladistic analyses (Chapter 3) of their evolutionary history. Rather than combine geographically distinct, but potentially interbreeding, populations with distinct evolutionary histories into one larger, variable—or polytypic—species, the Phylogenetic Species Concept would distinguish each one as a separate species unit for phylogenetic analysis. Each geographically distinct population of some bird, a Song Sparrow, for example, would be treated as a separate species. This approach gives greater weight to the recognition of the separate evolutionary histories of the isolated populations than to the fact that possibly they interbreed—hybridize—wherever they overlap.

Both the Biological and Phylogenetic Species Concepts are easily applied to the birds in a particular place, such as the White-crowned Sparrow and Song Sparrow in Seattle or the White Ibis and Glossy Ibis in South Carolina. Variation among populations in different places, however, increases the difficulties of application of both concepts. Geographical variation renders more difficult both the diagnosability of populations and judgments on whether populations interbreed. Variations may also be due to environmental rather than genetic differences among populations.

Geographical Variation

One-third of the species of North American birds show conspicuous geographical variation among distinct regional populations—or subspecies. Historically, ornithologists described subspecies when at least 75 percent of the individuals in a regional population were distinguishable, usually by their plumage or size, from other populations of the same species.

The subspecies of Song Sparrows, for example, range from sooty in the Pacific Northwest to pale brown in the deserts of California, and from medium-sized in Ohio to large, thrush-sized birds in the Aleutian islands. Early studies of geographical variation led to formal descriptions of sparrow subspecies with three Latin names, for example, *Melospiza melodia melodia*. But a thorough study of large samples gathered throughout a species range often reveals complex patterns of geographical variation in color and size that cannot be split logically into well-defined subspecies, and the practice is now declining (Barrowclough 1982; Gill 1982; Mayr 1982).

Geographical variation can evolve because different environments favor different attributes. The crests of Steller's Jays in the western United

Figure 22–2 Geographical variation in the Fox Sparrow. Local populations apparently diverge in bill dimensions as they adapt to local environments. (Adapted from Zink 1986)

States vary in length in relation to the openness of the vegetation in their habitats (see Figure 9–12). Populations of Fox Sparrows in the western United States differ in bill dimensions, which presumably reflect differences in their diets (Zink 1986) (Figure 22–2). Climatic adaptation is also a conspicuous feature of geographical divergence, which may take place in less than a century (Parker 1987; see also Chapter 6). We assume that size and color variation are genetically controlled and that they are not directly affected by the environment. This is a reasonable assumption, but it is not a simple or certain one.

Diet mediates adult size in Snow Geese. Long-term studies of Snow Geese in the Canadian Arctic show that they are getting smaller (Cooch and Cooke 1991). The large populations of Snow Geese at La Pérouse Bay, Manitoba, depleted the supplies of their favorite arctic tundra grasses, on which young goslings depend for their early growth (Box 21–1). The alternative, less nutritious grasses stunt their growth. As a result, the average weights of goslings at fledging and subsequent adult weights have declined 11 percent during the period 1978 to 1988.

In a pioneering study, Frances James (1983) demonstrated the effects of local environments on size features of Red-winged Blackbirds. Both their bill shapes and their wing lengths vary geographically. Some of this variation is attributable directly to the environment. When James transplanted eggs from the nests of one population to nests of another morphologically distinct population, the dimensions of fostered chicks grew to resemble those of their foster parents. Red-winged Blackbirds transplanted from the Everglades to Tallahassee, Florida, grew shorter, thicker bills, similar to those of Tallahassee Red-wings. Colorado Red-wings transplanted to Minnesota developed longer wings and toes (Figure 22–3). Thus, these young acquired some of the attributes of the host population. The morphological shifts by transplanted birds, however, were not complete, revealing a significant degree of genetic control.

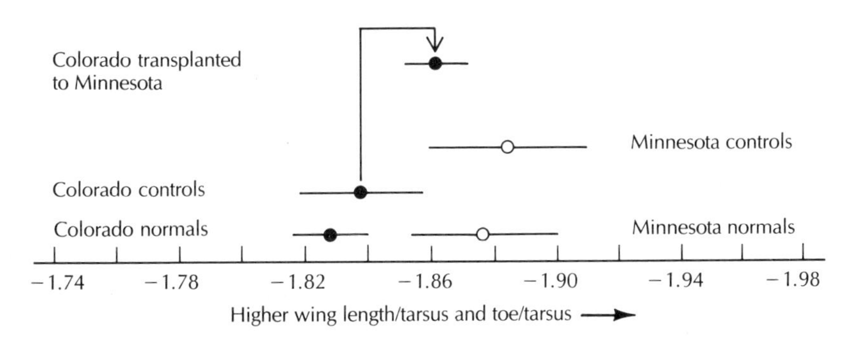

Figure 22–3 Environmental influence on the dimensions of nestling Red-winged Blackbirds. When transplanted to nests in Minnesota, eggs from nests in Colorado yielded nestlings that were shaped more like Minnesota Red-wings than were the controls in Colorado, which were those nestlings transplanted to new nests in the same locality. Nestling shape is here defined in terms of a discriminant function that relates wing length to size of the legs and feet. (From James 1983)

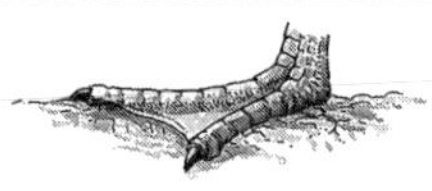

BOX 22–1

Anatomical characters vary in their heritability

The proportion of total observed variability that is controlled by the genes rather than by the environment is called the heritability (*H*) of a character. One-half to most of the size variation observed in bird species has a genetic basis. Body masses of chickens are moderately heritable (*H* = 0.53), whereas feathering traits, breast angle, body depth, keel length, and shank pigmentation have lower heritabilities (*H* = 0.25 to 0.40) (Kinney 1969). Studies of character heritability in wild birds, often difficult exercises in quantitative genetics, indicate moderate to high heritabilities: 0.43 to 0.95 (Boag and Noordwijk 1987). Such heritabilities expose characters to long-term genetic change by natural selection and also to short-term environmental modifications (because *H* is less than 1.0).

Patterns of Genetic Variation

The development of laboratory techniques in biochemical systematics has opened a new era of surveys of geographical variation of individual genotypes as well as size and color. Some bird populations in North America exhibit differences in genetically controlled protein markers—called allozymes (Corbin 1983). The frequencies of alternative alleles may change where two such populations come into contact. But as a rule, bird populations exhibit minor genetic differences, based on allozyme comparisons, and certainly less than small mammal populations (Barrowclough 1983).

Comparisons of the mitochondrial DNA (mtDNA) genome (which evolves faster than the chromosomal genes for proteins) reveal greater differences between closely related species and, in some cases, striking patterns of population structure not previously detected from morphology. Several widespread North American bird species, including Red-winged Blackbirds, Downy Woodpeckers, Mourning Doves, and Black-capped Chickadees, have nearly the same mtDNA genotypes from one side of the continent to another, a finding that reflects recent, postglacial expansions of their populations throughout the northern United States and Canada and not enough time for substantial genetic divergence (Ball et al. 1988; Ball and Avise 1992; Gill et al. 1993a).

Certain species in the southern United States, however, exhibit pronounced genetic divergences between regions. Carolina Chickadees are divided into eastern and western mtDNA genotypes that come together near the Mobile Basin in western Alabama (Gill et al. 1993a), and Seaside Sparrows are divided into Atlantic coast and Gulf coast genetic populations (Avise and Nelson 1989). Such southern species remained in place during the periods of Pleistocene glaciation, thereby allowing greater genetic divergence to take place. Similarly, bird species in tropical South America exhibit more genetic variation among populations than do Temperate Zone species (Capparella 1988, 1991).

BOX 22–2

Conservation biologists mixed the wrong populations of the Dusky Seaside Sparrow

Artificial flooding of marsh-grass habitat to control mosquitoes, plus conversion of the marshes of Brevard County, Florida, to pastures caused the decline and extinction of a blackish—melanic—form of the Seaside Sparrow, called the Dusky Seaside Sparrow. By 1980, only six individuals—all males—remained. To preserve the genes of this subspecies, conservation biologists captured five of the males and hybridized them with females of the presumed closely related "Scott's" Seaside Sparrow from the Gulf coast. By backcrossing the first generation (F1) hybrids to the male Dusky Seaside Sparrows, the proportion of Dusky genes could be increased to as high as 87.5 percent in some individuals, which would be used to start a reintroduction program to save the subspecies.

The well-intentioned conservation biologists did not know, however, that they had chosen the wrong population as a breeding stock (Avise and Nelson 1989). Subsequent analyses of geographical variation in mitochondrial DNA of Seaside Sparrows revealed not only that there was a major difference between Atlantic and Gulf coast populations but also that the nearly extinct Dusky Seaside Sparrow was derived from and genetically similar to the Atlantic coast birds, not the Gulf coast birds. Had the designers of the breeding program known, they could have manufactured a breeding stock that was much closer to the original Dusky Seaside Sparrow, the last pure member of which died on 16 June 1987.

Population Structure

The movement of a young bird from the site where it hatches to the site where it breeds—called dispersal—determines population structure. Natal dispersal is the permanent movement of young birds from their birth sites to their own breeding locations (Figure 22–4). The tendency to stay near one's birthplace—called philopatry—increases the probability of breeding with near relatives, even siblings, and thus increases the risk of inbreeding (Greenwood 1987). Conversely, dispersal promotes outbreeding. Colonial seabirds such as albatrosses, gulls, and terns return to their natal colonies; and songbirds such as the Great Tit, the European Pied Flycatcher, and the Song Sparrow stay within a few kilometers of their natal territories. Only a few individuals of philopatric species disperse widely.

Generally speaking, females of a species disperse farther than males, a pattern that is most extreme in cooperative breeders such as the Florida Jay, in which young females disperse and join new family groups while young males wait at home to inherit a portion of their family plot (see Chapter 19). Waterfowl, however, behave in the opposite way. Females return to their natal marsh or colony, and males disperse more widely, following whichever females accepted them on the wintering grounds. Approximately 50 percent of all Snow Geese breeding for the first time at the La Pérouse Bay colony are immigrant males from other colonies (Cooke et al. 1975).

Dispersal distance of a species defines the size of a local, reproductively cohesive population—or deme—in which gene exchange is theoreti-

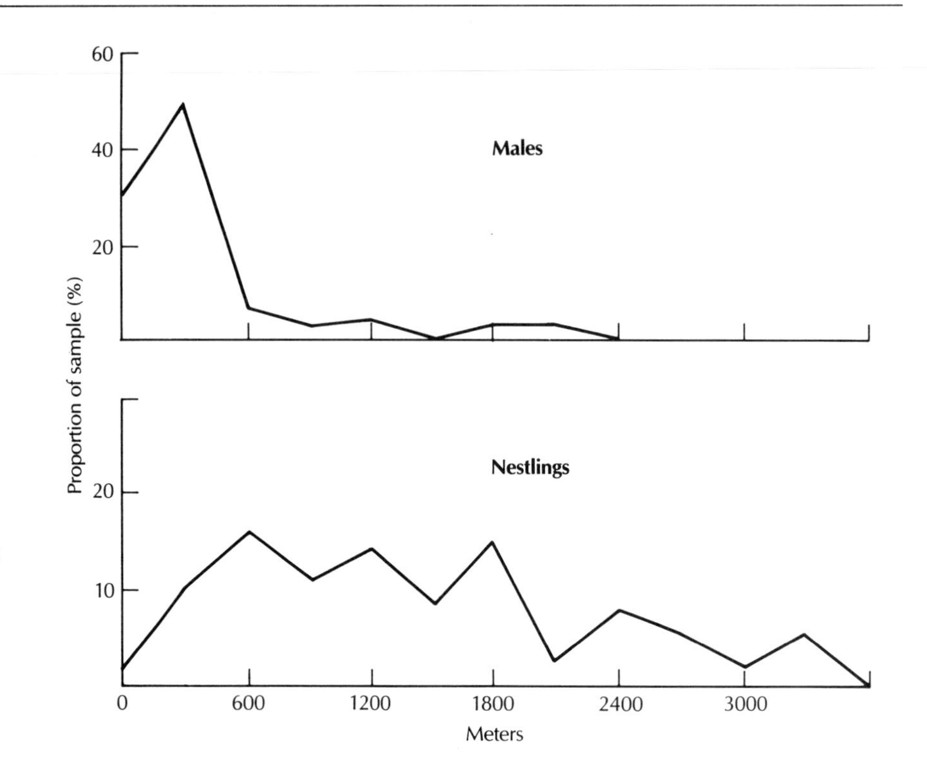

Figure 22–4 Dispersal of adult male (top) and nestling (bottom) House Wrens. Most adult males disperse over a small area, whereas young wrens disperse more widely. (Adapted from Barrowclough 1978)

cally a random process (see Rockwell and Barrowclough 1987; Payne 1990). The number of individuals in a deme is called the effective population size. The evolutionary potential of small demes differs from that of large demes in that small demes evolve faster and in directions that sometimes can be dictated by chance. Effective population size decreases with decreasing dispersal distances of juveniles (Findlay 1987; Greenwood 1987). The effective sizes of bird populations also decrease when small groups of colonists start new populations, when populations fragment into small isolates with limited dispersal, when populations are confined to isolated colonies on long, narrow coastlines, or when just a few individuals dominate nonmonogamous breeding systems, as in lekking manakins or grouse (Chapter 17).

Ornithologists estimate that noncolonial passerine birds disperse roughly 1 kilometer per year, with a range of 350 to 1700 meters per year (Barrowclough 1980a). This value indicates that the effective population sizes of such birds are quite large—roughly 175 to 7700 individuals—and that evolutionary change tends to be slow and adaptive except in very small, isolated groups of birds. It follows, also, that bird speciation usually results from gradual, adaptive divergence of large, fragmented populations or from rapid genetic reorganization in small, founder populations (Barrowclough 1983). For example, populations of Common Mynas introduced to Australia, New Zealand, Hawaii, Fiji, and South Africa from India differ genetically (allozymes) more from one another than do populations in Asia. Most of these introduced populations started as small founder populations. Subsequent reductions in population size and random changes in gene compositions in 100 to 120 years have promoted genetic shifts comparable to those between different subspecies of other birds (Baker and Moeed 1987).

Gene Flow and Clines

The evolution of geographical differences among bird populations depends on the relative strength of two opposing forces: natural selection and gene flow. Natural selection—the differential propagation of genotypes—promotes divergence by favoring one genetic attribute over another; gene flow—the movement and incorporation of alleles among local populations due to dispersal—opposes divergence by blending the differences among adjacent populations (Rockwell and Barrowclough 1987). Clines—gradients of character states, such as body size or feather color—are the expression of the opposing actions of divergent selection and blending gene flow in contiguous populations. Clinal variation is especially conspicuous in birds that have simple (Mendelian) genetic color phases, which parallel brown eyes versus blue eyes in humans (Buckley 1987). In one of many cases, the proportions of red- and gray-phase Eastern Screech-Owls change systematically with locality. Local populations change from mostly red-phase owls in Tennessee to mostly gray-phase owls in Maine and Florida (Owen 1963) (Figure 22–5). We do not

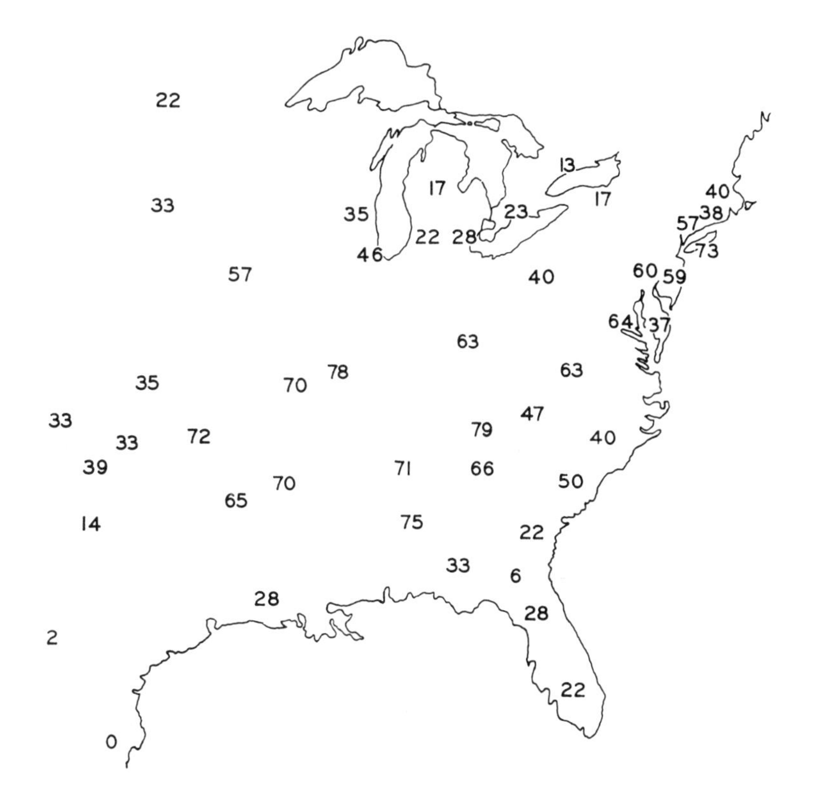

Figure 22–5 The proportions of red-phase Eastern Screech-Owls found in local populations decline from high values of 70 to 80 percent in the center of the range of this species to 30 percent or less at the edges of the range. (From Owen 1963)

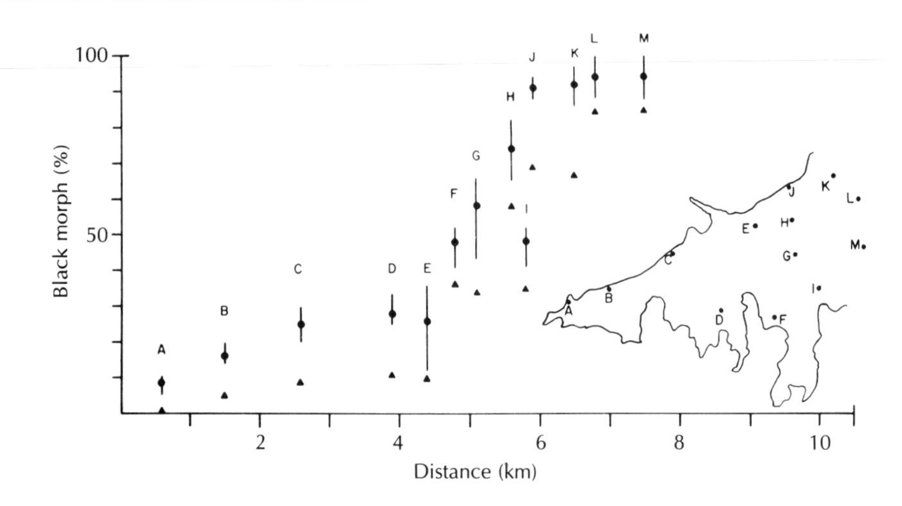

Figure 22–6 Yellow-and-black Bananaquits are common only in southwestern Grenada, the region of their recent colonization of this island. All-black birds (black morphs) occupy the rest of the island. Shown here are means and ranges of the proportions of the black morph at sampling sites for the interval 1974–1978, graphed in terms of distance from the westernmost point of the island. Triangles indicate the results of resampling the localities in 1981. (From Wunderle 1983)

know what advantages favor gray-phase screech-owls in the north and south, but quite likely, the advantages derive from protective coloration and exposure to predators such as the Great Horned Owl. As the type of forests changes from rich brown hardwoods in the center of their range to grayish conifers in the north and to pinelands in the far south, so does the concealing advantage of the gray color phase.

Clines may be either static or dynamic. Static clines are those in which an equilibrium between selection and gene flow has been established, and no changes in the composition of the populations are predicted. Dynamic clines change with time as a result of slow diffusion of neutral traits or as a result of a selective advantage of one trait over its alternatives. An example of a dynamic cline is found in Bananaquits, a small, tropical, warbler-like bird that feeds on nectar and fruit on the Caribbean island of Grenada (Wunderle 1981, 1983). The yellow-and-black color form of Bananaquits prevails throughout most of the Caribbean, but until the early 1900s, an all-black form of this species inhabited Grenada. About 80 years ago, yellow-and-black Bananaquits colonized the arid southwestern corner of Grenada, where they replaced the black Bananaquits. The relative numbers of the two color forms change rapidly in favor of black Bananaquits as one proceeds north and east. Yellow-and-black Bananaquits are advancing eastward at a rate of roughly 400 meters per year, mixing with and then replacing black Bananaquits—referred to as black morphs. This rate of advance suggests that yellow-and-black Bananaquits have a 17 percent selective advantage over black Bananaquits and will soon replace them throughout the island (Figure 22–6).

Local Variation

Sometimes bird populations evolve differences on a local scale. One extreme case of microgeographical variation occurs within the confines of small, rugged Réunion Island in the western Indian Ocean, the home of the Mascarene Gray White-eye, an Old World ecological equivalent of the Bananaquit. First described in terms of four subspecies, the complex patterns of local geographical variation in the Mascarene Gray White-eye feature both clinal variation of color phases and climatic adaptation (Gill 1973).

The Mascarene Gray White-eye has brown and gray color phases. The proportions of gray-phase white-eyes in local populations increase with altitude from none in coastal populations to over 90 percent in populations above 2000 meters. Whereas the proportions of color phases change gradually with latitude in the screech-owl, the clines in the white-eye change rapidly over distances of only a few kilometers on steep mountainsides. The advantage of gray-phase white-eyes at high altitudes is not known, but it must be substantial to maintain such a steep gradient.

Independent of the proportions of the two color phases, brown-phase white-eyes vary strikingly in color and size at different localities on the island. The variations involve three distinct populations, each with altitudinal clines in size and pigmentation. Occupying the coastal regions of Réunion Island are three distinct populations of brown-phase white-eyes: a gray-headed form, a brown-headed form, and a gray-crowned, brown-naped form. These three forms apparently evolved as small, isolated populations on different parts of the island before the arrival of humans. Following widespread cutting of the forests, the ranges of each color form expanded and now abut one another at major riverbeds and at a lava flow, and thus remain partially isolated. The gray-crowned, brown-naped form appears to be a result of hybridization of the other two forms (Figure 22–7).

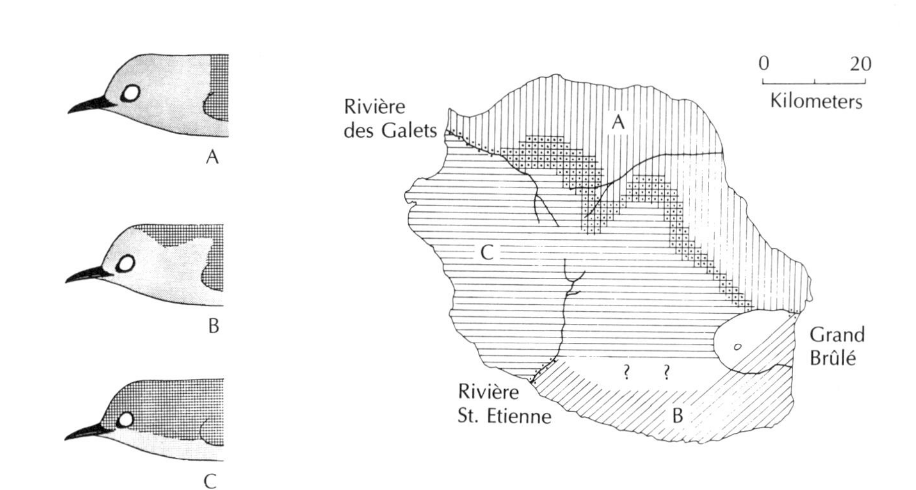

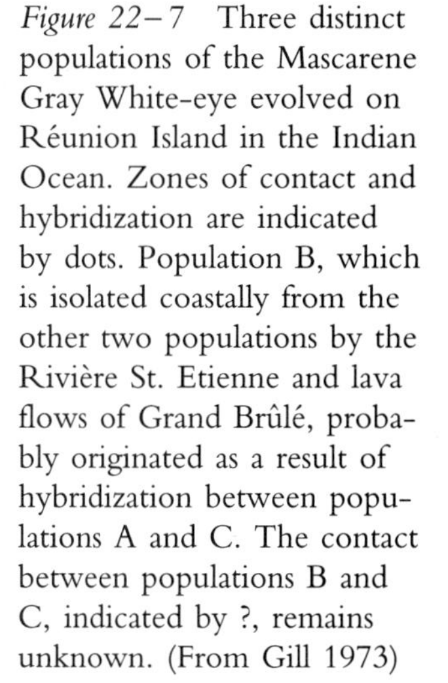
Figure 22–7 Three distinct populations of the Mascarene Gray White-eye evolved on Réunion Island in the Indian Ocean. Zones of contact and hybridization are indicated by dots. Population B, which is isolated coastally from the other two populations by the Rivière St. Etienne and lava flows of Grand Brûlé, probably originated as a result of hybridization between populations A and C. The contact between populations B and C, indicated by ?, remains unknown. (From Gill 1973)

Brown-phase white-eyes are larger and darker at higher elevations and in the cold, wet interior of the island. In contrast to the coastal brown-headed birds, which are small and pale with white bellies, those of the central highlands, only 30 kilometers away, are 6 to 8 percent larger and darkly colored with rufous-brown or charcoal-gray bellies. This aspect of geographical variation in the Mascarene Gray White-eye is similar to the variations recently evolved in House Sparrows in North America (Johnston and Selander 1971) and is widely characteristic of geographical variation in birds. However, the distances separating distinct forms of this island white-eye are extraordinarily short ones.

Geographical Isolation

Most species of birds evolve in geographical isolation—called allopatry—under conditions of minimal gene exchange with sister populations. Bird populations become isolated in two principal ways. First, individuals may colonize an oceanic island that is well separated from their main population on the mainland or other islands, as the Bananaquits and Mascarene White-eyes have done. Classic examples of major evolutionary

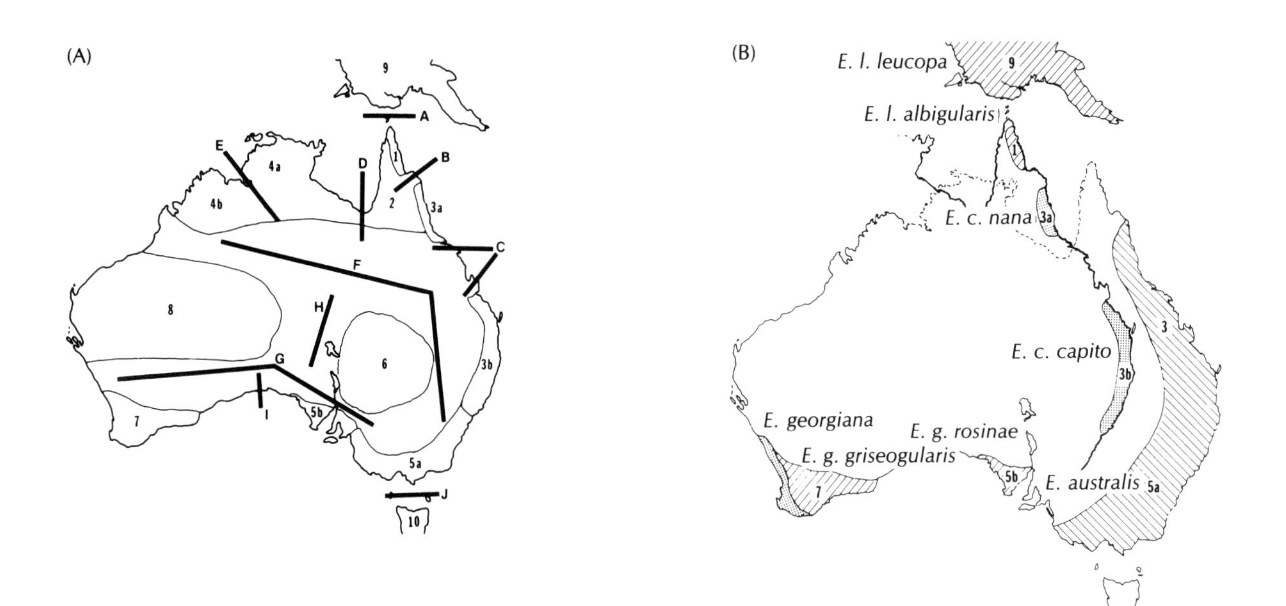

Figure 22–8 Fragmentations of Australian habitats were responsible for the geographical speciation patterns on that continent. (A) Ten primary ecological barriers (A–J) separate the geographical regions of endemism of Australian birds (1–10). (B) Distributions of Australian robins (*Eopsaltria*) in relation to the major areas of restricted distributions—endemism—of Australian birds. The eastern coast appears in a double image to show the different distributions of the two species that occur together there. (From Cracraft 1982a)

change come from remote islands such as the Galápagos and Hawaiian archipelagos. Comparison of the different species of mockingbirds (not the finches, named after him) on islands of the Galápagos archipelago led to Darwin's initial insights about the origin of species. Islands of special habitats, such as oases in deserts, set a similar stage for continental speciation by the populations that occupy them.

Historical separations of major habitats may fragment the ranges of whole communities. Habitat fragmentation was a regular consequence of the advance and retreat of the glaciers and corresponding climate changes during the last 3 million years. Pairs of sister species—or vicariants—may then evolve concordantly in taxa that were separated at the same time. For example, Joel Cracraft (1982a) postulated 10 specific fractures—or vicariance barriers—in the recent history of the Australian avifauna. Among these was a major separation of the northern and eastern avifaunas from the central and southern avifaunas. The birds that occupied wet habitats on the southern edge of Australia also split from those of the central arid region. Later the central arid avifaunas and southern mesic avifaunas split again. The Pleistocene deterioration of climates caused the two northern avifaunas to split repeatedly at different sites. Accompanying these separations was the divergence of sister taxa, such as the species and subspecies of Australian robins (Figure 22–8). These sister taxa came back into contact when their ranges expanded during favorable climatic periods.

Assortative Mating

Preferential pairing of like types—or positive assortative mating—maintains the cohesion of populations and the separation of coexisting species. Assortative mating also may occur among variants within a species, potentially changing the genetic composition of a population (Findlay 1987). The two best-studied cases are those involving light versus dark color phases of the Arctic Skua in the northern British Isles (called Parasitic Jaeger in North America) (O'Donald 1983, 1987) and the Snow Goose in the Canadian Arctic (Cooke 1987). In the first case, female Arctic Skuas studied on the Shetland islands have an innate preference for dark—or melanic—males, which consequently tend to breed earlier than pale-colored males (O'Donald 1987). Heterozygotes (birds of intermediate colors) are at a selective disadvantage, and the pale color phase persists in the population because of immigration from more northern, high colonies where the pale color phase is favored by natural selection.

In the Snow Goose, early imprinting by goslings on their parents determines later mate preferences and leads to assortative mating (Table 22–1; see also Chapter 9). At the La Pérouse Bay colony in northern Manitoba, only 15 to 18 percent of all pairs of Snow Goose were mixed, much less than the 35 to 41 percent expected if pairing were random with respect to color phase.

Box 22–3
DNA comparisons reveal the sequence of speciation in sparrows

The boldly marked, large sparrows of the genus *Zonotrichia* include some of the most widespread and best-studied species of North America. Genetic (mitochondrial DNA) studies of their phylogenetic relationships reveal the apparent history of speciation. These species split from one another in a hierarchical sequence that started with separation of the ancestor of the North American species from the (currently) widespread Rufous-collared Sparrow of Central and South America (Figure 21–6), perhaps 1 million years ago (Zink et al. 1991). This was followed by a separation of the ancestor of the crowned sparrows from the north central Harris's Sparrow, and was further followed by the separation of the White-throated Sparrow. The process concluded very recently with the separation of the Golden-crowned Sparrow of Alaska and the widespread White-crowned Sparrow (see figure). The complex songs of the dialect-prone White-crowned Sparrow appear to be derived from the simple, clear, whistled songs of the rest of the members of this group of closely related sparrows.

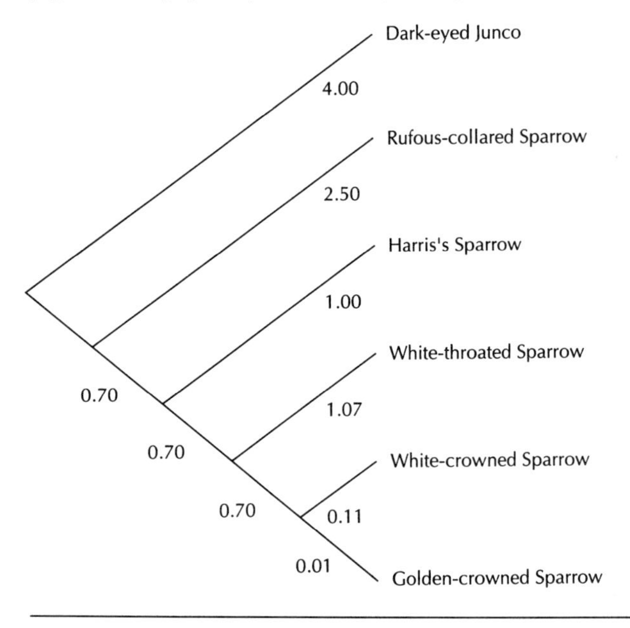

The sequence of speciation (from top to bottom) of modern species of *Zonotrichia* sparrows is indicated by this phylogenetic tree based on restriction fragment analysis of mitochondrial DNA (Box 3–1). Numbers indicate the amount of genetic divergence for each branch, in terms of percentage of nucleotide differentiation. The Dark-eyed Junco, a relative of *Zonotrichia* sparrows, was used as an external reference—or outgroup—to define the base of the speciation sequence. (After Zink et al. 1991)

Table 22–1
Mate selection by Snow Geese at La Pérouse Bay

Parent color	Mate color: White	Mate color: Blue
White pair	191	37
Mixed pair	37	27
Blue pair	9	57

After Cooke 1978.

Secondary Contact and Hybridization

Secondary contact—or the reuniting of previously isolated divergent populations—tests the reproductive compatibility of the two populations. Members of sister taxa in secondary contact face new options of mating with dissimilar individuals. The possible consequences of these matings range from free interbreeding and blending of the attributes of the sister taxa through limited hybridization to strict assortative mating and reproductive isolation and, thus, conformity to the definition of biological species. Over 10 percent of bird species are known to hybridize, a situation resulting in problems for phylogenetic reconstruction, formulation of species concepts, and conservation (Grant and Grant 1992).

The Yellow-rumped Warbler, now considered to be a single biological species, provides an example of free interbreeding between divergent, but reunited, eastern and western populations (Barrowclough 1980b). Separated by glaciers during the Wisconsin glaciation, eastern and western populations evolved into two phylogenetic units, the distinctive Myrtle Warbler and Audubon's Warbler, respectively (Figure 22–9).

The two divergent populations came into secondary contact about 7500 years ago when the forests reunited as the glaciers retreated. Myrtle Warblers and Audubon's Warblers now interbreed freely in mountain passes of the Canadian Rockies. Westward movement of Myrtle Warbler genes and eastward movement of Audubon's Warbler genes has extended the zone of intergradation—hybrid zone—far beyond the mountain passes where hybridization takes place. The 150-kilometer width of the

BOX 22–4

Western and Clark's Grebes mate assortatively

Until 1965, ornithologists recognized just one species of the large black-and-white grebes, genus *Aechmophorus*, of western North America—called the Western Grebe. This grebe, which is best known for its elaborate "rushing" courtship display, has two color forms. The light-phase bird has a pale back, an orange-yellow bill, and is white above the ruby red eyes; the dark-phase bird has a yellow-green bill with black extending below the red eyes. While studying the courtship behavior of these handsome water birds, Robert Storer (1965) tallied the compositions of mated pairs in northern Utah and discovered that light-phase grebes paired preferentially with each other, as did dark-phase birds. Censuses of large grebe populations in Utah, Oregon, and California proved that mixed pairs of light-phase and dark-phase grebes are rare, constituting less than 3 percent of all pairs. Subsequent study revealed differences in their advertising call, which they use to locate their mates, plus differences in foraging behavior, size, and DNA (Storer and Nuechterlein 1992).

The discovery of assortative mating indicated at least partial reproductive isolation between coexisting—or sympatric—populations. Consequently, the light and dark color phases are now considered to be separate species, called Clark's and Western Grebes, respectively.

zone is close to the theoretical prediction for a dynamic cline involving neutral—or nonadaptive—characters that has been widening slowly but steadily for the past 7500 years. Only the restriction of contact to the high, narrow mountain passes of Alberta, where populations are small, prevents a more rapid blurring of the differences between the two populations.

With time, two hybridizing populations may fuse into a new taxon with intermediate characters. The Gilded Flicker of the southwestern United States may be of hybrid origin (Short 1965) as is one population of Mascarene Gray White-eyes (Figure 22–7). Some local populations of towhees in Mexico consist only of hybrids with characteristics of both the Rufous-sided Towhee and the Collared Towhee (Sibley 1954). Quite possibly, hybridization is responsible for the modern characteristics of other bird species.

The opposite also happens. Hybridization may take place on initial contact of two species and then stop. One such case involves a distant relative of the Mascarene Gray White-eye, the Silver-eye of Australia (Gill 1970). This species colonized Norfolk Island in the South Pacific east of Australia at least three times, most recently in 1904. Shortly after the third invasion, some of the Silver-eyes hybridized with the descendants

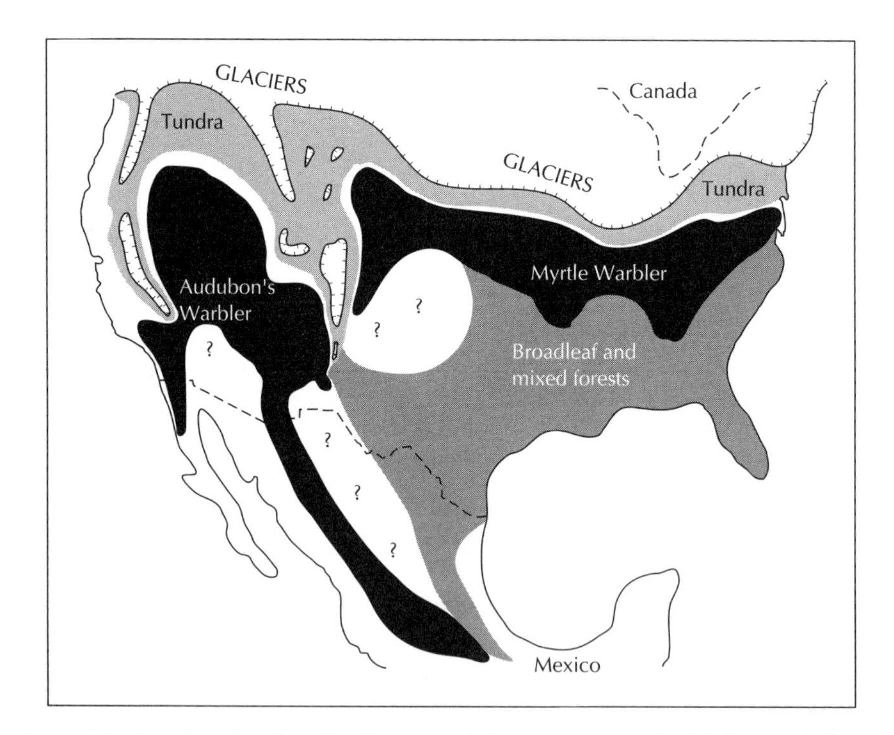

Figure 22–9 Model of the distributions of the eastern Myrtle Warbler and the western Audubon's Warbler separated during the Pleistocene epoch (Wisconsin glaciation), the time when they probably diverged from a common ancestor. Question marks indicate regions for which habitat type is unknown. (From Hubbard 1969)

of the previous invasion, which in the interim had evolved into the larger Slender-billed White-eye. But hybridization did not continue; the two white-eyes now coexist as distinct species on Norfolk Island without interbreeding.

Hybridization does not normally continue between populations with genetic, behavioral, or ecological incompatibilities when the hybrid offspring are inferior in some way. Like mules—the hybrids of female horses and male donkeys—almost all the hybrids of the Eastern Meadowlark and the Western Meadowlark are sterile (Lanyon 1979). The hybrids appear normal and healthy but produce infertile eggs when paired in captivity with an Eastern Meadowlark or a Western Meadowlark. Because there is no gene flow between them, the two meadowlarks remain distinct biological species, in contrast to Myrtle and Audubon's Warblers.

Hybrids of the two differentiated populations may be sterile both as males and as females, but there tends to be a higher degree of sterility among hybrid individuals of the heterogametic sex—that is, the sex with the ZW combination of sex chromosomes (Chapter 16)—than among hybrids of the homogametic sex (ZZ). This expectation—known as Haldane's rule because it was stated first by J.B.S. Haldane (1922)—is broadly supported by data from vertebrates and invertebrates. Among birds, females are the heterogametic sex and will be the first to exhibit sterility in interspecific matings. Haldane's prediction is supported by studies of hybridization between European Pied and Collared Flycatchers in northern Europe (Tegelström and Gelter 1990; Gelter et al. 1992). In the hybrid zone, 15 to 20 percent of the breeding pairs are mixed species pairs, and about 5 percent of all individuals are hybrids. Male hybrids are fertile, but female hybrids are sterile.

Hybrids may be able to produce viable sperm or fertile eggs, but the final test of their fertility comes later when the gene combinations carried by these gametes must function correctly during early development of the fertilized zygote. Blocks of genes of one parental species recombine with the genes of the other species for the first time during meiosis and gamete formation in the first generation (F1) hybrid. Incompatible gene combinations may then disrupt the delicate process of embryo development in second generation (F2) offspring. F2 breakdown, as this phenomenon is called, is a common result of hybridization. For example, in junglefowl a large proportion of the eggs produced by female F1 hybrids are fertile, but few of the eggs hatch (Morejohn 1968). The embryos perish before hatching as a result of developmental failure caused by incompatible sets of genes.

Hybrid inferiority may also be evident in intermediate plumage or displays that render the hybrid less effective in courtship. Hybrid crosses of Anna's Hummingbirds and Costa's Hummingbirds, for example, are intermediate in many details of plumage as well as in the circular courtship flight displays characteristic of these species (Wells et al. 1978). Similarly, a hybrid male of a cross between the Sharp-tailed Grouse and the Greater Prairie-Chicken was unable to perform bobs, bows, and foot stomps correctly and mated infrequently as a result (Evans 1966). Hybrid progeny of crosses between the Blue-winged Warbler and the Golden-winged Warbler take a few days longer than nonhybrid birds to secure mates (Ficken and Ficken 1968).

The proportions of hybrid and parental phenotypes in a zone of overlap can serve as criteria for judging whether two populations are the same species. If no hybrids are present, reproductive isolation is manifest and species status is warranted. If hybrids appear in low frequencies, interspecific pairings are infrequent; or if hybrids are less viable than the parental forms, taxonomic distinction as species is inferred. When hybrids are fairly common and blend with parental types, however, the taxonomic decision becomes more difficult.

Samples of individuals from a series of localities along a transect—a sample area, usually in the form of a long, continuous strip—through zones of overlap and hybridization provide evidence of the amount of hybridization required for taxonomic decisions. A variety of bird species found throughout eastern North America are replaced by closely related sister taxa in the western part of the continent. In the Great Plains alone, there are 14 such pairs of species, and of these, 11 pairs engage in hybridization (Rising 1983), including the Yellow-shafted Flicker and the Red-shafted Flicker, the Baltimore Oriole and the Bullock's Oriole (Figure 22–10). Farther north are additional cases of replacement with hybridization, such as the contact between Myrtle and Audubon's Warblers. As one proceeds east to west through the zones of hybridization between these taxa: The first samples include only the eastern representative of the pair; the samples from the hybrid zone consist of intermediate and variable phenotypes; and finally, the composition switches to include only the western representative. Most individuals at certain localities in the hybrid zone are intermediate in appearance, a finding that indicates reproductive cohesion. Following the Biological Species Concept, the eastern and western counterpart populations of the flickers, orioles, and warblers are now lumped into single species.

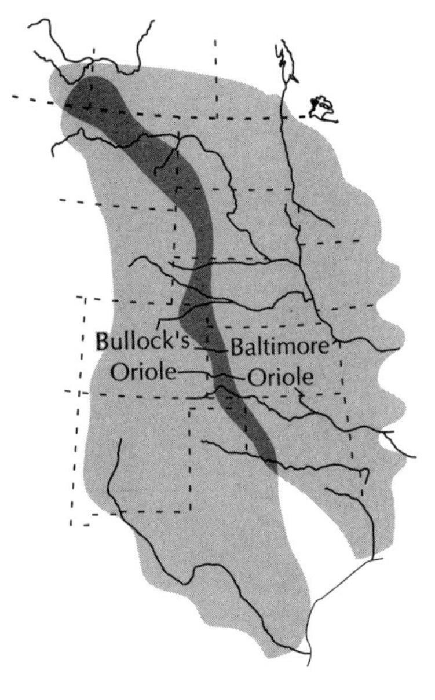

Figure 22–10 The eastern Baltimore Oriole and the western Bullock's Oriole interbreed in a narrow zone of overlap in the Great Plains. The extent of introgression of Bullock's characters eastward and of Baltimore characters westward is indicated by the lighter screen. (From Rising 1983)

In contrast, hybridization is limited between Lazuli Buntings and Indigo Buntings and between Rose-breasted Grosbeaks and Black-headed Grosbeaks (Rising 1983). The two buntings mostly pair assortatively, as do the grosbeaks (Emlen et al. 1975; Baker and Baker 1990). Hybrid grosbeaks, for example, never constitute more than 37 percent of local populations in South Dakota. Although no evidence of assortative mating was found, hybrid female grosbeaks laid smaller clutches than did pure females (Anderson and Daugherty 1974). In the buntings, mixed matings and hybrids were both quite uncommon, and little introgression was evident. Hybrids tended to be excluded from optimal habitats and may have reduced viability (Emlen et al. 1975; Kroodsma 1975). Many pure buntings and grosbeaks persist despite some hybridization, a finding that suggests greater reproductive isolation than is operating between the flickers or between the warblers. The two grosbeaks act as separate biological species despite their limited hybridization. So do the two buntings.

Stable Hybrid Zones

Hybridization in zones of secondary contact often persists and continues unabated for centuries without leading to assortative mating or population fusion. The narrow hybrid zone between the Hooded Crow and the Carrion Crow of Europe has not changed in width for at least 500 years (Mayr 1963). At least some of the hybrid zones in the Great Plains region of North America, including that of the Northern Flicker, are of ancient origin, dating back to expansion of isolated populations following the retreat of the glaciers 10,000 years ago.

Two theoretical models have been developed to explain the stability of hybrid zones. In one, the dynamic equilibrium model of Nicholas Barton and Godfrey Hewitt (1985), the hybrid zone is a population sink of inferior hybrids produced relentlessly by immigrants from the adjacent, large, pure populations. The clines of character states that change as a result of interbreeding and gene flow—also called introgression—tend to converge at particular locations, thereby forming sharp boundaries between hybridizing species.

The other model is the bounded superiority model of Moore (1977). This model, in which hybrid zones coincide with intermediate ecological or climatic conditions where hybrids are equally or better adapted than their parents, best explains the stable hybrid zones of both the crows in Europe and the flickers in North America. The stable hybrid zone of Hooded and Carrion Crows coincides with an ecological interface between alpine valleys and the intensively cultivated plains, where hybrid and nonhybrid crows are equally fit (Saino and Villa 1992). Hybrids between the all-black Carrion Crow and black-and-gray Hooded Crow are easily recognized by their variable color patterns (Figure 22–11). The hybrid zone between the western (red-shafted) and the eastern (yellow-shafted) populations of the Northern Flicker has not changed in width or

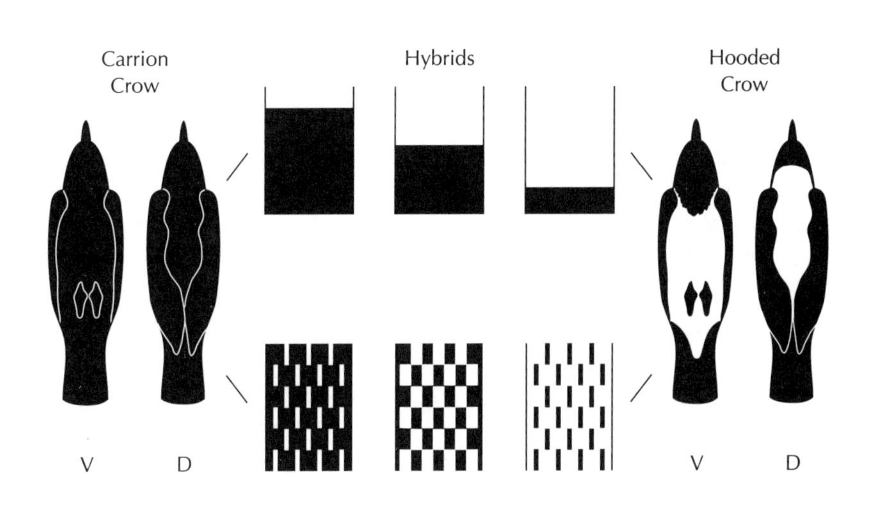

Figure 22–11 Color patterns of hybrid crows. Two main series of continuous variation in the distribution of black pigmentation characterize hybrid phenotypes. Boxes schematically represent the body. V, ventral view; D, dorsal view. The same amount of black pigmentation may be either uniformly scattered over gray parts of the "pure" Hooded Crow phenotype (series below) or may be concentrated toward the rear of the body (series above). (From Saino and Villa 1992)

location for at least 100 years and probably much longer (Moore and Buchanan 1985). The continued free interbreeding between the flickers suggests no disadvantages and perhaps even some advantages of hybrids in the zone of contact.

Ecology of Speciation

Although divergence and reproductive isolation are primary ingredients, ecological isolation is another ingredient of the speciation process. Incomplete reproductive isolation fosters hybridization and genetic fusion of divergent populations. Incomplete ecological isolation leads to competition and possible geographical replacement. Closely related but ecologically incompatible species tend to have contiguous—or parapatric—distributions, because they are too similar to coexist with each other. In time, ecological differences may evolve and permit expansion of one species into the geographical range of the other. An overlapping distribution of reproductively isolated and ecologically compatible species in local communities is the final stage of the speciation process.

If sister taxa evolved ecological differences while isolated, they may not compete at all when they first come into contact, and coexistence is immediate. If they have not diverged ecologically, however, they tend to compete, a situation that may lead to the extinction of one by the other or to open confrontation. The Eastern Meadowlark and the Western Meadowlark, for example, defend mutually exclusive territories where their ranges overlap (Lanyon 1957). Dusky Flycatchers and Gray Flycatchers also defend mutually exclusive territories in areas of recent secondary contact (Johnson 1963).

Such similar species may at first defend mutually exclusive territories, but with time one of the competitors is likely to exclude the other one. The current territorial relationships of the Sharp-tailed and LeConte's Sparrows seem to be based on just this kind of resolution of their ecological interactions (Murray 1969, 1971). LeConte's Sparrows are territorial; Sharp-tailed Sparrows are not. Territorial male LeConte's Sparrows attack Sharp-tails when they first arrive in the spring, but the Sharp-tails remain, nonaggressively sharing both marsh and song perches with the aggressive species.

The case of Blue-winged Warblers and Golden-winged Warblers combines both hybridization and ecological aspects of speciation. The Blue-winged Warbler was once an uncommon species of the south-central United States. Its sister species, the Golden-winged Warbler, occurred farther north and at higher altitudes in the Appalachians. The ranges of the two species were well separated (Short 1963). The clearing of forests and the increases in second-growth vegetation throughout the northeastern United States that occurred in the mid-1800s benefited both warblers, and the Blue-winged Warbler expanded northward into the range of the Golden-winged Warbler (Figure 22–12).

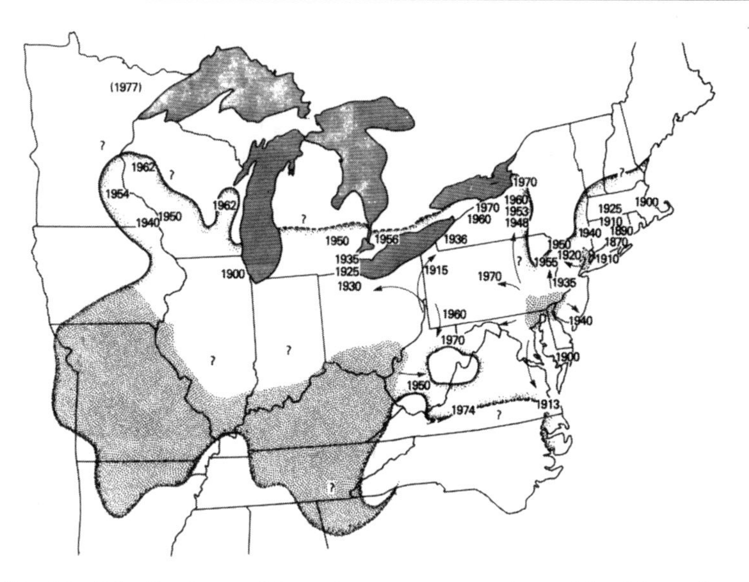

Figure 22–12 Distribution and breeding range expansion of the Blue-winged Warbler in the last century. Dates on the map indicate when Blue-wings first established themselves at that locality. Range boundaries and arrows indicating patterns of spread are hypotheses based on historical information. Stippled area indicates the approximate range of Blue-wings in the 1800s. (From Gill 1980)

The two warblers have strikingly different color patterns (Figure 22–13). Blue-winged Warblers are bright yellow with white wing bars and a narrow black line through the eye. Golden-winged Warblers are gray above, white below, with yellow wingbars and crown, and bold, black patches on the throat and the eyes. The contrasting facial color patterns of the two species have a simple genetic basis, analogous to the color phases of Bananaquits. The plain throat and narrow black eye line of the Blue-wing are dominant to the black throat and black eye patch of the Golden-wing. Other plumage color characteristics are controlled by several genes that supplement one another. First-generation hybrids, called Brewster's Warblers, inherit the genetically dominant Blue-wing face pattern but are intermediate between the two parental species in other aspects of plumage coloration. These fertile hybrids produce viable offspring when they mate with either Blue-wings, Golden-wings, or other hybrids. Blending of plumage colors coupled to either facial color pattern produces a variety of hybrid types, including Lawrence's Warbler, which looks like a yellow Blue-wing with a bold, black Golden-winged Warbler facial pattern.

Blue-winged Warblers generally replace Golden-winged Warblers within 50 years of local contact (Gill 1980). The Golden-winged Warbler faces widespread extinction as a result. Blue-winged Warblers use a broad range of habitats, including those of the Golden-winged Warbler, which usually inhabits abandoned fields that are in the early stages of conversion

Figure 22–13 Two species and two hybrids of warblers: (A) Blue-winged Warbler; (B) Golden-winged Warbler; (C) Lawrence's hybrid; (D) Brewster's hybrid. Stippling indicates bright yellow; slanted lines indicate olive. Gray, black, and white are as represented by those shades in the drawing. The Lawrence's hybrid resembles a Blue-winged Warbler with the lace pattern of a Golden-winged Warbler. (After Ficken and Ficken 1968)

to shrublands and young forest (Confer and Knapp 1981). Their territories often overlap. These two warblers apparently have not achieved the ecological compatibility needed for coexistence. Hybridization may hasten the extinction of the rarer Golden-wing species, but ecological forces apparently will determine its fate.

The steady replacement of Golden-winged by Blue-winged Warblers causes a predictable shift in the composition of local assemblages of these warblers. At first only the former are present, with perhaps an occasional pioneering Blue-winged Warbler or odd hybrid form. As Blue-winged Warblers move in, balanced proportions of the two extreme forms plus an assortment of intermediate hybrids develop. Finally, as the number of

Golden-winged Warblers declines and that of Blue-winged Warblers continues to increase, the distribution shifts in favor of a majority of the latter plus various hybrids that are the remnants of previous hybridizations. In the last phase, only Blue-winged Warblers remain, perhaps with a few individuals of mixed blood. Elsewhere in the last century, Mallard ducks expanded eastward through southern Canada, where they hybridized with and replaced American Black Ducks (Johnsgard and DiSilvestro 1976; Avise et al. 1990).

Behavior and Speciation

Returning now to the process of speciation, we ponder the features that catalyze speciation. Birds speciate continually, even though they are highly mobile, living in large populations connected by much gene flow. Some of the Darwin's finches, which live on oceanic islands of the Galápagos archipelago, have moderately large populations united by gene flow (Barrowclough 1983; Grant and Grant 1992). Nevertheless, they are prime examples of adaptive radiation of bill sizes, associated feeding habits, and behavioral innovations (see Chapter 7). Speciation apparently takes place when small colonizing founder populations undergo rapid but simple genetic changes followed by population growth and adaptive divergence. The evolutionary history of Galápagos finches has been marked by episodes of strong selection for changes in plumage or skeletal characteristics. The drought of 1976, for example, resulted in a severe, albeit temporary, shift as a result of the natural selection of bill sizes that enabled the finches to feed efficiently on the seeds that were available (Chapter 7). Periodic and stringent sorting of individuals from such populations with new behaviors and anatomical variations would promote the evolution of new species of finches and is precisely the process envisioned for birds generally.

The behavioral attributes of birds, particularly their capacity for new behavior and its cultural transmission, may be extraordinary advantages (Wyles et al. 1983; West-Eberhard 1983). As noted in Chapter 8, broad correlations between brain size, taxonomic diversity, and rates of molecular genetic change suggest that enhanced brain capacities and behavioral innovations may catalyze speciation and taxonomic diversification in both primates and songbirds (Wyles et al. 1983; Fitzpatrick 1988; Wilson 1991). Behavior, rather than the environment, can be the driving force of evolutionary change when individuals exploit the environment in new ways, followed by the rapid spread of the new habit through the population by cultural transmission, followed finally by the evolution of anatomical traits that enhance the effectiveness of individuals that practice the new habit. New behaviors that ultimately spawn anatomical change are more likely to arise in populations of individuals that have the intelligence to develop such behavior, such as songbirds in the Order Passeriformes.

Box 22–5
Brood parasites may undergo cultural speciation

Brood parasitic indigobirds lay their eggs in the nests of species of host estrildine finches in Africa (Chapter 19). Originally it was thought that the parasites and their hosts had speciated together through a normal process of geographical isolation. Genetic comparisons, however, reveal that the indigobirds speciated more recently than did their hosts and did so by switching to, or colonizing, a new host species through a behavioral process that could be called "cultural speciation" (Payne 1973b; Klein et al. 1993).

A model of cultural speciation in indigobirds is as follows. First, a female indigobird lays her eggs in the nest of a different (new) species of foster finch. The parasitic young survive despite the fact that they do not closely mimic the host's young, probably during a period of food abundance when hosts are less discriminating about which young they feed. The young parasites imprint on the songs of the new host. Males singing these new songs attract like-minded females, which proceed to lay their eggs in the nests of the new host, starting a new host–parasite relationship. Assortative mating of indigobirds imprinted on the same hosts effectively isolates them from other host races or "species."

New behavior, with its subsequent cultural transmission, is one catalyst of avian speciation. Social selection, which favors acquisition of new social signals, is another (West-Eberhard 1983). Charles Darwin (1871) recognized the importance of social selection, but only very recently have ornithologists started to appreciate its implications. Attributes associated with communication and competition among individuals for mates, space, or access to food enable some individuals to survive and reproduce while others are dying or failing to reproduce (see discussion of sexual selection in Chapter 9). Attributes that are favored in social competition, however, such as the extravagant display plumages of male birds-of-paradise, may not be adaptive in other contexts and, in fact, may prove to be a liability when they render the bird conspicuous to predators or desirable to humans. Thousands of birds-of-paradise are killed every year by New Guinea tribesmen for head ornaments.

Songs also are subject to elaboration through vocal contests. The interesting feature of the process of social selection with respect to speciation is that, through ritualization, these same attributes often enable pair formation, species recognition, and reproductive isolation.

Recall also that early visual and vocal imprinting affects subsequent reproductive behavior, including mate choice. Early imprinting can catalyze the process of social selection, leading initially to different ornamentation, social behavior, and ultimately to the evolution of new species. Thus, behavioral flexibility coupled to social selection in the form of contests or group coherence could be a force in the speciation process.

The development of song differences between Marsh Wrens of eastern and western North America provides a fitting conclusion to this chapter. Recall that western Marsh Wrens have innate capacities for larger song repertoires than do eastern Marsh Wrens (Chapter 8). Male Marsh Wrens duel vocally with one another to win the best territories and most females. The larger repertoires of the western Marsh Wrens reflect more intense competition for females in restricted pothole cattail marshes. This case illustrates how sexual selection and the behavior of countersinging have led to the elaboration of the brain nuclei that control singing behavior, associated differences in song-learning abilities, and perhaps mating preferences.

Generally separated by a 100-kilometer gap, eastern and western Marsh Wrens coexist in some marshes in the northern Great Plains (Kroodsma 1989). Intermediate song types and hybridization are not apparent. Not known, however, is whether female Marsh Wrens divide into two groups that pair assortatively with the differentially talented males and hence behave as two reproductively isolated species. Alternatively, western males with their larger repertoires may attract more females than their eastern rivals and inch their way eastward, progressively replacing eastern males, just as Blue-winged Warblers are replacing Golden-winged Warblers and Mallards replaced American Black Ducks.

Summary

Species are the primary units of systematic biology; they serve as the basis for describing and analyzing biological diversity. Of the many competing species concepts and definitions, the one that prevails in ornithology is called the Biological Species Concept, which states that a species comprises a set of populations that are capable of successfully interbreeding under natural conditions. The merits of an alternative concept, called the Phylogenetic Species Concept, are now under discussion. It stresses the historical origin of genetic differences and would recognize more distinct local populations than does the Biological Species Concept.

New species of birds evolve via the gradual divergence of large isolated populations adapting to different environments and via a rapid reorganization of some of the genes of birds in small, isolated populations. The evolution of geographical differences among natural bird populations depends on the relative strengths of two opposing forces: the intensity of natural selection favoring one genetic attribute over another and the rate of genetic blending as a result of interbreeding of individuals from different locations—gene flow.

Secondary contact—the reuniting of previously isolated populations—tests the ability of populations to interbreed. Once considered separate species, the Audubon's Warbler and the Myrtle Warbler interbreed freely where they come into contact in the Canadian Rockies. On this evidence they are now considered to be populations of the same biological species, the Yellow-rumped Warbler. The Phylogenetic Species Concept would continue to recognize both the Myrtle Warbler and the Audubon's Warbler. Many other species that interbreed do not produce viable hybrids. Hybrids resulting from interspecific breeding have one of several selective disadvantages: They may be unable to produce viable sperm and eggs; they may have incompatible blocks of genes that disrupt early embryo development; or they may suffer from intermediate plumage and display behaviors that reduce their reproductive success.

The capacities of birds to develop new, learned behaviors may contribute to the process of speciation. Behavior, rather than the environment, can be the driving force of evolutionary change if a new behavior is followed by the evolution of new anatomical traits that support the behavior.

Further Readings

Avise, J.C. 1994. Molecular Markers, Natural History and Evolution. New York: Chapman and Hall. *A synthesis of current evidence by a pioneer in the analysis of geographical variation in mitochondrial DNA.*

Cooke, F., and P.A. Buckley. 1987. Avian Genetics. New York: Academic Press. *A collection of significant papers relating genetics to the population ecology of birds, with implications for speciation.*

Cracraft, J. 1989. Speciation and its ontology: The empirical consequences of alternative species concepts for understanding patterns and processes of differentiation. *In* Speciation and Its Consequences, D. Otte and J.A. Endler, Eds., pp. 28–59. Sunderland, Mass.: Sinauer Associates. *A provocative challenge to the classic viewpoint.*

Findlay, C.S. 1987. Non-random mating: A theoretical and empirical overview with special reference to birds. *In* Avian Genetics, F. Cooke and P.A. Buckley, Eds., pp. 289–314. New York: Academic Press. *Reviews the potential roles of assortative mating in the evolution of bird populations.*

Grant, P.R. 1986. Ecology and Evolution of Darwin's Finches. Princeton, N.J.: Princeton University Press. *A synthesis of long-term studies of the evolutionary dynamics of an island species.*

Hailman, J.P. 1986. The heritability concept applied to wild birds. Current Ornithology 4: 71–95. *A critique of field measurements of character heritability in birds.*

Mayr, E. 1970. Populations, Species, and Evolution. Cambridge, Mass.: Belknap Press. *A classic by the leading proponent of geographical speciation in birds.*

Otte, D., and J.A. Endler, Eds. 1989. Speciation and Its Consequences. Sunderland, Mass.: Sinauer Associates. *A contemporary collection of important papers.*

Selander, R.K. 1971. Systematics and speciation in birds. Avian Biology 1: 57–147. *A detailed review of the old literature on avian speciation.*

West-Eberhard, M.J. 1983. Sexual selection, social competition, and speciation. Q. Rev. Biol. 58: 155–183. *A vital perspective.*

Zink, R.M., and J.V. Remsen, Jr. 1986. Evolutionary processes and patterns of geographical variation in birds. Current Ornithology 4: 1–69. *An excellent, modern overview of the analysis and interpretation of geographical variation.*

C H A P T E R 2 3

Communities

RESOURCES, SUCH AS FOOD and nest holes, determine not only the local population size for a species but also how many species can coexist locally in one habitat. Such coexisting groups of species are called communities. The sizes and diversities of communities increase regularly from temperate to tropical latitudes; several hundred species of birds coexist in a lowland tropical forest, whereas fewer than 50 species coexist in a northern temperate forest.

After a review of the general patterns of species diversity, including both spatial and seasonal components, this chapter shifts to the continuum of community structures that range from "open" communities with limited species interactions to "closed" communities that are organized by interspecific competition for limited resources. Examples of the effects of competition on reproduction and foraging by tits illustrate the consequences of coexistence of ecologically similar and closely related species. Patterns of ecological segregation and geographical replacement among similar species point to competition as a determinant of community structure. The chapter concludes with discussions of colonization and species turnover in island avifaunas, which illustrate the dynamics of community formation, including extinctions.

Species Diversity

The number of bird species in an area—called species richness or species diversity—increases from the Arctic to the Tropics: Greenland has 56 breeding bird species, New York State about 135, Honduras over 550, and Colombia over 1300. The variety of species also increases from high to low altitudes. In Colombia, 47 species reside above the timberline, 270 at altitudes that produce temperate conditions, over 480 at altitudes supporting subtropical conditions, and over 1000 in the tropical lowlands. Many more species are found in New World tropical forests than in comparable forests of the Old World. Community ecologists seek to understand why these differences exist and, in particular, why there are so many more species in tropical communities than in the temperate communities.

Spatial Components of Diversity

The distributions of most bird species are restricted both globally and locally. Penguins are limited to the Southern Hemisphere, auks to the Northern Hemisphere, curassows to tropical South America, turacos to Africa, and the Dodo (once upon a time) to the island of Mauritius. Such boundaries may reflect physical restrictions; the flightless Dodo was limited to one oceanic island. More often, however, the limits of the distribution of a bird species, even on continents, reflect intrinsic limits of population growth, competitive replacement by another species, availability of resources, physiological tolerance, or some combination of these factors. Even at the local level of a county or shire perhaps, few birds use the full variety of habitats available to them.

Each species usually has specific requirements—called its ecological niche (Grinnell 1917; Hutchinson 1959; MacArthur 1968). Thus, we expect to find a Pileated Woodpecker in a forest with large trees full of carpenter ants. Pileated Woodpeckers occupy a wide variety of forests with big, ant-ridden trees, but the Red-cockaded Woodpecker of the southeastern United States has a differently specialized, fateful niche. It requires old pine forests with trees 80 to 100 years old that have been infected by the red heart fungus (Box 21–2).

Species that coexist in seemingly homogeneous habitats, such as grasslands or spruce forests, may segregate their niches even more finely. In his classic study of niche partitioning by wood warblers in northern spruce forests, Robert MacArthur (1958) discovered that the Yellow-rumped Warbler fed mostly in the understory below 3 meters, the Black-throated Green Warbler in the middle story, and the Blackburnian Warbler at the tops of the same spruce trees. Sharing the middle part of the trees with the Black-throated Green Warbler, which explored the foliage for food, was the Cape May Warbler, which fed on insects attracted to sap on the tree trunk. Sharing the treetops with the Blackburnian Warbler, which fed on the outer twigs and sallied out after aerial insects, was the Bay-breasted Warbler, which searched for insects close to the trunk. In Europe, tits show parallel choices of their foraging stations. The analysis of overlap among such niches has played a central role in the development of models of avian community ecology (MacArthur 1972; Cody and Diamond 1975).

Local diversity—or alpha diversity—reflects the structural complexity of the habitat. The vertical distribution of vegetation provides a rough index to the variety of foraging opportunities and, hence, the variety of species that can occupy a habitat (Figure 23–1). The physical structure of habitats provides courtship and display stations, nest sites, protection from predators, shelter from climatic stresses, and, of course, food. Variations in

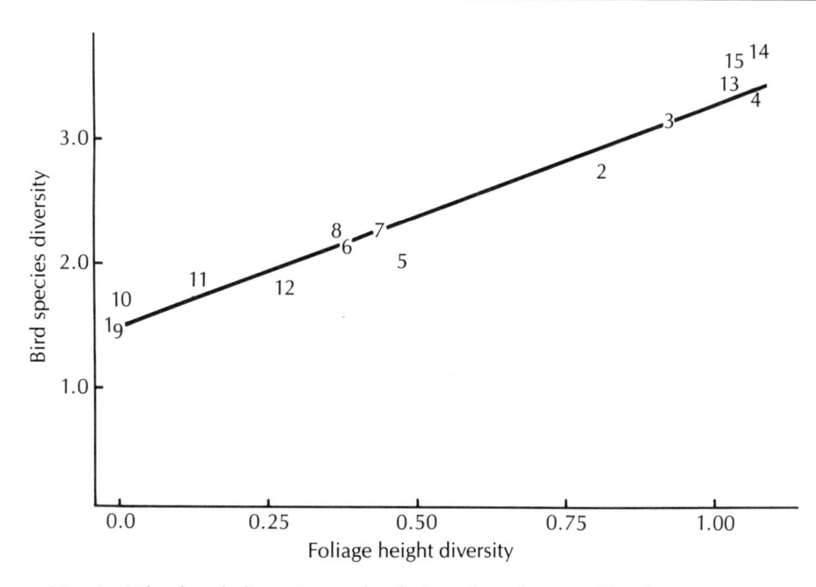

Figure 23–1 The local diversity and relative abundance of bird species are correlated with the relative height and diversity of the foliage, illustrated here for sites in Illinois (sites 1–4), Texas (sites 5–8), and Panama (sites 9–15). (Adapted from Karr and Roth 1971)

foraging behavior among insectivorous birds in the Hubbard Brook forests of New Hampshire are directly related to variations in foliage height (Holmes et al. 1979). Plant species and physical form, percentage of vegetative cover, and local variations in habitat structure all contribute to habitat complexity and influence the local diversity of birds (James 1971).

Subtle differences in habitat preferences are seen among birds that inhabit fields. In a 20-year study of birds associated with the conversion of old, abandoned fields to shrublands and then to forest—called old field succession—on Long Island, New York, Wesley Lanyon (1981) showed that nine species of birds established nesting territories in a sequence that corresponded to the availability of nest cover. Red-winged Blackbirds were the first to nest in a field, and Rufous-sided Towhees were the last. As the open field converts to shrubland and then to forest, the availability of nest-supporting vegetation and the amount of shade for nests determine the suitability of the habitat for breeding. In this example, the procession of species of overlapping tenures caused the average number of territorial species present to peak at 20 years after the field was last cultivated (Figure 23–2).

Thomas Martin (1988b, c) suggested that nest predation must be added as a major process to current theories of space use and coexistence of open nesting birds, such as those of old fields. Nest predation is commonly considered to be density dependent, and predators learn to specialize on particular nests if they are locally common. Nest predation, therefore, can determine local species compositions by favoring bird species that nest at different heights and in different microhabitats.

Box 23–1
Distance to cover defines the niches of sparrows

Distance to protective cover affects the variety of sparrows that can coexist in open, simply structured grassland habitats (Pulliam and Mills 1977). In southeastern Arizona, four species of sparrows inhabit open grasslands that have scattered mesquite trees, which provide some protection from predators such as Prairie Falcons. As far as the sparrows are concerned, this habitat offers concentric rings of increasing distance from the nearest cover. The Vesper Sparrow stays closest to the mesquite trees (within 4 meters), the Savannah Sparrow feeds farther out (4 to 16 meters), the Grasshopper Sparrow still farther out (8 to 32 meters), and the Chestnut-collared Longspur feeds far from the trees in the most open grassland. The behavior of these species when flushed reflects the risks of flying increasing distances to cover. Vesper Sparrows fly quickly to nearby cover, Savannah Sparrows fly to an exposed perch the first time they are flushed and then to full cover if flushed again. Rather than face the risks of a longer flight, Grasshopper Sparrows usually drop back into the grass when flushed, but they fly for cover if repeatedly flushed. Longspurs, however, either crouch to the ground to hide or fly off in tight flocks that help thwart predators.

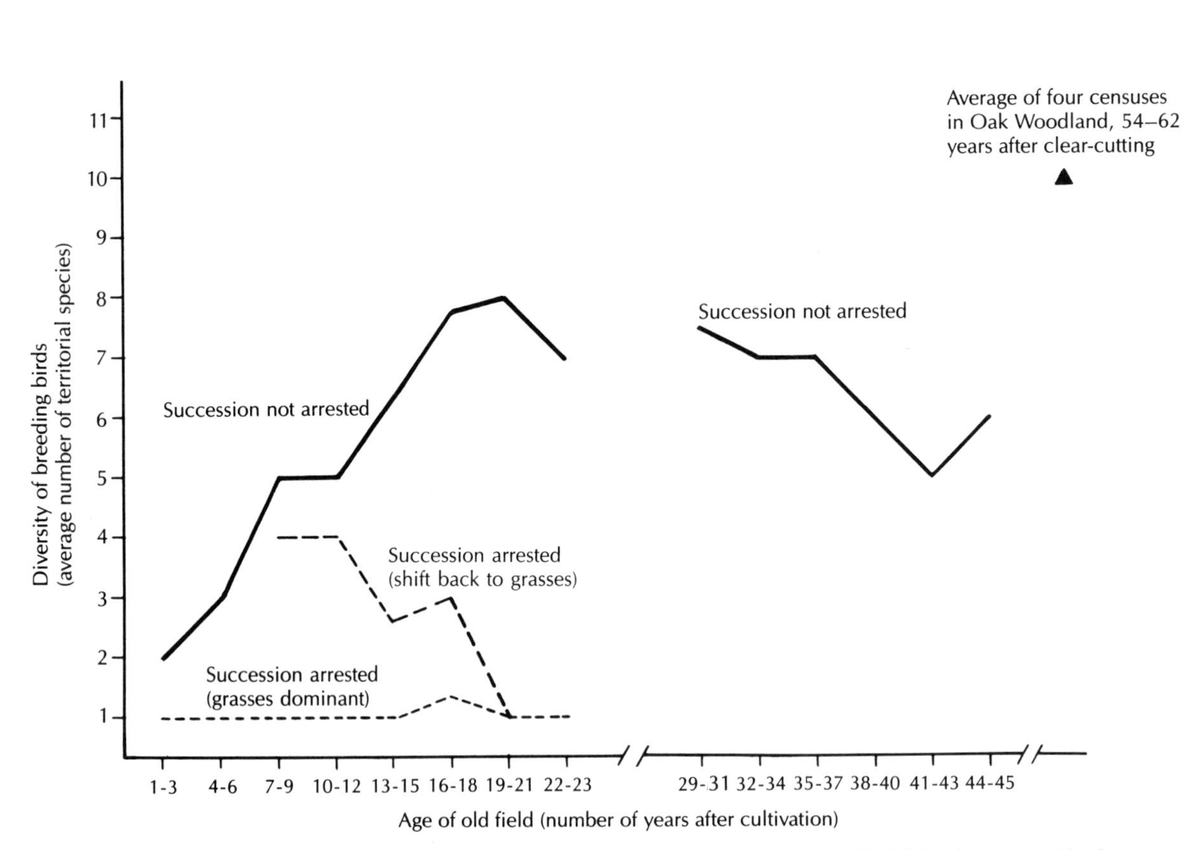

Figure 23–2 The diversity of species nesting in old fields changes with the age of the field and the successional change of the vegetation from grass, to bushes, and ultimately to trees. Successional change was arrested by cultivation. (From Lanyon 1981)

Habitat preferences cause the variety of birds encountered to increase as the number of distinct habitats sampled increases. Bird species diversity increases from that in a single local habitat (alpha diversity), to that in multiple local habitats (beta diversity), to large-scale regional diversity (gamma diversity). On a trip from the West Coast of the United States to the East Coast, the number of species peaks in the Sierra, Rocky, and Appalachian mountains, where markedly different habitats at various altitudes increase beta diversity. In South America, Colombia, Ecuador, and Peru owe their extraordinary variety of birds in part to the topographic and ecological diversity of the Andes.

Tropical Diversity

More species of birds live in the Tropics than in a comparable area in a temperate region; a 5-acre plot of forest in Panama has 2.5 times more species than a similar plot in Illinois. The density of species in an area increases as one proceeds southward from North America and peaks in the tropical forests of western Amazonia. A 400-mile square section (160,000 square miles in area) of the United States contains 120 to 150 species of breeding land birds, but the same area in Central America contains 500 to 600 (MacArthur 1969) (Figure 23–3). In western Amazonia, over 1000 species can be found in such an area, and over 535 species can occur locally in a 100-hectare site (Terborgh et al. 1990).

The greater diversity of species in the Tropics compared with diversity in the Temperate Zone is due in part to different and more varied food resources (Ricklefs and Travis 1980). For example, groups of fruit eaters—toucans, hornbills, barbets, trogons, cotingas, manakins, broadbills, and turacos—expand the dimensions of tropical communities. Large and small parrots consume a wide variety of seeds, fruits, and nectars that are not available in northern forests. Hummingbirds and tanagers, of which only a few species live in the north, abound in New World tropical forests. Some families of strictly tropical birds—puffbirds, motmots, antbirds, woodhoopoes, jacamars—depend on large insects and small reptiles that are not present in temperate ecosystems. The diversity of insect sizes is greater in the Tropics than in the Temperate Zone habitats, and the diversity of bill sizes of tropical birds increases accordingly (Schoener 1971) (Figure 23–4). Foraging specialists, such as ant followers and epiphyte probers, also add to the diversity of bird communities in tropical regions.

More specialized species, large and small, can exist in stable, benign climates than in variable cool climates. Tropical species tend to use a narrower range of habitats (Karr 1971; Lovejoy 1974), may be more specialized in their foraging behavior (Terborgh and Weske 1969; E.W. Stiles 1978), and may be less tolerant of climatic variation than their temperate counterparts. Greater ecological specialization leads to tighter packing of species in local communities and smaller geographical distributions.

Community diversity reflects regional and historical processes as well as local forces (Ricklefs 1987, 1989b). Diversity in tropical communities in particular reflects their long, stable histories of accumulation of specialized species (Moreau 1966; Pianka 1966; Mayr 1969). Ancient communities may be the most species rich of all. For example, the forest fauna of Panama is richer than that of Africa, but the Panamanian grasslands and savannas are not. The lowland forests in Africa were restricted in extent during the Pleistocene period, which seems to have prevented the development of rich forest avifaunas (Moreau 1966; Karr 1976). The man-made grasslands in Panama are quite young (15,000 years) relative to the ancient natural grasslands and savannas of Africa. As a result, grassland communities in Africa are species rich, whereas those in Central America are species poor.

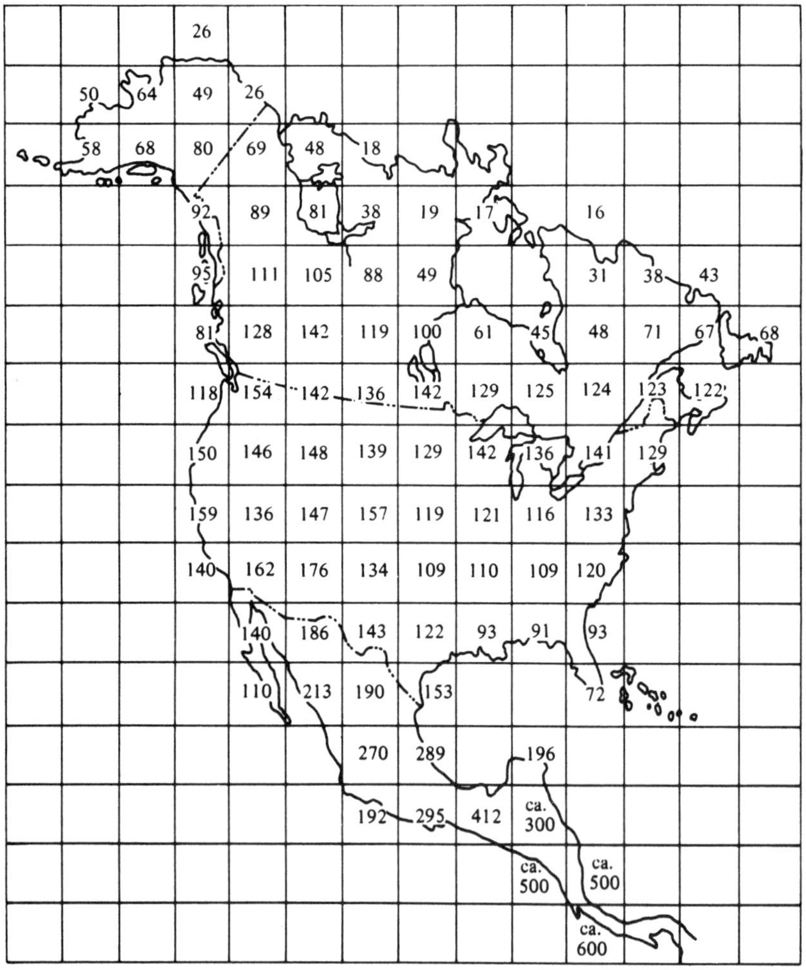

Figure 23–3 The number of land bird species that breed in geographical areas of 400 miles square in North and Central America decreases with increasing latitude. (From MacArthur 1969)

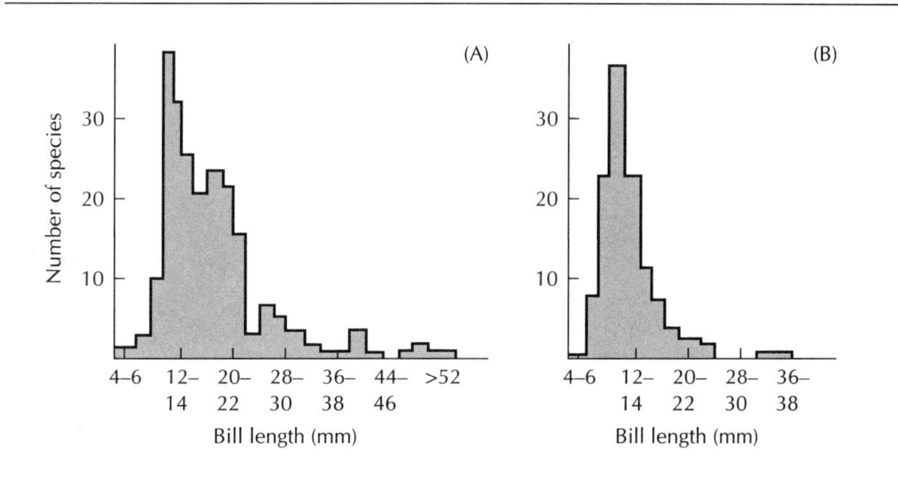

Figure 23–4 The addition of large-billed birds causes an increase in bird species diversity in the Tropics. Shown here are the bill lengths of insectivorous birds that breed in (A) tropical latitudes (8 to 10 degrees north latitude) and (B) temperate latitudes (42 to 44 degrees north latitude). (From Schoener 1971)

Temporal Components of Diversity

Local assemblages of species change in time as well as in space. Their composition fluctuates regularly with the season and irregularly with climate and resource availability. Disturbances due to deforestation or fire or to colonization or extinction keep many habitats in flux. Virtually all bird communities consist of both resident and nonresident species. Nonresidents are seasonal specialists, which take advantage of predictable periods of regional food abundance. The mobility of birds and the evolution of the migratory habit have made possible nonresidency and the opportunistic exploitation of variable environments (see Chapter 12).

Ephemeral resources attract opportunistic species. Temporary assemblages of highly mobile birds may last hours, weeks, or years. Aggregations of seabirds over a shoal of fish, for example, are brief and highly variable in species composition. Assemblages of sunbirds or hummingbirds at flowers feature high turnover of both individuals and species during the brief blooming periods (Figure 23–5). The regional diversity of small, short-billed hummingbirds depends on their ability to circulate among locally blooming flowers (Feinsinger 1980). Only two short-billed species, the Copper-rumped Hummingbird and the Ruby-topaz Hummingbird, inhabit the small island of Tobago, where they must coexist year-round in the principal nonforested habitat. Seven similar species coexist on the larger island of Trinidad, where the habitats are more diverse, thereby enabling more complicated seasonal patterns of local migration.

Seasonal residents form a major component of most bird communities. Migrant shorebirds often dominate the bird life of coastal wetlands. The influx of wintering migrants from the north triples the number of species found in the open pine forests of Grand Bahama Island and increases the density of individuals from 900 to 1600 per square kilometer (Emlen 1980) (Figure 23–6). Migrants comprise 1 out of every 16 birds on Barro Colorado Island on an annual basis and up to 1 out of 7 birds during the peak migration in October. In the tropical evergreen forests of western Mexico, the density of small foliage gleaners increases from an average of 2 to 64 per hectare with the arrival of the migrants (Hutto 1980). These extraordinary densities of wintering birds result from the compression of large populations into small areas. Migrant North American land birds from 16 million square kilometers of breeding range are compressed into 2 million square kilometers of winter range in northern Central America and the West Indies.

The migrants coexist only temporarily with resident species, however, and change their community membership with the seasons. The interactions between migrants and residents, therefore, pose difficult and still unanswered questions about the spatial scale of community structure. Do local assemblages belong to a greater global community? To what degree do competitive interactions in one season influence community structures in another season? In the extreme case, the local densities of migrants on the breeding grounds may be controlled primarily by interactions with a completely different set of species in another hemisphere in the opposite season.

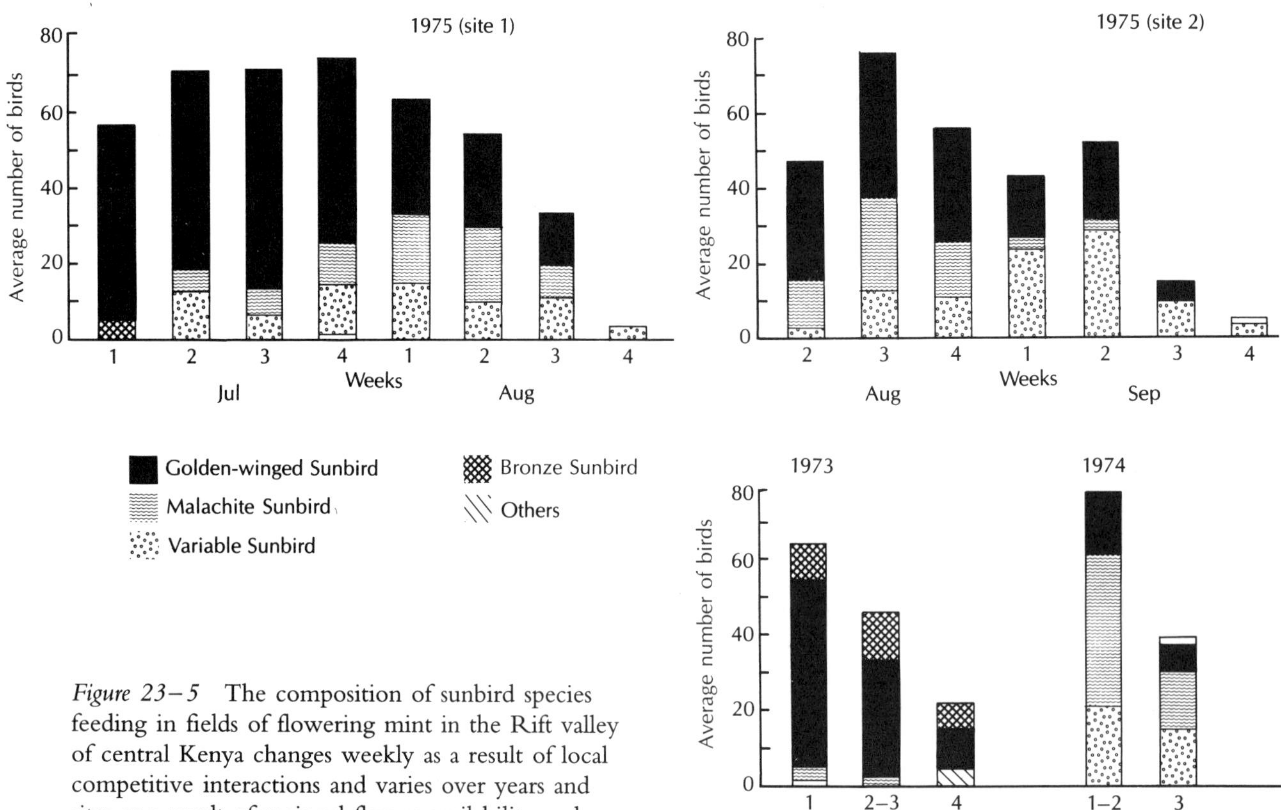

Figure 23–5 The composition of sunbird species feeding in fields of flowering mint in the Rift valley of central Kenya changes weekly as a result of local competitive interactions and varies over years and sites as a result of regional flower availability and colonization. (Adapted from Wolf and Gill 1980)

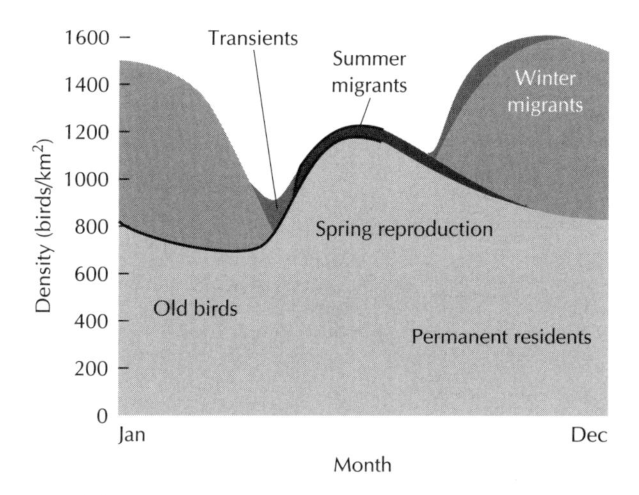

Figure 23–6 Model of seasonal composition of the pine forest bird community of Grand Bahama Island. Local numbers increase with the addition of young birds in the summer and also increase with the addition of wintering migrants, which leave in April. (Adapted from Emlen 1980)

Open versus Closed Communities

Evolution and resource availability both play major roles in the formation of communities. Species compositions change from epoch to epoch in accordance with evolution and from season to season in accordance with resource availability. Resource availability and evolution give rise to a third force: competition—the interactions among species through which one species may directly or indirectly preclude the presence of another. Competition can select those species best able to coexist from a large pool of potential candidates for a community (Figure 23–7).

The relative importance of these three forces in community ecology is a matter of continuing debate. Competition has long been assumed to be the main force in structuring bird communities, but irregular seasonal changes may keep communities in a state of dynamic flux. Habitat disturbance, fluctuating resources, colonization, and local extinctions impose short-term changes on avian communities. In addition, deforestation and rotation of farmlands have transformed woodland habitats and their avian communities into patchworks of chronically unstable and unsaturated habitats that invite opportunistic use by birds. Some communities, therefore, lack predictable structure and may simply be temporary and fortuitous collections of species.

Two polar views of communities derive from early botanical thought. According to one view, communities are dynamic, open systems in which each species aligns itself independently along environmental gradients according to its own ecological requirements (Gleason 1926, 1939; Wiens 1990). Thus, open communities are fortuitous assemblages of

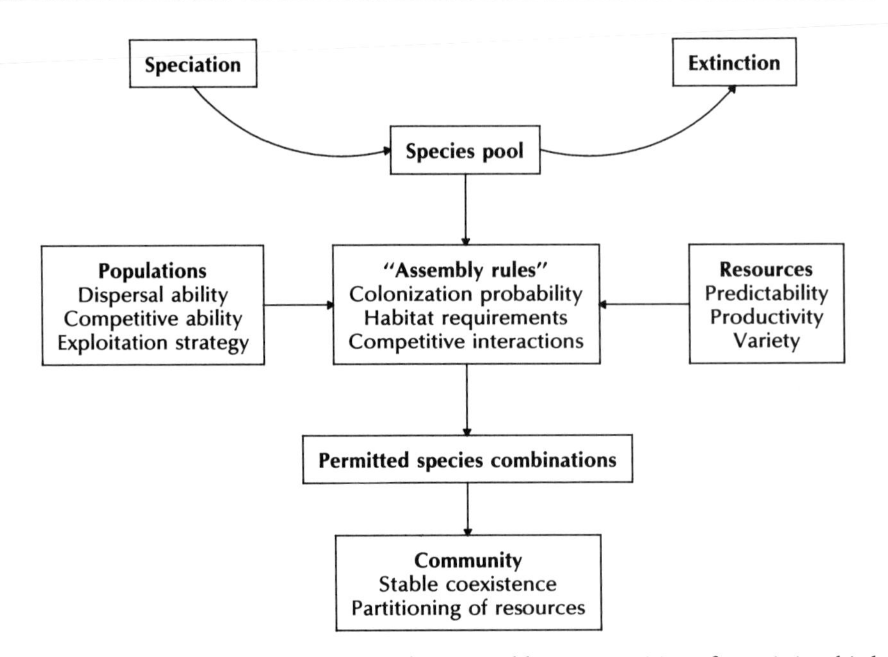

Figure 23–7 According to some ecologists, stable communities of coexisting bird species derive from a larger pool of species through the dynamics of population dispersal, through colonization in relation to habitat or other resources, and sometimes by competitive resolution of unstable species combinations. (Adapted from Wiens 1983)

noninteracting species. According to the other view, communities are closed, integrated sets of ecologically compatible species (Clements 1916, 1936; MacArthur 1972; Cody 1974). Closed communities include predictable sets of interacting species, whose distributions along an environmental gradient are determined by the presence or absence of competing species. These are extreme views, but evidence exists for both (Figure 23–8).

The birds that breed in upland hardwood stands in southern Wisconsin form open assemblages of species in habitats that range from open, dry, deciduous forests dominated by Black Oak trees to denser, moist forests dominated by Sugar Maple trees (Bond 1957). Certain bird species, such as the Red-eyed Vireo, are more common in wet-climate forests; others, such as the Black-capped Chickadee, are more common in dry-climate forests; and some, such as the American Redstart, are most common in forest types that are between these two. Each species has specific preferences or needs and chooses its habitat accordingly (Lack 1971; Lanyon 1981). These distributions suggest independent, ecologically related associations, not coincident relationships of coadapted species.

Few environments are stable. Series of unpredictable wet and dry years or severe and benign winters are the norm worldwide. Pronounced year-to-year variations are typical even of tropical rain forests, once thought to be the most stable of ecosystems. Fire, an extreme form of disturbance, naturally and periodically devastates chaparral communities in California—recall the fires of 1993—and the grasslands of both Africa and the western United States. Regrowth after a burn proceeds through regular patterns of plant succession and associated bird communities. Over the period of a year, localized burns and recovery create an ever-changing mosaic of unstable habitats. The communities that occupy them are dynamic rather than self-perpetuating systems at equilibrium. In the fire-controlled shrub-steppe bird communities of the Great Basin of western North America, birds respond opportunistically to abundant, nonlimiting resources. Interspecific competition does not seem to be an important force in these dynamic communities, whose component species may be limited by winter food availability rather than by summer food availability (Rotenberry and Wiens 1980; Dunning 1986).

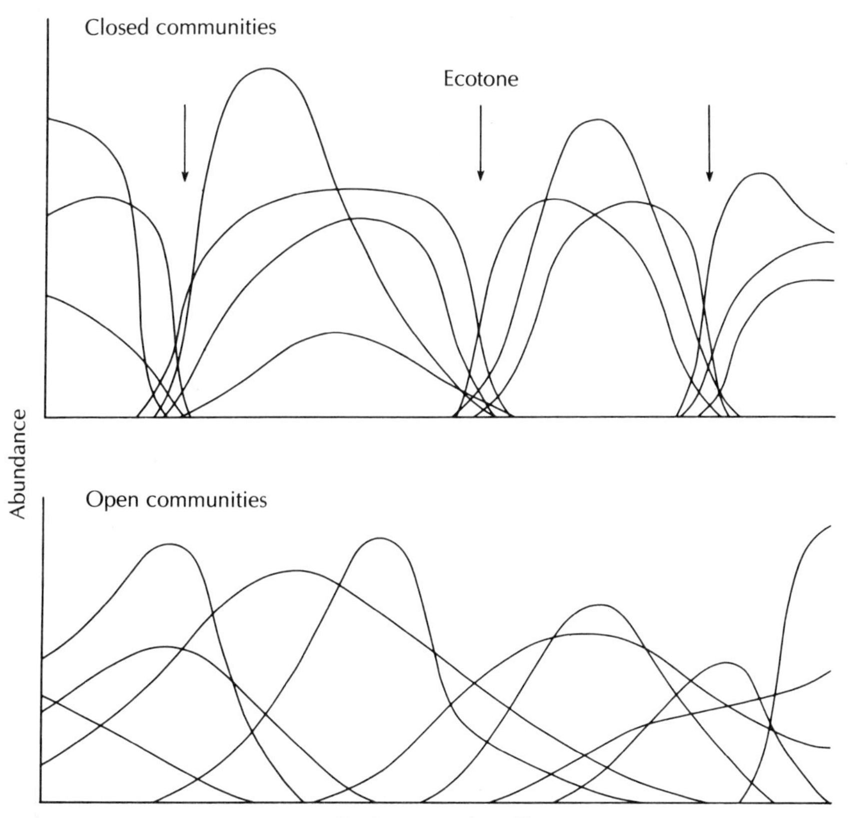

Figure 23–8 Open and closed communities are extreme forms of the continuum of possible community structures along environmental gradients, such as dry forest to wet forest. In open communities, species are arrayed independently in accord with their particular ecological needs; in closed communities, distinct sets of species occupy particular habitats with breaks at the interfaces between habitats—called ecotones (arrows). (From Ricklefs 1979a)

In comparison, nectar-feeding hummingbirds provide some of the best examples of closed communities based on competition for food (Wolf et al. 1976; Feinsinger and Colwell 1978). Coexisting hummingbirds organize themselves in predictable sets around one or two large, territorial species that control access to preferred flowers. In the highlands of Costa Rica, for example, the aggressive Fiery-throated Hummingbird and Green Violet-Ear control the clusters of flowers rich in nectar. Complementing such territorial species in each local set of species will be (1) a large, nonterritorial, long-billed species, which visits scattered flowers containing large nectar volumes; (2) a small, nonterritorial short-billed species, which visits small flowers containing small amounts of nectar; and (3) a small-sized, sharp-billed filcher, which steals nectar from undefended big flowers by piercing the base of the corolla.

Competition

The closed community concept dominates the modern study of bird communities (MacArthur 1971, 1972; Cody and Diamond 1975; Strong et al. 1984). In this view, stable combinations of species separate themselves from a pool of possible colonists and competitors and thereby become a community. Community efficiency, stability, and resistance to invasion by additional species increase with evolutionary adjustment among members of the community. Advocates of the closed community concept project competition among species for limited resources to be the single most important structuring force.

Competition occurs when use or defense of a resource by one individual reduces the availability of that resource to other individuals. Interspecific competition occurs when individuals of coexisting species require some of the same limited resources; the use or defense of those resources by individuals of one species reduces the availability of resources to individuals of another species. Recall that competition among individuals of one species reduces the rate of population growth in that species by limiting survival or reproduction. Competition among individuals of different species can have similar or more pronounced effects on one another's population growth.

The competitive exclusion principle—a fundamental concept of ecology that is also called Gause's law after G. F. Gause, a pioneering Russian ecologist—states that two species with identical ecological niches cannot coexist in the same environment. Laboratory results support this concept, because it has been observed that one species usually replaces another similar species when the two are forced to share the same environment in a laboratory.

Competition can be expressed as overt aggressive displacement of individuals—called interference competition—or as direct reduction of the fecundity and survival of one species by another. In the first case, large, dominant species of sunbirds and hummingbirds can exclude other species from the densest concentrations of flowers. Forced by dominant species to use other feeding grounds with fewer flowers, subordinate species quickly shift back to the best available feeding grounds whenever possible. Similarly, Golden-crowned Sparrows aggressively restrict juncos's use of foraging space near shrubs (Davis 1973). The juncos increase their use of sites closer to protective cover when Golden-crowned Sparrows are removed experimentally but revert to infrequent use when the sparrows return. Antbirds that gather at swarms of army ants exhibit similar behavior (see Chapter 14).

Rather than being manifested as overt aggression, most interspecific competition subtly depresses a species' survival or breeding success through reduction of critical resources—called exploitation competition. Some of the best evidence of the effects of one species on the fecundity, survival, and population recruitment of another comes from research on Great Tits and Blue Tits (Dhondt and Eyckerman 1980; Dhondt 1989). This research is an extension of the work on population regulation in the Great Tit, which is reviewed in Chapter 21.

Competition Between Great Tits and Blue Tits

Competitive interactions should be most intense within sets of ecologically similar species—called guilds—that are dependent on the same set of resources (Root 1967). Local assemblages of titmice compose such guilds, which have been the focus of intense research on the role of interspecific competition in bird communities. Reduction of the food supplies by tits can affect the reproductive success of species outside their guild, such as Collared Flycatchers in Sweden (Gustafsson 1987). However, the details of competitive interactions between Great Tits and Blue Tits have been of particular interest (Dhondt 1977, 1989). These two species negatively affect each other in a variety of ways, both trivial and consequential.

Recall that reproduction success—or fecundity—in Great Tits is sensitive to their population density, decreasing as population density increases. Reproduction of Blue Tits, however, is not density dependent, nor is it affected seriously by the local numbers of Great Tits, with one caveat—Great Tits may control nest boxes, if they are in short supply, and may even kill Blue Tits in the process (Löhrl 1977). This is an extreme form of interference competition.

In addition to being sensitive to the local densities of members of their own species, the fecundity of Great Tits is sensitive to the numbers of coexisting Blue Tits. High densities of Blue Tits during the breeding season reduce food availability and thereby reduce the reproductive output of Great Tits by increasing nestling mortality and by causing fewer Great Tits to attempt second broods. At least six different studies, four of which were experimental, corroborate this result. The effects of Blue Tits on fecundity of the Great Tit during the breeding season, however, are only temporary ones. Competition with Blue Tits does not cause a serious

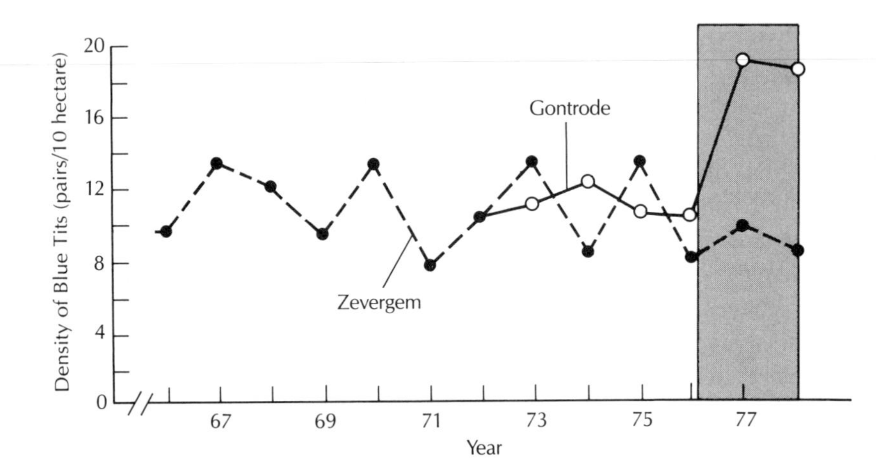

Figure 23–9 Experimental demonstration of interspecific competition. When Great Tits were excluded from nest boxes (from 1976 to 1978, screened area) in the experimental area at Gontrode, Belgium (white circles), more Blue Tits established themselves there than in a control area at Zevergem (black circles) that had the normal number of Great Tits. (Adapted from Dhondt and Eyckerman 1980)

reduction of the population density of Great Tits, which is controlled more severely by winter survival and recruitment of juveniles (see Chapter 21).

A reversal of competition between these two species outside the breeding season, however, is of serious consequence—Great Tits truly limit the number of Blue Tits in a woodlot by controlling the availability of roost holes. When the number of Great Tits in a population that depended on man-made boxes for roosting (and also for nesting) was halved (by narrowing the nest entrances from 32 to 26 millimeters and thereby excluding the larger Great Tits), many more juvenile male Blue Tits were recruited into the woodlot in the autumn and subsequently joined the breeding population in the following year (Figure 23–9).

Ecological Segregation

A corollary of the competitive exclusion principle is that competition should increase directly with overlap in the use of limited resources. Detrimental ecological overlap may foster the evolution of ecological differences that reduce competition. Observed ecological differences between related species, therefore, may be the "ghosts of competition past" (Connell 1980).

Local separation by habitat and feeding stations is typical of titmice (Table 23–1) (Lack 1971). In Europe, the Great Tit, Blue Tit, and Marsh Tit inhabit broadleaf forests. The Crested Tit and Coal Tit live primarily in coniferous forest used by the other three species only as a suboptimal habitat. The forest preferences of the Willow Tit vary geographically. The species that live together feed in different places: Great Tits on the

TABLE 23–1
Ecological segregation of titmice

Locality	Number of species	Number (percentage) of species pair combinations isolated by			
		Range[a]	No contact[b]	Habitat	Feeding station
Europe	9	3 (25)	6	11.5 (32)	15.5 (43)
North America	10	13 (85)	25	5 (11)	2 (4)
Africa	10	7 (67)	23	13 (29)	2 (4)
Asiatic mountains	14	2 (26)	22	51 (56)	16 (18)

a. Percentages in parentheses include No Contact data.
b. These data are included with species sharing the same range for the purpose of calculating percentages.
From Lack 1971.

ground, Marsh Tits on large branches, and Blue Tits on the smaller twigs. Differences among European titmice in their feeding locations are associated with adaptive differences in body mass and beak size. Larger species feed at a lower level and on larger insects and harder seeds than do the smaller species. Species that live in coniferous forests have longer and narrower beaks than those that live in broadleaf woods.

Each species of European tit has a counterpart in North America (Figure 23–10). However, only two of the North American species usually live together in the same habitat. In many areas, this may be a small chickadee that coexists with a large titmouse, which has different ecological requirements. Where two species of small chickadees occur together, they inhabit different habitats. In New England, the Boreal Chickadee inhabits dark conifer stands, whereas the Black-capped Chickadee inhabits more open, mixed deciduous and conifer forest. On the West Coast, Chestnut-backed Chickadees and Black-capped Chickadees segregate similarly by habitat.

If competition actually restricts a species, one would expect shifts in the distribution, habitat use, or foraging behavior of a species when it is not limited by a competitor. On the San Juan Islands of the Pacific Northwest, where there are no Black-capped Chickadees, the Chestnut-backed Chickadees inhabit broadleaf forests used elsewhere by the Black-capped Chickadees. Shifts in habitat use in the absence of other species are well documented among European tits (Lack 1971; Alatalo et al. 1986; Dhondt 1989). Marsh Tits inhabit pine plantations only in Denmark, where Willow Tits are absent from this habitat. In Ireland, Coal Tits feed regularly in the understory of evergreen forests in the absence of the Marsh Tits, Willow Tits, and Crested Tits that normally preempt this niche.

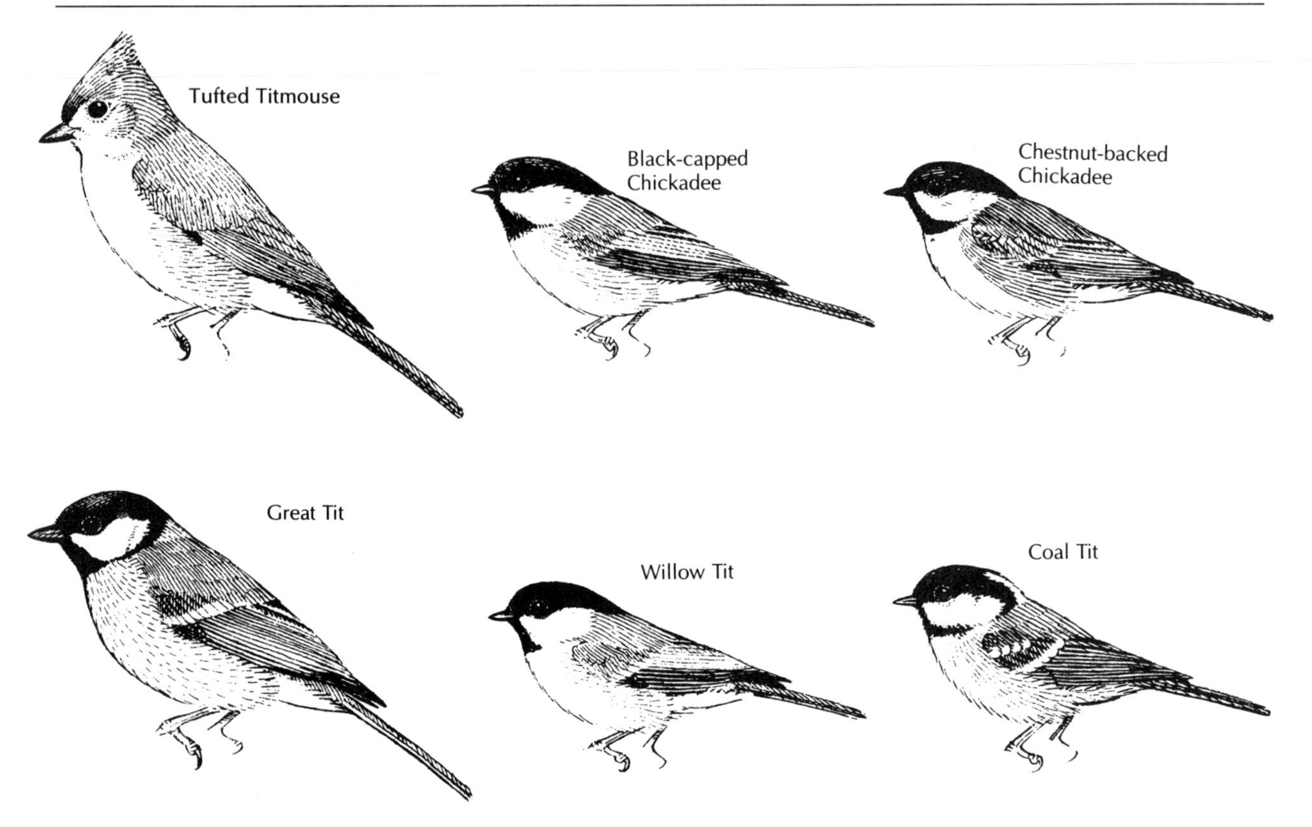

Figure 23–10 Certain species of North American (top) and European (bottom) chickadees and titmice act as ecological equivalents. (From Lack 1971)

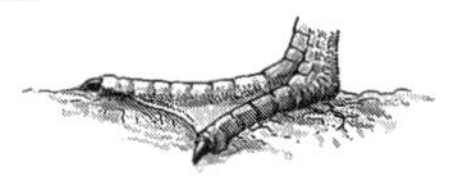

BOX 23–2

Competition affects use of foraging sites by tits

In the ornithological literature are many, often anecdotal, observations of apparent niche shifts in the absence of a competitor. Controlled experimental demonstrations using free-living birds in natural populations, however, are few. One exception is the study of foraging tits and Common Goldcrests in the coniferous forests of central Sweden (Alatalo et al. 1987). These small birds exploit nonrenewable insect and seed resources in their group territories during the long, cold winter. Two smaller and socially subordinate species, the Coal Tit and Common Goldcrest, forage on the outermost branches and needles, while two larger and dominant species, the Willow Tit and Crested Tit, forage inside the trees. In this experiment, the ornithologists removed the Coal Tits and Common Goldcrests from three of six flocks to test whether Willow Tits and Crested Tits would change their foraging behavior. They did. In late winter Crested Tits shifted farther out on the spruce branches in experimental flocks than in control flocks. Willow Tits did so in pine trees, but not in spruce trees. The Swedish team concluded that exploitation competition directly based on food depletion, without any interference, influences the use of foraging sites by tits that coexist in coniferous forests.

Character and Ecological Displacement

Simple ecological displacements as a result of competition should theoretically lead to evolutionary reinforcement in the form of morphological character displacement—or enhanced differences (in size, for example) where two species exist (Brown and Wilson 1956). On the Swedish island of Gotland in the absence of larger competitors—specifically, Crested Tits and Willow Tits—Coal Tits are larger than on the mainland (Alatalo et al. 1986). Their larger size on Gotland coincides with a shift in foraging niche from the outside of the tree and on needles, where small size is advantageous, to the inner parts of the pine trees, the converse of the experimental result described above (Box 23–2).

The Darwin's finches of the Galápagos archipelago provide a classic example of the apparent role of competitive exclusion and character displacement (Lack 1947a; Grant 1981; Schluter and Grant 1984). The adaptive radiation of these finches has promulgated species with a variety of bill sizes that relate directly to seed sizes (Abbott et al. 1977). Ground-finches and cactus-finches with distinctly different bill sizes inhabit every island. The differences in average bill size of coexisting species are consistent with the hypothesis of interspecific competition for food; species with similar-sized bills replace one another on various islands, and the bills of various species are more alike when they do not live together (Figure 23–11).

Ornithologists frequently discover that the ranges of closely related species are mutually exclusive. Each island in the West Indies, for example, is inhabited by only 2 or 3 hummingbird species although 15 hummingbird species inhabit the region (Lack 1971) (Figure 23–12). Only two resident species, a small one and a large one, inhabit low-lying islands. Mountainous islands are populated by three types of hummingbirds, a small, widespread species and two large ones, one species in the lowlands and another in the highlands. Competition can be inferred to be the process historically responsible for such replacement patterns of ecologically similar species.

Abrupt replacement of one species by another at various altitudes in the Andes and in New Guinea also suggests that competition from one species limits the distribution of another. New Guinea birds with well-defined altitudinal distributions replace each other abruptly at various elevations (Diamond 1973, 1975). The range of elevations occupied by a species seems to depend on the presence or absence of related species. For example, the Red-flanked Lorikeet is confined to low elevations in regions with one of the two highland species, either the Red-fronted Lorikeet or the Red-chinned Lorikeet. In regions that the Red-flanked Lorikeet inhabits alone, however, it ascends to high elevations. Conversely, in regions where either of the Red-fronted or Red-chinned Lorikeets occurs alone, it descends to sea level (Figure 23–13).

Similar patterns of altitudinal replacement are found in the mountains of eastern North America, where up to five species of thrushes nest in the high mountain forests of New England (Noon 1981). Veeries share low elevations with the Wood Thrush and are replaced at higher elevations by the Swainson's Thrush and the Bicknell's Thrush. In the Great Smoky mountains, where the Swainson's Thrush and the Bicknell's Thrush are absent, the Veery shifts to higher elevations and overlaps only slightly with the Wood Thrush at low elevations.

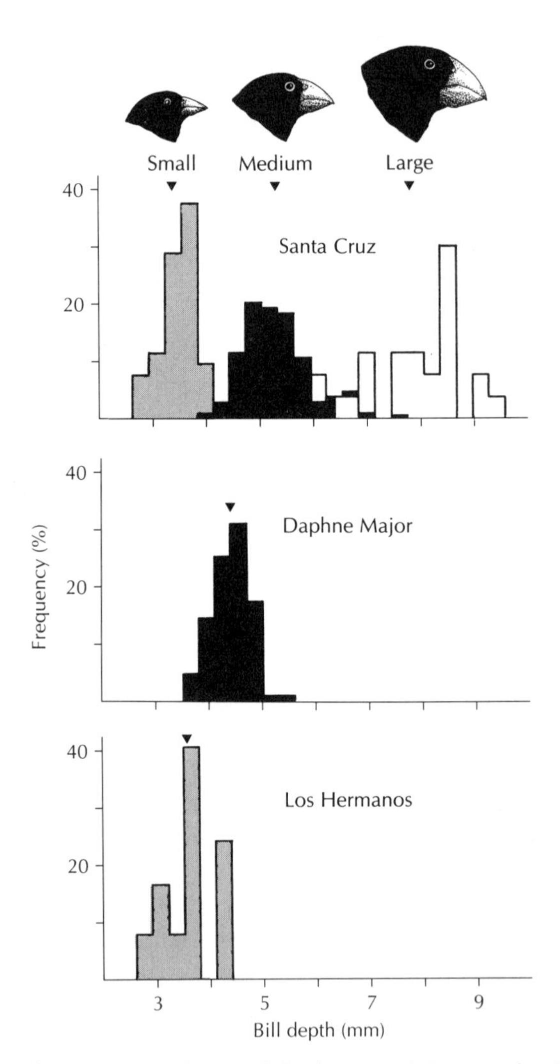

Figure 23–11 Three species of ground-finches coexisting on the Galápagos island of Santa Cruz have bills of different depths, which enable them to feed on different seeds. Only one species inhabits certain islands, such as Daphne Major and Los Hermanos. In the absence of other species, such solo populations evolve intermediate-sized bills. (Adapted from Grant 1986)

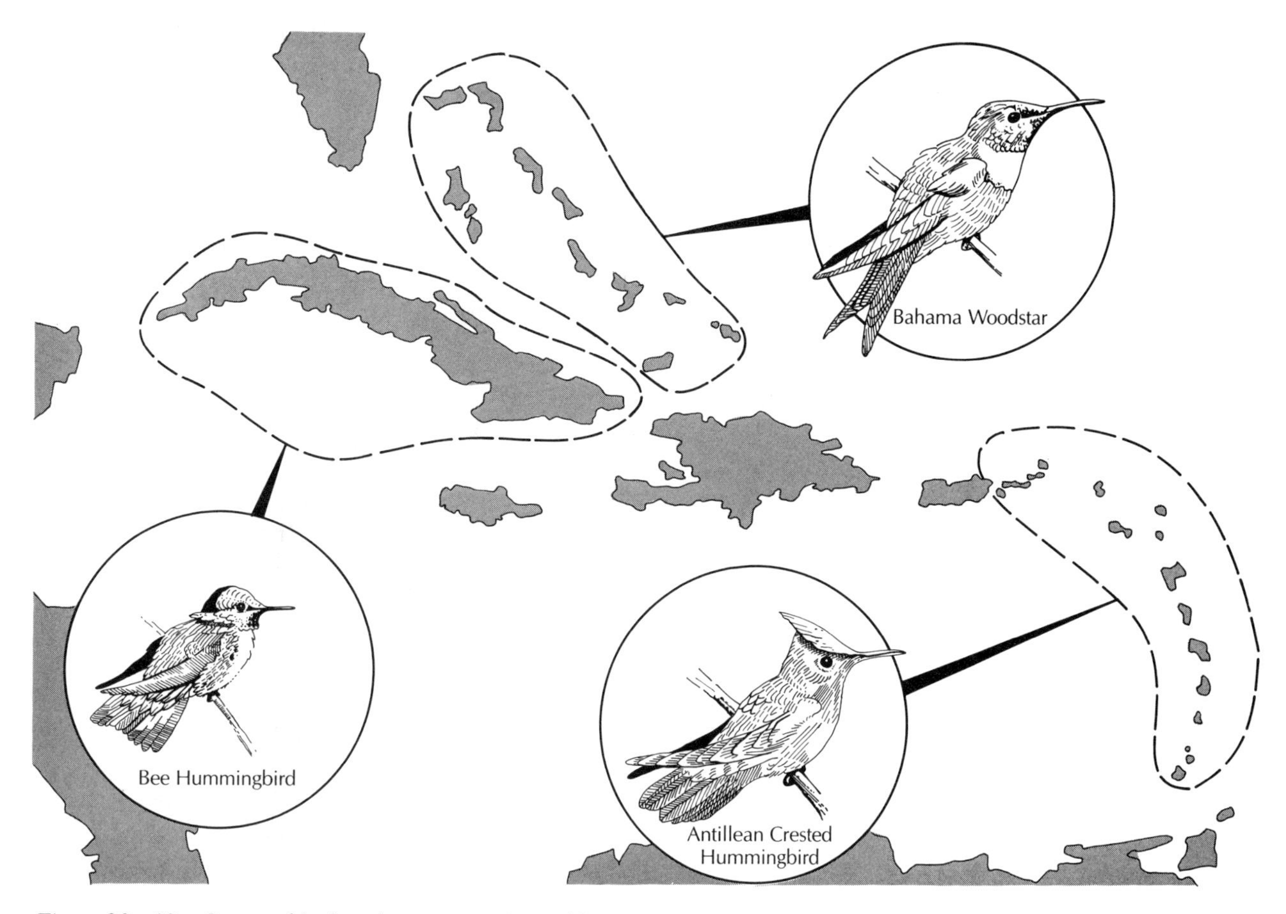

Figure 23–12 Geographical replacements of possible competitors. Different species of small hummingbirds do not coexist in the West Indies but have segregated distributions. (Adapted from Lack 1971)

The conclusion from such studies that competition limits altitudinal distributions is based on the unlikely assumption that the various locations are identical except for the presence or absence of the purported competitors (Wiens 1983). Thus, patterns of ecological isolation or replacement provide only weak evidence of structuring communities by competition (Newton 1980). At best, the process of competition must be inferred from an observed pattern. What may appear to be the direct, competitive replacement of older endemic species on islands by new colonists may only be coincidence. For example, the endemic Socorro Dove was replaced on Socorro Island, off Baja California, by Mourning Doves in the 1970s. It turns out that the extinction of the Socorro Dove and the colonization by the Mourning Dove were independent events related to the establishment of a garrison of the Mexican army, whose cats ate the tame endemic doves and whose new wells provided the drinking water required by the colonizing Mourning Doves (Jehl and Parkes 1983).

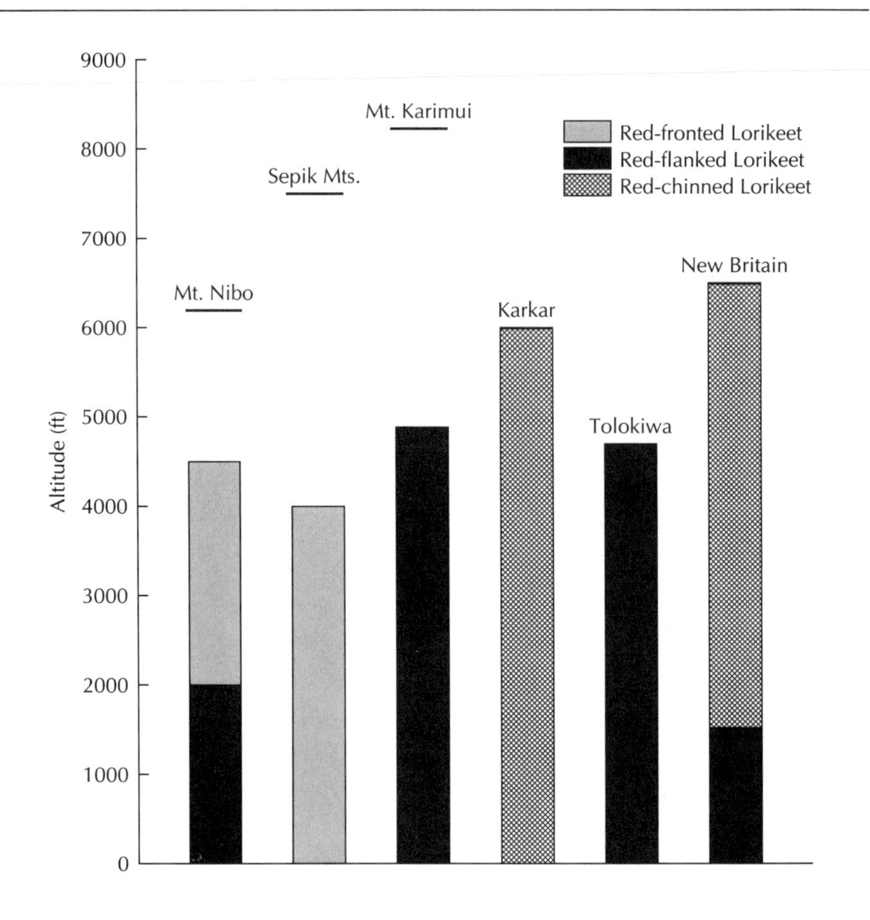

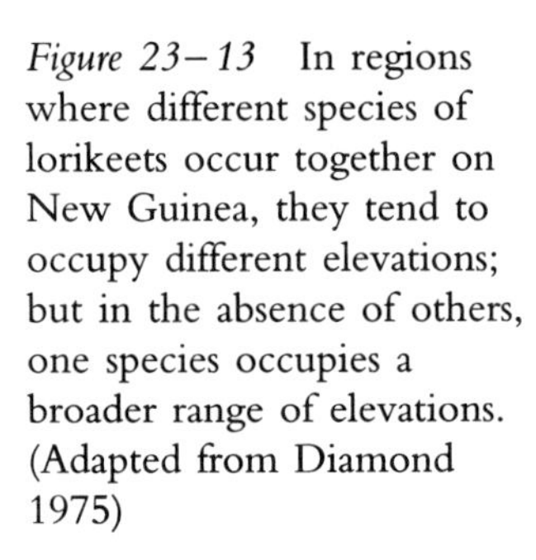
Figure 23–13 In regions where different species of lorikeets occur together on New Guinea, they tend to occupy different elevations; but in the absence of others, one species occupies a broader range of elevations. (Adapted from Diamond 1975)

The Dynamics of Local Avifaunas

Local communities are subsets of larger avifaunas of a particular geographical area. Reflecting regional history, avifaunas are the grand result of millions of years of evolution, adaptive radiation, immigration, dispersal, and extinction of bird taxa. Invasions and fusions of species from other regions supplement adaptive radiations of ecological types within a region. Turnover—the addition and loss of species—causes the compositions of avifaunas to change. The time scales range from the long-term, gradual replacements of ancient taxa by more recently evolved taxa to short-term, yearly changes. An equilibrium in the number of species may be established where new additions through colonization or speciation balance extinctions.

Colonization of Islands

Active dispersal and colonization of isolated places are trademarks of bird behavior. The dynamics of colonization are most apparent on oceanic islands, such as the West Indies, which receive periodic arrivals of new visitors dispersing over water from larger source areas. Water barriers favor colonization by highly mobile species that travel in small groups. In the West Indies, Bananaquits, and in the Indian and Pacific Oceans, white-eyes are superb island colonists—or "supertramps" (Diamond 1974). Their high reproductive potential and extraordinary dispersal abilities enable them to be the most predictable first colonists of newly formed islands. Successful colonization of one island may be followed by colonization of adjacent islands and continued spread throughout a region.

A colonist's ecological flexibility and its ability to fit into the local community determine its chances of establishing a population on a new island. Bananaquits and white-eyes, for example, are generalized opportunists, able to take advantage of local situations. They breed readily and repeatedly. Once established, their populations tend to thrive and grow rapidly in an environment with few specialized predators, competitors, diseases, or parasites. Population growth under such conditions of ecological release leads to large, dense populations and to the use of a wider variety of habitats than is the case on the mainland. Resident birds of the Pearl islands off western Panama, for example, achieve densities 20 to 40 percent higher than those reached on the adjacent mainland. They also forage over a greater vertical range and use more habitats than do their mainland counterparts (MacArthur et al. 1972). Comparison of the birds on small Caribbean islands such as St. Lucia and St. Kitts with the three- to sixfold larger species assemblages on Trinidad or mainland Panama

Box 23-3
Character displacement evolved rapidly in supertramps

Between the large, exotic islands of New Guinea and New Britain in the South Pacific lie several small islands, the Long group. On these little-known islands, Jared Diamond and his colleagues (1989) discovered an example of rapid evolution of character displacement between two species of small, nectar-feeding honeyeaters (genus *Myzomela*). The Long group's volcanic caldera collapsed 300 years ago in one of the largest known volcanic explosions, which undoubtedly eliminated all birdlife. The current avifauna consists mostly of small island specialists, or supertramps, that have recolonized the island. Among the now established supertramps are two similar *Myzomela* honeyeaters, the Ebony Myzomela from the northern Bismarck archipelago and Scarlet-bibbed Myzomela from the southern Bismarck archipelago. The two honeyeaters occur abundantly together all over the Long islands and often feed in the same flowering trees.

These are the only two bird populations on the Long islands that have evolved to different sizes since colonization. The Ebony Myzomela is now larger than its ancestral populations, and the Scarlet-bibbed Myzomela is now smaller. Their size difference has increased from 24–43 percent on the Bismarck archipelago to 52 percent on the Long islands. Elsewhere, closely related species of honeyeaters coexist only if they differ in size by 50 percent or more. Diamond and his colleagues conclude that the size shifts on the Long islands represent character displacement in response to each other's presence in the few centuries since Long's devastating eruption.

reveals conspicuous increases in the relative abundance per species of the island birds (Cox and Ricklefs 1977) (Table 23–2). Both the average number of habitats used by a species and the density of each species in a particular habitat may double on an island.

Populations on different islands diverge from one another as they adapt to unfilled local niches. Generalized colonists, such as white-eyes, may take over the specialized niches of species that are missing from island communities. The increased specialization of the first colonists then contributes to their ability, and to that of their descendants, to coexist with later arrivals. Unusually large species of white-eyes have evolved independently on 12 small Pacific islands that have few other species, and these large white-eyes often coexist with one or more other white-eyes (Lack 1971). Ponape and Belau (formerly Palau) both have an extremely large species of white-eye (Long-billed White-eye or Giant White-eye), the medium-sized Gray White-eye, and the small Bridled White-eye (Figure 23–14). On Réunion Island in the Indian Ocean, where there is no nectar-feeding sunbird, the Réunion Olive White-eye has evolved into a specialized nectar-feeding species in both bill morphology and behavior (Gill 1971). It coexists there with a second, generalized white-eye, the Mascarene Gray White-eye (see Chapter 22).

With time, established residents may become further restricted to specialized habitats because of competition with aggressive new colonists or changes in habitat distribution attributable to climatic shifts (Pregill and Olson 1981). Local extinctions then fragment the distribution of the species. Reproductive success and population size tend to decline as a result of increased susceptibility to parasites and predators. Declining population sizes and concomitant reductions in genetic variability increase the prob-

TABLE 23–2

Relative abundance and habitat distribution of birds in five tropical localities[a]

Locality	Number of species observed (regional diversity)	Average number of species per habitat (local diversity)	Habitats per species	Relative abundance per species per habitat[b] (density)	Relative abundance per species[b]	Relative abundance of all species[b]
Panama	135	30.2	2.01	2.95	5.93	800
Trinidad	108	28.2	2.35	3.31	7.78	840
Jamaica	56	21.4	3.43	4.97	17.05	955
St. Lucia	33	15.2	4.15	5.77	23.95	790
St. Kitts	20	11.9	5.35	5.88	31.45	629

a. Based on 10 counting periods in each of 9 habitats in each locality.

b. The relative abundance of each species in each habitat is the number of counting periods in which the species was seen (maximum 10); this, times number of habitats, gives relative abundance per species; this, times number of species, gives relative abundance of all species together.

From Cox and Ricklefs 1977.

Figure 23–14 White-eyes are excellent island colonists that often occupy unfilled niches on remote islands that have few other birds. Shown here are (A) a typical species, the Bridled White-eye, and (B) a large, thrushlike species, the Giant White-eye, which are found together on Belau (Palau) in the Caroline islands. (From Lack 1971)

ability of extinction of some populations. The House Wren, for example, has disappeared for no apparent reason from Guadeloupe and Martinique and is close to extinction on St. Vincent. Individual taxa seem to progress through taxon cycles, the three phases of which—colonization, differentiation, and local extinction—can be likened to the life cycles of individuals—youth, maturity, and senescence (Ricklefs and Cox 1972).

Equilibrium Theory

Although island avifaunas may change identity, the number of species present reflects a balance between losses due to extinction and gains due to immigration (MacArthur and Wilson 1967). The resulting number of species present is the equilibrium species number, which is the point of intersection between the extinction curve and the immigration curve (Figure 23–15). The rate of extinction increases with the number of species on an island because of the increased number of candidates for extinction, promoted by competition-reduced population sizes that are intrinsically prone to extinction. The rate of immigration, on the other hand, falls as the number of species on the island increases, for two reasons: additions of new species from source areas become fewer, and it becomes more difficult for a new species to colonize an island on which resources have been preempted by earlier immigrants.

Different rates of extinction and colonization are found on islands of different sizes and degrees of isolation. Colonization of islands close to the mainland occurs more readily than colonization of remote islands. The probability of extinction is greater on small islands with small populations of resident bird species than on large islands with large populations of many resident bird species. Thus, small, isolated islands have the lowest number of equilibrium species because of infrequent immigration and high extinction levels. Large islands near other landmasses (mainland or another large island) have frequent immigration and infrequent extinction and, therefore, large numbers of equilibrium species. The number of equilibrium species for large land-bridge islands, which were once part of a mainland with a full complement of species, is much greater (often three times) than that for large oceanic islands. Oceanic islands do not begin with a full complement of mainland species and depend solely on colonists that cross the seas. Whether a true equilibrium is established as a result of the balance between immigration and extinction (as envisioned by Robert MacArthur and Edward Wilson) is not yet certain (Lynch and Johnson 1974; Abbott and Grant 1976; Simberloff 1976). Nevertheless, the observed relations between the number of species and island size seem

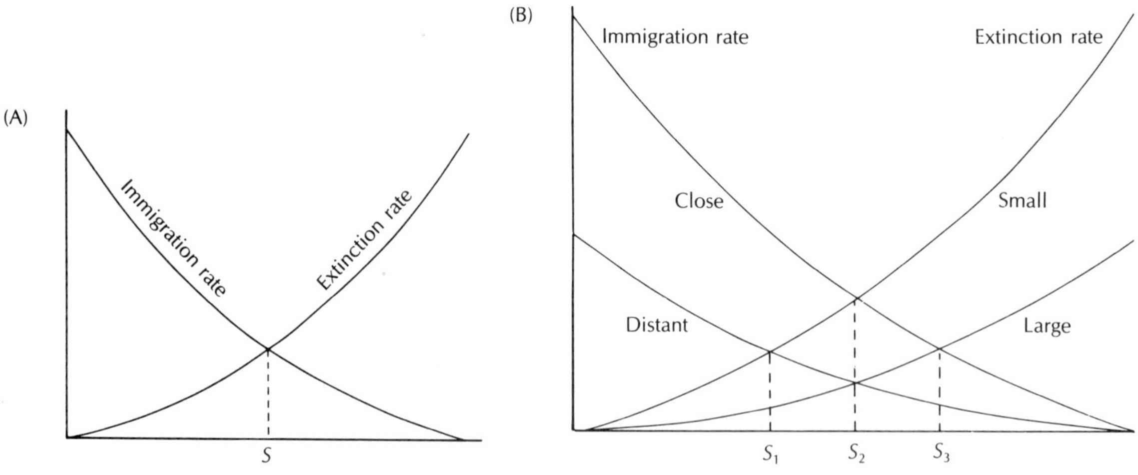

Figure 23–15 (A) The number of species found on an island reflects the balance between the rates of immigration (colonization) and extinction. (B) Immigration rates on islands that are distant from source areas are lower than rates on islands close to source areas. Extinction rates on large islands are lower than rates on small islands. Extinction rates increase as the number of species present on an island increases. The point of intersection of the two curves for any particular island defines the expected equilibrium number of species (S). (From Ricklefs 1976a)

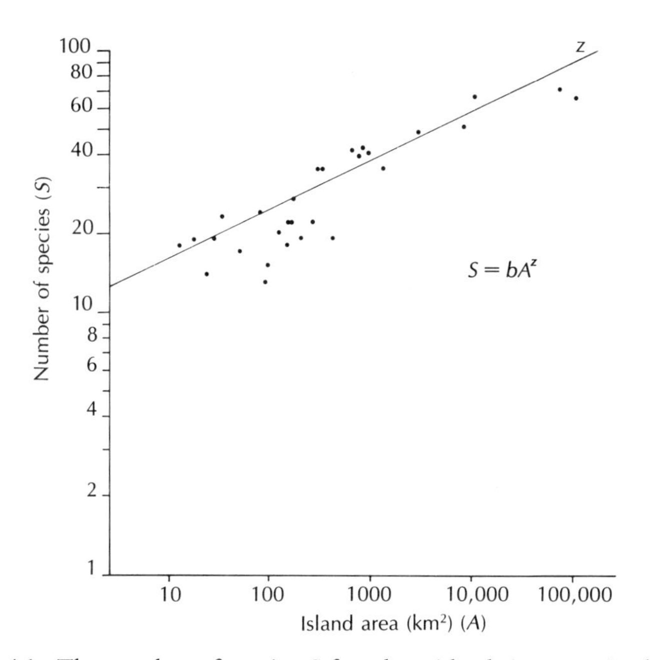

Figure 23–16 The number of species S found on islands increases in direct relation to island area A, according to the standard equation for a simple line, $S = bA^2$, where b is the intercept of the line with the ordinate. The slope z of this relationship represents the degree to which extinction rate varies with island size, assuming that immigration is constant; z is 0.22 on this graph plotted for the West Indies. (From Ricklefs and Cox 1972)

to be in general accord with the model (Figure 23–16), but they do not take into account massive prehistoric extinctions caused, for example, by Polynesian colonists on the Hawaiian islands and elsewhere in the South Pacific (Olson and James 1982).

Species Turnover

The turnover of species is a fundamental concept of geographical ecology. Species are subject to evolutionary change, multiplication through speciation, and extinction. Both island and mainland avifaunas are subject to conspicuous species turnover. Turnover rates on small, offshore islands, such as the Channel islands off the coast of California, are roughly 1 to 20 percent a year (Diamond 1980). Annual censuses of birds on a small British island, Calf of Man, revealed that Northern Wheatears and Common Stonechats repeatedly became extinct on the island, only to be replaced within a short time by new immigrants (Figure 23–17). Thus, a high turnover will not be detected if an island is censused infrequently, because regular immigrations may obscure regular extinctions. Jared Diamond (1969, 1980) estimated that censuses spaced more than a decade apart will substantially underestimate turnover.

The number of species present in northern Scandinavian bird communities varies by 15 percent from year to year (Järvinen 1980). Farther south, in the rich deciduous forests of Birdsong Valley, Sweden, annual turnover is about 10 percent. In both regions, repeated extinction and immigration of small numbers of rare species (one to two pairs total) are the primary causes of annual variations in community richness. Higher turnover in the north is caused by the greater unpredictability of weather during the breeding season and the greater proportion of rare species, which are more likely to become locally extinct.

Extinction can result simply from the inevitable fluctuations in the size of small populations. The probability of extinction, therefore, depends on population size and on the area containing the population. Small islands lose species more frequently than do large ones, an effect that is seen most clearly in the analysis of land-bridge islands such as Trinidad. Land-bridge islands have lost species steadily since they were isolated by rising sea levels at the end of the Pleistocene epoch (10,000 years ago), and small land-bridge islands have lost a greater proportion of their initial populations of birds than have large land-bridge islands of comparable age (Table 23–3).

Islands of a particular habitat, which also lose species because of their isolation, may be natural, such as bog and mountaintop forest habitats, or they may be the result of man's agricultural activities, including forest fragments surrounded by crops or housing developments. As on land-bridge islands, the loss of species from these habitat islands decreases with area. When isolated by the growth of sugarcane and coffee plantations in the last century, subtropical woodlots in southern Brazil supported about 220 bird species (Willis 1980). Today a large, isolated woodlot (1400 hectares) continues to support 202 species, whereas a medium-sized woodlot (250 hectares) and a small woodlot (21 hectares) have only 146

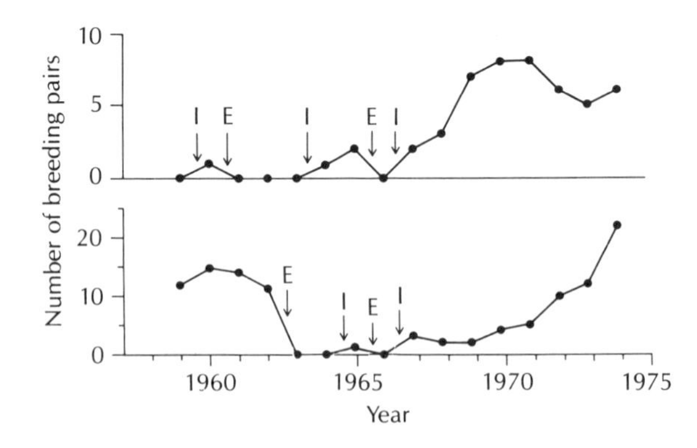

Figure 23–17 Turnover of Northern Wheatears (top) and Common Stonechats (bottom) as indicated by annual breeding bird censuses on the Calf of Man, a small British island. Regular extinctions and replacements of small populations take place. I, immigration; E, extinction. (From Diamond 1980)

TABLE 23–3
Present and probable past land bird faunas of five major land-bridge islands

Island	Area (km^2)	Number of species			Extinct species (%)
		Original	Present	Extinct	
Fernando Po	2,036	360	128	232	84
Trinidad	4,834	350	220	130	37
Hainan	33,710	198	123	75	38
Ceylon	65,688	239	171	68	28
Tasmania	67,978	180	88	92	51

From Terborgh and Winter 1980.

species and 93 species, respectively. The birds lost from the largest plots were mostly large species found in low densities, such as eagles, macaws, parrots, toucans, tinamous, a wood-quail, a pigeon, and a fruit-crow. The birds most likely to disappear from the small woodlots were primarily large, canopy, fruit-eating birds and large, ground, insect-eating birds. The birds lost from the small woodlot also included many microhabitat specialists, which were replaced by birds that thrived in edge and second growth.

Extinction is inevitable, but mushrooming human populations have hastened the process. Now too apparent to ignore is the unprecedented scale of habitat change due to human activities and the potential loss of global biodiversity. The international mandate for active conservation and wise stewardship of birds is the topic of the next and final chapter of this text.

Summary

The availability of resources, such as food and nest holes, determines not only the local population density of a given species but also the number and kinds of different species that can coexist in a given habitat. Two major theories attempt to explain the structures of avian communities. According to one theory, communities are open systems in which each species arrays itself independently along environmental gradients according to its specific ecological requirements. According to the other theory, communities are closed, discrete, and integrated sets of mutually compatible species.

The number of species found locally increases from the Arctic to the Tropics. The local availability of key resources determines which species can live in a particular habitat. Habitat heterogeneity contributes to the species diversity of a region, especially so in the Tropics and in mountainous regions. Interactions between residents and migrants in seasonally variable environments also affect species diversity.

Competition is the key structuring force in the closed-community concept that dominates modern ornithology. Theoretically, the degree to which species compete should relate directly to overlapping use of shared resources. This corollary to the competition theory has been illustrated by tit populations. Another corollary of the competition theory is that ecological displacements should lead to evolutionary reinforcement in the form of morphological character displacements. The differing bill sizes of Darwin's finches on the Galápagos islands are a classic example of character displacement. Competitive exclusion also is suggested by patterns of geographical replacement in species too similar to coexist. Replacement patterns are particularly well documented on the isolated islands of the West Indies and at different altitudes in the mountains of New Guinea and North America.

Turnover—the addition and loss of species—drives the changing compositions of avifaunas. The composition of island avifaunas reflects not only ancient history but an ongoing cycle of colonization and extinction, the frequency of which depends on the isolation and the size of the island. Small, isolated islands have the smallest equilibrium number of species, whereas large islands near continental source areas have the highest equilibrium number of species. Increasingly, mainland forests are reduced to small, island fragments subject to loss of species.

Further Readings

MacArthur, R.H. 1972. Geographical Ecology. New York: Harper & Row. *An elegant summary of the classic viewpoint.*

Martin, T.E. 1986. Competition in breeding birds: On the importance of considering processes at the level of the individual. Current Ornithology 4: 181–210. *A call for closer study of the competitive interactions among individuals, as the basis for larger scale patterns.*

Ricklefs, R.E., and D. Schluter. 1994. Species Diversity and Communities. Chicago: University of Chicago Press. *A rich collection of contemporary papers that integrate history, biogeography, and community ecology from a global perspective.*

Strong, D.R., Jr., D. Simberloff, L.G. Abele, and A.B. Thistle. 1984. Ecological Communities: Conceptual Issues and the Evidence. Princeton, N.J.: Princeton University Press. *A collection of advanced papers embracing polar views of controversies in community ecology.*

Wiens, J.A. 1990. The Ecology of Bird Communities. Vols. 1 and 2. New York: Cambridge University Press. *A detailed and critical synthesis of avian community ecology.*

CHAPTER 24

Conservation of Endangered Species

MAJOR EPISODES OF EXTINCTION have punctuated the long history of life on Earth, and the Class Aves did not escape these traumas. In one prolonged episode at the beginning of the Pleistocene epoch about 3 million years ago, climatic changes caused the extinction of at least 25 percent of the existing bird species (Brodkorb 1960). Past climatic changes also were responsible for habitat changes that affected the distributions and viabilities of bird populations. In the last century and a half, however, human beings have taken over as the primary force affecting the natural world. Birds now face a new episode of species extinctions as a result of a rapidly accelerating transformation of their habitats.

Terminal extinctions—the disappearances of lineages—may reflect the natural processes of biotic change, such as those during the Pleistocene. However, the numerous terminal extinctions of species due to human exploitation of the Earth's resources have sounded the international conservation alarms. Destruction of habitat, excessive human predation, introduced species, and other factors such as toxic chemicals and natural, uncontrollable events combine to threaten almost 1000 species worldwide (Temple 1986; Collar and Andrew 1988).

The plights of several majestic species—Whooping Cranes, California Condors, Ivory-billed Woodpeckers in North America, to mention a few—have been widely publicized. Such charismatic species are in serious trouble, and some may be past the point of no return. Ornithologists also recognize danger signs in many groups of birds worldwide: Neotropical migrants and grassland species, migrant shorebirds, the seabirds of the Southern Hemisphere, and rain forest species with restricted ranges. Many of the world's rarest birds persist only in small numbers on remote islands. On larger landmasses, just 10 countries support over 400 species of threatened birds with highly restricted geographical ranges (Figure 24–1).

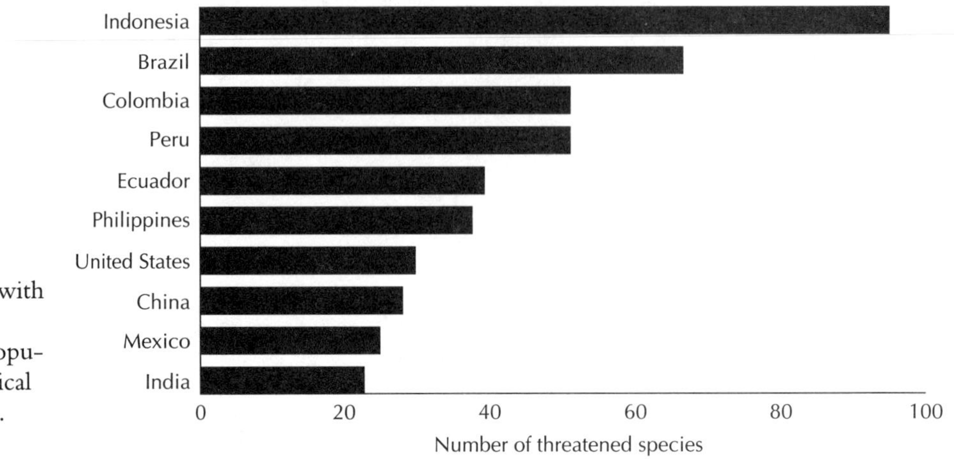

Figure 24–1 Countries with the highest numbers of threatened bird species populating restricted geographical ranges. (After Bibby et al. 1992)

Responding to such alarming statistics is the new field of conservation biology, which wrestles with the theoretical aspects of managing endangered species: defining critical population sizes, minimal habitat requirements, and the role of corridors between local sanctuaries (Soulé 1986). Conservation biologists also address the economics of species protection: How much does it cost to save one species? What will be the return to the local community of people concerned about their own survival? Belatedly, conservationists have recognized that the time to save species is while they are still common and that the wise management of healthy bird populations will avoid future costly, and politically volatile, rescue efforts (Senner 1988).

This final chapter brings together many of the elements of avian biology in a review of bird conservation efforts—past, present, and future. A discussion of the causes of avian extinction introduces the threats faced by birds, past and present, and some of the responses to these threats. Then the chapter shifts to the rise of conservation efforts in North America, including international efforts to save the Whooping Crane. Among the conservation success stories are innovative programs that promote protection of endangered species through local pride and that restore viable populations of species such as the California Condor and Peregrine Falcon. The growth of recreational birdwatching adds economic and political leverage to conservation programs and also provides volunteer manpower for the gathering of data needed to monitor bird populations and to manage them successfully.

Causes of Extinction

Populations of the most common birds number in the hundreds of millions of individuals, but those numbers do not guarantee continued existence. The estimated 2 billion Passenger Pigeons that flew over colonial

America were legendary (Schorger 1955), but during the late nineteenth century they declined rapidly into extinction (Figure 24–2). Advancing European colonists cut down the beech forests that provided abundant food for the pigeons and simultaneously slaughtered the pigeons themselves for human food (Blockstein and Tordoff 1985; Bucher 1992).

A host of other species that once existed in large numbers now hovers at the brink of extinction. Conservation biologists assign bird species in

Figure 24–2 Passenger Pigeon, a North American species that was abundant in the nineteenth century. Uncontrolled commercial hunting caused the extinction of this species. (Courtesy of S. Holt/VIREO)

trouble to one or two categories—endangered or threatened—as defined by the Endangered Species Act of 1973, As Amended. An endangered species is one that is in danger of extinction throughout all or a significant portion of its range. A threatened species is one that is likely to become an endangered species in the foreseeable future throughout all or a significant portion of its range.

Populations are likely to be listed as endangered or threatened when they decline to small and vulnerable sizes. Some short-term population declines simply reflect an unfavorable balance between reproduction and mortality (Chapter 21), but loss or fragmentation of specialized habitats, excessive hunting, and repeated nesting failures due to pesticides or predators can cause long-term and fatal population declines of bird species. Usually more than one such factor is responsible for fatal declines. Of 92 extinctions recorded from 1600 to 1980, 91 percent were due partly to introduced species (especially on islands), 32 percent were due partly to habitat changes, and 25 percent involved excessive human predation (King 1980; Temple 1986).

Of course, not all species suffer from human influences; some adapt well to man-made environments. Canada Geese are thriving along polluted streams and on golf courses. Ospreys nest on telephone poles, power lines, and channel buoys (Figure 24–3). American Crows, known for 200 years as shy, rural birds in the United States, are invading suburban backyards and city parks, as House Crows and Common Mynas did

Figure 24–3 Osprey, a species benefitting from human structures that provide safe nest sites. (Courtesy of C.H. Greenewalt/VIREO)

in Asia centuries ago. Equally impressive is the growing dominance of introduced species—called exotic species—throughout the world (Long 1981). Exotic species are the most common birds on Hawaii and Puerto Rico; at least 89 exotic species have been introduced into New Zealand. Rock Doves, European Starlings, and House Sparrows are human associates with prospects of widespread success. A redistribution of birds able to coexist with human societies is underway. The future of most species, however, will depend on the extent of the natural habitat that we can retain for them.

Small Island Effect

Losses of island birds accounted for 90 percent of bird extinctions during historical times (Moors 1985). Island-bound flightless birds and nesting seabirds are particularly vulnerable to rats, hogs, goats, and cats brought to once-safe islands by ships and sailors. The extermination of the Dodo and other flightless birds on the Mascarene islands in the late 1600s is a classic example of the eradication of vulnerable island birds (Box 24–1). Like the Dodo, the Great Auk of the North Atlantic was slaughtered to

Box 24–1
The Dodo is a symbol of extinction

The legendary Dodo is a symbol for the process of the extinction of vulnerable bird species by human beings (Figure A–15). Not just a whimsical character in Lewis Carroll's *Alice in Wonderland*, the Dodo was a real bird that once lived on the remote tropical island of Mauritius, one of three Mascarene islands in the western Indian Ocean. The Dodo was a large, flightless, turkey-sized pigeon, assigned to the Family Raphidae. Cohabiting the Mascarene islands with the Dodo was an amazing array of flightless pigeons, rails, parrots, waterfowl, and other birds, almost of all of which were exterminated in the seventeenth century (Hachisuka 1953). The Dodo ate fruit, became extremely fat, and was easily captured (hence the use of the name *Dodo* to indicate stupidity). It was prized as a readily available source of food. In the early 1600s, a few living Dodos were sent to Europe, where they captured public interest as a great curiosity. Few survived to the middle of the century, however. The last eyewitness account of wild Dodos comes from the journal of Volquard Iversen, who was shipwrecked and stranded on Mauritius for 5 days in 1662 before being rescued (see Cheke 1987). He found no Dodos on the mainland but discovered some on a small islet accessible by foot at low tide. Iversen's brief account was this:

> Amongst other birds were those which men in the Indies call *doddaerssen*; they were larger than geese but not able to fly. Instead of wings they had small flaps; but they could run very fast. (Cheke 1987, p. 38)

Perhaps the last Dodos learned to fear human hunters. But they did not run fast enough. Only fossils and a few preserved specimens remain as evidence of this odd species.

extinction (by 1840). Valued as food to resupply ships that had crossed the Atlantic, these flightless birds were easy to catch and kill. One enterprising crew even built a bridge of sail canvas from shore to ship and herded the helpless (and flightless) auks directly into the ship's cargo hold (Matthiessen 1959).

In one way or another, human beings have been responsible for the catastrophic destruction of most of the unique Hawaiian avifauna. Island birds often lose resistance to mainland diseases, in addition to losing their ability to fly and their fear of predators. Certain lowland populations of the Hawaiian honeycreepers were destroyed by bird pox and malaria when mosquitoes that carried these diseases were accidentally introduced in the eighteenth century (Warner 1968; van Riper et al. 1986). In addition, the early Polynesians destroyed most of the lowland forests after reaching the Hawaiian islands roughly 1500 years ago, long before Captain James Cook brought European civilization and mosquitoes to the islands. As a result of this destruction, the Polynesians eliminated at least 39 species of land birds, including 7 geese, 2 flightless ibises, 3 owls, 7 flightless rails, and 15 species of honeycreepers (Olson and James 1982). This is one example from the broad pattern of prehistorical destruction by early human colonists of island avifaunas in the South Pacific (Steadman and Olson 1985). Similar waves of extinctions followed the settlement of the Caribbean islands 3000 to 4500 years ago (Pregill and Olson 1981), and New Zealand 1000 years ago (Baker 1991).

Although small populations of birds are vulnerable, they can be resilient. Most small populations have an intrinsic potential to rebound from severe reductions (Pimm et al. 1988). One legendary case is that of the Short-tailed Albatross, which once nested in abundance in the western Pacific and congregated at the entrance to San Francisco Bay, when whale slaughtering produced abundant food there. Feather hunters reduced this species by 1929 to one population of 1400 birds that bred at Toroshima, an island refuge off southeastern Japan. Eruptions of the island's volcano in 1939 and 1941 destroyed this remaining albatross colony. The species was declared extinct when no birds returned to the island to breed from 1946 to 1949. Remaining to save the species, however, were some young birds that had been at sea. (Albatrosses wait 10 or more years before starting to breed.) In 1954 six pairs of these survivors returned to Toroshima and produced a total of three young. Today, despite their low fecundity and delayed maturity, Short-tailed Albatrosses are recovering slowly, and the world population is now estimated to be 300 to 400 birds.

Modern Threats to Bird Populations

Despite a growing conservation ethic, expanding human populations continue to threaten native bird populations. The threats range from direct exploitation by hunting, to habitat loss or degradation, to poisoning of food supplies with pesticides and other chemical contaminants, to new forms of exploitation such as the commercial pet trade. Added to these major threats is a steady annual attrition due to road kills, window strikes, and predation by pets (Figure 24–4).

Human activities are directly responsible for roughly 270 million bird deaths every year in the continental United States, about 2 percent of the 10 to 20 billion birds that inhabit the continental United States (Banks 1979). Such attrition adds to the primary losses due to destruction of breeding habitat and interference with reproduction. Nearly one-half of the annual deaths are the direct result of licensed hunting and pest control. Scientists collect a relatively small number of birds for research purposes, roughly 10,000 annually. Collisions with man-made objects, however, are a significant source of mortality. A minimum of 57 million birds are killed by vehicles in the United States every year, assuming 15 bird deaths per road mile per year (Hodson and Snow 1965). About 1 million migrating birds are killed annually in collisions with large, lighted buildings and television towers. Collisions with plate glass windows of homes and office buildings kill an estimated 80 million songbirds annually throughout the United States, nearly two-thirds the annual harvest of waterfowl and gamebirds by hunters. Roughly half of the birds that collide with windows die of skull fractures and intracranial hemorrhaging. Systematic monitoring over 1 year registered 61 collisions at a house in Illinois and 47 collisions at a house in New York (Klem 1989). These probably are minimal estimates because cats, raccoons, skunks, and opossums often move the carcasses before the investigators can count them. Efficient nocturnal scavenging removes what otherwise would be conspicuous littering of the landscape by dead birds.

Dwarfing these losses are those attributable to predation by pets. Domesticated cats in North America may kill 4 million songbirds every day, or perhaps over a billion birds each year (Stallcup 1991). Millions of hungrier, feral (wild) cats add to this toll, which is not included in the estimate of 270 million bird deaths each year.

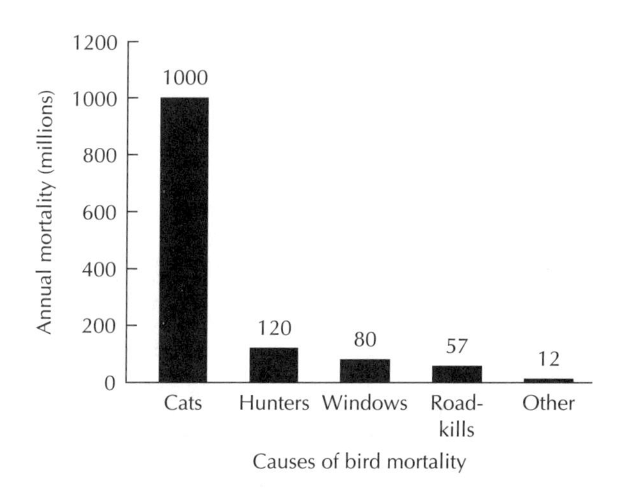

Figure 24–4 Human-related causes of bird deaths.

Environmental Poisons

Biologists and politicians both recognize that birds are sensitive indicators of environmental health (Morrison 1986)—and their recent declines bode ill for other species. Human poisoning of the environment has had a devastating impact on the fecundity of birds.

Pesticides

Accumulated pesticides, particularly DDT, not only kill birds directly but also interfere with eggshell production and thus cause nesting failure (Risebrough 1986). Peregrine Falcons and Ospreys in the eastern United States and Eurasian Sparrowhawks in Britain were nearly exterminated as a result of pesticide poisoning (Fyfe et al. 1976; Newton 1979). These raptors served as indicator species because, as predators at the top of the food chain, they concentrated the toxins. Eggs with shells 15 to 20 percent thinner than normal generally result in reproductive failure and population decline (Cooke 1975; Kiff et al. 1979). Reproduction of Bald Eagles in northwestern Ontario, for example, declined from an average of 1.26 young per breeding area (nest) in 1966 to a record low average of 0.46 in 1974, increasing to an average of 1.12 after DDT was banned (Grier 1982) (Figure 24–5).

Paralleling the case of the Bald Eagles is that of the Brown Pelican, one of the most familiar and abundant birds of the Gulf and West coasts of North America (Figure 24–6). This species faced extinction in the 1960s because of widespread reproductive failure (Schreiber 1980b).

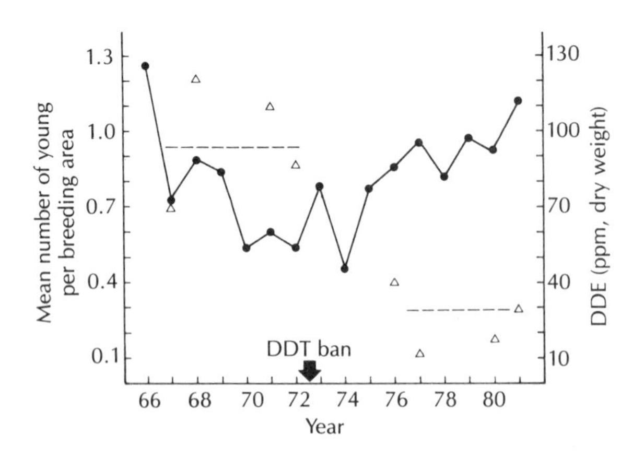

Figure 24–5 Reproduction in Bald Eagles (solid lines) improved following the ban on the use of the pesticide DDT; the ban resulted in a drop in chemical residues (DDE) in eggs (triangles). Dashed lines represent weighted mean concentrations of DDE before and after the ban. (From Grier 1982, copyright 1981 by the AAAS)

Figure 24–6 Brown Pelican, a species that experienced reduced reproduction and severe population declines due to thinning of eggshells by DDT. The pelican populations are now on the increase on both the East and West coasts of North America. (Courtesy of A.D. Cruickshank/VIREO)

Hydrocarbon pesticides in the marine food webs of coastal California, coastal Louisiana, and nearby Texas interfered with the production of normal eggshells, and the pelicans typically laid eggs with very thin or no shells. The fragile eggs were easily broken under the weight of an incubating parent. The lack of reproduction in Brown Pelicans in California, where eggshell thinning was most severe, and the alarming disappearance of pelicans from Louisiana and Texas resulted in placement of this bird on the endangered species list in 1973. Recent reductions of pesticides in the environment have enabled Brown Pelicans to nest successfully once again. No longer endangered, their populations are expanding rapidly in California and on both the Gulf and Atlantic coasts.

Xenobiotics: New, Insidious Environmental Poisons

Toxicologists now recognize subtle and sometimes unsubtle effects of a wide variety of chemicals on the endocrine systems of vertebrate animals, including birds. They discovered that certain chemicals disrupt the normal course of embryonic development, but often without obvious manifestations until adulthood. This class of chemicals—called xenobiotics—

includes fungicides, herbicides, and insecticides, plus assorted industrial chemicals, synthetic products including soy and pet food products, and some metals including cadmium, lead, and mercury.

There is recent and convincing evidence of the disturbing effects of these compounds on bird (as well as fish, mammal, and turtle) populations of North America, especially on the coast of southern California, Puget Sound, and the Great Lakes.

> We are certain of the following: A large number of man-made chemicals that have been released into the environment . . . have the potential to disrupt the endocrine systems of animals, including humans. (Colborn and Clement 1992, p. 1)

The effects include, but are not limited to, thyroid dysfunction, compromised immune systems, decreased fertility, decreased hatching success, gross birth deformities, metabolic and behavioral abnormalities, and sex reversal.

Western Gulls breeding in California and Herring Gulls breeding on Lakes Ontario and Michigan show some of the effects of these contaminants, including a high incidence of clutches with extra eggs, female–female pairings, and feminization and high mortality of males (Gilbertson et al. 1991; Fox 1992). Gulls in these colonies also suffered from embryonic and chick mortality, edema, growth retardation and deformities, and altered nest defense and incubation behavior, all of which severely reduced reproductive success.

Loss of Critical Habitat

Currently paramount among the negative forces of humankind on bird populations is the rapid destruction of the natural habitats of the world, including the replacement of virgin rain forest by pastures and coffee or banana plantations, the conversion of rich grasslands into agricultural monocultures or croplands, the draining of wetlands, and the consumption of diverse biological habitats by urban sprawl.

The accelerating destruction of tropical rain forests has the highest profile as a global conservation issue, because rain forests are the most diverse terrestrial ecosystem on the planet; they cover less than 7 percent of the Earth's landmass but contain 66 percent of all species. Originally, rain forests covered about 12 percent of the Earth's landmass, but commercial logging and expanding tropical civilizations have reduced their extent by nearly half in recent decades (Figure 24–7). Half again of the remaining rain forests on Earth will be gone by the year 2022, if their destruction continues at the present rate (50 million acres annually), extinguishing or dooming to extinction about 27,000 species each year, including many birds (Wilson 1992).

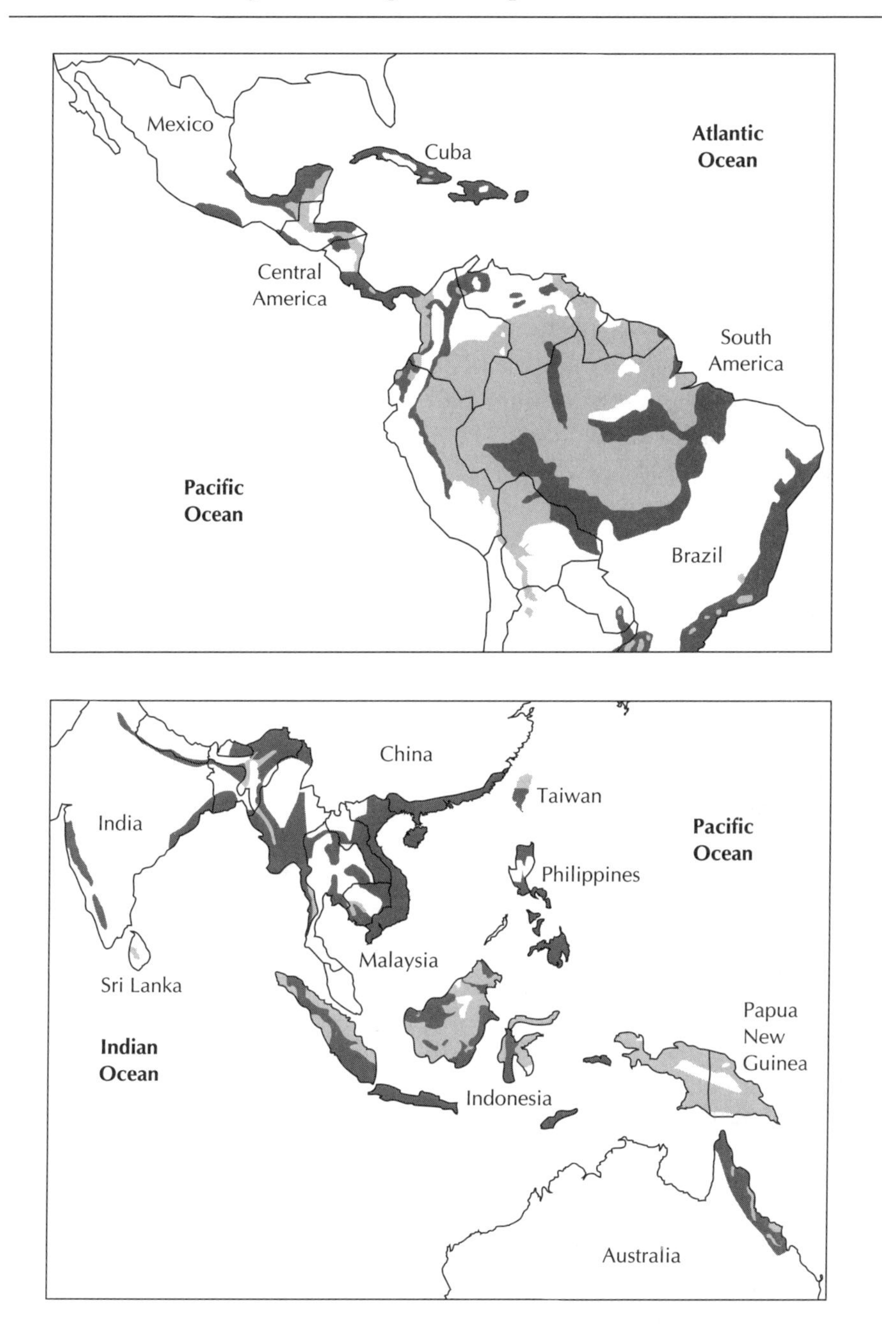

Figure 24–7 Tropical rain forest destruction in the New World Tropics and in Southeast Asia. Roughly half of the planet's rain forests have been removed or degraded in the last 50 years. Gray areas = extent of tropical forests in the late 1980s. Black areas = forest destroyed or seriously degraded since 1940. (After *The Philadelphia Inquirer*/Kirk Montgomery, May 11, 1992)

Loss of Old-growth Forests

The challenging politics of habitat loss, however, can be found not only in distant tropical settings but in the backyard as well. Preceding the cutting of tropical rain forests by over a century was the consumption of North American forests for fuel, lumber, and agriculture. The northeastern United States—east of the Finger Lakes—was virtually cleared by

1800. The bottomland forests of the Southeast, home to the Ivory-billed Woodpecker, and the giant forests of the West Coast, home to the Spotted Owl, were next. Other native habitats—scrublands, grasslands, and wetlands—were also exploited. The protection of endangered species has provided the principal legal basis for challenging commercial exploitation and development of declining natural habitats. The highly publicized political conflict over management of old-growth forests brought notoriety, for example, to the Red-cockaded Woodpecker of the Southeast and to its political counterpart in the Pacific Northwest, the Spotted Owl, both endangered species.

Intense logging in this century has reduced the old-growth forests in the Pacific Northwest to about 10 percent of their original extent. Much of the remaining old-growth forest is owned and managed by the federal government, more specifically, the U.S. Forest Service and the Bureau of Land Management. The Spotted Owl came front and center into the battle between environmentalists anxious to protect the remnant of what remained and loggers anxious to continue harvesting timber (Figure 24–8). This little known species symbolized the old-growth forest ecosystem and seemed destined to repeat the fates of the Passenger Pigeon and the Ivory-billed Woodpecker, losses that weighed on the consciences of some U.S. citizens (Wilcove 1987).

Typical Spotted Owl habitat in the Pacific Northwest consists of low- to mid-elevation old-growth forests dominated by Douglas fir trees. These forests have mixed age classes of trees, including some large ones that are over 200 years old, plus an abundance of dead trees and branches. The owls' preference for old-growth forests relates to the availability of large, old dead trees for nesting and of small mammalian prey and protection from predators. Depending on location, each pair requires 500 to 2000 hectares of mature forest.

Demographic data and life table analyses project declines in the populations of the Spotted Owl, which delays breeding until it is 3 years old, later than most medium-sized owls. Breeding success is good in some years but bad in others, depending on the availability of prey. Breeding success has been particularly poor recently, with high mortality of up to 82 percent of dispersing juveniles (Marcot and Holthausen 1987). In addition to raw loss of habitat due to commercial logging, Spotted Owls are affected by the fragmentation of old-growth forests. They also are sensitive to competition from larger and more aggressive Barred Owls, which are increasing in adjacent regrowth areas, and are subject to increased predation by Great Horned Owls, which frequent the forest edges and openings created by logging.

Required by law to ensure viable populations of all native vertebrate species in the national forests, the U.S. Forest Service developed and revised guidelines for the management of the Spotted Owl. Their plan in July 1986 was to set aside 500 Spotted Owl habitat areas in Washington and Oregon. The protected areas would vary in size—the average area containing 1000 hectares of old-growth forest—and would form a well-

Figure 24–8 Northern Spotted Owl, an endangered species that came to symbolize old-growth forests of the Pacific Northwest, now reduced to only 10 percent of its original extent. (Courtesy of A. and S. Carey/VIREO)

distributed network that allowed dispersal of young owls. A review of the minimum population requirements by a blue-ribbon advisory panel of ornithologists convened by the National Audubon Society raised the stakes. They recommended protection of a minimum of 1500 breeding pairs by setting aside habitat areas with 2100 hectares each in Washington, 1100 hectares in Oregon and northern California, and 650 hectares in the Sierra Nevadas (Dawson et al. 1986). Following more public hearings, President Bill Clinton proposed a new compromise plan in June 1993 that would reduce the rate of logging of the remaining old-growth forests, thereby protecting a minimum number of owls, and that also would help to retrain loggers for new jobs when there were no more big trees to cut. Whether the populations of the Spotted Owl and the loggers will stabilize and coexist remains to be seen.

Box 24–2
A seabird needs old-growth forest too

Few birds have been as mysterious and eluded study for as long as the Marbled Murrelet has; it was the last North American species to have its nest discovered. Now listed as a threatened species, the Marbled Murrelet is a quail-sized seabird that is vulnerable to coastal oil spills and high mortality in underwater fishing nets. Recent discoveries suggest that in the Pacific Northwest the murrelet depends on the disappearing old-growth forests, as does the Spotted Owl (Marshall 1988). By accident, a tree surgeon named Hoyt Foster discovered a moss nest containing one downy young murrelet 45 meters off the ground in a tall Douglas fir 16 kilometers from the sea in California's Santa Cruz Mountains. These small seabirds, it turns out, fly inland to nest in tall, old-growth forest. Now that ornithologists know where these murrelets nest, a major effort has been launched to define as precisely as possible their population requirements and especially their dependence on the old-growth forests from northern California to Alaska.

Scrubland Birds versus Suburbia

Like the owl and the murrelet (Box 24–2), other endangered species are the focus of conflicts between commercial development and preservation of natural habitats. The Golden-cheeked Warbler, and especially the Black-capped Vireo, are declining songbird residents of the scrub oak–juniper forests of the Edwards Plateau of southern Oklahoma and central Texas (Figure 24–9). Expanding suburban developments and the continued clearing of scrublands for cattle grazing now threaten the remaining populations, estimated to total fewer than 250 pairs of vireos and 5000 to 15,000 pairs of the warbler (Grzybowski et al. 1986; Wahl et al. 1990).

Like the Spotted Owl, the Golden-cheeked Warbler depends on mature trees; it peels the bark of 50- to 80-year-old Ashe Juniper trees for its nests. It is a strict habitat specialist that requires a mixture of junipers and oaks that is limited to only 10 percent of the vegetation in this area (Sexton 1992). The warbler and the vireo have two enemies, human beings and Brown-headed Cowbirds, both of which are thriving in suburban habitats. In recent years, nearly all of the vireo nests have been parasitized and therefore have failed to fledge young vireos. Direct removal of cowbird eggs and trapping and removal of cowbirds greatly improves nesting success, as it did for Kirtland's Warblers (Chapter 19).

After a highly publicized conflict, opposing forces agreed on a program of joint preventative action. The U.S. Fish and Wildlife Service plans to establish a new refuge, the Balcones Canyonlands National Wildlife Refuge, near Austin, Texas, to protect these two species. Enabled by generous gifts of critical land near Austin, Texas, the refuge started with 3239 hectares (8000 acres), including about 125 hectares of high-quality Black-capped Vireo habitat, but will expand to 16,600 hectares. Among the many acts certifying local interest, the citizens of Austin voted in 1992 to approve a $22 million bond issue to acquire lands for the preservation of the unique biodiversity of the region.

Figure 24–9 Golden-cheeked Warbler, a declining Texas species that requires the bark of old juniper trees for its nests. (Courtesy of H. Irby)

The Black-capped Vireo Recovery Plan (U.S. Fish and Wildlife Service 1991) has the goal of downlisting this species from endangered to threatened by the year 2020. Four criteria must be fulfilled before downlisting: (1) All existing populations are protected and maintained. (2) A minimum of six viable breeding populations of 500 to 1000 pairs will exist in Texas, Oklahoma, and Mexico. (3) Sufficient winter habitat will exist to support the priority breeding populations. (4) The designated breeding populations are maintained for at least 5 consecutive years with evidence of continued viability.

Habitat for Neotropical Migrants

Concern is growing over the apparent population declines of many species of birds that migrate between the New World Tropics and Temperate Zone latitudes (Terborgh 1989; Morton and Greenberg 1989; Robbins et al. 1989; Askins et al. 1990; Hagan and Johnston 1992), although there are valid debates about the statistical quality of the data used to document the long-term trends (James et al. 1992; Morton 1992). Even more challenging is the task of pinpointing the main threats to these populations. Either the rapid rate of habitat loss on the tropical wintering grounds, or recent fragmentation and degradation of northern forest

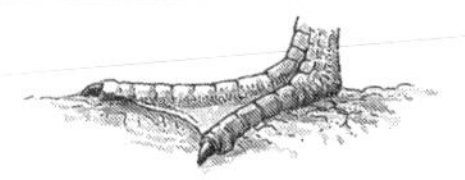

BOX 24–3

Elevation to species status puts California Gnatcatchers at odds with real estate developers

Before 1989, the tiny, 6-gram scrubland gnatcatchers of southern California were simply a subspecies of the widespread Black-tailed Gnatcatcher. Jonathan Atwood (1988) then discovered not only that the differences between California Gnatcatchers and Black-tailed Gnatcatchers elsewhere were greater than had been appreciated, but also that the California Gnatcatcher actually coexisted with Black-tailed Gnatcatchers in Baja California without interbreeding. On the basis of his detailed study, the Checklist Committee of the American Ornithologists' Union decided to recognize the California Gnatcatcher as a full species (*Polioptila californica*), one restricted to the scrublands, especially mesquite and creosote bush, in southwestern California and Baja California.

The population of California Gnatcatchers in the United States, now fewer than 2000 pairs, has been threatened over the years by the loss of critical habitat to housing developments. If added to the list of endangered species, as proposed by the U.S. Fish and Wildlife Service, the gnatcatcher would ultimately protect thousands of acres of sage scrubland in southern California from development. The California Fish and Game Commission, however, opposes listing the species as endangered, mostly for political reasons. The southern California building industry has publicly declared the gnatcatcher to be its worst enemy, escalating the conflict between economics and ecology (Atwood 1993). Because each pair of gnatcatchers may require a territory of 10 or more hectares in size, the estimated land values of $550,000 per hectare bring this cost to $5 million per territory! The political battle quickly turned to whether the California Gnatcatcher was truly a valid biological species, but Atwood's careful field research on the differences between the two species and their coexistence without interbreeding has held up against the legal challenges. The political battle continues, as does the destruction of prime gnatcatcher habitat.

---

habitats that the migrants visit to nest, or both, could be responsible. Forest fragmentation in North America clearly promotes increased nest predation and brood parasitism by Brown-headed Cowbirds, which can cause local reproductive failure (Robinson 1992). Most likely, Neotropical migrants are challenged increasingly on both wintering and breeding grounds, a situation requiring major commitments to habitat preservation in both regions to halt the decline.

One migrant species in trouble is the Cerulean Warbler, a treetop species with an azure blue back. The Cerulean Warbler showed the most precipitous decline from 1966 to 1987 of any North American wood warbler (3.4 percent per year) (Robbins et al. 1992). Like other Neotropical migrants, it has suffered extensive loss of breeding habitat during the past century, but it differs from many other Neotropical migrants because it distinctly prefers mature floodplain forest with tall trees. This particular habitat has become scarce over much of the Cerulean Warbler's nesting range. Sensitivity to fragmentation within the remaining suitable tracts of river-edge forest places this warbler at an additional disadvantage. Furthermore, Cerulean Warblers overwinter strictly in primary, humid evergreen forest along a narrow elevational zone at the base of the Andes in South America. This particular forest zone is among the most intensively logged and cultivated regions of the Neotropics. Unless steps are taken to protect large tracts of habitat of this ecologically specialized species, both in North America and in the foothills of the Andes, the Cerulean Warbler surely will disappear from our avifauna.

Wetlands and Water Birds

The protection and restoration of wetlands and their water birds are major conservation priorities (Figure 24–10). The goals of birdwatchers and duck hunters are converging as they discover a common cause—the conservation of wetlands. Wetlands have many ecological roles, including the recharging of groundwater supplies and purifying the drinking water, preventing flood damage, and serving as the spawning grounds for commercial fisheries. Wetlands of vital importance to hundreds of bird species range from the prairie marshes that produce most of the continent's ducks to the coastal salt marshes that host shorebirds, rails, colonial water birds, and waterfowl throughout the year. Before the settlement of North

Figure 24–10 Waterfowl populations in North America depend on wetlands, a habitat that is disappearing at the rate of 300 hectares daily. (Courtesy of A. Cruickshank/VIREO)

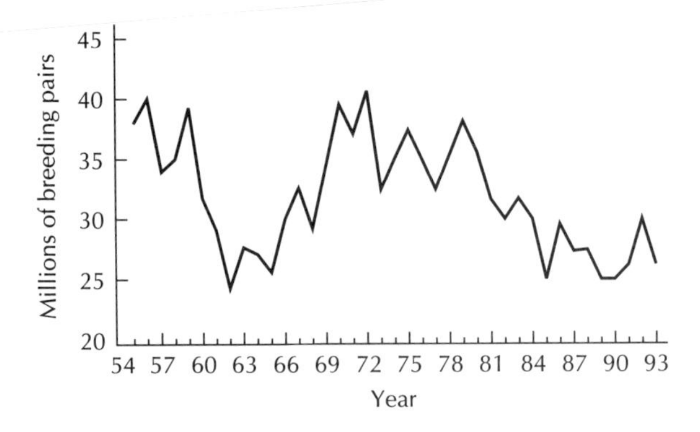

Figure 24–11 Population fluctuations and long-term decline of North American duck populations from 1954 to 1993. Numbers expressed as millions of breeding pairs. (From U.S. Fish and Wildlife Service 1993)

America by Europeans, there were roughly 100 million hectares of wetlands; now there are only 45 million hectares, a loss of more than half of the original acreage (Howe 1987). California has lost 90 percent of its wetlands, including 74,000 of an original 90,000 hectares of San Francisco Bay once used by hordes of wintering waterfowl and shorebirds. Despite regulations to protect wetlands, the destruction of wetlands in North America continues at a rate of 300 hectares daily.

Sharp declines of waterfowl and shorebird populations have accompanied the destruction of wetlands. Duck populations have declined to low levels of roughly 25 million breeding pairs, compared with 40 million pairs at peaks in the 1970s (Figure 24–11). The numbers of Northern Pintails have dropped from historical highs of 10 million pairs to record lows of 1.8 million pairs. With the exception of Gadwalls and Green-winged Teals, all duck species surveyed by the U.S. Fish and Wildlife Service in 1991 were well below their 40-year average population sizes.

Habitat Conservation

First and foremost, birds require specialized habitats—old-growth forests, natural scrublands, wetlands, and so on. Management of viable bird populations requires a commitment to set aside critical habitat.

With a commitment for protection of habitat for endangered species comes the challenge of designing these reserves, a fertile and controversial topic of conservation biology (Soulé 1986). Protection of adequate areas of gnatcatcher habitat in southern California (Box 24–3) requires a sound landscape design that takes into account how large the reserves must be and whether they should be connected by corridors of suitable habitat to allow dispersal, which enhances genetic variability and reduces the probability of extinction due to the small island effect.

Thomas Lovejoy's minimum-critical-size project in the Amazon rain forests of Brazil was launched in 1979 to address some of these questions for a tropical setting (Lovejoy et al. 1986; Bierregaard et al. 1992). The long-term study demonstrated dramatic losses—specifically in forest islands less than 10 hectares in size—both of birds that follow army ants to catch flushed prey and of birds that participate in mixed-species flocks. Three species of obligate army ant followers disappeared immediately upon isolation of small fragments. The majority of the members of mixed-species flocks disappeared from all small rain forest fragments in 1 to 2 years. Corridors 100 to 300 meters wide between blocks of forest, however, helped to maintain species diversity in forest fragments up to 100 hectares in area. These and other results of the study led to the recommendation that small forest fragments should be connected by corridors of rain forest and that local landscapes should include one or more forest tracts larger than 1000 hectares.

Species inhabiting tropical mountain forests, such as the Resplendent Quetzal of Central America, require corridors for seasonal altitudinal migration between separated living areas. The Resplendent Quetzal, a trogon (Trogoniformes) of the cloud forests of Central America, is one of the most dramatic of all tropical birds, iridescent green and scarlet in color and bearing lacy, 2-foot-long upper tail coverts. It feeds and breeds in mountain preserves, such as the popular 28,000-hectare Monteverde Cloud Forest Preserve in Costa Rica. However, ornithologists found that the quetzals also migrate downslope to find food during the nonbreeding seasons. By tracking the seasonal movements of quetzals wearing radio transmitters, George Powell and his colleagues from the RARE Center of Tropical Bird Conservation discovered that the slopes at altitudes between 615 and 1540 meters were critically important corridors and nonbreeding residences for the quetzal. Future plans for the conservation of the quetzal now include redesigning the preserve to provide the network of habitats the quetzal requires throughout the year.

In North America, a program for the protection and restoration of wetlands has begun. Improved legislation, including the Coastal Barrier Resources Act of 1982, which removes federal flood insurance and other subsidies for development of barrier islands and adjacent wetlands, and changing methods of mosquito control are helping to protect some of these fragile ecosystems (Bildstein et al. 1991). Migratory shorebird species, which require protection at key sites along their traditional migration routes, benefit from the formation of the Western Hemisphere Shorebird Reserve Network (Senner and Howe 1984; Myers et al. 1987; Hunter et al. 1991) (see Box 12–2). There is also growing interest in managing refuge water levels to accommodate both shorebirds and waterfowl (Helmers 1992).

In partnerships with government and private agencies, Ducks Unlimited has spearheaded the restoration and improvement of 2.5 million hectares of the nation's wetlands. Founded in 1937 by sportsmen concerned about declining waterfowl populations, Ducks Unlimited has contributed $640 million toward habitat conservation and has grown into a conservation force of a half a million members.

Faced with continuing wetland destruction and accompanying declines in waterfowl populations, the governments of the United States and Canada initiated in 1986 a new, intense effort to protect wetlands and associated wildlife. The North American Waterfowl Management Plan signed by representatives of both countries seeks to protect over 2.4 million hectares of important wetlands. Mexico recently joined this effort. With the participation of nearly 200 public and private organizations, including Ducks Unlimited, the North American Waterfowl Management Plan strives to increase waterfowl populations, with specific goals for each of 32 species of ducks, geese, and swans. The overall goal of the plan is to build waterfowl populations in the year 2000 to a total of 62 million breeding pairs that will support an annual fall flight of 100 million birds (including young of the year).

The Live Bird Trade

Few people are aware of the dimensions of the caged bird industry and its impact on the populations of certain wild birds (Beissinger et al. 1991). Two to five million birds move annually from tropical habitats to the living rooms of developed countries. The United States, presently the largest importer of exotic birds, legally imported nearly 2 million birds from 85 countries from 1986 to 1988 (Nilsson 1990). Over half of these (54 percent) were finches of various species, primarily from Africa. Forty-three percent were parrots, and the remaining 3 percent represented various other birds of the world—no less than 77 different taxonomic families.

Pet birds foster an appreciation and love of birds. Many cage birds are produced by aviculturists, who master the art of breeding a few species, such as Budgerigars and Common Canaries in captivity. Millions of other cage birds, however, are harvested from wild populations. Because the exotic cage bird trade is growing rapidly, it increasingly threatens some wild bird populations and even some species. Controlling the international pet trade is the Convention on International Trade in Endangered Species of Wild Fauna and Flora. In response to a list compiled by this organization, Congress passed the Wild Bird Conservation Act of 1992 in an effort to eliminate importation of endangered wild bird species.

The exotic pet bird trade is a multi-million-dollar industry, and parrots, particularly, command high prices. At the top of the price list are rare macaws, such as the Hyacinth and Little Blue (formerly Spix's) Macaws, which now sell for $10,000 to $50,000 apiece. About 1.8 million parrots were exported legally from tropical countries for trade from 1982 to 1988; most of these were captured from the wild. Such numbers are underestimates: Mortality before exportation plus illegal smuggling doubles or triples these levels of depletion of wild populations. Parrots and their nesting habitats are most sensitive to exploitation. Approximately 42 of the 140 species of parrots in the New World Tropics are currently endangered, 22 of them as a result of the pet trade (Collar and Juniper 1991). Most threatened are the Little Blue Macaw, Hyacinth Macaw, and Red-crowned Parrot. The extreme case is that of the beautiful Little Blue Macaw, of which only one free-living individual still persists in northeastern Brazil (Juniper and Yamashita 1991). About 40 exist in private aviaries. Conservationists hope to use some of them to establish a breeding program and to release a potential mate for the remaining wild male in Brazil before it is too late.

Excessive Human Predation

The roots of current conservation concerns in North America, as well as in other parts of the world, go deep into past practices of uncontrolled exploitation. The wholesale slaughter of wildlife in the United States was a national pastime of the expanding new nation (Matthiessen 1959).

The earliest settlers of the United States lived off the abundant game, severely depleting local stocks of turkey and deer. Larks, bobolinks, robins, and many other songbirds were also fair game. Full-scale market gunning took its toll later in the mid-1800s. First, the great bison herds and other large mammals of the Great Plains were exterminated. Flocks of waterfowl were decimated by cannons mounted on low boats that could approach closely without detection. Then the market gunners turned to the seemingly unlimited flocks of Passenger Pigeons, until the last wild Passenger Pigeon was killed in Ohio in 1900.

Shorebirds were hunted also, particularly the vast flocks of Lesser Golden-Plovers and Eskimo Curlews, which migrated north in the spring through the Great Plains and then south in the fall from maritime Canada to South America. John James Audubon reported millions of Lesser Golden-Plovers near New Orleans in the early nineteenth century and compared curlew flights with those of the Passenger Pigeon. Occasionally, southbound plover and curlew flocks appeared on the New England coast.

> On August 29, 1863, both curlew and plover appeared on Nantucket in such numbers as to "almost darken the sun"; seven or eight thousand were destroyed before the island's supply of powder and shot gave out. (Matthiessen 1959, p. 162)

By the turn of the century, Eskimo Curlews were virtually extinct. Continued but rare sightings of lone birds or pairs suggest that a tiny relict population still persists, perhaps breeding in the Northwest Territories of Canada (Faanes and Senner 1991).

As the flocks of shorebirds and pigeons declined in the late nineteenth century, another threat materialized—plume hunting for the millinery trade. The mounting of bird feathers and, indeed, even of whole birds on ladies' hats became the height of fashion in the 1870s and 1880s. An estimated 5 million birds were killed annually for this purpose alone. At first, the breeding plumes of large wading birds—egrets, herons, and spoonbills—were prized, with devastating impact on their nesting colonies. The millinery trade next turned to gulls and terns and then to a full array of species from brightly colored songbirds to crows. Drawing rave reviews was an entire crow—beak, feet, and all—seen on a hat in New York City in 1886. Frank M. Chapman, the distinguished ornithologist at the American Museum of Natural History, amused himself by identifying the species on hats during strolls through New York City; in one census, 542 out of 700 hats sported mounted birds of at least 20 species, including a Ruffed Grouse and a Green Heron (Matthiessen 1959, p. 168).

Rise of Conservation Efforts in North America

The vocal and increasingly organized opposition to the killing and exploitation of native American birds in the nineteenth century was inevitable. States passed laws limiting the hunting season or protecting particular species (Table 24–1). Additional local game laws were passed and then repealed in response to political pressure (Bean 1983). The work of the conservation committee of the American Ornithologists' Union, founded in 1883, led first to the formation of various state Audubon societies and soon thereafter to federal agencies charged with bird protection. In 1886, the American Ornithologists' Union proposed a Model Law that was adopted immediately by New York State and the Commonwealth of Pennsylvania and that eventually became the prototype for bird protection legislation throughout the country.

Effective federal legislation followed on the heels of growing public concern for birds. New sanctuaries protected colonial wading birds from the plume hunters. President Theodore Roosevelt established in 1908 the first National Wildlife Refuge in the United States at Pelican Island, a small island in the Indian River near Cape Canaveral, Florida. This tiny island is now part of the much larger Merritt Island National Wildlife Refuge. A seminal piece of conservation legislation, The Migratory Bird Treaty Act of 1918, made the management of migratory birds a federal responsibility. The killing of migratory songbirds became illegal. This law now implements migratory bird treaties involving Canada, Mexico, Japan, and Russia. Finally, the Endangered Species Act of 1973 (amended in 1978 and 1982) brought broad new powers to protect threatened birds, such as the Whooping Crane, whose survival required the protection of threatened habitats and well-coordinated rescue efforts (Box 24–4). A new environmental ethic has swept the country; conservation of biodiversity is now both a national and international priority. Government agencies, both state and federal, are beginning to devote increasing attention to the management of nongame bird populations (Senner 1986; Henderson 1988; Office of Migratory Bird Management 1990; Howe 1991).

Restoration Programs

Revival of severely threatened bird populations through the release of captive-bred individuals is a tool of conservation management. Gerald Durrell pioneered this approach by breeding endangered species in special facilities on the Isle of Jersey, off the coast of Great Britain. Among his special interests were the birds of Mauritius Island (where the Dodo once lived), such as the Pink Pigeon, the Mauritius Kestrel (the world's rarest raptor), and the Mauritius Parakeet (one of the world's rarest parrots). Restoration of Whooping Cranes and California Condors include captive propagation programs, as does Noel Snyder's current effort to restore Thick-billed Parrots to the Chiricahua Mountains in southeastern Arizona (see Box 14–7). These and other restoration programs must avoid imprinting hand-reared birds on their captors and must master the art of parenting young birds to independence, including learning how to forage and how to avoid predators.

Rescue of the California Condor

Restoration programs tend to be expensive and sometimes both politically and scientifically controversial. The program to save the California Condor illustrates some of the conflicts between using a species as symbol for habitat preservation and saving a species for its own sake.

The California Condor, North America's largest vulture, is a relic of the past (Figure 5–4). Other condor species, many of them even larger, once prospered along with the continent's prehistoric large-mammal fauna. Today, only two species remain, the Andean Condor of South America and the California Condor of North America.

The Californa Condor once roamed widely across the United States as far east as New York in search of carrion. Over the millennia, the large local populations shrank to a single remnant population in southern California. Estimates of only 60 remaining condors in the 1940s started a 50-year battle to save these special vultures one way or another—although in hindsight, there were probably many more than 60 birds left at the time the battle began. Two veterans, Noel and Helen Snyder, summarized the recent conclusion of that war:

> On Easter Sunday of 1987, the last California Condor (*Gymnogyps californianus*) known to exist in the wild was trapped for captive breeding, joining 26 others of his species at the San Diego and Los Angeles Zoos. A young male adult, he was a bird whose life had been followed closely for a number of years. His movements and interactions with other condors, his molting patterns and changes in coloration with age, as well as his pairing with an old female in late 1985, and his first breeding attempts in 1986 had all been documented in considerable detail. Like every other bird in the remnant population, he was known and had been studied as an individual. With his capture, an era of intensive investigations of condor natural history and ecology had come to a close. (Snyder and Snyder 1989, p. 175)

TABLE 24–1

A chronology of key legislation and events affecting the birds of North America

Year	Legislation
1616–1622	First New World wildlife protection: The government of Bermuda issues a proclamation protecting the cahow
1708	First closed season on birds, in certain New York counties, on heath hen, grouse, quail, and turkey
1710	Massachusetts prohibits use of camouflaged canoes or boats equipped with sails in pursuit of waterfowl
1782	Bald Eagle recognized as national emblem
1818	First law protecting non-game birds, establishing a closed season on larks and robins in Massachusetts
1838	First law against the use of batteries, or multiple guns, on waterfowl; later repealed (New York)
1846	First law against spring shooting (wood duck, black duck, woodcock, snipe); later repealed (Rhode Island)
1848	Massachusetts passes law protecting pigeon netters from molestation
1851	Non-game bird protection established in Vermont; subsequently in Massachusetts (1855) and 10 other states by 1864
1869	First Passenger Pigeon protection, in Michigan and later Pennsylvania (1878); no firearms to be discharged within 1 mile of roosts
1872	First law providing rest days for waterfowl, in Maryland
1875	First law prohibiting market hunting of waterfowl, in Arkansas
1877	Florida passes a plume-bird law prohibiting wanton destruction of eggs and young
1886	American Ornithologists' Union Model Law for states
1908	First U.S. National Wildlife Refuge established at Pelican Island, Florida
1916	Ratification of convention between United States and Great Britain for Protection of Migratory Game Birds in United States and Canada, including full protection for Band-tailed Pigeon, cranes, swans, most shorebirds
1918	Federal Migratory Bird Treaty Act prohibits spring shooting, awards to government the right to prescribe bag limits of migratory birds

TABLE 24–1 *(continued)*
A chronology of key legislation and events affecting the birds of North America

Year	Legislation
1934	Migratory Bird Hunting Stamp Act establishes federal hunting license in form of "duck stamp"
1939–1940	U.S. Fish and Wildlife Service (FWS) formed by merger of the Bureau of Biological Survey (Agriculture) and Bureau of Fisheries (Commerce)
1942	Convention on Nature Protection and Wildlife Preservation in the Western Hemisphere protects bird species that cross U.S. boundaries at any season; Bald Eagle receives full protection throughout the United States, also Alaska
1973	Endangered Species Act gives threatened taxa legal priority over commercial habitat destruction
1975	Convention on Wetlands of International Importance 1975 promotes international conservation of wetlands and waterfowl
1980	Fish and Wildlife Conservation Act (FWCA) encourages states to establish nongame programs
1985	Western Hemisphere Reserve System Network (WHSRN) established
1986	North American Waterfowl Management Plan (NAWMP) provides a cooperative framework for waterfowl conservation and management efforts by Canada and the United States through the year 2000
1988	Amendments to FWCA require FWS to monitor all species and populations of migratory birds
1989	North American Wetlands Conservation Act protects endangered species and migratory non-game birds as well as waterfowl
1989	Convention on International Trade in Endangered Species of Wild Fauna and Flora (CITES)
1992	Wild Bird Conservation Act bans importation into the United States of wild birds that are popular pets but threatened in the wild

After Matthiessen 1959; Senner and Howe 1984; Howe 1991.

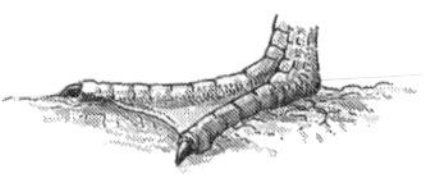

Box 24–4
Neighboring nations cooperate to save the Whooping Crane

Cooperation between the governments of the United States and Canada has restored hope for the future of the stately black-and-white Whooping Crane, an endangered species that inspired international concern and constructive action. The population of Whooping Cranes, which once nested widely in the

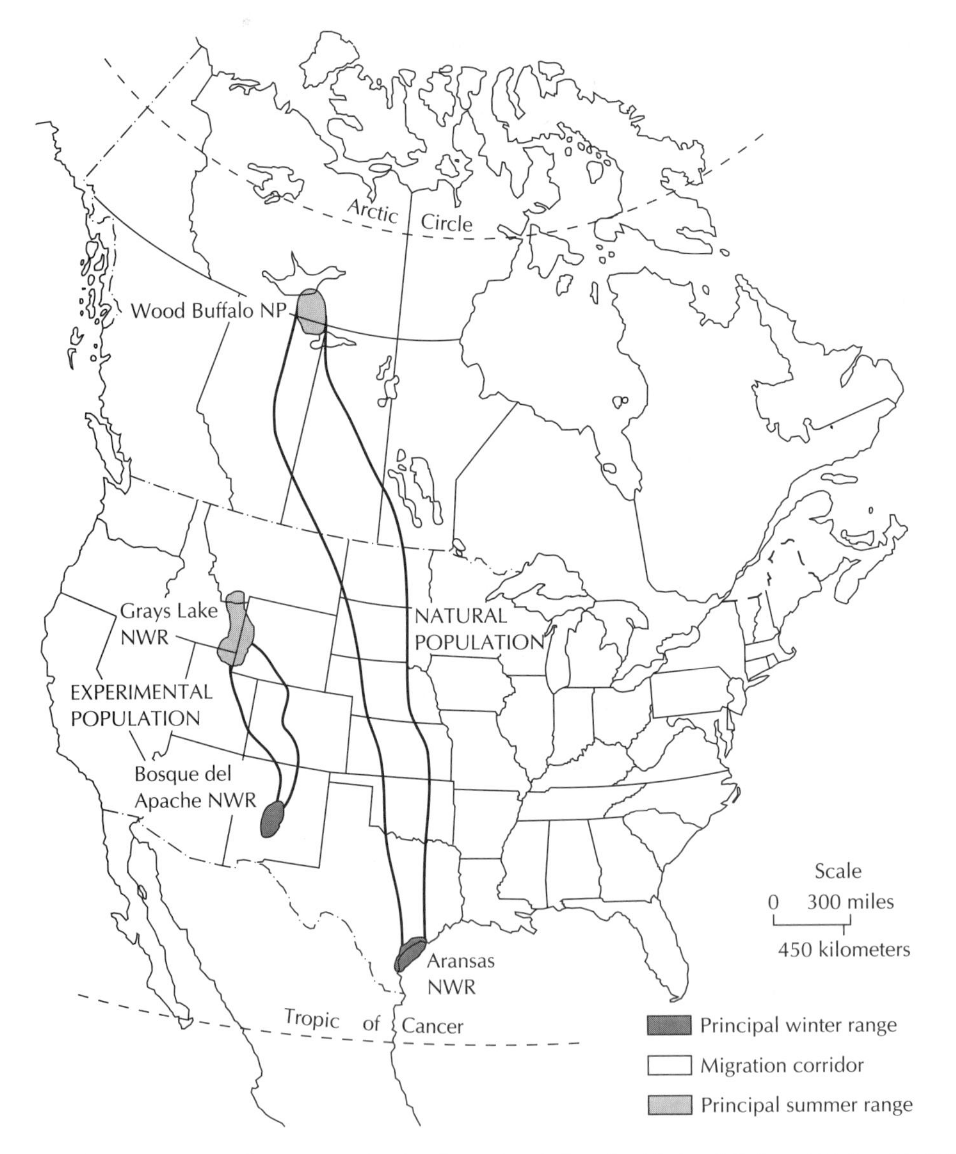

Current distribution of the Whooping Crane, a species that is slowly recovering from near extinction. NP and NWR are abbreviations for National Park and National Wildlife Refuge. (After McMillen 1988)

Box 24–4 *(continued)*

upper midwestern states and prairie provinces during the nineteenth century, declined to a low of only 18 individuals (in 1939) that wintered at Aransas National Wildlife Refuge on the Texas coast (U.S. Fish and Wildlife Service 1986; McMillen 1988) (see map). In addition, three nonmigratory cranes lived year-round in southwestern Louisiana, but they had not bred since 1939, when the conservation efforts began.

International concern about this endangered species impelled the governments of the United States and Canada to work together to prevent the extinction of the Whooping Crane. In 1945, the two nations combined forces in a multifaceted, international rescue program that included habitat preservation, captive propagation and release, and public awareness campaigns. The U.S. Fish and Wildlife Service and the Canadian Wildlife Service now jointly manage the rebounding Whooping Crane populations, according to a formal agreement signed in 1985. This agreement assigns the United States and Canada equal ownership of all eggs, birds, and specimens of Whooping Cranes. The goal has been to build the Wood Buffalo (Canada) population up to 40 breeding pairs and to create two other free-living populations of 25 breeding pairs each.

As of 1993, the main Wood Buffalo–Aransas (Texas) flock was up to 121 adults and 15 young from 1992; at least 44 pairs are expected to breed. The population of Whooping Cranes has increased dramatically to roughly 160 birds in two free-living populations, plus an additional 48 in captivity and experimental populations (see figure). By the end of the century, the cooperative and determined scientific effort should establish healthy, self-sustaining populations of Whooping Cranes.

An ambitious but not entirely successful component of the international program to rescue the Whooping Crane has been a combined effort of government and private organizations to build flocks through experimental propagation techniques. The two most important experimental flocks have been a captive one at the Patuxent, Maryland, facility of the U.S. Fish and Wildlife Service, and a set of young cranes raised by free-living Sandhill Cranes in Idaho. These experimental flocks were built largely through incubation of eggs removed from the nests of wild cranes. This clever technique works because cranes lay two eggs but fledge only one chick (see insurance eggs, Chapter 21).

The captive breeding program has had both successes and failures. Outbreaks of disease in the breeding facility at Patuxent, Maryland, have killed the much-needed productive adult females. Efforts to build a flock of young Whooping Cranes cross-fostered by parent Sandhill Cranes in Idaho have been hampered by high chick mortality, collisions with power lines along migration routes, and most recently, avian tuberculosis. Also, surviving chicks were not inclined to breed with one another, and at least one paired instead with a Sandhill Crane, producing a hybrid Sandhill Crane X Whooping Crane.

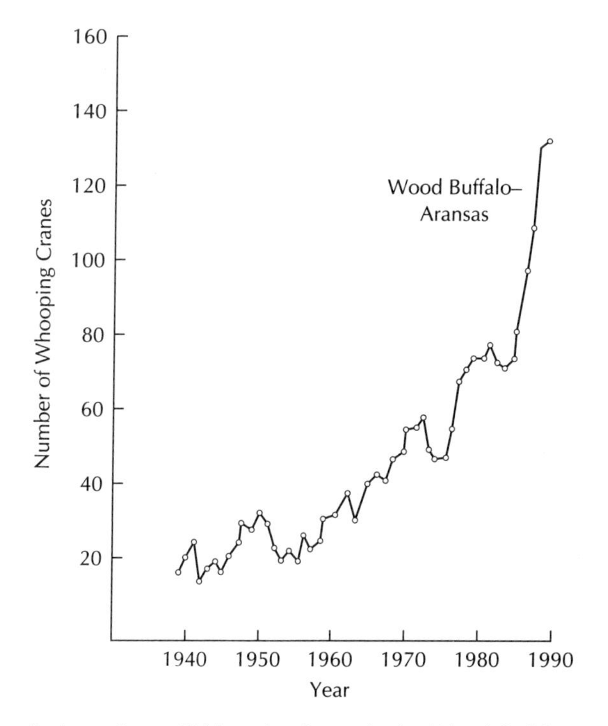

Peak numbers of Whooping Cranes in the Wood Buffalo–Aransas population from 1939 to 1993. (After McMillen 1988)

The initial efforts to save the California Condor polarized two principal factions (Snyder and Snyder 1989). In one political camp were those who considered the condors an untouchable symbol of wilderness and of the vast remaining expanses of southern California threatened by expanding populations of people. At worst, should this species become extinct, its noble death would conclude an era of Earth's history. In the other camp were those who believed that intervention was both warranted and essential to save the species, even if only as captives in zoos, because it was humankind that had brought the condors to this sorry state. Beneath these differences in philosophy, however, were other agendas of open land preservation, continued program funding, and a fear of failing. Contributing to the conflict was the assumption that the condor's population problem was one of reproduction, because they were wary, sensitive birds, intolerant of disturbance and incapable of attempting to raise a second brood if the first was lost. It turns out that this was a historical myth. Critical breakthroughs were finally achieved with the impartial assessments of alternative prescriptions by Noel and Helen Snyder and by a committee of experts appointed jointly by the American Ornithologists' Union and the National Audubon Society (Ricklefs 1978).

In the end, mortality of free-living condors due to illegal shooting and to lead poisoning from bullet fragments in deer carcasses depleted the remnant population to a few survivors. With the wrenching decision to capture the last free-living condor in 1987, their return to the skies of southern California now shifts to the release and successful rehabilitation of condors hatched and raised in captivity (Box 9–1). The first six young condors were released back into the wild in January 1992, but they have already suffered some casualties. One of the first ones released died after drinking water contaminated with antifreeze. The success of this risky and very expensive conservation effort will depend not only on teaching naive young condors to forage and survive on their own but also on the ability of conservation groups and government agencies to work constructively with one another and to reduce incidental mortality through public education.

Restoration of Peregrines

Restoration programs have the goal of reestablishing self-sustaining natural populations of a species and, if successful, may reach the point of calling a halt to a public-spirited campaign. This is the case of the successful effort to restore the Peregrine Falcon to eastern North America (Figure 24–12).

Peregrine populations in North America, particularly in the eastern United States and Canada, virtually disappeared in the 1950s and 1960s, primarily as a result of reproductive failure due to DDT pesticide poisoning (Hickey 1969). The ban on DDT for most uses in the United States removed the immediate problem and set the stage for a bold conservation

Figure 24–12 Peregrine Falcon, a raptor whose extinct populations have been replaced by local restoration programs using captive-raised young birds. (Courtesy of T. Pedersen)

initiative, namely, to rebuild a free-living population of eastern Peregrines by raising young falcons in captivity and then releasing them into the wild in a procedure called hacking. Private falconers joined the program led by Thomas Cade at the Cornell Laboratory of Ornithology to help breed the large numbers of young birds necessary for the success of the hacking effort.

Now that the program is a major success, with a substantial population of independent falcons, its managers must decide when to terminate it. This decision must take into account both biological projections and political interests (Tordoff 1992). To reestablish a self-sustaining breeding population in midwestern North America, one group of dedicated conservationists released 669 young falcons between 1982 and 1992 at a cost of about $2500 per bird, or a total investment of $1,675,000. This effort had produced 34 nesting pairs by 1992. Fecundity of the new midwestern Peregrines averages 1.5 fledged young per pair annually, the same as in healthy, wild Peregrine populations. Models of population growth project a stable final population size of 100+ adult Peregrines in the mid-1990s. From a genetic standpoint, the rebuilt population also appears to be strong. For example, Minnesota Peregrines are mixing with populations to the west, north, and east, fostering increased genetic variability.

From a theoretical standpoint at least, it soon may be time to direct such restoration efforts toward other species in trouble. Powerful political forces, however, favor continuation of the Peregrine program. Falcon breeders want to continue to produce young Peregrines for conservation and government wildlife programs are reluctant to give up the positive publicity and potential financial support.

The Hawk Mountain Story

Sixty years ago, in 1934, a dedicated New York conservationist named Rosalie Edge challenged local traditions; she put an end to hawk shooting on a mountaintop in southeastern Pennsylvania. Rosalie Edge believed that the time to save a species is while it is still common. Her private initiative—now a classic conservation story—started the sport of hawk watching as an alternative to hawk shooting and created a model for the conservation of migratory hawks and eagles worldwide (Senner 1989; Bildstein et al. 1993).

On their way south in the fall, migrating hawks hug the tops of mountain ridges, riding favorable, rising air currents. They pass key sites in great numbers on days of favorable winds, providing living targets to those who wished to eliminate predators. The annual toll of tens of thousands of Sharp-shinned Hawks and other species was staggering. Determined to stop the slaughter, Rosalie Edge raised the money to buy 567 hectares (1400 acres) on Hawk Mountain (Figure 24–13) and installed a brave young naturalist warden, Maurice Broun, to protect the hawks and to share his appreciation of raptors as beneficial rather than harmful members of natural ecosystems. Since its founding, the Hawk Mountain Sanctuary Association has played a key role in protecting North

Figure 24–13 The lookout at Hawk Mountain Sanctuary near Reading, Pennsylvania. (Courtesy of Hawk Mountain Sanctuary Association)

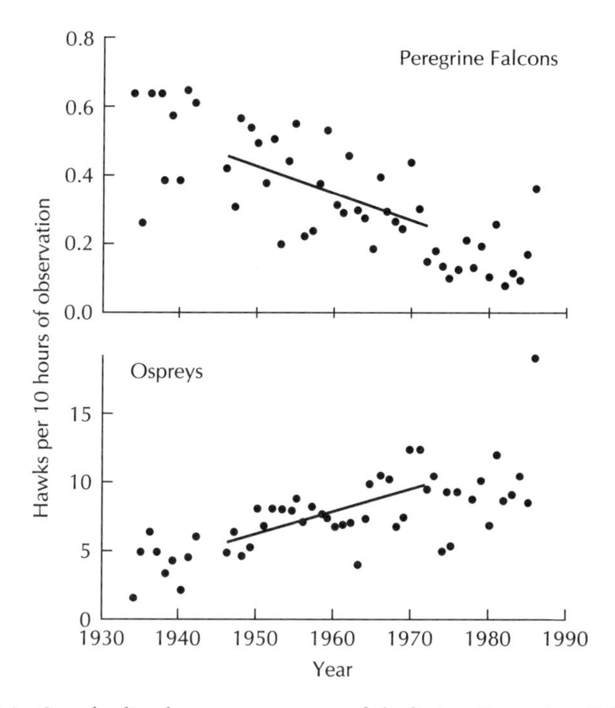

Figure 24–14 Standardized autumn counts of declining Peregrine Falcons and increasing Ospreys at Hawk Mountain Sanctuary, near Reading, Pennsylvania, from 1934 to 1986. (After Bednarz et al. 1990)

America's raptors and their essential habitats by developing grass-roots support for state and national legislation (Senner 1984). Each year, 70,000 visitors come to view the inspiring passage of hawks and eagles, and leave Hawk Mountain with a greater commitment to conservation.

Reflecting its long, pioneering history, Hawk Mountain also maintains the world's longest and most detailed record of raptor migration. The millionth raptor—an immature female Sharp-shinned Hawk—was officially logged at 12:41 P.M. on Thursday, October 8, 1992. The annual counts of hawks and eagles that migrate past Hawk Mountain have proved to be effective tools in assessing long-term trends in raptor populations throughout eastern North America (Figure 24–14) (Bednarz et al. 1990). This data base played a key role in exposing first generation organochlorine pesticides, such as DDT, as causative agents for the declines of several species of birds of prey, as well as measuring their population recoveries following decreased use of environmental pesticides.

Promoting Protection Through Pride

Public support of conservation programs is an essential ingredient for success. An unusual conservation program in the Caribbean demonstrates the power of fully engaging public participation. A populist revolution is sweeping through the islands there (Butler 1992). The revolution started on the island of St. Lucia, where Paul Butler, a British naturalist, piqued the local indifference about wildlife into a passionate embrace of an endangered parrot, the St. Lucia Parrot, known locally as the Jacquot. The population of this species had declined to a precariously small size as a result of being hunted for food and captured for the pet trade. The mountain forests where it lived had also been cut for firewood and for farming. Now the island's national bird and center of pride, the Jacquot is increasing in numbers, and as the island's conservation spokesman, the Jacquot exhorts fellow St. Lucians to save the island's forests, keep the water clean, protect their island's coral reefs, and engage in other environmental efforts.

> The parrot's colorful image now appears on billboards, bumper-stickers, T-shirts, and St. Lucia passports. A lively combination of classroom visits [featuring a person dressed up like a parrot], reggae songs, music videos, church sermons, and puppet shows has made saving the jacquot a cause célèbre with all age groups on the island. As a result, the parrot has not only stopped its slide toward extinction, but has nearly doubled its population to about 350 birds. (Nielsen 1993, p. 48)

Paul Butler's technique of using local pride is spreading throughout the Caribbean and Latin America, sponsored by the RARE Center for Tropical Bird Conservation, an action-oriented, Philadelphia-based membership organization. RARE Center's philosophy, like Butler's, is to get to the root of a conservation issue quickly and then to work with the local people, who can make a difference. Paul Butler's recipe for success can convert an island's human population into an effective conservation force in just 1 year. This conservation blitz heightens awareness everywhere, with special attention to schoolchildren, who educate their parents.

The Value of Birdwatchers

Recreational birdwatching is now an economically significant sport (Kerlinger and Wiedner 1990). In the United States, there are an estimated 61 million birdwatchers. Canadians are traditionally even more ardent, as are the British.

The majority of birdwatchers simply enjoy watching birds at the feeders in their backyards. In the United States, roughly 300,000 of these are seriously committed birders who can identify more than 100 species. The average committed American birder is well educated, earns an income well above the national average, belongs to at least three conservation or birding organizations, and spends $1884 annually directly on his or her favorite pastime, for travel, equipment, and books—which does not include birdseed and other backyard bird attractions (Table 24–2). Overall,

TABLE 24-2
Summary of economic expenditures by birders

Expenditure	Calculation	Total Dollars Spent per Birder per Year
Travel		
Airplane flights	0.59 flights × $250 per round trip	$ 148
Automobile	2,763 miles × $0.30 per mile	829
Car rental	0.50 rentals × $75 per rental	38
Hotel/motel	12.9 nights × 21.7% of birders × $20 per night	99
Campsite	12.9 nights × 38.3% of birders × $10 per night	28
Both campsite and motel/hotel	12.9 nights × 22.6% of birders × $15 per night	44
Meals	13 days × 82% of birders × $15 per day	160
Birding tour expenses	28% of birders × $150 to tour group	42
Miscellaneous items		
Books	4.2 books per birder × $18.40 per book	77
Magazines	1.6 magazines per birder × $20 per year	32
Conservation organizations	3.1 organizations × $20 per year	62
Artwork	Average per birder	56
Other paraphernalia	Average per birder	172
Optical equipment		
Binoculars	Average per birder	53
Scopes	Average per birder	30
Tripods	Average per birder	10
Christmas bird count (fee)	$4 per participant	4
Total annual expenditures per birder		$1884

Source: Kerlinger and Wiedner 1990.

the total spent by birders throughout North America in 1990 was more than $20 billion. In Great Britain, the presence of a nearby Royal Society for the Protection of Birds bird reserve substantially improves the economies of small rural towns by increasing sales of petrol, beer, and pub lunches.

The growth of birdwatching as a national pastime reflects the efforts of pioneering conservationists like Rosalie Edge. Before her, when he was not censusing birds on women's hats in New York City, Frank Chapman promoted winter birdwatching in the form of Christmas bird counts to replace the traditional end-of-the-year bird shooting parties. The ninety-fourth Christmas bird count organized by the National Audubon Society in 1993 boasted 43,000 participants, all volunteers and mostly skilled amateurs.

The evolution of lightweight, inexpensive field binoculars increased the reliability as well as the fun of field identifications and reduced the need to collect specimens. Above all, the publication in 1934 of Roger Tory Peterson's seminal *Field Guide to the Birds* and subsequent field guides catalyzed popular as well as professional interest in birdwatching. These and other sources of stimulation of the birdwatching "industry" also are responsible for a new political force for conservation, because each new birder and amateur ornithologist is likely to be a conservationist.

Personal Conservation Initiatives

The roots of current avian population declines in North America, as well as in other parts of the world, go deep into past practices of uncontrolled exploitation. A new environmental ethic has swept the North American continent and countries elsewhere in the world. Conservation of biodiversity is now both a national and international priority, with increasing legislative protection of wild bird species.

Each of us can help to reduce losses of local bird populations and prevent future extinctions. Here are eight simple steps toward being a constructive conservationist:

1. Keep your cats inside the house.
2. Enhance your property with plants and cover that attract and sustain birds.
3. Put a hawk silhouette on your large windows to reduce window kills.
4. Help to control human encroachment on the natural habitats of birds.
5. Recognize that conservation is a personal responsibility and that we must heed the warnings of our past mistakes.
6. Organize or join a local bird club and help to educate others in your community.
7. Support national and international bird conservation organizations, listed in the National Wildlife Federation's annually updated *Conservation Directory*.
8. Help to gather data on the status and trends of bird species as a volunteer participant in organized, professionally managed programs, advertised in the American Birding Association's "Opportunities for Volunteers."

Summary

Birds now face a new episode of species extinctions because of human encroachments onto natural habitats. Loss or fragmentation of specialized habitats, excessive hunting, and repeated nesting failures caused by pesticides or predators threaten 11 percent of all bird species with global extinction. Island species are particularly vulnerable. Human activities in North America cause an estimated 270 million bird deaths each year, principally as a result of hunting and pest control, but also from collisions with vehicles, television towers, and picture windows. Predation by domestic cats is one of the greatest causes of human-related deaths of wild birds. Pesticides have devastated some bird populations by interfering with eggshell formation. The exotic cage bird trade is growing rapidly, and increasingly threatens some wild bird populations and even some species.

Currently paramount among the negative forces of humankind on bird populations is the rapid destruction of the natural habitats of the world ranging from the tropical rain forests to a variety of Temperate Zone backyard habitats. Four cases of bird conservation illustrate the challenges of providing adequate habitat for endangered bird species: the conflict between the logging industry and the Spotted Owl in the Pacific Northwest; the endangerment of songbirds by the growth of suburbia in Texas and southern California; the decline of migrant songbirds throughout North America; and the loss of wetland habitats and associated water birds. With the commitment to set aside critical habitat for endangered species comes the challenge of designing these reserves, a fertile and controversial topic of the redefined field of conservation biology. Revival of severely threatened bird populations through the release of captive-bred individuals is a tool of conservation management.

Inevitably, tempering and sometimes aiding practical efforts are both the politics and economics of conservation. The economic value of birdwatchers as ecotourists and the cultivation of local pride are powerful forces behind successful conservation projects. In addition to public concern, the key ingredients for success of bird conservation programs worldwide are sound ornithological knowledge of a species' biology and the political will to help species prosper.

Further Readings

Askins, R.A., J.F. Lynch, and R. Greenberg. 1990. Population declines in migratory birds in eastern North America. Current Ornithology 7: 1–57. *A detailed review of current concerns.*

Beissinger, S.R., and N.F.R. Snyder, Eds. 1991. New World Parrots in Crisis. Washington, D.C.: Smithsonian Institution Press. *Key papers reviewing future strategies for managing parrot populations harvested for the pet trade.*

Collar, N.J., and P. Andrew. 1988. Birds to Watch: The ICBP Annotated Checklist of Threatened Birds of the World. Cambridge: International Council for Bird Preservation. *A succinct list of currently threatened bird species.*

Collar, N.J., L.P. Gonzaga, N. Krabbe, A. Madroño-Nieto, L.G. Naranjo, T.A. Parker III, and D.C. Wege. 1992. Threatened Birds of the Americas. Cambridge: International Council for Bird Preservation. *A primary reference.*

Hagan, J.M., III, and D.W. Johnston, Eds. 1992. Ecology and Conservation of Neotropical Migrant Landbirds. Washington, D.C.: Smithsonian Institution Press. *A major collection of important and timely papers on a problem of growing concern.*

Matthiessen, P. 1959. Wildlife in America. New York: Viking Press. *A classic flowing account of wildlife destruction and conservation in North America.*

Morrison, M.L. 1986. Bird populations as indicators of environmental change. Current Ornithology 3: 429–451. *Contains many examples of the values of birds to a society concerned about the environment.*

National Wildlife Federation. 1994. Conservation Directory. *The complete annual reference guide to environmental conservation organizations.*

Salathe, T., Ed. 1991. Conserving Migratory Birds. ICBP Technical Publication No. 12. Cambridge: International Council for Bird Preservation. *An important and timely set of papers on one of the most challenging global bird conservation problems.*

Soulé, M., Ed. 1986. Conservation Biology: The Science of Scarcity and Diversity. Sunderland, Mass.: Sinauer Associates. *One of the pioneering texts on a rapidly developing new science.*

Temple, S.A. 1986. The problem of avian extinctions. Current Ornithology 3: 453–485. *A survey of the patterns of endangerment of modern bird populations, and conservation management solutions.*

APPENDIX

Birds of the World

PART I (CHAPTERS 1–3) INTRODUCED the characteristics of birds and their evolution, as well as the principles of classification and phylogeny. This appendix introduces the major taxonomic groups—called orders—of the birds of the world. Included for each order are summaries of general characteristics and distinguishing features, component families with their geographical distributions, the number of genera and species in each family, plus illustrations by James Coe of some representative species. For students in advanced courses that focus on avian systematics, a section of taxonomic comments for each order points to the recent technical literature.

A detailed analysis of the evidence supporting avian relationships is beyond the scope of this book; see Charles Sibley and Jon Ahlquist (1990) for summaries of technical diagnoses. In most cases the relationships among orders are unknown; therefore, the sequence in which orders are listed is based largely on tradition. This presentation follows the established classification of Robert Storer (1971b), modified by recent advances, particularly the genetic analyses of evolutionary relationships that use biochemical techniques. Familiar, traditional names are standardized at the family level. Again, following Storer, loons (Gaviiformes) are placed after the Order Charadriiformes, a still controversial placement far from the grebes (Podicipediformes) at the beginning of the classification. Also controversial are the relationships of the Hoatzin (Opisthocomidae), which is included in the Order Cuculiformes, following Sibley and Ahlquist (1973, 1990), instead of in the Order Galliformes. The turacos (Musophagiformes), whose relationships remain unknown, appear as a separate order. Many groups of passerine birds appear as separate families, so that they can be regrouped in the future in accordance with new evidence. For example, the current subfamilies of the Family Muscicapidae are elevated to separate families—Old World warblers as Sylviidae, thrushes as Turdidae, and so on. Families of oscine passerine birds are grouped as recommended by Charles Sibley and Burt Monroe (1990), even though these relationships are still fresh and under discussion.

A well-established and familiar classification should be revised only when justified by evidence from several sources. Challenging traditional views is the revolutionary classification of the birds of the world proposed by Charles Sibley and Jon Ahlquist (1990). Based on comparisons of genomic DNA that were made using the techniques of DNA hybridization, this classification overhauls the nomenclature, hierarchical structure, and sequence of taxa included in the Class Aves. Except for the general groupings of passerine birds, however, I have not accepted their classification outright because it is still under critical review and may be subject to substantial revision (Gill and Sheldon 1991; Lanyon 1992b; Siegel-Causey 1992). Many independent tests of Sibley and Ahlquist's hypotheses and reanalyses of their data are under way. In time, the best of their monumental work will be carefully integrated with other evidence. Then it will be time to revise, perhaps radically, some parts of the classification presented here.

The species lists of Charles Sibley and Burt Monroe (1990, 1993) provide the most comprehensive and up-to-date definition of species limits and distributions of the birds of the world, even though they are structured according to the Sibley and Ahlquist (1990) classification. Consequently, I have updated the numbers of genera and species in each family to follow this presentation. Kenneth Parkes (1992) and Robert Storer (1992) review some of the limits of this work.

Birds: Class Aves

Birds are bipedal, tetrapod vertebrates with feathers. Feathers alone distinguish birds from all other vertebrates. Birds are derived from reptiles, some of which, for example, Crocodilia, are more closely related to the birds than to other lineages included in the Class Reptilia. The avian skeleton, which retains many reptilian features, is modified for flight. The bones and feathers of the forelimb, particularly, are transformed into wings capable of powerful, flapping flight. Birds maintain a high body temperature through metabolic heat production and have a large, four-chambered heart that supports the demands of sustained activity and high metabolism.

The Class Aves comprises two superorders: (1) Superorder Palaeognathae, which includes the ratites and tinamous, and (2) Superorder Neognathae, which includes all other modern birds. The complete classification of modern, living birds is a hierarchical arrangement of 29 orders and 187 families, which include 2029 genera and approximately 9600 species. The numbers of genera and species will continue to change as revisions and discoveries are made. Excluded from this classification are most extinct taxa known only from the fossil record. Following tradition, I include species such as the Carolina Parakeet and even the Dodo that have become extinct in historical times.

Ratites and Tinamous

The ratites and tinamous are an ancient group of bird families that share a unique configuration of bones between the nasal passages (the paleognathous palate). The ratites are flightless birds with reduced wing bones and a sternum that lacks a keel—the bony plate that serves as an anchor for flight muscles in most birds. Tinamous have a reduced sternal keel.

The Ostrich is the largest living bird; some individuals are 2.75 meters tall and 150 kilograms in weight. It and the other large ratites—rheas, Emus, and cassowaries—are long-necked birds with strong, muscular legs adapted for running. All are terrestrial, and the number of toes is reduced to two in the Ostrich and to three in the rheas, Emus, and cassowaries. The smaller kiwis and most tinamous still have four toes.

In the Ostrich, rheas, Emus, and cassowaries, the bill is relatively flat and the nostrils are oval; in the kiwis, the bill is long and slightly curved, and the nostrils are located at the tip; in the tinamous, the bill is more chickenlike, and the nostrils are covered with a fleshy plate called a cere.

Most members of the superorder Palaeognathae have distinct feather tracts, but the Ostrich has feathers distributed continuously over its body. The plumage is loose in all groups except the tinamous; it is hairlike in the kiwis. An aftershaft is strongly developed in Emus and cassowaries, absent in the Ostrich, rheas, and kiwis, and small to well developed in tinamous. Kiwis lay enormous eggs, and tinamous lay glossy, heavily pigmented, often brightly colored eggs. Males incubate the eggs and care for the young in unusual mating systems. The precocial chicks of both ratites and tinamous are clad in down at hatching.

Order/Family	Members	Distribution	Genera	Species
Tinamiformes				
Tinamidae	Tinamous	Neotropics	9	47
Rheiformes				
Rheidae	Rheas	Neotropics	1	2
Struthioniformes				
Struthionidae	Ostrich	Africa	1	1
Casuariiformes				
Dromiceidae	Emu	Australia	1	1
Casuariidae	Cassowaries	Australia, New Guinea	1	3
Dinornithiformes				
Apterygidae	Kiwis	New Zealand	1	3

1
2
3
4
5
♂
COE

Taxonomic Comments

The monophyly of the ratites now seems well established by morphological, biochemical, and chromosomal evidence (see Bock 1963; Parkes and Clark 1966; Sibley and Frelin 1972; Cracraft 1974a; Prager et al. 1976; de Boer 1980; Sibley and Ahlquist 1981a; Stapel et al. 1984). Some classifications treat the ratites as separate families in the Order Struthioniformes. Houde and Olson (1981) suggest that flying paleognathous birds of the early Tertiary of North America were the ancestors of modern birds in the Superorder Neognathae as well as the extant ratites. The paleognathous palate thus may represent a primitive character state.

Relations among the major groups of ratite birds remain controversial, including whether they originated in the Southern or Northern Hemisphere (see Chapter 3) (Houde 1986; Sibley and Ahlquist 1990; Gill and Sheldon 1991; Cooper et al. 1992). Tinamous probably are the sister group to flightless ratites. The extinct moas may represent a separate lineage not allied to kiwis.

Figure A–1 Ratites and tinamous: (1) Elegant Crested-Tinamou (Tinamidae); (2) Southern Cassowary (Casuariidae); (3) Brown Kiwi (Apterygidae); (4) Greater Rhea (Rheidae); (5) Ostrich (Struthionidae).

1
2
3
4
5

Grebes
Order Podicipediformes

Grebes are medium-sized, foot-propelled diving birds with stocky bodies, slender necks, and small heads. The toes are lobed and the tarsi are laterally compressed, thereby offering little resistance when drawn forward through the water. The legs are located far back on the body, thus making it difficult for the birds to move on land but giving them a powerful forward thrust and maneuverability during dives. The claws on the toes are unusual in being flattened like fingernails.

The dense, satiny plumage of grebes is waterproofed by secretions from a tufted oil gland above the base of the tail. The tail feathers are reduced, so grebes look tailless, a feature that contributes to the stocky appearance of some species. In most species, the bill is slender and pointed; but in some—for example, the Pied-billed Grebe of the New World—the bill is stouter, an adaptation to this bird's diet of hard-shelled crustaceans. Like many highly adapted diving birds, grebes have difficulty taking off from the water, and, with rapidly beating wings, they must run across the surface before becoming airborne.

The sexes are similar in appearance; some species acquire distinct breeding plumages. Elaborate courtship displays include fancy dives and rises out of the water, and prolonged rushes side by side. The young are clad in fine down at hatching. In most species, the down is handsomely striped in black and white; but in the Western and Clark's grebes of North America, the down is plain gray. Most grebes build floating platforms of aquatic weeds on which they mate and lay their eggs. They carry their young on their back, even when diving. Grebes eat their own feathers, which then trap fish bones in the stomach, holding the bones for prolonged digestion or regurgitation.

Family	Members	Distribution	Genera	Species
Podicipedidae	Grebes	Worldwide	6	21

Taxonomic Comments

The affinities of the grebes, which have no close relatives, remain uncertain (Storer 1960, 1986), although traditionally and, probably incorrectly, they have been placed next to loons. The fossil record suggests that they originated in southern South America perhaps 80 million years ago in the Cretaceous. Sibley and Ahlquist (1990) suggest that grebes share a common ancestor with the tropicbirds, cormorants, herons, and other members of a comprehensive water bird assemblage.

Species relationships are reviewed by Storer (1963, 1979) and Vlug and Fjeldså (1990).

Figure A–2 Grebes: (1) Eared Grebe; (2) Horned Grebe; (3) Great Crested Grebe; (4) Pied-billed Grebe; (5) Western Grebe.

1
2
3
4
CoE

Penguins
Order Sphenisciformes

The penguins form a distinctive order of flightless, marine, diving birds mostly of the southern oceans. They can dive down to 800 meters in search of food. The Emperor Penguin fasts for 3 months while incubating its one egg, which it balances on its feet.

The most striking adaptation of these birds involves their wings. The bones of the wing are flattened and somewhat fused, so the wing cannot be folded as in other birds; the result is a very efficient flipper, which is the principal means of locomotion under water. Because the wing is still used for locomotion, the keel of the sternum is well developed, unlike the keelless sternum of the flightless ratites. At the base of the wing is a complex network of blood vessels—called a vascular rete—in which cooled blood returning to the body from the wing absorbs heat from blood flowing outward into the wing. As a result, more heat is retained in the body and not lost during the inevitable chilling of the wings.

The legs are short, and the stout, webbed feet are located far back on the body, thereby enabling the birds to maintain an upright posture when they are on land. Unlike many other diving birds, penguins can walk when onshore, some clumsily, some with agility. The plumage of penguins is dense and waterproof; the feathers are distributed continuously over the body rather than in discrete tracts as in most birds. The feathers are shed and replaced in patches. The body is further insulated from cold water by a heavy layer of fat beneath the skin. The bill of most species is laterally compressed, but in the largest penguins, the King and Emperor, the bill is long and slender. Most penguins are clad in bluish black and white and often have distinctive adult head patterns. The young are clad in dense down at hatching.

The order is confined to the Southern Hemisphere, but reaches the equator in the Galápagos islands.

Family	Members	Distribution	Genera	Species
Spheniscidae	Penguins	Southern oceans	6	17

Taxonomic Comments

Penguins evolved early in the Tertiary period from flying, tube-nosed seabird (procellariiform) ancestors (Simpson 1946; Sibley and Ahlquist 1972; Ho et al. 1976). The presence of tubular nostrils in both fossil penguins and in the Little Penguin, considered the most primitive living member of the order, supports this relationship, as do DNA comparisons. The two groups also have similar bill-fencing courtship displays and methods of feeding their young.

Figure A–3 Penguins: (1) Chinstrap Penguin; (2) Rockhopper Penguin; (3) Jackass Penguin; (4) King Penguin, juvenile (left), adult (right).

1
2
3
COE
4

Tube-nosed Seabirds
Order Procellariiformes

The tube-nosed seabirds comprise several families of pelagic birds, most of which seek food from the surface of the sea. They range in size from the 6-inch Least Storm-Petrel to the huge Wandering Albatross, which has a wingspan of nearly 12 feet. All have tubular nostrils and a distinctly hooked bill. The plumage is dense and waterproof, and beneath the outer feathering is a dense coat of down. In the roof of the orbit is a large gland that concentrates and excretes salt in drops from the bill.

Most species have long wings, held stiffly and used for soaring and planing over the waves, but the storm-petrels flutter close to the surface on shorter wings, and the diving-petrels of the Southern Hemisphere have very short wings, which they use for locomotion under water, as do small shearwaters. Diving-petrels are convergent in morphology and ecology with the unrelated auklets of the Northern Hemisphere.

Tube-nosed seabirds are clad in black, white, brown, or gray and show little bright color except in the bill and feet of some species. Generally monogamous, males are larger than females. Small species nest in burrows; large albatrosses and some petrels nest on the surface. All lay only one egg and have long incubation and nestling periods. The young are densely downy. Tube-noses have a well-developed sense of smell, which some species use to find their nest burrows at night. Both parents and young squirt foul-smelling stomach oil at intruders.

The order is distributed throughout the oceans and seas of the world, but most species are found in the Southern Hemisphere.

Family	Members	Distribution	Genera	Species
Diomedeidae	Albatrosses	North Pacific, southern oceans	2	14
Procellariidae	Shearwaters, petrels, fulmars	All oceans	14	76
Hydrobatidae	Storm-petrels	All oceans	7	21
Pelecanoididae	Diving-petrels	Southern oceans	1	4

Taxonomic Comments

Members of the Orders Procellariiformes and Sphenisciformes are closely related (Simpson 1946; Ho et al. 1976). In turn, skeletal similarities between albatrosses and frigatebirds, the most primitive members of their respective orders, link the Orders Procellariiformes and Pelecaniformes (Cracraft 1981).

Figure A–4 Tube-nosed seabirds: (1) Wandering Albatross (Diomedeidae); (2) Cape Petrel (Procellariidae); (3) Wilson's Storm-Petrel (Hydrobatidae); (4) Short-tailed Shearwater (Procellariidae).

♂
3
2
1
♂
4
5
6
COE

Pelicans and Allies
Order Pelecaniformes

Belonging to the worldwide, perhaps artificial, Order Pelecaniformes are several groups of large or medium-sized, aquatic birds that eat fish or squid. They differ from all other birds in having totipalmate feet, with all four toes joined by webs. Most have a more or less distensible pouch of bare skin between the branches of the lower mandible; this gular pouch is absent in the tropicbirds and most highly developed in the pelicans, which use it to capture fish, and in the frigatebirds, in which it is inflated by males and used in courtship displays. The external nostrils of cormorants, boobies, and gannets are closed; the birds breathe instead through the mouth.

In all except the tropicbirds, the young are without down at hatching. The plumage of many members of the order is black and white, and others are clad in somber black or dark brown. The feet and bare areas of facial skin may be brightly colored.

The diversity of the members of the Order Pelecaniformes reflects the means the birds employ to obtain fish or squid. Boobies, gannets, and tropicbirds catch prey by diving from the air. Cormorants and anhingas pursue fish under water, cormorants seizing their prey with their hooked bills and anhingas spearing it. Pelicans dive and scoop fish up in their pouched bills, allowing water to drain out before swallowing their catch. Frigatebirds snatch food from other fish-eating birds.

Nearly all members of the order nest in colonies, those of the gannets numbering in the hundreds or thousands. One to three eggs are laid on bare ground, among rocks (tropicbirds, some boobies and cormorants) or in a nest of sticks, reeds, seaweed, or guano. Nests may be on the ground, cliff ledges, or trees. The initially helpless young are tended by both parents.

Family	Members	Distribution	Genera	Species
Phaethontidae	Tropicbirds	Tropical oceans	1	3
Fregatidae	Frigatebirds	Tropical oceans	1	5
Sulidae	Boobies, gannets	All oceans	3	9
Phalacrocoracidae	Cormorants	All continents and oceans	1	38
Anhingidae	Darters, anhingas	Tropical oceans	1	4
Pelecanidae	Pelicans	All continents	1	8

Figure A–5 Pelicans and allies: (1) Red-footed Booby (Sulidae); (2) Magnificent Frigatebird (Fregatidae); (3) White-tailed Tropicbird (Phaethontidae); (4) Anhinga (Anhingidae); (5) Brown Pelican (Pelecanidae); (6) Guanay Cormorant (Phalacrocoracidae).

Taxonomic Comments

Ornithologists debate whether these taxa form a natural unit (Cracraft 1985; Sibley and Ahlquist 1990). Probably related to the pelicans is the Shoebill (*Balaeniceps*), currently classified with the storks (Ciconiiformes).

♂
1
♂
2
3
♂
4
5
6
COE

Waterfowl
Order Anseriformes

The Order Anseriformes contains two distinctive families that share few external features. In both, the aftershaft is reduced or absent, the oil gland is feathered, and the precocial young are clad in down at hatching. Internally, they share characters of the skull, sternum, and syrinx. The ducks, geese, and swans are a diverse group of mainly aquatic birds that have webbed feet with a hind toe that is somewhat elevated, a flattened, blunt-tipped bill covered with a thin layer of skin and bearing a nail at the tip of the maxilla and fine lamellae along the margins of the maxilla and mandible, pointed wings, and a dense coat of firm, waterproof feathers, in distinct tracts, with a layer of down beneath. Males have a penis. Many of these birds are colorful; the sexes of Temperate Zone ducks are often patterned differently. Some species are flightless.

Waterfowl vary in ecology from terrestrial grazers to deepwater divers to agile riders of ocean surfs and mountain streams. Bill form varies with diet, from those with lateral lamellae for straining microscopic food from mud, to strong, broad bills for wrenching mollusks from rock moorings, to narrow bills with sharp toothlike serrations for capturing fish, to short, blunt bills for grazing on field plants.

Drakes typically do not assist in incubation or care of the young, yet they are monogamous. In northern temperate species, drakes pair with hens on the wintering grounds and then return to their mate's birthplace.

Screamers are loud-voiced, turkey-sized birds that have long, slender toes with only rudimentary webs, a hind toe on the same level as the front toes, a short, slightly hooked bill, rounded wings, and no defined feather tracts or apteria. They have stout spurs at the bend of the wings. The males lack a penis, and, unusual among birds, screamers lack uncinate processes. The most peculiar feature of the screamers is a skin filled with small bubbles of air about a quarter of an inch thick, which produce a crackling sound when pressed; the function of this layer of air bubbles is unknown.

Family	Members	Distribution	Genera	Species
Anhimidae	Screamers	South America	2	3
Anatidae	Swans, geese, ducks	Worldwide	46	158

Figure A–6 Waterfowl: (1) Musk Duck (Anatidae); (2) Smew (Anatidae); (3) Black-necked Swan (Anatidae); (4) Mallard (Anatidae); (5) Magpie Goose (Anatidae); (6) Horned Screamer (Anhimidae).

Taxonomic Comments

Waterfowl traditionally have been considered related to the Order Galliformes (Johnsgard 1968; Prager and Wilson 1976). Rather than retaining primitive, galliformlike features, as was once thought, the turkeylike screamers may be highly specialized waterfowl (Olson and Feduccia 1980b). The Pied Goose of Australia, however, retains primitive characters and represents an early lineage; and it may merit recognition as a separate family (Sibley and Monroe 1990).

Flamingos
Order Phoenicopteriformes

The flamingos are distinguished by their long necks with elongated vertebrae, long legs with webbed feet, generally pink coloration, and their specialized, filter-feeding bills, which are bent downward in the middle. The margins of the bill bear long lamellae for filtering small organisms out of mud or water (see Figure 7–4). The maxilla fits within the mandible, and during feeding the bill is placed in the water with the maxilla downward. The tongue is thick and fleshy and is used to circulate water between the lamellae.

Males are larger than females, but the sexes are alike in their pink plumage, which includes carotenoid pigments obtained from their food. Immatures are drab in color.

Flamingos mostly occupy large, shallow lagoons and lakes, some at extremely high altitudes in the Andes of South America. The salt concentrations and alkalinity of these lakes can be extreme, eliminating fish but allowing dense concentrations of small invertebrates and blue-green algae. Flamingos wander nomadically and breed erratically in dense, highly synchronized colonies.

Flamingos construct tall cone-shaped nests of mud, in which the female lays a single, chalky white egg. The young are clad in whitish down at hatching, and their bills gradually assume the specialized adult shape. Like pigeons, flamingos produce a milk rich in fat and protein, which they feed to young chicks.

Family	Members	Distribution	Genera	Species
Phoenicopteridae	Flamingos	Tropics, except Australia	1	5

Taxonomic Comments

Modern genera of flamingos were present in the fossil record 30 million years ago, with more primitive taxa dating back over 50 million years ago.

The relationships of flamingos, which are an ancient, distinct group of birds, remain controversial. Traditional debate has been whether they are closest to the Ciconiiformes or the Anseriformes, because flamingos seem to have characters of both orders (see Chapter 3). Olson and Feduccia (1980a) contend that the flamingos are related instead to the stilts (Recurvirostridae) and recommend placing them in the Order Charadriiformes. DNA studies point to the Order Ciconiiformes (Sibley and Ahlquist 1990). Bile acids analyses place flamingos in the Order Anseriformes, closest to geese (Hagey et al. 1990).

Figure A–7 Flamingos: Greater Flamingos.

1
2
3
4
5
6
COE

Herons, Storks, and Allies
Order Ciconiiformes

Most members of the Order Ciconiiformes are long-legged, long-necked birds that wade in shallow water or, in some cases, feed on open ground. Many species breed colonially and place their stick nests in trees. The various families differ in the shape of the bill: Herons have long, spear-shaped bills; storks have bills that are usually straight and sharp, but sometimes have a slight curve at the tip; ibises have long, curved bills; spoonbills have long bills with a broad, flattened, spoon-shaped tip. The Shoebill has a massive, hook-tipped, bulbous bill, which serves among other functions to bring water to its nestlings.

The herons have powderdowns (feathers that disintegrate and are used to condition the rest of the plumage), a comblike margin on the claw of the middle toe, and a modification of the vertebrae of the neck that provides a spearing mechanism and also allows the neck to fold into an S-shaped curve. The Shoebill also has a tract of powderdowns. The Hammerkop builds a huge, domed stick nest. Ibises and spoonbills, despite the difference in the shape of their bills, show their relationship in a number of features, the most noticeable of which is a pair of grooves that extend from the nostrils to the tip of the bill.

The storks lack a syrinx and clatter their bills. Members range in size from the Least Bitterns, about a foot tall, to the Goliath Heron of Africa and the Marabou Storks (*Leptoptilos*), the latter attaining a height of 5 feet.

Family	Members	Distribution	Genera	Species
Ardeidae	Herons, bitterns, egrets	Worldwide	20	65
Balaenicipitidae	Shoebill	Africa	1	1
Scopidae	Hammerkop	Africa, Madagascar	1	1
Threskiornithidae	Ibises, spoonbills	Tropics; some temperate regions	14	34
Ciconiidae	Storks	Tropics; temperate Eurasia	6	19

Figure A–8 Herons, storks, and allies: (1) Eurasian Spoonbill (Threskiornithidae); (2) White Stork (Ciconiidae); (3) Sacred Ibis (Threskiornithidae); (4) Great Blue Heron (Ardeidae); (5) Hammerkop (Scopidae); (6) Shoebill (Balaenicipitidae).

Taxonomic Comments

Members of the Order Ciconiiformes may not all be related. The Shoebill may belong in the Order Pelecaniformes (Cottam 1957; Saiff 1978; Sibley and Ahlquist 1990). Herons may belong in the Order Gruiformes (Olson 1978a). The storks may be more closely related to New World vultures (Falconiformes, Cathartidae) (Ligon 1967; Olson 1985; Sibley and Ahlquist 1990).

1
2
3
4
5
6
COE

Birds of Prey
Order Falconiformes

Most members of the Order Falconiformes are diurnal, raptorial birds with short, strongly hooked bills with a fleshy cere containing the imperforate nostrils, and sharp, curved talons (claws). The typical falcons (*Falco*) have a bony tubercle in the nostril. The legs are generally short but are very long in the largely terrestrial Secretary-bird of Africa. The wings vary considerably in shape: broad and rounded in eagles and many hawks, long and narrow in harriers (*Circus*), and pointed in falcons and many kites. The Osprey is distinguished by a reversible outer toe and sharp spicules on the underside of the foot, both adaptations for holding slippery fish.

Diets range from small insects to large vertebrates, carrion, and oily fruits. Falconiform birds are patterned in black, white, brown, rufous, and gray, and species have different color phases, often including an all-black form. Females are often larger than males, particularly so in bird-eating species. These birds range in size from the tiny Philippine Falconet, 13 centimeters long, to the magnificent Harpy Eagle and Great Philippine Eagle, both over a meter long. The Andean Condor is one of the largest flying birds in the world, with a wingspan of 9.5 feet.

Most species breed in trees, some on cliff ledges or in tree cavities. Clutches range from three to five eggs in small species to one to two in large species. The semiprecocial young hatch covered with down and their eyes open. After they leave the nest, young birds of prey depend on their parents for food for up to several months until they develop hunting skills.

Like storks and unlike most raptors, New World vultures have perforate nostrils, lack a syrinx, and have bare heads. New World vultures are convergent with the scavenging Old World vultures (Accipitridae).

Family	Members	Distribution	Genera	Species
Pandionidae	Osprey	Worldwide	1	1
Accipitridae	Hawks, harriers, eagles, kites, Old World vultures	Worldwide	64	239
Sagittariidae	Secretary-bird	Africa	1	1
Falconidae	Falcons, caracaras	Worldwide	10	63
Cathartidae	New World vultures	North and South America	5	7

Taxonomic Comments

The monophyly of the Order Falconiformes is still debated (Jollie 1953, 1976, 1977; Cracraft 1981), but has been reaffirmed in a recent cladistic analysis (Griffiths 1994). The New World vultures may be relatives of the storks (Ciconiiformes) rather than of birds of prey (Ligon 1967; Sibley and Ahlquist 1990).

Figure A–9 Birds of prey: (1) Cooper's Hawk (Accipitridae); (2) Osprey (Pandionidae); (3) Harpy Eagle (Accipitridae); (4) Peregrine Falcon (Falconidae); (5) Andean Condor (Cathartidae); (6) Secretary-bird (Sagittariidae).

1
2
♂
♂
3
4
5
♂
COE

Fowllike Birds
Order Galliformes

Members of the Order Galliformes are medium to large terrestrial birds with short, rounded wings, a well-developed keel, and sturdy legs with four toes. Among members of the Families Phasianidae and Numididae, the hind toe is elevated and not in contact with the ground, but among members of the Families Megapodiidae and Cracidae, the hind toe is on the same level as the front toes. The bill is short and more or less conical in most species, with an arched culmen and with the tip of the upper mandible overlapping the lower mandible. All members of the order have a large, muscular gizzard, a well-developed aftershaft, and large intestinal ceca. Many members of the Family Phasianidae have spurs on the tarsus; and in the grouse, the tarsi, and sometimes the toes, are feathered. The turkeys have bare heads, ornamented with wattles.

Most galliform birds are cryptically colored, patterned in black, gray, and brown, but among the true pheasants, the males and sometimes the females are clad in reds, yellows, silver, and other bright colors; in the peacocks, large members of the pheasant group, the upper tail coverts are greatly lengthened and bear large, iridescent eyespots; these great trains are erected during display. Intricate patterns are also found in some of the New World quails and in some of the small quails of the Old World.

Fowllike birds typically produce large clutches of eggs and precocial, downy young. Young of species in the Family Megapodiidae are fully feathered and capable of flight when they emerge from their mound nests, the result of unique incubation using heat from the sun, decaying compost, or volcanoes rather than a parent.

Family	Members	Distribution	Genera	Species
Cracidae	Curassows, guans, chachalacas	Neotropics	11	50
Megapodiidae	Moundbuilders	Australasia, Malaysia	6	19
Numididae	Guineafowl	Africa	4	6
Phasianidae	Pheasants, Old World quails, grouse, turkeys	Nearly worldwide	45	177
Odontophoridae	New World quails	North and South America	4	6

Figure A–10 Fowllike birds: (1) Lady Amherst's Pheasant (Phasianidae); (2) Red Junglefowl (Phasianidae); (3) Great Currasow (Cracidae); (4) Vulturine Guineafowl (Numididae); (5) Sage Grouse (Phasianidae).

Taxonomic Comments

The galliform birds are usually considered to be most closely related to birds in the Order Anseriformes and secondarily to birds in the Order Falconiformes (Cracraft 1981; Olson and Feduccia 1980b; Sibley and Ahlquist 1990). Moundbuilders and curassows may be sister taxa (Sibley and Ahlquist 1990).

The Hoatzin (Opisthocomidae), often placed in the Order Galliformes, herein has been assigned to the Order Cuculiformes (Sibley and Ahlquist 1973; but see Brush 1979; Cracraft 1981).

The New World quails (Odontophoridae) are an old South American lineage that merits family recognition. They are not closely related to the Old World quails (Phasianidae) (Sibley and Ahlquist 1990).

Cranes, Rails, and Allies
Order Gruiformes

The Order Gruiformes is an old, widely dispersed, and diverse group of small to large birds with few unifying characters. No member of the order has a crop, and most share certain skeletal and palatal features. An oil gland is present in most families but absent in the bustards. The nostrils are pervious in rails, sungrebes, the Sunbittern, cranes, seriemas, and bustards, and covered with an operculum flap in the Kagu.

Most gruiform birds are terrestrial, aquatic, or marsh-dwelling birds with strong, unwebbed or only slightly webbed toes, although the aquatic sungrebes (Family Heliornithidae) and the coots (Family Rallidae) have lobed toes. The condition of the hind toe varies: It is lacking, for example, in the terrestrial bustards and the buttonquails; elevated in rails, trumpeters, and roatelos of Madagascar, the seriemas of South America, and the cranes; and on the same level as the front toes in the Sunbittern and the Limpkin. Except for the large, stately cranes and the heavy (up to 18 kilograms) bustards of open country habitats, most gruiform birds are secretive and little known. Rails, which constitute the majority of species in the order, are heard more often than they are seen.

The downy young of all gruiform birds leave the nest soon after hatching; cranes and bustards have a protracted period of parental care. Some species are polyandrous, that is, females compete for and pair with multiple males, which incubate the eggs.

Family	Members	Distribution	Genera	Species
Rallidae	Rails, coots	Worldwide	34	142
Heliornithidae	Sungrebes	Pantropical	3	3
Rhynochetidae	Kagu	New Caledonia	1	1
Eurypygidae	Sunbittern	Neotropics	1	1
Mesoenatidae	Roatelos	Madagascar	2	3
Turnicidae	Buttonquails	Tropical and warm temperate parts of Old World	2	17
Gruidae	Cranes	All continents, except South America	2	15
Aramidae	Limpkin	Neotropics	1	1
Psophiidae	Trumpeters	South America	1	3
Cariamidae	Seriemas	South America	2	2
Otididae	Bustards	Old World	6	25

1
2
3
4
♂
5
COE

Taxonomic Comments

The taxonomic boundary between members of the Gruiformes and Charadriiformes, the most closely related order, is vague (Cracraft 1973b). The intermediate jacanas are now placed in the Order Charadriiformes. The Plains-wanderer (Pedionomidae), once included in this order, belongs in the Charadriiformes (Olson and Steadman 1981). Buttonquails also may not belong in this order (Sibley and Ahlquist 1990).

Cracraft (1973b, 1981) reviewed relationships among the gruiform families. Skeletal characters link the trumpeters, Sunbittern, and Kagu. DNA analysis suggests that the Kagu of New Caledonia is most closely allied to the seriemas of South America (Sibley and Ahlquist 1990). The relationships of the roatelos and buttonquail remain obscure.

Figure A–11 Cranes, rails, and allies: (1) Sunbittern (Eurypygidae); (2) Gray Crowned-Crane (Gruidae); (3) Water Rail (Rallidae); (4) Gray-winged Trumpeter (Psophiidae); (5) Great Bustard (Otididae).

1
2
3
4
5
6
7
COE

Shorebirds, Gulls, and Allies
Order Charadriiformes

The 18 families and more than 300 species contained in the Order Charadriiformes are all water birds or birds, like the terrestrial lapwings, that are clearly derived from water birds. The group is united by various characteristics of the skull, vertebral column, and syrinx, but it is so varied that it is divided into four suborders that bear little outward resemblance to one another.

Members of the Suborder Charadrii, which includes the sandpipers, plovers, and other birds that are collectively known as "shorebirds" or "waders," are small to medium-sized birds with slender, probing bills and rather long legs. The feet are webbed in only a few species, and the hind toe is well developed in all but a small number of species. Many are cryptically colored, but some are boldly patterned. In other ways, they vary greatly; they make up about two-thirds of the order and are the most varied of the three suborders.

The Suborder Lari comprises long-winged, rather short-legged, web-footed birds, often with a large salt-excreting gland located in the orbit above the eye and with the hind toe small or lacking. These birds are usually clad in white, gray and white, black and white, or brown. The gulls have a stout, somewhat hooked bill, the jaegers have a more strongly hooked bill, the bills of terns are slender and sharply pointed, and the bill of the skimmers is uniquely modified, with the lower mandible bladelike and longer than the upper.

The auks, murres, and puffins make up the Suborder Alcae. They are small to medium-sized, stocky marine birds, usually black and white, with webbed feet, no hind toe, and dense, waterproof plumage. In most members of this group, the feet are located so far back on the body that the birds have a distinctive upright stance and resemble the penguins of the Southern Hemisphere. Their bills are quite varied in shape, ranging from the flattened, triangular bills of the puffins to the slender bills of murres and guillemots. All modern species of auks retain the ability to fly; one flightless species, the Great Auk of the North Atlantic, became extinct in 1844, after years of predation by seafarers.

These three suborders reflect three distinct foraging life styles among these largely aquatic birds. Most members of the Suborder Charadrii feed by wading in shallow water or along the edge of the water, using their bills to probe in mud or sand or to pluck prey items from the surface of the ground. All members of the Order Charadriiformes that feed on open ground away from water are members of this suborder.

The Suborder Lari consists of birds that feed in flight, scavenging along shores in the case of gulls, diving for fish that swim near the surface in the case of terns, and "skimming" over the surface snapping up small fish in the case of skimmers. The jaegers and skuas are predators and pirates, stealing food from other members of the suborder, or preying on rodents and other birds. Many members of this suborder are colonial nesters.

Figure A–12 Shorebirds, gulls, and allies: (1) Pheasant-tailed Jacana (Jacanidae); (2) Snowy Sheathbill (Chionididae); (3) Eurasian Woodcock (Scolopacidae); (4) Atlantic Puffin (Alcidae); (5) Blacksmith Lapwing (Charadriidae); (6) Ring-billed Gull (Laridae); (7) Black Skimmer (Rynchopidae).

Figure A–13 Sandgrouse: Chestnut-bellied Sandgrouse (Pteroclidae).

The birds in Suborder Alcae are diving birds that forage below the surface, pursuing small fish and other marine animals by swimming rapidly with their short, paddlelike wings. Most species in the suborder are colonial nesters, although many nest in burrows, so their colonies are not as conspicuous as those of gulls and terns.

Most species in this order have precocial young, but there is a trend from nidicolous to nidifugous in the alcids—members of Suborder Alcae. Most lay patterned eggs, but the Crab Plover lays white eggs in burrows. Varied breeding systems exist within the order, including polyandry in phalaropes, jacanas, and various sandpipers. Most species nest on the ground, but others use floating vegetation, cliff ledges, burrows, and even trees. Most species also show age and seasonal plumage variation, and a few, for example, the Ruff, have sex-related differences.

Once separated as a distinct order are the sandgrouse (Suborder Pterocli), which are stocky, medium-sized, pigeonlike birds with short, feathered tarsi, small heads, a large crop, and chickenlike bills. The hind toe is elevated in *Pterocles* and absent in *Syrrhaptes*. Some species have long, pointed central tail feathers. Unlike the pigeons and doves, they lack a fleshy cere and have a well-developed oil gland and an aftershaft; although they have a crop, they do not produce "pigeon's milk." The birds are clad in brown, tan, buff, and rufous and often have breast bands. The sexes differ in color, with the females being more cryptically patterned than the males. They nest solitarily on the ground, as do plovers.

The young are downy at hatching, another feature in which these birds differ from the birds in the Order Columbiformes.

Sandgrouse, which are birds of the arid grasslands of the Old World, feed mostly on dry seeds. They visit remote desert water holes at dawn and dusk to drink. Some species carry water in soaked breast feathers up to 30 kilometers to their young. Sandgrouse, unlike pigeons, raise their heads to swallow water (Maclean 1968).

Family	Members	Distribution	Genera	Species
Suborder Charadrii				
Jacanidae	Jacanas	Pantropical	6	8
Rostratulidae	Painted-snipes	Tropics	1	2
Scolopacidae	Woodcocks, snipes, sandpipers, phalaropes, turnstones	Worldwide	21	88
Chionididae	Sheathbills	Subantarctic	1	2
Pluvianellidae	Magellanic Plover	Patagonia	1	1
Pedionomidae	Plains-wanderer	Australia	1	1
Thinocoridae	Seedsnipes	Temperate South America	2	4
Burhinidae	Thick-knees	Worldwide, excluding North America	1	9
Haematopodidae	Oystercatchers	Worldwide	1	11
Ibidorhynchidae	Ibisbill	Asia	1	1
Recurvirostridae	Avocets, stilts	All continents	3	10
Dromadidae	Crab Plover	Indian Ocean	1	1
Glareolidae	Pratincoles, coursers	Warm Old World	5	17
Charadriidae	Plovers, lapwings	Worldwide	10	66
Suborder Lari				
Laridae	Gulls, terns	Worldwide	13	95
Stercorariidae	Jaegers, skuas	Polar regions	2	8
Rynchopidae	Skimmers	Americas, Africa, Southeast Asia	1	3
Suborder Alcae				
Alcidae	Auks, murres, puffins, guillemots	Northern oceans	12	23
Suborder Pterocli				
Pteroclidae	Sandgrouse	Old World	2	16

Taxonomic Comments

Birds in the Order Charadriiformes are considered to be closely related to birds in the Order Gruiformes and possibly to birds in the Order Columbiformes. Storer (1971b) and Sibley and Ahlquist (1990) include sandgrouse as typical members of the Order Charadriiformes. There is now general agreement that sandgrouse are shorebirds, closer to the species making up the Family Charadriidae than to the species in the Family Scolopacidae (Maclean 1967, 1969; Fjeldså 1976; Zusi 1986).

Several important revisions of charadriiform birds have appeared in recent years (Jehl 1968; Strauch 1978; Stegmann 1978; Gochfeld et al. 1984). It is now believed that the Family Scolopacidae split off early in the evolution of the order. Jacanas and the painted-snipes are related. The Ibisbill probably is related to the stilts; the Crab Plover, to the pratincoles and coursers. The relations of the sheathbills and seedsnipes remain uncertain. Enigmatic genera are discussed in several papers: *Pluvianellus,* which may be related to seedsnipes (Jehl 1975); *Aechmorhynchus, Prosobonia,* and *Phegornis* (Zusi and Jehl 1970); and *Pedionomus* (Bock and McEvey 1969; Olson and Steadman 1981; Sibley and Ahlquist 1990).

Loons
Order Gaviiformes

Loons, known as divers in Europe and Asia, are a small, homogeneous group of large, heavy, foot-propelled diving birds with spear-shaped bills, stocky necks, streamlined bodies, webbed feet, and laterally compressed tarsi. The legs are positioned far back on the body as in many other diving birds, and the birds have difficulty moving on land. Among the best diving birds, loons can reach depths of 75 meters and stay under water for 8 minutes. Loons mostly eat fish.

Males and females look the same; both have different breeding and nonbreeding plumages. The breeding plumages are boldly patterned in black and white or with gray or chestnut, and the winter plumages are dark brownish or gray above and white below. The young are downy at hatching and wear a second coat of down before acquiring adult feathers.

Loons nest in masses of water vegetation, which they pile on small islands or close to the edge of open water. The young ride on their parents' backs when small; they depend on their parents for food until they are nearly full grown.

The loud, haunting, yodeling calls used by loons on their breeding territories are one of the most familiar sounds of the northern wilderness and were regarded as evil omens by some northern cultures. Loons are long-lived birds, which occasionally reach ages of 20 to 28 years. They pair monogamously, typically for life. Highly publicized local campaigns are under way to revitalize local populations threatened by human disturbance and lake pollution.

Family	Members	Distribution	Genera	Species
Gaviidae	Loons	Holarctic	1	5

Debate continues over the relationships of loons, which evolved in the early Tertiary (Cracraft 1982b; Olson 1985; del Hoyo et al. 1992). They have been associated with other foot-propelled diving birds, such as the grebes, because of similarities in hindlimb anatomy; these similarities are now believed to be due to convergence (Storer 1971a). Two species of primitive loons (*Colymboides*), known only as fossils from the late Oligocene, are not as highly specialized for diving as modern loons and resemble gulls in skeletal characters such as the canals of the hypotarsus and the structure of the coracoid (Storer 1956). Sibley and Ahlquist (1990) consider loons to be closer to tube-nosed seabirds than to grebes.

Figure A–14 Loons: Common Loon (Gaviidae).

1
2
3
4
5
COE

Pigeons and Doves
Order Columbiformes

Pigeons and doves are small, medium-sized, or large, plump birds with small heads, short legs covered with small, reticulate scales, and a fleshy cere at the base of the bill. Most have a muscular gizzard, and all have a large crop, the lining of which secretes a substance known as "pigeon's milk."

The plumage of columbiform birds is dense and is easily detached from the thin skin; there is little down under the outer feathering. The oil gland is small or absent, and the feathers have no aftershaft. The tail is commonly fan shaped but may be long and narrow or pointed. A large group, pigeons and doves exhibit a wide variety of colors and patterns, with soft, pastel grays and buff predominating; many species have patches of softly iridescent feathers on the sides of the neck, and others have dark bands on the nape. Some, such as the large crowned-pigeons (*Goura*) of New Guinea, bear elaborate crests.

Columbiform birds invariably lay two eggs. The young are nearly naked at hatching. Both parents attend their young, which they nourish initially with a rich milk produced by the lining of the crop.

Pigeons are unusual among birds in that they can drink by sucking water rather than having to tilt the head back to swallow it mouthful by mouthful.

The Dodo of Mauritius Island and the Rodriguez Solitaire of Rodriguez Island in the western Indian Ocean were large, flightless pigeons that were extirpated in the seventeenth century (Livezey 1993).

Family	Members	Distribution	Genera	Species
Columbidae	Pigeons, doves	Worldwide	40	310
Raphidae	Dodo, Rodriguez Solitaire	Indian Ocean islands	2	3

Taxonomic Comments

This is a distinct order, possibly related to the Order Charadriiformes. See Goodwin (1983) for a review of pigeons and doves of the world.

Figure A–15 Pigeons and doves: (1) Mourning Dove (Columbidae); (2) Common Wood-Pigeon (Columbidae); (3) Superb Fruit-Dove (Columbidae); (4) Dodo (Raphidae); (5) Tooth-billed Pigeon (Columbidae).

1
2
3
4
COE

Parrots
Order Psittaciformes

Parrots are a well-defined group of small to medium-sized birds with stout, hooked bills, in which the upper mandible is movable, being attached by a hingelike articulation to the skull. There is a fleshy cere. The tongue also is fleshy, and some species in the Australasian region have brush-tipped tongues for feeding on nectar. The neck and legs are short, and the toes are zygodactyl, adapted for perching and climbing; the scales on the legs and toes are small and granular. Parrots hold and manipulate food with their feet, and use their bill as well as their feet in climbing.

Psittaciform birds tend to be gregarious, vocal birds and can be destructive. The raptor like Kea of New Zealand pulls large nails from buildings, rips automobile upholstery, and kills sick sheep to feed on their kidney fat. The wings of parrots are generally pointed, and the keel is well-developed except in the flightless Owl Parrot, or Kakapo, of New Zealand. The plumage is sparse, and some species, most notably the cockatoos, are crested. The tail is most commonly fan-shaped but may be long and pointed. Many parrots are green or largely green, but others, especially in the Australasian region, are clad in a variety of brilliant colors as well as solid black.

Parrots and macaws generally nest in holes. The young are naked at hatching in some species or covered with down in other species. As a group, the parrots and macaws include the greatest proportion of threatened and endangered species of any large family of birds.

Family	Members	Distribution	Genera	Species
Psittacidae	Parrots, macaws	Pantropical, few temperate	62	281
Cacatuidae	Cockatoos	Australia	6	23
Loriidae	Lories	Australasian region, Pacific	12	54

Taxonomic Comments

The parrots are so distinctive that their affinities remain obscure. They may be distantly related either to the pigeons and doves or to the cuckoos.

See Forshaw (1989) for reviews and illustrations of the parrots of the world.

Figure A–16 Parrots: (1) Sulphur-crested Cockatoo (Cacatuidae); (2) Rainbow Lorikeet (Loriidae); (3) Chestnut-fronted Macaw (Psittacidae); (4) Kea (Psittacidae).

Mousebirds
Order Coliiformes

The mousebirds, or colies, are a distinctive group of small, crested, African birds with dense, gray or brown plumage and long, pointed tails. The first and fourth toes are reversible, so all four toes can be directed forward. Mousebirds lack bare, unfeathered areas of skin (apteria) and molt their primary wing feathers in irregular patterns.

Mousebirds build shallow, platformlike nests, which they ornament with fresh green leaves. Several females may lay eggs in the same nest. Both sexes incubate the eggs and feed the young by regurgitation.

Mousebirds inhabit savannah, woodland edge, and brushy country throughout Africa south of the Sahara. Some species are nomadic in arid regions. They travel in small, tight flocks during most of the year. They often hang upside down from branches and can scurry about in bushes like mice, a habit that explains their name. They feed mainly on fruit, buds, and flowers, and at times may damage fruit crops.

Family	Members	Distribution	Genera	Species
Coliidae	Mousebirds	Africa	2	6

Taxonomic Comments

Although they have no obvious allies, the mousebirds may be distantly related to parrots, which they resemble in the structure of the palate, heart, pelvis, intestines, and oil gland (Berman and Raikow 1982).

Schifter (1985) reviews the systematics and biology of mousebirds.

Figure A–17 Mousebirds: Speckled Mousebird (Coliidae).

Turacos
Order Musophagiformes

The African turacos and plaintain-eaters are medium-sized, long-tailed, chiefly arboreal birds with a foot on which the outer toe is reversible but not permanently reversed; therefore, their toes are not truly zygodactyl like the toes of cuckoos. The short, strong, chickenlike bill has serrate edges and often a keel. Nostrils vary from circular to slit-shaped. Turacos and plaintain-eaters are mostly vegetarian birds of the forests and woodlands.

Most species in the Order Musophagiformes are crested and have patches of bare facial skin. The hairy feathers of the head and breast have few barbules. A few species are gray or clad in blue or purple, but both sexes of most species are green, with a patch of bright red concealed in the flight feathers of the wing. Both of these colors are due to pigments that are unique to the Musophagidae: The green pigment is called turacoverdin and the red pigment is turacin.

Turacos and plaintain-eaters build flimsy, pigeonlike twig nests. The downy young leave the nest before they can fly and clamber about the nest tree, aided in some species by a well-developed wing claw.

Family	Members	Distribution	Genera	Species
Musophagidae	Turacos, plantain-eaters	Africa	5	23

Taxonomic Comments

The turacos and plaintain-eaters are a distinctive group not clearly related to species in any other order (Voous 1973; Sibley and Ahlquist 1990). In the past they have been associated with the Orders Cuculiformes and Galliformes.

Figure A–18 Turacos: Ross's Turaco (Musophagidae).

1
2
3
COE

Cuckoos
Order Cuculiformes

Cuckoos are small to medium-sized, slender, usually long-tailed birds with zygodactyl feet. Terrestrial species have sturdy legs, and arboreal species somewhat weaker ones, but in all species the foot is well adapted for perching. The bill is usually slender and slightly decurved but is laterally compressed in the anis, chickenlike in some African species (*Centropus*), and almost toucanlike in the Channel-billed Cuckoo. Many species have a colorful, fleshy eye-ring, and some are crested. The plumage colors of cuckoos range from streaked or solid brown to solid black, to brilliant, metallic green in the emerald cuckoos (*Chrysococcyx*).

All Old World species in the Order Cuculiformes are brood parasites, some of which produce eggs that mimic the eggs of their hosts. Anis of the New World Tropics have communal nests. Cuckoos are mostly insectivorous, often specializing on hairy caterpillars. The coucals of Africa and Asia often take small birds and eggs.

The Hoatzin is a large, distinctive bird (Figure 3–8), slender, with rounded wings, and with the keel shortened to accommodate a greatly enlarged crop, in which leaves—its primary food—ferment. The flight muscles are weakly developed, perhaps to make room for the large crop, so the Hoatzin can fly only short distances. It has a long, loose crest. The species's most peculiar feature is found in young birds. At hatching, the young have two functional claws on the second and third digits of the wing; for a few days, the young are capable of clambering about among the branches near the nest. After a few days, the claws regress, and the wing develops like that of any other bird.

Family	Members	Distribution	Genera	Species
Cuculidae	Cuckoos, anis, coucals, ground-cuckoos	Worldwide	29	142
Opisthocomidae	Hoatzin	South America	1	1

Taxonomic Comments

The Order Cuculiformes includes several ancient lineages with no clear living relatives. The Hoatzin is tentatively included here on the basis of biochemical evidence (Sibley and Ahlquist 1973, 1990), but Brush (1979) and Cracraft (1981) dispute this assignment.

Sibley and Monroe (1990) subdivide the Family Cuculidae into five separate families: Old World cuckoos (Cuculidae), coucals (Centropidae), American cuckoos (Coccyzidae), anis (Crotophagidae), and ground-cuckoos (Neomorphidae).

Figure A–19 Cuckoos: (1) Common Cuckoo (Cuculidae); (2) Smooth-billed Ani (Cuculidae); (3) Greater Roadrunner (Cuculidae); see also Figure 3–8 for Hoatzin (Opisthocomidae).

1
2
3
4
COE

Owls
Order Strigiformes

The owls are mainly nocturnal birds of prey with large, rounded heads and big eyes. Owls must turn their heads just to look sideways, to compensate for the socket-fixed positions of their forward-facing eyes. Some species can rotate their head 270 degrees to look behind them. The eyes are strengthened by bony plates like those of other birds, but in owls these plates form a lengthened cylinder that provides telescopic vision. Surrounding the eyes are large facial disks of feathers that concentrate sound and greatly increase hearing ability. In some species, the outer, bony portions of the two ears are differently shaped, creating a stereophonic effect that enables the birds to locate precisely sounds made by prey.

The bill is long and hooked, and bears a cere, but because it is directed downward, with its base covered by bristles, it appears short. The legs, and often the toes, are feathered, and the outer toe can be reversed. The barn-owls (Family Tytonidae) differ from typical owls of the Family Strigidae in a number of internal characters, among them a longer and narrower skull and a furcula fused to the sternum; also the legs of barn-owls are longer than those of most typical owls.

Owls feed predominantly on small rodents and shrews, but also take insects, earthworms, crabs, fish, frogs and salamanders, reptiles, and birds. The diets of owls are easy to decipher from the bones and hard parts packed into regurgitated pellets, which accumulate below daytime roosts.

Nearly all owls are cryptically colored, reducing discovery while they roost by day. Some species have different color phases—rufous versus gray. The dense plumage is soft, thus enabling the strigiform birds to fly silently. In both families, the young are clad in whitish down at hatching.

Rather than construct their own, owls usually use the nests of other birds, mammal burrows or natural cavities, and sometimes buildings. The numbers of round, chalky white eggs laid varies in relation to prey availability, and competition among siblings for food may be deadly.

Family	Members	Distribution	Genera	Species
Tytonidae	Barn-owls	Worldwide	2	17
Strigidae	Owls	Worldwide	23	161

Taxonomic Comments

Owls may be related to the diurnal birds of prey (Falconiformes, especially birds in the Family Falconidae), or to the nightjars and their allies (Caprimulgiformes), but both possibilities remain controversial (Jollie 1976, 1977; Cracraft 1981). Sibley and Ahlquist (1990) consider owls and nightjars to be close relatives.

Figure A–20 Owls: (1) Elf Owl (Strigidae); (2) Northern Hawk Owl (Strigidae); (3) Eurasian Eagle-Owl (Strigidae); (4) Barn Owl (Tytonidae).

1
2
3
COE

Nightjars and Allies
Order Caprimulgiformes

Members of the Order Caprimulgiformes are nocturnal or crepuscular birds with soft and cryptically patterned plumage and short legs. The true nightjars have small, weak feet, whereas the feet of birds in those families whose members dwell in trees have stronger toes, well suited to perching.

Most caprimulgiform birds have a small, weak bill with a very large mouth opening, enabling them to capture insects on the wing; the bill is often surrounded by long bristles, which aid in catching prey and protect the eyes from damage in an aerial encounter with an insect. The frogmouths have heavier bills but still have the wide gape of other members of the order. The Oilbird has a stronger, more hooked bill, with a smaller mouth opening. The cave-dwelling Oilbird, which feeds on oil-rich fruits, is one of the few birds known to navigate by means of echolocation. It also has a well-developed sense of smell.

Usually birds in these families lay one or two eggs in the open on the bare ground or on tree stumps, in tree cavities, or on cave ledges (oilbirds). Frogmouths build small nest pads. Both parents care for their downy, nidicolous young.

Family	Members	Distribution	Genera	Species
Steatornithidae	Oilbird	South America	1	1
Podargidae	Frogmouths	Orient, Australasia	2	14
Aegothelidae	Owlet-nightjars	Australasia	1	8
Nyctibiidae	Potoos	Neotropics	1	7
Caprimulgidae	Nightjars	Worldwide	15	83

Taxonomic Comments

Members of the Order Caprimulgiformes may be related to the owls, or perhaps instead to the Orders Apodiformes and Trogoniformes. The Oilbird resembles owls in some anatomical characters but has egg white proteins like those of nightjars and also resembles nightjars in characters of the skull, humerus, and sternum (Sibley and Ahlquist 1972; Cracraft 1981).

Nightjars of the Caribbean genus *Siphonorhis* may be primitive members of the order, related to the owlet-nightjars of Australasia (Olson 1978b).

Sibley and Monroe (1990) separate Australian frogmouths (Podargidae) from Asian frogmouths (Batrachostomidae), and also eared-nightjars (Eurostopodidae) from nighthawks and other nightjars (Caprimulgidae).

Figure A–21 Nightjars and allies: (1) Common Nighthawk (Caprimulgidae); (2) Tawny Frogmouth (Podargidae); (3) Eurasian Nightjar (Caprimulgidae).

1
2
3
4
COE

Swifts and Hummingbirds
Order Apodiformes

Swifts and hummingbirds are small or very small birds. The hummingbirds include the smallest bird in the world—the Cuban Bee Hummingbird is only 2.25 inches from bill tip to tail tip and weighs only 2 grams. Both swifts and hummingbirds have tiny feet, extremely short humeri, long bones in the outer portion of the wing, long, sturdy primaries, and short secondaries. These are all adaptations for the special flight modes of swifts and hummingbirds, which are perhaps the most accomplished fliers of all birds. Eurasian Swifts remain airborne for over 6 months in the breeding season, sleeping aloft.

In the swifts and treeswifts, the bill is short and the gape very broad to aid in capturing insects in flight. In the hummingbirds the gape is small and the bill is always slender but has various shapes adapted to feeding at flowers of diverse structures. All hummingbirds have a long and extensile tongue for reaching nectar. The nostrils are rounded and exposed in swifts and crested swifts, or slitlike and covered by an operculum in the hummingbirds.

In the true swifts (Apodidae), there is a small claw on the wing (hand); this is absent in the treeswifts and hummingbirds. A crop is lacking in adult apodiform birds but is present in nestling hummingbirds. An oil gland is present and unfeathered.

The swifts tend to be dull colored or patterned in blackish and white, with no differences between the sexes, whereas most hummingbirds are brilliantly iridescent and sexually dimorphic, with brightly colored throat patches or crests on the males.

Hummingbirds are promiscuous; females rear the young alone. Hatchling swifts and hummingbirds both are blind and helpless at hatching. Nestling swifts enter torpor and stop growing when bad weather prevents their parents from finding aerial insects.

Family	Members	Distribution	Genera	Species
Hemiprocnidae	Tree swifts	Southern Asia	1	4
Apodidae	Swifts	Worldwide	18	99
Trochilidae	Hummingbirds	New World	109	319

Taxonomic Comments

Members of the Order Apodiformes are related to birds in the Order Caprimulgiformes and perhaps to birds in the Order Passeriformes.

Although usually placed in the same order, the relationship of the swifts and hummingbirds is uncertain. Their similar, specialized wing morphology may reflect common ancestry, but it could be due to convergence (Cohn 1968; Zusi and Bentz 1982, 1984). They share a unique egg white protein (Kitto and Wilson 1966) and are usually classified together.

Figure A–22 Swifts and hummingbirds: (1) White-collared Swift (Apodidae); (2) Streamertail (Trochilidae); (3) Tufted Coquette (Trochilidae); (4) Black-chinned Hummingbird (Trochilidae).

♂
1
♂
2

Trogons
Order Trogoniformes

Trogons are a small group of tropical and subtropical, forest-dwelling birds with short bills, strongly arched culmens, and serrated edges on the upper mandibles. The feet, small and weak, differ from those of all other birds. Trogons have heterodactyl feet, with the first and second toes directed backward, and the third and fourth toes directed forward; all other birds with two toes in front and two behind have the first and fourth directed backward (zygodactyl feet).

Trogons have dense, lax plumage and a well-developed aftershaft; the feathers tear loose from the thin skin very easily. The wings are short and rounded. The tail is long, usually squared at the tip. The upper tail coverts are much longer than the other tail feathers of the quetzals of the New World Tropics. Most species have sexually dimorphic, iridescent green plumage; many have bright red or yellow underparts. The red pigment is unusual because it is unstable, fading quickly in museum specimens, and is maintained in life only as a result of feather replacement during molt.

Trogons eat fruits and large insects. They nest in cavities. The young are naked and blind at hatching.

Family	Members	Distribution	Genera	Species
Trogonidae	Trogons, quetzals	Pantropical, excluding Australasia	6	39

Taxonomic Comments

The trogons are an enigmatic group with no close, living relatives; perhaps they are most directly related to birds in the Order Coraciiformes (Maurer and Raikow 1981; Sibley and Ahlquist 1990).

Figure A–23 Trogons: (1) Resplendent Quetzal (Trogonidae); (2) Red-headed Trogon (Trogonidae).

1
2
3
4
5
6

Rollers, Kingfishers, and Allies
Order Coraciiformes

The rollers, kingfishers, and their allies are small to medium-sized, stocky birds with large heads and small feet. Characters unifying this diverse group largely involve the structure of the palate and leg muscles, and fusion of toes (syndactyly). The kingfishers, todies, motmots, bee-eaters, and, to a lesser extent, the other members of the order, have the anterior toes fused at the base.

Feather features peculiar to individual families include eyelashes in hornbills, a long, erectile crest in the Hoopoe, and spatulate tips to the central tail feathers in motmots. In addition to their varied patterns and often brilliant colors, the families differ greatly in the shape of the bill, ranging from the slender bills of bee-eaters, woodhoopoes, and the Hoopoe and the sturdier bills of kingfishers and rollers, to the huge, ornamented bills of the larger hornbills.

The animal prey of coraciiform birds include arthropods (particularly insects, spiders, and scorpions), fish and aquatic invertebrates including crabs and shellfish, small terrestrial vertebrates, and worms. Hornbills are omnivorous and consume many fruits as well as animals. Feeding techniques range from aerial sallies after flying insects (bee-eaters and rollers), to probing into deep crevices (hornbills and woodhoopoes), to hovering and plunge-diving by some kingfishers. Sit-and-wait predation is a common tactic.

Most coraciiform birds nest in cavities; hornbill females molt while sealed by the male into the nest. The young are typically helpless (altricial) and naked when they hatch. Cooperative breeding systems are quite common in these mostly tropical birds.

Family	Members	Distribution	Genera	Species
Alcedinidae	Kingfishers	Worldwide	18	94
Todidae	Todies	Greater Antilles	1	5
Momotidae	Motmots	Neotropics	6	9
Meropidae	Bee-eaters	Old World	3	26
Coraciidae	Rollers	Old World	2	12
Brachypteraciidae	Ground-rollers	Madagascar	3	5
Leptosomatidae	Courol (or Cuckoo Roller)	Madagascar	1	1
Upupidae	Hoopoe	Africa, warm Eurasia, Madagascar	1	2
Phoeniculidae	Woodhoopoes	Africa	2	8
Bucerotidae	Hornbills	Tropical Africa and Asia; East Indies	9	56

Figure A–24 Rollers, kingfishers, and allies: (1) Puerto Rican Tody (Todidae); (2) European Bee-eater (Meropidae); (3) Lilac-breasted Roller (Coraciidae); (4) Turquoise-browed Motmot (Momotidae); (5) Oriental Pied-Hornbill (Bucerotidae); (6) Pied Kingfisher (Alcedinidae).

Taxonomic Comments

The characteristics of the birds in the Order Coraciiformes link them to the Orders Piciformes and Passeriformes, but birds in the Orders Cuculiformes, Psittaciformes, Trogoniformes, and even Caprimulgiformes also have been suggested as possible relatives.

This order may not be monophyletic (Cracraft 1971, 1981; Feduccia 1977, 1980; Maurer and Raikow 1981). The Hoopoe and woodhoopoes are relatively closely related, as are the motmots and todies, and probably also the kingfishers and bee-eaters. The rollers, and particularly the ground-rollers and Courol—or Cuckoo Roller—of Madagascar, probably represent an ancient lineage.

Sibley and Ahlquist (1990) divide the kingfishers into three families: the little kingfishers of the Old World (Alcedinidae), the Australian kookaburras and their diverse relatives (Dacelonidae), and the Belted Kingfisher and its relatives (Cerylidae).

Woodpeckers and Allies
Order Piciformes

The Order Piciformes comprises eight varied families, of which the Picidae (woodpeckers) is the largest. All piciform birds have zygodactyl feet and a unique arrangement of tendons in the toes. Most have an aftershaft. The families differ most obviously in the shape of the bill: It is a strong, sturdy, and chisellike in woodpeckers, wrynecks, and piculets; long, slender, and sharply pointed in jacamars; stout, anteriorly compressed, and somewhat decurved at the tip in puffbirds; stout, more or less conical, and surrounded by bristles in barbets; small and slightly hooked in honeyguides; and very large and inflated in toucans.

The true woodpeckers (Subfamily Picinae) have stiffened tail feathers that serve as a prop when the bird is clinging to a tree. Color and pattern vary, from the dull greens of some honeyguides and the black-and-white patterns of woodpeckers to the brilliant reds, blues, and yellows of many barbets and toucans. The sexes are generally similar, but among the woodpeckers the males often have patches of red on the head that are reduced or lacking in females.

All species nest in holes, or cavities. The young are naked and blind at hatching.

The diversity of feeding specializations is a feature of this order. Woodpeckers hitch up vertical tree trunks and chip wood and bark with their chisellike bills. Long, barb-tipped tongues enable them to extract woodboring insects from tiny crevices. Jacamars catch butterflies on the wing. Puffbirds pluck large caterpillars from tropical foliage. Barbets and toucans consume much fruit, but toucans also snatch eggs and young from the nests of other birds. Some honeyguides eat beeswax and bee larvae, which they obtain by leading humans and honeybadgers to natural beehives. Honeyguides are also specialized brood parasites that lay their eggs in the nests of woodpeckers, barbets, and other birds.

Family	Members	Distribution	Genera	Species
Galbulidae	Jacamars	Neotropics	5	18
Bucconidae	Puffbirds	Neotropics	10	33
Indicatoridae	Honeyguides	Africa, tropical Asia	4	17
Picidae	Woodpeckers, wrynecks, piculets	Worldwide, except Australasia	28	215
Lybiidae	African barbets	Africa	7	42
Megalaimidae	Old World barbets	Asia and Africa	3	26
Capitonidae	New World barbets	Neotropics	3	18
Ramphastidae	Toucans	Neotropics	6	41

1
2
3
4
5
6
COE

Taxonomic Comments

Members of the Order Piciformes may be related to members of the Order Coraciiformes. Whether Piciformes constitutes a monophyletic group is currently a matter of debate (Swierczewski and Raikow 1981; Simpson and Cracraft 1981; Olson 1983; Raikow and Cracraft 1983). The main issue is whether the jacamars and puffbirds are perhaps closer to members of the Order Coraciiformes than to other members of the Order Piciformes.

New World barbets are more closely related to toucans than to Old World barbets, and hence the two groups of barbets are treated as three separate families (Prum 1988; Sibley and Monroe 1990; Lanyon and Hall 1994).

Figure A–25 Woodpeckers and allies: (1) Great Spotted Woodpecker (Picidae); (2) Greater Honeyguide (Indicatoridae); (3) Double-toothed Barbet (Capitonidae); (4) White-chinned Jacamar (Galbulidae); (5) White-eared Puffbird (Bucconidae); (6) Keel-billed Toucan (Ramphastidae).

1
2
3
4
5
COE

Perching Birds
Order Passeriformes

Perching birds—called passerines—constitute more than half of the species of birds of the world, presently comprising approximately 5700 species assigned to 1161 genera (Sibley and Monroe 1990). They represent a diverse, species-rich, monophyletic order of mostly small land birds. Defining characters include a distinctive bony palate structure; unique oil glands, spermatozoa, and forelimb and hindlimb muscles; and feet with an enlarged flexible hind toe (hallux). Many passerine species rub ants (with noxious fluids) on their feathers for protection against ectoparasites. The metabolism of perching birds tends to be higher than that of other birds of comparable size. They also have relatively large brains and superior learning abilities, particularly with respect to vocalizations.

The reasons for the apparent evolutionary success of this major group of birds have been a topic of recent interest and debate (Raikow 1986, 1988; Fitzpatrick 1988). Could any of their defining attributes have been "key adaptations" that gave passerine birds an advantage over nonpasserine competitors and that promoted speciation? The behavioral plasticity and experimental learning abilities of passerine birds are profound and might also have catalyzed adaptive radiation and speciation.

The classification of passerine bird families has a turbulent recent history (Voous 1985; Sibley and Ahlquist 1990). Fundamental differences in the anatomy of the syrinx distinguish two suborders of perching birds, the suboscines (Tyranni) and the oscines (Passeres). The boundaries between families of suboscines, however, remain poorly defined. Also, convergence in ecology and morphology has confused the family classifications of both suboscines and oscines. Passerine birds have repeatedly evolved into convergent ecological forms, such as flycatchers, thrushes, warblers, creepers, and seed eaters.

Perching Birds—Suboscines
Order Passeriformes
Suborder Tyranni

Suboscine passerine birds have few unifying characters other than arrangements of the syringeal muscles that are simpler than those distinguishing the oscine passeriform birds. Traditionally, suboscine families have included broadbills and pittas of Africa and Asia, asities of Madagascar, and the New Zealand wrens. Broadbills are sluggish, tropical, fruit-eating birds that are traditionally separated from other suboscine birds on the basis of distinct foot tendons (but see Olson 1971). Pittas are secretive, brightly colored, ground birds with long legs and short tails. The asities of Madagascar include two fruit-eating species and two nectar-feeding species that closely resemble sunbirds (Nectariniidae). Both DNA and morphological evidence suggest that the New Zealand wrens (Xenicidae) may be the most ancient lineage of passerine birds (Sibley et al. 1982; Raikow 1984).

Figure A–26 Perching birds—suboscines: (1) Scissor-tailed Flycatcher (Tyrannidae); (2) Guianan Cock-of-the-Rock (Cotingidae); (3) Banded Pitta (Pittidae); (4) Barred Woodcreeper (Dendrocolaptidae); (5) Great Antshrike (Thamnophilidae).

The suboscines also include two major radiations of South American birds: tyrant-flycatchers, cotingas, and manakins in one case, and woodcreepers, ovenbirds, antbirds, and tapaculos in the other. Although their body forms vary with feeding styles, tyrant-flycatchers are distinguished by cranial, syringeal, and tarsal characters. Cotingas are medium sized and large tropical fruit-eating birds with broad bills, rounded wings, and short legs. Manakins are small, stocky, tropical fruit-eating birds with short wings and tail and broad bills. Some cotingas and manakins are brightly colored birds with elaborate courtship behavior.

Woodcreepers, ovenbirds, antbirds, and tapaculos include many small to medium-sized, often sombre, insect-eating birds of tropical forests. The woodpeckerlike woodcreepers typically have powerful feet with sharp claws and stiff, bracing tails. Ovenbirds are a diverse group of small, brown birds. Antbirds and tapaculos, which mostly have thick, often hook-tipped bills and short, rounded wings, reside in deeply shaded, thick vegetation.

Family	Members	Distribution	Genera	Species
Xenicidae	New Zealand wrens	New Zealand	2	4
Pittidae	Pittas	Old World Tropics	1	31
Eurylaimidae	Broadbills	Africa, Southeast Asia	8	14
Philepittidae	Asities, false-sunbirds	Madagascar	2	4
Tyrannidae	Tyrant-flycatchers	New World	102	413
Cotingidae	Cotingas	Neotropical	26	65
Oxyruncidae	Sharpbill	Neotropical	1	1
Phytotomidae	Plantcutters	South America	1	3
Pipridae	Manakins	Neotropical	17	56
Furnariidae	Ovenbirds	Neotropical	34	218
Dendrocolaptidae	Woodcreepers	Neotropical	13	49
Thamnophilidae	Antbirds	Neotropical	45	188
Formicariidae	Ant-thrushes	Neotropical	7	56
Conopophagidae	Gnateaters	Neotropical	1	8
Rhinocryptidae	Tapaculos	Neotropical	12	28

Taxonomic Comments

The New Zealand wrens are relicts of an ancient lineage not clearly related to the other passerine birds and probably deserve a suborder of their own. Alan Feduccia and Storrs Olson (1982) suggest that the Australian lyrebirds and scrub-birds and the South American tapaculos are closely related relicts of an early radiation in the Southern Hemisphere of suboscine birds that gave rise to the rest of the Order Passeriformes. The pittas, broadbills, and asities of the Old World are more closely related to one another than any is to New World suboscines (Feduccia 1974; Olson 1971; Raikow 1987; Sibley and Ahlquist 1990; Prum 1993). Within this clade, asities and broadbills appear to be sister taxa. The typical antbirds (Thamnophilidae) are separated from the ant-thrushes (Formicariidae), which are related to gnateaters and tapaculos (Sibley and Ahlquist 1990; Ridgely and Tudor 1994).

1
2
3
4
♂
5
♂
6

Perching Birds—Oscines
Order Passeriformes
Suborder Passeres

Oscine passerine birds have more complex syringeal musculature than do suboscines and, unlike most birds, tend to learn their vocalizations from natural tutors rather than inheriting innate vocalizations. Diverse in habits and form, they constitute the majority of passerine birds and include over 4500 species. Sibley and Monroe (1990) group the oscine passerines into two groups called parvorders.

Perching Birds—Crow Relatives

Crow relatives include most of the families of Australo-Papuan birds as well as crows, jays, drongos, shrikes, vangas, birds-of-paradise, and vireos. Considered to be among the most "intelligent" and perhaps evolutionarily advanced birds (Mayr 1958), crows, jays, and their relatives the nutcrackers (*Nucifraga*) cache seeds for later recovery based on exceptional spatial memory, facilitated by an enlarged hippocampus.

The corvoid assemblage includes several of the best-known adaptive radiations of all songbirds. Foremost among the endemic birds of Madagascar is the radiation of vangid shrikes, which parallel the Hawaiian honeycreepers in the diversity of ecological forms and bill types. Elsewhere, the variety of ecological types of Australasian warblers, robins, honeyeaters, and shrikes rivals the adaptive radiations of marsupial mammals and eucalyptus trees in that region. Many of these songbirds of Australasia were once squeezed conveniently by ornithologists into northern Temperate Zone bird families, which they resemble convergently in ecology and morphology.

A diversity of breeding systems also is seen among the corvoid birds. Well known for their extraordinary displays and promiscuous mating systems are the lyrebirds, bowerbirds, and birds-of-paradise (Figures 9–1, 9–19, and 9–20). Male lyrebirds display their lyre-shaped tail feathers like a peacock in the dark forests of Australia. Excellent mimics, they include the roar of chain saws or church bells in their vocal repertoires. Contrasting with the highly competitive mating systems of bowerbirds and birds-of-paradise are the highly social, communal breeding systems of fairywrens, Australo-Papuan babblers, wood-swallows, helmet-shrikes, as well as crows and jays. The cooperative social behavior of certain New World jays—the Florida Jay and the Mexican Jay of Arizona—are especially well studied.

Figure A–27 Perching birds—crow relatives: (1) Fork-tailed Drongo (Dicruridae); (2) Eurasian Jay (Corvidae); (3) Common Fiscal (Laniidae); (4) New Holland Honeyeater (Meliphagidae); (5) Superb Fairywren (Maluridae); (6) Spotted Quail-Thrush (Cinclosomatidae).

Family	Members	Distribution	Genera	Species
Climacteridae	Australo-Papuan treecreepers	Australia, New Guinea	2	7
Menuridae	Lyrebirds	Australia	1	2
Atrichornithidae	Scrub-birds	Australia	1	2
Ptilonorhynchidae	Bowerbirds	Australasia	7	20
Maluridae	Fairywrens, emuwrens	Australia	5	26
Meliphagidae	Honeyeaters	Australasia	42	182
Pardalotidae	Pardalotes, bristlebirds	Australasia	2	7
Acanthizidae	Australian warblers	Australasia	14	61
Eopsaltriidae	Australasian robins	Australasia	14	46
Irenidae	Fairy-bluebirds, leafbirds	Oriental	2	10
Aegithinidae	Ioras	Oriental	1	4
Orthonychidae	Logrunners, chowchillas	Australasia	1	2
Pomatostomatidae	Australo-Papuan babblers	Australasia	1	5
Laniidae	True shrikes	North America, Africa, Eurasia	3	38
Vireonidae	Vireos, peppershrikes	New World	4	51
Cinclosomatidae	Quail-thrushes, whipbirds	Australia	6	15
Corcoracidae	Australian Chough, Apostlebird	Australia	2	2
Pachycephalidae	Whistlers, sitellas	Australasia	14	59
Corvidae	Crows, jays, magpies	Worldwide	25	117
Paradisaeidae	Birds-of-paradise	Australasia	17	45
Artamidae	Wood-swallows, currawongs	Orient, Australasia	6	24
Oriolidae	Old World orioles, figbirds	Warm Old World	1	27
Campephagidae	Cuckoo-shrikes	Old World Tropics	7	84
Rhipiduridae	Fantails	Australasia	1	42
Dicruridae	Drongos	Old World Tropics	2	24

Family	Members	Distribution	Genera	Species
Monarchidae	Monarch-flycatchers, magpie-larks	Southeastern Asia, Australasia	18	98
Malaconotidae	Bushshrikes	Africa	8	48
Prionopidae	Helmet-shrikes, batises	Africa	7	44
Vangidae	Vangas	Madagascar	12	14
Callaeatidae	New Zealand wattlebirds	New Zealand	3	3
Picathartidae	Chaetops, rockfowl	New Guinea, Africa	2	4

Taxonomic Comments

Included among the corvoid birds are several of the world's most bizarre species, such as the Bornean Bristlehead. Its scientific name, *Pityriasis,* is also the name of a human skin disease and was assigned in reference to the yellow waxy tubercles, actually modified feathers, that cover its bald head. It also has stiff bristly ear coverts and a scarlet and black plumage of short, round, silky feathers. Also intriguing are the two species of rockfowl (*Picathartes*), a bare-headed ground bird that lives mostly in caves in tropical West Africa. Both the Bristlehead and rockfowl defied taxonomic classification until recent DNA studies linked them to the crow assemblage.

1
2
3
4
5
COE

Perching Birds—Thrush Relatives

This group includes assorted insect- and berry-eating songbirds such as mockingbirds and starlings, waxwings, chats, dippers, and the familiar thrushes of backyards such as the American Robin and the Mistle Thrush. Among the thrushlike birds are many excellent vocalists, including the Hermit and Wood Thrushes of North America, the nightingales of Europe and Asia, and the nightingale thrushes of tropical America. Both starlings and the mimic thrushes, especially the Northern Mockingbird, have large vocal repertoires made up of the imitated songs and calls of other species. The Hill Myna of Southeast Asia is a popular cage bird because it imitates human speech. Most familiar, perhaps, are the American Robin and Eurasian Blackbird of North America and Europe, respectively, but similar species in the genus *Turdus* occur throughout the world. Thrushes and their relatives feed on fruit and insects, often forming large flocks during the nonbreeding season.

Family	Members	Distribution	Genera	Species
Bombycillidae	Waxwings, silky-flycatchers	Holarctic	5	8
Dulidae	Palmchat	Hispaniola	1	1
Cinclidae	Dippers	Holarctic, South America	1	5
Turdidae	Thrushes	Worldwide	27	179
Muscicapidae	Old World flycatchers and chats	Eurasia	47	270
Sturnidae	Starlings, mynas	Old World	27	114
Mimidae	Mimic thrushes, thrashers	New World	11	34

Taxonomic Comments

Both the New World silky flycatchers and the enigmatic *Hypocolius* of the Middle East are tentatively included in the Bombycillidae as relatives of the waxwings.

The Australasian robins, including *Drymodes* (Eopsaltriidae), formerly placed in or close to the Turdidae, are now included in the Parvorder Corvida. Chats (*Oenanthe*) are now thought to be "ground-dwelling" Old World flycatchers, Muscicapidae (Sibley and Ahlquist 1990).

Starlings (Sturnidae) of the Old World and mimic thrushes of the New World may be related (Sibley and Ahlquist 1984).

Figure A–28 Perching birds—thrush relatives: (1) Northern Mockingbird (Mimidae); (2) European Pied Flycatcher (Muscicapidae); (3) Bohemian Waxwing (Bombycillidae); (4) Brahminy Starling (Sturnidae); (5) Mistle Thrush (Turdidae).

1
2
3
4
5
6

Perching Birds—Old World Insect Eaters

This group includes some of our most familiar backyard birds, such as nuthatches, titmice, wrens, and swallows, as well as the Old World warblers and the babblers. Excluded are convergent Australasian birds, most of which are classified in their own families as crow relatives.

Nuthatches are specialized for clinging upside down on tree trunks. Like nutcrackers and also most titmice, they recover cached seeds a task requiring exceptional spatial memory. Creepers and wrens use their thin, curved bills for extracting insects from narrow crevices. Only one species of wren occurs in Eurasia; the rest are restricted to the New World. Wrens with polygynous mating systems also have elaborate vocal repertoires, which are used incompetitive song duels to establish dominance among territorial males. Titmice, especially European species, are among the most intensively studied of all land birds. Models of population regulation, ecological coexistence of similar species, and the evolution of clutch size are based on long-term studies of banded tits.

Swallows are aerial insectivores. Some species migrate extreme distances between northern breeding grounds and wintering places. Their nests, which vary from natural holes to burrows dug in sand banks to elaborate mud constructions, reflect their phylogenetic relationships that have been defined on the basis of DNA comparisons.

Figure A–29 Perching birds—insect eaters: (1) Broad-ringed White-eye (Zosteropidae); (2) Barn Swallow (Hirundinidae); (3) Sardinian Warbler (Sylviidae); (4) Long-tailed Tit (Aegithalidae); (5) White-breasted Nuthatch (Sittidae); (6) White-crested Laughingthrush (Sylviidae).

Family	Members	Distribution	Genera	Species
Sittidae	Nuthatches, wallcreepers	Widespread, not South America	2	25
Certhiidae	Creepers	Holarctic, Africa, India	2	7
Troglodytidae	Wrens	New World; one species Holarctic	16	75
Polioptilidae	Gnatcatchers, Verdin	New World	4	15
Paridae	Titmice	Eurasia, Africa, North America	3	53
Remizidae	Penduline-tits	Eurasia, Africa, western North America	4	12
Aegithalidae	Long-tailed tits	Eurasia, western North America	3	8
Hirundinidae	Swallows, martins	Worldwide	14	89
Regulidae	Kinglets	Palearctic	1	6
Pycnonotidae	Bulbuls	Africa, Southern Asia	15	123

Family	Members	Distribution	Genera	Species
Hypocoliidae	Hypocolius	Southwest Asia	1	1
Cisticolidae	African warblers	Africa	14	119
Zosteropidae	White-eyes	Old World Tropics and subtropics	11	83
Sylviidae	Old World warblers, babblers, Wrentit, Laughingthrush	Mostly Old World	101	552

Taxonomic Comments

Sibley and Ahlquist (1983, 1990) restructured the classical babbler family Timaliidae and suggested further that traditional distinctions between babblers and Old World warblers may be confused by convergence of ecological types that cross traditional taxonomic boundaries. Members of the genus *Sylvia,* for example, may be closer to the babblers than to other Old World warblers. According to Sibley and Ahlquist (1990), the Family Sylviidae comprises the leaf-warblers (Subfamily Acrocephalinae), grass-warblers (Subfamily Megalurinae), laughingthrushes (Subfamily Garrulacinae), plus the assemblages of true babblers, Wrentit, and *Sylvia* warblers as equal-ranked taxa.

Perching Birds—Weaver Relatives

This group comprises major radiations of Old World granivorous and insectivorous birds, including weaverbirds and estrildine finches; the ground-living wagtails, pipits, and accentors; the nectar-feeding sunbirds and flowerpeckers; sugarbirds and a few Australasian taxa; and the New World wood warblers, tanagers, and blackbirds diagnosed by a strongly reduced tenth primary. Larks appear to be the sister family of these radiations. DNA studies suggest that a variety of seed-eating and nectar-feeding species belong to the ecological radiation of tanager relatives in South America. Seed-eating passerines (sparrows, buntings, weavers, waxbills, siskins, finches, grosbeaks) are assigned to at least five different families. The commonly used English names of seed-eating birds, such as finch or sparrow, do not relate in a simple way to the formal classification but reflect traditional, conflicting uses.

Weavers, which are diverse in their ecology and behavior, craft elaborate, globular grass nests with stitches that range from simple knots to slip knots and overhand knots. Different species use different knots. Their diverse social systems, which include complex social colonies, harems and promiscuity, and isolated monogamous pairs, parallel the social systems of North American blackbirds with similar ecological correlations. Certain weavers are major agricultural pests in Africa. Estrildine finches include some of the most popular small cage birds, such as the Zebra Finch, which breeds prolifically in rapid response to unpredictable rains in its native Australia.

Featured in the avifaunas of the New World is a radiation of a diverse, colorful species that includes the wood warblers and tanagers as well as the cardinal-grosbeaks, buntings, and blackbirds. Many of these species migrate from the Tropics to the United States and Canada to nest in June and July. Apparent population declines of Neotropical migrants are now a major concern.

Wood warblers are among North America's most boldly colored and popular small birds. Field and laboratory studies of wood warblers have contributed to advances in community ecology, speciation, and feeding behavior. Several species, for example, the Golden-cheeked Warbler in eastern Texas and the Kirtland's Warbler in Michigan, are endangered and play central roles in conservation programs. Tanagers are mostly small, brightly colored, fruit-eating birds of the New World Tropics. They form large mixed-species flocks, some of which have distinctive color combinations. The cardinals, grosbeaks, and saltators of the New World are medium-sized, seed and bud eaters with large powerful bills. Also in this clade are the brightly colored *Passerina* buntings, including the Painted and Indigo Buntings. Sparrows are the most familiar buntings in North America. Most species inhabit brushy habitats, but the longspurs are flocking birds of open, short-grass plains. Several species, such as the White-crowned Sparrow and the Song Sparrow, have been the focal subjects of key research on vocal learning and physiology of annual cycles

♂
♂
2
1
♂
3
4
♂
5

in birds. The New World oriole and blackbird assemblage includes medium-sized birds as diverse as oropendolas, cowbirds, and meadowlarks. Their equally diverse social systems include dense colonies, harem formation, and obligate brood parasitism. Several species in North America are major agricultural pests.

Family	Members	Distribution	Genera	Species
Alaudidae	Larks	Worldwide	17	91
Promeropidae	Sugarbirds	South Africa	1	2
Dicaeidae	Flowerpeckers	Oriental	2	44
Nectariniidae	Sunbirds	Old World Tropics	5	123
Melanocharitidae	Berrypeckers, longbills	New Guinea	5	12
Passeridae	Old World sparrows, rock-sparrows	Old World	4	36
Motacillidae	Wagtails, pipits	Nearly worldwide	5	65
Prunellidae	Accentors	Eurasia	1	13
Ploceidae	Weavers	Eurasia, Africa	17	117
Estrildidae	Estrildine finches, whydahs	Old World Tropics	30	155
Fringillidae	Carduelline finches, Chaffinch, Olive "Warbler"	Worldwide, except Australasia	22	140
Drepanididae	Hawaiian honeycreepers	Hawaiian islands	18	30
Emberizidae	Buntings, longspurs, sparrows, towhees, brush-finches	Worldwide, except Australasia	32	156
Parulidae	New World warblers, Wrenthrush	New World	25	115
Thraupidae	Tanagers, seedeaters, honeycreepers, and allies	New World	104	413
Cardinalidae	Cardinal-grosbeaks and allies	New World	13	42
Icteridae	New World orioles, blackbirds, and allies	New World	26	97

Figure A–30 Perching birds—weaver relatives: (1) Village Weaver (Ploceidae); (2) Beautiful Sunbird (Nectariniidae); (3) Pine Grosbeak (Fringillidae); (4) Eurasian Skylark (Alaudidae); (5) Alpine Accentor (Prunellidae).

1
♂
2
♂
3
4
♂
5

Taxonomic Comments

Historical debates about the evolutionary relationships of the sugarbirds (*Promerops*) centered on whether they were African relatives of the Australian honeyeaters or highly modified sunbirds. Initial DNA studies suggested that they were nectar-feeding starlings, but Sibley and Ahlquist (1990) have supported the sunbird hypothesis. Bock (1985) continues to advocate the honeyeater hypothesis. Sunbirds and flowerpeckers are probably sister taxa (Sibley and Ahlquist 1981b).

The Passeridae are characterized by a unique character state of the preglossal skeleton structure of the tongue (Bock and Morony 1971).

The Olive Warbler of the pine forests of Mexico and southern Arizona was long assumed to be a wood warbler until George (1962) discovered fundamental differences in the syringeal musculature. Proposed affinities include thrushes, Old World warblers, and most recently from DNA studies, Old World finches such as the Chaffinch. Sibley and Ahlquist (1990) suggest that the Olive Warbler is the sister species of the New World nine-primaried oscine clade.

Hawaiian honeycreepers evolved from carduelline finches (Beecher 1953; Tordoff 1954; Raikow 1976; Sibley and Ahlquist 1982).

The sixth edition of the American Ornithologists' Union *Checklist of North American Birds* treats the Emberizidae as a broadly inclusive family that encompasses the Emberizinae, Parulinae, Thraupinae, Cardinalinae, and Icterinae as subfamilies.

The Wrenthrush (*Zeledonia*) has been shown to be a thrushlike wood warbler rather than a relative of the Turdidae.

Sibley and Ahlquist's (1990) and Bledsoe's (1988) DNA studies revealed that the concept of tanagers should be expanded to include a radiation of ecological types, including the nectarivorous honeycreepers, the swallowlike *Tersina,* and a variety of tanager-finches (*Sicalis, Oryzoborus*), some of which were once considered to be emberizid buntings.

Recent monographs include ones on the biology of wood warblers (Morse 1989), tanagers (Isler and Isler 1987), and blackbirds (Orians 1980).

Figure A–31 Perching birds—New World nine-primaried oscines: (1) Yellow-rumped Cacique (Icteridae); (2) Magnolia Warbler (Parulidae); (3) Rose-breasted Grosbeak (Cardinalidae); (4) Magpie Tanager (Thraupidae); (5) Lapland Longspur (Emberizidae).

Bibliography

Abbott, I., and P.R. Grant. 1976. Nonequilibrial bird faunas on islands. Am. Nat. 110: 507–528. **[576]**

Abbott, I., L.K. Abbott, and P.R. Grant. 1977. Comparative ecology of Galápagos ground finches (*Geospiza* Gould): Evaluation of the importance of floristic diversity and interspecific competition. Ecol. Monogr. 47: 151–184. **[569]**

Able, K.P. 1982. Skylight polarization patterns at dusk influence migratory orientation in birds. Nature London 299: 550–551. **[323]**

Ainley, D.G., and D.P. DeMaster. 1980. Survival and mortality in a population of Adelie Penguins. Ecology 61: 522–530. **[494]**

Ainley, D.G., R.E. LeRosche, and W.J.L. Sladen. 1983. Breeding Biology of the Adelie Penguin. Berkeley: University of California Press. **[407, 494, 505]**

Åkerman, B. 1966a, b. Behavioural effects of electrical stimulation in the forebrain of the pigeon, I and II. Behaviour 26: 323–338, 339–350. **[179]**

Alatalo, R.V., and A. Lundberg. 1984. Polyterritorial behavior in the Pied Flycatcher *Ficedula hypoleuca*—Evidence for the deception hypothesis. Ann. Zool. Fenn. 21: 217–228. **[414]**

Alatalo, R.V., D. Ericksson, L. Gustafsson, and K. Larsson 1987. Exploitation competition influences the use of foraging sites by tits: Experimental evidence. Ecology 68: 284–290. **[568]**

Alatalo, R.V., L. Gustafsson, and A. Lundberg. 1986. Interspecific competition and niche shifts in tits *Parus*: Evaluation of non-experimental data. Am. Nat. 127: 819–834. **[567, 569]**

Alcock, J. 1993. Animal Behavior, 5th ed. Sunderland, Mass.: Sinauer Associates. **[232]**

Aldrich, J.W., and F.C. James. 1991. Ecogeographic variation in the American Robin (*Turdus migratorius*). Auk 108: 230–249. **[129]**

Alerstam, T. 1990. Bird Migration. Cambridge: Cambridge University Press. **[307]**

Alexander, R.M. 1992. Exploring Biomechanics. New York: Scientific American Library. **[101, 102, 104, 108, 113]**

Alvarez-Buyilla, A., J.R. Kirn, and F. Nottebohm. 1990. Birth of projection neurons in adult avian brain may be related to perceptual or motor learning. Science 249: 1444–1445. **[184]**

Alvarez del Toro, M. 1971. On the biology of the American Finfoot in southern Mexico. Living Bird 10: 79–88. **[434]**

Amadon, D. 1980. Varying proportions between young and old raptors. Proc. Pan-Afr. Ornithol. Congr. 4: 327–331. **[109]**

American Ornithologists' Union. 1983. Checklist of North American Birds, 6th ed. Lawrence, Kan.: American Ornithologists' Union. **[xvi]**

Ames, P.L. 1975. The application of syringeal morphology to the classification of the Old World insect-eaters. Bonn. Zool. Beitr. 26: 107–134. **[57]**

Anderson, B.W., and R.J. Daugherty. 1974. Characteristics and reproductive biology of grosbeaks (*Pheucticus*) in the hybrid zone in South Dakota. Wilson Bull. 86: 1–11. **[544]**

Anderson, D.J. 1989. The role of hatching asynchrony in siblicidal brood reduction of two booby species. Behav. Ecol. Sociobiol. 25: 363–368. **[492]**

Anderson, D.J. 1990. Evolution of obligate siblicide in boobies. I. A test of the insurance egg hypothesis. Am. Nat. 135: 334–351. **[492]**

Anderson, D.J., N.C. Stoyan, and R.E. Ricklefs. 1987. Why are there no viviparous birds? Am. Nat. 130: 941–947. **[361]**

Andersson, M. 1976. Social behaviour and communication in the Great Skua. Behaviour 58: 40–77. **[222]**

Andersson, M. 1978. Optimal egg shape in waders. Ornis Fenn. 55: 105–109. **[370]**

Andersson, M. 1982. Female choice selects for extreme tail length in a widowbird. Nature London 299: 818–820. **[225]**

Andersson, M., and M.O.G. Eriksson. 1982. Nest parasitism in Goldeneyes *Bucephala clangula*: Some evolutionary aspects. Am. Nat. 120: 1–16. **[458]**

Andrewartha, H.G., and L.C. Birch. 1954. The Distribution

Note: The boldfaced bracketed numbers are the text page numbers on which references or figure or table citations occur.

and Abundance of Animals. Chicago: University of Chicago Press. **[510]**

Ankney, C.D., and C.D. MacInnes. 1978. Nutrient reserves and reproductive performance of female Lesser Snow Geese. Auk 95: 459–471. **[365, 366, 398]**

Ankney, C.D., and D.M. Scott. 1980. Changes in nutrient reserves and diet of breeding Brown-headed Cowbirds. Auk 97: 684–696. **[365]**

Armstrong, E.A. 1942. Bird Display. Cambridge: Cambridge University Press. **[419]**

Armstrong, E.A. 1963. A Study of Bird Song. New York: Academic Press. **[242]**

Arnold, A.P. 1980. Anatomical and electrophysiological studies of sexual dimorphism in a passerine vocal control system, pp. 648–652. *In* Acta XVII Congressus Internationalis Ornithologici (R. Nöhring, Ed.). Berlin: Verlag der Deutschen Ornithologen-Gesellschaft. **[184]**

Arnold, A.P., and A. Saltiel. 1979. Sexual difference in pattern of hormone accumulation in the brain of a songbird. Science 205: 702–705. **[184]**

Aschoff, J. 1980. Biological clocks in birds, pp. 113-136. *In* Acta XVII Congressus Internationalis Ornithologici (R. Nöhring, Ed.). Berlin: Verlag der Deutschen Ornithologen-Gesellschaft. **[268]**

Ashmole, N.P. 1963a. The biology of the Wideawake or Sooty Tern *Sterna fuscata* on Ascension Island. Ibis 103b: 297–364. **[282]**

Ashmole, N.P. 1963b. The regulation of numbers of tropical oceanic birds. Ibis 103b: 458–473. **[500]**

Ashmole, N.P. 1965. Adaptive variation in the breeding regime of a tropical sea bird. Proc. Natl. Acad. Sci. USA 53: 311–318. **[279, 283]**

Ashmole, N.P. 1968. Breeding and molt in the White Tern (*Gygis alba*) on Christmas Island, Pacific Ocean. Condor 70: 35–55. **[281]**

Askins, R.A., J.F. Lynch, and R. Greenberg. 1990. Population declines in migratory birds in eastern North America. Current Ornithology 7: 1–57. **[287, 595, 615]**

Atwood, J.L. 1988. Separation and geographic variation in Black-tailed Gnatcatchers. Ornithol. Monogr. No. 42. **[596]**

Atwood, J.L. 1993. The politics of protection. Living Bird 12: 14–21. **[596]**

Aulie, A. 1976. The pectoral muscles and the development of thermoregulation in chicks of Willow Ptarmigan (*Lagopus lagopus*). Comp. Biochem. Physiol. 53A: 343–346. **[437]**

Austin, G.T. 1976. Behavioral adaptations of the Verdin to the desert. Auk 93: 245–262. **[390]**

Austin, O.L., Jr., and A. Singer. 1985. Families of Birds. New York: Golden Press. **[381]**

Avise, J.C. 1994. Molecular Markers, Natural History and Evolution. New York: Chapman and Hall. **[552]**

Avise, J.C., and W.S. Nelson. 1989. Molecular genetic relationships of the extinct Dusky Seaside Sparrow. Science 243: 646–648. **[532, 533]**

Avise, J.C., C.D. Ankney, and W.S. Nelson. 1990. Mitochondrial gene trees and the evolutionary relationship of Mallard and Black ducks. Evolution 44: 1109–1119. **[549]**

Avise, J.C., R.M. Ball, and J. Arnold. 1988. Current versus historical population sizes in vertebrate species with high gene flow: A comparison based on mitochondrial DNA lineages and inbreeding theory for neutral mutations. Mol. Biol. Evol. 5: 331–344. **[524]**

Axelrod, R., and W.D. Hamilton. 1981. The evolution of cooperation. Science 211: 1390–1396. **[468]**

Badgerow, J.P. 1988. An analysis of function in the formation flight of Canada Geese. Auk 105: 749–755. **[101]**

Bagg, A.M., W.W.H. Gunn, D.S. Miller, J.T. Nichols, W. Smith, and F.P. Wolfarth. 1950. Barometric pressure patterns and spring bird migration. Wilson Bull. 62: 5–19. **[304]**

Bailey, R.E. 1952. The incubation patch of passerine birds. Condor 54: 121–136. **[392]**

Baker, A.J. 1980. Morphometric differentiation in New Zealand populations of the House Sparrow (*Passer domesticus*). Evolution 34: 638–653. **[130]**

Baker, A.J. 1991. A review of New Zealand ornithology. Current Ornithology 8: 1–67. **[586]**

Baker, A.J, and A. Moeed. 1987. Rapid genetic differentiation and founder effect in colonizing populations of common mynas (*Acridotheres tristis*). Evolution 41: 525–538. **[535]**

Baker, M.C. 1982. Genetic population structure and vocal dialects in *Zonotrichia* (Emberizidae), pp. 209–235. *In* Acoustic Communication in Birds, Vol. 2 (D.E. Kroodsma and E.H. Miller, Eds.). New York: Academic Press. **[258]**

Baker, M.C., and A.E.M. Baker. 1990. Reproductive behavior of female buntings: Isolation mechanisms in a hybridizing pair of species. Evolution 44: 332–338. **[544]**

Baker, M.C., and S.F. Fox. 1978. Dominance, survival, and enzyme polymorphism in Dark-eyed Juncos, *Junco hyemalis*. Evolution 32: 697–711. **[334]**

Bakker, R.T. 1975. Dinosaur renaissance. Sci. Am. 232(4): 58–78. **[31]**

Balda, R.P., and A.C. Kamil. 1989. A comparative study of cache recovery by three corvid species. Anim. Behav. 38: 486–495. **[185]**

Ball, R.M., and J.C. Avise. 1992. Mitochondrial DNA phylogeographic differentiation among avian populations and the evolutionary significance of subspecies. Auk 109: 626–636. **[532]**

Ball, R.M., S. Freeman, F.C. James, E. Bermingham, and J.C.Avise. 1988. Phylogeographic population structure of Red-winged Blackbirds assessed by mitochondrial DNA. Proc. Natl. Acad. Sci. USA 85: 1558–1562. **[532]**

Balthazart, J., and Schoffeniels, E. 1979. Pheromones are involved in the control of sexual behavior in birds. Naturwissenschaften 66: 55–56. **[200]**

Bancroft, G.T., and G.E. Woolfenden. 1982. The molt of Scrub Jays and Blue Jays in Florida. Ornithol. Monogr. No. 29. **[279]**

Bang, B.G., and B.M. Wenzel. 1986. Nasal cavity and olfactory system, pp. 195–225. *In* Form and Function in Birds, Vol. 3 (A.S. King and J. McLelland, Eds.). New York: Academic Press. **[199]**

Banks, R.C. 1979. Human related mortality of birds in the United States. U.S. Fish Wildl. Serv. Spec. Sci. Rep. Wildl. No. 215. **[587]**

Barreto, C., R.M. Albrecht, D.E. Bjorling, J.R. Horner, and N.J. Wilsman. 1993. Evidence of the growth plate and the growth of long bones in juvenile dinosaurs. Science 262: 2020–2023. **[32]**

Barrowclough, G.F. 1978. Sampling bias in dispersal studies based on finite area. Bird-Banding 49: 333–341. **[534]**

Barrowclough, G.F. 1980a. Gene flow, effective population sizes, and genetic variance components in birds. Evolution 34: 789–798. **[534]**

Barrowclough, G.F. 1980b. Genetic and phenotypic differentiation in a wood warbler (genus *Dendroica*) hybrid zone. Auk 97: 655–668. **[541]**

Barrowclough, G.F. 1982. Geographic variation, predictiveness, and subspecies. Auk 99: 601–603. **[529]**

Barrowclough, G.F. 1983. Biochemical studies of microevolutionary processes, pp. 223–261. *In* Perspectives in Ornithology (A.H. Brush and G.A. Clark, Jr., Eds.). Cambridge: Cambridge University Press. **[57, 532, 534, 549]**

Barrowclough, G. 1992. Biochemical studies of the higher level systematics of birds. Bull. Br. Ornithol. Club 112A: 39–52. **[61]**

Barrowclough, G.F., and K.W. Corbin. 1978. Genetic variation and differentiation in the Parulidae. Auk 95: 691–702. **[57]**

Bartholomew, G.A. 1942. The fishing activities of Double-crested Cormorants on San Francisco Bay. Condor 44: 13–21. **[337]**

Bartholomew, G.A. 1982. Body temperature and energy metabolism, pp. 333–406. *In* Animal Physiology: Principles and Adaptations, 4th ed. (M.S. Gordon, Ed.). New York: Macmillan. **[126]**

Bartholomew, G.A., and T.J. Cade. 1963. The water economy of landbirds. Auk 80: 504–539. **[138, 139]**

Barton, N.H., and G.M. Hewitt. 1985. Analysis of hybrid zones. Annu. Rev. Ecol. Syst. 16: 113–148. **[545]**

Bateson, P.P.G. 1976. Specificity and the origins of behavior, pp. 1–20. *In* Advances in the Study of Behavior, Vol. 6 (J. Rosenblatt, R.A. Hinde, and C. Beer, Eds.). New York: Academic Press. **[430]**

Batten, L.A., and J.H. Marchant. 1977. Bird population changes for the years 1974–75. Bird Study 24: 55–61. **[513]**

Baxter, M., and M.D. Trotter. 1969. The effect of fatty materials extracted from keratins on the growth of fungi, with particular reference to the free fatty acid content. Sabouraudia 7: 199–206. **[77]**

Baylis, J.R. 1982. Avian vocal mimicry: Its function and evolution, pp. 51–83. *In* Acoustic Communication in Birds, Vol. 2 (D.E. Kroodsma and E.H. Miller, Eds.). New York: Academic Press. **[252]**

Bean, M. 1983. The Evolution of Natural Wildlife Law. Urbana, Ill.: Praeger. **[602]**

Becker, P.H. 1976. Artkennzeichnende Gesangsmerkmale bei Winter- und Sommergoldhänchen (*Regulus regulus, R. ignicapillus*). Z. Tierpsychol. 42: 411–437. **[244]**

Becker, P.H. 1982. The coding of species-specific characteristics in bird sounds, pp. 213–252. *In* Acoustic Communication in Birds, Vol. 1 (D.E. Kroodsma and E.H. Miller, Eds.). New York: Academic Press. **[244, 245]**

Becking, J.H. 1975. The ultrastructure of the avian eggshell. Ibis 117: 143–151. **[371]**

Bednarz, J.C. 1988. Cooperative hunting in Harris' Hawks (*Parabuteo unicinctus*). Science 239: 1525–1527. **[339]**

Bednarz, J., D. Klem, L.J. Goodrich, and S.E. Senner. 1990. Migration counts at Hawk Mountain, Pennsylvania, as indicators of population trends. Auk 107: 96–109. **[611]**

Beecher, M.D. 1982. Signature systems and kin recognition. Am. Zool. 22: 477–490. **[431]**

Beecher, W.J. 1953. A phylogeny of the oscines. Auk 70: 270–333. **[691]**

Beehler, B. 1983. Lek behavior of the Lesser Bird of Paradise. Auk 100: 992–995. **[420]**

Beehler, B. 1985. Adaptive significance of monogamy in the Trumpet Manucode *Manucodia keraudrenii* (Aves: Paradisaeidae). Ornithol. Monogr. No. 37. **[412]**

Beehler, B.M. 1989. The Birds of Paradise. Sci. Am. 261 (Dec): 117–123. **[412]**

Beehler, B.M., and M.S. Foster. 1988. Hotshots, hotspots, and female preference in the organization of lek mating systems. Am. Nat. 131: 203–219. **[420]**

Beehler, B., and S.G. Pruett-Jones. 1983. Display dispersion and diet of birds of paradise: A comparison of nine species. Behav. Ecol. Sociobiol. 13: 229–238. **[412, 420, 421]**

Beer, C.G. 1970. Individual recognition of voice in the social behaviour of birds. Adv. Study Behav. 3: 27–74. **[247]**

Beer, C.G. 1979. Vocal communication between Laughing Gull parents and chicks. Behaviour 70: 118–146. **[247]**

Beissinger, S.R., and N.F.R. Snyder, Eds. 1991. New World Parrots in Crisis. Washington, D.C.: Smithsonian Institution Press. **[434, 615]**

Beissinger, S.R., N.F.R. Snyder, S.R. Derrickson, F.C. James, and S.M. Lanyon. 1991. International trade in live, exotic birds creates a vast movement that must be halted. Auk 108: 982–984. **[600]**

Bellairs, R. 1960. Development of birds, pp. 127–188. *In* Biology and Comparative Physiology of Birds, Vol. 1 (A.J. Marshall, Ed.). New York: Academic Press. **[369]**

Bellrose, F.C. 1967. Orientation in waterfowl migration. Proc. Annu. Biol. Colloq. (Oregon State Univ.) 27: 73–79. **[302]**

Benedict, F.G., W. Landauer, and E.L. Fox. 1932. The physiology of normal and Frizzle fowl, with special reference to the basal metabolism. Storrs (Conn.) Agr. Expt. Sta. Bull. 177: 12–101. **[127]**

Bennett, A.F. 1980. The metabolic foundations of vertebrate behavior. BioScience 30: 452–456. **[116]**

Bent, A.C. 1926. Life histories of North American marsh birds. U.S. Natl. Mus. Bull. No. 135. **[434]**

Bent, A.C. 1939. Life histories of North American woodpeckers. U.S. Natl. Mus. Bull. No. 174. **[365]**

Bentz, G.D. 1983. Myology and histology of the phalloid organ of the buffalo weaver. Auk 100: 501–504. **[359]**

Berger, M., and J.S. Hart. 1974. Physiology and energetics of flight. Avian Biology 4: 415–477. **[126, 137, 301, 302]**

Berman, S.L., and R.J. Raikow. 1982. The hindlimb musculature of the mousebirds (Coliiformes). Auk 99: 41–57. **[454]**

Berthold, P. 1975. Migration: Control and metabolic physiology. Avian Biology 5: 77–128. **[299, 302]**

Berthold, P. 1978. Circannuale Rhythmik: Freilaufende

selbsterregte Periodik mit lebenslanger Wirksamkeit bei Vögeln. Naturwissenschaften 65: 546–547. **[269]**

Berthold, P. 1988. Evolutionary aspects of migratory behavior in European warblers. J. Evol. Biol. 1: 195–209. **[321]**

Berthold, P., Ed. 1991. Orientation in Birds. Basel: Birkhauser Verlag. **[325]**

Berthold, P., A.J. Helbig, G. Mohr, and U. Querner. 1992. Rapid microevolution of migratory behavior in a wild bird species. Nature London 360: 668–670. **[321]**

Berthold, P., and U. Querner. 1981. Genetic basis of migratory behavior in European warblers. Science 212: 77–79. **[321]**

Bertram, B.C.R. 1980. Vigilance and group size in ostriches. Anim. Behav. 28: 278–286. **[341]**

Bibby, C.J., N.J. Collar, M.J. Crosby. M.F. Heath, Ch. Imboden, T.H. Johnson, A.J. Long, A.J. Stattersfield, and S.J. Thirgood. 1992. Putting biodiversity on the map: Priority areas for global conservation. Cambridge: International Council for Bird Preservation. **[582]**

Biebach, H. 1983. Genetic determination of partial migration in the European Robin (*Erithacus rubecula*). Auk 100: 601–606. **[292]**

Bierregaard, R.O. Jr., T.E. Lovejoy, V. Kapos, A. Augusto dos Santos, and R.W. Hutchings. 1992. The biological dynamics of tropical rain-forest fragments. BioScience 42: 859–866. **[599]**

Bildstein, K.L., G.T. Bancroft, P.J. Dugan, D.H. Gordon, R.M. Erwin, E. Nol, L.X. Payne, and S.E. Senner. 1991. Approaches to the conservation of coastal wetlands in the western hemisphere. Wilson Bull. 103: 218–254. **[599]**

Bildstein, K.L., J. Brett, L. Goodrich, and C. Viverette. 1993. Shooting Galleries. Am. Birds 47: 38–43. **[287, 610]**

Bingman, V.P. 1988. The avian hippocampus: Its role in the neural optimization of the spatial behavior of homing pigeons, Vol. 2, pp. 2075–2093. *In* Acta XIX Congressus Internationalis Ornithologici (H. Ouellet, Ed.). Ottawa: University of Ottawa Press. **[185]**

Birkhead, T.R. 1988. Behavioral aspects of sperm competition in birds. Behaviour 101: 101–138. **[360]**

Birkhead, T.R. 1993. Avian mating systems: The Aquatic Warbler is unique. TREE 8: 390–391. **[353]**

Birkhead, T.R., and A.P. Møller. 1992. Sperm Competition in Birds. New York: Academic Press. **[360, 374, 424]**

Blackburn, D.G., and H.E. Evans. 1986. Why are there no vivaparous birds? Am. Nat. 128: 165–190. **[360]**

Blank, J.L., and V. Nolan, Jr. 1983. Offspring sex ratio in red-winged blackbirds is dependent on maternal age. Proc. Natl. Acad. Sci. USA 80: 6141–6145. **[504]**

Bledsoe, A.H. 1988. A phylogenetic analysis of postcranial skeletal characters of the ratite birds. Ann. Carnegie Mus. 57: 73–90. **[52, 691]**

Bledsoe, A.H., and R.J. Raikow. 1990. A quantitative assessment of congruence between molecular and nonmolecular estimates of phylogeny. J. Mol. Evol. 30: 247–259. **[56]**

Blem, C.R. 1980 The energetics of migration, pp. 175–224. *In* Animal Migration, Orientation, and Navigation (S.A. Gauthreaux, Ed.). Orlando, Fla.: Academic Press. **[307]**

Blem, C.R. 1990. Avian energy storage. Current Ornithology 7: 59–114. **[170, 365]**

Blockstein, D.E., and H.B. Tordoff. 1985. Gone forever—A contemporary look at the extinction of the Passenger Pigeon. Am. Birds 39: 845–851. **[583]**

Bluhm, C.K. 1988. Temporal patterns of pair formation and reproduction in annual cycles and associated endocrinology in waterfowl. Current Ornithology 5: 123–185. **[264, 278, 284]**

Boag, P.T., and P.R. Grant. 1981. Intense natural selection in a population of Darwin's finches (Geospizinae) in the Galápagos. Science 214: 82–85. **[152]**

Boag, P.T., and A.J. van Noordwijk. 1987. Quantitative genetics, pp. 45–78. *In* Avian Genetics (F. Cooke and P.A. Buckley, Eds.). New York: Academic Press. **[364, 492]**

Board, R.G., and G. Love. 1980. Magnesium distribution in avian eggshells. Comp. Biochem. Physiol. 66A: 667–672. **[363]**

Board, R.G., and V.D. Scott. 1980. Porosity of the avian eggshell. Am. Zool. 20: 339–349. **[371]**

Bock, C.E. 1980. Winter bird population trends: Scientific evaluation of Christmas Bird count data. Atl. Nat. 33: 28–37. **[515]**

Bock, C.E., and L.W. Lepthien. 1976a. Population growth in the Cattle Egret. Auk 93: 164–166. **[509]**

Bock, C.E., and L.W. Lepthien. 1976b. Growth in the eastern House Finch population, 1962–1971. Am. Birds 30: 791–792. **[509, 510]**

Bock, C.E., and L.W. Lepthien. 1976c. Synchronous eruptions of boreal seed-eating birds. Am. Nat. 110: 559–571. **[514]**

Bock, W.J. 1961. Salivary glands in the gray jays (*Perisoreus*). Auk 78: 355–365. **[155]**

Bock, W.J. 1963. The cranial evidence for ratite affinities, pp. 39–54. *In* Proc. XIII Int. Ornithol. Congr. (C.G. Sibley, Ed.). Lawrence, Kans.: Allen Press. **[621]**

Bock, W.J. 1965. The role of adaptive mechanisms in the origin of higher levels of organization. Syst. Zool. 14: 272–287. **[32, 33]**

Bock, W.J. 1966. An approach to the functional analysis of bill shape. Auk 83: 10–51. **[149]**

Bock, W.J. 1973. Philosophical foundations of classical evolutionary classification. Syst. Zool. 22: 375–392. **[50]**

Bock, W.J. 1985. Relationships of the sugarbird (Promerops: Passeriformes. ?Meliphagidae), pp. 349–374. In Proc. Int. Congr. African Vertebrates (K.-L. Schuchmann, Ed.). Bonn: Mus. Koenig. **[691]**

Bock, W.J., and J. Farrand, Jr. 1980. The number of species and genera of recent birds: A contribution to comparative systematics. Am. Mus. Novit. No. 2703. **[16]**

Bock, W.J., and R.S. Hikida. 1968. An analysis of twitch and tonus fibers in the hatching muscle. Condor 70: 211–222. **[427]**

Bock, W.J., and A. McEvey. 1969. Osteology of *Pedionomus torquatus* (Aves: Pedionomidae) and its allies. Proc. R. Soc. Victoria 82: 187–232. **[648]**

Bock, W.J., and W.D. Miller. 1959. The scansorial foot of the woodpeckers, with comments on the evolution of perching and climbing feet in birds. Am. Mus. Novit. No. 1931. **[54]**

Bock, W.J., and J. Morony. 1971. The preglossale of *Passer* (Aves)—A skeletal neomorph. Am. Zool. 11: 705. **[463]**

Boer, M.H. den. 1971. A colour-polymorphism in caterpillars of *Bupalus piniarius* (L.) (Lepidoptera: Geometridae). Neth. J. Zool. 21: 61–116. **[165]**

Bolze, G. 1968. Anordnung und Bau der Herbstschen Körperchen in Limicolenschnäbeln im Zusammenhang mit der Nahrungsfindung. Zool. Anz. 181: 313–355. **[197]**

Bond, J. 1948. Origin of the bird fauna of the West Indies. Wilson Bull. 60: 207–229. **[37]**

Bond, J. 1963. Derivation of the Antillean avifauna. Proc. Acad. Nat. Sci. Philadelphia 115: 79–98. **[37]**

Bond, R.R. 1957. Ecological distribution of breeding birds in the upland forests of southern Wisconsin. Ecol. Monogr. 27: 351–384. **[562]**

Booth, D.T., D.H. Clayton, and B.A. Block. 1994. Experimental demonstration of the energetic cost of parasitism in wild hosts. Proc. R. Soc. London Ser. B 253(1337): 125-129. **[77]**

Borgia, G. 1985a. Bower quality, number of decorations and mating success of male Satin Bowerbirds (*Ptilonorhynchus violaceus*): An experimental analysis. Anim. Behav. 33: 266–271. **[228]**

Borgia, G. 1985b. Bowers as markers of male quality: Test of a hypothesis. Anim. Behav. 35: 266–271. **[228]**

Borgia, G. 1986. Sexual selection in bowerbirds. Sci. Am. 254(6): 92–100. **[226]**

Borgia, G., and M.A. Gore. 1986. Sexual competition by feather stealing in the Satin Bowerbird (*Ptilonorhynchus violaceus*). Anim. Behav. 34: 727–738. **[228]**

Borgia, G., S.G. Pruett-Jones, and M.A. Pruett-Jones. 1985. The evolution of bower-building and the assessment of male quality. Z. Tierpsychol. 67: 225–236. **[226, 228]**

Boswall, J. 1977. Tool-using by birds and related behavior. Avic. Mag. 83: 88–97, 146–159, 220–228. **[164]**

Boswall, J. 1983. Tool-using and related behaviour in birds: More notes. Avic. Mag. 89: 94–108. **[164]**

Bottjer, S.W., E.A. Miesner, and A.P. Arnold. 1984. Forebrain lesions disrupt development but not maintenance of song in passerine birds. Science 224: 901–903. **[255]**

Boughey, M.J., and N.S. Thompson. 1976. Species specificity and individual variation in the songs of the Brown Thrasher (*Toxostoma rufum*) and Catbird (*Dumetella carolinensis*). Behaviour 57: 64–90. **[246]**

Bourne, W.R.P. 1955. The birds of the Cape Verde Islands. Ibis 97: 508–556. **[275]**

Bowen, B.S., R.R. Koford, and S.L. Vehrencamp. 1989. Dispersal in the communally breeding Groove-billed Ani (*Crotophaga sulcirostris*). Condor 91: 62–65. **[474]**

Bowman, R.I., and S.L. Billeb. 1965. Blood-eating in a Galápagos finch. Living Bird 4: 29–44. **[164]**

Boyce, M.S., and C.M. Perrins. 1987. Optimizing Great Tit clutch size in a fluctuating environment. Ecology 68: 142–153. **[497, 498, 499]**

Brackenbury, J.H. 1982. The structural basis of voice production and its relationship to sound characteristics, pp. 53–73. *In* Acoustic Communication in Birds, Vol. 1 (D.E. Kroodsma and E.H. Miller, Eds.). New York: Academic Press. **[239, 241]**

Braun, E.J. 1982. Renal function. Comp. Biochem. Physiol. 71A: 511–517. **[141]**

Bray, O.E., J.J. Kennelly, and J.L. Guarino. 1975. Fertility of eggs produced on territories of vasectomized Red-winged Blackbirds. Wilson Bull. 87: 187–195. **[409]**

Breitwisch, R. 1989. Mortality patterns, sex ratios, and parental investment in monogamous birds. Current Ornithology 6: 1–50. **[487]**

Breitwisch, R., and G.H. Whitesides. 1987. Directionality of singing and non-singing behavior of mated and unmated northern mockingbirds. Anim. Behav. 35: 331–339. **[253]**

Briskie, J.V. 1992. Copulation patterns and sperm competition in the polygynandrous Smith's Longspur. Auk 109: 563–575. **[359, 360]**

Briskie, J.V. 1993. Smith's Longspur. *In* The Birds of North America, No. 34. (A. Poole, P. Stettenheim, and F. Gill, Eds.). Philadelphia: Academy of Natural Sciences; Washington, D.C.: American Ornithologists' Union. **[359]**

Briskie, J.V., S.G. Sealy, and K.A. Hobson. 1992. Behavioral defenses against avian brood parasitism in sympatric and allopatric host populations. Evolution 46: 341–352. **[464]**

Brodkorb, P. 1955. Number of feathers and weights of various systems in a Bald Eagle. Wilson Bull. 67: 142. **[74]**

Brodkorb, P. 1960. How many species of birds have existed? Bull. Florida State Mus. 4: 349–371. **[581]**

Brooks, A.S., P.E. Hare, J.E. Kokis, G.H. Miller, R.D. Ernst, and F. Wendorf. 1990. Dating Pleistocene archeological sites by protein diagenesis in ostrich eggshell. Science 248: 60–64. **[371]**

Brooks, W.S. 1978. Avian prehatching behavior: Functional aspects of the tucking pattern. Condor 80: 442–444. **[427]**

Broom, R. 1913. On the South-African pseudosuchian *Euparkeria* and allied genera. Proc. Zool. Soc. London 1913: 619–633. **[30]**

Brower, L.P., B.S. Alpert, and S.C. Glazier. 1970. Observational learning in the feeding behavior of Blue Jays (*Cyanocitta cristata* Oberholser, Fam. Corvidae). Am. Zool. 10: 475–476. **[178]**

Brown, C.R. 1984. Laying eggs in a neighbor's nest: Benefit and cost of colonial nesting in swallows. Science 224: 518–519. **[458]**

Brown, C.R. 1986. Cliff Swallow colonies as information centers. Science 234: 83–85. **[340]**

Brown, C.R. 1988. Enhanced foraging efficiency through information centers: A benefit of coloniality in Cliff Swallows. Ecology 69: 602–613. **[384]**

Brown, C.R., and M.B. Brown. 1986. Ectoparasitism as a cost of coloniality in Cliff Swallows (*Hirundo pyrrhonotoa*). Ecology 67: 1206–1218. **[385]**

Brown, J.L. 1964a. The integration of agonistic behavior in the Steller's Jay, *Cyanocitta stelleri* (Gmelin). Univ. Calif. Publ. Zool. No. 60. **[217]**

Brown, J.L. 1964b. The evolution of diversity in avian territorial systems. Wilson Bull. 76: 160–169. **[330]**

Brown, J.L. 1969. Territorial behaviour and population regulation in birds: A review and re-evaluation. Wilson Bull. 81: 293–329. **[518]**

Brown, J.L. 1974. Alternate routes to sociality in jays—with a theory for the evolution of altruism and communal breeding. Am. Zool. 14: 63–80. **[468]**

Brown, J.L. 1975. The Evolution of Behavior. New York: Norton. **[333, 334]**

Brown, J.L. 1987. Helping and Communal Breeding in Birds:

Ecology and Evolution. Princeton, N.J.: Princeton University Press. **[471, 477]**

Brown, J.L., and E.R. Brown. 1981a. Extended family system in a communal bird. Science 211: 959–960. **[468]**

Brown, J.L., and E.R. Brown. 1981b. Kin selection and individual selection in babblers, pp. 244–256. *In* Natural Selection and Social Behavior: Recent Research and New Theory (R.D. Alexander and D.W. Tinkle, Eds.). New York: Chiron Press. **[469, 470]**

Brown, J.L., E.R. Brown, S.D. Brown, and D.D. Dow. 1982. Helpers: Effects of experimental removal on reproductive success. Science 215: 421–422. **[468, 469]**

Brown, J.L., and G.H. Orians. 1970. Spacing patterns in mobile animals. Annu. Rev. Ecol. Syst. 1: 239–262. **[329]**

Brown, L.H., V. Gargett, and P. Steyn. 1977. Breeding success in some African eagles related to theories about sibling aggression and its effects. Ostrich 48: 65–71. **[442]**

Brown, L.H., and E.K. Urban. 1969. The breeding biology of the Great White Pelican *Pelecanus onocrotalus roseus* at Lake Shala, Ethiopia. Ibis 111: 199–237. **[385]**

Brown, R.G.B. 1962. The aggressive and distraction behavior of the Western Sandpiper *Ereunetes mauri.* Ibis 104: 1–12. **[383]**

Brown, W.L., Jr., and E.O. Wilson. 1956. Character displacement. Syst. Zool. 5: 49–64. **[571]**

Brush, A.H. 1969. On the nature of "cotingin." Condor 71: 431–433. **[90]**

Brush, A.H. 1972. Review of structure and spectral reflectance of green and blue feathers of the Rose-faced Lovebird (*Agapornis roseicollis*), by J. Dyke, 1971, Biol. Skr. 18: 1–67. Auk 89: 679–681. **[89]**

Brush, A.H. 1979. Comparison of egg-white proteins: Effect of electrophoretic conditions. Biochem. Syst. Ecol. 7: 155–165. **[640, 659]**

Brush, A.H. 1993. The evolution of feathers: A novel approach. Avian Biology 9: 121–162. **[65,91]**

Bryant, D.M. 1975. Breeding biology of House Martins *Delichon urbica* in relation to aerial insect abundance. Ibis 117: 180–216. **[441]**

Bryant, D.M. 1978a. Environmental influences on growth and survival of nestling House Martins *Delichon urbica.* Ibis 120: 271–283. **[441–442]**

Bryant, D.M. 1978b. Establishment of weight hierarchies in the broods of House Martins *Delichon urbica.* Ibis 120: 16–26. **[441–442]**

Bucher, E.H. 1992. The causes of extinction of the Passenger Pigeon. Current Ornithology 9: 1–36. **[583]**

Buckley, F.G., and P.A. Buckley. 1974. Comparative feeding ecology of wintering adult and juvenile Royal Terns (Aves: Laridae, Sterninae). Ecology 55: 1053–1063. **[329, 449]**

Buckley, P.A. 1987. Mendelian genes, pp. 1–44. *In* Avian Genetics (F. Cooke and P.A. Buckley, Eds.). New York: Academic Press. **[90, 352, 535]**

Buckley, P.A., and F.G. Buckley. 1977. Hexagonal packing of Royal Tern nests. Auk 94: 36–43. **[329]**

Bulmer, M.G., and C.M. Perrins. 1973. Mortality in the Great Tit *Parus major.* Ibis 115: 277–281. **[487]**

Burger, J. 1981. A model for the evolution of mixed-species colonies of ciconiiformes. Q. Rev. Biol. 56: 143–167. **[345]**

Burley, N. 1980. Clutch overlap and clutch size: Alternative and complementary reproductive tactics. Am. Nat. 115: 223–246. **[490]**

Burley, N., and N. Moran. 1979. The significance of age and reproductive experience in the mate preferences of feral pigeons, *Columba livia.* Anim. Behav. 27: 686–698. **[407]**

Burtt, E.H., Jr. 1979. Tips on wings and other things, pp. 75–110. *In* The Behavioral Significance of Color (E.H. Burtt, Jr., Ed.). New York: Garland STPM Press. **[86]**

Burtt, E.H., Jr., and J.P. Hailman. 1978. Head-scratching among North American wood-warblers (Parulidae). Ibis 120: 153–170. **[77, 79]**

Buskirk, W.H. 1976. Social systems in a tropical forest avifauna. Am. Nat. 110: 293–310. **[344]**

Buskirk, W.H. 1980. Influence of meteorological patterns and trans-gulf migration on the calendars of latitudinal migrants, pp. 485–491. *In* Migrant Birds in the Neotropics (A. Keast and E.S. Morton, Eds.). Washington, D.C.: Smithsonian Institution Press. **[304]**

Butcher, G.S., and S. Rohwer. 1989. The evolution of conspicuous and distinctive coloration for communication in birds. Current Ornithology 6: 51–108. **[91, 232]**

Butler, P.J. 1992. Parrots, pressure, people, and pride, pp. 25–46. *In* New World Parrots in Crisis (S.R. Beissinger and N.F.R. Snyder, Eds.). Washington, D.C.: Smithsonian Institution Press. **[611]**

Buttemer, W.A. 1992. Differential overnight survival by Bumpus' House Sparrows: An alternative interpretation. Condor 94: 944–954. **[130]**

Cade, T.J. 1953. Sub-nival feeding of the Redpoll in interior Alaska: A possible adaptation to the northern winter. Condor 55: 43–44. **[132]**

Cade, T.J., and G.L. Maclean. 1967. Transport of water by adult sandgrouse to their young. Condor 69: 323–343. **[68]**

Cairns, D.K. 1992. Population regulation of seabird colonies. Current Ornithology 9: 37–61. **[514, 525]**

Calder, W.A. 1971. Temperature relationships and nesting of the Calliope Hummingbird. Condor 73: 314–321. **[390]**

Calder, W.A. 1974. Consequences of body size for avian energetics, pp. 86–151. *In* Avian Energetics (R.A. Paynter, Jr., Ed.). Publ. Nuttall Ornithol. Club No. 15. **[171, 390]**

Calder, W.A. 1984. Size, Function, and Life History. Cambridge, Mass.: Harvard University Press. **[145]**

Calder, W.A., and J.R. King. 1974. Thermal and caloric relations of birds. Avian Biology 4: 259–413. **[127, 131, 132, 133, 145]**

Calford, M.B., and Piddington, R.W. 1988. Avian interaural canal enhances interaural delay. J. Comp. Physiol. 162: 503–510. **[196]**

Canady, R.A., D.E. Kroodsma, and F. Nottebohm. 1984. Population differences in complexity of a learned skill are correlated with the brain space involved. Proc. Natl. Acad. Sci. USA 81: 6232–6234. **[184]**

Caple, G., R.P. Balda, and W.R. Willis. 1983. The physics of leaping animals and the evolution of preflight. Am. Nat. 121: 455–476. **[33, 109]**

Caple, G., R.P. Balda, and W.R. Willis. 1984. Flap about flight. Anim. Kingdom 87: 33–38. **[33, 109]**

Capparella, A. 1988. Genetic variation in Neotropical birds: Implications for the speciation process, pp. 1658–1664. *In* Acta XIX Congressus Internationalis Ornithologici (H. Ouellet, Ed.). Ottawa: University of Ottawa Press. **[532]**

Capparella, A. 1991. Neotropical avian diversity and riverine barriers, pp. 307–316. *In* Acta XX Congressus Internationalis Ornithologici (B.D. Ball, Ed.). Wellington, N.Z.: Ornithological Congress Trust Board. **[532]**

Caraco, T. 1979. Time budgeting and group size: A test of theory. Ecology 60: 618–627. **[342]**

Caraco, T., S. Martindale, and H.R. Pulliam. 1980. Avian flocking in the presence of a predator. Nature London 285: 400–401. **[342]**

Carey, C. 1980. Adaptation of the avian egg to high altitude. Am. Zool. 20: 449–459. **[373]**

Carey, C. 1983. Structure and function of avian eggs. Current Ornithology 1: 69–103. **[371, 372, 374, 401]**

Carey, C. 1991. Respiration of avian embryos at high altitudes, pp. 265–278. *In* Acta XX Congressus Internationalis Ornithologici (B.D. Ball, Ed.). Wellington, N.Z.: Ornithological Congress Trust Board. **[372, 373]**

Carey, C., W.R. Dawson, L.C. Maxwell, and J.A. Faulkner. 1983. Seasonal acclimatization to temperature in carduelline finches. II. Changes in body composition and mass in relation to season and acute cold stress. J. Comp. Physiol. 125: 101–103. **[133]**

Carey, M., and V. Nolan, Jr. 1975. Polygyny in Indigo Buntings: A hypothesis tested. Science 190: 1296–1297. **[413]**

Carlson, A., and J. Moreno. 1982. The loading effect in central place foraging Wheatears (*Oenanthe oenanthe* L.). Behav. Ecol. Sociobiol. 11: 173–183. **[167, 168]**

Carr, C.E. 1992. Evolution of the central auditory system in reptiles and birds, pp. 511–544. *In* Evolutionary Biology of Hearing (D.R. Webster, R.R. Fay, and A.N. Popper, Eds.). New York: Springer-Verlag. **[193, 196, 202]**

Carrick, R. 1963. Ecological significance of territory in the Australian Magpie, *Gymnorhina tibicen*, pp. 740–753. *In* Proc. XIII Int. Ornithol. Congr. (C.G. Sibley, Ed.). Lawrence, Kans.: Allen Press. **[518]**

Caryl, P.G. 1979. Communication by agonistic displays: What can games theory contribute to ethology? Behaviour 68: 136–169. **[222]**

Cavé, A.J. 1968. The breeding of the Kestrel, *Falco tinnunculus* L., in the reclaimed area Oostelijk Flevoland. Neth. J. Zool. 18: 313–407. **[398, 512]**

Chappuis, C. 1971. Un exemple de l'influence du milieu sur les émissions vocales des oiseaux: l'évolution des chants en foret équatoriale. Terre Vie 25: 183–202. **[237]**

Chatterchee, S. 1991. Cranial anatomy and relationships of a new Triassic bird from Texas. Philos. Trans. R. Soc. London Ser B 1265: 277–342. **[30]**

Cheke, A.S. 1987. An ecological history of the Mascarene Islands, with particular reference to extinctions and introductions of land vertebrates, pp. 5–89. *In* Studies of Mascarene Island Birds (A.W. Diamond, Ed.). Cambridge: Cambridge University Press. **[585]**

Chen, D.-M., J.S. Collins, and T.H. Goldsmith. 1984. The ultraviolet receptor of bird retinas. Science 225: 337–340. **[191]**

Clara, M. 1925. Ueber den Bau des Schnabels der Waldschnepfe. Z. Mikrosk.-Anat. Forsch. Leipzig 3: 1–108. **[197]**

Clark, A.B., and D.S. Wilson. 1981. Avian breeding adaptations: Hatching asynchrony, brood reduction, and nest failure. Q. Rev. Biol. 56: 253–277. **[428]**

Clark, G.A., Jr. 1961. Occurrence and timing of egg teeth in birds. Wilson Bull. 73: 268–278. **[428]**

Clark, G.A., Jr., and J.B. de Cruz. 1989. Functional interpretation of protruding filoplumes in oscines. Condor 91: 962–965. **[73]**

Clark, K.L., and R.J. Robertson. 1981. Cowbird parasitism and evolution of anti-parasite strategies in the Yellow Warbler. Wilson Bull. 93: 249–258. **[464]**

Clark, L. 1983. The development of effective homeothermy and endothermy by nestling starlings. Comp. Biochem. Physiol. 73A: 253–260. **[429, 436]**

Clark, L. 1991. The nest protection hypothesis: The adaptive use of plant secondary compounds by European starlings, pp. 205–220. *In* Bird–Parasite Interactions: Ecology, Evolution, and Behaviour (J.E. Loye and M. Zuk, Eds.). New York: Oxford University Press. **[380]**

Clark, L., and J.R. Mason. 1985. Use of nest material as insecticidal and anti-pathogenic agents by the European Starling. Oecologia Berlin 67: 169–176. **[378]**

Clark, L., and J.R. Mason. 1993. Interaction between sensory and postingestional repellents in starlings: Methyl anthranilate and sucrose. Ecol. Appl. 3: 262–270. **[161]**

Clark, L., and P.S. Shah. 1992. Information content of prey odor plumes: What do foraging Leach's Storm-Petrels know?, pp. 231–238. *In* Chemical Signals in Vertebrates, Vol 6 (R. Doty and H. Meuller-Schwarze, Eds.). New York: Plenum Press. **[200]**

Clark, L., K.V. Avilova, and N.J. Bean. 1992. Odor thresholds in passerines. Comp. Biochem. Physiol. 104A: 305–312. **[199, 200]**

Clayton, D.H. 1990. Mate choice in experimentally parasitized Rock Doves: Lousy males lose. Am. Zool. 30: 251–262. **[77, 78]**

Clayton, D.H. 1991. Coevolution of avian grooming and ectoparasite avoidance, pp. 258–289. *In* Bird–Parasite Interactions (J.E. Loye and M. Zuk, Eds.). New York: Oxford University Press. **[77]**

Clements, F.E. 1916. Plant succession: An analysis of the development of vegetation. Carnegie Inst. Wash. Publ. No. 242. **[562]**

Clements, F.E. 1936. Nature and structure of the climax. J. Ecol. 24: 252–284. **[562]**

Clements, J.F. 1991. Birds of the World: A Checklist. Vista, Calif.: Ibis Publishing Co. **[13]**

Clench, M.H. 1978. Tracheal elongation in birds-of-paradise. Condor 80: 423–430. **[241]**

Clobert, J., and J.-D. Lebreton. 1991. Estimation of demographic parameters in bird populations, pp. 75–104. *In* Bird Population Studies (C.M. Perrins, J.-D. Lebreton, and G.J.M. Hirons, Eds.). New York: Oxford University Press. **[488, 505]**

Clutton-Brock, T.H. 1991. The Evolution of Parental Care. Princeton, N.J.: Princeton University Press. **[417]**

Cochran, W.W., G.G. Montgomery, and R.R. Graber. 1967.

Migratory flights of *Hylocichla* thrushes in spring: A radiotelemetry study. Living Bird 6: 213–225. **[313]**

Cody, M.L. 1966. A general theory of clutch size. Evolution 20: 174–184. **[499]**

Cody, M.L. 1974. Competition and the Structure of Bird Communities. Princeton, N.J.: Princeton University Press. **[562]**

Cody, M.L., and J.M. Diamond, Eds. 1975. Ecology and Evolution of Communities. Cambridge, Mass.: Harvard University Press. **[554, 564]**

Cohn, J.M.W. 1968. The convergent flight mechanism of swifts (*Apodi*) and hummingbirds (*Trochili*) (Aves). Doctoral dissertation, University of Michigan, Ann Arbor, Mich. **[665]**

Colbert, E.H. 1955. Evolution of Vertebrates. New York: Wiley. **[8, 26]**

Colborn, T., and C. Clement. 1992. Chemically induced alterations in sexual and functional development: The wildlife/human connection. Princeton, N.J.: Princeton Science Publishing Co. **[590]**

Collar, N.J., and P. Andrew. 1988. Birds to Watch: The ICBP Annotated Checklist of Threatened Birds of the World. ICBP Tech. Publ. No. 8. Cambridge: International Council for Bird Preservation. **[579, 615]**

Collar, N.J., and A.T. Juniper. 1991. Dimensions and causes of the parrot crisis, pp. 1–24. *In* New World Parrots in Crisis: Solutions from Conservation Biology (S.R. Beissinger and N.F.R. Snyder, Eds.). Washington, D.C.: Smithsonian Institution Press. **[601]**

Collar, N.J., L.P. Gonzaga, N. Krabbe, A. Madroño-Nieto, L.G. Naranjo, T.A. Parker, III, and D.C. Wege. 1992. Threatened Birds of the Americas. Cambridge: International Council for Bird Preservation. **[616]**

Collias, N.E. 1952. The development of social behavior in birds. Auk 69: 127–159. **[430]**

Collias, N.E., and E.C. Collias. 1964. Evolution of nest-building in the weaverbirds (Ploceidae). Univ. Calif. Publ. Zool. No. 73. **[388, 389]**

Collias, N.E., and E.C. Collias. 1971. Some observations on behavioral energetics in the Village Weaverbird. I. Comparison of colonies from two subspecies in nature. Auk 88: 124–133. **[390]**

Collias, N.E., and E.C. Collias. 1984. Nest Building and Bird Behavior. Princeton, N.J.: Princeton University Press. **[378, 386, 401]**

Confer, J.L., and K. Knapp. 1981. Golden-winged Warblers and Blue-winged Warblers: The relative success of a habitat specialist and a generalist. Auk 98: 108–114. **[548]**

Connell, J.H. 1980. Diversity and the coevolution of competitors, or the ghost of competition past. Oikos 35: 131–138. **[566]**

Cooch, E.G., and F. Cooke. 1991. Demographic changes in a Snow Goose population: Biological and management implications, pp. 168–189. *In* Bird Population Studies (C.M. Perrins, J-D. Lebreton, and G.J.M. Hirons, Eds.). New York: Oxford University Press. **[511, 531]**

Cooke, A.S. 1975. Pesticides and eggshell formation. Symp. Zool. Soc. London 35: 339–361. **[363, 588]**

Cooke, F. 1978. Early learning and its effect on population structure. Studies of a wild population of Snow Geese. Z. Tierpsychol. 46: 344–358. **[209, 540]**

Cooke, F. 1987. Lesser Snow Goose: A long-term population study, pp. 407–432. *In* Avian Genetics (F. Cooke and P.A. Buckley, Eds.). New York: Academic Press. **[539]**

Cooke, F., and P.A. Buckley, Eds. 1987. Avian Genetics. New York: Academic Press. **[525, 552]**

Cooke, F., C.D. MacInnes, and J.P. Prevett. 1975. Gene flow between breeding populations of Lesser Snow Geese. Auk 92: 493–510. **[533]**

Cooke, F., P.D. Taylor, C.M. Francis, and R.F. Rockwell. 1990. Directional selection and clutch size in birds. Am. Nat. 136: 261–267. **[492]**

Cooper, A., C. Mourer-Chauviré, G.K. Chambers, A. von Haeseler, A.C. Wilson, and S. Pääbo. 1992. Independent origins of New Zealand moas and kiwis. Proc. Natl. Acad. Sci. USA 89: 8741–8744. **[52, 621]**

Corbin, K.W. 1983. Genetic structure and avian systematics. Current Ornithology 1: 211–292. **[532]**

Cortopassi, A.J., and L.R. Mewaldt. 1965. The circumannual distribution of White-crowned Sparrows. Bird-Banding 36: 141–169. **[265, 266]**

Corwin, J.T., and D.A. Cotanche. 1988. Regeneration of sensory hair cells after acoustic trauma. Science 240: 1772–1774. **[xxvii]**

Costa, R., and R.E.F. Escano. 1989. Red-cockaded Woodpecker status and management in the southern region. U.S.D.A. Forest Service Technical Publication R8-TP12. Washington, D.C.: USDA. **[512]**

Cottam, P.A. 1957. The pelecaniform characters of the skeleton of the Shoe-bill Stork, *Balaeniceps rex*. Bull. Br. Mus. Nat. Hist. Zool. 5: 49–72. **[635]**

Cox, G.W. 1968. The role of competition in the evolution of migration. Evolution 22: 180–192. **[293]**

Cox, G.W. 1985. The evolution of avian migration systems between temperate and tropical regions of the New World. Am. Nat. 126: 451–474. **[293]**

Cox, G.W., and R.E. Ricklefs. 1977. Species diversity and ecological release in Caribbean land bird faunas. Oikos 28: 113–122. **[574]**

Cracraft, J. 1971. The relationships and evolution of the rollers: Families Coraciidae, Brachypteraciidae, and Leptosomatidae. Auk 88: 723–752. **[670]**

Cracraft, J. 1973a. Continental drift, paleoclimatology, and the evolution and biogeography of birds. J. Zool. 169: 455–545. **[36, 37, 43]**

Cracraft, J. 1973b. Systematics and evolution of the Gruiformes (Class Aves). 3. Phylogeny of the suborder Grues. Bull. Am. Mus. Nat. Hist. 151: 1–128. **[643]**

Cracraft, J. 1974a. Phylogeny and evolution of the ratite birds. Ibis 116: 494–521. **[52, 621]**

Cracraft, J. 1974b. Continental drift and vertebrate distribution. Annu. Rev. Ecol. Syst. 5: 215–261. **[36]**

Cracraft, J. 1981. Toward a phylogenetic classification of the recent birds of the world (Class Aves). Auk 98: 681–714. **[627ff.]**

Cracraft, J. 1982a. Geographic differentiation, cladistics, and vicariance biogeography: Reconstructing the tempo and mode of evolution. Am. Zool. 22: 411–424. **[538, 539]**

Cracraft, J. 1982b. Phylogenetic relationships and monophyly of loons, grebes, and hesperornithiform birds, with comments on the early history of birds. Syst. Zool. 31: 35–56. **[649]**

Cracraft, J. 1985. Monophyly and phylogenetic relationships of the Pelecaniformes: A numerical cladistic analysis. Auk 102: 834–853. **[629]**

Cracraft, J. 1986. The origin and early diversification of birds. Paleobiology 12: 383–399. **[43, 443]**

Cracraft, J. 1989. Speciation and its ontology: The empirical consequences of alternative species concepts for understanding patterns and processes of differentiation, pp. 28–59. *In* Speciation and Its Consequences (D. Otte and J. Endler, Eds.). Sunderland, Mass.: Sinauer Associates. **[46, 529, 552]**

Craig, J.V., L.L. Ortman, and A.M. Guhl. 1965. Genetic selection for social dominance ability in chickens. Anim. Behav. 13: 114–131. **[334]**

Cronin, E.W., Jr., and P.W. Sherman. 1976. A resource-based mating system: The Orange-rumped Honeyguide. Living Bird 15: 5–32. **[414]**

Crook, J.H. 1964. The evolution of social organization and visual communication in the weaver birds (Ploceinae). Behav. Suppl. No. 10. **[411]**

Crowe, T.M., and P.C. Withers. 1979. Brain temperature regulation in Helmeted Guineafowl. S. Afr. J. Sci. 75: 362–365. **[137]**

Csada, R.D., and R.M. Brigham. 1992. Common Poorwill. *In* The Birds of North America, No. 32 (A. Poole, P. Stettenheim, and F. Gill, Eds.). Philadelphia: Academy of Natural Sciences; Washington, D.C.: American Ornithologists' Union. **[134]**

Cullen, J.M. 1966. Reduction of ambiguity through ritualization. Philos. Trans. R. Soc. London Ser B 251: 363–374. **[216, 222]**

Curio, E. 1959. Beiträge zur Populationsökologie des Trauerschnäppers (*Ficedula h. hypoleuca* Pallas). Zool. Jahrb. Abt. Syst. Oekol. Geogr. Tiere 87: 185–230. **[408]**

Curio, E. 1978. The adaptive significance of avian mobbing. I. Teleonomic hypotheses and predictions. Z. Tierpsychol. 48: 175–183. **[342]**

Curio, E., U. Ernst, and W. Vieth. 1978. The adaptive significance of avian mobbing. II. Cultural transmission of enemy recognition in blackbirds: Effectiveness and some constraints. Z. Tierpsychol. 48: 184–202. **[342, 429]**

Curry, R.L., and P.R. Grant. 1990. Galápagos Mockingbirds: Territorial cooperative breeding ina climatically variable environment, pp. 291–331. *In* Cooperative Breeding in Birds (P.B. Stacey and W.D. Koenig, Eds.). Cambridge: Cambridge University Press. **[476]**

Dane, B., C. Walcott, and W.H. Drury. 1959. The form and duration of the display actions of the Goldeneye (*Bucephala clangula*). Behaviour 14: 265–281. **[216]**

Darwin, C. 1859. On the Origin of Species by Means of Natural Selection. London: J. Murray. **[11–12, 27, 46–47]**

Darwin, C. 1871. The Descent of Man, and Selection in Relation to Sex. New York: D. Appleton and Company. **[550]**

Davidson, N.C. 1983. Identification of refuelling sites by studies of weight changes and fat deposition, pp. 68–78. *In* Shorebirds and Large Waterbirds Conservation (P.R. Evans, H. Hafner, and P. L'Hermite, Eds.). Brussels: Commission of the European Communities. **[299, 302]**

Davies, N.B. 1976a. Parental care and the transition to independent feeding in the young spotted flycatcher (*Muscicapa striata*). Behaviour 59: 280–295. **[169]**

Davies, N.B. 1976b. Food, flocking, and territorial behavior of the pied wagtail (*Motacilla alba yarrellii* Gould) in winter. J. Anim. Ecol. 45: 235–253. **[336]**

Davies, N.B. 1977. Prey selection and social behaviour in wagtails (Aves: Motacillidae). J. Anim. Ecol. 46: 37–57. **[166]**

Davies, N.B. 1978. Ecological questions about territorial behaviour, pp. 317–350. *In* Behavioural Ecology: An Evolutionary Approach (J.R. Krebs and N.B. Davies, Eds.). Sunderland, Mass.: Sinauer Associates. **[331]**

Davies, N.B. 1983. Polyandry, cloaca-pecking and sperm competition in dunnocks. Nature London 302: 334–336. **[360]**

Davies, N.B. 1992. Dunnock Behavior and Social Evolution. New York: Oxford University Press. **[418, 424]**

Davies, N.B., and R.E. Green. 1976. The development and ecological significance of feeding techniques in the Reed Warbler (*Acrocephalus scirpaceus*). Anim. Behav. 24: 213–229. **[169]**

Davis, D.E. 1950. The growth of Starling, *Sturnus vulgaris*, populations. Auk 67: 460–465. **[509]**

Davis, J. 1973. Habitat preferences and competition of wintering juncos and Golden-crowned Sparrows. Ecology 54: 174–180. **[565]**

Davis, S.D., J.B. Williams, W.J. Adams, and S.L. Brown. 1984. The effect of egg temperature on attentiveness in the Belding's Savannah Sparrow. Auk 101: 556–566. **[394]**

Dawson, W.R. 1984. Physiological studies of desert birds: Present and future considerations. J. Arid Environ. 7: 133–155. **[135, 146, 392]**

Dawson, W.R., and C. Carey. 1976. Seasonal acclimatization to temperature in carduelline finches. J. Comp. Physiol. 112: 317–333. **[133]**

Dawson, W.R., C. Carey, and T.J. Van't Hof. 1992. Metabolic aspects of shivering thermogenesis in passerines during winter. Ornis Scand. 23: 381–387. **[133]**

Dawson, W.R., J.D. Ligon, J.R. Murphy, J.P. Myers, D. Simberloff, and J. Verner. 1986. Report of the advisory panel on the Spotted Owl. Audubon Conservation Report No. 7. New York, N.Y.: National Audubon Society. **[593]**

Dawson, W.R., J.D. Ligon, J.R. Murphy, J.P. Myers, D. Simberloff, and J. Verner. 1987. Report of the scientific advisory panel on the spotted owl. Condor 89: 205–229. **[512]**

Dawson, W.R., R.L. Marsh, and M.E. Yacoe. 1983. Metabolic adjustments of small passerine birds for migration and cold. Am. J. Physiol. 245: R755–R767. **[132, 298]**

DeBenedictis, P.A. 1966. The bill-brace feeding behavior of the Galápagos finch *Geospiza conirostris*. Condor 68: 206–208. **[164]**

DeBenedictis, P.A., F.B. Gill, F.R. Hainsworth, G.H. Pyke,

and L.L. Wolf. 1978. Optimal meal size in hummingbirds. Am. Nat. 112: 301–316. **[167]**

de Boer, L.E.M. 1980. Do the chromosomes of the kiwi provide evidence for a monophyletic origin of the ratites? Nature London 287: 84–85. **[621]**

del Hoyo, J., A. Elliott, and J. Sargatal, Eds. 1992. Handbook of the Birds of the World, Vol. 1. Barcelona: Lynx Edicions. **[171, 283, 492, 649]**

Delius, J.D. 1965. A population study of Skylarks *Alauda arvensis*. Ibis 107: 466–492. **[274]**

Derrickson, K.C., and R. Breitwisch. 1992. Northern Mockingbird. *In* The Birds of North America, No. 7 (A. Poole, P. Stettenheim, and F. Gill, Eds.). Philadelphia: Academy of Natural Sciences; Washington, D.C.: American Ornithologists' Union. **[252]**

De Sante, D.F. 1983. Annual variability in the abundance of migrant landbirds on southeast Farallon Island, California. Auk 100: 826–852. **[319]**

De Steven, D. 1980. Clutch size, breeding success, and parental survival in the Tree Swallow (*Iridoprocne bicolor*). Evolution 34: 278–291. **[499]**

Desselberger, H. 1931. Der Verdauungskanal der Dicaeiden nach Gestalt und Funktion. J. Ornithol. 79: 353–370. **[161]**

Dhondt, A.A. 1977. Interspecific competition between Great and Blue Tit. Nature London 268: 521–523. **[565]**

Dhondt, A.A. 1989. Ecological and evolutionary effects of interspecific competition in tits. Wilson Bull. 101: 198–216. **[565, 567]**

Dhondt, A.A., and R. Eyckerman. 1980. Competition and the regulation of numbers in Great and Blue Tit. Ardea 68: 121–132. **[565, 566]**

Diamond, A.W., and A.R. Place. 1988. Wax digestion in Black-throated Honeyguides. Ibis 130: 558–561. **[163]**

Diamond, J.M. 1969. Avifaunal equilibria and species turnover rates on the Channel Islands of California. Proc. Natl. Acad. Sci. USA 64: 57–63. **[577]**

Diamond, J.M. 1973. Distributional ecology of New Guinea birds. Science 179: 759–769. **[40, 569]**

Diamond, J.M. 1974. Colonization of exploded volcanic islands by birds: The supertramp strategy. Science 184: 803–806. **[573]**

Diamond, J.M. 1975. Assembly of species communities, pp. 342–444. *In* Ecology and Evolution of Communities (M.L. Cody and J.M. Diamond, Eds.). Cambridge, Mass.: Harvard University Press. **[40, 42, 569, 572]**

Diamond, J.M. 1980. Species turnover in island bird communities, pp. 777–782. *In* Acta XVII Congressus Internationalis Ornithologici (R. Nöhring, Ed.). Berlin: Verlag der Deutschen Ornithologen-Gesellschaft. **[577, 578]**

Diamond, J. 1986. Biology of birds-of-paradise and bowerbirds. Annu. Rev. Ecol. Syst. 17: 17–37. **[226]**

Diamond, J.M., S.L. Pimm, M.E. Gilpin, and M. LeCroy. 1989. Rapid evolution of character displacement in myzomelid honeyeaters. Am. Nat. 134: 675–708. **[575]**

Dickerson, J.W.T., and R.A. McCance. 1960. Severe undernutrition in growing and adult animals. III. Avian skeletal muscle. Br. J. Nutr. 14: 331–338. **[442]**

Dilger, W. 1962. The behavior of lovebirds. Sci. Am. 206(1): 88–98. **[386]**

Dixon, E.S. 1848. Ornamental and Domestic Poultry. London: The Gardeners' Chronicle. **[128]**

Doak, D. 1989. Spotted Owls and old growth logging in the Pacific Northwest. Conserv. Biol. 3: 389–396. **[512]**

Dobson, A. 1990. Survival rates and their relationship to life-history traits in some common British birds. Current Ornithology 7: 115–146. **[488]**

Dobson, C.W., and R.E. Lemon. 1975. Re-examination of monotony threshold hypothesis in bird song. Nature London 257: 126–128. **[251]**

Dobson, C.W., and R.E. Lemon. 1977. Markovian versus rhomboidal patterning in the song of Swainson's Thrush. Behaviour 62: 277–297. **[250]**

Docters van Leeuwen, W.M. 1954. On the biology of some Javanese Loranthaceae and the role birds play in their life-historie. Beaufortia 4: 105–207. **[160]**

Dolnik, V.R., and V.M. Gavrilov. 1979. Bioenergetics of molt in the Chaffinch (*Fringilla coelebs*). Auk 96: 253–264. **[279]**

Dooling, R.J. 1982. Auditory perception in birds, pp. 95–130. *In* Acoustic Communication in Birds, Vol. 1 (D.E. Kroodsma and E.H. Miller, Eds.). New York: Academic Press. **[193, 194, 195, 202]**

Dorst, J. 1962. The Migrations of Birds. Boston: Houghton Mifflin. **[288, 289]**

Dow, D.D. 1965. The role of saliva in food storage by the Gray Jay. Auk 82: 139–154. **[155]**

Downing, R.L. 1959. Significance of ground nesting by Mourning Doves in northwestern Oklahoma. J. Wildl. Manage. 23: 117–118. **[381]**

Dowsett-Lemaire, F. 1979. The imitative range of the song of the Marsh Warbler *Acrocephalus palustris*, with special reference to imitations of African birds. Ibis 121: 453–468. **[253]**

Drent, R.H. 1970. Functional aspects of incubation in the Herring Gull. Behav. Suppl. 17: 1–132. **[396]**

Drent, R.H. 1972. Adaptive aspects of the physiology of incubation, pp. 255–280. *In* Proc. XV Int. Ornithol. Congr. (K.H. Voous, Ed.). Leiden: E.J. Brill. **[135, 394]**

Drent, R.H. 1975. Incubation. Avian Biology 5: 333–419. **[370, 372, 391, 393, 399, 401]**

Drent, R.H., and S. Daan. 1980. The prudent parent: Energetic adjustments in avian breeding. Ardea 68: 225–252. **[366, 401]**

Driver, P.M. 1967. Notes on the clicking of avian egg-young, with comments on its mechanism and function. Ibis 109: 434–437. **[428]**

Drobney, R.D. 1980. Reproductive bioenergetics of Wood Ducks. Auk 97: 480–490. **[365]**

Duffy, D. 1983. The ecology of tick parasitism on densely nesting Peruvian seabirds. Ecology 64: 110–119. **[385]**

Dumbacher, J.P., B.M. Beehler, T.F. Spande, H.M. Garraffo, and J.W. Daly. 1992. Homobatrachotoxin in the genus *Pitohui*: Chemical defense in birds? Science 258: 799–801. **[77]**

Dunne, P., Ed. 1989. New Jersey at the Crossroads of Migration. Franklin Lakes, N.J.: New Jersey Audubon Society. **[307]**

Dunning, J.B., Jr. 1986. Shrub-steppe bird assemblages revisited: Implications for community theory. Am. Nat. 128: 82–98. **[563]**

Dunning, J.B., Jr., and J.H. Brown. 1982. Summer rainfall and winter sparrow densities: A test of the food limitation hypothesis. Auk 99: 123–129. **[516]**

DuPlessis, M.A. 1990. The influence of roost-cavity availability on flock size in Red-billed Woodhoopoes. Ostrich Suppl. 14: 97–104. **[472]**

Dürrer, H., and W. Villiger. 1966. Schillerfarben der Trogoniden. J. Ornithol. 107: 1–26. **[90]**

Dwight, J., Jr. 1900. The sequence of plumages and moults of the passerine birds of New York. Ann. N.Y. Acad. Sci. 13: 73–360. **[264]**

Dwight, J., Jr. 1907. Sequence in moults and plumages, with an explanation of plumage-cycles. Proc. Int. Ornithol. Congr. 4: 513–518. **[83]**

Dyck, J. 1971. Structure and spectral reflectance of green and blue feathers of the Rose-faced Lovebird (*Agapornis roseicollis*). Biol. Srk. No. 18(2): 1–67. **[89]**

Dyck, J. 1992. Reflectance spectra of plumage areas colored by green feather pigments. Auk 109: 293–301. **[87]**

Dyrcz, A. 1977. Polygamy and breeding success among Great Reed Warblers *Acrocephalus arundinaceus* at Milicz, Poland. Ibis 119: 73–77. **[414]**

East, M. 1981. Aspects of courtship and parental care of the European Robin *Erithacus rubecula*. Ornis Scand. 12: 230–239. **[409]**

Einarsen, A.S. 1942. Specific results from Ring-necked Pheasant studies in the Pacific Northwest. Trans. N. Am. Wildl. Nat. Resour. Conf. 7: 130–145. **[510]**

Einarsen, A.S. 1945. Some factors affecting Ring-necked Pheasant population density. Murrelet 26: 39–44. **[510]**

Eldredge, N., and J. Cracraft. 1980. Phylogenetic Patterns and the Evolutionary Process. New York: Columbia University Press. **[62]**

Emlen, J.T. 1980. Interactions of migrant and resident land birds in Florida and Bahama pinelands, pp. 133–144. *In* Migrant Birds in the Neotropics (A. Keast and E.S. Morton, Eds.). Washington, D.C.: Smithsonian Institution Press. **[559, 561]**

Emlen, J.T., and R.L. Penney. 1964. Distance navigation in the Adelie Penguin. Ibis 106: 417–431. **[313]**

Emlen, S.T. 1967a. Migratory orientation in the Indigo Bunting, *Passerina cyanea*. Part I: Evidence for use of celestial cues. Auk 84: 309–342. **[314]**

Emlen, S.T. 1967b. Migratory orientation in the Indigo Bunting, *Passerina cyanea*. Part II: Mechanism of celestial orientation. Auk 84: 463–489. **[314]**

Emlen, S.T. 1969. Bird migration: Influence of physiological state upon celestial orientation. Science 165: 716–718. **[273, 315]**

Emlen, S.T. 1970. Celestial rotation: Its importance in the development of migratory orientation. Science 170: 1198–1201. **[322]**

Emlen, S.T. 1972. An experimental analysis of the parameters of bird song eliciting species recognition. Behaviour 41: 130–171. **[246]**

Emlen, S.T. 1975a. Migration: Orientation and navigation. Avian Biology 5: 129–219. **[312, 313, 320, 324]**

Emlen, S.T. 1975b. The stellar-orientation system of a migratory bird. Sci. Am. 233(2): 102–111. **[314, 315, 322]**

Emlen, S.T. 1978. The evolution of cooperative breeding in birds. *In* Behavioural Ecology: An Evolutionary Approach (J.R. Krebs and N.B. Davies, Eds.). pp. 245–281. Sunderland, Mass.: Sinauer Associates. **[468]**

Emlen, S.T. 1981. Altruism, kinship, and reciprocity in the White-fronted Bee-eater, pp. 217–230. *In* Natural Selection and Social Behavior: Recent Research and New Theory (R.D. Alexander and D.W. Tinkle, Eds.). New York: Chiron Press. **[474]**

Emlen, S.T. 1982a. The evolution of helping. II: The role of behavioral conflict. Am. Nat. 119: 40–53. **[472]**

Emlen, S.T. 1982b. The evolution of helping. I: An ecological constraints model. Am. Nat. 119: 29–39. **[470, 472]**

Emlen, S.T. 1984. Cooperative breeding in birds and mammals, pp. 305–339. *In* Behavioural Ecology: An Evolutionary Approach, 2nd ed. (J.R. Krebs and N.B. Davies, Eds.). Sunderland, Mass.: Sinauer Associates. **[470, 471]**

Emlen, S.T. 1990. White-fronted bee-eaters: Helping in a colonially nesting species, pp. 489–525. *In* Cooperative Breeding in Birds (P.B. Stacey and W.D. Koenig, Eds.). Cambridge: Cambridge University Press. **[474]**

Emlen, S.T. 1991. Cooperative breeding in birds and mammals, pp. 301–337. *In* Behavioral Ecology: An Evolutionary Approach, 3rd ed. (J.R. Krebs and N.B. Davies, Eds.). Sunderland, Mass.: Sinauer Associates. **[477]**

Emlen, S.T., and H.W. Ambrose, III. 1970. Feeding interactions of Snowy Egrets and Red-breasted Mergansers. Auk 87: 164–165. **[337]**

Emlen, S.T., and J.T. Emlen. 1966. A technique for recording migratory orientation of captive birds. Auk 83: 361–367. **[314]**

Emlen, S.T., and L.W. Oring. 1977. Ecology, sexual selection, and the evolution of mating systems. Science 197: 215–223. **[414]**

Emlen, S.T., and P.H. Wrege. 1988. The role of kinship in helping decisions among white-fronted bee-eaters. Behav. Ecol. Sociobiol. 23: 305–315. **[474]**

Emlen, S.T., N.J. Demong, and D.J. Emlen. 1989. Experimental induction of infanticide in female Wattled Jacanas. Auk 106: 1–7. **[457]**

Emlen, S.T., J.D. Rising, and W.L. Thompson. 1975. A behavioral and morphological study of sympatry in the Indigo and Lazuli Buntings of the Great Plains. Wilson Bull. 87: 145–179. **[544]**

Emmanuel, V. 1993. Fallout! Unpublished manuscript. **[297]**

Erickson, C.J., and P.G. Zenone. 1978. Courtship differences in male Ring Doves: Avoidance of cuckoldry? Science 192: 1353–1354. **[409]**

Ettinger, A.O., and J.R. King. 1980. Time and energy budgets of the Willow Flycatcher (*Empidonax traillii*) during the breeding season. Auk 97: 533–546. **[365]**

Evans, K. 1966. Observations on a hybrid between the Sharp-tailed Grouse and the Greater Prairie Chicken. Auk 83: 128–129. **[544]**

Evans, P.G.H. 1988. Intraspecific nest parasitism in the European Starling *Sturnus vulgaris*. Anim. Behav. 36: 1282–1294. **[458]**

Evarts, S., and C.J. Williams. 1987. Multiple paternity in wild populations of Mallards. Auk 104: 597–602. **[409]**

Faanes, C.A., and S.E. Senner. 1991. Status and conservation of the Eskimo Curlew. Am. Birds 45: 237–239. **[601]**

Fagen, R. 1981. Animal Play Behavior. New York: Oxford University Press. **[449]**

Falls, J.B. 1982. Individual recognition by sounds in birds, pp. 237–278. *In* Acoustic Communication in Birds, Vol. 2 (D.E. Kroodsma and E.H. Miller, Eds.). New York: Academic Press. **[245, 247]**

Farabaugh, S.M. 1982. The ecological and social significance of duetting, pp. 85–124. *In* Acoustic Communication in Birds, Vol. 2 (D.E. Kroodsma and E.H. Miller, Eds.). New York: Academic Press. **[249]**

Farner, D.S. 1967. The control of avian reproductive cycles, pp. 107–133. *In* Proc. XIV Int. Ornithol. Congr. (D.H. Snow, Ed.). Oxford: Blackwell. **[284]**

Farner, D.S. 1970. Some glimpses of comparative avian physiology. Fed. Proc. 29: 1649–1663. **[126]**

Farner, D.S. 1980a. Endogenous periodic functions in the control of reproductive cycles, pp. 123–138. *In* Biological Rhythms in Birds: Neural and Endocrine Aspects (Y. Tanabe, T. Hirano, and M. Wada, Eds.). Tokyo: Japan Scientific Societies Press. **[271]**

Farner, D.S. 1980b. Evolution of the control of reproductive cycles in birds, pp. 185–191. *In* Hormones, Adaptation and Evolution (S. Ischii, Ed.). Tokyo: Japan Scientific Societies Press. **[268]**

Farner, D.S. 1980c. The regulation of the annual cycle of the White-crowned Sparrow, *Zonotrichia leucophrys gambelii*, pp. 71–82. Acta XVII Congressus Internationalis Ornithologici (R. Nöhring, Ed.). Berlin: Verlag der Deutschen Ornithologen-Gesellschaft. **[268, 270, 273, 284]**

Farner, D.S., and R.A. Lewis. 1971. Photoperiodism and reproductive cycles in birds. Photophysiology 6: 325–370. **[303]**

Farner, D.S., and L.R. Mewaldt. 1952. The relative roles of photoperiod and temperature in gonadal recrudescence in male *Zonotrichia leucophrys gambelii*. Anat. Rec. 113: 612–613. **[274]**

Fay, R.R. 1988. Hearing in vertebrates: A psychophysics databook. Winnetka, Ill.: Hill-Fay Associates. **[193]**

Feduccia, A. 1974. Morphology of the bony stapes in New and Old World suboscines: New evidence for common ancestry. Auk 91: 427–429. **[677]**

Feduccia, A. 1977. A model for the evolution of perching birds. Syst. Zool. 26: 19–31. **[51, 192, 670]**

Feduccia, A. 1980. The Age of Birds. Cambridge, Mass.: Harvard University Press. **[24, 27, 32, 33, 43, 670]**

Feduccia, A. 1993a. Evidence from claw geometry indicating arboreal habits of *Archaeopteryx*. Science 259: 790–793. **[26–27]**

Feduccia, A. 1993b. Aerodynamic model for the early evolution of feathers provided by *Propithecus* (Primates, Lemuridae). J. Theor. Biol. 160: 159–164. **[28]**

Feduccia, A., and S.L. Olson. 1982. Morphological similarities between the Menurae and the Rhinocryptidae, relict passerine birds of the southern hemisphere. Smithson. Contrib. Zool. No. 366. **[677]**

Feduccia, A., and H.B. Tordoff. 1979. Feathers of *Archaeopteryx*: Asymmetric vanes indicate aerodynamic function. Science 203: 1021–1022. **[27]**

Feinsinger, P. 1980. Asynchronous migration patterns and the coexistence of tropical hummingbirds, pp. 411–419. *In* Migrant Birds in the Neotropics (A. Keast and E.S. Morton, Eds.). Washington, D.C.: Smithsonian Institution Press. **[559]**

Feinsinger, P., and R.K. Colwell. 1978. Community organization among Neotropical nectar-feeding birds. Am. Zool. 18: 779–795. **[564]**

Fenna, L., and D.A. Boag. 1974. Adaptive significance of the caeca in Japanese Quail and Spruce Grouse (Galliformes). Can. J. Zool. 52: 1577–1584. **[162]**

Ferns, P.N. 1978. Individual differences in the head and neck plumage of Ruddy Turnstones (*Arenaria interpres*) during the breeding season. Auk 95: 753–755. **[210]**

Ficken, M.S. 1977. Avian play. Auk 94: 573–582. **[449]**

Ficken, M.S., and R.W. Ficken. 1968. Courtship of Blue-winged Warblers, Golden-winged Warblers, and their hybrids. Wilson Bull. 80: 161–172. **[544, 548]**

Findlay, C.S. 1987. Non-random mating: A theoretical and empirical overview with special reference to birds, pp. 289–314. *In* Avian Genetics (F. Cooke and P.A. Buckley, Eds.). New York: Academic Press. **[534, 539, 552]**

Fisher, A.C., Jr. 1979. Mysteries of bird migration. National Geographic 156(2): 154–193. **[xxi]**

Fisher, C.D., E. Lindgren, and W.R. Dawson. 1972. Drinking patterns and behavior of Australian desert birds in relation to their ecology and abundance. Condor 74: 111–136. **[139]**

Fisher, H.I. 1971. The Laysan Albatross: Its incubation, hatching, and associated behaviors. Living Bird. 10: 19–78. **[398]**

Fisher, H. 1972. The nutrition of birds. Avian Biology 2: 431–469. **[171, 439]**

Fisher, J., and R.A. Hinde. 1949. The opening of milk bottles by birds. Br. Birds 42: 347–357. **[164]**

Fisher, R.A. 1930. The Genetical Theory of Natural Selection. Oxford: Clarendon Press. **[230]**

Fitzpatrick, J.W. 1988. Why so many passerine birds? A response to Raikow. Syst. Zool. 37: 71–76. **[549, 675]**

Fjeldså, J. 1976. The systematic affinities of sandgrouses, Pteroclidae Vidensk. Medd. Dan. Naturhist. Foren. 139: 179–243. **[648]**

Fleischer, R.C., and R.F. Johnston. 1982. Natural selection on body size and proportions in house sparrows. Nature London 298: 747–749. **[130]**

Flower, S.S. 1938. Further notes on the duration of life in animals. IV. Birds. Proc. Zool. Soc. London Ser A 108: 195–235. **[489]**

Fogden, M.P.L. 1972. The seasonality and population dynamics of equatorial forest birds in Sarawak. Ibis 114: 307–342. **[263, 265]**

Fogden, M.P.L., and P.M. Fogden. 1979. The role of fat and protein reserves in the annual cycle of the Grey-backed Camaroptera in Uganda (Aves: Sylviidae). J. Zool. 189: 233–258. **[281]**

Ford, N.L. 1983. Variation in mate fidelity in monogamous birds. Current Ornithology 1: 329–356. **[413]**

Forshaw, J. 1989. Parrots of the World, 2nd ed. Melbourne: Lansdowne Press. **[653]**

Forsythe, D.M. 1971. Clicking in the egg-young of the Long-billed Curlew. Wilson Bull. 83: 441–442. **[428]**

Foster, M.S. 1974. A model to explain molt–breeding overlap

and clutch size in some tropical birds. Evolution 28: 182–190. **[500]**

Foster, M.S. 1975. The overlap of molting and breeding in some tropical birds. Condor 77: 304–314. **[281]**

Foster, M.S. 1977. Odd couples in manakins: A study of social organization and cooperative breeding in *Chiroxiphia linearis*. Am. Nat. 111: 845–853. **[421]**

Foster, M.S. 1978. Total frugivory in tropical passerines: A reappraisal. Trop. Ecol. 19: 131–154. **[438]**

Foster, M.S. 1981. Cooperative behavior and social organization of the Swallow-tailed Manakin (*Chiroxiphia caudata*). Behav. Ecol. Sociobiol. 9: 167–177. **[421, 423]**

Foster, M.S. 1983. Disruption, dispersion, and dominance in lek-breeding birds. Am. Nat. 122: 53–72. **[421]**

Fox, G.A. 1976. Eggshell quality: Its ecological and physiological significance in a DDE-contaminated Common Tern population. Wilson Bull. 88: 459–477. **[363, 369]**

Fox, G.A. 1992. Epidemiological and pathobiological evidence of contaminant-induced alterations in sexual development in free-living wildlife, pp. 147–158. *In* Chemically-Induced Alterations in Sexual and Functional Development: The Wildlife/Human Connection (T. Colburn and C. Clement, Eds). Princeton, N.J.: Princeton Science Publishing Co. **[590]**

Franks, E.C. 1967. The response of incubating Ringed Turtle Doves (*Streptopelia risoria*) to manipulated egg temperatures. Condor 69: 268–276. **[394]**

Fretwell, S. 1968. Habitat distribution and survival in the Field Sparrow (*Spizella pusilla*). Bird-Banding 39: 293–306. **[334]**

Friedmann, H., and J. Kern. 1956. The problem of cerophagy or wax-eating in the Honey-guides. Q. Rev. Biol. 31: 19–30. **163**

Frith, H.J. 1962. The Mallee-Fowl. Sydney: Angus and Robertson. **[395]**

Fürbringer, M. 1888. Bijdragen tot de Dierkunde. Vol. 15. Untersuchungen zur Morphologie und Systematik der Vögel. Amsterdam: Tj. Van Holkema. **[30]**

Fyfe, R.W., S.A. Temple, and T.J. Cade. 1976. The 1975 North American Peregrine Falcon survey. Can. Field-Nat. 90: 228–273. **[588]**

Gadow, H. 1892. On the classification of birds. Proc. Zool. Soc. London 1892: 229–256. **[47]**

Gadow, H. 1893. Dr. H.G. Bronn's Klassen und Ordnungen des Thier-Reichs. Vol. 6, Pt. 4. Vogel, II: Systematischer Theil. Leipzig: C.F. Winter. **[47, 54, 55, 61]**

Gales, R., and B. Green. 1990. The annual energetics cycle of Little Penguins (*Eudyptula minor*). Ecology 71: 2297–2312. **[454]**

Gardner, L.L. 1925. The adaptive modifications and the taxonomic value of the tongue in birds. Proc. U.S. Natl. Mus. 67(19): 1–49. **[157]**

Gargett, V. 1978. Sibling aggression in the Black Eagle in the Matopos, Rhodesia. Ostrich 49: 57–63. **[442]**

Garrod, A.H. 1876. Notes on the anatomy of *Plotus anhinga*. Proc. Zool. Soc. London 1876: 335–345. **[159]**

Gasaway, W.C. 1976a. Seasonal variation in diet, volatile fatty acid production and size of the cecum of Rock Ptarmigan. Comp. Biochem. Physiol. 53A: 109–114. **[162]**

Gasaway, W.C. 1976b. Volatile fatty acids and metabolizable energy derived from cecal fermentation in the Willow Ptarmigan. Comp. Biochem. Physiol. 53A: 115–121. **[162]**

Gaston, A.J. 1978. The evolution of group territorial behavior and cooperative breeding. Am. Nat. 112: 1091–1100. **[468]**

Gaunt, A.S., and M.K. Wells. 1973. Models of syringeal mechanisms. Am. Zool. 13: 1227–1247. **[237]**

Gaunt, A.S., and S.L.L. Gaunt. 1985. Syringeal structure and avian phonation. Current Ornithology 2: 213–245. **[259]**

Gauthier, J. 1986. Saurischian monophyly and the origin of birds, pp. 1–55. *In* The Origin of Birds and the Evolution of Flight (K. Padian, Ed.). Memoirs of the California Academy of Sciences, No. 8. **[43]**

Gauthreaux, S.A., Jr. 1971. A radar and direct visual study of passerine spring migration in southern Louisiana. Auk 88: 343–365. **[304]**

Gauthreaux, S.A., Jr. 1972. Behavioral responses of migrating birds to daylight and darkness: A radar and direct visual study. Wilson Bull. 84: 136–148. **[306]**

Gauthreaux, S.A., Jr., Ed. 1980. Animal Migration, Orientation, and Navigation. Orlando, Fla.: Academic Press. **[307]**

Gauthreaux, S.A., Jr. 1982. The ecology and evolution of avian migration systems. Avian Biology 6: 93–168. **[289, 307]**

Gauthreaux, S.A., Jr. 1992. The use of weather radar to monitor long-term patterns of trans-Gulf migration in spring, pp. 96–100. *In* Ecology and Conservation of Neotropical Migrant Landbirds (J.M. Hagen III and D.W. Johnston, Eds.). Washington, D.C.: Smithsonian Institution Press. **[296]**

Gehlbach, F.R. 1989. Screech-owl, pp. 315–326. *In* Lifetime Reproduction in Birds (I. Newton, Ed.). New York: Academic Press. **[485]**

Gelter, H.P., H. Tegelström, and L. Gustafsson. 1992. Evidence from hatching success and DNA fingerprinting for the fertility of hybrid Pied × Collared Flycatchers *Ficedula hypoleuca* × *albicollis*. Ibis 134: 62–68. **[543]**

George, F.W., J.F. Noble, and J.D. Wilson. 1981. Female feathering in Sebright cocks is due to conversion of testosterone to estradiol in skin. Science 213: 557–559. **[352]**

George, J.C., and A.J. Berger. 1966. Avian Myology. New York: Academic Press. **[96, 297]**

George, W.G. 1962. The classification of the olive warbler, *Peucedramus taeniatus*. Amer. Mus. Novit. 2103: 1–41. **[691]**

Geramita, J.M., F. Cooke, and R.F. Rockwell. 1982. Assortative mating and gene flow in the Lesser Snow Goose: A modelling approach. Theor. Popul. Biol. 22: 177–203. **[209]**

Gessaman, J.A. 1972. Bioenergetics of the Snowy Owl (*Nyctea scandiaca*). Arct. Alp. Res. 4: 223–238. **[129]**

Gibb, J.A. 1960. Populations of tits and goldcrests and their food supply in pine plantations. Ibis 102: 163–208. **[170]**

Gibbs, H.L., P.J. Weatherhead, P.T. Boag, B.N. White, L.M. Tabak, and D.J. Hoysak. 1990. Realized reproductive success of polygynous red-winged blackbirds revealed by DNA markers. Science 250: 1394–1397. **[409]**

Gibson, R. 1993. [Review of] Dunnock Behavior and Social Evolution, by N.B. Davies. 1992. New York: Oxford University Press. Science 260: 374–375. **[418]**

Gilbert, A.B. 1979. Female genital organs, pp. 237–360. *In* Form and Function in Birds, Vol. 1 (A.S. King and J. McLelland, Eds.). New York: Academic Press. **[374]**

Gilbertson, M., T. Kubiak, J. Ludwig, and G. Fox. 1991. Great Lakes embryo mortality, edema, and deformities syndrome (GLEMEDS) in colonial fish-eating birds: Similarity to chick edema disease. J. Toxicol. Environ. Health 33: 455–520. **[590]**

Gill, F.B. 1970. Hybridization in Norfolk Island white-eyes (*Zosterops*). Condor 72: 481–482. **[542]**

Gill, F.B. 1971. Ecology and evolution of the sympatric Mascarene white-eyes, *Zosterops borbonica* and *Zosterops olivacea.* Auk 88: 35–60. **[574]**

Gill, F.B. 1973. Intra-island variation in the Mascarene White-eye, *Zosterops borbonica.* Ornithol. Monogr. No. 12. **[87, 537]**

Gill, F.B. 1980. Historical aspects of hybridization between Blue-winged and Golden-winged Warblers. Auk 97: 1–18. **[547]**

Gill, F.B. 1982. Might there be a resurrection of the subspecies? Auk 99: 598–599. **[529]**

Gill, F.B. 1985. Hummingbird flight speeds. Auk 102: 97–101. **[100]**

Gill, F.B. 1988. Trapline foraging by hermit hummingbirds: Competition for an undefended, renewable resource. Ecology 69: 1933–1942. **[168]**

Gill, F.B., and F.H. Sheldon. 1991. The birds reclassified. Science 252: 1003–1005. **[57, 618, 621]**

Gill, F.B., and B. Slikas. 1992. Patterns of mitochondrial DNA divergence in North American crested titmice. Condor 94: 20–28. **[60]**

Gill, F.B., and L.L. Wolf. 1975. Economics of feeding territoriality in the Golden-winged Sunbird. Ecology 56: 333–345. **[329, 332]**

Gill, F.B., and L.L. Wolf. 1978. Comparative foraging efficiencies of some montane sunbirds in Kenya. Condor 80: 391–400. **[154]**

Gill, F.B., and L.L. Wolf. 1979. Nectar loss by Golden-winged Sunbirds to competitors. Auk 96: 448–461. **[171, 332]**

Gill, F.B., A.M. Mostrom, and A.L. Mack. 1993a. Speciation in North American chickadees. I. Patterns of mtDNA genetic divergence. Evolution 47: 195–212. **[524, 532]**

Gilliard, E.T. 1956. Bower ornamentation versus plumage characters in bower-birds. Auk 73: 450–451. **[226]**

Gilliard, E.T. 1958. Living Birds of the World. Garden City, N.Y.: Doubleday. **[387]**

Gilliard, E.T. 1969. Birds of paradise and Bowerbirds. Garden City, N.Y.: Natural History Press. **[226, 228]**

Gjershaug, J.O., T. Järvi, and E. Roskaft. 1989. Marriage entrapment by solitary mothers: A study in male deception by female pied flycatchers. Am. Nat. 133: 273–276. **[414]**

Gleason, H.A. 1926. The individualistic concept of the plant association. Bull. Torrey Bot. Club 53: 7–26. **[561]**

Gleason, H.A. 1939. The individualistic concept of the plant association. Am. Midl. Nat. 21: 92–110. **[561]**

Gochfeld, M., J. Burger, and J.R. Jehl, Jr. 1984. The classification of the shorebirds of the world, pp. 1–15. *In* Behavior of Marine Animals, Vol. 5, Shorebirds: Breeding Behavior and Populations (J. Burger and B.L. Olla, Eds.). New York: Plenum Press. **[648]**

Goldsmith, A.R. 1991. Prolactin and avian reproduction strategies, pp. 2063–2071. *In* Acta XX Congressus Internationalis Ornithologici, Vol. 4 (B.D. Ball, Ed.). Wellington, N.Z.: Ornithological Congress Trust Board. **[391]**

Goldsmith, T.H. 1980. Hummingbirds see near-ultraviolet light. Science 207: 786–788. **[190, 191]**

Goldstein, D.L. 1984. The thermal environment and its constraint on activity of desert quail in summer. Auk 101: 542–550. **[144]**

Goldstein, D.L., and K.A. Nagy. 1985. Resource utilization by desert quail: Time and energy, food and water. Ecology 66: 378–387. **[144]**

Goodwin, D. 1983. Pigeons and Doves of the World, 3rd ed. Ithaca, N.Y.: Cornell University Press. **[651]**

Goss-Custard, J.D. 1975. Beach Feast. Birds (Sept/Oct): 23–26. **[149]**

Götmark, F. 1992. The effects of investigator disturbance on nesting birds. Current Ornithology 9: 63–104. **[396]**

Gottlander, K. 1987. Parental feeding behavior and sibling competition in the pied flycatcher (*Ficedula hypoleuca*). Ornis Scand. 18: 269–276. **[444]**

Gottlieb, G. 1968. Prenatal behavior of birds. Q. Rev. Biol. 43: 148–174. **[430]**

Gottlieb, G. 1971. Development of Species Identification in Birds. Chicago: University of Chicago Press. **[430]**

Gould, J.L., and P. Marler. 1987. Learning by instinct. Sci. Am. 255(1): 74–85. **[203]**

Gould, S. 1985. A clock of evolution. Nat. Hist. 94(4): 12–25. **[424]**

Grant, B.R., and P.R. Grant. 1989. Evolutionary Dynamics of a Natural Population. Princeton, N.J.: Princeton University Press. **[525]**

Grant, G.S. 1982. Avian incubation: Egg temperature, nest humidity, and behavioral thermoregulation in a hot environment. Ornithol. Monogr. No. 30. **[396]**

Grant, P.R. 1968. Polyhedral territories of animals. Am. Nat. 102: 75–80. **[329]**

Grant, P.R. 1972. Centripetal selection and the House Sparrow. Syst. Zool. 21: 23–30. **[130]**

Grant, P.R. 1981. Speciation and adaptive radiation of Darwin's finches. Am. Sci. 69: 653–663. **[569]**

Grant, P.R. 1986. Ecology and Evolution of Darwin's Finches. Princeton, N.J.: Princeton University Press. **[515, 552, 570]**

Grant, P.R., and B.R. Grant. 1980. Annual variation in finch numbers, foraging and food supply on Isla Daphne Major, Galápagos. Oecologia Berlin 46: 55–62. **[515]**

Grant, P.R., and B.R. Grant. 1992. Hybridization of bird species. Science 256: 193–197. **[541, 549]**

Grassé, P.-P., Ed. 1950. Traité de Zoologie. Vol. XV. Oiseaux. Paris: Masson et Cie. **[188, 241]**

Grau, C.R. 1976. Ring structure of avian egg yolk. Poult. Sci. 55: 1418–1422. **[357]**

Grau, C.R. 1982. Egg formation in Fiordland Crested Penguins (*Eudyptes pachyrhynchus*). Condor 84: 172–177. **[356, 364]**

Green, C. 1972. Use of tool by Orange-winged Sitella. Emu 72: 185–186. **[164]**

Greenberg, R. 1983. The role of neophobia in determining the degree of foraging specialization of some migrant warblers. Am. Nat. 122: 444–453. **[169]**

Greenberg, R. 1984. Neophobia in the foraging-site selection of a Neotropical migrant bird: An experimental study. Proc. Natl. Acad. Sci. USA 81: 3778–3780. **[169]**

Greenberg, R., and J. Gradwohl. 1983. Sexual roles in the Dot-winged Antwren (*Microrhopias quixensis*), a tropical forest passerine. Auk 100: 920–925. **[343, 408]**

Greenewalt, C.H. 1960a. Hummingbirds. Garden City, N.Y.: Doubleday. **[89, 90, 106, 107]**

Greenewalt, C.H. 1960b. The wings of insects and birds as mechanical oscillators. Proc. Am. Philos. Soc. 104: 605–611. **[108]**

Greenewalt, C.H. 1968. Bird Song: Acoustics and Physiology. Washington, D.C.: Smithsonian Institution Press. **[235, 236, 237, 238, 240, 241, 259]**

Greenewalt, C.H. 1969. How birds sing. Sci. Am. 221(5): 126–139. **[237]**

Greenewalt, C.H. 1975. The flight of birds. Trans. Am. Philos. Soc. New Ser 65(4): 1–67. **[103, 302]**

Greenewalt, C.H., W. Brandt, and D.D. Friel. 1960. Iridescent colors of hummingbird feathers. J. Opt. Soc. Am. 50: 1005–1013. **[90]**

Greenwood, P.G. 1987. Inbreeding, philopatry, and optimal outbreeding in birds, pp. 207–222. *In* Avian Genetics (F. Cooke and P.A. Buckley, Eds.). New York: Academic Press. **[491, 533, 534]**

Greenwood, P.J., P.H. Harvey, and C.M. Perrins. 1979. The role of dispersal in the Great Tit (*Parus major*): The causes, consequences and heritability of natal dispersal. J. Anim. Ecol. 48: 123–142. **[517]**

Grier, J.W. 1982. Ban of DDT and subsequent recovery of reproduction in Bald Eagles. Science 218: 1232–1235. **[588]**

Griffin, D.R. 1974. Bird Migration. New York: Dover. **[324]**

Griffiths, C.S. 1994. Monophyly of the Falconiformes based on syringeal morphology. Auk 111: 787–805. **[637]**

Grinnell, J. 1917. The niche relationships of the California thrasher. Auk 34: 427–433. **[554]**

Grubb, T.C., Jr. 1972. Smell and foraging in shearwaters and petrels. Nature London 237: 404–405. **[200]**

Grubb, T.C., Jr. 1974. Olfactory navigation to the nesting burrow in Leach's Petrel (*Oceanodroma leucorrhoa*). Anim. Behav. 22: 192–202. **[200, 316]**

Grubb, B., J.M. Colacino, and K. Schmidt-Nielsen. 1978. Cerebral blood flow in birds: Effect of hypoxia. Am. J. Physiol. 234(3): H230–H234. **[120]**

Grubb, B., J.H. Jones, and K. Schmidt-Nielsen. 1979. Avian cerebral blood flow: Influence of the Bohr effect on oxygen supply. Am. J. Physiol. 236(5): H744–H749. **[120]**

Grzybowski, J.A., R.B. Clapp, and J.T. Marshall, Jr. 1986. History and current population status of the Black-capped Vireo in Oklahoma. Am. Birds 40: 1151–1161. **[594]**

Guhl, A.M. 1968. Social inertia and social stability in chickens. Anim. Behav. 16: 219–232. **[336]**

Gurney, M. 1988. Songbirds, neuroleukin, and AIDS-dementia. Medicine on the Midway 41(3): 2–6. **[184]**

Gurney, M.E., and M. Konishi. 1980. Hormone-induced sexual differentiation of brain and behavior in Zebra Finches. Science 208: 1380–1383. **[184]**

Gustafsson, L. 1987. Interspecific competition lowers fitness in Collared Flycatchers *Ficedula albicollis*: An experimental demonstration. Ecology 68: 291–296. **[565]**

Gwinner, E. 1966. Ueber einige Bewegungsspiele des Kolkraben (*Corvus corax* L.). Z. Tierpsychol. 23: 28–36. **[449]**

Gwinner, E. 1975. Circadian and circannual rhythms in birds. Avian Biology 5: 221–285. **[267]**

Gwinner, E. 1977. Circannual rhythms in bird migration. Annu. Rev. Ecol. Syst. 8: 381–405. **[269, 285, 319, 323]**

Gwinner, E., Ed. 1990. Bird Migration. Berlin: Springer-Verlag. **[307]**

Haartman, L. von. 1951. Der Trauerfliegenschnapper. II. Populationsproblem. Acta Zool. Fenn. 67: 1–60. **[512]**

Haartman, L. von. 1953. Was reizt den Trauerfliegenschnapper (*Muscicapa hypoleuca*) zu füttern? Vogelwarte 16: 157–164. **[443]**

Haartman, L. von. 1956. Der Einfluss der Temperatur auf den Brutrhythmus experimentell nachgeweisen. Ornis Fenn. 33: 100–107. **[393]**

Haartman, L. von. 1958. The incubation rhythm of the female Pied Flycatcher (*Ficedula hypoleuca*) in the presence and absence of the male. Ornis Fenn. 35: 71–76. **[408]**

Haartman, L. von. 1971. Population dynamics. Avian Biology 1: 391–459. **[490]**

Haartman, L. von. 1973. Changes in the breeding bird fauna of north Europe, pp. 448–481. *In* Breeding Biology of Birds (D.S. Farner, Ed.). Washington, D.C.: National Academy of Sciences. **[508]**

Hachisuka, M. 1953. The Dodo and kindred birds, or the extinct birds of the Mascarene Islands. London: H.F. and G. Witherby. **[585]**

Häcker, V. 1900. Der Gesang der Vogel. Jena: Gustav Fischer. **[238]**

Haffer, J. 1969. Speciation in Amazonian forest birds. Science 165: 131–137. **[40]**

Haffer, J. 1974. Avian speciation in tropical South America. Publ. Nuttall Ornithol. Club No. 14. **[40, 41]**

Haftorn, S. 1959. The proportion of spruce seeds removed by the tits in a Norwegian spruce forest in 1954–55. Det Kgl. Norsk Vidensk. Selsk. Forh. 32: 121–125. **[185]**

Haftorn, S. 1978a. Egg-laying and regulation of egg temperature during incubation in the Goldcrest *Regulus regulus*. Ornis Scand. 9: 2–21. **[393]**

Haftorn, S. 1978b. Energetics of incubation by the Goldcrest *Regulus regulus* in relation to ambient air temperatures and the geographical distribution of the species. Ornis Scand. 9: 22–30. **[393]**

Haftorn, S. 1978c. Cooperation between the male and female Goldcrest *Regulus regulus* when rearing overlapping double broods. Ornis Scand. 9: 124–129. **[491]**

Hagan, J.M., III, and D.W. Johnston, Eds. 1992. Ecology and Conservation of Neotropical Migrant Landbirds. Washington, D.C.: Smithsonian Institution Press. **[287, 307, 595, 616]**

Hagey, L.R., C.D. Schteingart, H.-T. Ton-Nu, S.S. Rossi, D. Odell, and A.F. Hofmann. 1990. Beta-phocacholic acid in bile: Biochemical evidence that the flamingo is related to an ancient goose. Condor 92: 593–597. **[633]**

Hahn, T.P., J. Swingle, J.C. Wingfield, and M. Ramenofsky. 1992. Adjustments of the prebasic molt schedule in birds. Ornis Scand. 23: 314–321. **[274]**

Hailman, J.P. 1967. The ontogeny of an instinct. Behav. Suppl. No. 15. **[429, 430]**

Hailman, J.P. 1969. How an instinct is learned. Sci. Am. 221(6): 98–106. **[430]**

Hailman, J.P. 1977. Optical Signals: Animal Communication and Light. Bloomington: Indiana University Press. **[360]**

Hailman, J.P. 1986. The heritability concept applied to wild birds. Current Ornithology 4: 71–95. **[552]**

Hainsworth, F.R., and L.L. Wolf. 1970. Regulation of oxygen comsumption and body temperature during torpor in a hummingbird, *Eulampis jugularis*. Science 168: 368–369. **[134]**

Haldane, J.B.S. 1922. Sex ratio and unisexual sterility in hybrid animals. J. Genet. 12: 101–109. **[543]**

Hamilton, W.D. 1964. The genetical evolution of social behaviour I, II. J. Theor. Biol. 7: 1–52. **[468]**

Hamilton, W.D. 1971. Geometry for the selfish herd. J. Theor. Biol. 31: 295–311. **[340]**

Hamilton, W.D., and M. Zuk. 1982. Heritable true fitness and bright birds: A role for parasites? Science 218: 384–387. **[229]**

Hamilton, W.J., III. 1973. Life's Color Code. New York: McGraw-Hill. **[345]**

Hamilton, W.J., III, and F.H. Heppner. 1967. Radiant solar energy and the function of black homeotherm pigmentation: An hypothesis. Science 155: 196–197. **[128]**

Harris, M.P. 1969. The biology of storm petrels in the Galápagos Islands. Proc. Calif. Acad. Sci. 37: 95–166. **[282]**

Harris, M.P., and S. Wanless. 1991. Population studies and conservation of Puffins *Fratercula arctica*, pp. 230–248. *In* Bird Population Studies (C.M. Perrins, J.-D. Lebreton, and G.J.M. Hirons, Eds.). Oxford: Oxford University Press. **[521]**

Hartshorne, C. 1973. Born to Sing: An Interpretation and World Survey of Bird Song. Bloomington: University of Indiana Press. **[251]**

Hartwig, H.-G. 1993. The central nervous system of birds: A study of functional morphology. Avian Biology 9: 1–119. **[202]**

Healy, S., and T. Guilford. 1990. Olfactory-bulb size and nocturnality in birds. Evolution 44: 339–346. **[200]**

Hecht, M.K., J.R. Ostrom, G. Viohl, and P. Wellenhofer. 1985. The Beginnings of Birds. Willibaldsburg: Freunde des Jura-Museums Eichstätt. **[43]**

Hegner, R.E. 1985. Dominance and anti-predator behavior in blue tits (*Parus caeruleus*). Anim. Behav. 33: 762–768. **[335]**

Hegner, R.E., S.T. Emlen, and N.J. Demong. 1982. Spatial organization of the White-fronted Bee-eater. Nature London 298: 264–266. **[474, 475]**

Hegner, R.E., and J.C. Wingfield. 1986. Social modulation of gonadal development and circulating hormone levels during autumn and winter, pp. 109–117. *In* Behavioural Rhythms (Y. Quéinnec and N. Delvolvé, Eds.). Toulouse: Privat. I.E.C. **[265]**

Heilmann, G. 1927. The Origin of Birds. New York: D. Appleton and Company. **[30, 31, 35]**

Heinroth, O. 1922. Die Beziehungen zwischen Vogelgewicht, Eigewicht, Gelegegewicht und Brutdauer. J. Ornithol. 70: 172–285. **[363]**

Heinsohn, R.G. 1991. Slow learning of foraging skills and extended parental care in cooperatively breeding White-winged Choughs. Am. Nat. 137: 864–881. **[472]**

Helbig, A.J. 1991. Inheritance of migratory direction in a bird species: A cross-breeding experiment with SE- and SW-migrating blackcaps (*Sylvia atricapilla*). Behav. Ecol. Sociobiol. 28: 9–12. **[321]**

Helmers, D.L. 1992. Shorebird management manual. Manomet, Mass.: Western Hemisphere Shorebird Reserve Network. **[599]**

Henderson, C.L. 1988. Nongame bird conservation. Current Ornithology 5: 297–312. **[602]**

Henley, C., A. Feduccia, and D.P. Costello. 1978. Oscine spermatozoa: A light- and electron-microscopy study. Condor 80: 41–48. **[354]**

Hensley, M.M., and J.B. Cope. 1951. Further data on removal and repopulation of the breeding birds in a spruce–fir forest community. Auk 68: 483–493. **[519]**

Heppner, F. 1970. The metabolic significance of differential absorption of radiant energy by black and white birds. Condor 72: 50–59. **[128]**

Herrick, F.H. 1932. Daily life of the American Eagle: Early phase. Auk 49: 307–323. **[378]**

Hess, E.H. 1959a. Imprinting. Science 130: 133–141. **[430]**

Hess, E.H. 1959b. Two conditions limiting critical age for imprinting. J. Comp. Physiol. Psychol. 52: 515–518. **[430]**

Hess, E.H. 1973. Imprinting. New York: Van Nostrand Reinhold. **[430]**

Hickey, J.J., Ed. 1969. Peregrine Falcon Populations: Their Biology and Decline. Madison: University of Wisconsin Press. **[608]**

Higginson, T.W. 1863. Outdoor Papers. Boston: Lee and Shepard. **[366]**

Hill, G.E. 1988. Age, plumage brightness, territory quality, and reproductive success in the Black-headed Grosbeak. Condor 90: 379–388. **[518]**

Hill, G.E. 1990. Female house finches prefer colourful males: Sexual selection for a condition-dependent trait. Anim. Behav. 40: 563–572. **[88]**

Hill, G.E. 1991. Plumage coloration is a sexually selected indicator of male quality. Nature London 350: 337–339. **[88]**

Hill, G.E. 1992. Proximate basis of variation in carotenoid pigmentation in male House Finches. Auk 109: 1–12. **[88]**

Hill, G.E. 1993. Geographic variation in the carotenoid plumage pigmentation of male house finches (*Carpodacus mexicanus*). Biol. J. Linn. Soc. 49: 63–86. **[88]**

Hillis, D.M., and C. Moritz. 1990. Molecular Systematics. Sunderland, Mass.: Sinauer Associates. **[62]**

Hinde, R.A. 1956. The biological significance of territories of birds. Ibis 98: 340–369. **[328]**

Hindwood, K.A. 1959. The nesting of birds in the nests of social insects. Emu 59: 1–36. **[383]**

Hitchcock, C.L., and D.F. Sherry. 1990. Long-term memory

for cache sites in the black-capped chickadee. Anim. Behav. 40: 701–712. **[185]**

Ho, C.Y.-K., E.M. Prager, A.C. Wilson, D.T. Osuga, and R.E. Feeney. 1976. Penguin evolution: Protein comparisons demonstrate phylogenetic relationship to flying aquatic birds. J. Mol. Evol. 8: 271–282. **[625, 627]**

Hockey, P.A.R., R.A. Navarro, B. Kalejta, and C.R. Velasquez. 1992. The riddle of the sands: Why are shorebird densities so high in southern estuaries. Am. Nat. 140: 961–979. **[293]**

Hodson, N.L., and D.W. Snow. 1965. The road deaths enquiry, 1960–61. Bird Study 12: 90–99. **[587]**

Hoffman, K. 1954. Versuche zu der im Richtungsfinden der Vögel enthaltenen Zeitschätzung. Z. Tierpsychol. 11: 453–475. **[313]**

Hogan-Warburg, A.J. 1966. Social behavior of the Ruff, *Philomachus pugnax* (L.). Ardea 54: 109–229. **[210]**

Hogstad, O. 1967. Seasonal fluctuation in bird populations within a forest area near Oslo (southern Norway) in 1966–67. Nytt Mag. Zool. Oslo 15: 81–96. **[170]**

Högstedt, G. 1980. Evolution of clutch size in birds: Adaptive variation in relation to territory quality. Science 210: 1148–1150. **[498]**

Höhn, E.O. 1961. Endocrine glands, thymus, and pineal body, pp. 87–114. *In* Biology and Comparative Physiology of Birds, Vol. 2, (A.J. Marshall, Ed.). New York: Academic Press. **[272]**

Holmes, R.T. 1970. Differences in population density, territoriality, and food supply of Dunlin on arctic and subarctic tundra, pp. 303–319. *In* Animal Populations in Relation to Their Food Resources (A. Watson, Ed.). Oxford and Edinburgh: Blackwell Scientific Publications. **[518]**

Holmes, R.T., R.E. Bonney, Jr., and S.W. Pacala. 1979. Guild structure of the Hubbard Brook bird community: A multivariate approach. Ecology 60: 512–520. **[165, 557]**

Houck, M.A., J.A. Gauthier, and R.E. Strauss. 1990. Allometric scaling in the earliest fossil bird, *Archaeopteryx lithographica*. Science 247: 195–198. **[26]**

Houde, P. 1986. Ostrich ancestors found in the Northern Hemisphere suggest new hypothesis of ratite origins. Nature London 324: 563–565. **[621]**

Houde, P., and S.L. Olson. 1981. Paleognathous carinate birds from the early Tertiary of North America. Science 214: 1236–1237. **[52, 621]**

Howard, H.E. 1920. Territory in Bird Life. London: John Murray. **[329]**

Howard, R., and A. Moore. 1991. A Complete Checklist of the Birds of the World, 3rd ed. San Diego, Calif.: Academic Press. **[13]**

Howe, H.F. 1978. Initial investment, clutch size, and brood reduction in the Common Grackle (*Quiscalus quiscula* L.). Ecology 59: 1109–1122. **[503]**

Howe, M.A. 1987. Wetlands and waterbird conservation. Am. Birds 41: 204–209. **[598]**

Howe, M.A. 1991. Federal research on the conservation of migratory nongame birds in the United States, pp. 225–258. *In* Conserving Migratory Birds (T. Salathe, Ed.). Cambridge: International Council for Bird Preservation. **[602, 605]**

Howell, T.R. 1959. A field study of temperature regulation in young Least Terns and Common Nighthawks. Wilson Bull. 71: 19–32. **[437]**

Howell, T.R. 1979. Breeding biology of the Egyptian Plover, *Pluvianus aegyptius*. Univ. Calif. Publ. Zool. No. 113. **[396]**

Howell, T.R., and G.A. Bartholomew. 1962. Temperature regulation in the Sooty Tern *Sterna fuscata*. Ibis 104: 98–105. **[437]**

Howell, T.R., B. Araya, and W.R. Millie. 1974. Breeding biology of the Gray Gull, *Larus modestus*. Univ. Calif. Publ. Zool. No. 104. **[396]**

Hubbard, J.P. 1969. The relationships and evolution of the *Dendroica coronata* complex. Auk 86: 393–432. **[542]**

Hudson, J.W., and M.H. Bernstein. 1981. Temperature regulation and heat balance in flying white-necked ravens, *Corvus cryptoleucus*. J. Exp. Biol. 90: 267–282. **[137]**

Hudson, P.J., and A.P. Dobson. 1991a. Control of parasites in natural populations: Nematode and virus infections of Red Grouse, pp. 413–432. *In* Bird Population Studies (C.M. Perrins, J.-D. Lebreton, and G.J.M. Hirons, Eds.). Oxford: Oxford University Press. **[517]**

Hudson, P.J., and A.P. Dobson. 1991b. The direct and indirect effects of the caecal nematode *Trichostrongylus tenuis* on red grouse, pp. 49–68. *In* Bird–Parasite Interactions (J.E. Loye and M. Zuk, Eds.). Oxford: Oxford University Press. **[517]**

Hudson, W.H. 1920. Birds of La Plata. London: J.M. Dent and Sons Ltd. **[466]**

Humphrey, P.S., and K.C. Parkes. 1959. An approach to the study of molts and plumages. Auk 76: 1–31. **[82]**

Hunt, J.H. 1971. A field study of the Wrenthrush, *Zeledonia coronata*. Auk 88: 1–20. **[57]**

Hunter, L., P. Canevari, J.P. Myers, and L.X. Payne. 1991. Shorebird and wetland conservation in the western hemisphere, pp. 279–290. *In* Conserving Migratory Birds (T. Salathé, Ed.). Cambridge: International Council for Bird Preservation. **[599]**

Hussell, D.J.T. 1969. Weight loss of birds during nocturnal migration. Auk 86: 75–83. **[301]**

Hussell, D.J.T. 1972. Factors affecting clutch size in Arctic passerines. Ecol. Monogr. 42: 317–364. **[497, 498, 499]**

Hussell, D.J.T., and A.B. Lambert. 1980. New estimates of weight loss in birds during nocturnal migration. Auk 97: 547–558. **[301]**

Hutchinson, G.E. 1959. Homage to Santa Rosalia, or why are there so many kinds of animals. Am. Nat. 93: 145–159. **[554]**

Hutchison, L.V., and B.M. Wenzel. 1980. Olfactory guidance in foraging by Procellariiformes. Condor 82: 314–319. **[200]**

Hutto, R.L. 1980. Winter habitat distribution of migratory land birds in western Mexico, with special reference to small foliage-gleaning insectivores, pp. 181–204. *In* Migrant Birds in the Neotropics (A. Keast and E.S. Morton, Eds.). Washington, D.C.: Smithsonian Institution Press. **[560]**

Huxley, T.H. 1867. On the classification of birds and on the taxonomic value of the modifications of certain of the cranial bones observable in that class. Proc. Zool. Soc. London 1867: 415–472. **[22, 49]**

Huxley, T.H. 1868. On the animals which are most nearly intermediate between birds and reptiles. Ann. Mag. Nat. Hist. 4th Series 2: 66–75. **[30]**

Idyll, C.P. 1973. The anchovy crisis. Sci. Am. 228(6): 23–29. **[513]**

Immelmann, K. 1972a. The influence of early experience upon the development of social behavior in estrildine finches, pp. 316–338. *In* Proceedings of the XV International Ornithlogical Congress, Den Haag 1970 (K.H. Voous, Ed.). Leiden: E.J. Brill. **[210]**

Immelmann, K. 1972b. Sexual and other long-term aspects of imprinting in birds and other species. Adv. Study Anim. Behav. 4: 147–174. **[210]**

Immelmann, K. 1975. Ecological significance of imprinting and early learning. Annu. Rev. Ecol. Syst. 6: 15–37. **[430]**

Ingolfsson, A. 1969. Behaviour of gulls robbing eiders. Bird Study 16: 45–52. **[335]**

Ingram, W.J. 1907. On the display of the King Bird-of-Paradise. Ibis 1 (Ninth Series): 225–229. **[204]**

Isack, H.A., and H.-U. Reyer. 1989. Honeyguides and honeygatherers: Interspecific communication in a symbiotic relationship. Science 243: 1343–1346. **[163]**

Isler, M.L., and P.R. Isler. 1987. The Tanagers. Washington, D.C.: Smithsonian Institution Press. **[691]**

Jackson, J.A. 1986. Biopolitics, management of Federal Lands, and the conservation of the Red-cockaded Woodpecker. Am. Birds 40: 1162–1168.

Jackson, J.A. 1994. Red-cockaded Woodpecker. *In* Birds of North America, No. 85 (A. Poole and F.B. Gill, Eds.). Philadelphia: Academy of Natural Sciences; Washington, D.C.: American Ornithologists' Union. **[512]**

Jacob, J. 1978. Uropygial gland secretions and feather waxes. Chem. Zool. 10: 165–211. **[77]**

Jacob, J., and V. Ziswiler. 1982. The uropygial gland. Avian Biology 6: 199–324. **[76, 77, 91]**

James, F.C. 1970. Geographic size variation in birds and its relationship to climate. Ecology 51: 365–390. **[129]**

James, F.C. 1971. Ordinations of habitat relationships among breeding birds. Wilson Bull. 83: 215–236. **[555]**

James, F.C. 1983. Environmental component of morphological differentiation in birds. Science 221: 184–186. **[531]**

James, F.C., and H.H. Shugart, Jr. 1974. Robin phenology study. Condor 76: 159–168. **[275]**

James, F.C., D.A. Wiedenfield, and C.E. McCulloch. 1992. Trends in breeding populations of warblers: Declines in the southern highlands and increases in the lowlands, pp. 43–56. *In* Ecology and Conservation of Neotropical Migrant Landbirds (J.M. Hagan, III, and D.W. Johnston, Eds.). Washington, D.C.: Smithsonian Institution Press. **[595]**

Järvinen, O. 1980. Dynamics of North European bird communities, pp. 770–776. *In* Acta XVII Congressus Internationalis Ornithologici (R. Nöhring, Ed.). Berlin: Verlag der Deutschen Ornithologen-Gesellschaft. **[508, 578]**

Jehl, J.R. 1968. Relationships in the Charadrii (shorebirds): A taxonomic study based on color patterns of the downy young. Mem. San Diego Soc. Nat. Hist. 3: 1–54. **[49, 648]**

Jehl, J.R., Jr. 1975. *Pluvianellus socialis*: Biology, ecology, and relationships of an enigmatic Patagonian shorebird. Trans. San Diego Soc. Nat. Hist. 18: 25–74. **[648]**

Jehl, J.R., Jr., and K.C. Parkes. 1983. "Replacements" of landbird species on Socorro Island, Mexico. Auk 100: 551–559. **[571]**

Jehl, J.R., Jr., and B.G. Murray, Jr. 1986. The evolution of normal and reverse sexual size dimorphism in shorebirds and other birds. Current Ornithology 3: 1–86. **[407]**

Jenkin, P.M. 1957. The filter-feeding and food of flamingoes (Phoenicopteri). Philos. Trans. R. Soc. London Ser B 240: 401–493. **[150, 151]**

Jenkins, F.A., K.P. Dial, and G.E. Goslow. 1988. A cineradiographic analysis of bird flight: The wishbone in starlings is a spring. Science 241: 1495–1498. **[95, 117]**

Jenni, D.A. 1974. Evolution of polyandry in birds. Am. Zool. 14: 129–144. **[417]**

Jensen, R.A.C. 1980. Cuckoo egg identification by chromosome analysis. Proc. Pan-Afr. Ornithol. Congr. 4: 23–25. **[461]**

Johnsgard, P.A. 1967. Animal Behavior. Dubuque, Iowa: Wm. C. Brown. **[230]**

Johnsgard, P.A. 1968. Waterfowl: Their Biology and Natural History. Lincoln: University of Nebraska Press. **[631]**

Johnsgard, P.A., and R. DiSilvestro. 1976. Seventy-five years of changes in Mallard–Black duck ratios in eastern North America. Am. Birds 30: 905–908. **[549]**

Johnsgard, P.A., and J. Kear. 1968. A review of parental carrying of young by waterfowl. Living Bird 7: 89–102. **[447]**

Johnson, L.L., and M.S. Boyce. 1991. Female choice of males with low parasite loads in sage grouse, pp. 177–388. *In* Bird–Parasite Interactions (J.E. Loye and M. Zuk, Eds.). New York: Oxford University Press. **[420]**

Johnson, N.K. 1963. Biosystematics of sibling species of flycatchers in the *Empidonax hammondii-oberholseri-wrightii* complex. Univ. Calif. Publ. Zool. 66: 79–238. **[546]**

Johnson, R.A. 1969. Hatching behavior of the Bobwhite. Wilson Bull. 81: 79–86. **[428]**

Johnson, R.F., Jr., and N.F. Sloan. 1978. White Pelican production and survival of young at Chase Lake National Wildlife Refuge, North Dakota. Wilson Bull. 90: 346–352. **[385]**

Johnson, S.R. 1971. Thermal adaptability of Sturnidae introduced into North America. Unpublished thesis, University of British Columbia. **[394, 395]**

Johnston, D.W. 1988. A morphological atlas of the avian uropygial gland. Bull. Br. Mus. Nat. Hist. Zool. 54(5): 199–259. **[77]**

Johnston, R.F., and R.K. Selander. 1971. Evolution in the house sparrow. II. Adaptive differentiation in North American populations. Evolution. 25: 1–28. **[130, 538]**

Johnston, R.F., D.M. Niles, and S.A. Rohwer. 1972. Herman Bumpus and natural selection in the House Sparrow *Passer domesticus*. Evolution 26: 20–31. **[130]**

Jollie, M. 1953. Are the Falconiformes a monophyletic group? Ibis 95: 369–371. **[637]**

Jollie, M. 1976, 1977. A contribution to the morphology and phylogeny of the Falconiformes. Evol. Theory 1: 285–298, 2: 115–300, 3: 1–141. **[637, 661]**

Jones, D.R., and K. Johansen. 1972. The blood vascular system of birds. Avian Biology 2: 157–285. **[122, 146]**

Jones, P.J., and P. Ward. 1976. The level of reserve protein as the proximate factor controlling the timing of breeding and clutch-size in the Red-billed Quelea *Quelea quelea.* Ibis 118: 547–574. **[365]**

Jones, P.J., and P. Ward. 1979. A physiological basis for colony desertion by Red-billed Queleas (*Quelea quelea*). J. Zool. London 189: 1–19. **[385]**

Jönsson, P.E., and T. Alerstam. 1990. The adaptive significance of parental role division and sexual size dimorphism in breeding shorebirds. Biol. J. Linn. Soc. 41: 301–314. **[407]**

Jouventin, P., M. Guillotin, and A. Cornet. 1979. Le chant du Manchot empereur et sa signification adaptative. Behaviour 70: 231–250. **[247]**

Joyce, F.J. 1993. Nesting success of rufous-naped wrens (*Campylorhynchus rufinucha*) is greater near wasp nests. Behav. Ecol. Sociobiol. 32: 71–77. **[383]**

Juniper, A.T., and C. Yamashita. 1991. The habitat and status of Spix's Macaw *Cyanopsitta spixii.* Bird Conserv. Int. 1: 1–9. **[601]**

Kahl, M.P., Jr. 1963. Thermoregulation in the Wood Stork with special reference to the role of the legs. Physiol. Zool. 36: 141–151. **[137]**

Källander, H., and H.G. Smith. 1990. Food storage in birds: An evolutionary perspective. Current Ornithology 7: 147–208. **[171, 173]**

Kamil, A.C. 1985. The evolution of higher learning abilities in birds, pp. 109–119. *In* Proc. XVIII Int. Ornithol. Congr. (V.D. Ilyicheve and V.M. Gavrilov, Eds.). Moscow: Academy of Sciences of the USSR. **[177]**

Kamil, A.C. 1988. A synthetic approach to the study of animal intelligence, pp. 257–308. *In* Nebraska Symposium on Motivation: Vol. 35. Comparative Perspectives in Modern Psychology (D.W. Leger, Ed.). Lincoln: University of Nebraska Press. **[177, 202]**

Kamil, A.C., and R.P. Balda. 1985. Cache recovery and spatial memory in Clark's nutcrackers (*Nucifraga columbiana*). J. Exp. Psychol. Anim. Behav. Process. 11: 95–111. **[185]**

Kamil, A.C., T.B. Jones, A. Pietrewicz, and J.E. Mauldin. 1977. Positive transfer from successive reversal training to learning set in blue jays. J. Comp. Physiol. Psychol. 91: 79–86. **[178]**

Karasov, W.H., D. Phan, J.M. Diamond, and F.L. Carpenter. 1986. Food passage and intestinal nutrient absorption in hummingbirds. Auk 103: 453–464. **[161]**

Karr, J.R. 1971. Structure of avian communities in selected Panama and Illinois habitats. Ecol. Monogr. 41: 207–233. **[557]**

Karr, J.R. 1976. Within- and between-habitat avian diversity in African and Neotropical lowland habitats. Ecol. Monogr. 46: 457–481. **[558]**

Karr, J.R., J.D. Nichols, M.K. Klimkiewicz, and J.D. Brawn. 1990. Survival rates of birds of tropical and temperate forests: Will the dogma survive? Am. Nat. 136: 277–291. **[488–489]**

Karr, J.R., and R.R. Roth. 1971. Vegetation structure and avian diversity in several New World areas. Am. Nat. 105: 423–435. **[555]**

Kear, J. 1963. Parental feeding in the Magpie Goose. Ibis 105: 428. **[433]**

Keast, A. 1972. Ecological opportunities and dominant families, as illustrated by the Neotropical Tyrannidae (Aves). Evol. Biol. 5: 229–277. **[53]**

Keeton, W.T. 1971. Magnets interfere with pigeon homing. Proc. Natl. Acad. Sci. USA 68: 102–106. **[316]**

Keeton, W.T. 1972. Effects of magnets on pigeon homing. NASA Spec. Publ. 262: 579–594. **[316]**

Keeton, W.T. 1974. The mystery of pigeon homing. Sci. Am. 231(6): 96–107. **[317, 318]**

Keeton, W.T. 1980. Avian orientation and navigation: New developments in an old mystery, pp. 137–158. *In* Acta XVII Congressus Internationalis Ornithologici, Vol 1 (R. Nöhring, Ed.). **[325]**

Keeton, W.T., and A. Gobert. 1970. Orientation by untrained pigeons requires the sun. Proc. Natl. Acad. Sci. USA 65: 853–856. **[323]**

Keeton, W.T., M.L. Kreithen, and K.L. Hermayer. 1977. Orientation by pigeons deprived of olfaction by nasal tubes. J. Comp. Physiol. 114: 289–299. **[316]**

Kendeigh, S.C. 1949. Effect of temperature and season on energy resources of the English Sparrow. Auk 66: 113–127. **[171]**

Kendeigh, S.C. 1952. Parental care and its evolution in birds. Ill. Biol. Monogr. No. 22. **[393]**

Kendeigh, S.C. 1976. Latitudinal trends in the metabolic adjustments of the House Sparrow. Ecology 57: 509–519. **[133]**

Kennedy, E.D. 1991. Predicting clutch size of the House Wren with the Murray-Nolan equation. Auk 108: 728–731. **[502]**

Kenward, R.E. 1978. Hawks and doves: Factors affecting success and selection in goshawk attacks on wood pigeons. J. Anim. Ecol. 47: 449–460. **[341]**

Kerlinger, P., and F.R. Moore. 1989. Atmospheric structure and avian migration. Current Ornithology 6: 109–142. **[304, 305, 306, 307]**

Kerlinger, P., and D. Wiedner. 1990. Economics of birding: A national survey of active birders. Am. Birds 44: 209–213. **[612, 613]**

Ketterson, E.D. 1977. Male Prairie Warbler dies during courtship. Auk 94: 393. **[123]**

Ketterson, E.D., and V. Nolan, Jr. 1982. The role of migration and winter mortality in the life history of a temperate-zone migrant, the Dark-eyed Junco, as determined from demographic analyses of winter populations. Auk 99: 243–259. **[293]**

Ketterson, E.D., and V. Nolan, Jr. 1983. The evolution of differential bird migration. Current Ornithology 1: 357–402. **[293, 294]**

Kiff, L.F., D.B. Peakall, and S.R. Wilbur. 1979. Recent changes in California Condor eggshells. Condor 81: 166–172. **[588]**

King, A.P., and M.J. West. 1977. Species identification in the North American cowbird: Appropriate responses to abnormal song. Science 195: 1002–1004. **[246, 463]**

King, A.S. 1981. Phallus, pp. 107–147. *In* Form and Function in Birds. Vol 2 (A.S. King and J. McLelland, Eds.). New York: Academic Press. **[359, 374]**

King, A.S., and J. McLelland, Eds. 1980–1988. Form and Function. *In* Birds, Vol. 1–4. London: Academic Press. **[146]**

King, J.R. 1972. Adaptive periodic fat storage by birds, pp. 200–217. *In* Proc. XV Int. Ornithol. Congr. (K.H. Voous, Ed.). Leiden: E.J. Brill. **[303]**

King, J.R. 1973. Energetics of reproduction in birds, pp. 78–107. *In* Breeding Biology of Birds (D.S. Farner, Ed.). Washington, D.C.: National Academy of Sciences. **[356, 364, 365]**

King, J.R. 1974. Seasonal allocation of time and energy resources in birds, pp. 4–85. *In* Avian Energetics. Publ. Nuttall Ornithol. Club No. 15. **[268, 278, 285]**

King, J.R., and D.A. Farner. 1965. Studies of fat deposition in migratory birds. Ann. N.Y. Acad. Sci. 131: 422–440. **[297, 298]**

King, J.R., and M.E. Murphy. 1985. Periods of nutritional stress in the annual cycles of endotherms: Fact or fiction? Am. Zool. 25: 955–964. **[169]**

King, S.C., and C.R. Henderson. 1954. Variance components in heritability studies. Poult. Sci. 33: 147–154. **[364]**

King, W.B. 1980. Ecological basis of extinctions of birds, pp. 905–911. *In* Acta XVII Congr. Int. Ornithol. (R. Nöhring, Ed.). Berlin: Verlag der Deutschen Ornithologen-Gesellschaft. **[584]**

Kinney, T.B., Jr. 1969. A summary of reported estimates of heritabilities and of genetic and phenotypic correlations for traits of chickens. U.S. Dept. Agric. Handb. 363. **[364, 532]**

Kinsky, F.C. 1971. The consistent presence of paired ovaries in the Kiwi (*Apteryx*) with some discussion of this condition in other birds. J. Ornithol. 112: 334–357. **[350]**

Kitto, G.B., and A.C. Wilson. 1966. Evolution of malate dehydrogenase in birds. Science 153: 1408–1410. **[665]**

Klein, N.K., R.B. Payne, and M.E.D. Nhlane. 1992. A molecular genetic perspective on speciation in the brood parasite Vidua finches. Proc. Pan-Afr. Congr. 8: 29-39. **[550]**

Klem, D. 1989. Bird–window collisions. Wilson Bull. 101: 606–620. **[587]**

Klomp, H. 1970. The determination of clutch-size in birds. Ardea 58: 1–124. **[442]**

Klomp, H. 1980. Fluctuations and stability in Great Tit populations. Ardea 68: 205–224. **[521, 522, 523, 524]**

Klopfer, P. 1963. Behavioral aspects of habitat selection: The role of early experience. Wilson Bull. 75: 15–22. **[430]**

Kluijver, H.N. 1935. Waarnemingen over de levenswijze van den Spreeuw (*Sturnus v. vulgaris* L.) met behulp van geringde individuen. Ardea 24: 133–166. **[493]**

Kluijver, H.N. 1950. Daily routines of the Great Tit, *Parus m. major* L. Ardea 38: 99–135. **[394]**

Kluijver, H.N. 1951. The population ecology of the Great Tit, *Parus m. major* L. Ardea 39: 1–135. **[507, 523]**

Kluijver, H.N. 1966. Regulation of a bird population. Ostrich Suppl. 6: 389–396. **[522]**

Klump, G.M., W. Windt, and E. Curio. 1986. The great tit's (*Parus major*) auditory resolution in azimuth. J. Comp. Physiol. 158: 383–390. **[196]**

Knorr, O.A. 1957. Communal roosting of the Pygmy Nuthatch. Condor 59: 398. **[132]**

Knudsen, E.I. 1981. The hearing of the Barn Owl. Sci. Am. 245(6): 112–125. **[196]**

Kodric-Brown, A., and J.H. Brown. 1978. Influence of economics, interspecific competition, and sexual dimorphism on territoriality in migrant hummingbirds. Ecology 59: 285–296. **[329]**

Kodric-Brown, A., and J. Brown 1984. Truth in advertising: The kinds of traits favored by sexual selection. Am. Nat. 124: 309–323. **[408]**

Koehler, O. 1950. The ability of birds to "count". Bull. Anim. Behav. 9: 41–45. **[178]**

Koenig, W.D. 1981. Space competition in the Acorn Woodpecker: Power struggles in a cooperative breeder. Anim. Behav. 29: 396–409. **[333]**

Koenig, W.D. 1984. Geographic variation in clutch size in the Northern Flicker (*Colaptes auratus*): Support for Ashmole's hypothesis. Auk 101: 698–706. **[501, 502]**

Koenig, W.D., and R.L. Mumme. 1987. Population Ecology of Cooperatively Breeding Acorn Woodpeckers. Princeton, N.J.: Princeton University Press. **[477]**

Koenig, W.D., F.A. Pitelka, W.J. Carmen, R.L. Mumme, and M.T. Stanback. 1992. The evolution of delayed dispersal in cooperative breeders. Q. Rev. Biol. 67: 111–150. **[470, 477]**

Komdeur, J. 1991. Influence of territory quality and habitat saturation on dispersal options in the Seychelles warbler: An experimental test of the habitat saturation hypothesis for cooperative breeding, pp. 1325–1332. *In* Acta Congressus Internationalis Ornithologici (B.D. Ball, Ed.). Wellington, N.Z.: Ornithological Congress Trust Board. **[471]**

Konishi, M. 1963. The role of auditory feedback in the vocal behavior of the domestic fowl. Z. Tierpsychol. 20: 349–367. **[253]**

Konishi, M., and E.I. Knudsen. 1979. The Oilbird: Hearing and echolocation. Science 204: 425–427. **[195]**

Konishi, M., and F. Nottebohm. 1969. Experimental studies in the ontogeny of avian vocalisations, pp. 29–48. *In* Bird Vocalisations (R.A. Hinde, Ed.). London: Cambridge University Press. **[255]**

Konishi, M., S.T. Emlen, R.E. Ricklefs, and J.C. Wingfield. 1989. Contribution of bird studies to biology. Science 246: 465–471. **[xxvii]**

Korhonen, K. 1981. Temperature in the nocturnal shelters of the Redpoll (*Acanthis flammea* L.) and the Siberian Tit (*Parus cinctus* Budd.) in winter. Ann. Zool. Fenn. 18: 165–168. **[132]**

Koskimies, J. 1948. On temperature regulation and metabolism in the Swift, *Microapus a. apus* L., during fasting. Experientia Basel 4: 274–276. **[442]**

Koskimies, J. 1950. The life of the swift, *Micropus apus* (L.), in relation to the weather. Ann. Acad. Sci. Fenn. Ser. A IV Biol. No. 15. **[171]**

Kramer, G. 1950. Orientierte Zugaktivität gekäfigter Singvögel. Naturwissenschaften 37: 188. **[312]**

Kramer, G. 1951. Eine neue Methode zur Erforschung der Zugorientierung und die bisher damit erzielten Ergebnisse, pp. 269–280. *In* Proc. X Int. Ornithol. Congr. (S. Hörstadius, Ed.). Uppsala: Almqvist and Wiksell. **[312]**

Kramer, G. 1952. Experiments on bird orientation. Ibis 94: 265–285. **[312]**

Krebs, J.R. 1970. The efficiency of courtship feeding in the Blue Tit *Parus caeruleus*. Ibis 112: 108–110. **[409]**

Krebs, J.R. 1973. Social learning and the significance of mixed-species flocks of chickadees (*Parus* spp.). Can. J. Zool. 51: 1275–1288. **[339, 344]**

Krebs, J.R. 1977. The significance of song repertoires: The Beau Geste hypothesis. Anim. Behav. 25: 475–478. **[243, 250–251]**

Krebs, J.R. 1978. Optimal foraging: Decision rules for predators, pp. 23–63. *In* Behavioural Ecology: An Evolutionary Approach (J.R. Krebs and N.B. Davies, Eds.). Sunderland, Mass.: Sinauer Associates. **[167]**

Krebs, J.R., and R. Dawkins. 1984. Animal signals: Mind-reading and manipulation, pp. 380–402. *In* Behavioural Ecology: An Evolutionary Approach, 2nd ed. (J.R. Krebs and N.B. Davies, Eds.). Sunderland, Mass.: Sinauer Associates. **[223]**

Krebs, J.R., and N.B. Davies. 1984. Behavioral Ecology, 2nd ed. Sunderland, Mass.: Sinauer Associates. **[173, 346]**

Krebs, J.R., and D.E. Kroodsma. 1980. Repertoires and geographical variation in bird song. Adv. Study Behav. 11: 143–177. **[250]**

Krebs, J.R., J.T. Erichsen, M.I. Webber, and E.L. Charnov. 1977. Optimal prey selection in the Great Tit (*Parus major*). Anim. Behav. 25: 30–38. **[166]**

Krebs, J.R., M.H. MacRoberts, and J.M. Cullen. 1972. Flocking and feeding in the Great Tit *Parus major*—An experimental study. Ibis 114: 507–530. **[339]**

Krebs, J.R., D.F. Sherry, S.D.Healy, V.H. Perry, and A.L. Vaccarino. 1989. Hippocampal specialization of food-storing birds. Proc. Natl. Acad. Sci. USA 86: 1388–1392. **[185]**

Kreithen, M.L., and T. Eisner. 1978. Ultraviolet light detection by the homing pigeon. Nature London 272: 347–348. **[191]**

Kreithen, M.L., and W.T. Keeton. 1974. Detection of changes in atmospheric pressure by the homing pigeon, *Columba livia*. J. Comp. Physiol. 89: 73–82. **[198]**

Kreithen, M.L., and D.B. Quine. 1979. Infrasound detection by the homing pigeon: A behavioral audiogram. J. Comp. Physiol. A 129: 1–4. **[194, 304]**

Kroodsma, D.E. 1977. Correlates of song organization among North American wrens. Am. Nat. 111: 995–1008. **[250]**

Kroodsma, D.E. 1978. Continuity and versatility in bird song: Support for the monotony threshold hypothesis. Nature London 274: 681–683. **[251]**

Kroodsma, D.E. 1979. Vocal dueling among male Marsh Wrens: Evidence for ritualized expressions of dominance/subordinance. Auk 96: 506–515. **[251]**

Kroodsma, D.E. 1980. Winter Wren singing behavior: A pinnacle of song complexity. Condor 82: 357–365. **[250]**

Kroodsma, D.E. 1982. Song repertoires: Problems in their definition and use, pp. 125–146. *In* Acoustic Communication in Birds, Vol. 2 (D.E. Kroodsma and E.H. Miller, Eds.). New York: Academic Press. **[251, 257, 258]**

Kroodsma, D.E. 1984. Songs of the Alder Flycatcher (*Empidonax alnorum*) and Willow Flycatcher (*Empidonax traillii*) are innate. Auk 101: 13–24. **[244, 253]**

Kroodsma, D.E. 1989. Two North American song populations of the Marsh Wren reach distributional limits in the Great Plains. Condor 91: 332–340. **[551]**

Kroodsma, D.E., and R.A. Canady. 1985. Differences in repertoire size, singing behavior, and associated neuroanatomy among Marsh Wren populations have a genetic basis. Auk 102: 439–446. **[184]**

Kroodsma, D.E., and E.H. Miller, Eds. 1982. Acoustic Communication in Birds. New York: Academic Press. **[260, 451]**

Kroodsma, D.E., and R. Pickert. 1980. Environmentally dependent sensitive periods for avian vocal learning. Nature London 288: 477–479. **[254]**

Kroodsma, D.E., M.C. Baker, and L.F. Baptista. 1985. Vocal "dialects" in Nuttall's White-crowned Sparrow. Current Ornithology 2: 103–133. **[257, 258]**

Kroodsma, R.L. 1975. Hybridization in buntings (*Passerina*) in North Dakota and eastern Montana. Auk 92: 66–80. **[544]**

Kruijt, J.P., G.J. de Vos, and I. Bossema. 1972. The arena system of Black Grouse, pp. 399–423. *In* Proc. XV Int. Ornithol. Congr. (K.H. Voous, Ed.). Leiden, E.J. Brill. **[420]**

Kruuk, H. 1964. Predators and anti-predator behavior of the black-headed gull (*Larus ridibundus* L.). Behaviour Suppl. 11: 1–130. **[384]**

Kumerloeve, H. 1987. Le gynandromorphisme chez les oiseaux–recapitulation des donnees connues. Alauda 55: 1–9. **[352]**

Kusmierski, R., G. Borgia, R.H. Crozier, and B.H.Y. Chan. 1993. Molecular information on bowerbird phylogeny and the evolution of exaggerated male characteristics. J. Evol. Biol. 6: 737–752. **[226]**

Lack, D. 1947a. Darwin's Finches. Cambridge: Cambridge University Press. **[569]**

Lack, D. 1947b. The significance of clutch-size, I–II. Ibis 89: 302–352. **[496, 498]**

Lack, D. 1948. The significance of clutch-size, III. Ibis 90: 25–45. **[496]**

Lack, D. 1954. The Natural Regulation of Animal Numbers. Oxford: Clarendon Press. **[380, 442, 510, 513, 525]**

Lack, D. 1956. Swifts in a Tower. London: Methuen. **[441]**

Lack, D. 1958. The significance of the colour of turdine eggs. Ibis 100: 145–166. **[371]**

Lack, D. 1963. Cuckoo hosts in England. (With an appendix on cuckoo hosts in Japan by T. Royama). Bird Study 10: 185–202. **[460]**

Lack, D. 1966. Population Studies of Birds. Oxford: Clarendon Press. **[507. 508, 510, 513]**

Lack, D. 1968. Ecological Adaptations for Breeding in Birds. London: Methuen. **[364, 407, 419, 424, 440, 460, 461, 462, 477]**

Lack, D. 1971. Ecological Isolation in Birds. Cambridge, Mass.: Harvard University Press. **[562, 566, 567, 568, 569, 571, 574, 575]**

Lack, D., and R.E. Moreau. 1965. Clutch size in tropical passerine birds of forest and savanna. Oiseau Rev. Fr. Ornithol. 35 (No. Special): 76–89. **[501]**

Lande, R. 1981. Models of speciation by sexual selection on polygenic traits. Proc. Natl. Acad. Sci. USA 78: 3721–3725. **[230]**

Lane, B.A., and D. Parish. 1991. A review of the Asian–Australasian bird migration system. *In* ICBP Technical Publication No. 12. Cambridge: International Council for Bird Preservation. **[288, 307]**

Lank, D.B., and C.M. Smith. 1987. Conditional lekking in ruff (*Philomachus pugnax*). Behav. Ecol. Sociobiol. 30: 323–329. **[421]**

Lank, D.B., and C.M. Smith. 1992. Females prefer larger leks: Field experiments with ruffs (*Philomachus pugnax*). Behav. Ecol. Sociobiol. 30: 323–329. **[421]**

Lank, D.B, L.W. Oring, and S.J. Maxson. 1985. Mate and nutrient limitation of egg-laying in a polyandrous shorebird. Ecology 66: 1513–1524. **[415–416]**

Lanyon, S.M. 1992a. [A review of] Phylogeny and classification of birds: A study in molecular evolution, by C.G. Sibley and J.E. Ahlquist. Condor 94: 304–310. **[57]**

Lanyon, S.M. 1992b. Interspecific brood parasitism in blackbirds (Icterinae): A phylogenetic perspective. Science 225: 77–79. **[618]**

Lanyon, S.M., and J.G. Hall. 1994. Reexamination of barbet monophyly using mitochondrial-DNA sequence data. Auk 111: 389–397. **[61, 673]**

Lanyon, W.E. 1957. The comparative biology of meadowlarks (*Sturnella*) in Wisconsin. Publ. Nuttall Ornithol. Club No. 1. **[253, 546]**

Lanyon, W.E. 1967. Revision and probable evolution of the *Myiarchus* flycatchers of the West Indies. Bull. Am. Mus. Nat. Hist. 136: 329–370. **[39, 40, 244]**

Lanyon, W.E. 1978. Revision of the *Myiarchus* flycatchers of South America. Bull. Am. Mus. Nat. Hist. 161: 427–628. **[244]**

Lanyon, W.E. 1979. Hybrid sterility in meadowlarks. Nature London 279: 557–558. **[543]**

Lanyon, W.E. 1981. Breeding birds and old field succession on fallow Long Island farmland. Bull. Am. Mus. Nat. Hist. 168: 1–60. **[555, 556, 562]**

Lasiewski, R.C. 1962. The energetics of migrating hummingbirds. Condor 64: 324. **[302]**

Lasiewski, R.C. 1972. Respiratory function in birds, pp. 287–342. *In* Avian Biology, Vol. 2 (D.S. Farner, J.R. King, and K.C. Parkes, Eds.). New York: Academic Press. **[118, 136]**

Lasiewski, R.C., W.W. Weathers, and M.H. Bernstein. 1967. Physiological responses of the giant hummingbird, *Patagona gigas*. Comp. Biochem. Physiol. 23: 797–813. **[123]**

Lawick-Goodall, J. van. 1968. Tool-using bird: The Egyptian vulture. Natl. Geogr. Mag. 133: 630–641. **[164]**

Lawton, M.F., and R.O. Lawton. 1986. Heterochrony, deferred breeding, and avian sociality. Current Ornithology 3: 187–222. **[495]**

Laybourne, R.C. 1967. Bilateral gynandrism in an Evening Grosbeak. Auk 84: 267–272. **[353]**

Leask, M.J.M. 1977. A physicochemical mechanism for magnetic field detection by migrating birds and homing pigeons. Nature (London) 267: 144–145. **[192]**

Lebreton, J.-D., K.P. Burnham, J. Clobert, and D.R. Anderson. 1992. Modeling survival and testing biological hypotheses using marked animals: A unified approach with case studies. Ecol. Monogr. 62: 67–118. **[488]**

Lein, M.R. 1978. Song variation in a population of Chestnut-sided Warblers (*Dendroica pennsylvanica*): Its nature and suggested significance. Can. J. Zool. 56: 1266–1283. **[243]**

Lemon, R.E. 1977. Bird song: An acoustic flag. BioScience 27: 402–408. **[250]**

Lemon, R.E., and C. Chatfield. 1971. Organization of song in cardinals. Anim. Behav. 19: 1–17. **[250]**

Lennartz, M.R., and B.G. Henry. 1985. Endangered species recovery plan: Red-cockaded Woodpecker (*Picoides borealis*). Atlanta, Ga.: U.S. Fish and Wildlife Service. **[512]**

Leopold, F. 1951. A study of nesting Wood Ducks in Iowa. Condor 63: 209–220. **[447]**

Lessells, C.M. 1986. Brood size in Canada Geese: A manipulation experiment. J. Anim. Ecol. 55: 669–689. **[498]**

Levey, D.J., and F.G. Stiles. 1992. Evolutionary precursors of long distance migration: Resource availability and movement patterns in Neotropical landbirds. Am. Nat. 140: 447–476. **[291]**

Lifjeld, J.T., P.O. Dunn, R.J. Robertson, and P.T. Boag. 1993. Extra pair paternity in monogamous Tree Swallows. Anim. Behav. 45: 213–229. **[409]**

Ligon, J.D. 1967. Relationships of the cathartid vultures. Occ. Papers Mus. Zool. Univ. Mich. No. 651. **[57, 635, 637]**

Ligon, J.D. 1970. Still more responses of the Poor-will to low temperatures. Condor 72: 496–498. **[134–135]**

Ligon, J.D. 1974. Green cones of the piñon pine stimulate late summer breeding in the piñon jay. Nature London 250: 80–82. **[274]**

Ligon, J.D. 1981. Demographic patterns and communal breeding in the Green Woodhoopoe, *Phoeniculus purpureus*, pp. 231–243. *In* Natural Selection and Social Behavior: Recent Research and New Theory (R.D. Alexander and D.W. Tinkle, Eds.). New York: Chiron Press. **[472]**

Ligon, J.D. 1993. The role of phylogenetic history in the evolution of contemporary avian mating and parental care systems. Current Ornithology 10: 1–46. **[415, 417, 424, 453, 500]**

Ligon, J.D., and S.H. Ligon. 1978. The communal social system of the Green Woodhoopoe in Kenya. Living Bird 17: 159–197. **[472]**

Ligon, J.D., and S.H. Ligon. 1983. Reciprocity in the Green Woodhoopoe (*Phoeniculus purpureus*). Anim. Behav. 31: 480–489. **[473]**

Ligon, J.D., C. Carey, and S.H. Ligon. 1988. Cavity roosting and cooperative breeding in the Green Woodhoopoe reflect a physiological trait. Auk 105: 123–127. **[472]**

Ligon, J.D., R. Thornhill, M. Zuk, and K. Johnson. 1990. Male–male competition, ornamentation, and the role of testosterone in sexual selection in red jungle fowl. Anim. Behav. 40: 367–373. **[229]**

Lill, A. 1974a. Sexual behavior of the lek-forming White-bearded Manakin (*Manacus manacus trinitatis* Hartert). Z. Tierpsychol. 36: 1–36. **[420]**

Lill, A. 1974b. Social organization and space utilization in the lek-forming White-bearded Manakin (*Manacus manacus trinitatis* Hartert). Z. Tierpyschol. 36: 1–36. **[420]**

Lill, A. 1974c. The evolution of clutch size and male "chauvinism" in the White-bearded Manakin. Living Bird 13: 211–231. **[500]**

Lind, E. 1964. Nistzeitliche Gesselligkeit der Mehlschwalbe, *Delichon u. urbica* (L.). Ann. Zool. Fenn. 1: 7–43. **[386]**

Livezey, B.C. 1988. Morphometrics of flightlessness in the Alcidae. Auk 105: 681–698. **[110]**

Livezey, B.C. 1989a. Flightlessness in grebes (Aves, Podicipedidae): Its independent evolution in three genera. Evolution 43: 29–54. **[110]**

Livezey, B.C. 1989b. Morphometric patterns in Recent and fossil penguins (Aves, Sphenisciformes). J. Zool. London 219: 269–307. **[110]**

Livezey, B.C. 1993. An ecomorphological review of the Dodo (*Raphus cucullatus*) and Solitaire (*Pezophaps solitaria*), flightless Columbiformes of the Mascarene Islands. J. Zool. 230: 247–292. **[110, 651]**

Lockwood, W.B. 1984. The Oxford book of British bird names. Oxford: Oxford University Press. **[287]**

Lofts, B., and R.K. Murton. 1973. Reproduction in birds. Avian Biology 3: 1–108. **[374]**

Löhrl, H. 1977. Nistökologische und etholgische Anpassungserscheinungen bei Hohlenbrutern. Vogelwarte 29: 92–101. **[565]**

Löhrl, H. 1978. Beiträge zur Ethologie und Gewichtsentwicklung beim Wendehals *Jynx torquilla*. Ornithol. Beob. 75: 193–201. **[438]**

Lombardo, M.P., H.W. Power, P.C. Stouffer, L.C. Romagnano, and A.S. Hoffenberg. 1989. Egg removal and intraspecific brood parasitism in the European Starling (*Sturnus vulgaris*). Behav. Ecol. Sociobiol. 24: 217–223. **[458]**

Long, J.L. 1981. Introduced Birds of the World. London: David and Charles. **[585]**

Lorenz, K. 1941. Vergleichende Bewegungsstudien an Anatinen. J. Ornithol. 89: 194–294 (see translations in Avic. Mag. 1951–1953). **[216]**

Lorenz, K. 1965. Evolution and Modification of Behavior. Chicago: University of Chicago Press. **[232]**

Lorenz, K. 1969. Innate bases of learning, pp. 13–93. *In* On the Biology of Learning (K.H. Pribram, Ed.). New York: Harcourt Brace and World. **[388]**

Lorenz, K. 1981. The Foundations of Ethology. New York: Springer-Verlag. **[214]**

Lövei, G.L. 1989. Passerine migration between the Palearctic and Africa. Current Ornithology 6: 143–174. **[288, 307, 309]**

Lovejoy, T.E. 1974. Bird diversity and abundance in Amazon forest communities. Living Bird 13: 127–191. **[557]**

Lovejoy, T.E., R.O. Bierregaard, Jr., A.B. Rylands, J.R. Malcolm, C.E. Quintela, L.H. Haraper, K.S. Brown, Jr., A.H. Powell, G.V.N. Powell, H.O.R. Schubart, and M.B. Hays. 1986. Edge and other effects of isolation on Amazon forest fragments, pp. 257–285. *In* Conservation Biology (M.E. Soulé, Ed.). Sunderland, Mass.: Sinauer Associates. **[599]**

Lovell, H.B. 1958. Baiting of fish by a Green Heron. Wilson Bull. 70: 280–281. **[164]**

Lowther, P.E. 1977. Bilateral size dimorphism in House Sparrow gynandromorphs. Auk 94: 377–380. **[352]**

Lowther, P.E. 1993. Brown-headed Cowbird (*Molothrus ater*). *In* The Birds of North America, No. 47 (A. Poole and F. Gill, Eds.). Philadelphia: Academy of Natural Sciences; Washington, D.C.: American Ornithologists' Union. **[460]**

Lowther, P.E., and C.L. Cink. 1992. House Sparrow. *In* The Birds of North America, No. 12 (A. Poole, P. Stettenheim, and F. Gill, Eds.). Philadelphia: Academy of Natural Sciences; Washington, D.C.: American Ornithologists' Union. **[130]**

Loye, J.E., and M. Zuk, Eds. 1991. Bird–Parasite Interactions. New York: Oxford University Press. **[232, 517, 525]**

Lucas, A.M., and P.R. Stettenheim. 1972. Avian Anatomy: Integument. Washington, D.C.: U.S. Government Printing Office. **[69, 76, 91]**

Lumley, W.F., Ed. 1895. Fulton's Book of Pigeons, rev. ed. London: Cassell and Company, Ltd. **[309]**

Lustick, S. Energy requirements of molt in cowbirds. Auk 87: 742–746. **[279]**

Lynch, J.F., and N.K. Johnson. 1974. Turnover and equilibria in insular avifaunas, with special reference to the California Channel Islands. Condor 76: 370–384. **[576]**

Lyon, B.E., and R.D. Montgomerie. 1987. Ecological correlates of incubation feeding: A comparative study of high arctic finches. Ecology 68: 713–722. **[455]**

MacArthur, R.H. 1958. Population ecology of some warblers of northeastern coniferous forests. Ecology 39: 599–619. **[554]**

MacArthur, R. 1968. The Theory of the Niche: Population Biology and Evolution. Syracuse, N.Y.: Syracuse University Press. **[554]**

MacArthur, R. 1969. Patterns of communities in the tropics. Biol. J. Linn. Soc. 1: 19–30. **[557, 558]**

MacArthur, R. 1971. Patterns of terrestrial bird communities. Avian Biology 1: 189–221. **[564]**

MacArthur, R. 1972. Geographical Ecology. New York: Harper & Row. **[554, 562, 564, 580]**

MacArthur, R.H., and J.W. MacArthur. 1961. On bird species diversity. Ecology 42: 594–598. **[166]**

MacArthur, R.H., and E.O. Wilson. 1967. The Theory of Island Biogeography. Princeton, N.J.: Princeton University Press. **[575, 576]**

MacArthur, R.H., J.M. Diamond, and J.R. Karr. 1972. Density compensation in island faunas. Ecology 53: 330–342. **[573]**

Maclean, G.L. 1967. Die systematische Stellung der Flughühner (Pteroclididae). J. Ornithol. 108: 203–217. **[648]**

Maclean, G.L. 1968. Field studies on the sandgrouse of the Kalahari desert. Living Bird 7: 209–235. **[647]**

Maclean, G.L. 1969. The sandgrouse—doves or plovers? J. Ornithol. 110: 104–107. **[648]**

MacMillen, R.E. 1986. Energetic patterns and lifestyles in the Meliphagidae. N. Z. J. Zool. 12: 623–629. **[124]**

MacRoberts, M.H., and B.R. MacRoberts. 1976. Social organization and behavior of the Acorn Woodpeckers in central coastal California. Ornithol. Monogr. No. 21. **[333]**

Maher, W.J. 1962. Breeding biology of the Snow Petrel near Cape Hallett, Antarctica. Condor 64: 488–499. **[280]**

Maher, W.J. 1970. The Pomarine Jaeger as a brown lemming predator in northern Alaska. Wilson Bull. 82: 130–157. **[329]**

Manwell, C., and C.M.A. Baker. 1975. Molecular genetics of avian proteins. XIII. Protein polymorphism in three species of Australian passerines. Aust. J. Biol. Sci. 28: 545–557. **[458]**

Marchant, S. 1960. The breeding of some S.W. Ecuadorian birds. Ibis 102: 349–382. **[275, 501]**

Marcot, B.G., and R. Holthauser. 1987. Analyzing populations variability of the spotted owl in the Pacific Northwest. Trans. North Amer. Wildlife Natural Resource Conference No. 52: 33–347. **[592]**

Marder, J., Y. Arieli, and J. Ben-Asher. 1989. Defense strategies against environmental heat stress in birds. Israel J. Zool. 36: 61–75. **[136]**

Marler, P. 1955. Characteristics of some animal calls. Nature (London) 176: 6–8. **[236]**

Marler, P. 1956. The voice of the Chaffinch and its function as a language. Ibis 98: 231–261. **[242]**

Marler, P. 1967. Animal communication signals. Science 157: 769–774. **[236]**

Marler, P. 1969. Tonal quality of bird sounds, pp. 5–18. *In* Bird Vocalisations (R.A. Hinde, Ed.). Cambridge: Cambridge University Press. **[236, 237]**

Marler, P. 1981. Birdsong: The acquisition of a learned motor skill. Trend Neurosci. 4: 88–94. **[233, 254, 257]**

Marler, P. 1983. Some ethological implications for neuroethology: The ontogeny of birdsong, pp. 21–52. *In* Advances in Vertebrate Neuroethology (J.P. Ewert, R.R. Capranica, and D.J. Ingle, Eds.). New York: Plenum Press. **[254]**

Marler, P., and W.J. Hamilton, III. 1966. Mechanisms of Animal Behavior. New York: John Wiley & Sons. **[250]**

Marler, P., and P. Mundinger. 1971. Vocal learning in birds, pp. 389–450. *In* The Ontogeny of Vertebrate Behavior (H. Moltz, Ed.). New York: Academic Press. **[256]**

Marler, P., and S. Peters. 1977. Selective vocal learning in a sparrow. Science 198: 519–521. **[257]**

Marler, P., and S. Peters. 1981. Sparrows learn adult song and more from memory. Science 213: 780–782. **[255, 257]**

Marler, P., and S. Peters. 1982a. Subsong and plastic song: Their role in the vocal learning process, pp. 25–50. *In* Acoustic Communication in Birds, Vol. 2 (D.E. Kroodsma and E.H. Miller, Eds.). New York: Academic Press. **[256, 257]**

Marler, P., and S. Peters. 1982b. Structural changes in song ontogeny in the Swamp Sparrow, *Melospiza georgiana*. Auk 99: 446–458. **[255]**

Marsh, O.C. 1877. Introduction and succession of vertebrate life in America. Am. J. Sci. 3rd Ser. 14: 337–378. **[27]**

Marsh, R.L. W.R. Dawson, J.J. Camilliere, and J.M. Olson. 1990. Regulation of glycolysis in the pectoralis muscles of seasonally acclimatized American goldfinches exposed to cold. Am. J. Physiol. 258: R711–R717. **[133]**

Marshall, A.J. 1954. Bowerbirds. Cambridge: Oxford University Press. **[226, 228]**

Marshall, A.J. 1961. Reproduction, pp. 169–213. *In* Biology and Comparative Physiology of Birds. Vol. 2 (A.J. Marshall, Ed.). New York: Academic Press. **[354]**

Marshall, D.B. 1988. The Marbled Murrelet joins the old growth forest conflict. Am. Birds 42: 202–212. **[594]**

Martella, M.B., and E.H. Bucher. 1984. Nesting of the Spot-winged Falconet in Monk Parakeet nests. Auk 101: 614–615. **[378]**

Martin, K., and F. Cooke. 1987. Bi-parental care in willow ptarmigan: A luxury? Anim. Behav. 35: 369–379. **[455]**

Martin, L.D. 1983. The origin and early radiation of birds, pp. 291–338. *In* Perspectives in Ornithology (A.H. Brush and G.A. Clark, Jr., Eds.). Oxford: Oxford University Press. **[30]**

Martin, T.E. 1986. Competition in breeding birds: On the importance of considering processes at the level of the individual. Current Ornithology 4: 181–210. **[580]**

Martin, T.E. 1988a. Nest placement: Implications for selected life-history traits, with special reference to clutch size. Am. Nat. 132: 900–910. **[380]**

Martin, T.E. 1988b. Processes organizing open-nesting bird assemblages: Competition or nest predation. Evol. Ecol. 2: 37–50. **[380, 555]**

Martin, T.E. 1988c. On the advantage of being different: Nest predation and the coexistence of bird species. Proc. Natl. Acad. Sci. USA 85: 2196–2199. **[555]**

Martinez, M.M. 1983. Nidification de *Hirundo rustica erythrogaster* (Boddaert) en la Argentina (Aves, Hirundinidae). Neotropica La Plata 29: 83–86. **[291]**

Martinez del Rio, C., and B.R. Stevens. 1989. Physiological constraint on feeding behavior: Intestinal membrane disaccharidase of the starling. Science 243: 794–796. **[161]**

Mather, M.H., and R.J. Robertson 1991. Honest advertisement in flight displays of Bobolinks (*Dolichonyx oryzivorus*). Auk 109: 869–873. **[408]**

Mathiu, P.M., W.R. Dawson, and G.C. Whittow. 1991. Development of thermoregulation in Hawaiian Brown Noddies (*Anous stolidus pileatus*). J. Therm. Biol. 16: 317–325. **[437]**

Matthews, G.V.T. 1951. The experimental investigation of navigation in homing pigeons. J. Exp. Biol. 28: 508–536. **[312]**

Matthews, G.V.T. 1953. Sun navigation in homing pigeons. J. Exp. Biol. 30: 243–267. **[312]**

Matthews, G.V.T. 1968. Bird Navigation, 2nd ed. London: Cambridge University Press. **[312, 324]**

Matthiessen, P. 1959. Wildlife in America. New York: Viking Press. **[586, 601, 602, 605, 616]**

Maurer, D.R., and R.J. Raikow. 1981. Appendicular myology, phylogeny, and classification of the avian order Coraciiformes (including Trogoniformes). Ann. Carnegie Mus. 50: 417–434. **[667, 670]**

May, R.M., and S.K. Robinson. 1985. Population dynamics of avian brood parasitism. Am. Nat. 126: 475–494. **[477]**

Mayer, L., S. Lustick, and B. Battersby. 1982. The importance of cavity roosting and hypothermia to the energy balance of the winter acclimatized Carolina Chickadee. Int. J. Biometeorol. 26: 231–238. **[132]**

Mayfield, H.F. 1960. The Kirtland's Warbler. Bloomfield Hills, Mich.: Cranbrook Institute of Science. **[463]**

Maynard Smith, J. 1977. Parental investment: A prospective analysis. Anim. Behav. 25: 1–9. **[417]**

Mayr, E. 1926. Die Ausbreitung des Girlitz (*Serinus canaria serinus* L.). J. Ornithol. 74: 571–671. **[291]**

Mayr, E. 1958. The sequence of songbird families. Condor 60: 194–195. **[679]**

Mayr, E. 1963. Animal Species and Evolution. Cambridge, Mass.: Belknap Press. **[545]**

Mayr, E. 1965. What is a fauna? Zool. Jarhb. Abt. Syst. Oekol. Geogr. Tiere 92: 473–486. **[572]**

Mayr, E. 1969. Bird speciation in the tropics. Biol. J. Linn. Soc. 1: 1–17. **[558]**

Mayr, E. 1970. Population, Species, and Evolution. Cambridge, Mass.: Belknap Press. **[45, 527, 552]**

Mayr, E. 1982. Of what use are subspecies? Auk 99: 593–595. **[529]**

Mayr, E. 1984. The contributions of ornithology to biology. BioScience 34: 250–255. **[xxvii]**

Mayr, E., and D. Amadon. 1951. A classification of Recent birds. Am. Mus. Novit. 1496: 1–42. **[47]**

Mayr, E., and P.D. Ashlock. 1991. Principles of Systematic Zoology, 2nd ed. New York: McGraw-Hill. **[62]**

Mayr, E., and R.J. O'Hara. 1986. The biogeographic evidence supporting the Pleistocene forest refuge hypothesis. Evolution 40: 55–67. **[37]**

Mazzeo, R. 1953. Homing of the Manx Shearwater. Auk 70: 200–201. **[309]**

McCamant, R.E., and E.G. Bolen. 1979. A 12-year study of nest box utilization by Black-bellied Whistling Ducks. J. Wildl. Manage. 43: 936–943. **[458]**

McCance, R.A. 1960. Severe undernutrition in growing and adult animals. I. Production and general effects. Br. J. Nutr. 14: 59–73. **[442]**

McCleery, R.H., and C.M. Perrins. 1991. Effects of predation on the numbers of Great Tits *Parus major*, pp. 129–147. *In* Bird Population Studies (C.M. Perrins, J.-D. Lebreton, and G.J.M. Hirons, Eds.). Oxford: Oxford University Press. **[521, 522]**

McDonald, D.B. 1989. Cooperation under sexual selection: Age graded changes in a lekking bird. Am. Nat. 134: 709–730. **[423]**

McDonald, D.B. 1993. Delayed plumage maturation and orderly queues for status: A manakin mannequin experiment. Ethology 94: 31–45. **[423]**

McDonald, D.B., and H. Caswell. 1993. Matrix methods for avian demography. Current Ornithology 10: 139–185. **[487, 505]**

McDonald, D.B., J.W. Fitzpatrick, and G.E. Woolfenden. 1994. Senescent mortality in the Florida Scrub Jay. Manuscript. **[487]**

McFarland, D.J., and E. Baher. 1968. Feathers affecting feather posture in the Barbary Dove. Anim. Behav. 16: 171–177. **[129]**

McFarlane, R.W. 1963. The taxonomic significance of avian sperm, pp. 91–102. Proc. XIII Int. Ornithol. Congr. (C.G. Sibley, Ed.). Lawrence, Kans.: Allen Press. **[354, 355]**

McKinney, F., S.R. Derrickson, and P. Mineau. 1983. Forced copulations in waterfowl. Behaviour 86: 250–294. **[409]**

McKitrick, M.C., and R.M. Zink. 1988. Species concepts in ornithology. Condor 90: 1–14. **[46, 529]**

McLaughlin, C.L., and D. Grice. 1952. The effectiveness of large-scale erection of Wood Duck boxes as a management procedure. Trans. N. Am. Wildl. Nat. Resour. Conf. 17: 242–259. **[512]**

McLaughlin, R.L., and R.D. Montgomerie. 1990. Flight speeds of parent birds feeding nestlings: Maximization of foraging efficiency or food delivery rate? Can. J. Zool. 68: 2269–2274. **[100]**

McLean, I.G., and G. Rhodes. 1991. Enemy recognition and response in birds. Current Ornithology 8: 173–211. **[342, 346]**

McLelland, J. 1975. Aves digestive system, pp. 1857–1882. *In* Sisson and Grossman's The Anatomy of the Domestic Animals, 5th ed., Vol. 2 (R. Getty, Ed.). Philadelphia: Saunders. **[159]**

McLelland, J. 1979. Digestive system, pp. 69-181. *In* Form and Function in Birds, Vol. 1 (A.S. King and J. McLelland, Eds.). New York: Academic Press. **[162, 173]**

McMillen, J.L. 1988. Conservation of North American cranes. Am. Birds 42: 1212–1221. **[606, 607]**

Medway, Lord, and J.D. Pye. 1977. Echolocation and the systematics of swiftlets, pp. 225–238. *In* Evolutionary Ecology (B. Stonehouse and C. Perrins, Eds.). Baltimore, Md.: University Park Press. **[195]**

Meier, A.H., and A.C. Russo. 1985. Circadian organization of the avian annual cycle. Current Ornithology 2: 303–343. **[267, 273, 285]**

Mengel, R.M. 1970. The North American central plains as an isolating agent in bird speciation. Univ. Kansas Spec. Publ. 3: 279–340. **[40]**

Merkel, F.W., and W. Wiltschko. 1965. Magnetismus und Richtungsfinden zugunruhiger Rotkehlchen (*Erithacus rubecula*). Vogelwarte 23: 71–77. **[316]**

Mertens, J.A.L. 1969. The influence of brood size on the energy metabolism and water loss of nestling Great Tits *Parus major major*. Ibis 111: 11–16. **[442]**

Mewaldt, L.R. 1964. California sparrows return from displacement to Maryland. Science 146: 941–942. **[310, 311]**

Mewaldt, L.R., and J.R. King. 1978. Latitudinal variation of postnuptial molt in Pacific coast White-crowned Sparrows. Auk 95: 168–179. **[265, 266]**

Meyburg, B. 1974. Sibling aggression and mortality among nestling eagles. Ibis 116: 224–228. **[442]**

Michener, M.C., and C. Walcott. 1967. Homing of single pigeons—Analysis of tracks. J. Exp. Biol. 47: 99–131. **[312]**

Miller, A.H. 1959. Response to experimental light increments by Andean Sparrows from an equatorial area. Condor 61: 344–347. **[273]**

Miller, A.H. 1962. Bimodal occurrence of breeding in an equatorial sparrow. Proc. Natl. Acad. Sci. USA 48: 396–400. **[282]**

Miller, D.B. 1977. Two-voice phenomenon in birds: Further evidence. Auk 94: 567–572. **[240]**

Miller, E.H. 1988. Description of bird behavior for comparative purposes. Current Ornithology 5: 347–394. **[232]**

Millikan, G.C., and R.I. Bowman. 1967. Observations on Galápagos tool-using finches in captivity. Living Bird 6:23–41. **[164]**

Minton, C.D.T. 1968. Pairing and breeding of Mute Swans. Wildfowl 19: 41–60. **[407]**

Mock, D.W. 1975. Social behavior of the Boat-billed Heron. Living Bird 14: 185–214. **[215]**

Mock, D.W. 1976. Pair-formation displays of the Great Blue Heron. Wilson Bull. 88: 185-230. **[218, 219]**

Mock, D.W. 1978. Pair-formation displays of the Great Egret. Condor 80: 159–172. **[219]**

Mock, D.W. 1980. Communication strategies of Great Blue Herons and Great Egrets. Behaviour 72: 156–170. **[219]**

Mock, D.W. 1984a. Siblicidal aggression and resource monopolization in birds. Science 225: 731–733. **[442,443]**

Mock, D.W. 1984b. Infanticide, siblicide, and avian nestling mortality, pp. 3–30. *In* Infanticide: Comparative and Evolutionary Perspectives (G. Hausfater and S.B. Hrdy, Eds.). New York: Aldine. **[442]**

Møller, A.P. 1990. Effects of a haematophagous mite on the barn swallow (*Hirundo rustica*): A test of the Hamilton and Zuk hypothesis. Evolution 44: 771–784. **[229]**

Møller, A.P. 1991. Parasites, sexual ornaments, and mate choice in barn swallows, pp. 328–343. *In* Bird–Parasite Interactions (J.E. Loye and M. Zuk, Eds.). New York: Oxford University Press. **[229]**

Monastersky, R. 1991. The lonely bird. Science News 140: 104–105. **[30]**

Montgomerie, R.D., and P.J. Weatherhead. 1988. Risks and rewards of nest defense by parent birds. Q. Rev. Biol. 63: 167–187. **[342]**

Moore, F.R. 1977. Geomagnetic disturbance and the orientation of nocturnally migrating birds. Science 196: 682–684. **[319]**

Moore, F.R. 1978. Interspecific agression: Toward whom should a mockingbird be aggressive? Behav. Ecol. Sociobiol. 3: 173–176. **[333]**

Moore, F.R. 1982. Sunset and the orientation of a nocturnal bird migrant: A mirror experiment. Behav. Ecol. Sociobiol. 10: 153–155. **[323]**

Moore, W.S. 1977. An analysis of narrow hybrid zones in vertebrates. Q. Rev. Biol. 52: 263–278. **[545]**

Moore, W.S., and D.B. Buchanan. 1985. Stability of the Northern Flicker hybrid zone in historical times: Implications for adaptive speciation theory. Evolution 39: 135–151. **[546]**

Moors, P.J. 1985. Conservation of Island Birds. Cambridge: International Council for Bird Preservation. **[585]**

Moreau, R.E. 1961. Problems of Mediterranean-Saharan migration. Ibis 92: 223–267. **[296]**

Moreau, R.E. 1966. The Bird Faunas of Africa and Its Islands. London: Academic Press. **[291, 558]**

Moreau, R.E. 1972. The Palaearctic-African Bird Migration Systems. London: Academic Press. **[287, 288, 296, 309]**

Morejohn, C.V. 1968. Breakdown of isolation mechanisms in two species of captive junglefowl (*Gallus gallus* and *Gallus sonneratii*). Evolution 22: 576–582. **[543]**

Moreno, J., A. Carlson, and R.V. Alatalo. 1988. Winter energetics of coniferous forest tits Paridae in the north: The implications of body size. Funct. Ecol. 2: 163–170. **[133]**

Morris, D. 1956. The feather postures of birds and the problem of the origin of social signals. Behaviour 9: 75–113. **[216]**

Morrison, M.L. 1986. Bird populations as indicators of environmental change. Current Ornithology 3: 429–451. **[588, 616]**

Morse, D.H. 1970. Ecological aspects of adaptive radiation in birds. Biol. Rev. Cambridge Philos. Soc. 50: 167–214. **[340]**

Morse, D.H. 1978. Structure and foraging patterns of tit flocks in an English woodland. Ibis 120: 298–312. **[340]**

Morse, D.H. 1980. Behavioral Mechanisms in Ecology. Cambridge, Mass.: Harvard University Press. **[164, 173, 344, 346]**

Morse D.H. 1989. American Warblers. Cambridge, Mass.: Harvard Univ. Press. **[691]**

Morton, E.S. 1973. On the evolutionary advantages and disadvantages of fruit eating in tropical birds. Am. Nat. 107: 8–22. **[438]**

Morton, E.S. 1975. Ecological sources of selection on avian sounds. Am. Nat. 109: 17–34. **[237]**

Morton, E.S. 1976. Vocal mimicry in the Thick-billed Euphonia. Wilson Bull. 88: 485–487. **[253]**

Morton, E.S. 1982. Grading, discreteness, redundancy, and motivation-structural rules, pp. 183–212. *In* Acoustic Communication in Birds, Vol. 1 (D.E. Kroodsma and E.H. Miller, Eds.). New York: Academic Press. **[251]**

Morton, E.S. 1986. Predictions from the ranging hypothesis for the evolution of long distance signals in birds. Behaviour 99: 65–86. **[243]**

Morton, E.S. 1991. Cuckoldry in the Condo. Zoogoer, May–June 1991: 11–15. **[410]**

Morton, E.S. 1992. What do we know about the future of migrant landbirds, pp. 579–589. *In* Ecology and Conservation of Neotropical Migrant Landbirds (J.M. Hagan, III, and D.W. Johnston, Eds). Washington, D.C.: Smithsonian Institution Press. **[595]**

Morton, E.S., and R. Greenberg. 1989. The outlook for migratory songbirds: "Future shock" for birders. Am. Birds 43: 178–183. **[595]**

Morton, E.S., L. Forman, and M. Braun. 1990. Extra pair fertilizations and the evolution of colonial breeding in purple martins. Auk 107: 275–283. **[410]**

Morton, M.L. 1979. Fecal sac ingestion in the Mountain White-crowned Sparrow. Condor 81: 72–77. **[446]**

Morton, M.L., and L.R. Mewaldt. 1962. Some effects of castration on a migratory sparrow (*Zonotrichia atricapilla*). Physiol. Zool. 35: 237–247. **[303]**

Moseley, L.J. 1979. Individual auditory recognition in the Least Tern (*Sterna albifrons*). Auk 96: 31–39. **[248]**

Moss, R. 1972. Food selection by Red Grouse (*Lagopus lagopus scoticus* (Lath.)) in relation to chemical composition. J. Anim. Ecol. 41: 411–428. **[169]**

Moss, R., G.R. Miller, and S.E. Allen. 1972. Selection of heather by captive red grouse in relation to the age of the plant. J. Appl. Ecol. 9: 771–781. **[169]**

Moynihan, M. 1955. Some aspects of reproductive behavior in the Black-headed Gull (*Larus ridibundus ridibundus* L.) and related species. Behav. Suppl. 4: 1–201. **[220]**

Moyhihan, M. 1962. The organization and probable evolution of some mixed species flocks of neotropical birds. Smithson. Misc. Collect. 143: 1–140. **[344]**

Moynihan, M. 1968. Social mimicry: Character convergence versus character displacement. Evolution 22: 315–331. **[345]**

Mueller, A.J. 1992. Inca Dove. *In* The Birds of North America, No. 28 (A. Poole, P. Stettenheim, and F. Gill, Eds.). Philadelphia: Academy of Natural Sciences; Washington, D.C.: American Ornithologists' Union. **[132]**

Mugaas, J.N., and J.R. King. 1981. Annual variation of daily energy expenditure by the Black-billed Magpie: A study of thermal and behavioral energetics. Studies in Avian Biology No. 5. **[365]**

Mulligan, J.A. 1966. Singing behavior and its development in the Song Sparrow, *Melospiza melodia.* Univ. Calif. Publ. Zool. No. 81. **[250]**

Mumme, R.L. 1992. Do helpers increase reproductive success? An experimental analysis in the Florida Scrub Jay. Behav. Ecol. Sociobiol. 31: 319–328. **[468]**

Munn, C.A., and J.W. Terborgh. 1979. Multi-species territoriality in Neotropical foraging flocks. Condor 81: 338–347. **[343, 344]**

Murphy, E.C. 1985. Bergmann's rule, seasonality, and geographic variation in body size of House Sparrows. Evolution 39: 1318–1327. **[130]**

Murphy, E.C., and E. Haukioja. 1986. Clutch size in nidicolous birds. Current Ornithology 4: 141–180. **[496, 505]**

Murphy, M.E., and J.R. King. 1989. Sparrows discriminate between diets differing in valine or lysine concentrations. Physiol. Behav. 45: 423–430. **[170]**

Murphy, M.E., and J.R. King. 1991. Nutritional aspects of avian molt. Acta Congr. Int. Ornithol. XX,1: 2186–2194. **[279]**

Murphy, M.E., and J.R. King. 1992. Energy and nutrient use during moult by White-crowned Sparrows *Zonotrichia leucophrys gambelii.* Ornis Scand. 23: 304–313. **[279, 280, 285]**

Murray, B.G., Jr. 1969. A comparative study of the LeConte's and Sharp-tailed Sparrows. Auk 86: 199–231. **[546]**

Murray, B.G., Jr. 1971. The ecological consequences of interspecific territorial behavior in birds. Ecology 52: 414–423. **[333, 546]**

Murray, B.G., Jr. 1979. Population Dynamics. New York: Academic Press. **[510]**

Murray, B.G., Jr. 1989. A critical review of the transoceanic migration of the Blackpoll Warbler. Auk 106: 8–17. **[294]**

Murray, B.G., Jr. 1992a. Sir Isaac Newton and the evolution of clutch size in birds: A defense of the hypothetico-deductive method in ecology and evolutionary biology, pp. 143–180. *In* Beyond Belief: Randomness, Prediction, and Explanation in Science (J.L. Casti and A. Karlqvist, Eds.). Boca Raton, Fla.: CRC Press. **[496, 505]**

Murray, B.G., Jr. 1992b. The evolution of clutch size: A reply to Wootton, Young, and Winkler. Evolution 46: 1584–1587. **[496, 502]**

Murray, B.G., Jr., and V. Nolan, Jr. 1989. The evolution of clutch size. I. An equation for predicting clutch size. Evolution 43: 1699–1705. **[496, 502]**

Murray, B.G., Jr., J.W. Fitzpatrick, and G.E. Woolfenden. 1989. The evolution of clutch size. II. A test of the Murray-Nolan equation. Evolution 43: 1706–1711. **[502]**

Murton, R.K. 1965. The Wood Pigeon. London: Collins. **[398]**

Murton, R.K. 1967. The significance of endocrine stress in population control. Ibis 109: 622–623. **[334]**

Murton, R.K. 1971a. The significance of a specific search image in the feeding behaviour of the wood-pigeon. Behaviour 40: 10–42. **[165]**

Murton, R.K. 1971b. Why do some bird species feed in flocks? Ibis 113: 534–536. **[339]**

Murton, R.K., A.J. Isaacson, and N.J. Westwood. 1971. The significance of gregarious feeding behavior aand adrenal stress in a population of wood pigeons *Columba palumbus.* J. Zool. London 165: 53–84. **[334]**

Myers, J.P. 1981a. Cross-seasonal interactions in the evolution of sandpiper social systems. Behav. Ecol. Sociobiol. 8: 195–202. **[405]**

Myers, J.P. 1981b. A test of three hypotheses for latitudinal segregation of the sexes in wintering birds. Can. J. Zool. 59: 1527–1534. **[405]**

Myers, J.P., P.G. Connors, and F.A. Pitelka. 1979. Territory size in wintering Sanderlings: The effects of prey abundance and intruder density. Auk 96: 551–561. **[331, 335]**

Myers, J.P., R.I.G. Morrison, P.Z. Antas, B.A. Harrington, T.E. Lovejoy, M. Sallaberry, S.E. Senner, and A. Tarak. 1987. Conservation strategy for migratory species. Am. Sci. 75: 19–26. **[599]**

Myers, J.P., J.L. Maron, and M. Sallaberry. 1985. Going to extremes: Why do Sanderlings migrate to the Neotropics? Ornithol. Monogr. 36: 520–532. **[289, 331]**

Nagy, K.A. 1989. Field bioenergetics: Accuracy of models and methods. Physiol. Zool. 62: 237–252. **[143]**

National Wildlife Federation. 1994. Conservation Directory. **[614, 616]**

Nelson, J.B. 1964. Factors influencing clutch-size and chick growth in the North Atlantic Gannet *Sula bassana.* Ibis 106: 63–77. **[499]**

Nelson, J.B. 1969. The breeding ecology of the Red-footed Booby in the Galápagos. J. Anim. Ecol. 38: 181–198. **[398]**

Newton, I. 1972. Finches. London: Collins. **[335]**

Newton, I. 1979. Population Ecology of Raptors. Berkhamsted, England: T. and A.D. Poyser. **[588]**

Newton, I. 1980. The role of food in limiting bird numbers. Ardea 68: 11–30. **[513, 571]**

Newton, I., Ed. 1989. Lifetime Reproduction in Birds. New York: Academic Press. **[490]**

Nice, M.M. 1943. Studies in the life history of the Song Sparrow, II. Trans. Linn. Soc. N.Y. No. 6. **[449]**

Nice, M.M. 1962. Development of behavior in precocial birds. Trans. Linn. Soc. N.Y. 8: 1–211. **[432]**

Nicolai, J. 1964. Der Brutparasitismus der Viduinae als ethologisches Problem: Pragungsphaenomene als Faktoren der Rassen- und Artbuildung. Z. Tierpsychol. 21(2): 129–204. **[462]**

Nicolai, J. 1974. Mimicry in parasitic birds. Sci. Am. 231(4): 92–98. **[462]**

Nielsen, B. 1993. The Parrot Man. Caribbean Travel and Life, Am. Birds 47 (March–April): 48–51. **[612]**

Nilsson, G. 1990. Importation of Birds into the United States in 1986–1988. Washington, D.C.: Animal Welfare Institute. **[600]**

Nisbet, I.C.T. 1973. Courtship-feeding, egg-size and breeding success in Common Terns. Nature London 241: 141–142. **[408]**

Nishiyama, H. 1955. Studies of the accessory reproductive organs in the cock. Reprinted from J. Fac. Agric. Kyushu Univ. 10 (3). **[359]**

Nixon, K.G., and Q.D. Wheeler. 1990. An amplification of

the phylogenetic species concept. Cladistics 6: 211–223. **[529]**

Nolan, V., Jr. 1963. Reproductive success of birds in a deciduous scrub habitat. Ecology 44: 305–313. **[380]**

Nolan, V., Jr., and C.F. Thompson. 1975. The occurrence and significance of anomalous reproductive activities in two North American non-parasitic cuckoos *Coccyzus* spp. Ibis 117: 496–503. **[459]**

Noon, B.R. 1981. The distribution of an avian guild along a temperate elevational gradient: The importance and expression of competition. Ecol. Monogr. 51: 105–124. **[570]**

Norberg, R.A. 1981. Optimal flight speed in birds when feeding young. J. Anim. Ecol. 50: 473–477. **[100]**

Norman, D.M., J.R. Mason, and L. Clark. 1992. Capsaicin effects on consumption of food by Cedar Waxwings and House Finches. Wilson Bull. 104: 549–551. **[199]**

Norton-Griffiths, M. 1969. The organization, control and development of parental feeding in the oystercatcher (*Haematopus ostralegus*). Behaviour 34: 55–114. **[169]**

Nottebohm, F. 1967. The role of sensory feedback in the development of avian vocalizations, pp. 265–280. *In* Proc. XIV Int. Ornithol. Congr. (D.W. Snow, Ed.). Oxford: Blackwell. **[255]**

Nottebohm, F. 1975. Vocal behavior in birds, pp. 287–332. *In* Avian Biology, Vol. 5 (D.S. Farner, J.R. King, and K.C. Parkes, Eds.). New York: Academic Press. **[255]**

Nottebohm, F. 1980. Neural pathways for song control: A good place to study sexual dimorphism, hormonal influences, hemispheric dominance and learning, pp. 642–647. *In* Acta XVII Congressus Internationalis Ornithologici (R. Nöhring, Ed.). Berlin: Verlag der Deutschen Ornithologen-Gesellschaft. **[183]**

Nottebohm, F., and M.E. Nottebohm. 1971. Vocalizations and breeding behaviour of surgically deafened Ring Doves (*Streptopelia risoria*). Anim. Behav. 19: 313–327. **[253]**

Nottebohm, F., and M.E. Nottebohm. 1978. Relationship between song repertoire and age in the canary *Serinus canarius*. Z. Tierpsychol. 46. 298–305. **[250]**

Nottebohm, F., S. Kasparian, and C. Pandazis. 1981. Brain space for a learned task. Brain Res. 213: 99–110. **[184]**

Nowicki, S. 1987. Vocal tract resonances in oscine bird sound production: Evidence from birdsongs in a helium atmosphere. Nature London 325: 53–55. **[240, 241]**

Nowicki, S., and R. Capranica. 1986. Bilateral syringeal interaction in vocal production of an oscine bird sound. Science 231: 1297–1299. **[240]**

O'Connor, R.J. 1975. The influence of brood size upon metabolic rate and body temperature in nestling Blue Tits *Parus caeruleus* and House Sparrows *Passer domesticus*. J. Zool. London 175: 391–403. **[442]**

O'Connor, R.J. 1977. Growth strategies in nestling passerines. Living Bird 16: 209–238. **[440]**

O'Connor, R.J. 1978. Brood reduction in birds: Selection for fratricide, infanticide and suicide? Anim. Behav. 26: 79–96. **[442, 503]**

O'Connor, R. 1984. The Growth and Development of Birds. Chichester: John Wiley & Sons. **[451]**

O'Connor, R.J., and A. Cawthorne. 1982. How Britain's birds survived the winter. New Sci. 93: 786–788. **[514]**

O'Donald, P. 1983. The Arctic Skua: A Study of the Ecology and Evolution of a Seabird. Cambridge: Cambridge University Press. **[539]**

O'Donald, P. 1987. Polymorphism and sexual selection in the Arctic Skua, pp. 433–450. *In* Avian Genetics (F. Cooke and P.A. Buckley, Eds.). New York: Academic Press. **[539]**

O'Hara, R.J. 1993. Systematic generalization, historical fate, and the species problem. Syst. Biol. 43: 231–246. **[528]**

Oatley, T.B., and L.G. Underhill. 1993. pp. 77–90. *In* Marked Individuals in the Study of Bird Populations (J.D. Lebreton and P.M. North, Eds.). Boston: Birkhauser Verlag. **[488]**

Odum, E.P., and C.E. Connell. 1956. Lipid levels in migrating birds. Science 123: 892–894. **[297, 302]**

Odum, E.P., and J.D. Perkinson, Jr. 1951. Relation of lipid metabolism to migration in birds: Seasonal variation in body lipids of the migratory White-throated Sparrow. Physiol. Zool. 24: 216–230. **[299]**

Office of Migratory Bird Management. 1990. Conservation of avian diversity in North America. Washington, D.C.: U.S. Fish and Wildlife Service. **[602]**

Ohmart, R.D., and R.C. Lasiewski. 1971. Roadrunners: Energy conservation by hypothermia and absorption of sunlight. Science 172: 67–69. **[127]**

Olsen, M.W. 1960. Performance record of a parthenogenetic turkey male. Science 132: 1661. **[360]**

Olson, J.M., W.R. Dawson, and J.J. Camilliere. 1988. Fat from Black-capped Chickadees: Avian brown adipose tissue? Condor 90: 529–537. **[132]**

Olson, S.L. 1971. Taxonomic comments on the Eurylaimidae. Ibis 113: 507–516. **[675, 677]**

Olson, S.L. 1973. Evolution of the rails of the South Atlantic islands (Aves: Rallidae). Smithson. Contrib. Zool. No. 152. **[110]**

Olson, S.L. 1976. Oligocene fossils bearing on the origins of the Todidae and the Momotidae (Aves: Coraciiformes). Smithson. Contrib. Paleobiol. 27: 111–119. **[40]**

Olson, S.L. 1978a. Multiple origins of the Ciconiiformes. Proc. Colonial Waterbird Group 1978: 165–170. **[635]**

Olson, S.L. 1978b. A paleontological perspective of West Indian birds and mammals, pp. 99–117. *In* Zoogeography in the Caribbean (F.B. Gill, Ed.). Spec. Publ. Acad. Nat. Sci. Philadelphia No. 13. **[663]**

Olson, S.L. 1983. Evidence for a polyphyletic origin of the Piciformes. Auk 100: 126–133. **[673]**

Olson, S.L. 1985. The fossil record of birds. Avian Biology 5: 79–238. **[40, 43, 586, 635, 649]**

Olson, S.L., and A. Feduccia. 1979. Flight capability and the pectoral girdle of *Archaeopteryx*. Nature London 278(5701): 247–248. **[27]**

Olson, S.L., and A. Feduccia. 1980a. Relationships and evolution of flamingos. Smithson. Contrib. Zool. No. 316. **[54, 55, 663]**

Olson, S.L., and A. Feduccia. 1980b. *Presbyornis* and the origin of the Anseriformes (Aves: Charadriomorphae). Smithson. Contrib. Zool. No. 323. **[631, 640]**

Olson, S.L., and H.F. James. 1982. Fossil birds from the Hawaiian Islands: Evidence for wholesale extinction by man before western contact. Science 217: 633–635. **[577, 586]**

Olson, S.L., and D.W. Steadman. 1981. The relationships of the Pedionomidae (Aves: Charadriiformes). Smithson. Contrib. Zool. No. 337. **[643, 648]**

Orians, G.H. 1980. Some Adaptations of Marsh-nesting Blackbirds. Princeton, N.J.: Princeton Univ. Press **[691]**

Orians, G.H., and G.M. Christman. 1968. A comparative study of the behavior of Red-winged, Tri-colored, and Yellow-headed Blackbirds. Univ. Calif. Publ. Zool. 84: 1–81. **[224]**

Orians, G.H., and M.F. Willson. 1964. Interspecific territories of birds. Ecology 45: 736–745. **[333]**

Oring, L.W. 1982. Avian mating systems. Avian Biology 6: 1–91. **[404, 417, 424]**

Oring, L.W. 1986. Avian polyandry. Current Ornithology 3: 309–351. **[415, 416, 418, 424]**

Oring, L.W., and A.J. Fivizzani. 1991. Reproductive endocrinology of sex-role reversal. Acta Congr. Int. Congr. 2072–2081. **[416, 417]**

Oring, L.W., and M.L. Knudson. 1972. Monogamy and polyandry in the Spotted Sandpiper. Living Bird 11: 59–73. **[415]**

Österlöf, S. 1966. Kungsfågelns (*Regulus regulus*) flyttning. Vår Fågelvärld 25: 49–56. **[170]**

Ostrom, J.H. 1975. The origin of birds. Annu. Rev. Earth Planet Sci. 3: 55–77. **[29, 30]**

Ostrom, J.H. 1976. *Archaeopteryx* and the origin of birds. Biol. J. Linn. Soc. 8: 91–182. **[27]**

Otte, D., and J.A. Endler, Eds. 1989. Speciation and Its Consequences. Sunderland, Mass.: Sinauer Associates. **[552]**

Owen, D.F. 1963. Polymorphism in the Screech Owl in eastern North America. Wilson Bull. 75: 183–190. **[535]**

Packard, G.C., and M.J. Packard. 1980. Evolution of the cleidoic egg among reptilian antecedents of birds. Am. Zool. 20: 351–362. **[368]**

Page, G., and D.F. Whitacre. 1975. Raptor predation on wintering shorebirds. Condor 77: 73–83. **[340]**

Page, J., and E.S. Morton. 1989. Lords of the Air. Washington, D.C.: Smithsonian Books. **[20]**

Papi, F., and H.G. Wallraff, Eds. 1982. Avian Navigation. Berlin: Springer-Verlag. **[325]**

Papi, F., L. Fiore, V. Fiaschi, and S. Benvenuti. 1971. The influence of olfactory nerve section on the homing capacity of carrier pigeons. Monit. Zool. Ital. (N.S.) 5: 265–267. **[316]**

Papi, F., L. Fiore, V. Fiaschi, and S. Benvenuti. 1972. Olfaction and homing in pigeons. Monit. Zool. Ital. (N.S.) 6: 85–95. **[316]**

Parker, D.T. 1987. Evolutionary genetics of House Sparrows, pp. 381–406. *In* Avian Genetics (F. Cooke and P.A. Buckley, Eds.). New York: Academic Press. **[531]**

Parkes, K.C. 1966. Speculations on the origin of feathers. Living Bird 5: 77–86. **[28]**

Parkes, K.C. 1992. [Review of] Distribution and taxonomy of birds of the world (C.G. Sibley and B.L. Monroe, Jr., 1990). J. Field Ornithology 63: 228–235. **[618]**

Parkes, K.C. 1993. Note on taxonomy. Avian Biology 9: xxi–xxiii. **[xvi, 617]**

Parkes, K.C., and G.A. Clark, Jr. 1966. An additional character linking ratites and tinamous, and an interpretation of their monophyly. Condor 68: 459–471. **[621]**

Parmelee, D.F. 1992. Snowy Owl. *In* The Birds of North America, No. 10 (A. Poole, P. Stettenheim, and F. Gill, Eds.). Philadelphia: Academy of Natural Sciences; Washington, D.C.: American Ornithologists' Union. **[514]**

Parrish, J.W., J.A. Ptacek, and K.L. Will. 1984. The detection of near-ultraviolet light by nonmigratory and migratory birds. Auk 101: 53–58. **[191]**

Pasquier, R.F. 1983. The diversity of birdlife, pp. 18–48. *In* The Wonder of Birds (R.M. Poole, Ed.). Washington, D.C.: National Geographic Society. **[5]**

Patten, M.A. 1993. A probable gynandromorphic Black-throated Blue Warbler. Wilson Bull. 105: 695–698. **[352]**

Payne, R.B. 1969a. Overlap of breeding and molting schedules in a collection of African birds. Condor 71: 140–145. **[280]**

Payne, R.B. 1969b. Breeding seasons and reproductive physiology of Tricolored Blackbirds and Red-winged Blackbirds. Univ. Calif. Publ. Zool. 90: 1–115. **[385]**

Payne, R.B. 1972. Mechanisms and control of molt. Avian Biology 2: 103–155. **[85, 285]**

Payne, R.B. 1973a. Individual laying histories and the clutch size and numbers of eggs of parasitic cuckoos. Condor 75: 414–438. **[460]**

Payne, R.B. 1973b. Behavior, mimetic songs and song dialects, and relationships of the parasitic indigobirds (*Vidua*) of Africa. Ornithol. Monogr. No. 11. **[550]**

Payne, R.B. 1977a. The ecology of brood parasitism in birds. Annu. Rev. Ecol. Syst. 8: 1–28. **[459, 477]**

Payne, R.B. 1977b. Clutch size, egg size, and the consequences of single vs. multiple parasitism in parasitic finches. Ecology 58: 500–513. **[460]**

Payne, R.B. 1979. Sexual selection and intersexual differences in variance of breeding success. Am. Nat. 114: 447–452. **[224, 405]**

Payne, R.B. 1981a. Population structure and social behavior: Models for testing the ecological significance of song dialects in birds, pp. 108–120. *In* Natural Selection and Social Behavior: Recent Research and New Theory (R.D. Alexander and D.W. Tinkle, Eds.). New York: Chiron Press. **[257]**

Payne, R.B. 1981b. Song learning and social interaction in Indigo Buntings. Anim. Behav. 29: 688–697. **[257]**

Payne, R.B. 1982a. Species limits in the indigobirds (Ploceidae, *Vidua*) of West Africa: Mouth mimicry, song mimicry, and description of new species. Misc. Publ. Mus. Zool. Univ. Mich. No. 162. **[462, 463]**

Payne, R.B. 1982b. Ecological consequences of song matching: Breeding success and intraspecific song mimicry in Indigo Buntings. Ecology 63: 401–411. **[258]**

Payne, R.B. 1983. The social context of song mimicry: Song-matching dialects in Indigo Buntings (*Passerina cyanea*). Anim. Behav. 31: 788–805. **[257]**

Payne, R.B. 1986. Bird songs and avian systematics. Current Ornithology 3: 87–126. **[244, 246, 260]**

Payne, R.B. 1990. Natal dispersal, area effects, and effective population size. J. Field Ornithol. 61: 396–403. **[534]**

Payne, R.B., and K. Payne. 1977. Social organization and mating success in local song populations of Village Indigobirds, *Vidua chalybeata*. Z. Tierpsychol. 45: 113–173. **[258]**

Payne, R.B., and D.F. Westneat. 1988. A genetic and behavioral analysis of mate choice and song neighborhoods in Indigo Buntings. Evolution 42: 935–947. **[258]**

Payne, R.B., W.L. Thompson, K.L. Fiala, and L.L. Sweany. 1981. Local song traditions in Indigo Buntings: Cultural transmission of behavior patterns across generations. Behaviour 77: 199–221. **[257]**

Payne, R.S. 1971. Acoustic location of prey by Barn Owls (*Tyto alba*). J. Exp. Biol. 54: 535–573. **[195]**

Peaker, M., and J.L. Linzell. 1975. Salt Glands in Birds and Reptiles. Cambridge: Cambridge University Press. **[143]**

Pearson, O.P. 1950. The metabolism of hummingbirds. Condor 52: 145–152. **[302]**

Pearson, R. 1972. The Avian Brain. London: Academic Press. **[187]**

Pendergast, B.A., and D.A. Boag. 1971. Nutritional aspects of the diet of Spruce Grouse in central Alberta. Condor 73: 437–443. **[161]**

Pennycuick, C.J. 1960. The physical basis of astronavigation in birds: Theoretical considerations. J. Exp. Biol. 37: 573–593. **[324]**

Pennycuick, C.J. 1972. Soaring behaviour and performance of some East African birds, observed from a motor glider. Ibis 114: 178–218. **[104]**

Pennycuick, C.J. 1973. The soaring flight of vultures, pp. 38–45. *In* Birds (B.W. Wilson, Ed.). San Francisco: W.H. Freeman. **[104]**

Pennycuick, C.J. 1989. Bird Flight Performance. Oxford: Oxford University Press. **[113, 302]**

Pepperberg, I.M. 1981. Functional vocalizations by an African Grey Parrot (*Psittacus erithacus*). Z. Tierpsychol. 55: 139–160. **[179]**

Pepperberg, I.M. 1987. Acquisition of the same/different concept by an African Grey parrot (*Psittacus erithacus*): Learning with respect to categories of color, shape, and material. Anim. Learn. Behav. 15: 423–432. **[179]**

Pepperberg, I.M. 1988. Comprehension of "absence" by an African Grey Parrot: Learning with respect to questions of same/different. J. Exp. Anal. Behav. 50: 553–564. **[179]**

Pepperberg, I.M. 1991. Learning to communicate: The effects of social interaction. Perspect. Ethol. 9: 119–164. **[260, 449]**

Pepperberg, I.M. 1992. Proficient performance of a conjunctive, recursive task by an African Gray Parrot (*Psittacus erithacus*). J. Comp. Psychol. 106: 295–305. **[179]**

Perdeck, A.C. 1958. Two types of orientation in migrating Starlings *Sturnus vulgaris* L. and Chaffinches *Fringilla coelebs* L., as revealed by displacement experiments. Ardea 46: 1–37. **[320]**

Perdeck, A.C. 1967. Orientation of Starlings after displacement to Spain. Ardea 55: 194–202. **[320]**

Pernkopf, E., and J. Lehner. 1937. Vorderdarm. Vergleichende Beschreibung des Vorderdarmes bei den einzelnen Klassen der Kranioten, pp. 349–476. *In* Handbuch der Vergleichenden Anatomie der Wirbeltiere, Vol. 3 (L. Bolk, E. Göppert, E. Kallius, and W. Lubosch, Eds.). Berlin and Vienna: Urban and Schwarzenberg. **[158, 159]**

Perrins, C.M. 1965. Population fluctuations and clutch-size in the Great Tit, *Parus major* L. J. Anim. Ecol. 34: 601–647. **[365]**

Perrins, C.M. 1977. The role of predation in the evolution of clutch size, pp. 181–191. *In* Evolutionary Ecology (B. Stonehouse and C. Perrins, Eds.). Baltimore: University Park Press. **[500]**

Perrins, C.M. 1979. British Tits. London: Collins. **[521]**

Perrins, C.M. 1980. Survival of young Great Tits, *Parus major*, pp. 159–174. *In* Acta XVII Congressus Internationalis Ornithologici (R. Nöhring, Ed.). Berlin: Verlag der Deutschen Ornithologen-Gesellschaft. **[487]**

Perrins, C.M., and T.R. Birkhead. 1983. Avian Ecology. New York: Chapman and Hall. **[486]**

Perrins, C.M., and T.A. Geer. 1980. The effect of Sparrowhawks on tit populations. Ardea 68: 133–142. **[487]**

Perrins, C.M., and P.J. Jones. 1974. The inheritance of clutch size in the Great Tit (*Parus major* L.). Condor 76: 225–229. **[364]**

Perrins, C.M., and A.L.A. Middleton. 1985. The Encyclopedia of Birds. New York: Facts on File. **[20]**

Perrins, C.M., and D. Moss. 1975. Reproductive rates in the Great Tit. J. Anim. Ecol. 44: 695–706. **[497]**

Perrins, C.M., J.-D. Lebreton, and G.J.M. Hirons, Eds. 1991. Bird Population Studies. Oxford: Oxford University Press. **[373]**

Peters, J.L. 1931–1986. Check-list of Birds of the World. Cambridge, Mass.: Museum Comparative Zoology. **[47]**

Peterson, R.T. 1980. A Field Guide to the Birds. Boston: Houghton Mifflin (first edition, 1934). **[614]**

Pettingill, O.S. 1984. Ornithology, 4th ed. New York: Academic Press. **[234]**

Phillips, J.B., and S.C. Borland. 1992. Behavioural evidence for use of a light-dependent magnetoreception mechanism by a vertebrate. Nature London 359: 142–144. **[192]**

Phillips, R.E., and O.M. Youngren. 1981. Effects of denervation of the tracheo-syringeal muscles on frequency control in vocalizations in chicks. Auk 98: 299–306. **[241]**

Pianka, E.R. 1966. Latitudinal gradients in species diversity: A review of concepts. Am. Nat. 100: 33–46. **[558]**

Pierotti, R. 1987. Isolating mechanisms in seabirds. Evolution 41: 559–570. **[208]**

Pietrewicz, A.T., and A.C. Kamil. 1979. Search image formation in the Blue Jay (*Cyanocitta cristata*). Science 204: 1332–1333. **[165]**

Pimm, S.L., H.L Jones, and J. Diamond. 1988. On the risk of extinction. Am. Nat. 132: 757–785. **[586]**

Pinkowski, B.C. 1977. Breeding adaptations in the Eastern Bluebird. Condor 79: 289–302. **[499]**

Pitelka, F.A., R.T. Holmes, and S.F. MacLean, Jr. 1974. Ecology and evolution of social organization in Arctic sandpipers. Am. Zool. 14: 185–204. **[405]**

Pitelka, F.A., P.Q. Tomich, and G.W. Treichel. 1955. Ecological relations of jaegers and owls as lemming predators near Barrow, Alaska. Ecol. Monogr. 25: 85–117. **[329]**

Place, A.R. 1991. The avian digestive system—an optimally designed plug–flow chemical reactor with recycle? pp. 913–919. *In* Acta XX Congressus Internationalis Ornithologici (B.D. Ball, Ed.). Wellington, N.Z.: Ornithological Congress Trust Board. **[154, 158, 162]**

Place, A.R., and E.W. Stiles. 1992. Living off the wax of the land: Bayberries and Yellow-rumped Warblers. Auk 109: 334–345. **[163]**

Pleszczynska, W.K., and R.I.C. Hansell. 1980. Polygyny and decision theory: Testing of a model in Lark Buntings (*Calamospiza melanocorys*). Am. Nat. 116: 821–830. **[414]**

Pohlman, A.G. 1921. The position and functional interpretation of the elastic ligaments in the middle-ear region of *Gallus*. J. Morphol. 35: 229–262. **[193]**

Pool, R. 1988. Wishbones on display. Science 241: 1430–1431. **[95]**

Popper, K.R. 1972. Objective Knowledge: An Evolutionary Approach. Oxford: Oxford University Press. **[502]**

Portmann, A. 1961. Sensory organs: Skin, taste and olfaction, pp. 37–48. *In* Biology and Comparative Physiology of Birds, Vol. 2 (A.J. Marshall, Ed.). New York: Academic Press. **[118, 197]**

Portmann, A., and W. Stingelin. 1961. The central nervous system, pp. 1–36. *In* Biology and Comparative Physiology of Birds, Vol. 2 (A.J. Marshall, Ed.). New York: Academic Press. **[180]**

Post, W., and F. Enders. 1970. The occurrence of *Mallophaga* on two bird species occupying the same habitat. Ibis 112: 539–540. **[84]**

Post, W., and J.W. Wiley. 1977. Reproductive interactions of the Shiny Cowbird and the Yellow-shouldered Blackbird. Condor 79: 176–184. **[463]**

Powell, G.V.N. 1979. Structure and dynamics of interspecific flocks in a Neotropical mid-elevation forest. Auk 96: 375–390. **[343]**

Powell, G.V.N. 1985. Sociobiology and adaptive significance of interspecific foraging flocks in the Neotropics. Ornithol. Monogr. No. 36. **[345]**

Power, H.W., E.D. Kennedy, L.C. Romagnano, M.P. Lombardo, A.S. Hoffenberg, P.C. Stouffer, and T.R. McGuire. 1989. The parasitism insurance hypothesis: Why starlings leave space for parasitic eggs. Condor 91: 753–765. **[458]**

Prager, E.M., and A.C. Wilson. 1976. Congruency of phylogenies derived from different proteins. J. Mol. Evol. 9: 45–57. **[631]**

Prager, E.M., A.C. Wilson, D.T. Osuga, and R.E. Feeney. 1976. Evolution of flightless land birds on southern continents: Transferrin comparison shows monophyletic origin of ratites. J. Mol. Evol. 8: 283–294. **[621]**

Prange, H.D., and K. Schmidt-Nielsen. 1970. The metabolic cost of swimming in ducks. J. Exp. Biol. 53: 763–777. **[126]**

Pregill, G.K., and S.L. Olson. 1981. Zoogeography of West Indian vertebrates in relation to Pleistocene climatic cycles. Annu. Rev. Ecol. Syst. 12: 75–98. **[574, 586]**

Prinzinger, R., and I. Hänssler. 1980. Metabolism–weight relations in some small nonpasserine birds. Experientia 36: 1299–1300. **[124]**

Procter, D.L.C. 1975. The problem of chick loss in the South Polar Skua *Catharacta maccormicki*. Ibis 117: 452–459. **[442]**

Proctor-Gray, E., and R.T. Holmes. 1981. Adaptive significance of delayed attainment of plumage in male American Redstarts: Tests of two hypotheses. Evolution 35: 742–751. **[495]**

Pruett-Jones, S.G., M.A. Pruett-Jones, and H.I. Jones. 1991. Parasites and sexual selection in a New Guinea avifauna. Current Ornithology 8: 213–245. **[229, 232]**

Prum, R.O. 1988. Phylogenetic interrelationships of the barbets (Aves: Capitonidae) and toucans (Aves: Ramphastidae) based on morphology with comparisons to DNA × DNA hybridization. Zool. J. Linn. Soc. 92: 313–343. **[673]**

Prum, R.O. 1990. Phylogenetic analysis of the evolution of display behavior in the Neotropical manakins (Aves: Pipridae). Ethology 84: 202–231. **[214]**

Prum, R.O. 1993. Phylogeny, biogeography, and evolution of the broadbills (Eurylaimidae) and asities (Philepittidae), based on morphology. Auk 110: 304–324. **[677]**

Prum, R.O., R.L. Morrison, and G.R. Ten Eyck. 1994. Structural color production by constructive reflection from ordered collagen arrays in a bird (*Philepitta castanea*: Eurylaimidae). J. Morphol. in press. **[89]**

Prys-Jones, O.E. 1973. Interactions between gulls and eiders in St. Andrews Bay, Fife. Bird Study 20: 311–313. **[335]**

Pugesek, B.H. 1981. Increased reproductive effort with age in the California Gull (*Larus californicus*). Science 212: 822–823. **[493, 494]**

Pugh, G.J.F., and M.D. Evans. 1970. Keratinophilic fungi associated with birds. II: Physiological studies. Trans. Br. Mycol. Soc. 54: 241–250. **[77]**

Pulliam, H.R., and G.S. Mills. 1977. The use of space by wintering sparrows. Ecology 58: 1393–1399. **[556]**

Pulliam, H.R., and T.H. Parker. 1979. Population regulation of sparrows. Fortschr. Zool. 25: 137–147. **[516]**

Pumphrey, R.J. 1961. Sensory organs: Hearing, pp. 69–86. *In* Biology and Comparative Physiology of Birds, Vol. 2 (A.J. Marshall, Ed.). New York: Academic Press. **[198]**

Pyke, G.H. 1981. Optimal travel speeds of animals. Am. Nat. 118: 475–487. **[100]**

Quay, W.B. 1985. Cloacal sperm in spring migrants: Occurrence and interpretation. Condor 87: 273–280. **[410]**

Quay, W.B. 1989. Insemination of Tennessee warblers during spring migration. Condor 91: 660–670. **[410]**

Quinn, T.W., and B.N. White. 1987. Analysis of DNA sequence variation, pp. 163–198. *In* Avian Genetics (F. Cooke and P.A. Buckley, Eds.). New York: Academic Press. **[57]**

Rahn, H., and A. Ar. 1974. The avian egg: Incubation time and water loss. Condor 76: 147–152. **[399]**

Rahn, H., and A. Ar. 1980. Gas exchange of the avian egg: Time, structure, and function. Am. Zool. 20: 477–484. **[372]**

Rahn, H., A. Ar, and C.V. Paganelli. 1979. How bird eggs breathe. Sci. Am. 240(2): 46–55. **[372, 427]**

Rahn, H., T. Ledoux, C.V. Paganelli, and A.H. Smith. 1982. Changes in eggshell conductance after transfer of hens from an altitude of 3800 m to 1200 m. J. Appl. Physiol. 53: 1429–1431. **[373]**

Raikow, R.J. 1976. The origin and evolution of the Hawaiian honeycreepers (Drepanididae). Living Bird 15: 75–117. **[9]**

Raikow, R.J. 1978. Appendicular myology and relationships of the New World nine-primaried oscines (Aves: Passeriformes). Bull. Carnegie Mus. Nat. Hist. No. 7. **[57]**

Raikow, R.J. 1982. Monophyly of the Passeriformes: Test of a phylogenetic hypothesis. Auk 99: 431–445. **[50]**

Raikow, R.J. 1984. Hindlimb myology and phylogenetic position of the New Zealand wrens. Am. Zool. 24: 446. **[675]**

Raikow, R.J. 1985. Problems in avian classification. Current Ornithology 2: 187–212. **[62]**

Raikow, R.J. 1986. Why are there so many kinds of passerine birds? Syst. Zool. 35: 255–259. **[675]**

Raikow, R.J. 1987. Hindlimb myology and evolution of the Old World suboscine passerine birds (Acanthisittidae, Pittidae, Philepittidae, Eurylaimidae). Ornithol. Monogr. No. 41. **[677]**

Raikow, R.J. 1988. The analysis of evolutionary success. Syst. Zool. 37: 76–79. **[675]**

Raikow, R.J., and J. Cracraft. 1983. Monophyly of the Piciformes: A reply to Olson. Auk 100: 134–138. **[673]**

Rappole, J.H., and D.W. Warner. 1980. Ecological aspects of migrant bird behavior in Veracruz, Mexico, pp. 353–393. *In* Migrant Birds in the Neotropics (A. Keast and E.S. Morton, Eds.). Washington, D.C.: Smithsonian Institution Press. **[309]**

Rayner, J.M.V. 1985. Flight, speeds of, pp. 224–226. *In* A Dictionary of Birds (B. Campbell and E. Lack, Eds.). Staffordshire, England: Poyser. **[100]**

Rayner, J.M.V. 1988. Form and function in avian flight. Current Ornithology 5: 1–66. **[113]**

Raynor, G.S. 1956. Meteorological variables and the northward movement of nocturnal landbird migrants. Auk 73: 153–175. **[304]**

Rayner, J.M.V. 1988. Form and function in avian flight. Current Ornithology 5: 1–66. **[109]**

Reed, J.M., J.R. Walters, T.E. Emigh, and D.E. Seaman. 1993. An effective population size in Red-cockaded Woodpeckers: Populations and model differences. Conservation Biology 7: 302–308. **[308]**

Regal, P.J. 1975. The evolutionary origin of feathers. Q. Rev. Biol. 50: 35–66. **[28]**

Regal, P.J. 1977. Ecology and evolution of flowering plant dominance. Science 196: 622–629. **[36]**

Reinecke, K.J. 1979. Feeding ecology and development of juvenile Black Ducks in Maine. Auk 96: 737–745. **[438]**

Rensch, B. 1947. Neure probleme der Abstammungslehne. Stuttgart: Ferdinand Enke Verlag. **[462]**

Reyer, H.-U. 1980. Flexible helper structure as an ecological adaptation in the Pied Kingfisher (*Ceryle rudis rudis* L.). Behav. Ecol. Sociobiol. 6: 219–227. **[469]**

Rhijn, J.G. van. 1973. Behavioural dimorphism in male Ruffs, *Philomachus pugnax* (L.). Behaviour 47: 153–229. **[210]**

Rice, W.R. 1982. Acoustical location of prey by the Marsh Hawk: Adaptation to concealed prey. Auk 99: 403–413. **[193]**

Richards, D.G. 1981. Estimation of distance of singing conspecifics by the Carolina Wren. Auk 98: 127–133. **[251]**

Richardson, W.J. 1978. Timing and amount of bird migration in relation to weather: A review. Oikos 30: 224–272. **[302, 304]**

Richdale, L.E. 1951. Sexual Behavior of Penguins. Lawrence: University of Kansas Press. **[399]**

Richdale, L.E. 1957. A Population Study of Penguins. Oxford: Clarendon Press. **[495]**

Ricklefs, R.E. 1968. Patterns of growth in birds. Ibis 110: 419–451. **[439]**

Ricklefs, R.E. 1969a. The nesting cycle of songbirds in tropical and temperate regions. Living Bird 8: 165–175. **[491]**

Ricklefs, R.E. 1969b. An analysis of nesting mortality in birds. Smithson. Contrib. Zool. 9: 1–48. **[491]**

Ricklefs, R.E. 1973a. Patterns of growth in birds. II: Growth rate and mode of development. Ibis 115: 177–201.

Ricklefs, R.E. 1973b. Fecundity, mortality, and avian demography, pp. 366–435. *In* Breeding Biology of Birds (D.S. Farner, Ed.). Washington, D.C.: National Academy of Sciences. **[485, 486, 509]**

Ricklefs, R.E. 1974. Energetics of reproduction in birds, pp. 152–292. *In* Avian Energetics (R.A. Paynter, Ed.). Publ. Nuttall Ornithol. Soc. No. 15. **[161, 278, 364, 365, 374, 433, 454]**

Ricklefs, R.E. 1975. Dwarf eggs laid by a Starling. Bird-Banding 46: 169. **[363]**

Ricklefs, R.E. 1976a (1st ed.), 1993 (new ed.). The Economy of Nature. New York: W.H. Freeman. **[576]**

Ricklefs, R.E. 1976b. Growth rates of birds in the humid New World Tropics. Ibis 118: 179–207. **[438]**

Ricklefs, R.E. 1977. A note on the evolution of clutch size in altricial birds, pp. 193–214. *In* Evolutionary Ecology (B. Stonehouse and C. Perrins, Eds.). London: Macmillan. **[497, 499, 500]**

Ricklefs, R.E., Ed. 1978. Report of the advisory panel on the California Condor. Natl. Audubon Soc. Conserv. Rep. 6: 1–27. **[608]**

Ricklefs, R.E. 1979a. Adaptation, constraint, and compromise in avian postnatal development. Biol. Rev. Cambridge Philos. Soc. 54: 269–290. **[434, 440, 563]**

Ricklefs, R.E. 1979b. Patterns of growth in birds. V. A comparative study of development in the Starling, Common Tern, and Japanese Quail. Auk 96: 10–30. **[436, 440, 441]**

Ricklefs, R.E. 1979c (2nd ed.), 1990 (3rd ed.). Ecology. New York: Chiron Press and W.H. Freeman. **[136, 464, 510]**

Ricklefs, R.E. 1980a. To the editor. Condor 82: 476–477. **[412]**

Ricklefs, R.E. 1980b. Geographical variation in clutch size among passerine birds: Ashmole's hypothesis. Auk 97: 38–49. **[500, 501]**

Ricklefs, R.E. 1983a. Avian postnatal development. Avian Biology 7: 1–83. **[432, 433, 434, 440, 451]**

Ricklefs, R.E. 1983b. Comparative avian demography. Current Ornithology 1: 1–32. **[506]**

Ricklefs, R.E. 1987. Community diversity: Relative roles of local and regional processes. Science 235: 167–171. **[558]**

Ricklefs, R.E. 1989a. Nest predation and the species diversity of birds. TREE 4: 184–186. **[380]**

Ricklefs, R.E. 1989b. Speciation and diversity: The integration of local and regional processes, pp. 599–622. *In* Speciation and Its Consequences (D. Otte and J.A. Endler, Eds.). Sunderland, Mass.: Sinauer Associates. **[380, 558]**

Ricklefs, R.E. 1993. Sibling competition, hatching asynchrony, incubation period, and life span in altricial birds. Current Ornithology 11: 199–276. **[399, 442, 482, 506]**

Ricklefs, R.E., and G.W. Cox. 1972. Taxon cycles in the West Indian avifauna. Am. Nat. 106: 195–219. **[575, 577]**

Ricklefs, R.E., and F.R. Hainsworth. 1969. Temperature regulation in nestling Cactus Wrens: The nest environment. Condor 71: 32–37. **[390]**

Ricklefs, R.E., and D. Schluter. 1994. Species Diversity and Communities. Chicago: University of Chicago Press. **[580]**

Ricklefs, R.E., and J. Travis. 1980. A morphological approach to the study of avian community organization. Auk 97: 321–338. **[557]**

Ricklefs, R.E., and S.C. White. 1981. Growth and energetics of chicks of the Sooty Tern (*Sterna fuscata*) and Common Tern (*S. hirundo*). Auk 98: 361–378. **[438]**

Ricklefs, R.E., S. White, and J. Cullen. 1980. Postnatal development of Leach's Storm-Petrel. Auk 97: 768–781. **[434, 437, 438, 440]**

Ridgely, R.S., and G. Tudor. 1994. Birds of South America, Vol 2, Suboscine Passerine Birds. Austin: University of Texas Press. **[677]**

Ridpath, M.G. 1972. The Tasmanian Native Hen, *Tribonyx mortierii*. Commonwealth Scientific and Industrial Research Organization. Wildlife Research 17: 53–90, 91–118. **[384]**

Ridley, M. 1992. Swallows and scorpionflies find symmetry is beautiful. Science 257(5068): 327–328. **[230]**

Ripley, S.D. 1957. Notes on the Horned Coot, *Fulica cornuta* Bonaparte. Postilla No. 30. **[382]**

Risebrough, R.W. 1986. Pesticides and bird populations. Current Ornithology 3: 397–427. **[588]**

Rising, J.D. 1983. The Great Plains hybrid zones. Current Ornithology 1: 131–157. **[544]**

Robbins, C.S., J.W. Fitzpatrick, and P.B. Hamel. 1992. A warbler in trouble: *Dendroica caerulea*, pp. 549–562. *In* Ecology and conservation of Neotropical migrant landbirds (J.M. Hagan, III, and D.W. Johnston, Eds.). Washington, D.C.: Smithsonian Institution Press. **[596]**

Robbins, C.S., J.R. Sauer, R.S. Greenberg, and S. Droege. 1989. Population declines in North American birds that migrate to the Neotropics. Proc. Natl. Acad. Sci. USA 86: 7658–7662. **[595]**

Robertson, R.J., and B.J. Stutchbury. 1988. Experimental evidence for sexually selected infanticide in tree swallows. Anim. Behav. 36: 749–753. **[457]**

Robinson, S.K. 1985. Coloniality in the Yellow-rumped Cacique as a defense against nest predators. Auk 102: 506–519. **[385]**

Robinson, S.K. 1986. Benefits, costs, and determinants of dominance in a polygynous oriole. Anim. Behav. 34: 241–255. **[410]**

Robinson, S.K. 1992. Population dynamics of breeding Neotropical migrants in a fragmented Illinois landscape, pp. 408–418. *In* Ecology and Conservation of Neotropical Migrant Landbirds (J.M. Hagan III and D.W. Johnston, Eds.). Washington, D.C.: Smithsonian Institution Press. **[463, 596]**

Roby, D., A.R. Place, and R.R. Ricklefs. 1986. Assimilation and deposition of wax esters in planktivorous seabirds. J. Exp. Zool. 239: 29–41. **[162]**

Rochon-Duvigneaud, A. 1950. Les yeux et la vision, pp. 221–242. *In* Oiseaux, Traité de Zoologie, Vol. 15 (P.-P. Grassé, Ed.). Paris: Masson et Cie. **[189]**

Rockwell, R.F., and G.F. Barrowclough. 1987. Gene flow and the genetic structure of populations, pp. 223–255. *In* Avian Genetics (F. Cooke and P.A. Buckley, Eds.). New York: Academic Press. **[534, 535]**

Rockwell, R.F., C.S. Findlay, F. Cooke, and J.A. Smith. 1985. Life history studies of the Lesser Snow Goose (*Anser caerulescens caerulescens*). IV. The selective value of plumage polymorphism: Net viability, the timing of maturation, and breeding propensity. Evolution 39: 178–189. **[209]**

Rohwer, F.C., and M.G. Anderson. 1988. Female-biased philopatry, monogamy, and the timing of pair formation in migratory waterfowl. Current Ornithology 5: 187–221. **[413]**

Rohwer, S. 1975. The social significance of avian winter plumage variability. Evolution 29: 593–610. **[338]**

Rohwer, S. 1977. Status signaling in Harris Sparrows: Some experiments in deception. Behaviour 61: 107–129. **[338]**

Rohwer, S. 1982. The evolution of reliable and unreliable badges of fighting ability. Am. Zool. 22: 531–546. **[338]**

Rohwer, S. 1986. Selection for adoption versus infanticide by replacement "mates" in birds. Current Ornithology 3: 353–395. **[457]**

Rohwer, S., and G.S. Butcher. 1988. Winter versus summer explanations of delayed plumage maturation in temperate passerine birds. Am. Nat. 131: 556–572. **[495]**

Romanoff, A.L., and A.J. Romanoff. 1949. The Avian Egg. New York: John Wiley & Sons. **[356, 357, 363, 367, 374]**

Root, R.B. 1967. The niche exploitation pattern of the Blue-gray Gnatcatcher. Ecol. Monogr. 37: 317–350. **[565]**

Root, T. 1988. Energy constraints on avian distributions and abundances. Ecology 69: 330–339. **[131]**

Rotenberry, J.T., and J.A. Wiens. 1980. Temporal variation in habitat structure and shrubsteppe bird dynamics. Oecologia Berlin 47: 1–9. **[563]**

Rothstein, S.I. 1975. An experimental and teleonomic investigation of avian brood parasitism. Condor 77: 250–271. **[464]**

Rothstein, S.I. 1982. Successes and failures in avian egg and nestling recognition with comments on the utility of optimality reasoning. Am. Zool. 22: 547–560. **[464]**

Roudybush, T.E., C.R. Grau, M.R. Petersen, D.G. Ainley, K.V. Hirsch, A.P. Gilman, and S.M. Patten. 1979. Yolk formation in some charadriiform birds. Condor 81: 293–298. **[356, 357]**

Rowan, W. 1929. Experiments in bird migration. I. Manipulation of the reproductive cycle: Seasonal histological changes in the gonads. Proc. Boston Soc. Nat. Hist. 39: 151–208. **[269, 270]**

Rowley, I. 1965. The life history of the Superb Blue Wren, *Malurus cyaneus*. Emu 64: 251–297. **[468]**

Rowley, I., and E. Russell. 1990. Demography of passerines in the temperate southern hemisphere, pp. 22–44. *In* Bird Population Studies (C.M. Perrins, J.-D. Lebreton, and G.J.M. Hirons, Eds.). New York: Oxford University Press.

Royama, T. 1966a. A re-interpretation of courtship feeding. Bird Study 13: 116–129. **[409]**

Royama, T. 1966b. Factors governing feeding rate, food requirement and brood size of nestling Great Tits *Parus major*. Ibis 108: 313–347. **[442]**

Royama, T. 1969. A model for the global variation of clutch size in birds. Oikos 20: 562–567. **[497]**

Ruben, J. 1991. Reptilian physiology and the flight capacity of *Archaeopteryx*. Evolution 45:1–17. **[26]**

Rüppell, G. 1977. Bird Flight. New York: Van Nostrand Reinhold. **[98, 113]**

Ryan, P.G., R.P. Wilson, and J. Cooper. 1987. Intraspecific mimicry and status signals in juvenile African Penguins. Behav. Ecol. Sociobiol. 20: 69–76. **[338]**

Saarela, S., R. Hissa, A. Pyörnilä, R. Harjula, M. Ojanen, and M. Orell. 1989. Do birds possess brown adipose tissue? Comp. Biochem. Physiol. 92A: 219–228. **[132]**

Safriel, U.N. 1975. On the significance of clutch size in nidifugous birds. Ecology 56: 703–708. **[433, 500]**

Sagitov, A.K. 1964. The vestibular apparatus and the degree of mobility of gallinaceous birds. Tr. Samark. Gos. Univ. 137: 5–38. **[198]**

Saiff, E.J. 1978. The middle ear of the skull of birds: The Pelecaniformes and Ciconiiformes. Zool. J. Linn. Soc. 63: 315–370. **[635]**

Saino, N., and S. Villa. 1992. Pair composition and reproductive success across a hybrid zone of Carrion Crows and Hooded Crows. Auk 109: 543–555. **[545]**

Saither, B.-E. 1990. Age-specific variation in reproductive performance of birds. Current Ornithology 7: 251–283. **[493]**

Salathe, T., Ed. 1991. Conserving Migratory Birds. ICBP Technical Publication No. 12. Cambridge: International Council for Bird Preservation. **[616]**

Samson, F.B. 1976. Territory, breeding density, and fall departure in Cassin's Finch. Auk 93: 477–497. **[335]**

Sargent, T.D. 1965. The role of experience in the nest building of the Zebra Finch. Auk 82: 48–61. **[430]**

Sauer, E.G.F. 1957. Die Sternenorientierung nächtlich ziehender Grasmücken (*Sylvia atricapilla, borin* und *curruca*). Z. Tierpsychol. 14: 29–70. **[313]**

Sauer, E.G.F. 1958. Celestial navigation by birds. Sci. Am. 199: 42–47. **[313]**

Sauer, E.G.F., and E.M. Sauer. 1966. The behavior and ecology of the South Africa Ostrich. Living Bird 5: 45–75. **[428]**

Saunders, A.A. 1959. Forty years of spring migration in southern Connecticut. Wilson Bull. 71: 208–219. **[303]**

Schardien, B.J., and J.A. Jackson. 1979. Belly-soaking as a thermoregulatory mechanism in nesting Killdeers. Auk 96: 604–606. **[396]**

Scheid, P. 1982. Respiration and control of breathing. Avian Biology 6: 406–453. **[146]**

Schenkel, R. 1956. Zur Deutung der Balzleistungen einiger Phasianiden und Tetraoniden. Ornithol. Beob. 53: 182–201. **[216]**

Schermuly, L., and R. Klinke. 1990. Infrasound sensitive neurones in the pigeon cochlear ganglion. J. Comp. Physiol. 166: 355–363. **[194]**

Schifter, H. 1985. Systematics and distribution of mousebirds (Coliidae), pp. 325–347. *In* Proc. Intl. Symp. African Vertebrates (K.-L. Schuchmann, Ed.). Bonn: Museum Koenig. **[655]**

Schjeldrup-Ebbe, T. 1935. Social behavior of birds, pp. 947–973. *In* A Handbook of Social Psychology (C.A. Murchison, Ed.). Worcester, Mass.: Clark University Press. **[336]**

Schlichte, H.-J. 1973. Untersuchungen über die Bedeutung optischer Parameter für das Heimkehrverhalten der Brieftaube. Z. Tierpsychol. 32: 257–280. **[312]**

Schlichte, H.-J., and K. Schmidt-Koenig. 1971. Zum Heimfindevermögen der Brieftaube bei erschwerter optischer Wahrnehmung. Naturwissenschaften 58: 329–330. **[312]**

Schluter, D., and P.R. Grant. 1984. Determinants of morphological patterns in communities of Darwin's finches. Am. Nat. 123: 175–196. **[569]**

Schmidt, I., and W. Rautenberg. 1975. Instrumental thermoregulatory behavior in pigeons. J. Comp. Physiol. A, 101: 225–235. **[135]**

Schmidt-Nielsen, K. 1983. Animal Physiology: Adaptation and Environment, 3rd ed. Cambridge: Cambridge University Press. **[121, 142, 143]**

Schneider, K.J. 1984. Dominance, predation, and optimal foraging in White-throated Sparrow flocks. Ecology 65; 1820–1827. **[334]**

Schoener, T.W. 1968. Sizes of feeding territories among birds. Ecology 49: 123–141. **[330]**

Schoener, T.W. 1971. Large-billed insectivorous birds: A precipitous diversity gradient. Condor 73: 154–161. **[330, 557, 559]**

Scholander, P.F., R. Hock, V. Walters, F. Johnson, and L. Irving. 1950. Heat regulation in some arctic and tropical mammals and birds. Biol. Bull. Woods Hole Mass. 99: 237–258. **[131]**

Schönwetter, M. 1960–1980. Handbuch der Oologie. Berlin: Akademie-Verlag. **[363]**

Schorger, A.W. 1955. The Passenger Pigeon, Its Natural History and Extinction. Norman: University of Oklahoma Press. **[583]**

Schreiber, R.W. 1980a. Nesting chronology of the Eastern Brown Pelican. Auk 97: 491–508. **[276, 277]**

Schreiber, R.W. 1980b. The Brown Pelican: An endangered species? BioScience 30: 742–747. **[588]**

Schreiber, R.W., and N.P. Ashmole. 1970. Sea-bird breeding seasons on Christmas Island, Pacific Ocean. Ibis 112: 363–394. **[276, 277]**

Schreiber, R.W., and E.A. Schreiber. 1984. Central Pacific seabirds and the El Niño Southern Oscillation: 1982 to 1983 perspectives. Science 225: 713–716. **[276]**

Schwartz, P. 1964. The Northern Waterthrush in Venezuela. Living Bird 3: 169–184. **[309]**

Schwartzkopff, J. 1973. Mechanoreception. Avian Biology 3: 417–477. **[193, 197]**

Scott, D.M., and C.D. Ankney. 1983. The laying cycle of Brown-headed Cowbirds: Passerine chickens? Auk 97: 677–683. **[460]**

Searcy, W.A., and P. Marler. 1981. A test for responsiveness to song structure and programming in female sparrows. Science 213: 926–928. **[246, 250]**

Searcy, W.A., and K. Yasukawa. 1983. Sexual selection and Red-winged Blackbirds. Am. Sci. 71: 166–174. **[404, 406, 413]**

Searcy, W.A., and K. Yasukawa. 1989. Alternative models of teriitorial polygyny in birds. Am. Nat. 134: 323–343. **[424]**

Searcy, W.A., P. Marler, and S. Peters. 1981. Species song discrimination in adult female song and swamp sparrows. Anim. Behav. 29: 997–1003. **[247]**

Seastedt, T.R., and S.F. Maclean, Jr. 1977. Calcium supplements in the diet of nestling Lapland Longspurs (*Calcarius lapponicus*) near Barrow, Alaska. Ibis 119: 531–533. **[438]**

Seebohm, H. 1885. A history of British birds with coloured figures of their eggs. London: Published for the author by R.H. Porter. **[367]**

Selander, R.K. 1971. Systematics and speciation in birds. Avian Biology 1: 57–147. **[552]**

Selander, R.K., and L.L. Kuich. 1963. Hormonal control and development of the incubation patch in icterids, with notes on behavior of cowbirds. Condor 65: 73–90. **[392]**

Semm, P., and C. Demaine. 1986. Neurophysiological properties of magnetic cells in the pigeon's visual system. J. Comp. Physiol. 59: 619–625. **[191]**

Senner, S. 1984. The model hawk law—1934 to 1972. Hawk Mountain News 62: 29–36. **[611]**

Senner, S.E. 1986. Federal research on migratory nongame birds: Is the United States Fish and Wildlife Service doing its job? Am. Birds 40: 413–417. **[602]**

Senner, S.E. 1988. Saving birds while they are still common: An historical perspective. Endangered Species UPDATE 5: 1–3. **[582]**

Senner, S.E. 1989. Hawk Mountain Sanctuary Association, Pennsylvania. Am. Birds 43: 248–253. **[610]**

Senner, S.E., and M.A. Howe. 1984. Conservation of Nearctic shorebirds, pp. 379–421. *In* Shorebirds: Breeding Behavior and Populations (J. Burger and B.L. Olla, Eds.). New York: Plenum. **[599, 605]**

Sereno, P.C., and R. Chenggang. 1992. Early evolution of avian flight and perching: New evidence from the lower Cretaceous of China. Science 255: 845–848. **[33, 34]**

Sexton, C. 1992. The Golden-cheeked Warbler. Birding 24: 373–376. **[594]**

Sheldon, F.H. 1987a. Phylogeny of herons estimated from DNA–DNA hybridization data. Auk 104: 97–108. **[59]**

Sheldon, F.H. 1987b. Rates of single-copy DNA evolution in herons. Mol. Biol. Evol. 4: 56–69. **[59]**

Sheldon, F.H., and A.H. Bledsoe. 1993. Avian molecular systematics, 1970s to 1990s. Annu. Rev. Ecol. Syst. 24: 243–278. **[61, 62]**

Shepard, J.M. 1975. Factors influencing female choice in the lek mating system of the Ruff. Living Bird 14: 87–111. **[210]**

Sherman, P.W., and M.L. Morton. 1988. Extra-pair fertilizations in mountain White-crowned Sparrows. Behav. Ecol. Sociobiol. 22: 413–420. **[409]**

Sherry, D.F. 1989. Food storing in the Paridae. Wilson Bull. 101: 289–293. **[185]**

Sherry, D.F. 1990. Evolutionary modification of memory and the hippocampus, pp. 401–421. *In* The Biology of Memory (L.R. Squire and E. Lindenlaub, Eds.). Stuttgart: F.K. Schattauer Verlag. **[185]**

Sherry, D.F. 1992. Memory, the hippocampus, and natural selection: studies of food-storing birds, pp. 521–532. *In* Neuropsychology of Memory, 2nd ed. (L.R. Squire and N. Butters, Eds.). New York: Guilford Press. **[185]**

Sherry, D.F., L.F. Jacobs, and S.J.C. Gaulin. 1992. Spatial memory and adaptive specialization of the hippocampus. TINS 15: 298–302. **[185]**

Sherry, D.F., A.L. Vaccarino, K. Buckenham, and R.S. Herz. 1989. The hippocampal complex of food-storing birds. Brain Behav. Evol. 34: 308–317. **[185]**

Shields, G.F. 1987. Chromosomal variation, pp. 79–104. *In* Avian Genetics (F. Cooke and P.A. Buckley, Eds.). New York: Academic Press. **[352, 374]**

Shields, G.F., and K.M. Helm-Bychowski. 1988. Mitochondrial DNA of birds. Current Ornithology 5: 273–295. **[60]**

Shilov, I.A. 1973. Heat Regulation in Birds. An Ecological–Physiological Outline. New Delhi: Amerind Publishing Co. **[437]**

Short, L.L., Jr. 1963. Hybridization in the wood warblers *Vermivora pinus* and *V. chrysoptera*, pp. 147–160. *In* Proc. XIII Int. Ornithol. Congr. (C.G. Sibley, Ed.). Lawrence, Kans.: Allen Press. **[546]**

Short, L.L., Jr. 1965. Hybridization in the flickers (*Colaptes*) of North America. Bull. Am. Mus. Nat. Hist. 129: 307–428. **[542]**

Shugart, G.W. 1988. Uterovaginal sperm-storage glands in sixteen species with comments on morphological differences. Auk 105: 379–385. **[360]**

Sibley, C.G. 1954. Hybridization in the red-eyed towhees of Mexico. Evolution 8: 252–290. **[542]**

Sibley, C.G. 1968. The relationships of the "wren-thrush", *Zeledonia coronata* Ridgway. Postilla No. 125. **[57]**

Sibley, C.G. 1970. A comparative study of the egg-white proteins of passerine birds. Peabody Mus. Nat. Hist. Yale Univ. Bull. No. 32. **[56]**

Sibley, C.G., and J.E. Ahlquist. 1972. A comparative study of the egg white proteins of non-passerine birds. Peabody Mus. Nat. Hist. Yale Univ. Bull. No. 39. **[625, 663]**

Sibley, C.G., and J.E. Ahlquist. 1973. The relationships of the Hoatzin. Auk 90: 1–13. **[57, 617, 618, 640, 649]**

Sibley, C.G., and J.E. Ahlquist. 1981a. The phylogeny and relationships of the ratite birds as indicated by DNA-DNA hybridization. Proc. Second Int. Congr. Syst. Evol. Biol.: 301–335. **[621]**

Sibley, C.G., and J.E. Ahlquist. 1981b. The relationships of the wagtails and pipits (Motacillidae) as indicated by DNA–DNA hybridization. Oiseau Rev. Fr. Ornithol. 51: 189–199. **[691]**

Sibley, C.G., and J.E. Ahlquist. 1982. The relationships of the Hawaiian Honeycreepers (*Drepaninini*) as indicated by DNA–DNA hybridization. Auk 99: 130–140. **[691]**

Sibley, C.G., and J.E. Ahlquist. 1983. Phylogeny and classification of birds based on the data of DNA–DNA hybridization. Current Ornithology 1: 245–292. **[686]**

Sibley, C.G., and J.E. Ahlquist. 1984. The relationships of the starlings (Sturnidae: *Sturnini*) and the mockingbirds (Sturnidae: *Mimini*). Auk 101: 230–243. **[683]**

Sibley, C.G., and J.E. Ahlquist. 1985. The relationships of some groups of African birds, based on comparisons of the genetic material, DNA, pp. 115–161. *In* Proceedings of the International Symposium on African Vertebrates: Systematics, Phylogeny and Evolutionary Biology (K.-L. Schuchmann, Ed.). Bonn: Zoologisches Forschungsinstitut und Museum Alexander Koenig. **[61]**

Sibley, C.G., and J.E. Ahlquist. 1990. Phylogeny and Classification of Birds. New Haven, Conn.: Yale University Press. **[47, 57, 62, 617ff.]**

Sibley, C.G., and C. Frelin. 1972. The egg white protein evidence for ratite affinities. Ibis 114: 377–387. **[444]**

Sibley, C.G., and B.L. Monroe, Jr. 1990. Distribution and Taxonomy of Birds of the World. New Haven, Conn.: Yale University Press. **[xvi, 13, 16, 46, 62, 617ff.]**

Sibley, C.G., and B.L. Monroe, Jr. 1993. A Supplement to Distribution and Taxonomy of Birds of the World. New Haven, Conn.: Yale University Press. **[618]**

Sibley, C.G., G.R. Williams, and J.E. Ahlquist. 1982. The relationships of the New Zealand Wrens (Acanthisittidae) as indicated by DNA–DNA hybridization. Notornis 29: 113–130. **[675]**

Sick, H. 1964. Hoatzin. A New Dictionary of Birds (A.L. Thomson, Ed.). New York: McGraw-Hill. **[158]**

Sick, H. 1967. Courtship behavior in the manakins (Pipridae): A review. Living Bird. 6: 5–22. **[421–422]**

Siegel-Causey, D. 1992. [Review of] Distribution and taxonomy of birds of the world. Auk 109: 939–944. **[618]**

Siegel-Causey, D., and S.P. Kharitonov. 1990. The evolution of coloniality. Current Ornithology 7: 285–330. **[384, 401]**

Sileo, L, P.R. Sievert, and M.D. Samuel. 1990. Causes of mortality of albatross chicks at Midway Atoll. J. Wildl. Dis. 26: 329–338. **[437]**

Sillman, A.J. 1973. Avian vision. Avian Biology 3: 349–387. **[190]**

Silver, R., and G.F. Ball. 1989. Brain, hormone and behavior interactions in avian reproductive status and prospectus. Condor 91: 966–978. **[285]**

Silver, R., H. Andrews, and G.F. Ball. 1985. Parental care in an ecological perspective: A quantitative analysis of avian subfamilies. Am. Zool. 25: 823–840. **[453]**

Simberloff, D. 1976. Species turnover and equilibrium island biogeography. Science 194: 572–578. **[576]**

Simmons, K.E.L. 1952. The nature of the predator-reactions of breeding birds. Behaviour 4: 161–171. **[383]**

Simmons, K.E.L. 1955. The nature of the predator-reactions of waders towards humans; with special reference to the role of the aggressive-, escape-, and brooding-drives. Behaviour 8: 130–173. **[383]**

Simmons, K.E.L. 1957. The taxonomic significance of the head-scratching methods of birds. Ibis 99: 178–181. **[77]**

Simmons, K.E.L. 1964. Feather maintenance, pp. 278–286. *In* A New Dictionary of Birds (A.L. Thomson, Ed.). New York: McGraw-Hill. **[77]**

Simpson, G.G. 1946. Fossil penguins. Bull. Am. Mus. Nat. Hist. 87: 1–100. **[625, 627]**

Simpson, S.F., and J. Cracraft. 1981. The phylogenetic relationships of the Piciformes (Class Aves). Auk 98: 481–494. **[673]**

Skadhauge, E. 1981. Osmoregulation in Birds. Berlin: Springer-Verlag. **[146]**

Skowron, C., and M. Kern. 1980. The insulation in nests of selected North American songbirds. Auk 97: 816–824. **[390]**

Skutch, A.F. 1949. Do tropical birds rear as many young as they can nourish? Ibis 91: 430–455. **[500]**

Skutch, A.F. 1960. Life histories of Central American birds. II. Pacific Coast Avifauna No. 34. Berkeley, Calif.: Cooper Ornithological Society. **[387]**

Skutch, A.F. 1961. Helpers among birds. Condor 63: 198–226. **[467]**

Skutch, A.F. 1976. Parent Birds and Their Young. Austin: University of Texas Press. **[380, 387, 388, 399, 401, 447, 448, 451]**

Skutch, A.F. 1985. Clutch size, nesting success, and predation on nests of neotropical birds, reviewed, pp. 575–594. *In* Neotropical Ornithology, Ornithol. Monogr. No. 6. Lawrence, Kans.: Allen Press. **[380]**

Slagsvold, T. 1982. Clutch size variation in passerine birds: The nest predation hypothesis. Oecologia Berlin 54:159–169. **[500]**

Slagsvold, T. 1990. Fisher's sex ratio theory may explain hatching patterns in birds. Evolution 44: 1009–1017. **[503]**

Smith, H.G., and R. Montgomerie. 1991. Sexual selection and the tail ornaments of North American barn swallows. Behav. Ecol. Sociobiol. 28: 195–201. **[229]**

Smith, H.G., R. Montgomerie, T. Poldmaa, B.N. White, and P.T. Boag. 1991. DNA fingerprinting reveals relation between tail ornaments and cuckoldry in barn swallows, *Hirundo rustica*. Behav. Ecol. 2: 90–98. **[229]**

Smith, J.N.M. 1974. The food searching behaviour of two European thrushes. II. The adaptiveness of the search patterns. Behaviour 49: 1–61. **[165]**

Smith, J.N.M., and H.P.A. Sweatman. 1974. Food-searching behavior of titmice in patchy environments. Ecology 55: 1216–1232. **[167]**

Smith, N.G. 1968. The advantage of being parasitized. Nature London 219: 690–694. **[464]**

Smith, S.A., and Paselk, R.A. 1986. Olfactory sensitivity of the Turkey Vulture (*Cathartes aura*) to three carrion-associated odorants. Auk 103: 586–592. **[200]**

Smith, S.M. 1972. The ontogeny of impaling behaviour in the Loggerhead Shrike, *Lanius ludovicianus* L. Behaviour 42: 232–247. **[430]**

Smith, S.M. 1975. Innate recognition of coral snake pattern by a possible avian predator. Science 187: 759–760. **[429]**

Smith, S.M. 1977. Coral-snake pattern recognition and stimulus generalisation by naive great kiskadees (Aves: Tyrannidae). Nature London 265: 535–536. **[429]**

Smith, S.M. 1978. The "underworld" in a territorial sparrow: Adaptive strategy for floaters. Am. Nat. 112: 571–582. **[518, 519]**

Smith, S.M. 1983. The ontogeny of avian behavior. Avian Biology 7: 85–159. **[430, 449, 451]**

Smith, S.M. 1991. The Black-capped Chickadee. Ithaca, N.Y.: Cornell University Press. **[333]**

Smith, T.B. 1990a. Resource use by bill morphs of an African finch: Evidence for intraspecfic competition. Ecology 71: 1246–1257. **[153]**

Smith, T.B. 1990b. Natural selection on bill characters in the

two bill morphs of the African finch *Pyrenestes ostrinus*. Evolution 44: 832–842. **[153]**
Smith, W.J. 1969. Messages of vertebrate communication. Science 165: 145–150. **[218]**
Smith, W.J. 1977. The Behavior of Communicating. Cambridge, Mass.: Harvard University Press. **[218]**
Smith, W.J., Pawlukiewicz, J., and S.T. Smith. 1978. Kinds of activities correlated with singing patterns of the Yellow-throated Vireo. Anim. Behav. 26: 862–884. **[243]**
Smith, W.K., S.W. Roberts, and P.C. Miller. 1974. Calculating the nocturnal energy expenditure of an incubating Anna's Hummingbird. Condor 76: 176–183. **[390]**
Snow, D.W. 1962. A field study of the Black and White Manakin, *Manacus manacus*, in Trinidad. Zoologica N.Y. 47: 65–104. **[380, 490]**
Snow, D.W. 1976. The Web of Adaptation: Bird Studies in the American Tropics. New York: Quadrangle, N,Y. Times Books. **[213, 424]**
Snow, D.W. 1978. The nest as a factor determining clutch-size in tropical birds. J. Ornithol. 119: 227–230. **[500]**
Snow, D.W., and B.K. Snow. 1964. Breeding seasons and annual cycles of Trinidad land-birds. Zoologica New York 49: 1-39. **[276]**
Snyder, N.F.R., and H.A. Snyder. 1989. Biology and conservation of the California Condor. Current Ornithology 6: 175–267. **[603, 608]**
Snyder, N.F., H.A. Snyder, and T.B. Johnson. 1989. Parrots return to the Arizona skies. Birds Int. 1(2): 41–52. **[343, 608]**
Soulé, M., Ed. 1986. Conservation Biology: The Science of Scarcity and Diversity. Sunderland, Mass.: Sinauer Associates. **[582, 598]**
Southern, W.E. 1971. Gull orientation by magnetic cues: A hypothesis revisited. Ann. N.Y. Acad. Sci. 188: 295–311. **[319]**
Southern, W.E. 1972. Influence of disturbances in the earth's magnetic field on Ring-billed Gull orientation. Condor 74: 102–105. **[319]**
Southwick, E.E., and D.M. Gates. 1975. Energetics of occupied hummingbird nests, pp. 417–430. *In* Perspectives of Biophysical Ecology (D.M. Gates and R.B. Schmerl, Eds.). New York: Springer-Verlag. **[390]**
Spector, D.A. 1992. Wood-warbler song systems: A review of paruline singing behaviors. Current Ornithology 9: 199-238. **[260]**
Spellerberg, I.F. 1971. Aspects of McCormick Skua breeding biology. Ibis 113: 357–363. **[442]**
Spitzer, G. 1972. Jahreszeitliche Aspekte der Biologie der Bartmeise (*Panurus biarmicus*). J. Ornithol. 113: 241–275. **[160]**
Spurrier, M.F., M.S. Boyce, and B.F.J. Manly. 1991. Effects of parasites on mate choice by captive sage grouse, pp. 389–398. *In* Bird–Parasite Interactions (J.E. Loye and M. Zuk, Eds.). New York: Oxford University Press. **[420]**
Stacey, P.B., and W.D. Koenig, Eds. 1990. Cooperative Breeding in Birds. Cambridge: Cambridge University Press. **[467]**
Stacey, P.B., and J.D. Ligon. 1991. The benefits-of-philopatry hypothesis for the evolution of cooperative breeding: Variation in territory quality and group size effects. Am. Nat. 137: 831–846. **[470, 477]**
Stager, K.E. 1964. The role of olfaction in food location by the Turkey Vulture (*Cathartes aura*). Los Angeles County Museum Contribution Science. Sci. No. 81. **[200]**
Stager, K.E. 1967. Avian olfaction. Am. Zool. 7: 415–419. **[200]**
Stallcup, R. 1991. A reversible catastrophe. Observer 91 (Spring/Summer): 18–29. **[587]**
Stamps, J.A., A. Clark, P. Arrowood, and B. Kus. 1985. Parent offspring conflict in budgerigars. Behaviour 94: 1–39. **[444]**
Stapel, S.O., J.A.M. Leunissen, M. Versteeg, J. Wattel, and W.W. de Jong. 1984. Ratites as the oldest offshoot of avian stem—Evidence from alpha-crystallin A sequences. Nature London 311: 257–259. **[621]**
Starck, J.M. 1993. Evolution of avian ontogenies. Current Ornithology 10: 275–366. **[426, 432, 436, 451]**
Stegmann, B.K. 1978. Relationships of the superorders Alectoromorphae and Charadriomorphae (Aves): A comparative study of the avian hand. Publ. Nuttall Ornithol. Club No. 17. **[648]**
Stein, R.C. 1963. Isolating mechanisms between populations of Traill's Flycatchers. Proc. Am. Philos. Soc. 107: 21–50. **[244]**
Sternberg, H. 1972. The origin and age composition of newly formed populations of Pied Flycatchers (*Ficedula hypoleuca*), pp. 690–691. *In* Proc. XV Int. Ornithol. Congr. (K.H. Voous, Ed.). Leiden: E.J. Brill. **[512]**
Stettenheim, P. 1973. The bristles of birds. Living Bird 12: 201–234. **[92]**
Stettenheim, P. 1976. Structural adaptations in feathers, pp. 385–401. *In* Proc. XVI Int. Ornithol. Congr. (H.J. Frith and J.H. Calaby, Eds.). Canberra: Australian Academy of Science. **[71, 92]**
Stettner, L.J., and K.A. Matyniak. 1968. The brain of birds. Sci. Am. 218: 64–76. **[178, 181]**
Stewart, R.E., and J.W. Aldrich. 1951. Removal and repopulation of breeding birds in a spruce–fir forest community. Auk 68: 471–482. **[519]**
Stiles, E.W. 1978. Avian communities in temperate and tropical alder forests. Condor 80: 276–284. **[557]**
Stiles, F.G. 1978. Possible specialization for hummingbird-hunting in the Tiny Hawk. Auk 95: 550–553. **[420]**
Stiles, F.G. 1980. Evolutionary implications of habitat relations between permanent and winter resident landbirds in Costa Rica, pp. 421–435. *In* Migrant Birds in the Neotropics (A. Keast and E.S. Morton, Eds.). Washington, D.C.: Smithsonian Institution Press. **[291]**
Stinson, C.H. 1979. On the selective advantage of fratricide in raptors. Evolution 33: 1219–1225. **[442]**
Stokes, A.W. 1960. Nest-site selection and courtship behaviour of the Blue Tit *Parus caeruleus*. Ibis 102: 507–519. **[222]**
Storer, R.W. 1956. The fossil loon, *Colymbus minutus*. Condor 58: 413–426. **[649]**
Storer, R.W. 1960. Evolution in the diving birds, pp. 694–707. *In* Proc. XII Int. Ornithol. Congr. **[112, 623]**
Storer, R.W. 1963. Courtship and mating behavior and the

phylogeny of the grebes, pp. 562–569. *In* Proc. XIII Int. Ornithol. Congr. (C.G. Sibley, Ed.). Lawrence, Kans.: Allen Press. **[623]**

Storer, R.W. 1965. The color phases of the Western Grebe. Living Bird 4: 59–63. **[541]**

Storer, R.W. 1971a. Adaptive radiation of birds. Avian Biology 1: 149–188. **[6, 20, 649]**

Storer, R.W. 1971b. Classification of birds. Avian Biology 1: 1–18. **[47, 54, 55, 617, 648]**

Storer, R.W. 1979. Podicipediformes, pp. 140–155. *In* Check-list of Birds of the World. Vol. 1, 2nd ed. Cambridge, Mass.: Museum of Comparative Zoology. **[623]**

Storer, R.W. 1986. Birds: Podicipediformes (grebes), pp. 14–16. *In* Encyclopaedia Britannica, 15th ed., Macropaedia, Vol. 15. Chicago: Encyclopaedia Britannica, Inc. **[623]**

Storer, R.W. 1992. [Review of] Distribution and taxonomy of birds of the world, by Charles G. Sibley and Burt L. Monroe, Jr. Wilson Bull. 104: 554–569. **[617, 618]**

Storer, R.W., and G.L. Nuechterlein. 1992. Western and Clark's Grebes. *In* The Birds of North America, No. 26 (A. Poole, P. Stettenheim, and F. Gill, Eds.). Philadelphia: Academy of Natural Sciences; Washington, D.C.: American Ornithologists' Union. **[541]**

Strauch, J.G., Jr. 1978. The phylogeny of the Charadriiformes (Aves): A new estimate using the method of character compatibility analysis. Trans. Zool. Soc. London 34: 269–345. **[648]**

Stresemann, E. 1927–1934. Aves, pp. 729–853. *In* Handbuch Zoologie, Vol. VII B (W. Kükenthal, Ed.). Berlin: W. de Gruyter. **[47]**

Stresemann, E. 1959. The status of avian systematics and its unsolved problems. Auk 76: 269–280. **[55]**

Stresemann, E. 1967. Inheritance and adaptation in moult, pp. 75–80. *In* Proc. XIV Int. Ornithol. Congr. (D.W. Snow, Ed.). Oxford: Blackwell. **[84]**

Stresemann, E., and V. Stresemann. 1966. Die Mauser der Vögel. J. Ornithol. 107: 1–445. **[83, 92]**

Strong, D.R., Jr., D. Simberloff, L.G. Abele, and A.B. Thistle. 1984. Ecological Communities: Conceptual Issues and the Evidence. Princeton, N.J.: Princeton University Press. **[564, 580]**

Studd, M.V., and R. J. Robertson. 1985. Life span, competition, and delayed plumage maturation in male passerines: The breeding threshold hypothesis. Am. Nat. 126: 101–115. **[495]**

Sturkie, P.D. 1976 (3rd ed.), 1986 (4th ed.). Avian Physiology. New York: Springer-Verlag. **[146, 358, 359, 360, 374]**

Stutchbury, B.J., and R.J. Robertson. 1987. Signalling subordinate and female status: Two hypotheses for the adaptive significance of subadult plumage in female tree swallows. Auk 104: 717–723. **[496]**

Sulkava, S. 1969. On small birds spending the night in the snow. Aquilo Ser Zool. 7: 33–37. **[132]**

Sullivan, K.A. 1984a. The advantages of social foraging in Downy Woodpeckers. Anim. Behav. 32: 16–22. **[344]**

Sullivan, K.A. 1984b. Information exploitation by Downy Woodpeckers in mixed-species flocks. Behaviour 91: 294–311. **[344]**

Sullivan, K.A. 1988. Ontogeny of time-budgets in Yellow-eyed Juncos: Adaptation to ecological constraints. Ecology 69: 118–124. **[449]**

Sullivan, K.A. 1989. Predation and starvation: Age-specific mortality in juvenile juncos. J. Anim. Ecol. 58: 275–286. **[486]**

Summers, K.R., and R.H. Drent. 1979. Breeding biology and twinning experiments of Rhinoceros Auklets on Cleland Island, British Columbia. Murrelet 60: 16–22. **[441]**

Summers, R.W., and M. Waltner. 1979. Seasonal variations in the mass of waders in southern Africa, with special reference to migration. Ostrich 50: 21–37. **[302]**

Swennen, C. 1968. Nest protection of Eiderducks and Shovelers by means of faeces. Ardea 56: 248–258. **[383]**

Swierczewski, E.V., and R.J. Raikow. 1981. Hind limb morphology, phylogeny, and classification of the Piciformes. Auk 98: 466–480. **[673]**

Sy, M. 1936. Funktionell-anatomische Untersuchungen am Vogelflügel. J. Ornithol. 84: 199–296. **[96]**

Szumowski, P., and M. Theret. 1965. Causes possibles de la faible fertilité des oies et des difficultés de son amelioration. Recl. Méd. Vét Ec. Alfort 141: 583. **[349]**

Tarsitano, S., and M.K. Hecht. 1980. A reconsideration of the reptilian relationships of *Archaeopteryx*. Zool. J. Linn. Soc. 69: 149–182. **[30]**

Tasker, C.R., and J.A. Mills. 1981. A functional analysis of courtship feeding in the Red-billed Gull, *Larus novaehollandiae scopulinus*. Behaviour 77: 222–241. **[365]**

Taylor, T.G. 1970. How an eggshell is made. Sci. Am. 222(3): 88–95. **[362]**

Tegelström, H., and H.P. Gelter. 1990. Haldane's Rule and sex-biassed gene flow between two hybridizing flycatcher species (*Ficedula albicollis* and *F. hypoleuca*, Aves: Muscicapidae). Evolution 44: 2012–2021. **[543]**

Temple, S.A. 1977. The status and conservation of endemic kestrels on Indian Ocean Islands, pp. 74–92. *In* Proc. ICBP World Conference on Birds of Prey (R.D. Chancellor, Ed.). London: International Council for Bird Preservation. **[388–389]**

Temple, S.A. 1986. The problem of avian extinctions. Current Ornithology 3: 453–485. **[581, 584]**

ten Cate, C. 1987. Sexual preferences in zebra finch males raised by two species. II. The internal representation resulting from double imprinting. Anim. Behav. 35: 321–330. **[210]**

ten Cate, C., and P. Bateson. 1988. Sexual selection: The evolution of conspicuous characteristics in birds by means of imprinting. Evolution 42: 1355–1358. **[231]**

Terborgh, J.W. 1980. The conservation status of Neotropical migrants: Present and future, pp. 21–30. *In* Migrant Birds in the Neotropics (A. Keast and E.S. Morton, Eds.). Washington, D.C.: Smithsonian Institution Press. **[513]**

Terborgh, J.W. 1989. Where Have All the Birds Gone? Princeton, N.J.: Princeton University Press. **[287, 525, 595]**

Terborgh, J., S.K. Robinson, T.A. Parker, III, C. Munn, and N. Pierpont. 1990. Structure and organization of an Amazonian forest bird community. Ecol. Monogr. 60: 213–238. **[557]**

Terborgh, J.W., and J.S. Weske. 1969. Colonization of secondary habitats by Peruvian birds. Ecology 50: 765–782. **[557]**

Terborgh, J.W., and B. Winter. 1980. Some causes of extinction, pp. 119–133. *In* Conservation Biology: An Evolutionary and Ecological Perspective (M.E. Soule and B.A. Wilcox, Eds.). Sunderland, Mass.: Sinauer Associates. **[579]**

Thayer, G.H. 1909. Concealing-coloration in the Animal Kingdom. New York: Macmillan. **[206]**

Thielcke, G. 1961. Stammegeschichte und geographische Variation des Gesanges unserer Baumlaufer (*Certhia familiaris* L. und *C. brachydactyla* Brehm). Z. Tierpsychol. 18: 188–204. **[257]**

Thielcke, G. 1969. Geographic variation in bird vocalizations, pp. 311–339. *In* Bird Vocalisations (R.A. Hinde, Ed.). Cambridge: Cambridge University Press. **[257]**

Thomas, D.H. 1982. Salt and water excretion by birds: The lower intestine as an integrator of renal and intestinal excretion. Comp. Biochem. Physiol. 71A: 527–535. **[141]**

Thompson, A.L., Ed. 1964. A New Dictionary of Birds. New York: McGraw-Hill. **[17]**

Thompson, C.W. 1991. The sequence of molts and plumages in painted buntings and implications for theories of delayed plumage maturation. Condor 93: 209–235. **[495]**

Thompson, W.A., I. Vertinsky, and J.R. Krebs. 1974. The survival value of flocking in birds: A simulation model. J. Anim. Ecol. 43: 785–820. **[340]**

Thorpe, W.H. 1958. The learning of song patterns by birds, with especial reference to the song of the Chaffinch *Fringilla coelebs*. Ibis 100: 535–570. **[254]**

Thorpe, W.H. 1961. Bird-Song. London: Cambridge University Press. **[242, 260]**

Thorpe, W.H. 1963. Antiphonal singing in birds as evidence for avian auditory reaction time. Nature London 197: 774–776. **[249]**

Thorpe, W.H., and M.E.W. North. 1966. Vocal imitation in the tropical Bou-bou Shrike *Laniarus aethiopicus major* as a means of establishing and maintaining social bonds. Ibis 108: 432–435. **[249]**

Tinbergen, N. 1951. The Study of Instinct. London: Oxford University Press. **[341]**

Tinbergen, N. 1952. "Derived" activities: Their causation, biological significance, origin, and emancipation during evolution. Q. Rev. Biol. 27: 1–32. **[216]**

Tinbergen, N. 1959. Comparative studies of the behaviour of gulls (Laridae): A progress report. Behaviour 15: 1–70. **[221]**

Tinbergen, N. 1963. The shell menace. Nat. Hist. 72(7): 28–35. **[428]**

Tinbergen, N., and A.C. Perdeck. 1950. On the stimulus situation releasing the begging response in the newly hatched Herring Gull chick (*Larus argentatus argentatus* Pont.). Behaviour 3: 1–39. **[429]**

Tordoff, H.B. 1954. Relationships in the New World nine-primaried oscines. Auk 71: 273–284. **[691]**

Tordoff, H.B. 1992. Plethora of Peregrines? Unpublished essay. **[609]**

Trail, P.W. 1985. Courtship disruption modifies mate choice in a lek-breeding bird. Science 227: 778–780. **[421]**

Trainer, J.M. 1985. Changes in song dialect distributions and microgeographic variation in song of White-crowned Sparrows (*Zonotrichia leucophrys nuttalli*). Auk 100: 568–582. **[257]**

Trillmich, F. 1976. Learning experiments on individual recognition in budgerigars (*Melopsittacus undulatus*). Z. Tierpsychol. 41: 372–395. **[210]**

Trivers, R.L. 1971. The evolution of reciprocal altruism. Q. Rev. Biol. 46: 35–57. **[468]**

Trivers, R.L. 1974. Parent-offspring conflict. Am. Zool. 14: 249–264. **[425, 456]**

Tschanz, B. 1968. Trottellummen. Z. Tierpsychol. Beiheft 4: 1–103. **[430]**

Tschanz, B., P. Ingold, and H. Lengacher. 1969. Eiform und Bruterfolg bei Trottellummen *Uria aalge aalge* Pont. Ornithol. Beob. 66: 25–42. **[370]**

Tullett, S.G., and R.G. Board. 1977. Determinants of avian eggshell porosity. J. Zool. London 183: 203–211. **[372]**

Turner, E.L. 1924. Broadland Birds. London: Country Life. **[434]**

Tyler, C., and K. Simkiss. 1959. A study of the egg shells of ratite birds. Proc. Zool. Soc. London 133: 201–243. **[372]**

Tyrrell, R.A., and E.Q. Tyrrell. 1985. Hummingbirds: Their Life and Behavior. New York: Crown. **[94, 119, 155]**

U.S. Fish and Wildlife Service. 1986. Whooping Crane recovery plan. Albuquerque, N.M.: U.S. Fish and Wildlife Service. **[607]**

U.S. Fish and Wildlife Service. 1991. Black-capped Vireo Recovery Plan. Albuquerque, N.M.: U.S. Fish and Wildlife Service, Region 2. **[595, 598]**

U.S. Fish and Wildlife Service. 1993. Waterfowl. Status and Fall Flight Forecast 1993. Washington, D.C.: U.S. Fish and Wildlife Service. **[598]**

Van Balen, J.H. 1980. Population fluctuations of the Great Tit and feeding conditions in winter. Ardea 68:143–164. **[520]**

Vander Wall, S.B. 1982. An experimental analysis of cache recovery in Clark's Nutcracker. Anim. Behav. 30: 84–94. **[185]**

Vander Wall, S.B. 1990. Food hoarding in animals. Chicago: University of Chicago Press. **[171]**

Vander Werf, E. 1992. Lack's clutch size hypothesis: An examination of the evidence using meta-analysis. Ecology 73: 1699–1705. **[498]**

Van Iersel, J.J.A., and A.C.A. Bol. 1958. Preening of two tern species: A study on displacement activities. Behaviour 13: 1–88. **[216]**

van Noordwijk, A.J., and W. Scharloo. 1981. Inbreeding in an island population of the Great Tit. Evolution 35: 674–688. **[492]**

van Riper, C., III, S.G. van Riper, M.L. Goff, and M. Laird. 1986. The epizootiology and ecological significance of malaria in Hawaiian land birds. Ecol. Monogr. 56: 327–344. **[586]**

Van Tets, G.F. 1965. A comparative study of some social communication patterns in the Pelecaniformes. Ornithol. Monogr. No. 2. **[211, 212]**

Van Tyne, J., and A.J. Berger. 1976. Fundamentals of Ornithology, 2nd ed. New York: Wiley. **[48, 75, 239]**

Vehrencamp, S.L. 1978. The adaptive significance of communal nesting in Groove-billed Anis (*Crotophaga sulcirostris*). Behav. Ecol. Sociobiol. 4: 1–33. **[473]**

Vehrencamp, S.L., and J.W. Bradbury. 1984. Mating systems and ecology, pp. 251–278. *In* Behavioral Ecology: An Evolutionary Approach, 2nd ed. (J.R. Krebs and N.B. Davies, Eds.). Sunderland Mass.: Sinauer Associates. **[420]**

Vehrencamp, S.L., B.S. Bowen, and R.R. Koford. 1986. Breeding roles and pairing patterns within communal groups of Groove-billed Anis. Anim. Behav. 34: 347–366. **[473]**

Verbeek, N.A.M. 1972. Daily and annual time budget of the Yellow-billed Magpie. Auk 89: 567–582. **[398]**

Verbeek, N.A.M. 1973. The exploitation system of the Yellow-billed Magpie. Univ. Calif. Publ. Zool. 99: 1–58. **[336]**

Verner, J. 1964. Evolution of polygamy in the Long-billed Marsh Wren. Evolution 18: 252–261. **[413]**

Verner, J., and G.H. Engelson. 1970. Territories, multiple nest building, and polygyny in the Long-billed Marsh Wren. Auk 87: 557–567. **[386]**

Verner, J., and M.F. Willson. 1969. Mating systems, sexual dimorphism, and the role of male North American passerine birds in the nesting cycle. Ornithol. Monogr. No. 9. **[386, 407, 413, 455]**

Vernon, C.J. 1973. Vocal imitation by southern African birds. Ostrich 44: 23–30. **[252]**

Vince, M.A. 1969. How quail embryos communicate. Ibis 111: 441. **[428]**

Vlug, J.J., and J. Fjeldså. 1990. Working bibliography of grebes of the world with summaries of current taxonomy and of distributional status. Copenhagen: Zoological Museum, University of Copenhagen. **[623]**

Voous, K.H. 1957. Studies on the Fauna of Curacao and other Caribbean Islands, Vol. 7. The Birds of Aruba, Curacao and Bonaire. The Hague: Martinus Nijhoff. **[296]**

Voous, K.H. 1973. List of recent Holarctic bird species: Nonpasserines. Ibis 115: 612–638. **[657]**

Voous, K.H. 1985. Passeriformes, pp. 440–441. *In* A Dictionary of Birds (B. Campbell and E. Lack, Eds.). Calton: T. & A.D. Poyser. **[675]**

Vuilleumier, F. 1975. Zoogeography. Avian Biology 5: 421–495. **[43]**

Wahl, R.R., D.D. Diamond, and D. Shaw. 1990. The Golden-cheeked Warbler: A status review. Final report submitted to Office of Endangered Species, U.S. Fish and Wildlife Service, Albuquerque, NM. **[594]**

Walcott, C., and R.P. Green. 1974. Orientation of homing pigeons altered by a change in the direction of an applied magnetic field. Science 184: 180–182. **[316, 318]**

Walcott, C., J.L. Gould, and J.L. Kirschvink. 1979. Pigeons have magnets. Science 205: 1027–1028. **[191]**

Waldvogel, J.A. 1989. Olfactory orientation by birds. Current Ornithology 6: 269–322. **[199, 316]**

Walkinshaw, L.H. 1963. Some life history studies of the Stanley Crane, pp. 344–353. *In* Proc. XIII Int. Ornithol. Congr. (C.G. Sibley, Ed.). Lawrence, Kans.: Allen Press. **[438]**

Walkinshaw, L.H. 1972. Kirtland's Warbler—Endangered. Am. Birds 26: 3–9. **[463]**

Wallace, A.R. 1874. Migration of Birds. Nature London 10: 459. **[289]**

Walls, G.L. 1942. The Vertebrate Eye and Its Adaptive Radiation. Cranbrook Inst. Sci. Bull. No. 19. Bloomfield Hills, Mich.: Cranbrook Institute of Science. **[187, 189, 190, 202]**

Walsberg, G.E. 1975. Digestive adaptations of *Phainopepla nitens* associated with the eating of mistletoe berries. Condor 77: 169–174. **[160]**

Walsberg, G.E. 1978. Brood size and the use of time and energy by the Phainopepla. Ecology 59: 147–153. **[453, 454]**

Walsberg, G.E. 1983. Avian ecological energetics. Avian Biology 7: 161–220. **[124, 125, 278, 285, 364, 444, 454]**

Walsberg, G.E., and J.R. King. 1978. The energetic consequences of incubation for two passerine species. Auk 95: 644–655. **[390, 397]**

Walters, J.R. 1982. Parental behavior in lapwings (Charadriidae) and its relationships with clutch sizes and mating systems. Evolution 36: 1030–1040. **[433]**

Walters, J.R. 1984. The evolution of parental behavior and clutch size in shorebirds, pp. 243–287. *In* Behavior of Marine Animals, Vol. 5 (J. Burger and B.L. Olla, Eds.). New York: Plenum Press. **[433, 455, 500]**

Walters, J.R. 1991. Application of ecological principles to the management of endangered species: The case of the Red-cockaded Woodpecker. Annu. Rev. Ecol. Syst. 22: 505–523. **[512]**

Wangensteen, O.D., D. Wilson, and H. Rahn. 1970. Diffusion of gases across the shell of the hen's egg. Respir. Physiol. 11: 16–30. **[372]**

Ward, P. 1965. The breeding biology of the Black-faced Dioch *Quelea quelea* in Nigeria. Ibis 107: 326–349. **[412]**

Ward, P. 1969. The annual cycle of the Yellow-vented Bulbul *Pycnonotus goiavier* in a humid equatorial environment. J. Zool. London 157: 24–45. **[171]**

Warner, R.E. 1968. The role of introduced diseases in the extinction of the endemic Hawaiian avifauna. Condor 70: 101–120. **[516, 586]**

Wasserman, F.E. 1977. Mate attraction function of song in the White-throated Sparrow. Condor 79: 125–127. **[243]**

Waterman, A.J. 1977. The integumentary system. *In* Chordate Structure and Function, 2nd ed. (A. Kluge, Ed.). New York: Macmillan. **[28]**

Watson, G.E. 1963. The mechanism of feather replacement during natural molt. Auk 80: 486–495. **[80]**

Weathers, W.W. 1979. Climatic adaptation in avian standard metabolic rate. Oecologia Berlin 42: 81–89. **[132]**

Weathers, W.W., W.A. Buttemer, A.M. Hayworth, and K.A. Nagy. 1984. An evaluation of time-budget estimates of daily energy expenditure in birds. Auk 101: 459–472. **[127, 143]**

Webb, D.R. 1987. Thermal tolerance of avian embryos: A review. Condor 89: 874–898. **[401]**

Weeden, J.S. 1965. Territorial behavior of the Tree Sparrow. Condor 67: 193–209. **[329]**

Wegge, P. 1980. Distorted sex ratio among small broods in a declining Capercaille population. Ornis Scand. 11: 106–109. **[406]**

Weimerskirch, H., J.C. Stahl, and P. Jouventin. 1992. The breeding biology and population dynamics of King Penguins *Aptendodytes patagonica* on the Crozet Islands. Ibis 134: 107–117. **[283]**

Wellnhofer, P. 1988. A new specimen of *Archaeopteryx*. Science 240: 1790–1792. **[25]**

Wells, S., R.A. Bradley, and L.F. Baptista. 1978. Hybridization in *Calypte* hummingbirds. Auk 95: 537–549. **[543]**

Welsh, D.A. 1975. Savannah Sparrow breeding and territoriality on a Nova Scotia dune beach. Auk 92: 235–251. **[412]**

Welty, J.C. 1982 (3rd ed.), 1988 (4th ed). The Life of Birds. Philadelphia: W.B. Saunders. **[336, 486]**

Wenzel, B.M. 1968. Olfactory prowess of the kiwi. Nature London 220: 1133–1134. **[201]**

Wenzel, B.M. 1971. Olfactory sensation in the kiwi and other birds. Ann. N.Y. Acad. Sci. 188: 183–193. **[201]**

Wenzel, B.M. 1973. Chemoreception. Avian Biology 3: 389–415. **[199, 202]**

Werner, C.F. 1958. Der Canaliculus (*Aquaeductus*) cochleae und seine Beziehungen zu den Kanälen des IX. und X. Hirnnerven bei den Vögeln. Zool. Jahr. Abt. Anat. Ontog. Tiere 77: 1–8. **[198]**

West, M.J., and A.P. King. 1980. Enriching cowbird song by social deprivation. J. Comp. Physiol. Psychol. 94: 263–270. **[247]**

West, M.J., A.P. King, and D.H. Eastzer. 1981. The cowbird: Reflections on development from an unlikely source. Am. Sci. 69: 56–66. **[247]**

West-Eberhard, M.J. 1975. The evolution of social behavior by kin selection. Q. Rev. Biol. 50: 1–33. **[468]**

West-Eberhard, M.J. 1983. Sexual selection, social competition, and speciation. Q. Rev. Biol. 58: 155–183. **[549, 550, 552]**

Westneat, D.F., P.C. Frederick, and R.H. Wiley. 1987. The use of genetic markers to estimate the frequency of successful alternative tactics. Behav. Ecol. Sociobiol. 21: 35–45. **[409]**

Westneat, D.F., P.W. Sherman, and M.L. Morton. 1990. The ecology and evolution of extra-pair copulations in birds. Current Ornithology 7: 331–369. **[409]**

Wetherbee, D.K., and L.M. Bartlett. 1962. Egg teeth and shell rupture of the American Woodcock. Auk 79: 117. **[428]**

Wetmore, A. 1936. The number of contour feathers in passeriform and related birds. Auk 53: 159–169. **[74]**

Wetmore, A. 1960. A classification for the birds of the world. Smithson. Misc. Collect. 139 (11). **[47]**

White, F.N., and J.L. Kinney. 1974. Avian incubation. Science 186: 107–115. **[390]**

White, F.N., G.A. Bartholomew, and T.R. Howell. 1975. The thermal significance of the nest of the Sociable Weaver *Philetairus socius*: Winter observations. Ibis 117: 171–179. **[390]**

White, F.N., G.A. Bartholomew, and J.L. Kinney. 1978. Physiological and ecological correlates of tunnel nesting in the European Bee-eater, *Merops apiaster*. Physiol. Zool. 51: 140–154. **[390]**

White, S.C. 1974. Ecological aspects of growth and nutrition in tropical fruit-eating birds. Doctoral dissertation, University of Pennsylvania, Philadelphia. **[438]**

White, S.J. 1971. Selective responsiveness by the Gannet (*Sula bassana*) to played-back calls. Anim. Behav. 19: 125–131. **[247]**

Whittow, G.C., P.D. Sturkie, and G. Stein, Jr. 1964. Cardiovascular changes associated with thermal polypnea in the chicken. Am. J. Physiol. 207: 1349–1353. **[137]**

Wiens, J.A. 1983. Avian community ecology: An iconoclastic view, pp. 355–403. *In* Perspectives in Ornithology (A.H. Brush and G.A. Clark, Jr., Eds.). Cambridge: Cambridge University Press. **[562, 571]**

Wiens, J.A. 1990. The Ecology of Bird Communities, Vols. 1 and 2. New York: Cambridge University Press. **[561, 580]**

Wilcove, D.S. 1985. Nest predation in forest tracts and the decline of migratory songbirds. Ecology 66: 1211–1214. **[383]**

Wilcove, D.S. 1987. Public lands management and the fate of the Spotted Owl. Am. Birds 41: 361–367. **[592]**

Wiley, R.H. 1974. Evolution of social organization and life history patterns among grouse (Aves: Tetraonidae). Q. Rev. Biol. 49: 201–227. **[420]**

Wiley, R.H., and D.G. Richards. 1982. Adaptations for acoustic communication in birds: Sound transmission and signal detection, pp. 132–181. *In* Acoustic Communication in Birds, Vol. 1 (D.E. Kroodsma and E.H. Miller, Eds.). New York and London: Academic Press. **[237]**

Williams, A.J. 1980. Offspring reduction in Macaroni and Rockhopper penguins. Auk 97: 754–759. **[492]**

Williams, G.C. 1966. Natural selection, the costs of reproduction, and a refinement of Lack's principle. Am. Nat. 100: 687–690. **[499]**

Williams, H. 1990. Bird song, pp. 77–126. *In* Neurobiology of Comparative Cognition (R. Kesner and D.S. Olson, Eds.). Hillsdale, N.J.: Erlbaum. **[183]**

Williams, H.W., A.W. Stokes, and J.C. Wallen. 1968. The food call and display of the Bobwhite Quail (*Colinus virginianus*). Auk 85: 464–476. **[216]**

Williams, J.B., and K.A. Nagy. 1984. Daily energy expenditure of Savannah Sparrows: Comparison of time-energy budget and doubly-labeled water estimates. Auk 101: 221–229. **[144]**

Williams, J.B., M.A. DuPlessis, and W.R. Siegfried. 1991. Green Woodhoopoes (*Phoeniculus purpureus*) and obligate cavity roosting provide a test of the thermoregulatory insufficiency hypothesis. Auk 108: 285–293. **[472]**

Williams, T.C., and J.M. Williams. 1978. An oceanic mass migration of land birds. Sci. Am. 239(4): 166–176. **[294, 295, 296]**

Willis, E.O. 1972. The behavior of Spotted Antbirds. Ornithol. Monogr. No. 10. **[341]**

Willis, E.O. 1980. Species reduction in remanescent woodlots in southern Brazil, pp. 783–786. *In* Acta XVII Congressus Internationalis Ornithologici (R. Nöhring, Ed.). Berlin: Verlag der Deutschen Ornithologen-Gesellschaft. **[578]**

Willis, E.O., and Y. Oniki. 1978. Birds and army ants. Annu. Rev. Ecol. Syst. 9: 243–263. **[336, 337]**

Willoughby, F., and J. Ray. 1676. Ornithologiae libri tres. London: Royal Society, 307 pp (English translation 1678). **[13, 46]**

Willson, M.F. 1971. Seed selection in some North American finches. Condor 73: 415–429. **[167]**

Willson, M.F., and J.C. Harmeson. 1973. Seed preferences and digestive efficiency of Cardinals and Song Sparrows. Condor 75: 225–234. **[150]**

Wilson, A.C. 1991. From molecular evolution to body and brain evolution, pp. 331–340. *In* Perspectives on Cellular Regulation: From Bacteria to Cancer (M. Inouye, J. Campisi, D. Cunningham, and M. Riley, Eds.). New York: Wiley-Liss. **[549]**

Wilson, B.W. 1980. Birds. New York: W.H. Freeman. **[7, 11, 189]**

Wilson, E.O. 1975. Sociobiology. Cambridge, Mass.: Belknap Press. **[346]**

Wilson, E.O. 1992. The Diversity of Life. Cambridge, Mass.: Belknap Press. **[590]**

Wilson, R.P., P.G. Ryan, A. James, and M.-P. Wilson. 1987. Conspicuous coloration may enhance prey capture in some piscivores. Anim. Behav. 35: 1558–1560. **[338]**

Wiltschko, W., and R. Wiltschko. 1988. Magnetic orientation in birds. Current Ornithology 5: 67–121. **[191, 317, 322, 323, 325]**

Wiltschko, W., R. Wiltschko, W.T. Keeton, and R. Madden. 1983. Growing up in an altered magnetic field affects the initial orientation of young homing pigeons. Behav. Ecol. Sociobiol. 12: 135–142. **[323]**

Wingfield, J.C. 1991. Mating systems and hormone-behavior interactions, pp. 2055–2062. *In* Acta XX Congresus Internationalis Ornithologici, Vol. 4 (B.D. Ball, Ed.). Wellington, N.Z.: Ornithological Congress Trust Board. **[391]**

Wingfield, J.C., H. Schwabl, and P.W. Mattocks, Jr. 1990. Endocrine mechanisms of migration, pp. 232–256. *In* Bird Migration (E. Gwinner, Ed.). Berlin: Springer-Verlag. **[303]**

Winker, K., D.W. Warner, and A.R. Weisbrod. 1992a. Migration of woodland birds at a fragmented inland stopover site. Wilson Bull. 104: 580–598. **[299]**

Winker, K., D.W. Warner, and A.R. Weisbrod. 1992b. Daily mass gains among woodland migrants at an inland stopover site. Auk 109: 853–862. **[299]**

Winkler, D.W. 1985. Factors determining a clutch size reduction in California Gulls (*Larus californicus*). Evolution 39: 667–677. **[496]**

Winkler, D.W. 1991. Parental investment decision rules in tree swallows: Parental defense, abandonment, and the so-called Concorde Fallacy. Behav. Ecol. 2: 133–142. **[455–456]**

Winkler, D.W., and F.H. Sheldon. 1993. Evolution of nest construction in swallows (Hirundinidae): A molecular phylogenetic perspective. Proc. Natl. Acad. Sci. USA 90: 5705–5707. **[379]**

Winkler, D.W., and J.R. Walters. 1983. The determination of clutch size in precocial birds. Current Ornithology 1: 33–68. **[365, 366, 496, 500, 506]**

Winkler, D.W., and G.S. Wilkinson. 1988. Parental effort in birds and mammals: Theory and measurement. Oxford Surveys in Evolutionary Biology 5: 185–214. **[456]**

Winstanley, D., R. Spencer, and K. Williamson. 1974. Where have all the white-throats gone? Bird Study 21: 1–14. **[513]**

Winter, P. 1963. Vergleichende qualitative und quantitative Untersuchungen an der Hörbahr von Vögeln. Z. Morphol. Oekol. Tiere. 52: 365–400. **[193]**

Winterbottom, J.M. 1971. Priest's Eggs of Southern African Birds. Johannesburg: Winchester Press. **[367]**

Wittenberger, J.F., and G.L. Hunt, Jr. 1985. The adaptive significance of coloniality in birds. Avian Biology 8: 1–78. **[340, 346, 401]**

Woldhek, S. 1980. Bird killing in the Mediterranean. Zeist, Netherlands: European Committee for the Prevention of Mass Destruction of Migratory Birds. **[291]**

Wolf, L.L. 1969. Breeding and molting periods in a Costa Rican population of the Andean Sparrow. Condor 71: 212–219. **[282]**

Wolf, L.L. 1978. Aggressive social organization in nectarivorous birds. Am. Zool. 18: 765–778. **[335]**

Wolf, L.L., and F.B. Gill. 1980. Resource gradients and community organization of nectarivorous birds, pp. 1105–1113. *In* Acta XVII Congressus Internationalis Ornithologici (R. Nöhring, Ed.). Berlin: Verlag der Deutschen Ornithologen-Gesellschaft. **[560]**

Wolf, L.L., F.R. Hainsworth, and F.B. Gill. 1975. Foraging efficiencies and time budgets in nectar-feeding birds. Ecology 56: 117–128. **[126]**

Wolf, L.L., E.D. Ketterson, and V. Nolan, Jr. 1990. Behavioural response of female dark-eyed juncos to the experimental removal of their mates: implications for the evolution of male parental care. Anim. Behav. 39: 125–134. **[454]**

Wolf, L.L., F.G. Stiles, and F.R. Hainsworth. 1976. Ecological organization of a tropical, highland hummingbird community. J. Anim. Ecol. 45: 349–379. **[153, 564]**

Wolfson, A. 1942. Regulation of spring migration in juncos. Condor 44: 237–263. **[303]**

Wolfson, A. 1954. Sperm storage at lower-than-body temperature outside the body cavity in some passerine birds. Science 120: 68–71. **[354]**

Woolfenden, G.E. 1981. Selfish behavior by Florida Scrub Jay helpers, pp. 257–260. *In* Natural Selection and Social Behavior: Recent Research and New Theory (R.D. Alexander and D.W. Tinkle, Eds.) New York: Chiron Press. **[469]**

Woolfenden, G.E., and J.W. Fitzpatrick. 1984. The Florida Scrub Jay: Demography of a Cooperative-Breeding Bird. Princeton, N.J.: Princeton University Press. **[466, 468, 477, 487, 488, 506]**

Wooller, R.D. 1978. Individual vocal recognition in the Kittiwake Gull, *Rissa tridactyla* (L.). Z. Tierpsychol. 48: 68–86. **[247]**

Wootton, J.T., B.E. Young, and D.W. Winkler. 1991. Ecological versus evolutionary hypotheses: Demographic stasis and the Murray-Nolan clutch size equation. Evolution 45: 1947–1950. **[502]**

Wunderle, J.M., Jr. 1981. An analysis of a morph ratio cline in the Bananaquit (*Coereba flaveola*) on Grenada, West Indies. Evolution 35: 333–344. **[87, 536]**

Wunderle, J.M., Jr. 1983. A shift in the morph ratio cline in the Bananaquit of Grenada, West Indies. Condor 85: 365–367. **[536]**

Wunderle, J.M., Jr. 1991. Age-specific foraging efficiency in birds. Current Ornithology 8: 273–324. **[168, 173]**

Wunderle, J.M., Jr., and K.H. Pollock. 1985. The bananaquit-wasp nesting association and a random choice model. Neotropical Ornithology. Ornithol. Monogr. No. 36. Lawrence, Kans.: Allen Press. **[383]**

Würdinger, I. 1979. Olfaction and feeding behavior in juvenile geese (*Anser a. anser* and *Anser domesticus*). Z. Tierpsychol. 49: 132–135. **[200]**

Wyles, J.S., J.G. Kunkel, and A.C. Wilson. 1983. Birds, behavior, and anatomical evolution. Proc. Natl. Acad. Sci. USA 80: 4394–4397. **[549]**

Wynne-Edwards, V.C. 1962. Animal Dispersion in Relation to Social Behavior. Edinburgh: Oliver and Boyd. **[467]**

Yasukawa, K. 1981. Song repertoires in the Red-winged Blackbird (*Agelaius phoeniceus*): A test of the Beau Geste hypothesis. Anim. Behav. 29: 114–125. **[250, 251]**

Ydenberg, R.C. 1989. Growth-mortality trade-offs and the evolution of juvenile life histories in the Alcidae. Evolution 70: 1494–1506. **[448]**

Yeagley, H.L. 1947. A preliminary study of a physical basis of bird navigation. J. Appl. Phys. 18: 1035–1063. **[316]**

Yokoyama, K., and D.S. Farner. 1978. Induction of *Zugunruhe* by photostimulation of encephalic receptors in White-crowned Sparrows. Science 201: 76–79. **[272]**

Zach, R. 1979. Shell dropping, decision making and optimal foraging in Northwestern Crows. Behaviour 68: 106–117. **[166]**

Zach, R., and J.B. Falls. 1977. Influences of capturing prey on subsequent search in the ovenbird (Aves: Parulidae). Can. J. Zool. 55: 1958–1969. **[165]**

Zahavi, A. 1971. The social behavior of the White Wagtail *Motacilla alba alba* wintering in Israel. Ibis 113: 203–211. **[336]**

Zahavi, A. 1974. Communal nesting by the Arabian Babbler: A case of individual selection. Ibis 116: 84–87. **[472]**

Zahavi, A. 1975. Mate selection—A selection for a handicap. J. Theor. Biol. 53: 205–214. **[229]**

Zahavi, A. 1977. Reliability in communication systems and the evolution of altruism, pp. 253–259. *In* Evolutionary Ecology (B. Stonehouse and C. Perrins, Eds.). Baltimore, Md.: University Park Press. **[216]**

Zahavi, A. 1980. Ritualisation and the evolution of movement signals. Behaviour 72: 77–81. **[217]**

Zenone, P.G., M.E. Sims, and C.J. Erickson. 1979. Male Ring Dove behavior and the defense of genetic paternity. Am. Nat. 114: 615–626. **[409]**

Ziegler, H.P. 1964. Displacement activity and motivational theory: A case study in the history of ethology. Psychol. Bull. 61: 362–376. **[216]**

Ziegler, H.P., and H.-J. Bischof, Eds. 1993. Vision, Brain, and Behavior in Birds. Cambridge, Mass.: MIT Press. **[202]**

Zimmer, J.T. 1926. Catalogue of the Edward E. Ayer ornithological library. Field Mus. Nat. Hist. Publ. Zool. Ser. No. 16. **[46]**

Zink, R.M. 1986. Patterns and evolutionary significance of geographic variation in the shistacea group of the fox sparrows (*Passerella iliaca*). Ornithol. Monogr. No. 40. **[530, 531]**

Zink, R.M., and J.C. Avise. 1990. Patterns of mitochondrial DNA and allozyme evolution in the avian genus *Ammodramus*. Syst. Zool. 39: 148–161. **57**

Zink, R.M., D.L. Dittmann, and W.L. Rootes. 1991. Mitochondrial DNA variation and the phylogeny of *Zonotrichia*. Auk 108: 578–584. **[540]**

Zink, R.M., and J.V. Remsen, Jr. 1986. Evolutionary processes and patterns of geographical variation in birds. Current Ornithology 4: 1–69. **[129, 552]**

Ziswiler, V., and D.S. Farner. 1972. Digestion and the digestive system. Avian Biology 2: 343–430. **[437]**

Zuk, M. 1991. Parasites and bright birds: New data and a new prediction, pp. 318–327. *In* Bird–Parasite Interactions (J.E. Loye and M. Zuk, Eds.). New York: Oxford University Press. **[229]**

Zuk, M., R. Thornhill, J.D. Ligon, and K. Johnson. 1990. Parasites and mate choice in red jungle fowl. Am. Zool. 30: 235–244. **[229]**

Zuk, M., R. Thornhill, J.D. Ligon, K. Johnson, S. Austad, S.H. Ligon, N. Thornhill, and C. Costin. 1990. The role of male ornaments and courtship behavior in female mate choice of red jungle fowl. Am. Nat. 136: 459–473. **[229]**

Zusi, R.L. 1984. A functional and evolutionary analysis of rhynchokinesis in birds. Smithson. Contrib. Zool. No. 385. **[148]**

Zusi, R.L. 1986. Birds: Charadriiformes (plovers, sandpipers, gulls, terns, auks), pp. 54–63. *In* Encyclopaedia Britannica, 15th ed., Macropaedia, Vol. 15. Chicago: Encyclopaedia Britannica, Inc. **[648]**

Zusi, R.L., and G.D. Bentz. 1982. Variation of a muscle in hummingbirds and swifts and its systematic implications. Proc. Biol. Soc. Washington 95: 412–420. **[665]**

Zusi, R.L., and G.D. Bentz. 1984. Myology of the Purple-throated Carib (*Eulampis jugularis*) and other hummingbirds (Aves: Trochilidae). Smithson. Contrib. Zool. No. 385. **[665]**

Zusi, R.L., and D. Bridge. 1981. On the slit pupil of the Black Skimmer (*Rynchops niger*). J. Field Ornithol. 52: 338–340. **[188]**

Zusi, R.L., and J.R. Jehl. 1970. The systematic relationships of *Aechmorhynchus*, *Prosobonia*, and *Phegornis* (Charadriiformes; Charadrii). Auk 87: 760–780. **[648]**

Index

Note: Boldfaced folio indicates page on which term is defined.

thrushes, 53, 198, 570, 683